S0-ANR-052

MECHANICS OF SOLIDS

PWS SERIES IN ENGINEERING

Anderson, *Thermodynamics*
Askeland, *The Science and Engineering of Materials*, Third Edition
Borse, *FORTRAN 77 and Numerical Methods for Engineers*, Second Edition
Bolluyt/Stewart/Oladipupo, *Modeling for Design Using SilverScreen*
Clements, *68000 Family Assembly Language*
Clements, *Microprocessor Systems Design*, Second Edition
Clements, *Principles of Computer Hardware*, Second Edition
Das, *Principles of Foundation Engineering*, Third Edition
Das, *Principles of Geotechnical Engineering*, Third Edition
Das, *Principles of Soil Dynamics*
Duff/Ross, *Freehand Sketching for Engineering Design*
El-Wakil/Askeland, *Materials Science and Engineering Lab Manual*
Fleischer, *Introduction to Engineering Economy*
Gere/Timoshenko, *Mechanics of Materials*, Third Edition
Glover/Sarma, *Power System Analysis and Design*, Second Edition
Janna, *Design of Fluid Thermal Systems*
Janna, *Introduction to Fluid Mechanics*, Third Edition
Kassimali, *Structural Analysis*
Keedy, *An Introduction to CAD Using CADKEY 5 and 6*, Third Edition
Keedy/Teske, *Engineering Design Using CADKEY 5 and 6*
Knight, *The Finite Element Method in Mechanical Design*
Knight, *A Finite Element Method Primer for Mechanical Design*
Logan, *A First Course in the Finite Element Method*, Second Edition
McDonald, *Continuum Mechanics*
McGill/King, *Engineering Mechanics: Statics*, Third Edition
McGill/King, *Engineering Mechanics: An Introduction to Dynamics*, Third Edition
McGill/King, *Engineering Mechanics: Statics and an Introduction to Dynamics*, Third Edition
Meissner, *Fortran 90*
Raines, *Software for Mechanics of Materials*
Ray, *Environmental Engineering*
Reed-Hill/Abbaschian, *Physical Metallurgy Principles*, Third Edition
Reynolds, *Unit Operations and Processes in Environmental Engineering*
Russ, *CD-ROM for Materials Science*
Schmidt/Wong, *Fundamentals of Surveying*, Third Edition
Segui, *Fundamentals of Structural Steel Design*
Segui, *LRFD Steel Design*
Shen/Kong, *Applied Electromagnetism*, Second Edition
Sule, *Manufacturing Facilities*, Second Edition
Vardeman, *Statistics for Engineering Problem Solving*
Weinman, *VAX FORTRAN*, Second Edition
Weinman, *FORTRAN for Scientists and Engineers*
Wempner, *Mechanics of Solids*
Wolff, *Spreadsheet Applications in Geotechnical Engineering*
Zirkel/Berlinger, *Understanding FORTRAN 77 and 90*

MECHANICS OF SOLIDS

GERALD WEMPNER

Georgia Institute of Technology

PWS Publishing Company

I(T)P **An International Thomson Publishing Company**

Boston • Albany • Bonn • Cincinnati • Detroit • London
Madrid • Melbourne • Mexico City • New York • Paris
San Francisco • Singapore • Tokyo • Toronto • Washington

International Thomson Publishing
The trademark ITP is used under license.

For more information, contact:

PWS Publishing Co.
20 Park Plaza
Boston, MA 02116

International Thomson Publishing Europe
Berkshire House I68-I73
High Holborn
London WC1V 7AA
England

Thomas Nelson Australia
102 Dodds Street
South Melbourne, 3205
Victoria, Australia

Nelson Canada
1120 Birchmont Road
Scarborough, Ontario
Canada M1K 5G4

International Thomson Editores
Campos Eliseos 385, Piso 7
Col. Polanco
11560 Mexico D.F., Mexico

International Thomson Publishing GmbH
Konigswinterer Strasse 418
53227 Bonn, Germany

International Thomson Publishing Asia
221 Henderson Road
#05-10 Henderson Building
Singapore 0315

International Thomson Publishing Japan
Hirakawacho Kyowa Building, 31
2-2-1 Hirakawacho
Chiyoda Ku, Tokyo 102
Japan

Library of Congress Cataloging-in-Publication Data

Wempner, Gerald.
Mechanics of solids / Gerald Wempner.
p. cm.
Includes index.
ISBN 0-534-92739-4
1. Strength of materials. I. Title.
TA405.W46 1994 94-25417
620.1′12—dc20 CIP

Sponsoring Editor: Jonathan Plant
Editorial Assistant: Cynthia Harris
Developmental Editor: Mary Thomas
Production Coordinator: Robine Andrau
Marketing Manager: Marianne C. P. Rutter
Manufacturing Coordinator: Marcia Locke
Interior Designer: Jean Hammond
Cover Designer: Julia Gecha
Interior Illustrator: Carl Brown
Cover Art: Georgia O'Keefe's "Brooklyn Bridge"; courtesy of the Brooklyn Museum
Typesetter: Asco Trade Typesetting Ltd
Cover Printer: Henry N. Sawyer Company, Inc.
Text Printer/Binder: Quebecor Printing/Martinsburg

Printed and bound in the United States of America
94 95 96 97 98—10 9 8 7 6 5 4 3 2 1

To my greatest teachers,
Paul and Thekla Wempner

Contents

Preface

Mechanics of Solids is intended for the introductory-level mechanics of materials, strength of materials, or mechanics of solids course. The text's approach reflects new attitudes and methods in engineering study and practice; in particular, it focuses on the underlying concepts of mechanics and the logical extension of these concepts to advanced methods and modern technologies. Too often textbooks for these courses have taken the approach of providing procedural techniques and equations for specific problems. The foundations of the subject are most important and must be firmly established. Indeed, the time-tested foundations are much needed in contemporary engineering practice; they provide the basis for most finite element models and many widely used computational programs.

Emphasis on these major concepts will provide readers with a solid foundation for their subsequent studies. Civil engineers of the 1990s need to deal with complex structures, such as domelike stadiums and cable-suspended roofs; books that offer only specific formulas limited to beams and framelike structures are no longer adequate. Mechanical and aeronautical engineers' design work is not limited to simple shafts and beams; witness the integral frames and shells that make up the bodies of modern vehicles. In its presentation of the subject—the mechanics of continuous solids—this book is a refined and powerful tool for the analysis of structures and machines.

Mechanics of Solids has been carefully structured to provide many instructional options. It can be used for a standard introductory mechanics of materials course, for a mixed engineering majors course, or for a more advanced mechanics course. Suggested programs are presented in the accompanying Instructor's Manual.

In *Mechanics of Solids* the three basic aspects—dynamics-statics, kinematics, and behavior of materials—are presented separately in Chapters 2, 3, and 4, respectively, thus introducing one fundamental notion at a time. Definitions have been chosen deliberately for their simplicity and physical appeal rather than for mathematical elegance. Some advanced treatments (e.g., a precise analysis of the strain transformation) are set in optional sections, marked by an asterisk, so that they need not intrude upon a simpler approach.

Chapter 5 demonstrates the application of the fundamentals via the simplest problems (e.g., extension, torsion of circular shafts, and simple bending). These are the so-called exact solutions in the theory of small deformations. With due attention to Saint-Venant's principle, these fundamental solutions apply also to systems of axially loaded members (e.g., trusses) and shafts. To reinforce the fundamentals and to give a sense of practicality, such applications follow immediately; for example, applications to trusses follows the study of simple extension. Of course, this order need not be followed; in a course for students of mechanical engineering, for example, the segment on trusses might readily be omitted.

Many illustrative examples are given; each application follows a common theme, that is, the essential roles of *statics*, *kinematics*, and *behavior of materials*. A conscious effort has been made to avoid special formulas for special problems.

The analysis of beams is set apart in Chapter 6, since the theory requires specific approximations and differential equations and the solutions require specific integrations. Chapter 7 presents the notion of superposition and applications to combined loadings. It serves to demonstrate the utility of the prior solutions and to reinforce the earlier concepts.

Chapter 8 presents the basic methods of work and energy. Here, the most fundamental principle of virtual work and the principles of stationary and minimum potential are couched in general terms, that is, they are not restricted to small deformations or quadratic forms. Only subsequent applications of Castigliano's theorem are subject to such limitations. This chapter provides a basis for those who will pursue courses in the modeling of structures and machines via finite elements.

Though Chapter 8 addresses the energy criteria for stability of equilibrium, the treatment is not essential to Chapter 9's coverage of stability. The latter chapter emphasizes the phenomena and method of determining critical loads but also treats columns as an instructive and practical application.

PWS Publishing Company

Acknowledgments

The author owes a great debt of gratitude to the coauthors of an earlier work, from which *Mechanics of Solids* was created. Additionally, much is owed to those who have long encouraged our approach: Maciej P. Bieniek, who offered his advice and support; the late Robert W. Shreeves; and Wilton W. King.

Thanks also to Christopher Pionke from the University of Tennessee at Knoxville for his diligent checking of problems and solutions for the text and Instructor's Manual and to Linda Tucker for her patience and care in the typing of the entire manuscript. I also wish to thank the following reviewers:

Royce Beckett
Auburn University

M. P. Bieniek
Columbia University

William L. Bingham
North Carolina State University

Art Boresi
University of Wyoming

John Dickerson
University of South Carolina

Stephen M. Heinrich
Marquette University

Robert Miller
University of Illinois at Urbana—Champaign

Panayiotis Papadopoulos
University of California—Berkeley

J. Lyell Sanders, Jr.
Harvard University

Gerald Wempner

1 Introduction

1.1 Raison d'Être

Every structure or machine is designed for a specific purpose. A bridge, for example, serves to carry traffic over an obstacle. Various occurrences might impair the ability of the structure to fulfill its intended purpose: The bridge might sag or sway excessively or even collapse. The suspension bridge across the Tacoma Narrows in Tacoma, Washington, was dramatically destroyed by uncontrollable oscillations induced by winds. When a structure can no longer serve its purpose, it has failed. Any change that prohibits the intended usage constitutes failure. In some instances the cause cannot be anticipated, for example an automobile being struck by a meteorite. In most cases the conditions are foreseeable and the designer, engineer, or architect must be able to predict the response of the proposed system before it is fabricated and placed in use. This ability is founded upon the mechanics of deformable solids, which, in turn, rests upon statics or dynamics, kinematics (a study of deformation), and the physical properties of the material.

1.2 Brief Historical Remark

Attempts to investigate and explain the rupture of a solid can be traced to Galileo's inquiry[1] (1638) into the strength of a horizontal beam that is fixed at one end and subject to transverse load, such as its weight. By historical accounts[2] Galileo did not consider the deformation, that is, the changes in geometry, that invariably accompany

---

1 Galileo Galilei. *Discorsi e dimostrazioni matematiche.* Leiden, 1638.

2 Todhunter, I., and Pearson, K. *A History of the Theory of Elasticity and of the Strength of Materials*, Vol. 1. New York: Cambridge University Press, 1886. Reprint. Mineola, N.Y.: Dover Publications, 1960.

the loading. Subsequent investigators recognized that internal forces, which might cause fracture, were intimately related to deformation. Indeed, an accurate characterization of localized deformation (strain) preceded a similar characterization of internal force (stress), or actions between neighboring elements of a body. Hooke[3] (1678) implied the proportionality of the extension and force upon an elastic body in an anagram, *ceiiinosssttuu* (*ut tensio sic vis*). The linear relationship between stress and strain is a foundation of linear (hookean) elasticity. The interested reader can find a concise historical account of our subject in *The Mathematical Theory of Elasticity*[4] by A. E. H. Love and an extensive account in *A History of the Theory of Elasticity and the Strength of Materials* by Todhunter and Pearson.[5]

In retrospect, we know that the kinematics of deformation are common to all continuous solids, whether elastic or inelastic. That aspect of our subject is entirely geometrical. For that reason, most treatises on this subject address deformation, questions of geometrical changes (kinematics), prior to other considerations. Our treatment is, however, an introduction for those who have some familiarity with concepts of force and equilibrium. Consequently, these aspects are addressed first (Chapter 2) and geometrical concepts second (Chapter 3). We note that essential concepts of internal force (stress) and deformation (strain) are also independent of the material properties. Both studies (stress and strain) are prerequisite to an exposition of the behavior of material (Chapter 4).

1.3 Introductory Example

Our subject rests on the foundations of statics (Chapter 2), kinematics (Chapter 3), and material properties (Chapter 4). To appreciate the need for each and the roles that they play, let us consider the plane truss depicted in Fig. 1.1(a). Each of the members is supposedly joined at the ends by ideal pins. Each is a "two-force" member; two opposing forces are exerted on each member by the pins at their ends. The members are pinned to fixed surfaces at B, C, and D; the known static load W is applied at A, coaxial with member AD.

This simple structure might fail in various ways, but let us address the question of determining the force upon each member. (An excessive force might cause the rupture of a member.) Since the system is at rest, *equilibrium* is a consideration. From the free-body diagram of Fig. 1.1(b), we have the two equations of equilibrium.

$$\sum F_y = F_D + \frac{F_B}{\sqrt{2}} - W = 0 \tag{1.1a}$$

$$\sum F_x = -F_C - \frac{F_B}{\sqrt{2}} = 0 \tag{1.1b}$$

3 Hooke, Robert. *De potentia restitutiva*. London, 1678.

4 Love, A. E. H., *A Treatise on the Mathematical Theory of Elasticity*. New York: Cambridge University Press, 1927. Reprint. Mineola, N.Y.: Dover Publications, 1944.

5 Todhunter, I., and Pearson, K. *History of the Theory of Elasticity*.

Fig. 1.1

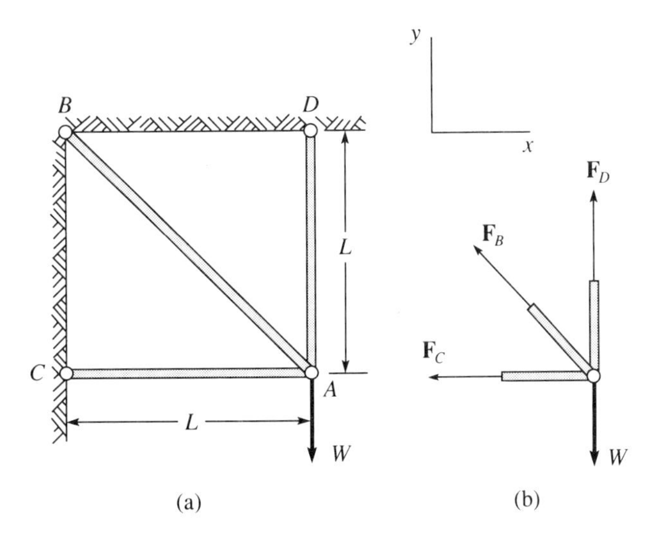

In these equations, we have our first lesson in the mechanics of deformable bodies (structures, machines, etc.). These two equations are both necessary and sufficient for equilibrium, but there are three unknowns (F_B, F_C, and F_D). *Statics alone is not sufficient.*

To obtain any meaningful solution we must consider the *deformation* of the structure. All real bodies deform under the application of force. In the present case the joint A must displace when force W is applied. Because the truss is symmetrical about the xy plane, the motion occurs in that plane. Therefore, the two components u and v in Fig. 1.2 describe the displacement of joint A, which moves to point A^*. The reader may think of the members as springs, which either stretch or contract as joint A displaces. We denote the elongation of a member by the letter e; our solution may yield a negative value, a contraction. It is now a *geometric* exercise to express the elongation of each bar in terms of the two components (u, v). By the Pythagorean theorem,

Fig. 1.2

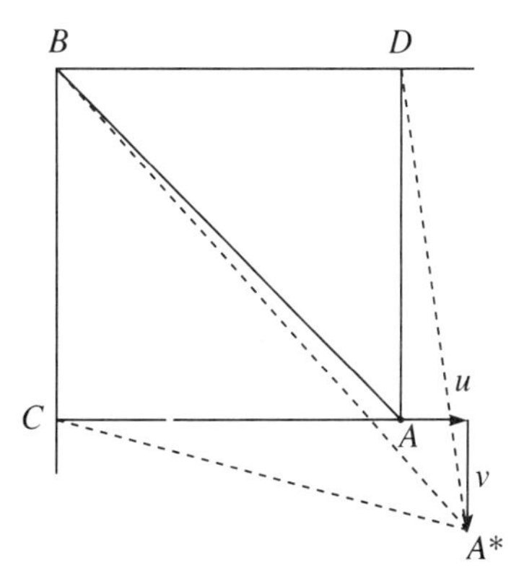

$$
\begin{aligned}
e_D &\equiv DA^* - DA \\
&= \sqrt{(L+v)^2 + u^2} - L \\
&= L\sqrt{(1+v/L)^2 + (u/L)^2} - L && \text{(1.2a)} \\
e_B &= L\sqrt{(1+v/L)^2 + (1+u/L)^2} - \sqrt{2}L && \text{(1.2b)} \\
e_C &= L\sqrt{(v/L)^2 + (1+u/L)^2} - L && \text{(1.2c)}
\end{aligned}
$$

The reader may wonder what has been gained by introducing more unknowns (u, v, e_D, e_B, and e_C) and three more equations. Indeed, the solution requires yet another essential consideration: The behavior of the individual members—but, more importantly, the *behavior of the material(s)*—must be introduced. Clearly, the relation between the force upon a member and the elongation depends upon the composition. If the materials were rubber, the elongations would be greater than if the materials were steel. Let us assume that the members are composed of a hookean material, such that the elongation is proportional to the applied force. Suppose that they are made of structural steel, which is nearly hookean if the forces are not excessive. Then,

$$F_D = K_D e_D \tag{1.3a}$$

$$F_B = K_B e_B \tag{1.3b}$$

$$F_C = K_C e_C \tag{1.3c}$$

Here the constants (K_D, K_B, K_C) depend upon the *properties* of the steel and also the size of each bar. In Chapter 5 the behavior of an individual bar and then trusses are considered in greater detail. The constants are then determined. Presently it is important to realize that the behavior of the material is embodied in these equations (1.3a, b, c).

Equilibrium, *kinematics* of deformation, and *behavior of the material* are individually accommodated by Eq. (1.1a, b), (1.2a, b, c), and (1.3a, b, c), respectively. These constitute a system of eight equations in eight unknowns ($F_D, F_B, F_C, e_D, e_B, e_C, u$, and v). Solution of this mathematical problem would be difficult because of the geometrically nonlinear equations (1.2a, b, c). Fortunately, we are concerned with a practical solution of an engineering problem, which admits realistic approximations. Indeed, we tacitly introduced approximations in Eq. (1.1a, b), since we neglected deformation when we imposed those conditions of equilibrium. In reality the imposition of load W causes deformation, as depicted in Fig. 1.2, *but* these steel bars deform very little ($u/L \ll 1$, $v/L \ll 1$). It is consistent now to make similar approximations in Eq. (1.2a, b, c). Each of these equations has a similar form. Consider Eq. (1.2a):

$$e_D = L\left[1 + 2\frac{v}{L} + \left(\frac{v}{L}\right)^2 + \left(\frac{u}{L}\right)^2\right]^{1/2} - L$$

Since $v/L \ll 1$ and $u/L \ll 1$, we can neglect the products and use the approximation

$$e_D \doteq L\left[1 + 2\frac{v}{L}\right]^{1/2} - L$$

By the binomial theorem,

$$(1 + \varepsilon)^n = 1 + n\varepsilon + \tfrac{1}{2}n(n - 1)\varepsilon^2 + \cdots \tag{1.4}$$

Again, the higher powers of a small number ($\varepsilon \ll 1$) can be neglected. Here, $\varepsilon = 2v/L$ and $n = \frac{1}{2}$, so that we have the first-order (linear) approximation

$$e_D \doteq v \tag{1.5a}$$

Likewise,

$$e_B \doteq \frac{u}{\sqrt{2}} + \frac{v}{\sqrt{2}}, \qquad e_C \doteq u \tag{1.5b, c}$$

Approximations such as (1.5a, b, c) are often justifiably employed because the relative displacements are small. It is first-order, or linear, in the small quantities. Now our problem is governed by a linear system of *approximations*, Eq. (1.1a, b), (1.3a, b, c), and (1.5a, b, c). In this case the solution is simple because these equations are not all coupled. Equations (1.5a, b, c) can be substituted into (1.3a, b, c), which can then be substituted into Eq. (1.1a, b) to obtain

$$(K_D + \tfrac{1}{2}K_B)v + \tfrac{1}{2}K_B u = W \tag{1.6a}$$

$$\tfrac{1}{2}K_B v + (K_C + \tfrac{1}{2}K_B)u = 0 \tag{1.6b}$$

The solution depends on the stiffnesses (K_D, K_B, and K_C), which depend upon a property of the steel and sizes of the bars, but, in any case, the *small* displacements (u and v) are proportional to the load (W). This is a consequence of the linear approximations of the geometrical equations (1.2a, b, c), consistent approximations in the equilibrium equations (1.1a, b), and the linear (hookean) material.

It is instructive to obtain the approximations in a direct, though less rigorous, manner. Consider a blowup of the view at the joint A, as depicted in Fig. 1.3. Because the displacement is relatively small, the rotation of member BA is nearly imperceptible. Indeed, a very small rotation about B does not stretch the bar. Taken in succession, the small rotation carries A to A' (nearly perpendicular to BA) and the elongation carries A' to A^* (nearly parallel to BA); hence, the first-order approximation of e_B is the projection of the displacement upon the line BA^* (parallel to BA in the first-order approximation), namely, Eq. (1.6b).

Fig. 1.3

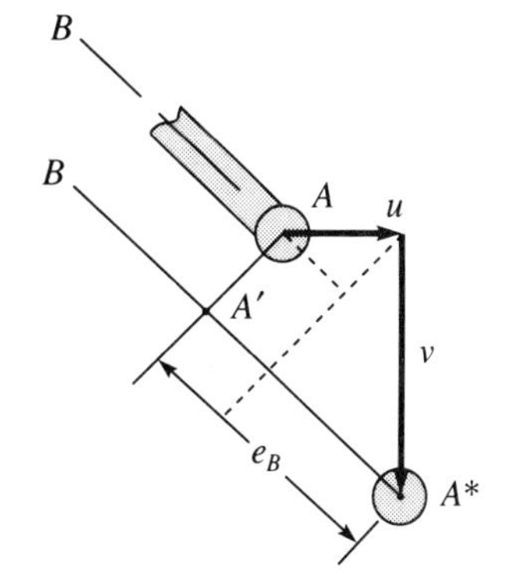

The foregoing example serves to illustrate essential aspects of our subject. Reviewing them, we see that internal and external forces are related by the conditions of *equilibrium* [Eq. (1.1a, b)], that elongations (changes of geometry) are related to displacement by *geometric* requirements [Eq. (1.2a, b, c)], and that the elongations are related to internal forces by the *resistive properties of the material* [Eq. (1.3a, b, c)]. These three aspects of our subject are distinct, but each enters into the solution. They can be described generally as follows.

Kinematics (Geometrical Changes)

A member of a structure or machine is subject to many external agencies. These include forces, changes of temperature, exposure to corrosive environments, and also the passage of time. The member may respond in various ways, but it invariably exhibits some changes in size and shape, for example, stretching, twisting, or bending.

A change in the shape or size of a structure or any part is called a *deformation.* Any study of deformation is an exercise in geometry, and any quantitative description requires only geometrical terms. These geometric aspects are called the *kinematics* of deformation.

Dynamics

When external forces or temperature changes are applied to a member, they create internal forces, which are governed by Newton's laws of motion or equivalent dynamic principles. Such considerations involve only forces, masses, and accelerations. These aspects of a solution constitute the *dynamics*, or *statics.*

Resistive Behavior

When members are subjected to forces or other environmental actions, they resist deformation to a varying degree that depends upon many factors. Size and shape are certainly important, but certain physical attributes of the material must also be considered. For example, a steel bar and a rubber bar of the same size and shape respond quite differently. Those properties that characterize only the material, independently of size or shape, describe the *resistive behavior*. Equations that describe the behavior of the material in a cause-effect manner are sometimes called *constitutive relations.*

The foundations of our subject, *kinematics*, *dynamics*, and *resistive behavior*, are examined separately in the next three chapters.

1.4 Distribution of Forces

In the mechanics of *ideal rigid* bodies, it is possible to consider only certain attributes of a distribution of force: If the resultant is a force, then only the line of action—and not the point of application—is needed. If the resultant is a couple, then the site of that action plays no role in the motion of the rigid body. In the mechanics of deformable bodies, we *cannot* take such liberties with the description of forces, external or internal. The following examples help to dispel any preconceived notions.

EXAMPLE 1

Deformable blocks are depicted in Fig. 1.4. Imagine two identical blocks of rubber bonded to the rigid base. In Fig. 1.4(a), a very concentrated force acts upon the left side; in Fig. 1.4(b), the same force acts along the same line, *but* it acts upon the right side. The difference in the deformations is evident; internal forces are also different. ◆

Fig. 1.4

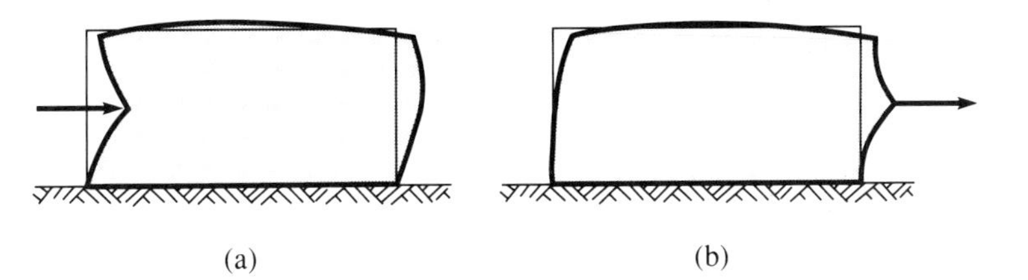

EXAMPLE 2

The identical blocks in Fig. 1.5 are subject to the same resultant, a force W that acts along the centerline. In the case of Fig. 1.5(a) the force is applied upon a very rigid plate, which distributes that force over the upper surface. In the case of Fig. 1.5(c) the

Fig. 1.5

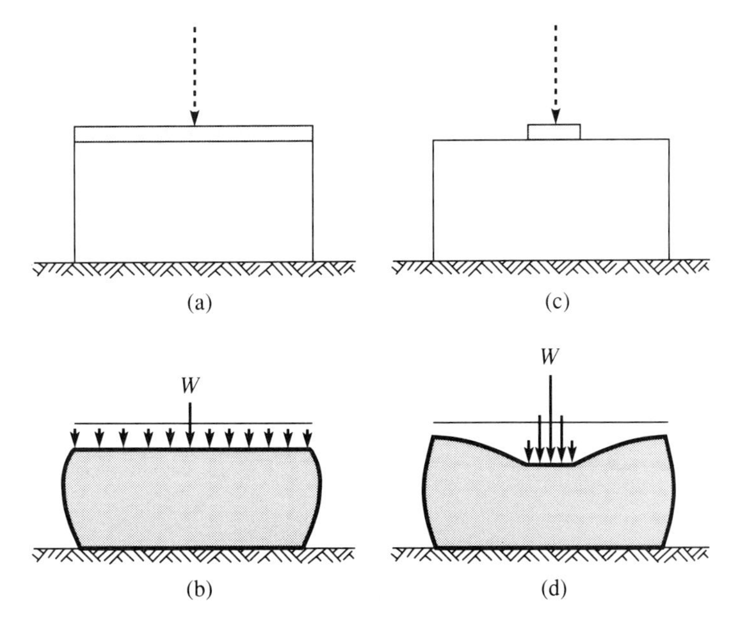

force is applied upon a very small plate, which then imposes a concentrated force on the block. Again, the difference in the deformations is evident. ◆

As a general rule, one *cannot* alter the distribution of actual loading upon a deformable body. In our subsequent studies, we examine certain circumstances and effects that can be treated with imprecise descriptions of the loading (Sec. 5.2). Otherwise, abide by the rule! Do not alter the loads!

1.5 Requisite Mathematics

The mechanics of deformable bodies requires the mathematics common to classical mechanics. Specifically, the reader must be fluent in geometry, trigonometry, and the rudiments of vector analysis. Additionally, the kinematics of deformation often involves quantities that are relatively small, as the displacements of the joint A in the truss of Figs. 1.1 and 1.2, where $u/L \ll 1$ and $v/L \ll 1$. Then good approximations and much simplification are possible but only with careful attention to the errors entailed.

Next we recount some useful mathematical relations and concepts.

Trigonometric Formulas

Some of the analyses of kinematics and statics are facilitated by trigonometric relations. Among these are the *Pythagorean identities*:

$$\cos^2\theta + \sin^2\theta = 1 \tag{1.7a}$$

$$1 + \tan^2\theta = \sec^2\theta \tag{1.7b}$$

$$1 + \cot^2\theta = \csc^2\theta \tag{1.7c}$$

Trigonometric relations for the sum of angles $(\alpha + \beta)$ are

$$\sin(\alpha + \beta) = \sin\alpha\cos\beta + \cos\alpha\sin\beta \tag{1.8a}$$

$$\cos(\alpha + \beta) = \cos\alpha\cos\beta - \sin\alpha\sin\beta \tag{1.8b}$$

$$\tan(\alpha + \beta) = \frac{\tan\alpha + \tan\beta}{1 - \tan\alpha\tan\beta} \tag{1.8c}$$

Finally, if $\alpha = \beta = \theta$, we obtain

$$\sin 2\theta = 2\sin\theta\cos\theta \tag{1.9a}$$

$$\cos 2\theta = \cos^2\theta - \sin^2\theta \tag{1.9b}$$

$$\tan 2\theta = \frac{2\tan\theta}{1 - \tan^2\theta} \tag{1.9c}$$

Directed Line Segments

A directed line segment is a line connecting an initial point P to a final point Q. It is denoted by PQ. Each directed line segment determines a rectangular parallelopiped whose edges are parallel to the coordinate axes, as shown in Fig. 1.6. The coordinates of point P and Q are (x_1, y_1, z_1) and (x_2, y_2, z_2), respectively. The differences $(x_2 - x_1)$, $(y_2 - y_1)$, and $(z_2 - z_1)$ are the projections PA, PB, and PC in Fig. 1.6.

Fig. 1.6

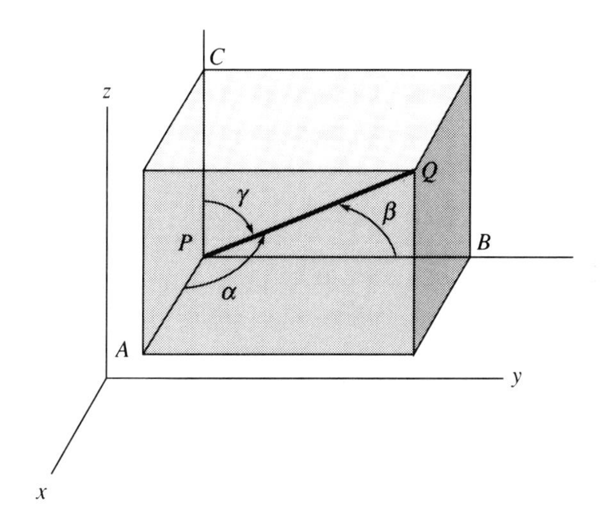

The length of PQ, denoted by L, is

$$L = \sqrt{(x_2 - x_1)^2 + (y_2 - y_1)^2 + (z_2 - z_1)^2} \tag{1.10}$$

The direction of PQ is specified by the angles PQ makes with the positive directions of the x-, y-, and z-axes. These angles are denoted by α, β, and γ, respectively, and are called *direction angles*. Sometimes it is more convenient to deal with the cosines of these angles. To compute $\cos\alpha$, for example, note that PAQ is a right triangle with the right angle at A. Thus,

$$\cos\alpha = \frac{x_2 - x_1}{L}$$

Similarly,

$$\cos\beta = \frac{y_2 - y_1}{L}, \qquad \cos\gamma = \frac{z_2 - z_1}{L}$$

Now

$$\cos^2\alpha + \cos^2\beta + \cos^2\gamma = \frac{(x_2 - x_1)^2 + (y_2 - y_1)^2 + (z_2 - z_1)^2}{L^2} = 1 \tag{1.11}$$

If $\cos\alpha = l$, $\cos\beta = m$, and $\cos\gamma = n$, Eq. (1.11) becomes

$$l^2 + m^2 + n^2 = 1 \tag{1.12}$$

The angle between two directed line segments is found as follows: PQ and PR are two directed segments, as shown in Fig. 1.7. Together, they form a triangle QPR. The angle θ and the lengths of the sides of this triangle are related by the law of cosines. Thus,

$$\cos\theta = \frac{L_2^2 + L_1^2 - L_3^2}{2L_1L_2}$$

which reduces to

$$\cos\theta = \frac{(x_2 - x_1)(x_3 - x_1) + (y_2 - y_1)(y_3 - y_1) + (z_2 - z_1)(z_3 - z_1)}{L_1L_2} \tag{1.13}$$

Fig. 1.7

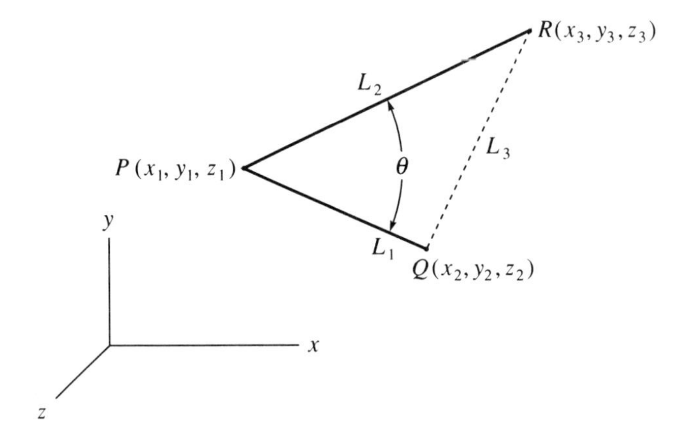

If l_1, m_1, and n_1 are the direction cosines of PQ and l_2, m_2, and n_2 are the direction cosines of PR, then Eq. (1.13) becomes

$$\cos\theta = l_1l_2 + m_1m_2 + n_1n_2 \tag{1.14}$$

If $\hat{\mathbf{q}}$ and $\hat{\mathbf{r}}$ denote *unit* vectors in the directions of PQ and PR and $\hat{\mathbf{i}}$, $\hat{\mathbf{j}}$, and $\hat{\mathbf{k}}$ are the *unit* vectors in the directions of the x-, y-, and z-axis, respectively, then

$$\hat{\mathbf{q}} = l_1\hat{\mathbf{i}} + m_1\hat{\mathbf{j}} + n_1\hat{\mathbf{k}}$$
$$\hat{\mathbf{r}} = l_2\hat{\mathbf{i}} + m_2\hat{\mathbf{j}} + n_2\hat{\mathbf{k}}$$

The scalar product is again Eq. (1.14):

$$\hat{\mathbf{q}}\cdot\hat{\mathbf{r}} = \cos\theta = l_1l_2 + m_1m_2 + n_1n_2$$

Vectors

Many of the physical quantities in mechanics can be represented by vectors; these include displacements, forces, moments, and couples. They can be represented by directed line segments and added according to the parallelogram law. The vector is depicted by a line segment with an arrow to indicate the sense. The length of the line segment represents the magnitude. In writing, the vector is denoted by a boldface letter, such as **a**.

The parallelogram law for the addition of two vectors **a** and **b** is illustrated in Fig. 1.8. The sum of the vectors is commutative

$$\mathbf{c} = \mathbf{a} + \mathbf{b} = \mathbf{b} + \mathbf{a}$$

and associative

$$\mathbf{a} + (\mathbf{b} + \mathbf{c}) = (\mathbf{a} + \mathbf{b}) + \mathbf{c} = \mathbf{a} + \mathbf{b} + \mathbf{c}$$

The negative of a vector is one of the same magnitude but opposite direction; thus

$$\mathbf{a} + (-\mathbf{a}) = 0$$

Subtraction of two vectors is defined as

$$\mathbf{a} - \mathbf{b} = \mathbf{a} + (-\mathbf{b})$$

Fig. 1.8

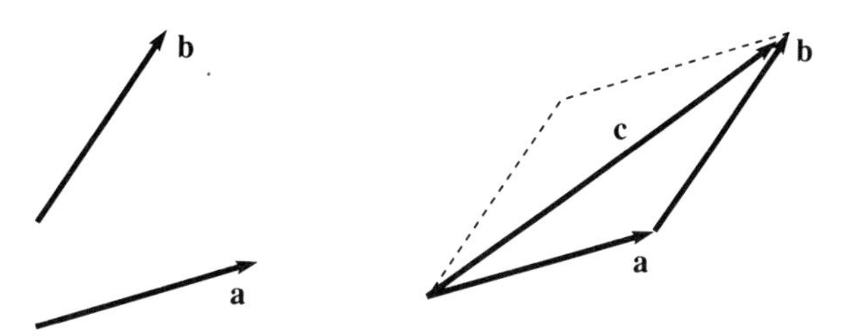

It is usually convenient to refer a vector to a coordinate system. In Fig. 1.9, the directed line segment PQ represents vector **V**. The magnitude of **V** is denoted by $|\mathbf{V}|$, or simply by V. The components of **V** in the xyz-coordinate system are its projections on the three coordinate axes. Thus,

$$V_x = V\cos\alpha, \qquad V_y = V\cos\beta, \qquad V_z = V\cos\gamma \tag{1.15}$$

Also

$$V^2 = V_x^2 + V_y^2 + V_z^2 \tag{1.16}$$

which determines the magnitude of a vector when its components are known. To find the direction cosines from the components, we have, from Eq. (1.15),

$$\cos\alpha = \frac{V_x}{V}, \qquad \cos\beta = \frac{V_y}{V}, \qquad \cos\gamma = \frac{V_z}{V} \tag{1.17}$$

Fig. 1.9

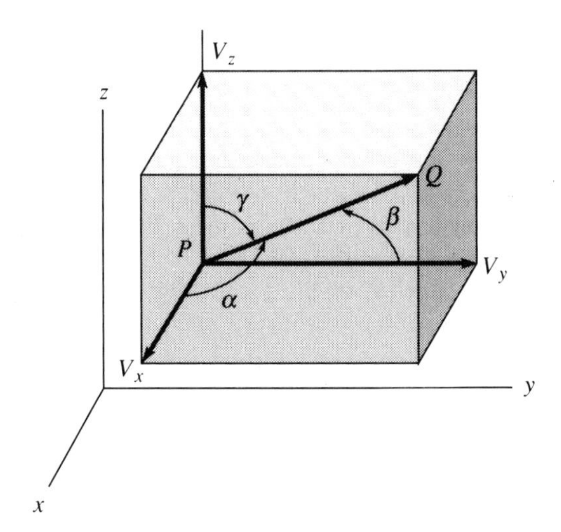

Let **U** be any other vector. The sum of **U** and **V** is the vector **W** = **U** + **V**. The components of **W** are

$$W_x = U_x + V_x, \qquad W_y = U_y + V_y, \qquad W_z = U_z + V_z \tag{1.18}$$

where the addition is now of the algebraic variety. This greatly simplifies the process of vector addition.

A vector is completely defined by its components. *This is not necessarily true for the physical quantity that the vector might represent.* As an example, the effects of a force on a deformable body depend upon the point of application, as demonstrated in Fig. 1.4. When specifying vectorial quantities, such as forces, we must provide such additional information as needed to determine the effects uniquely.

Approximation of Small Quantities

In the mechanics of deformable bodies, we frequently encounter quantities that are *relatively* small. The displacement of the joint in the truss of Figs. 1.1 and 1.2 is typical; there the displacements u and v are small relative to the length L. Often the important consequence is the smallness of rotation; for instance, the rotations of the members in a truss are imperceptible without careful measurements. Indeed, with little forethought we tend to examine the equilibrium with disregard for these rotations. Of course, real engineering problems do arise wherein the deformations and rotations are not inconsequential; rubberlike mountings, membranes, and thin flexible bodies demand our attention to geometric changes. However, the vast majority of engineering structures and machines entail only small changes in geometry. Then approximations are admissible. These approximations always hinge on a comparison of terms and the neglect of those that are relatively small. To illustrate, let us return to the small rotation θ. Usually the functions of the quantity can be expanded in a power series; for example,

$$\sin\theta = \theta - \frac{\theta^3}{3!} + \frac{\theta^5}{5!} - \cdots$$

For small values of θ, that is, $\theta \ll 1$,

$$\sin\theta \doteq \theta$$

In many structural problems a rotation $\theta = 0.1$ (6°) is large; the error is about 0.2%. Similarly, for small values of θ

$$\cos\theta = 1 - \frac{\theta^2}{2!} + \frac{\theta^4}{4!} - \cdots \doteq 1$$

The student is reminded that *mathematical analyses* require that angles be given in radians (arc/radius). Recall also the expansion by the binomial theorem [Eq. (1.4)].

Some specific examples follow ($x \ll 1$):

$$(1 + x)^n \doteq 1 + nx$$
$$(1 + x)^2 \doteq 1 + 2x$$
$$\sqrt{1 + x} \doteq 1 + \tfrac{1}{2}x$$
$$(1 + x)^{-1} \doteq 1 - x$$
$$(1 + x)(1 + y) \doteq 1 + x + y, \qquad (x \ll 1, y \ll 1)$$

Accuracy

For the sake of brevity, only nominal values are given in most problems. For example, a shaft might be described as one with a length of 3 ft and a diameter of 1 in. In actual practice, the length of a shaft might be held to tolerances of ± 0.01 in. and the diameter, to tolerances of ± 0.001 in. In such applications one has accuracies of four decimal places. The length and thickness of a structural steel beam are less precisely specified, probably ± 0.04 and ± 0.01 in., respectively. Properties of materials are typically measured to accuracies of three decimal places. In short, the accuracy of the data is limited by the particular application, capabilities and costs of machining and/or forming parts, and, ultimately, by cost-effective means of measurement and control. The student of engineering must appreciate the practical significance of numerical values as they relate to the specific problem.

Often one loses accuracy in computation. For example, a wire is known to have a length $L = 2$ m, ± 0.1 cm; the accuracy is $\pm 0.05\%$. This wire is subjected to an axial load, which causes it to elongate to a deformed length of $L^* = 200.4$ cm; again, the ability to measure is limited by instrumentation to ± 0.1 cm. Suppose that the important result is the change in length:

$$\Delta L = L^* - L = 200.4 - 200.0 = 0.4 \text{ cm}$$

Both measurements are subject to error of ± 0.1 cm. Conceivably, the numerical value $\Delta L = 0.4$ cm could have an error of ± 0.2 cm, or 50%. Beware! In an era of electronic computation, we can easily punch 6, 8, or 10 keys; however, we must recognize the accuracy of our results, which are limited by practical considerations, as stated previously.

PROBLEMS

1.1 In a Cartesian coordinate system, a directed line segment goes from the point $P(1, 2, 4)$ to the point $Q(4, 7, -1)$. What is its length? What angle does PQ make with the x-axis?

1.2 A quadrilateral has its vertices at the points $A(1, 1, 1)$, $B(2, 3, 2)$, $C(0, 4, 1)$, $D(-2, 2, 0)$. Six lines connect the four vertices. Which is the longest of the six lines? Which is most nearly parallel to the y-axis?

1.3 In Prob. 1.2, compute the cosines of the angles between pairs of the three lines emanating from A.

1.4 Four points are given by their coordinates: $P(0, 0, 0)$, $A(3, 4, 7)$, $B(1, -2, 5)$, $Q(0, 0, 6)$. Which is the shorter path, $PABQ$ or $PBAQ$?

1.5 In Prob. 1.4, which of the four segments PA, PB, AQ, or BQ is most nearly parallel to PQ?

1.6 A triangle has vertices at the points $A(2, 1, 4)$, $B(6, -2, 1)$, $C(5, 2, 3)$. Compute the length of the triangle's perimeter and find the cosine of the angle between the sides BA and BC.

1.7 Let **a** and **b** be arbitrary vectors whose directions differ by the angle θ. Denote their sum by **c** and let ϕ be the angle between the directions of **c** and **b**. Then show that

$$c^2 = a^2 + b^2 + 2ab\cos\theta, \qquad \tan\phi = \frac{a\sin\theta}{b + a\cos\theta}$$

1.8 Let **u** and **v** be the vectors shown in Fig. P1.8, with $u = \sqrt{3}$ and $v = \sqrt{2}$. Find the magnitude and direction cosines of the sum $\mathbf{u} + \mathbf{v}$.

Fig. P1.8

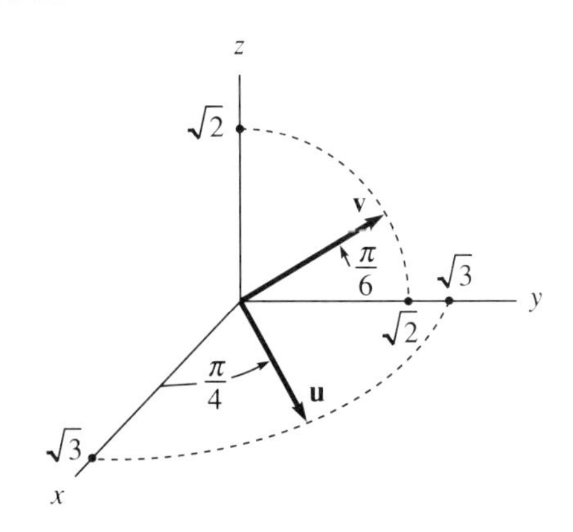

1.9 The vectors **a**, **b**, and **c** are specified by their Cartesian components: $\mathbf{a} = (3, -3, 2)$, $\mathbf{b} = (-1, 7, 4)$, $\mathbf{c} = (2, 1, 1)$. Find the magnitude and direction cosines of the sum $\mathbf{a} + \mathbf{b} + \mathbf{c}$.

1.10 A particle is acted upon by three forces; their magnitudes and direction numbers are

$\mathbf{F}_{\mathrm{I}}$	1 lb in the direction $(1, 3, 7)$
$\mathbf{F}_{\mathrm{II}}$	3 lb in the direction $(2, 1, -1)$
$\mathbf{F}_{\mathrm{III}}$	2 lb in the direction $(-1, 2, 2)$

Obtain the magnitude and direction cosines of the resultant force on the particle.

1.11 In the xy-rectangular coordinate system, vector **a** has components (2,2). What are the components of **a** with respect to the $x'y'$ rectangular coordinate system shown in Fig. P1.11? What are the components of **a** with respect to the $x''y''$ rectangular coordinate system?

Fig. P1.11

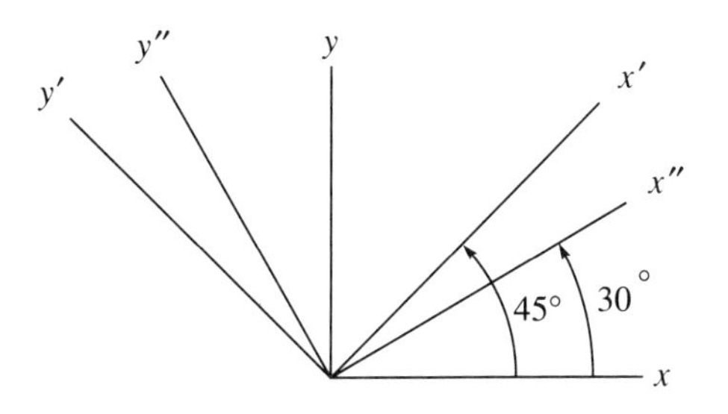

1.12 The vector **V** lies in the xy-plane shown in Fig. P1.12 and has components V_x and V_y. At the arbitrary point P defined by the polar coordinates (r, θ), a radial-tangential coordinate system is constructed. Show that the components of **V** in this coordinate system are given by

$$V_r = V_x \cos\theta + V_y \sin\theta$$
$$V_\theta = -V_x \sin\theta + V_y \cos\theta$$

Fig. P1.12

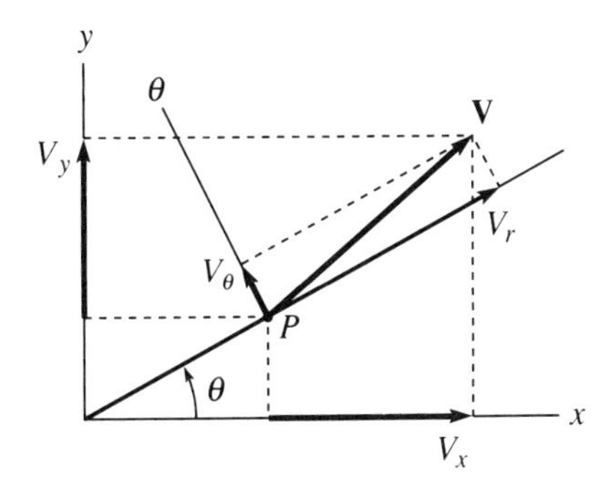

1.13 The unit vectors **u**, **v**, and **w** are perpendicular to each other. In the xyz coordinate system, **u** has the components $(1/\sqrt{3}, 1/\sqrt{3}, 1/\sqrt{3})$. If the vector **v** is parallel to the yz-plane, what is a possible set of direction cosines for **w**?

1.14 The unit vectors **u**, **v**, and **w** are perpendicular to each other. In the xyz-coordinate system, **u** has the direction cosines $(1/\sqrt{5}, 2/\sqrt{5}, 0)$ and **v** has the direction cosines $(2/\sqrt{30}, -1/\sqrt{30}, 5/\sqrt{30})$. What is a possible set of direction cosines for **w**?

1.15 A vector **U** has components U_x, U_y, U_z with respect to the xyz coordinate system shown in Fig. P1.15. At the arbitrary point P defined by the spherical polar coordinates r, ϕ, θ, another coordinate system is constructed with axes tangent to the lines of increasing r, ϕ, θ, respectively. Show that the components of **U** in this coordinate system are

$$U_r = U_x \cos\theta \sin\phi + U_y \sin\theta \sin\phi + U_z \cos\phi$$
$$U_\phi = U_x \cos\theta \cos\phi + U_y \sin\theta \cos\phi - U_z \sin\phi$$
$$U_\theta = -U_x \sin\theta + U_y \cos\theta$$

Fig. P1.15

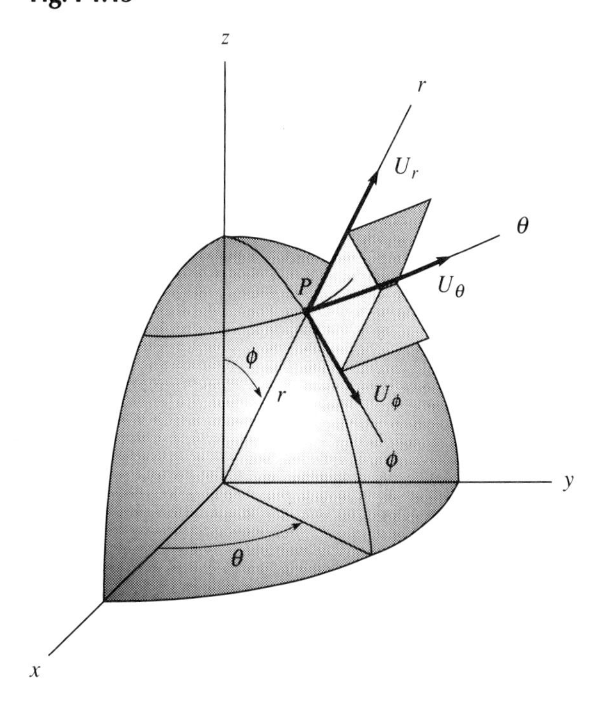

1.16 Treat θ, γ, and ε as small quantities. Find first-order approximations for the following expressions:

a. $\sin(a + \theta)$

b. $\dfrac{1}{k + \varepsilon}$

c. $\sqrt{(1 + \varepsilon)^2 + (1 - \theta)^2}$

d. $\dfrac{(1 + 2\varepsilon)^2}{1 - \gamma}$

e. $\dfrac{(3 + 2\theta)(1 - 2\theta)}{2 + 4\gamma}$

f. $\cos^2\left(\dfrac{\pi}{4} - \theta\right)$

g. $\dfrac{1}{(K+\varepsilon)^{3/2}}$

h. $\dfrac{\sin(\alpha+\gamma)}{\cos(\alpha-\gamma)}$

i. $\tan(\phi+\theta)\cot(\phi-\theta)$

j. $\sqrt{\sin^2\psi+\cos^2(\psi+\varepsilon)}$

k. $[(1+\varepsilon)^p+\gamma]^q$

l. $\dfrac{(1+\varepsilon)^p+\gamma}{(1+\varepsilon)^q-\gamma}$

1.17 For the triangle shown in Fig. P1.17, the right angle at C is increased by the small angle γ. If the lengths of a and b remain unchanged, compute the approximate increase in length of side c.

Fig. P1.17

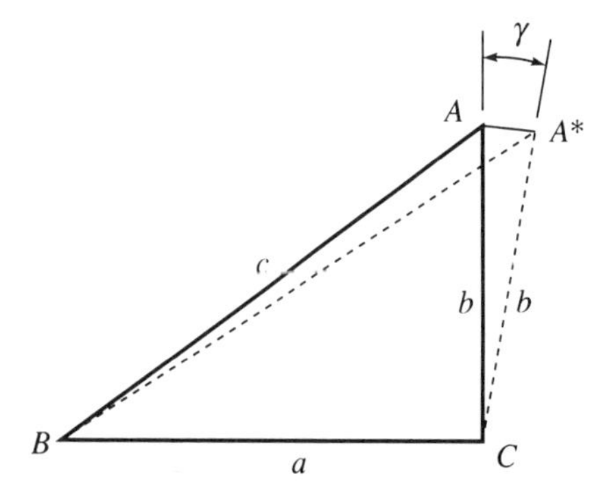

1.18. In Prob. 1.17, only side b is unchanged, but the angle at C is increased the small amount γ and the length of a is increased the small amount ε. Compute the approximate increase in length of side c.

1.19 A square has sides of length l. If the square is distorted into a rhombus (no change in length of the sides) by changing the right angles by the small amount γ, compute the approximate changes in the lengths of the diagonals. What is the first-order approximation to the change in area?

1.20 A rectangle has sides with lengths a and b. If these are increased by the small amounts Δa and Δb, respectively, what is the approximate length of a diagonal of the new rectangle? Compute the change in angle between the diagonal and the side originally of length a.

1.21 A line segment connects the points $P(x_1, y_1)$ and $Q(x_2, y_2)$. If P is held fixed but Q is moved by the amount $(\Delta x, \Delta y)$, compute the approximate change in slope of PQ.

1.22 A gauge for measuring curvature is shown in Fig. P1.22. The two outer points are fixed; the center contact is movable, and its displacement is recorded by a dial. The gauge is zeroed on a perfectly flat surface and then placed on the surface of a sphere with the gauge normal to the surface. If the dial indicates that the center contact was moved 0.0005 in. (from the flat reading), what is the approximate radius of the sphere?

Fig. P.1.22

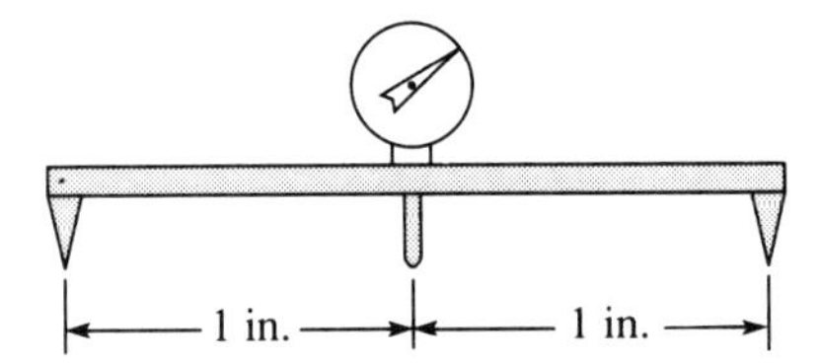

1.23 Two strings, AO and BO, are tied together as indicated in Fig. P1.23. Under a force at O, string AO elongates 0.1 in. and BO stretches 0.15 in. What are the approximate vertical and horizontal displacements of point O?

Fig. P.1.23

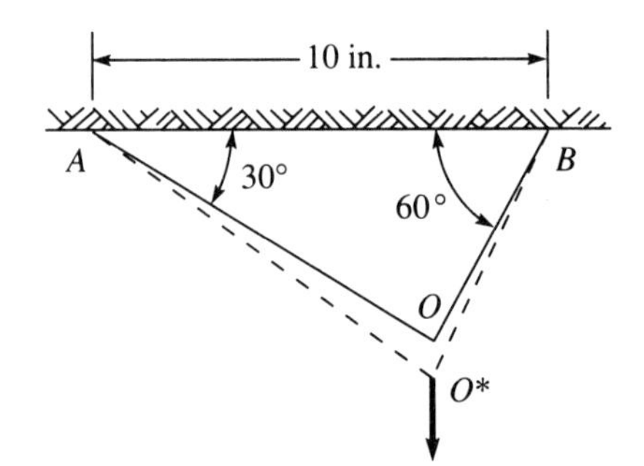

2 Force, Stress, and Equilibrium

In our introductory chapter, we identified three distinct facets in the analysis of deformable bodies: dynamics, kinematics, and behavior of the material(s). To lay a firm foundation and to avoid misconceptions, the dynamic and kinematic aspects are treated independently—the former in the present chapter, the latter in the next chapter. Both are prerequisite to a description of the behavior of a material.

Here the treatment of dynamics is confined to the Newtonian concepts, methods, and equations that enforce equilibrium. The reader has encountered the basic notions in previous studies of particles and "rigid" bodies via courses in physics and engineering. These ideas include the concepts of force, moment, and couple, the use of the free-body diagram, and the equations of equilibrium. Now, some extensions are needed to treat the mechanics of a *deformable* body.

Because each elemental portion of a body is subject to different actions and undergoes different deformations, it is necessary to examine the actions upon internal surfaces. Hence, we pass surfaces *through* components of machines and structures and isolate the separate parts as free bodies. Those parts are subject to forces that are internal to the entire component. Such *internal* forces require our special attention.

To describe accurately the effects of forces upon elements of *arbitrary* size, another concept is needed. That concept is the concept of *stress*, which serves to define the intensity and the nature of force upon surfaces of arbitrary size and orientation and, specifically, surfaces of infinitesimal size. The definitions and analyses of *stress* and the *components* of stress are *essential* to our subsequent treatment of the behavior of materials and, ultimately, to our analysis of each deformable body.

2.1 Classification of External Forces

Body and Surface Forces

The external forces that act on a body may be classified in different ways. For example, since every force is ultimately the action of one body on another, the force can arise in one of two ways. In one case the force results from physical contact between two bodies. Thus, a book applies a force to a desk on which it rests. If the book is raised so that there is no physical contact between book and desk, then the force vanishes. Such forces are called *contact forces*. In the other case, one body exerts forces on another body even though there is no contact between them. The gravitational force called weight, which the earth applies to both the book and desk of the previous example, is such a force. The electromagnetic force that keeps a compass needle pointing north-south is another example. To distinguish these forces from contact forces, we call them *actions at a distance*.

Since contact is not an essential element of actions at a distance, it follows that all parts of the body are involved; that is, the force is distributed throughout the body. Therefore, actions at a distance are usually referred to as *body forces*. Contact forces, however, can act only on those portions of a body that are in direct contact with the body exerting the force; that is, the force is distributed over a portion of the surface of the body. Hence, contact forces are often called *surface forces*. In many cases a surface force is distributed over a relatively small region compared with the other dimensions of the body. It is sometimes possible to idealize such forces and treat them as if they acted at a point. Such an idealization is called a *concentrated force*. Also, if a surface force is distributed over a relatively narrow region, it might be idealized as a *line force*—that is, a force distributed along a line. Concentrated and line forces do not exist in reality, but are convenient idealizations. For instance, in some cases it might be possible to regard the weight of a truck on a road as a concentrated force, while in other cases the actual pressure distribution under the truck tires must be considered. The forces which a floor joist exerts on the floor it supports can generally be treated as a line force.

We depict a concentrated force with an arrow in the same way as we usually represent a vectorial quantity. The length is proportional to the magnitude; the direction is indicated by the arrowhead. The point of application of a concentrated force may be indicated by the point to *or* from which the arrow is drawn. The simplified sketches of Fig. 2.1 depict concentrated forces acting upon a plate (a), a beam (b), a cable (c), and a rod (d).

A line force is usually sketched as a line of arrows, as shown in the examples of Fig. 2.2. The intensity of the line force at any point is proportional to the length of the arrow to that point. The units of line forces are [force]/[length]. To find the total force, we multiply the intensity by a length. Thus, for example, the total force acting on the bar of Fig. 2.2(a) is wL, whereas the total force on the body of Fig. 2.2(b) is $\int_0^L w(x)\,dx$.

Body forces are sometimes represented by their resultant, although this practice is dangerous. For example, the weight of the cylinder shown in Fig. 2.3(a) is indicated by a single vector acting at the center of gravity. *This vector must not be confused with a concentrated force.* If we consider only half the cylinder, as shown in Fig. 2.3(b), the body force is only one-half the previous resultant. This, of course, would not be the case if the force W were actually a concentrated force.

Fig. 2.1

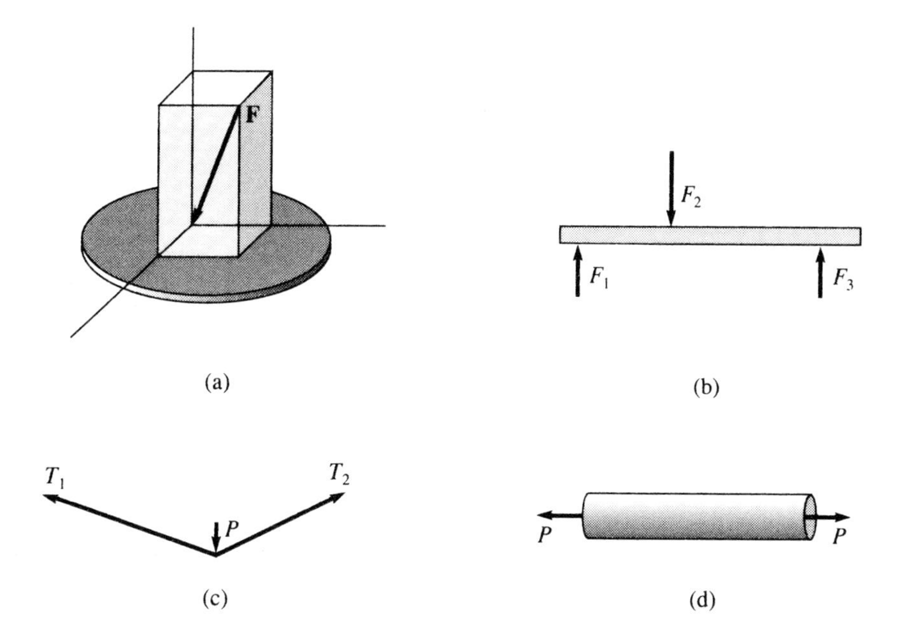

Fig. 2.2

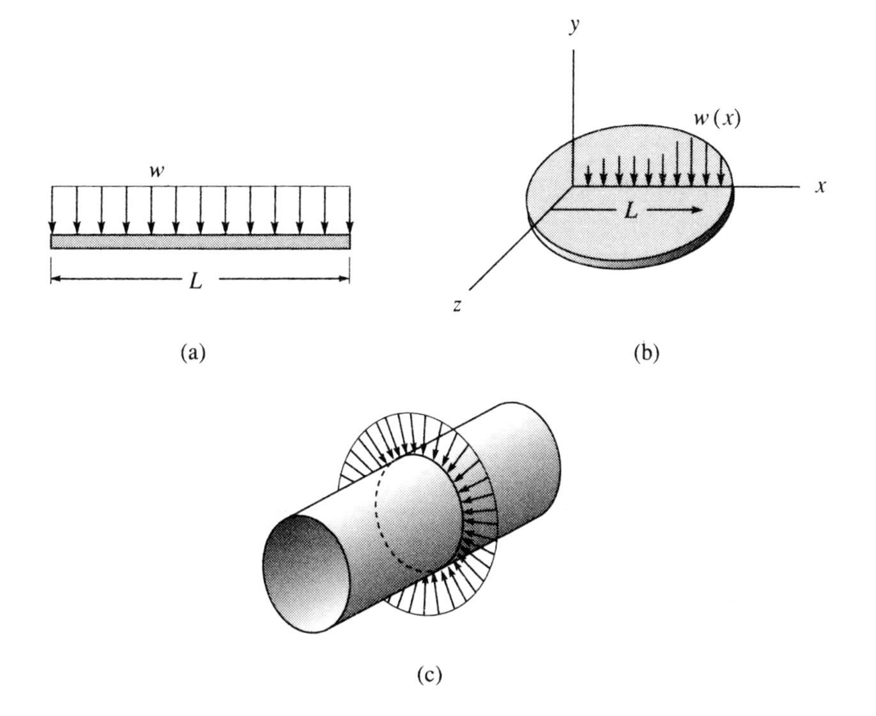

Fig. 2.3

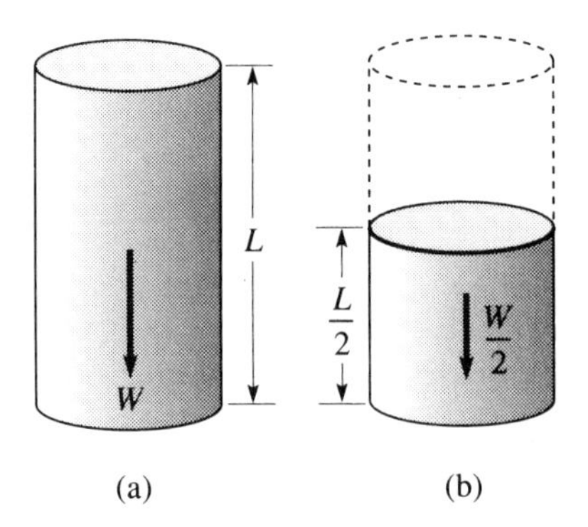

Loads and Reactions

Some of the external forces that act on a body arise from the connections that support the body. Let us return to the example of a book resting on a desk. The force that the desk exerts on the book is exactly enough to support it. If something else is placed on top of the book, the supporting force of the desk adjusts so that the book remains where it is. As another example, let us consider a door on hinges. The supporting forces, which the hinges exert on the door, will allow the door to open or close but will restrict any other motion. The essential element of supporting forces is that they arise in order to restrict the displacement of the body that they support. Such supporting forces are called *reactions.* Any other external forces acting on a body are called *loads.* Thus, the weight of a person on a floor is a load on the floor, but the supporting forces exerted by the floor joists are reactions.

The actual reactions (forces and/or couples) exerted by a support and also the restraints provided are often quite complicated. In many circumstances approximations, or idealizations, are admissible. Some simple idealizations are tabulated in Fig. 2.4, together with the motions, components of translation (→) and rotation (→→) that are or are not permitted, and the components of force (→) and couple (→→) that are provided. An inadmissible motion or action is indicated by a stroke (↛ or →↛). Notice that the restraint against a translation (↛) requires a force (→); a restraint against a rotation (→↛) requires a couple (→→). Conversely, if the action (force or couple) is absent, then the corresponding motion occurs.

The effect of a rocker or roller (4) is to restrain displacement in one direction while leaving the body free to move in the perpendicular direction and/or rotate. This reaction is always a force perpendicular to the direction of allowable motion. Such supports are common in bridge construction, since it is desirable to support the span while leaving it free to expand and contract with temperature changes.

The effect of a rocker or roller (4) is to restrain displacement in one direction while leaving the body free to move in the perpendicular direction and/or to rotate. This points of the span.

A *clamped*, or *fixed*, support (1) does not permit displacement or rotation. These supports are required for overhanging structures. The reaction at a clamped support consists, in general, of a force to restrict displacement and a couple to restrict rotation.

Fig. 2.4

Support	Depiction	Motion Translation–Rotation	Action Force–Couple
1. Fixed no translation, no rotation	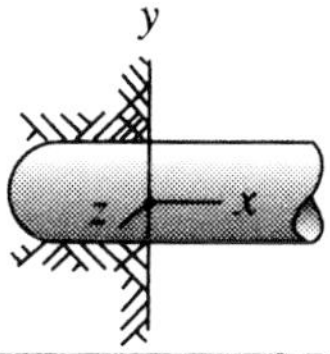	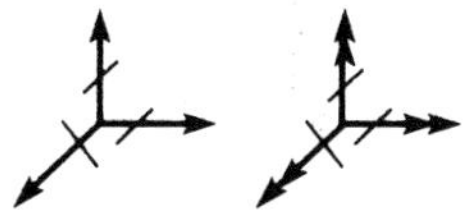	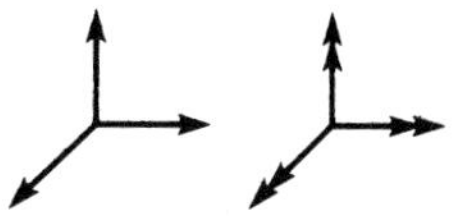
2. Ball and Socket no translation, no couple	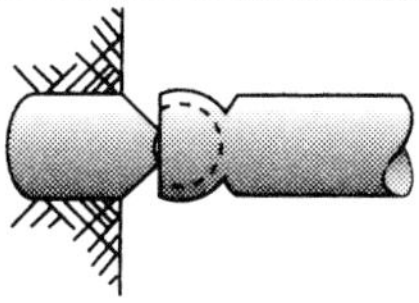	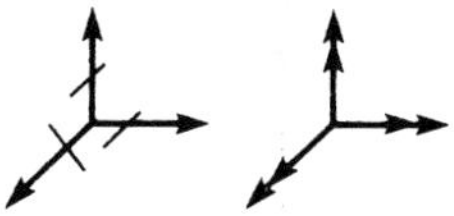	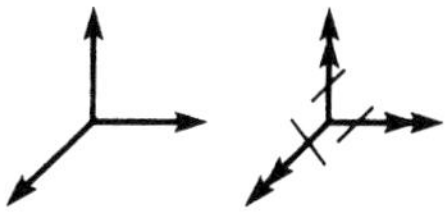
3. Pinned (2-D)		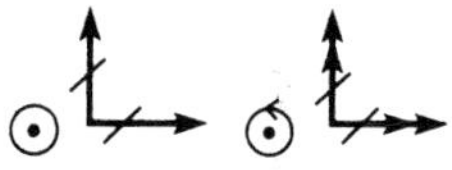	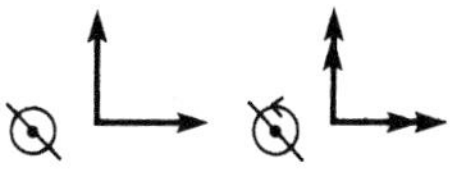
4. Simple (2-D)	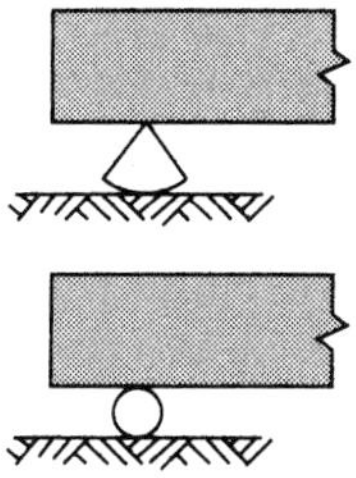	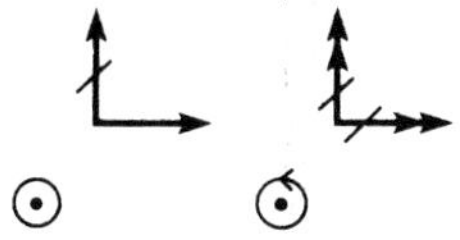	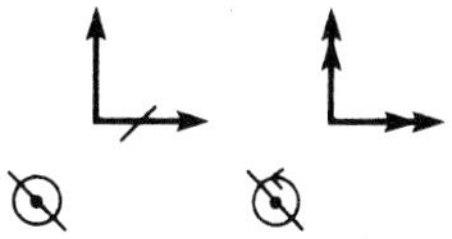
5. Free no force, no couple	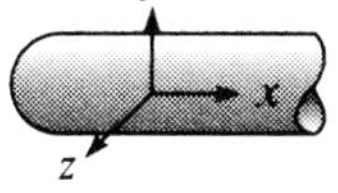	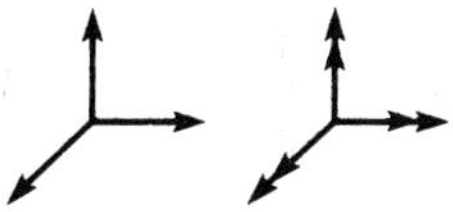	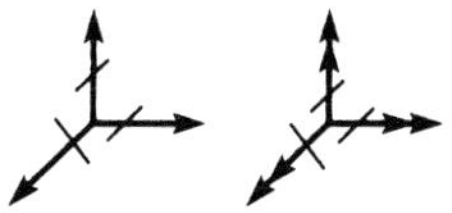

Static and Dynamic Loads

The effect of a load on a body might depend on how the load is applied. If a load is steady—that is, if the load does not change intensity, direction, or location with time—it is called a *static* load. Loads that change with time are called *dynamic* loads. Sometimes the changes in a dynamic load are so small or so gradual that the load may be treated as static. The weight of a building on its foundation as people enter and leave and the fluid pressure at any instant in the bottom of a slowly filling tank are usually taken as static loads. Dynamic loads may be further subdivided: A truck being driven over a bridge applies a *moving load* to the structure; a connecting rod in a reciprocating engine is subjected to *repeated loading*; a hammer striking the head of a nail applies an *impact load* to it.

The different classifications of external forces discussed in this section are very useful for descriptive purposes. The terms should, therefore, be treated as engineering definitions with specific meanings.

2.2 Equilibrium of Deformable Bodies

Newton's laws of motion deal with the motion of a particle that is acted upon by forces. The second law can be expressed by the (vector) equation

$$\sum \mathbf{F} = m\mathbf{a} \tag{2.1}$$

where $\sum \mathbf{F}$ represents the (vector) sum of all of the forces acting on a particle, m is the mass of the particle, and $\mathbf{a}$ is the (vector) acceleration of the particle with respect to a fixed coordinate system. The first law states that if no force acts on a particle, the acceleration of the particle is zero. The third law states that if one particle exerts a force on a second particle, the second exerts a force of the same magnitude but of opposite direction on the first.

If the sum of all the forces acting on a particle is zero; that is, if

$$\sum \mathbf{F} = \mathbf{0} \tag{2.2}$$

then the particle is said to be in equilibrium. We may think of a solid body as many particles held together in some way. If all the particles are in equilibrium, the body itself is said to be in equilibrium. The subsequent analyses are restricted to bodies in equilibrium, although this is not an essential requirement.[1]

Equilibrium of a rigid body is defined by the two (vector) equations

$$\sum \mathbf{F} = \mathbf{0} \tag{2.3}$$

$$\sum \mathbf{M} = \mathbf{0} \tag{2.4}$$

In Eqs. (2.3) and (2.4), $\sum \mathbf{F}$ represents the (vector) sum of all of the external forces acting on the body, and $\sum \mathbf{M}$ is the (vector) sum of the moments of all these forces about any point. For a deformable body in equilibrium, Eqs. (2.3) and (2.4) also hold, not only for the body as a whole but for any portion of it.

---

1 If necessary, the restriction can be removed at any point of the analysis by making use of D'Alembert's principle. For instance, if a particle is accelerating, we can define the so-called inertia force

$$\mathbf{F}' = -m\mathbf{a}$$

Equation (2.1) can then be written

$$\sum \mathbf{F} + \mathbf{F}' = \mathbf{0}$$

which is now in the same form as Eq. (2.2), the equation that defines equilibrium.

In any study of deformable bodies, it is necessary to examine the forces that act between adjacent parts of the body. With this in mind, consider the body depicted in Fig. 2.5. It is acted upon by the system of external forces $\mathbf{F}_1$, $\mathbf{F}_2$, $\mathbf{F}_3$, $\mathbf{F}_4$. We suppose that the body is in equilibrium under the action of this force system. Imagine the body to be cut by a plane that divides it into two parts, I and II, with the common surface *S*. Let us examine one of these parts, for instance portion II, and the forces that act on it. A free-body diagram of portion II is shown in Fig. 2.6. The forces $\mathbf{F}_3$ and $\mathbf{F}_4$ are not the only forces that act on it. In addition, we also have the forces that the material of portion I exert on the material in portion II. These forces are internal forces for the

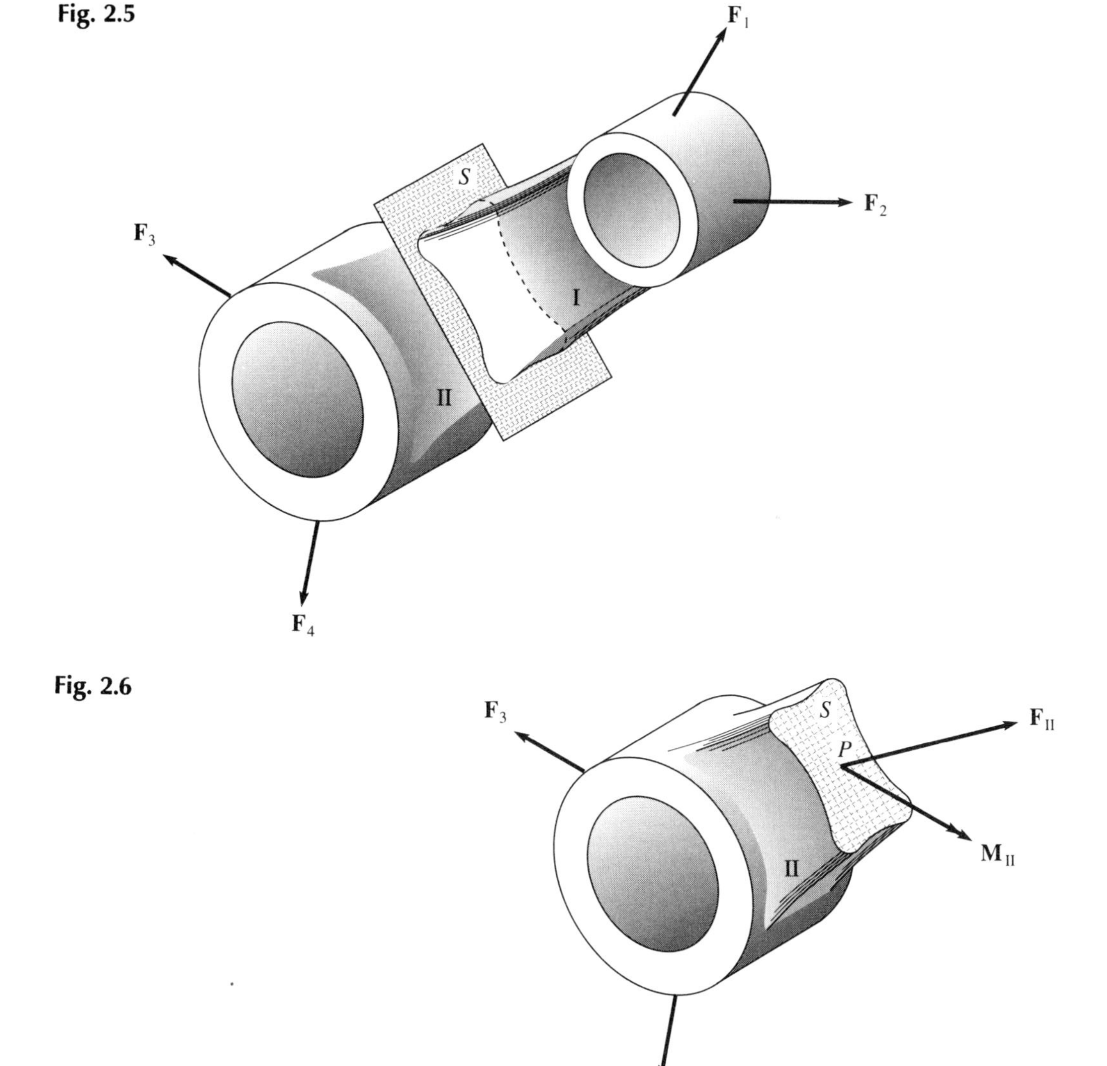

Fig. 2.5

Fig. 2.6

body as a whole but are external to portion II. Their distribution over the surface S is entirely unknown at present. We can, however, discuss the (vector) sum of these forces, which is denoted by $\mathbf{F}_{II}$, and the (vector) sum of the moments of these forces about a point P in S, which is denoted by $\mathbf{M}_{II}$. The force system consisting of the force $\mathbf{F}_{II}$ acting at P and the couple $\mathbf{M}_{II}$ is *statically equivalent* to the system of distributed forces on the surface S. This means that in any calculation involving *only* the equilibrium of portion II, the actual distribution of forces on S can be replaced by $\mathbf{F}_{II}$ acting at P and $\mathbf{M}_{II}$. We shall refer to $\mathbf{F}_{II}$ and $\mathbf{M}_{II}$ as the *net internal force system* on S.

Since portion II is in equilibrium, the force system consisting of $\mathbf{F}_3$, $\mathbf{F}_4$, $\mathbf{F}_{II}$, and $\mathbf{M}_{II}$ must satisfy the equilibrium conditions, Eqs. (2.3) and (2.4). From these equations, the net internal force system on S is

$$\mathbf{F}_{II} = -[\mathbf{F}_3 + \mathbf{F}_4], \qquad \mathbf{M}_{II} = -[\mathbf{M}_3 + \mathbf{M}_4]$$

where $\mathbf{M}_3$ and $\mathbf{M}_4$ are the moments of the forces $\mathbf{F}_3$ and $\mathbf{F}_4$ *about the point P*. The point P is the assumed point of application of the net internal force $\mathbf{F}_{II}$. It plays a passive but important role in the use of Eq. (2.4) that cannot be overlooked. If we change the location of P, we will also, in general, change the value of $\mathbf{M}_{II}$. Consequently, *we must always identify the point P* in such representations of the net internal system.

If we were to use a free-body diagram of portion I, as in Fig. 2.7, instead of portion II, then the net internal force system on S would be $\mathbf{F}_I$ and $\mathbf{M}_I$. As a direct consequence of Newton's third law,

$$\mathbf{F}_I = -\mathbf{F}_{II}, \qquad \mathbf{M}_I = -\mathbf{M}_{II}$$

We will, therefore, drop the subscript and use only $\mathbf{F}$ and $\mathbf{M}$ to represent the net internal force system on S.

Fig. 2.7

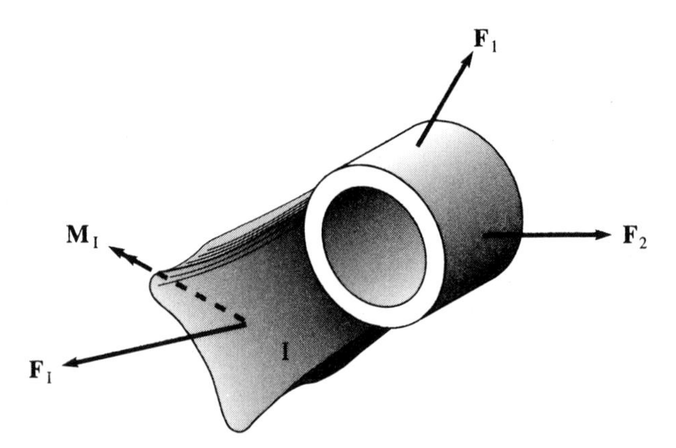

It is usually more convenient to use components of the net internal system rather than the net internal force and couple themselves. One component is taken normal to the plane of the sectioning surface, and the other component is taken tangent to the plane. This decomposition is shown in Fig. 2.8, where the n-axis is normal to the plane

Fig. 2.8

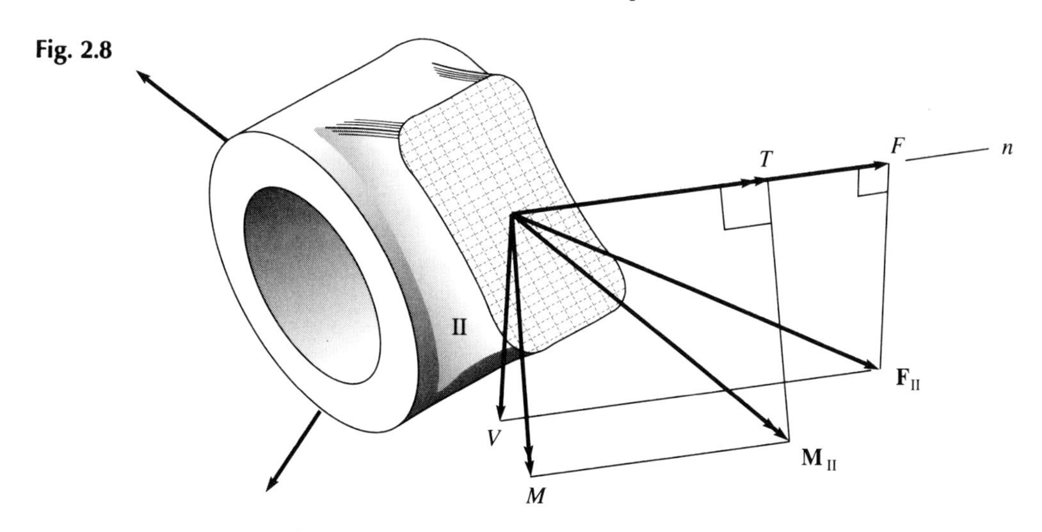

of S. The normal component of the net internal force, F, is called the *normal force* on the sectioning surface. The component V in the plane is called the *shear force*. The component of the net internal couple, T, perpendicular to the plane is called the *twisting couple*, and M, the component in the plane, is called the *bending couple*. The direction of the normal force F and twisting couple T are determined if the orientation of the sectioning plane is prescribed. However, the shear force V and bending couple M can have any direction in the plane of S. To be explicit, we can identify two directions tangent to surface S and decompose the shear force and bending couple into components in the chosen directions. Most conveniently, we identify orthogonal directions, s and t, tangent to the surface, as depicted in Fig. 2.9. Here, V_s and V_t are the components of V in the s- and t-directions. Similarly, M_s and M_t are the components of M. Forces and couples are shown separately in Fig. 2.9(a) and 2.9(b), respectively, simply to avoid confusion. These six components of the net internal force

Fig. 2.9

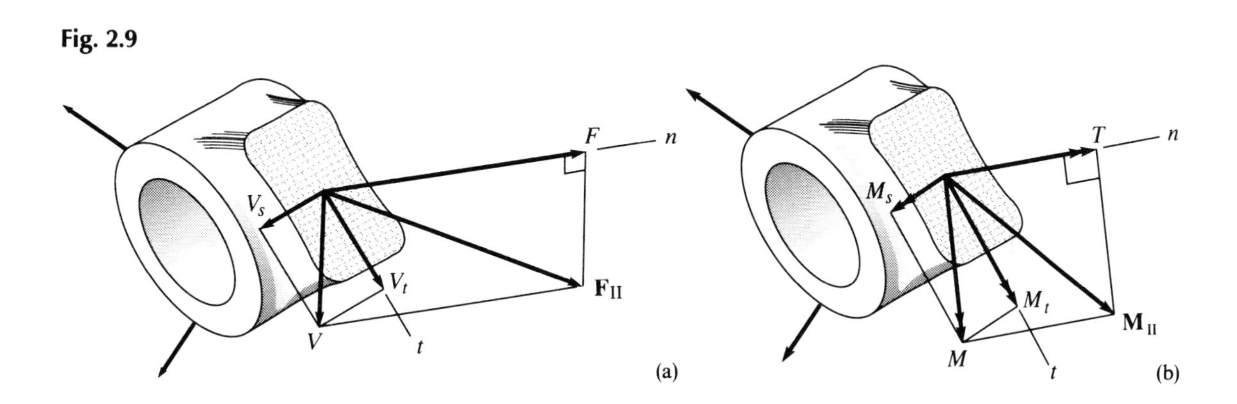

system can now be determined from the scalar form of the equilibrium conditions:

$$\begin{aligned} \sum F_s = 0, \quad & \sum F_t = 0, \quad \sum F_n = 0 \\ \sum M_s = 0, \quad & \sum M_t = 0, \quad \sum M_n = 0 \end{aligned} \tag{2.5}$$

Of course, if these equations are used, the external forces (in this case $\mathbf{F}_3$ and $\mathbf{F}_4$) must first be resolved into their components in the s-, t-, and n-directions.

A normal force on a cross section arises when there is a tendency on the part of the external forces to compress or pull apart the two portions of the body on either side of the sectioning plane. A shear force comes about when the external forces tend to slide one portion over the other. When the external forces tend to twist one portion with respect to the other, a twisting couple arises. If the two portions are bent by the external forces, a bending couple results. These four situations are depicted in Fig. 2.10, which depicts a cylinder stretched or compressed by a normal force F in (a), sheared by a force V in (b), twisted by a couple T in (c), and bent by a couple M in (d).

Fig. 2.10

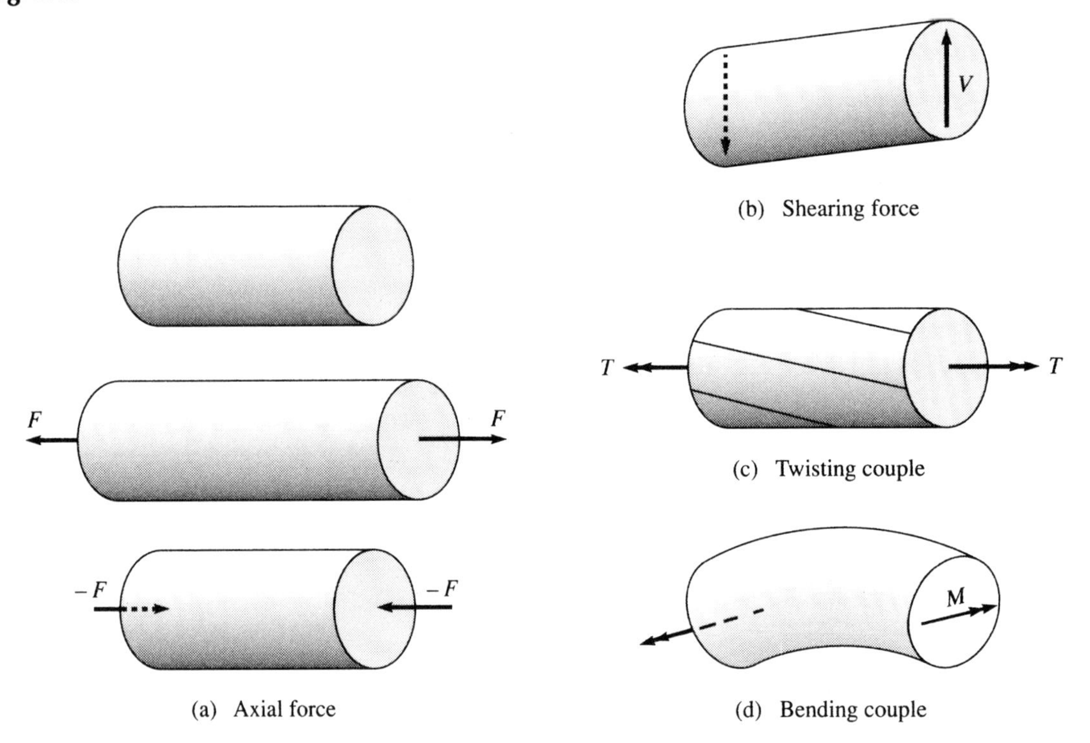

(a) Axial force (b) Shearing force (c) Twisting couple (d) Bending couple

To illustrate the application of these ideas, we compute the components of the net internal force system in the following examples.

EXAMPLE 1

Consider a rectangular bar, loaded as shown in Fig. 2.11(a). Let us determine the net internal force system on the plane at A that bisects the horizontal portion of the bar. In Fig. 2.11(b), the part of the bar to the left of the plane has been shown as a free body. Since none of the external forces acting on the bar have components perpendicular to the plane of the figure, the net internal force system has only three components: a normal force F, a shear force V, and a bending couple M. An xyz-coordinate system can be placed in the sectioning surface at A, as indicated in Fig. 2.11(b). Then the equilibrium equations for that portion of the bar are

$$\sum F_x = 0; \qquad F + 200 = 0, \qquad F = -200 \text{ lb} \quad [-900 \text{ N}]$$

$$\sum F_y = 0; \qquad V + 100 = 0, \qquad V = -100 \text{ lb} \quad [-450 \text{ N}]$$

$$\sum M_z = 0; \qquad M - (10)(100) = 0, \qquad M = 1000 \text{ in.-lb} \quad [113 \text{ N}\cdot\text{m}]$$

These three equations give the normal force, shear force, and bending couple on the cross section at A. The negative values for F and V merely indicate that the directions in which these forces act are opposite to the ones assumed in Fig. 2.11(b). ◆

Fig. 2.11

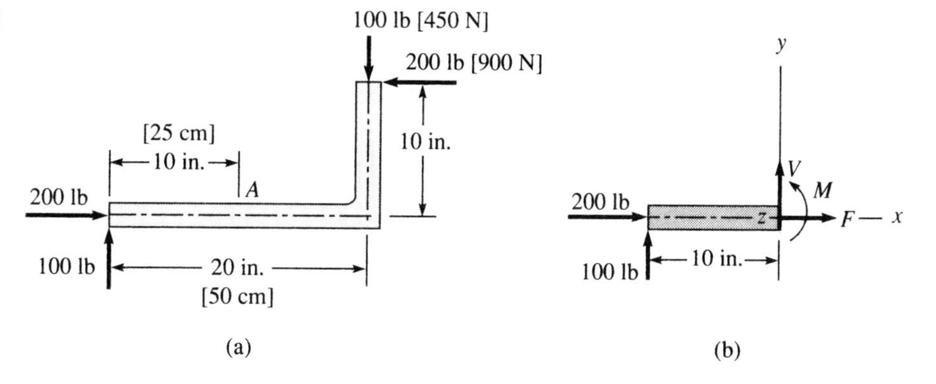

EXAMPLE 2

Consider a circular cylinder hanging under its own weight W, as shown in Fig. 2.12(a). The axis of the cylinder makes an angle θ with the vertical. Let us find the components of the net internal force system on a cross section one-third of the way from the free end of the cylinder.

In Fig. 2.12(b) a free-body diagram of the lower third of the cylinder is shown with an xyz-coordinate system indicated in the sectioning surface. The weight of this portion of the cylinder is $W/3$. Equilibrium equations yield

$$\sum F_y = 0; \qquad F - \frac{W}{3}\cos\theta = 0, \qquad F = \frac{W}{3}\cos\theta$$

$$\sum F_x = 0; \qquad V - \frac{W}{3}\sin\theta = 0, \qquad V = \frac{W}{3}\sin\theta$$

$$\sum M_z = 0; \qquad M - \frac{W}{3}\cdot\frac{L}{6}\sin\theta = 0, \qquad M = \frac{WL}{18}\sin\theta \ ◆$$

Fig. 2.12

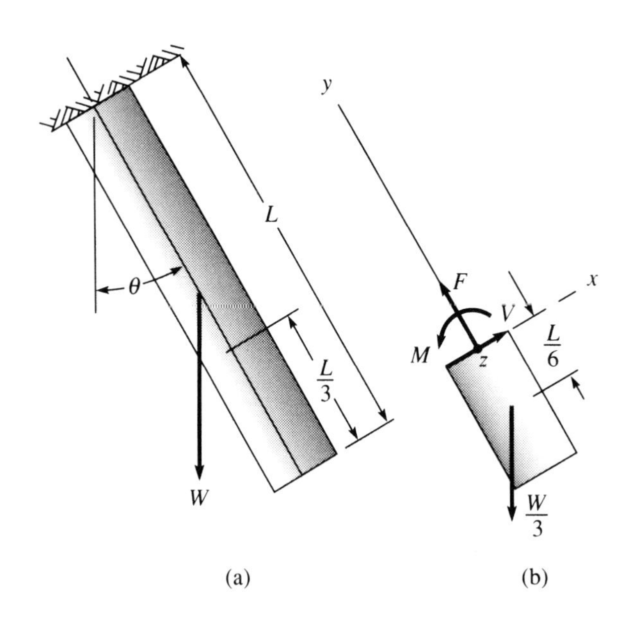

EXAMPLE 3

Let us determine the shear force and bending couple on the center cross section of the simply supported beam shown in Fig. 2.13(a). Regardless of which half of the beam we take as a free body, an unknown reactive force at A or B will appear. We therefore consider a free-body diagram of the entire beam first [Fig. 2.13(b)] to determine these forces:

$$\sum M_z = 0; \qquad 3R_B - (2.4)(2250) - (1.5)(150)(3) = 0$$

$$R_B = 2025 \text{ N} \quad [455 \text{ lb}]$$

$$\sum F_y = 0; \qquad R_A + R_B - 2250 - (3)(150) = 0$$

$$R_A = 675 \text{ N} \quad [150 \text{ lb}]$$

Fig. 2.13

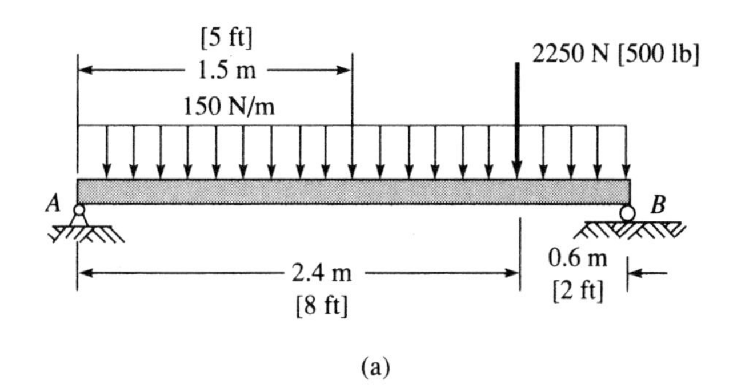

(a)

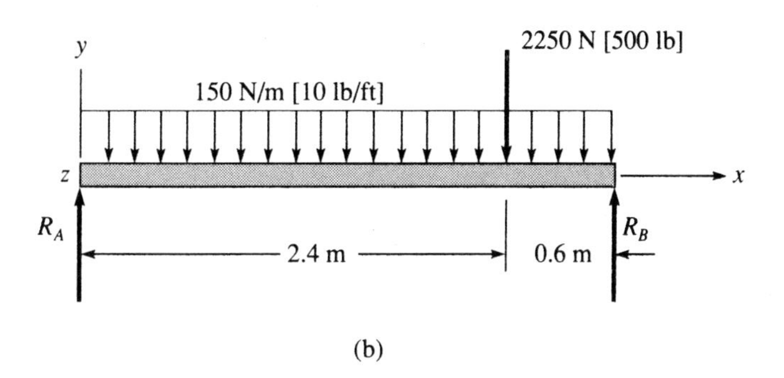

(b)

A free-body diagram of the left half of the beam is shown in Fig. 2.14, with the known value of R_A indicated. The equilibrium conditions for this portion enable us to determine V and M:

$$\sum F_y = 0; \qquad V + 675 - (1.5)(150) = 0,$$

$$V = -450 \text{ N} \quad [-100 \text{ lb}]$$

$$\sum M_{z'} = 0; \qquad M - (1.5)(675) + (0.75)(150)(1.5) = 0$$

$$M = 844 \text{ N}\cdot\text{m} \quad [622 \text{ ft-lb}] \quad \blacklozenge$$

Fig. 2.14

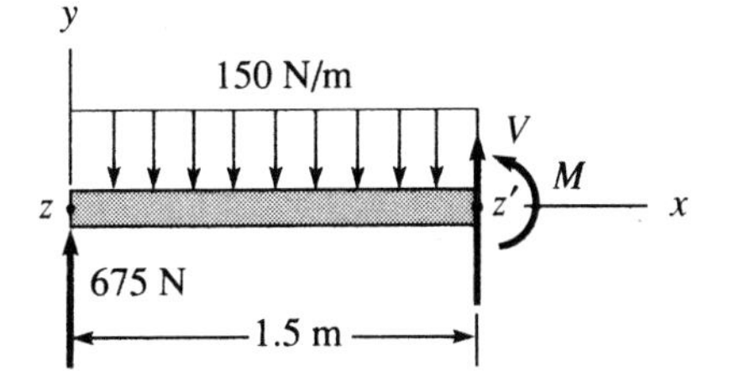

PROBLEMS

2.1 Two bars pinned at each end support a force P, as shown in Fig. P2.1. Find the components of the net internal force system on the cross sections at the midpoint of each bar.

Fig. P2.1

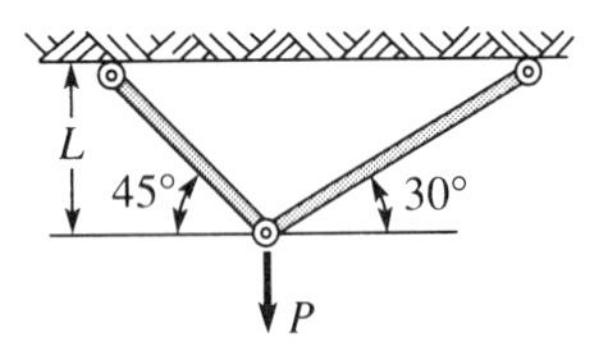

2.2 Calculate the normal force on cross sections of each member of the pin-connected truss shown in Fig. P2.2.

Fig. P2.2

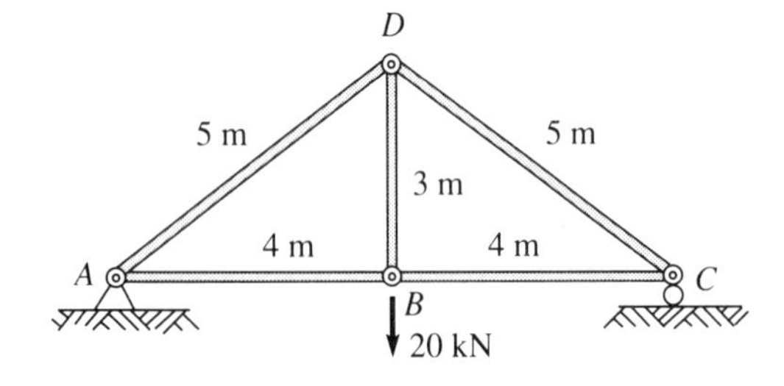

2.3 Each member of the truss shown in Fig. P2.3 is 4 ft long and is pinned at each end. Find the normal force on the cross sections of each member.

Fig. P2.3

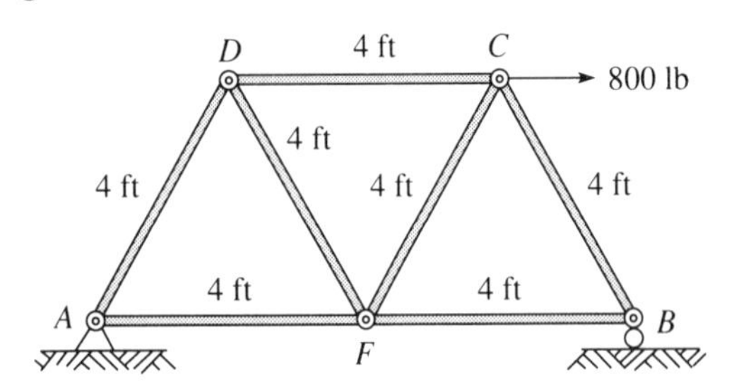

2.4 The cylindrical bar shown in Fig. P2.4 is subjected to an axial force P. Determine the normal force and shear force on a plane section whose normal makes an angle θ with the axis of the cylinder.

Fig. P2.4

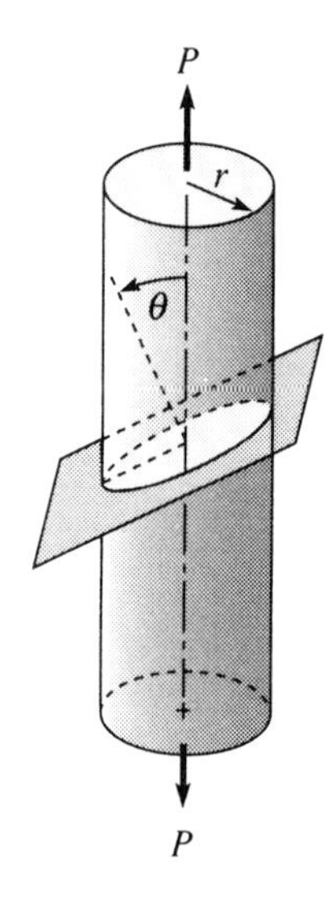

2.5 A simply supported beam is 8 ft long, with a 1000-lb load applied 3 ft from one end, as shown in Fig. P2.5. Find the shear force and bending couple on the cross sections 1 ft on either side of the load.

Fig. P2.5

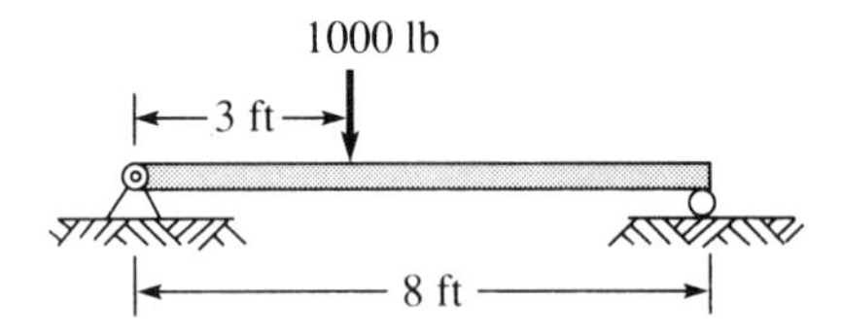

2.6 A beam 6 m long is supported by a pin at one end and a roller 4 m from that end. The beam supports a uniformly distributed load of 30 kN/m on the 2-m overhang. Find the components of the net internal force system on a cross section midway between the supports.

2.7 A bar 10 ft long, suspended by two wires, supports a load of 8000 lb at its midpoint, as shown in Fig. P2.7. What are the components of the net internal force system on a cross section midway between the load and either end of the bar?

Fig. P2.7

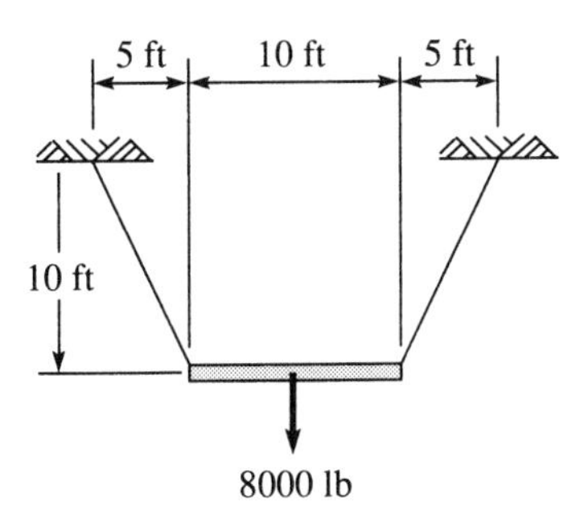

2.8 Members AB and AC are pinned at each end and member AB is loaded as shown in Fig. P2.8. Find the components of the net internal force system on the cross section at the midpoint of member AB.

Fig. P2.8

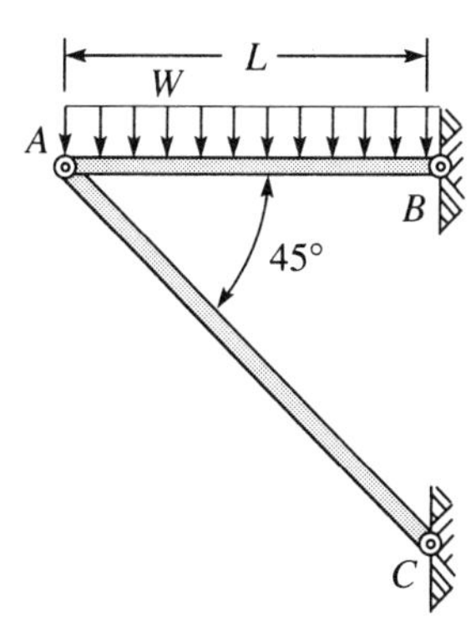

2.9 The bent bar shown in Fig. P2.9 is supported by a pin at B and a roller at A. Determine the normal force, shear force, and bending couple on the cross section at C.

Fig. P2.9

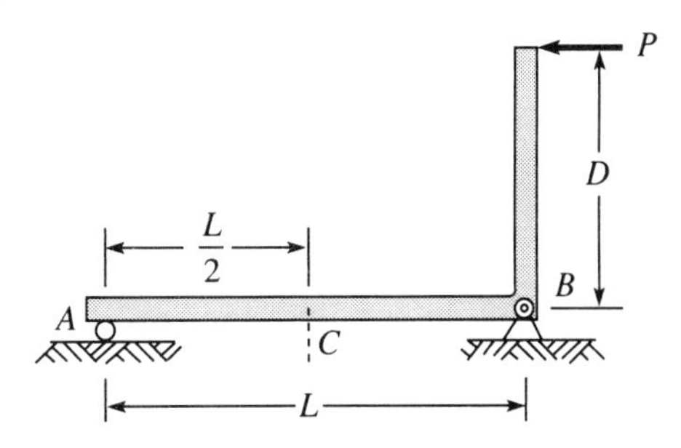

2.10 The bent bar shown in Fig. P2.10 is supported by a pin at C and a roller at D. Find the components of the net internal force system on planes A and B.

Fig. P2.10

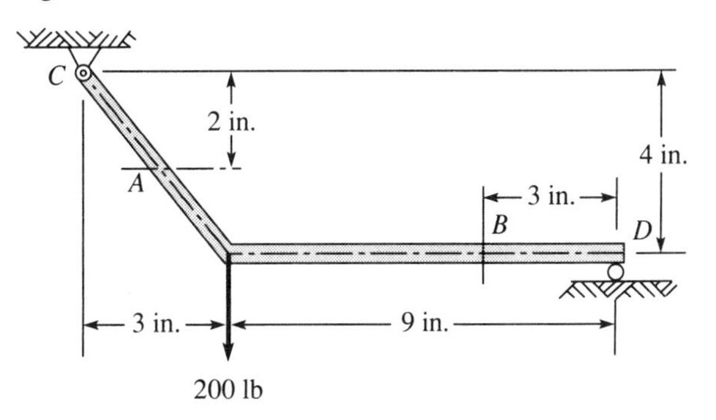

2.11 Repeat Prob. 2.10 with the 200-lb load shifted 4.5 in. to the right so that it acts at the midpoint of the horizontal position.

2.12 Compute the normal force, shear force, and bending couple on the cross section at the center of BC for the bent bar shown in Fig. P2.12.

Fig. P2.12

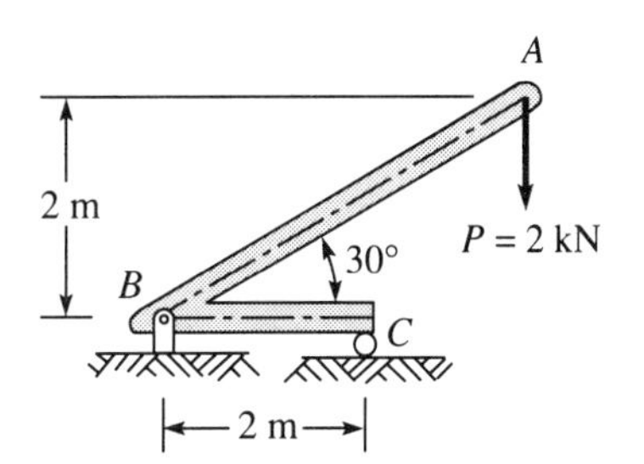

2.13 The bent bar shown in Fig. P2.13 is supported by a pin at A and a wire at B. Compute the components of the net internal force system on the cross section at N. Give your answer in customary and SI units.

Fig. P2.13

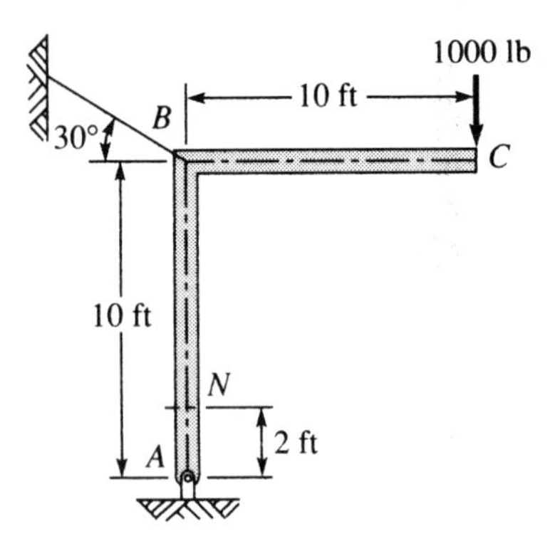

2.14 The arm shown in Fig. P2.14, firmly attached at its base, supports a load of 1200 lb. Find the normal force, shear force, and bending couple on the cross section at A.

Fig. P2.14

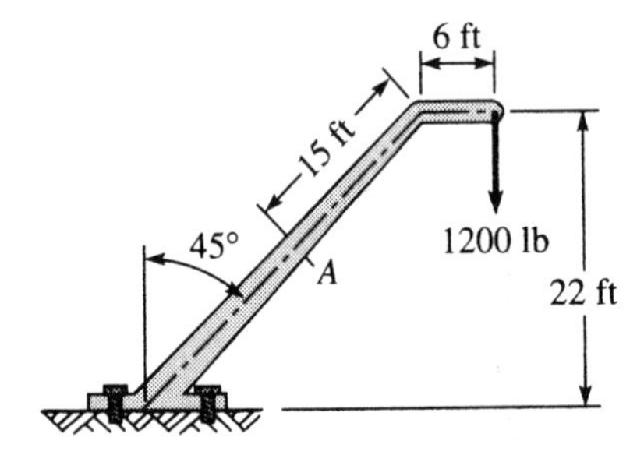

2.15 The lift force on a helicopter rotor blade is distributed as shown in Fig. P2.15. Compute the shear force and bending couple on the cross section at O resulting from the lift force.

Fig. P2.15

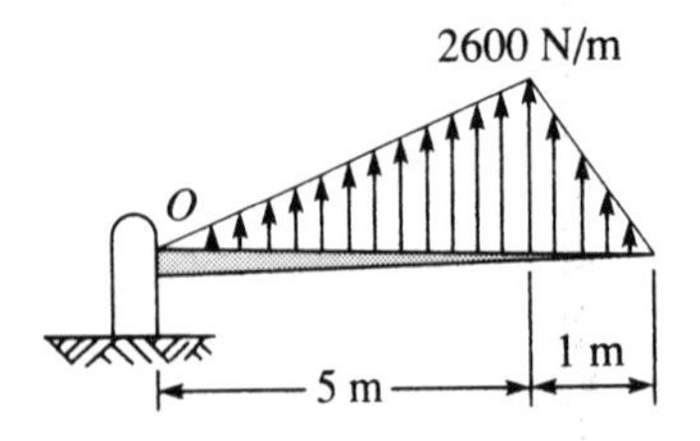

2.16 The intensity of the lift force on the wing of the airplane shown in Fig. P2.16 is given by

$$w_x = w_0 \sin \frac{\pi x}{2L}$$

Determine the shear force and bending couple on the cross section adjacent to the fuselage.

Fig. P2.16

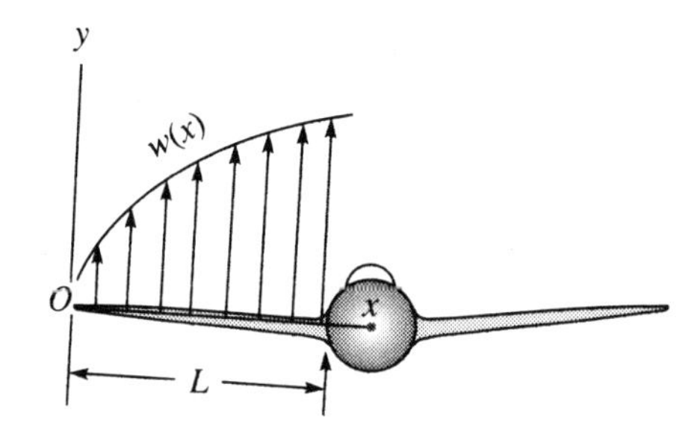

2.17 A pipe 18 in. long is clamped at one end and twisted by a wrench at the other end, as shown in Fig. P2.17. Compute the components of the net internal force system on a cross section 6 in. from the wall.

Fig. P2.17

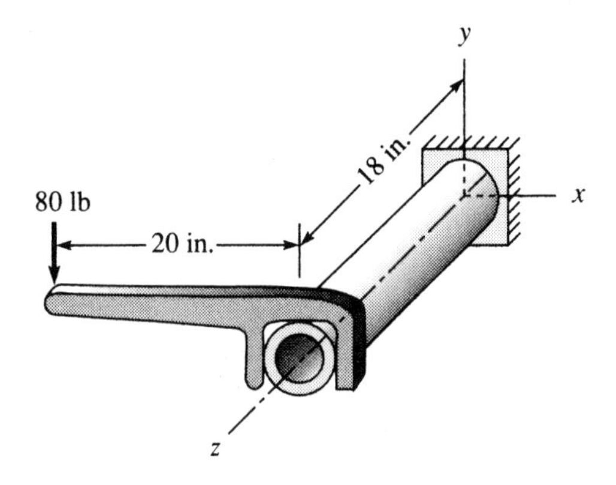

2.18 The bent bar shown in Fig. P2.18 consists of three straight portions joined consecutively at right angles. The first portion, OA, firmly clamped at its base O, has a length Y. The next portion, AB (parallel to the

z-axis), of length Z, is connected to a portion BC of length X (parallel to the x-axis). At the free end C a force having components F_x, F_y, and F_z is applied. Compute the components of the net internal force system on the cross section at the center of each straight portion of the bar.

Fig. P2.18

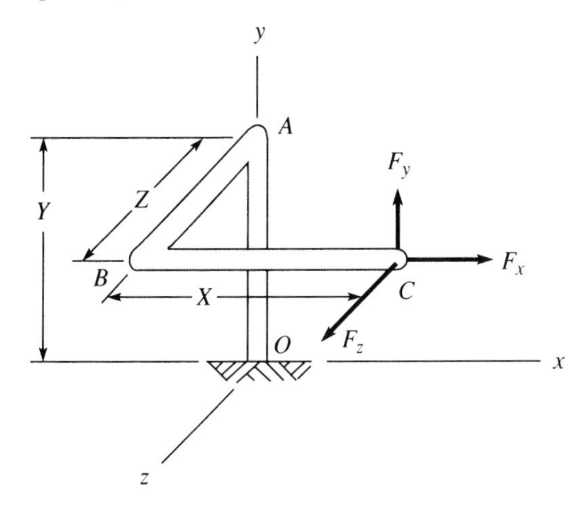

2.19 Bars AB and BC are pinned at each end, as shown in Fig. P2.19. They are uniform and weigh 30 N/m. Compute the components of the net internal force system on the cross section at the center of each bar.

Fig. P2.19

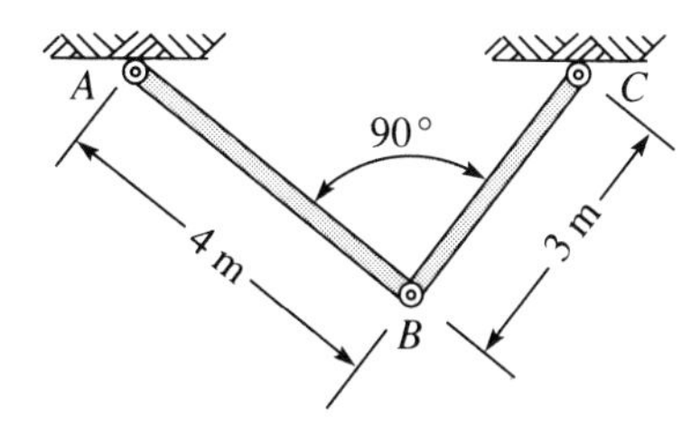

2.20 A thin triangular plate hangs under its own weight as shown in Fig. P2.20. If the plate weighs $\sqrt{3}$ lb for each square foot of area, compute the components of the net internal force system on the cross section at A.

Fig. P2.20

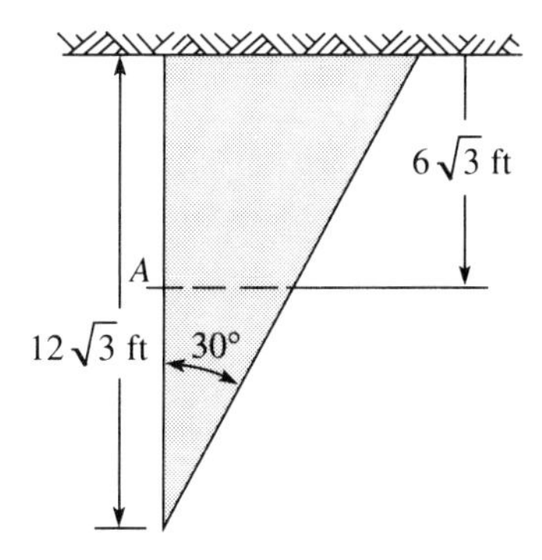

2.21 A bar in the form of a quarter circle is fixed at one end and a load of 600 N is applied to the free end, as shown in Fig. P2.21. What are the components of the net internal force system on the cross sections at A and B?

Fig. P2.21

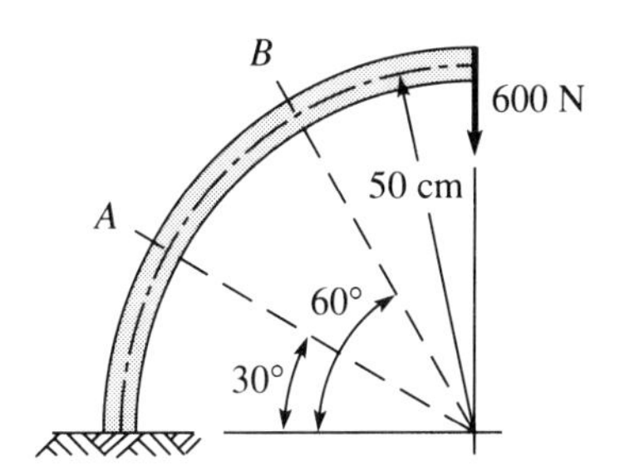

2.22 Repeat Prob. 2.21 with the load replaced by a 300-N · m bending couple, as shown in Fig. P2.22.

Fig. P2.22

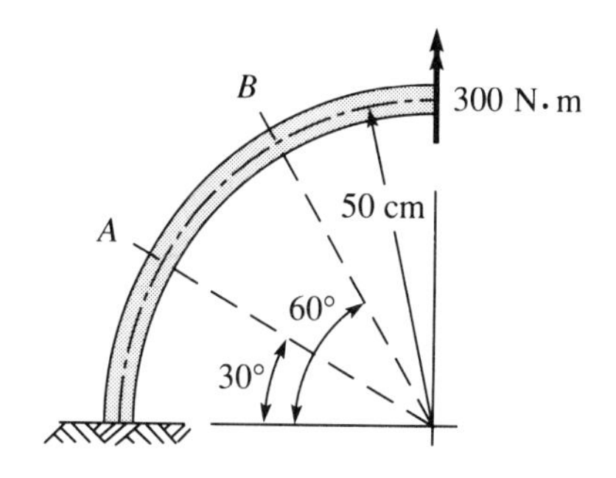

2.23 Repeat Prob. 2.21 with the force replaced by a 300-N·m twisting couple, as shown in Fig. P2.23.

Fig. P2.23

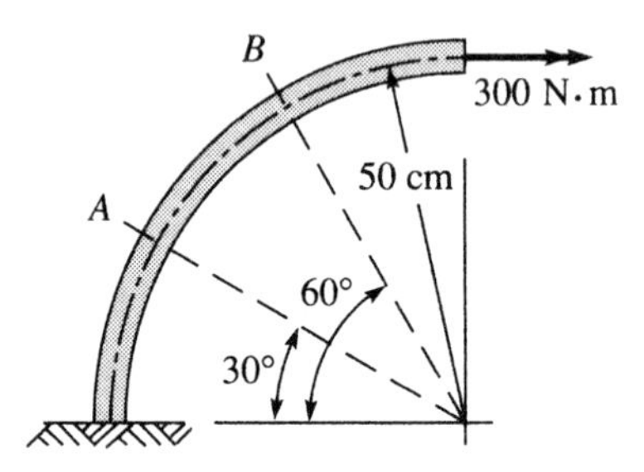

2.24 The semicircular bar shown in Fig. P2.24 is pinned at A. Determine the components of the net internal force system on a cross section as a function of the angle θ to that cross section. For what values of θ does each of these components take on its largest magnitude?

Fig. P2.24

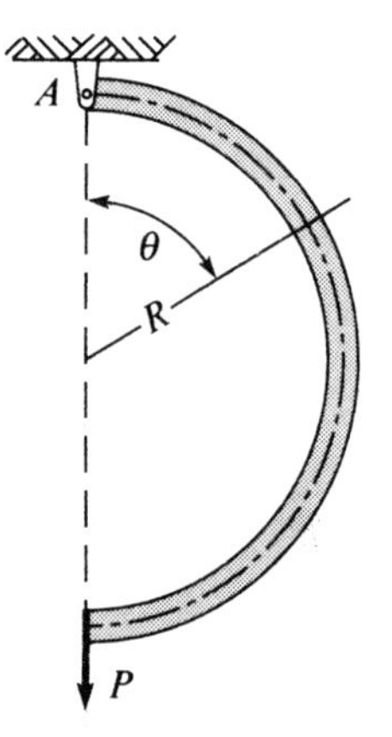

2.25 A solid cone is suspended by a pin at its vertex, as shown in Fig. P2.25. The density of the cone is 0.1 lb/in^3. Determine the normal force on a cross section x inches from the vertex.

Fig. P2.25

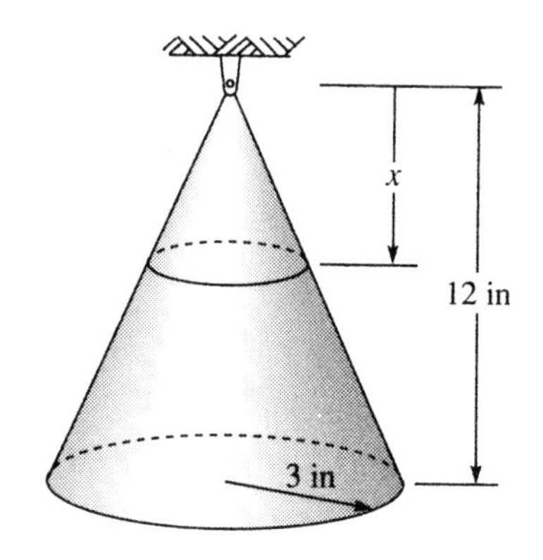

2.3 | Dependence of Internal Forces on Location

The net internal force system acting on a plane section of a body depends on the location and orientation of this plane. In general, as the position of the sectioning plane is changed, the normal and shear forces, as well as the twisting and bending couples, change. These components of the net internal force system can be related to the variables that specify the location of the sectioning plane. This gives an indication of how the internal forces vary throughout the body. Some examples serve to illustrate these remarks.

EXAMPLE 4

A cylindrical bar hanging under its own weight W is shown in Fig. 2.15(a). We can specify the location of a cross section by its distance from the fixed end. Let us denote this distance as x. A free-body diagram of the portion of the cylinder below the cross section at x is shown in Fig. 2.15(b). The net internal force system on this cross section consists only of a normal force F. For equilibrium, all other components must be zero. The weight of the portion below the cross section at x is

$$W\frac{(L-x)}{L} = W\left(1-\frac{x}{L}\right)$$

Fig. 2.15

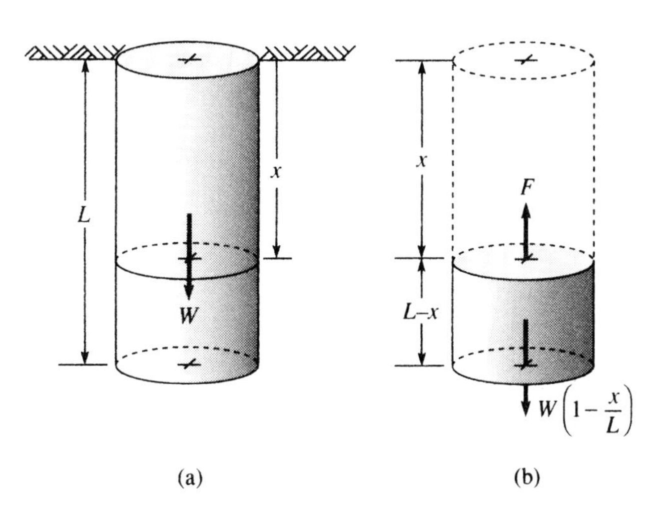

For equilibrium of this portion, we must have

$$\sum F_x = 0; \qquad F - W\left(1-\frac{x}{L}\right) = 0, \qquad F = W\left(1-\frac{x}{L}\right)$$

This result can be illustrated graphically by a plot of the normal force F versus the coordinate x of the sectioning plane, as shown in Fig. 2.16. This graph is called a *normal force diagram.*

Fig. 2.16

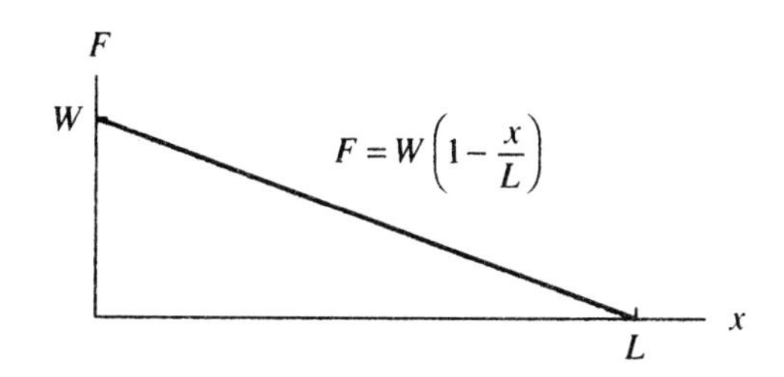

Similar graphs can be made for the other components of the net internal force system. A plot of the shear force versus x is called a *shear force diagram*, and a plot of the bending or twisting couple versus x is called a *bending couple diagram*, or a *twisting couple diagram*. ◆

EXAMPLE 5

Consider a cantilevered beam loaded as shown in Fig. 2.17(a). Again, an arbitrary cross section of the beam can be located by the one coordinate x, the distance of this cross section from the free end of the beam. To find the components of the net internal force system on this cross section, we take, as a free body, the portion of the beam to the left of this cross section [Fig. 2.17(b)]. Since the loading is coplanar, the only nonzero components of the net internal force system are a normal force, a shear force, and a bending couple, as indicated in the figure. Then

$$\sum F_x = 0; \qquad F + P\cos 45^\circ = 0, \qquad F = -\frac{P}{\sqrt{2}} \tag{2.6}$$

$$\sum F_y = 0; \qquad V - P\sin 45^\circ - wx = 0, \qquad V = \frac{P}{\sqrt{2}} + wx \tag{2.7}$$

$$\sum M_z = 0; \qquad M + (P\sin 45^\circ)x + wx\frac{x}{2} = 0, \qquad M = -\frac{Px}{\sqrt{2}} - \frac{wx^2}{2} \tag{2.8}$$

Fig. 2.17

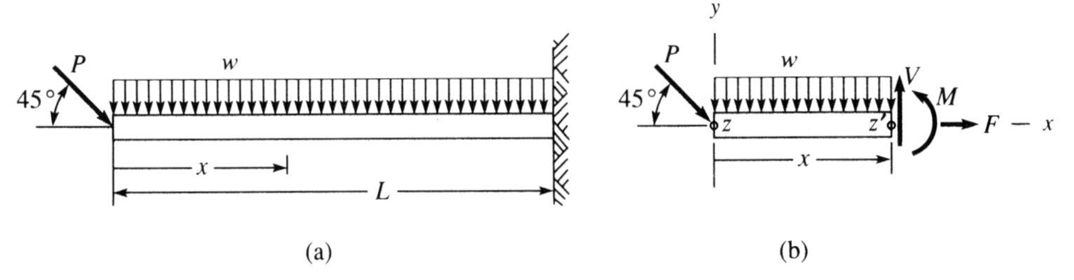

Equations (2.6), (2.7), and (2.8) are valid for every cross section defined by a value of x between $x = 0$ and $x = L$, because the same free-body diagram, Fig. 2.17(b), would result for any x in this range. Therefore, these equations give the components of the net internal force system on every cross section of the beam. They are plotted as the normal force diagram, the shear force diagram, and the bending couple diagram, in that order, in Fig. 2.18. (These diagrams give only the general shapes of the curves, since we do not have numerical values of P, L, and w).

Fig. 2.18

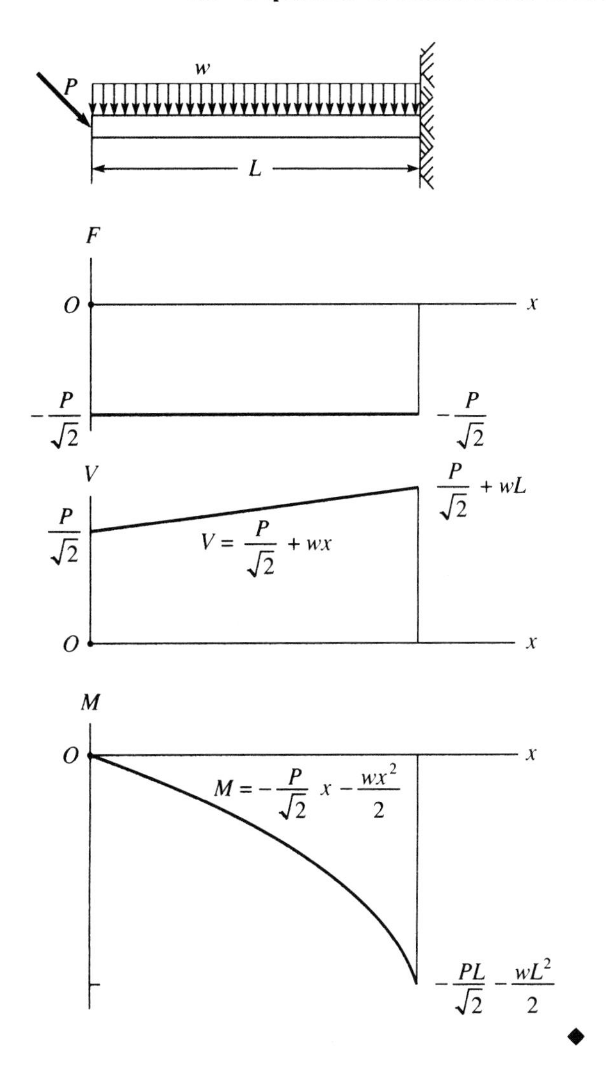

◆

EXAMPLE 6

Consider the beam shown in Fig. 2.19. An arbitrary cross section can again be located by x, the distance of the cross section from the pinned end of the beam. But when a free-body diagram is drawn of the portion to the left of the cross section, three different situations arise. These are shown in Fig. 2.19(c), 2.19(d), or 2.19(e), depending on whether $0 < x < 4$ ft, 4 ft $< x <$ 8 ft, or 8 ft $< x <$ 10 ft. Each of these free bodies is acted upon by a different system of external loading. Each must, therefore, be considered separately.

Fig. 2.19

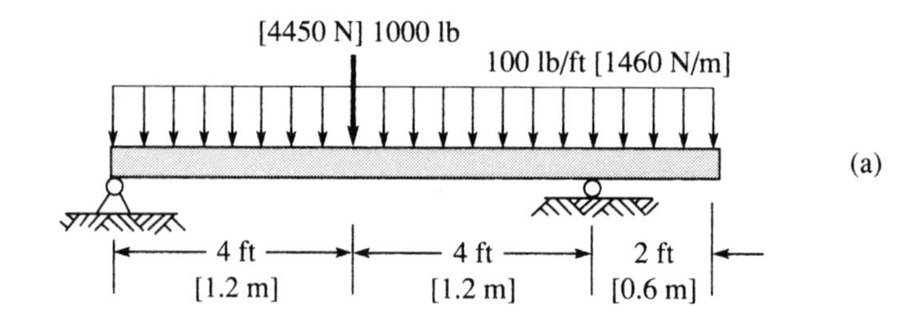

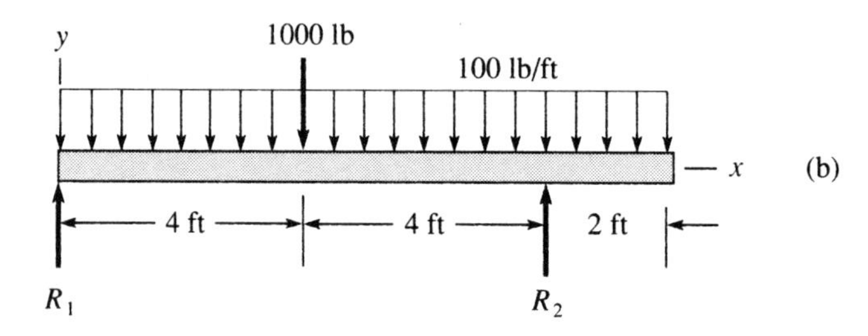

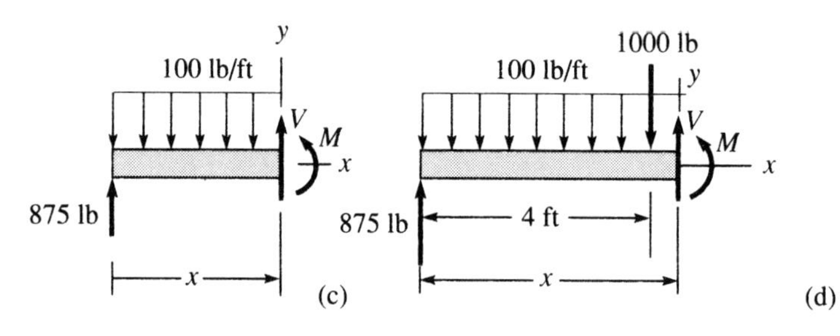

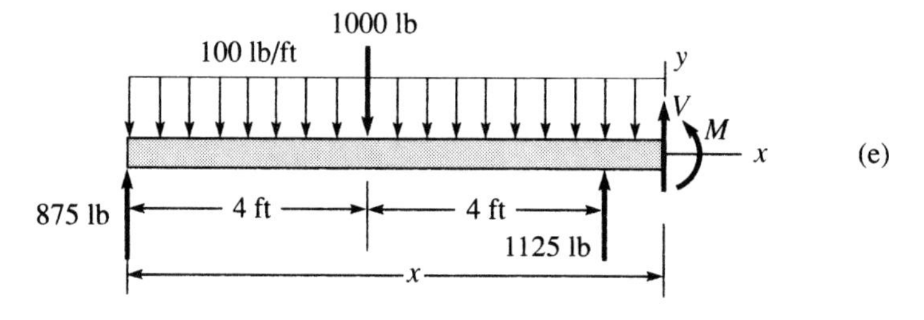

We first determine the reactions R_1 and R_2 from the equations of equilibrium applied to the beam as a whole. The free-body diagram is shown in Fig. 2.19(b).

$$\sum M_z = 0; \qquad R_2 = 1125 \text{ lb} \quad [500 \text{ N}]$$

$$\sum F_y = 0; \qquad R_1 = 875 \text{ lb} \quad [3900 \text{ N}]$$

Then, from Fig. 2.19(c), for $0 < x < 4$ ft,

$$\sum F_y = 0; \qquad V = -875 + 100x \tag{2.9}$$

$$\sum M_z = 0; \qquad M = 875x - 50x^2 \tag{2.10}$$

From Fig. 2.19(d), for 4 ft $< x <$ 8 ft,

$$\sum F_y = 0; \qquad V = 125 + 100x \tag{2.11}$$

$$\sum M_z = 0; \qquad M = 4000 - 125x - 50x^3 \tag{2.12}$$

From Fig. 2.19(e), for 8 ft $< x <$ 10 ft,

$$\sum F_y = 0; \qquad V = -1000 + 100x \tag{2.13}$$

$$\sum M_z = 0; \qquad M = -5000 + 1000x - 50x^2 \tag{2.14}$$

Equations (2.9) through (2.14) can be plotted in the regions in which they are valid to give the shear force diagram and the bending couple diagram of Fig. 2.20.

Fig. 2.20

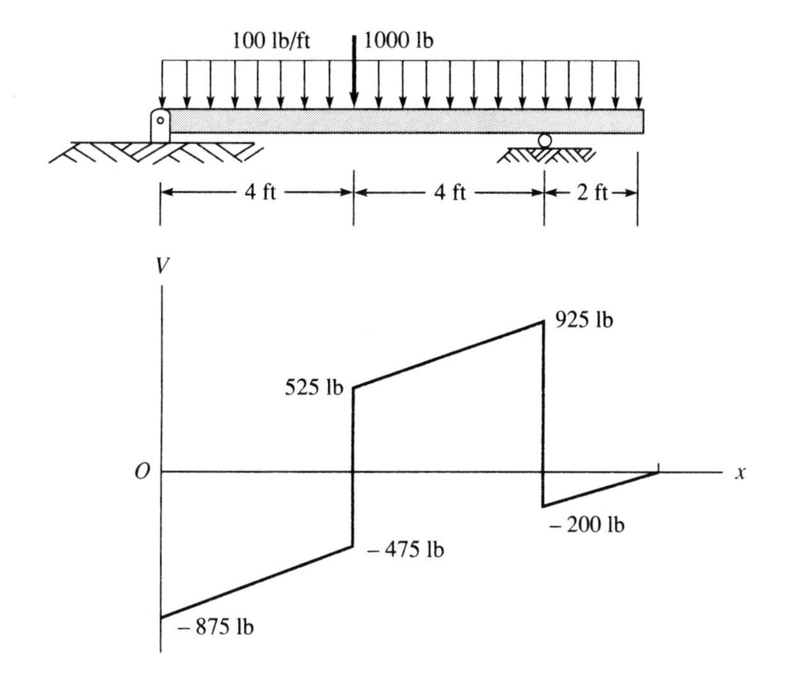

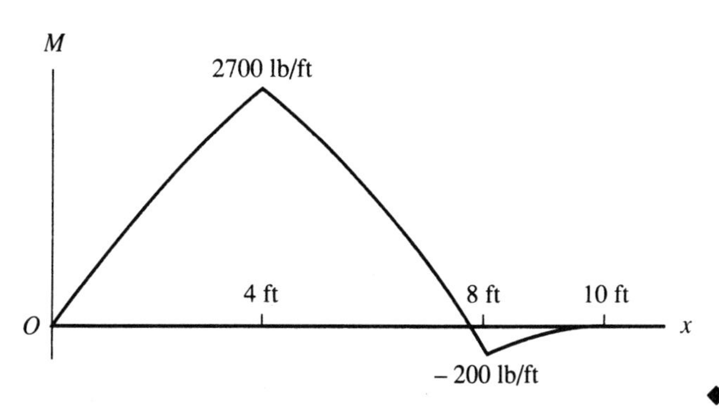

◆

EXAMPLE 7

As another example, let us consider the curved bar shown in Fig. 2.21(a). It is a quarter of a circular bar, fixed at one end, with a 500-N load applied to the free end perpendicular to the plane of the bar. In this case, we can locate an arbitrary cross section by the angle θ shown in Fig. 2.21.

Fig. 2.21

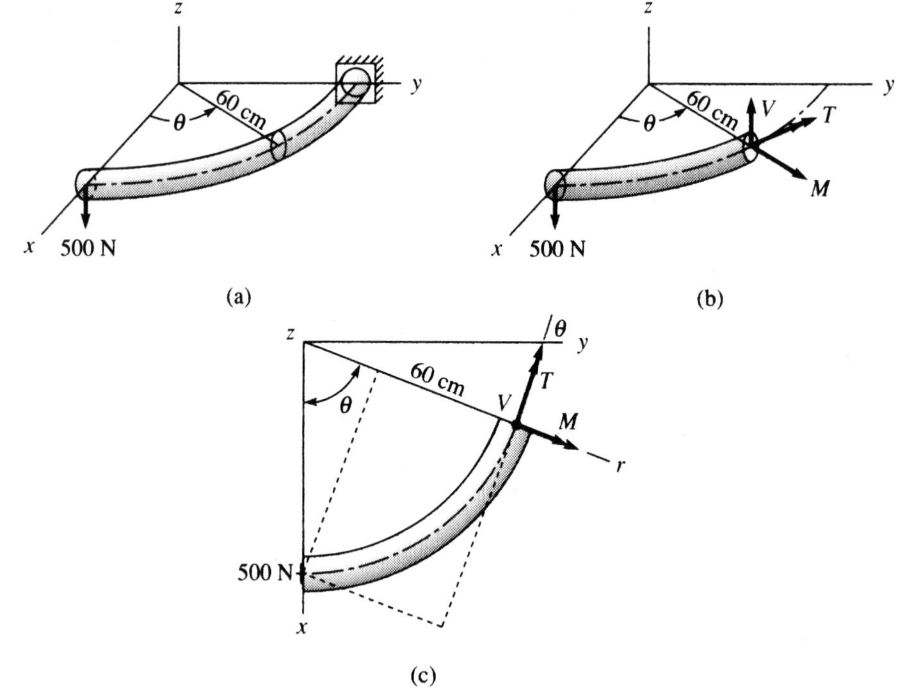

To find the components of the net internal force system on such a cross section, we consider as a free body the portion of the bar shown in Fig. 2.21(b) and 2.21(c). For equilibrium of this portion of the bar, we have

$$\sum F_z = 0; \qquad V = 500 \text{ N}$$

$$\sum M_r = 0; \qquad M = -300 \sin\theta \text{ N}\cdot\text{m}$$

$$\sum M_\theta = 0; \qquad T = 300(1 - \cos\theta) \text{ N}\cdot\text{m}$$

The resulting shear force, bending couple, and twisting couple diagrams are shown in Fig. 2.22. In this case the coordinate specifying the location of the sectioning plane is an angle rather than a distance. ◆

Fig. 2.22

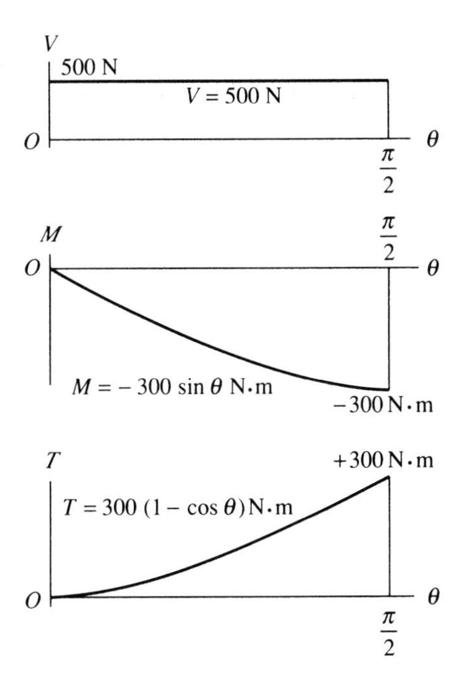

EXAMPLE 8

In each of the preceding examples, the loads were fixed in direction. Now the circular ring of Fig. 2.23(a) is subjected to pressure q, which acts normal to the surface; on each element of length $R\,d\theta$, the force $qR\,d\theta$ is directed radially as shown. Often domes of shell-like structures are subject to such loading, such as the pressure of water upon a submerged hull.

The free body of Fig. 2.23(b) is obtained by cutting the ring at an arbitrary angle θ. The external load upon a typical element at angle β is $qR\,d\beta$. Components of normal force F and shear force V act upon the section at A in directions n and t, respectively; additionally, a bending couple M acts upon the cross section. *Note* that the direction of the pressure changes with angle β, always normal to the surface of the ring. The component in direction n is $qR\,d\beta \sin(\theta - \beta)$, and the net component in direction n is the integral

$$\int_0^\theta qR \sin(\theta - \beta)\,d\beta = qR(1 - \cos\theta)$$

Equilibrium requires that

$$\sum F_n = F + \int_0^\theta qR \sin(\theta - \beta)\,d\beta = 0$$

Fig. 2.23

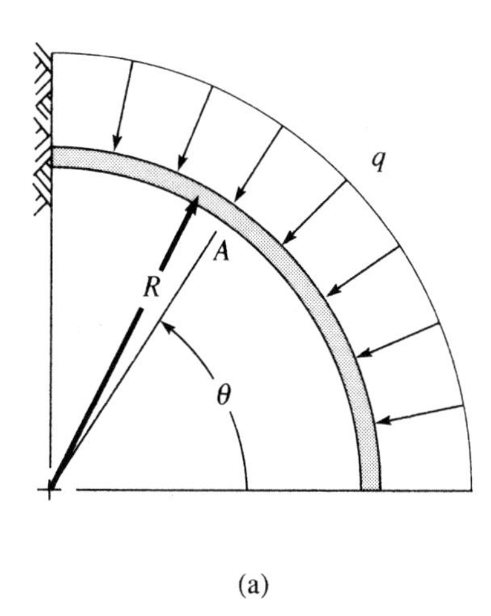

(a)

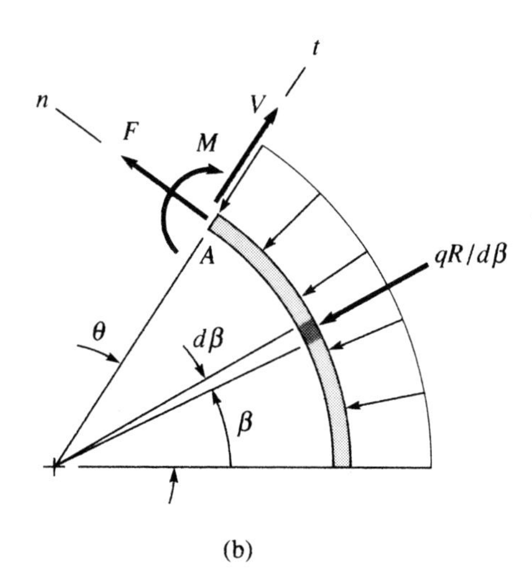

(b)

so that

$$F = -qR(1 - \cos\theta)$$

Additionally,

$$\sum F_t = V - \int_0^\theta \cos(\theta - \beta)\, d\beta = 0$$

$$V = qR\sin\theta$$

$$\sum M_A = M + \int_0^\theta qR^2 \sin(\theta - \beta)\, d\beta = 0$$

$$M = -qR^2(1 - \cos\theta)$$

The components of force F, V, and couple M can be plotted versus the variable θ, as in the previous example. It is most important, however, to note that we must duly account for the changing direction of the loading, as illustrated here. ◆

2.26 Write expressions for the shear force and bending couple on cross sections of the beams shown in Fig. P2.26(a) through P2.26(v). Also sketch the appropriate force and couple diagrams.

Fig. P2.26

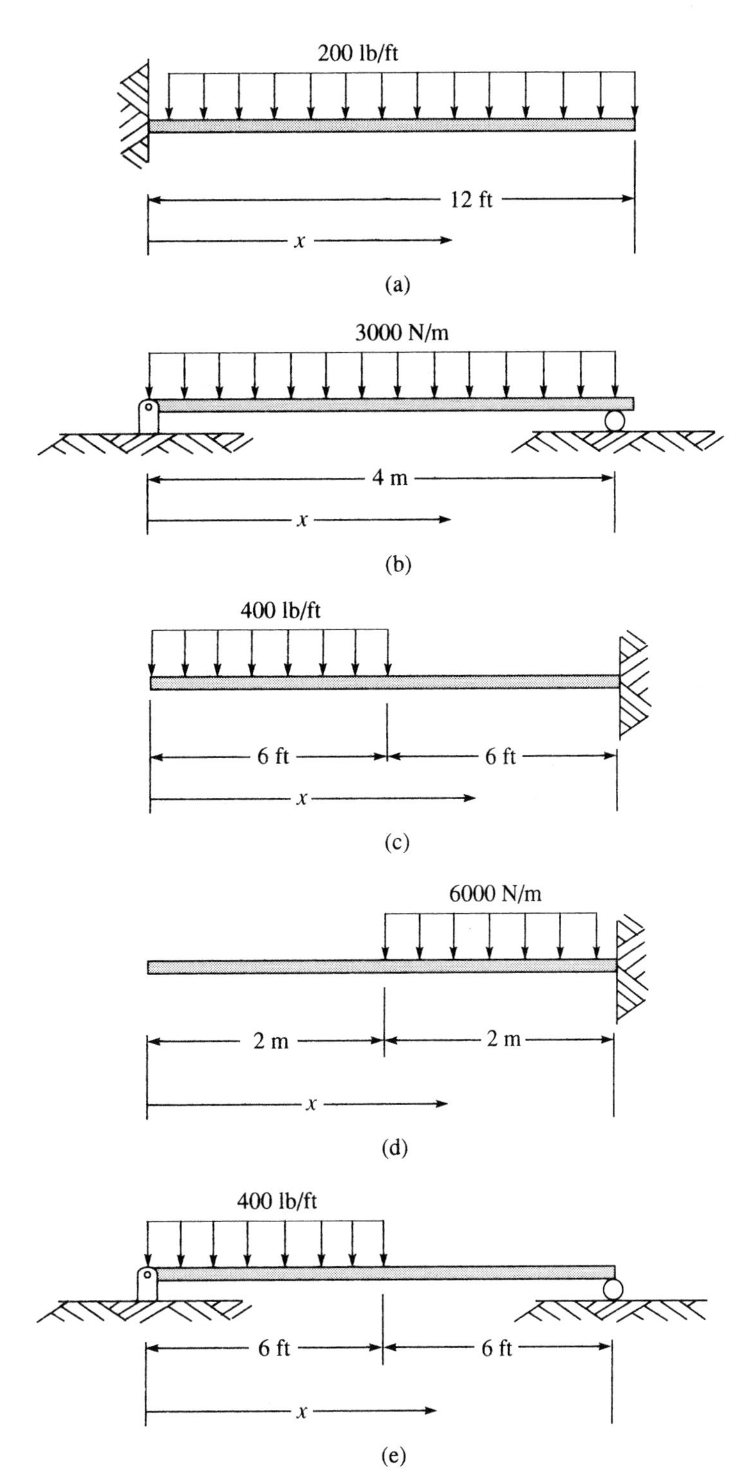
200 lb/ft
12 ft
x
(a)
3000 N/m
4 m
x
(b)
400 lb/ft
6 ft
6 ft
x
(c)
6000 N/m
2 m
2 m
x
(d)
400 lb/ft
6 ft
6 ft
x
(e)

Fig. P2.26
(cont.)

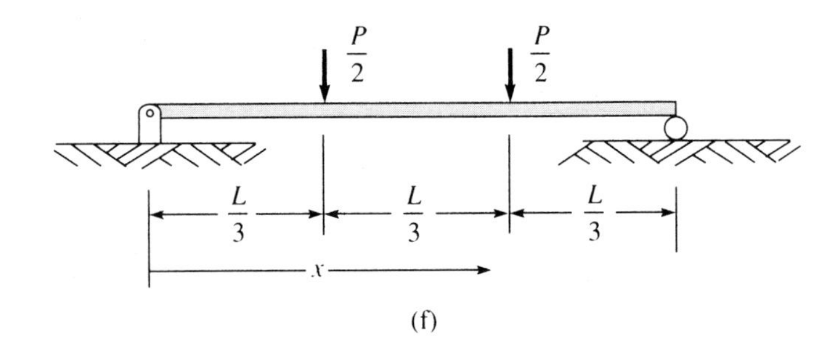

(f)

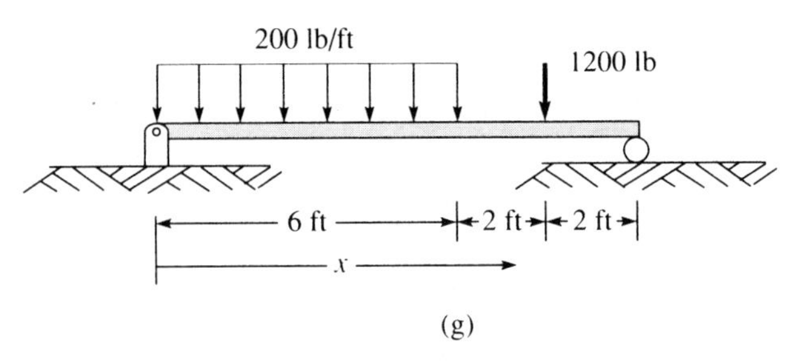

(g)

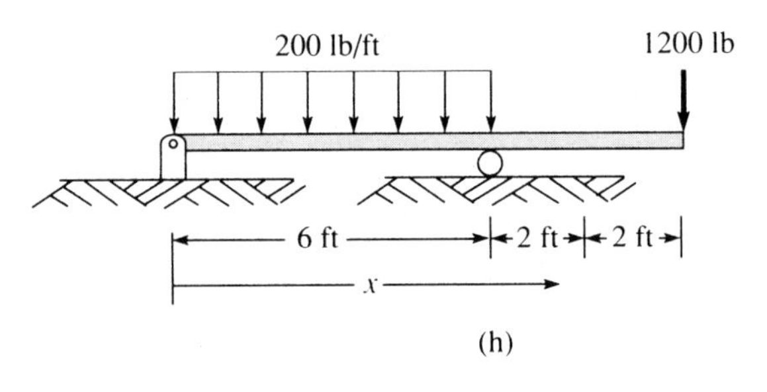

(h)

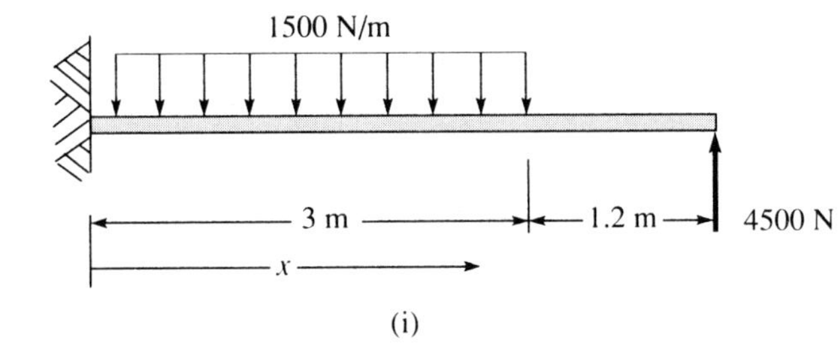

(i)

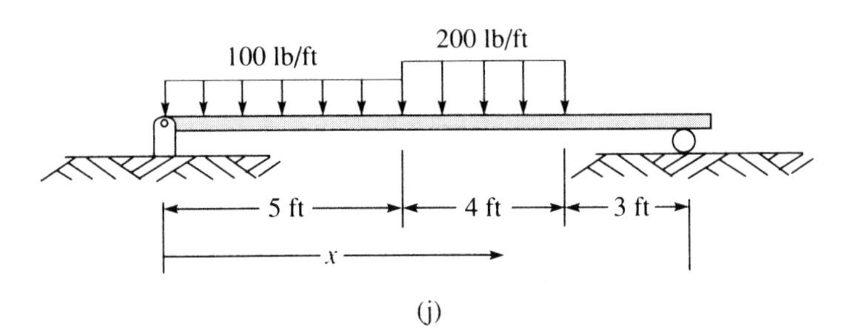

(j)

Fig. P2.26
(*cont.*)

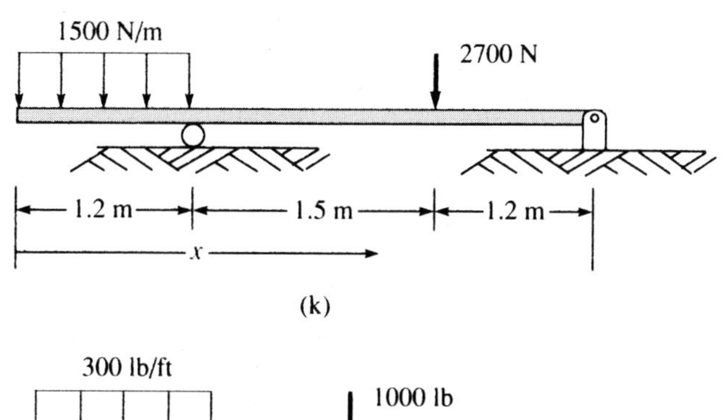

(k)

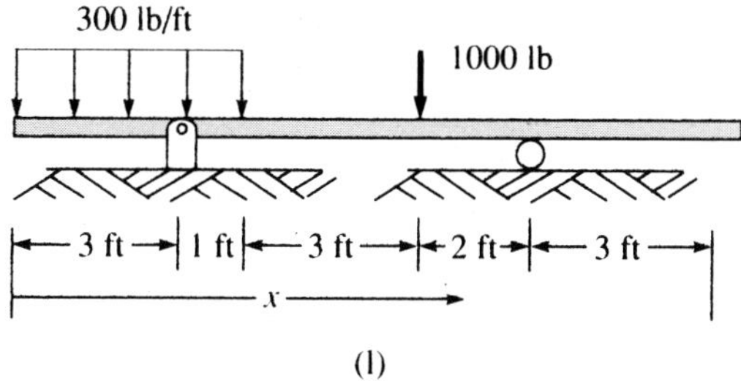

(l)

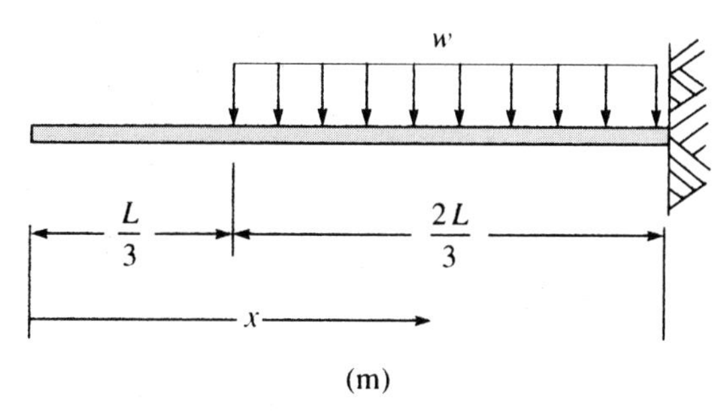

(m)

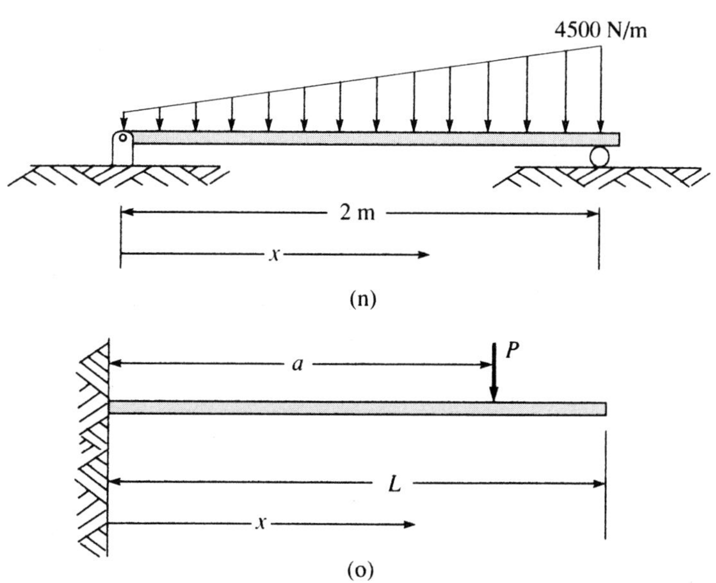

(n)

(o)

Fig. P2.26
(*cont.*)

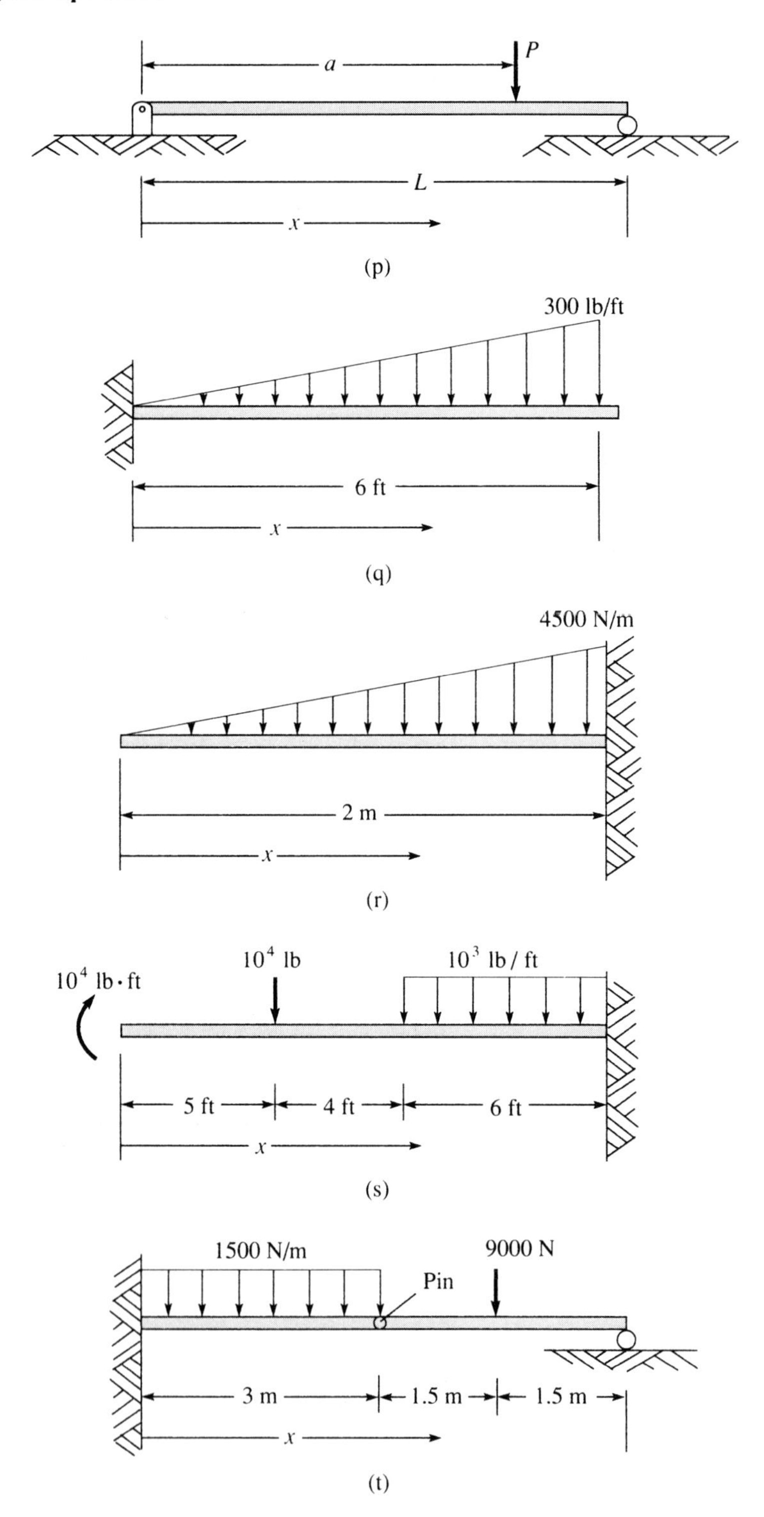

Fig. P2.26
(*cont.*)

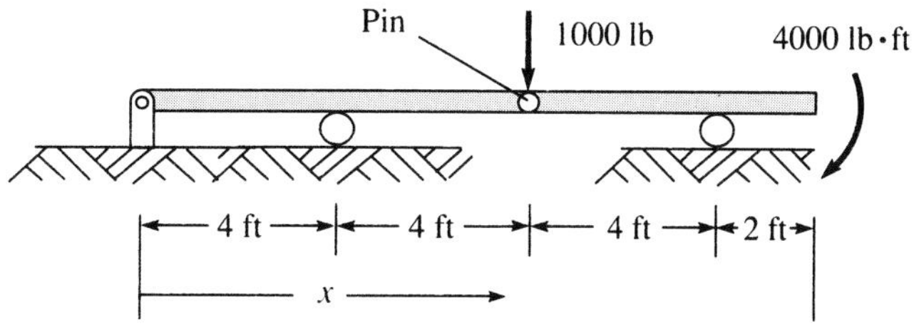

Pin
1000 lb
4000 lb·ft
4 ft
4 ft
4 ft
2 ft
x

(u)

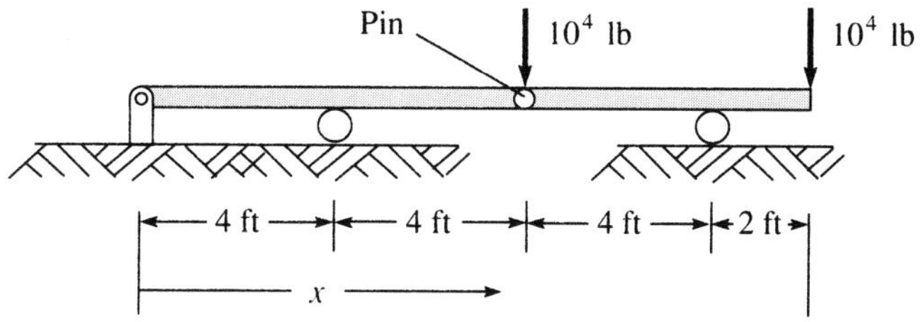

Pin
10^4 lb
10^4 lb
4 ft
4 ft
4 ft
2 ft
x

(v)

2.27 Write expressions for the normal force, shear force, and bending couple on cross sections of the member AB shown in Fig. P2.27(a) through P2.27(h). Also sketch the appropriate diagrams.

Fig. P2.27

200 lb
A
C
2 ft
B
8 ft
6 ft
x

(a)

220 N
C
A
0.9 m
B
2.4 m
1.2 m
x

(b)

x
6 ft
4 ft
A
B
30 °
C
1 ft
400 lb

(c)

C
4 kN
1.5 m
30 kN
B
A
1.5 m
1.5 m
1.5 m
x

(d)

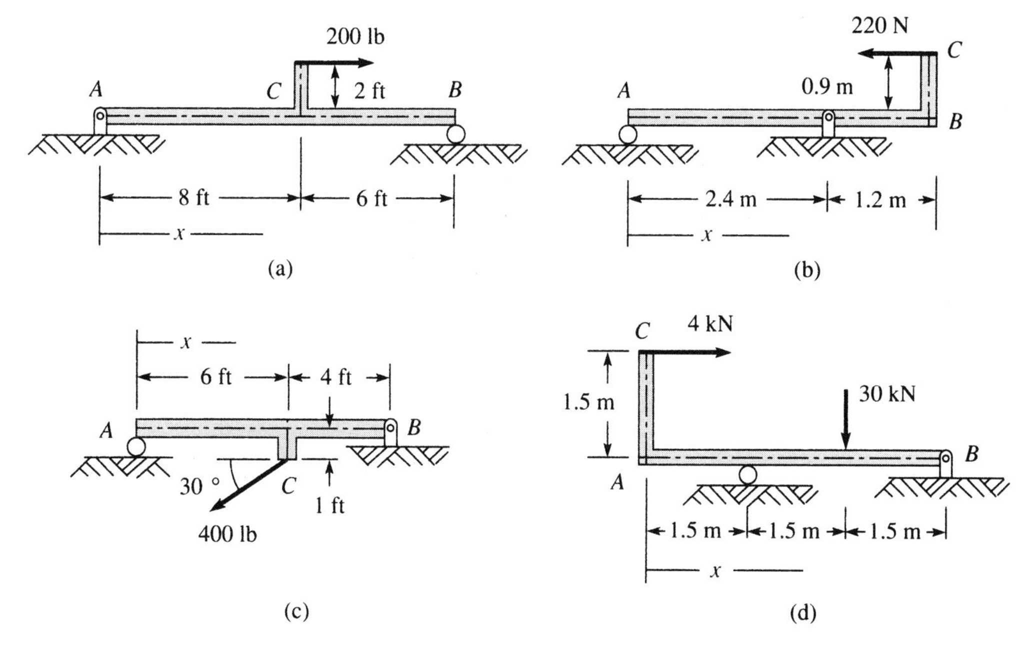

Fig. P2.27
(cont.)

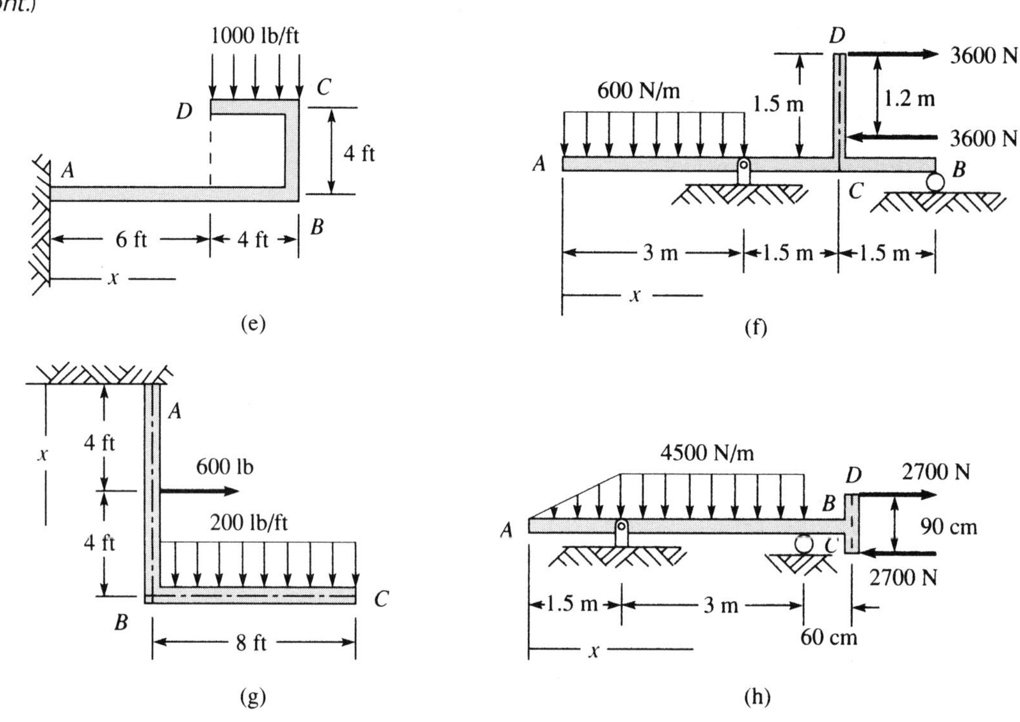

2.28 The homogeneous cylindrical bar shown in Fig. P2.28 hangs under its own weight. Sketch the normal force, shear force, and bending couple diagrams. Use the distance x from the built-in end to locate the cross sections.

Fig. P2.28

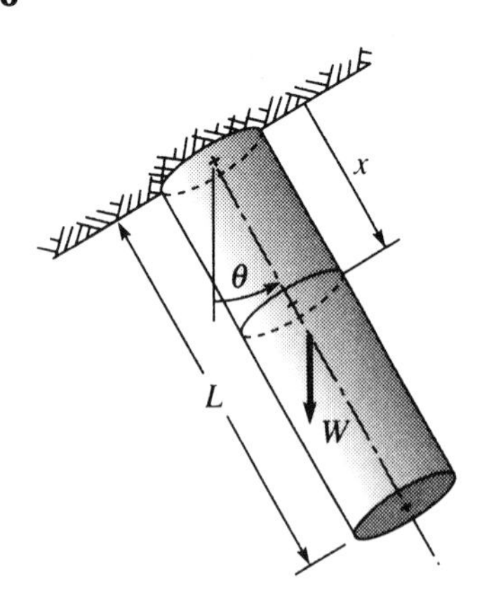

2.29 The cylinder shown in Fig. P2.29 is subjected to the eccentric load of 100 lb acting parallel to the z-axis. Sketch the bending and twisting couple diagrams for the cylinder.

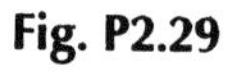

Fig. P2.29

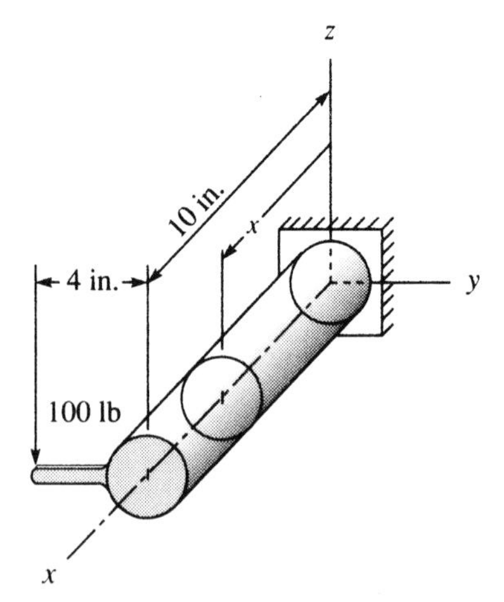

2.30 The bars AB and BC are pinned at each end, as shown in Fig. P2.30. They are uniform and weigh 0.2 lb for each inch of length. Determine the normal force, shear force, and bending couple on any cross section of the bar AB, and sketch the appropriate force and couple diagrams.

Fig. P2.30

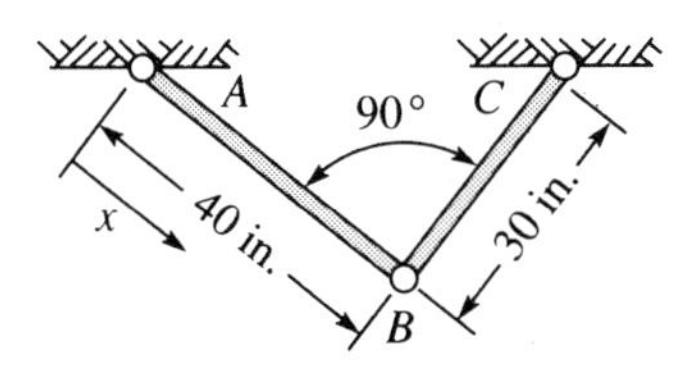

2.31 A member is clamped at one end and bent (as shown in Fig. P2.31) so that an initial portion OA of length X is parallel to the x-axis, the next portion AB of length Y is parallel to the y-axis, and the final portion BC of length Z is parallel to the z-axis. At the free end, the member is subjected to a force that has components F_x, F_y, F_z. Determine the components of the net internal force system on any cross section of the member (defined by x, y, or z).

Fig. P2.31

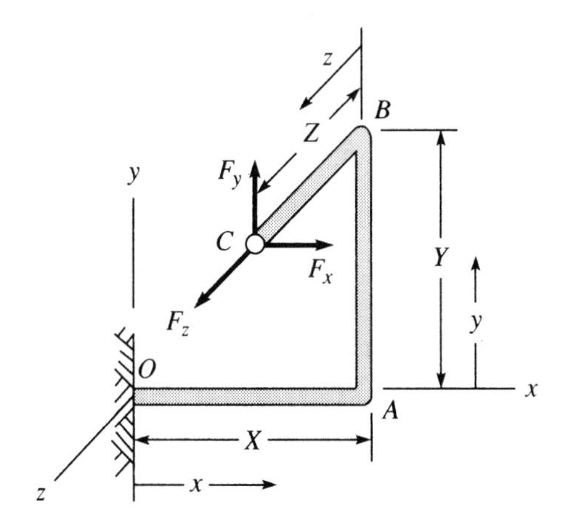

2.32 A bar in the form of a quarter circle is clamped at one end and loaded by a normal force at the other end, as shown in Fig. P2.32. Compute the components of the net internal force system on an arbitrary cross section (at θ) and plot the appropriate force and couple diagrams.

Fig. P2.32

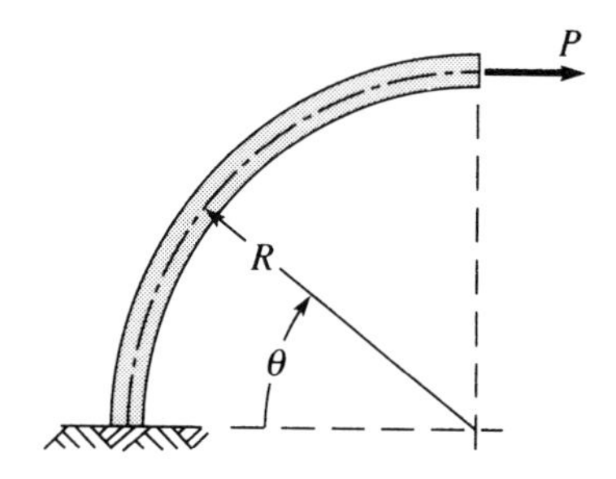

2.33 Repeat Prob. 2.32 for the bar loaded by a shear force, as shown in Fig. P2.33.

Fig. P2.33

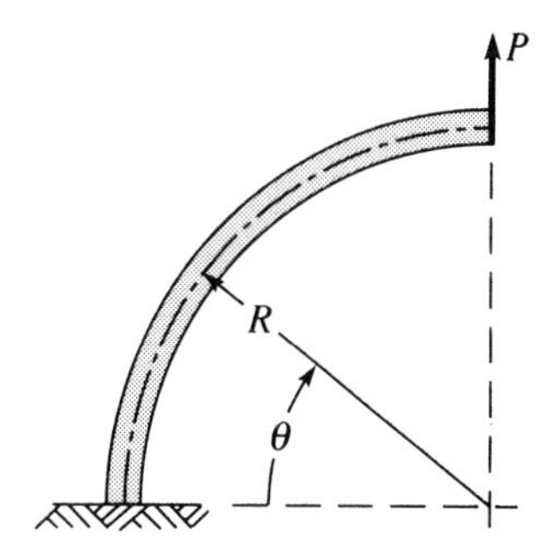

2.34 Repeat Prob. 2.32 for the bar loaded by a twisting couple, as shown in Fig. P2.34.

Fig. P2.34

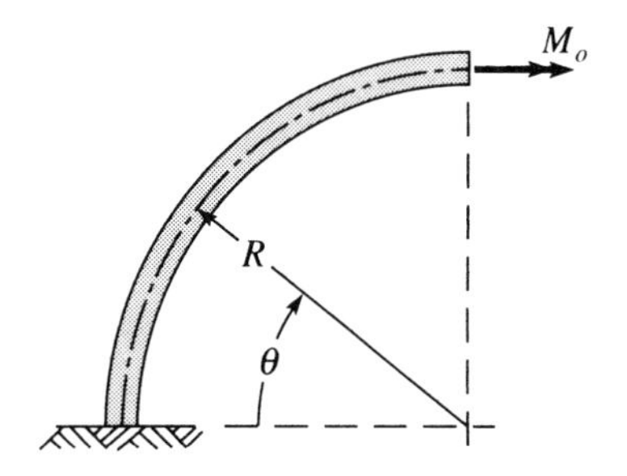

2.35 Repeat Prob. 2.32 for the bar loaded by a bending couple, as shown in Fig. P2.35.

Fig. P2.35

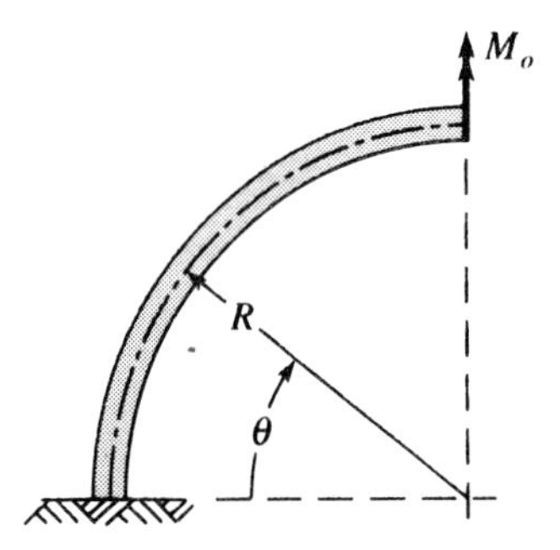

2.36 A thin triangular plate hangs under its own weight, as shown in Fig. P2.36. If the plate weighs $\sqrt{3}$ lb per square foot of area, compute the normal force and bending couple on any cross section N a distance x from the top of the plate. Sketch the appropriate force and couple diagrams.

Fig. P2.36

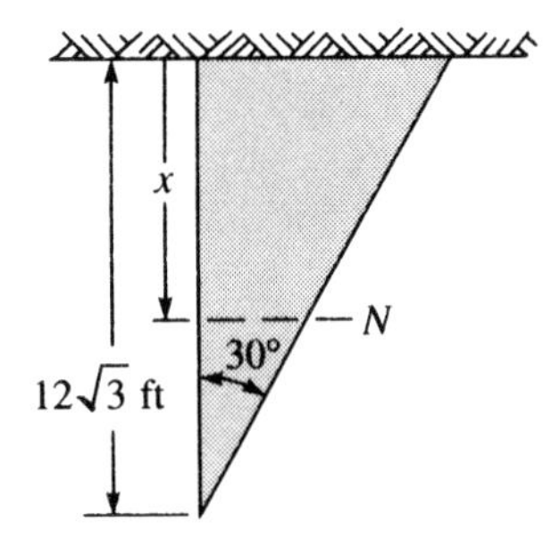

2.37 A uniform semicircular ring hangs under its own weight of 100π N, as shown in Fig. P2.37. Compute the components of the net internal force system on any cross section defined by $0 < \theta < \pi$. Sketch the appropriate force and couple diagrams.

Fig. P2.37

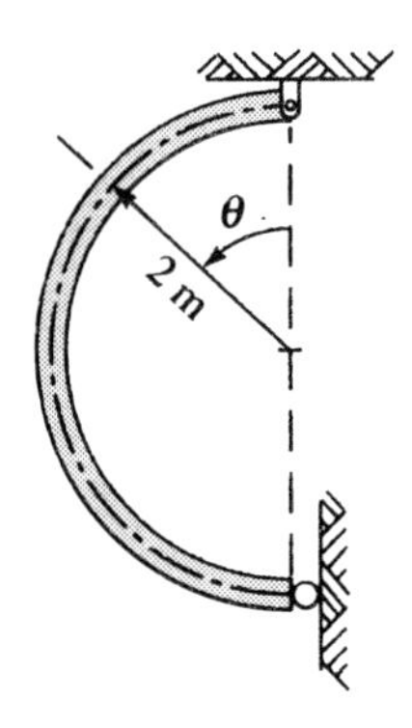

2.38 A beam of length L is built in at one end and supported by a rod at the other end, as shown in Fig. P2.38. The beam is subjected to a uniform load intensity w, and the tension in the rod is adjusted so that the largest (absolute value) bending couple in the beam is minimum. Determine this tension in terms of w and L.

Fig. P2.38

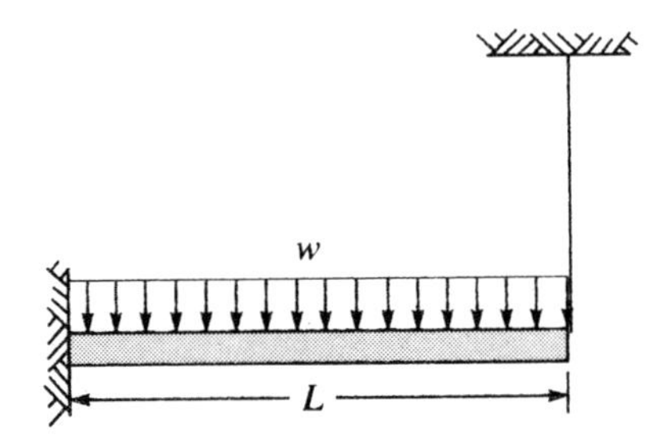

2.39 A beam of length L is pinned at one end and supported by a wire at another point at distance a from the pin, as shown in Fig. P2.39. The beam is subjected to a uniform load intensity w, and the distance a is chosen so that the largest (absolute value) bending couple in the beam is a minimum. Compute a/L.

Fig. P2.39

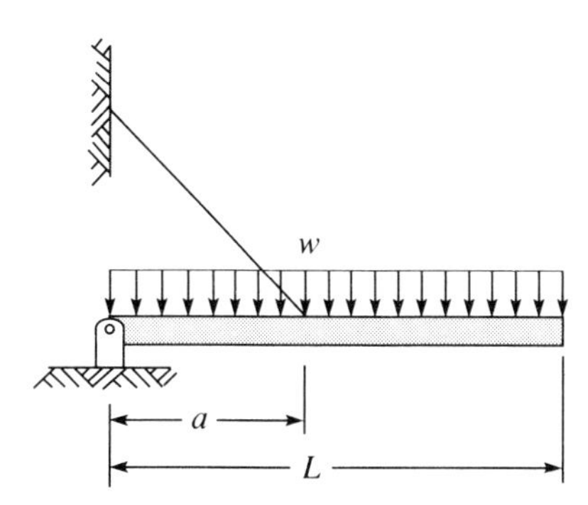

*2.40 The cylindrical ring in Fig. P2.40 is hinged at O and rests upon the smooth surface at A (neglect friction at hinge O and surface A). The convex (outer) surface is subjected to a uniform pressure (normal to the surface) q (force/length). Express the normal force, shear force, and bending couple on a cross section (M) in terms of the parameters (q, R) and position $\theta[F(\theta), V(\theta), M(\theta)]$.

Fig. P2.40

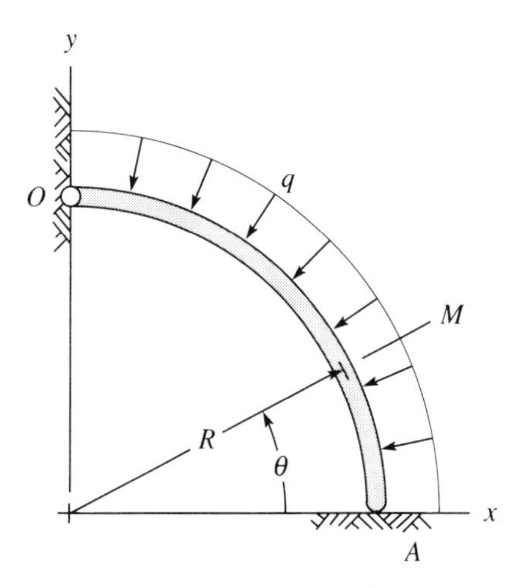

*2.41 Repeat Prob. 2.40 if the uniform pressure is replaced by a pressure that varies according to the equation $q = p(1 - \sin\theta)$. This corresponds to a hydrostatic pressure, wherein a fluid is contained by the upper surface from A to the level O.

2.4 Stress

The preceding sections were concerned with the net internal force system (the resultant) acting upon a plane section of a body. Such a resultant is sometimes said to be statically equivalent to the actual distribution of internal forces upon the surface. Here it is very important to appreciate the implications of the term *statically equivalent*. The term means only that two *different* distributions of forces (or couples) have the same resultant. It means only that the systems have the same effect on a *rigid* body. Two different distributions are *never* equivalent in their effect on a *deformable* body. To appreciate this fact, consider a deformable block, say rubber, subjected to the two different distributions, as depicted in Fig. 2.24. The resultants are the same, a central normal force; the effects are very different. In general, we must *never* replace one distribution by another, although *statically* equivalent, and expect the same

Fig. 2.24

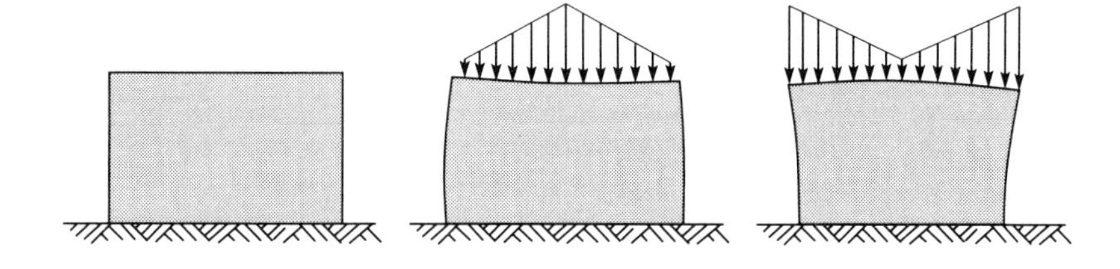

effects. The student of the mechanics of deformable bodies must adopt this as an inviolable rule.

Since we wish eventually to relate the deformational behavior of a body to the internal forces, the net internal force system alone will not adequately describe these internal forces. We need quantities that describe the internal forces in the immediate vicinity of a point. Such quantities can be defined in the following manner.

Consider a member subjected to some external equilibrated loads, as depicted in Fig. 2.25. The body may be divided by a surface S, as in the body of Fig. 2.5. The

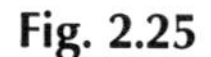
Fig. 2.25

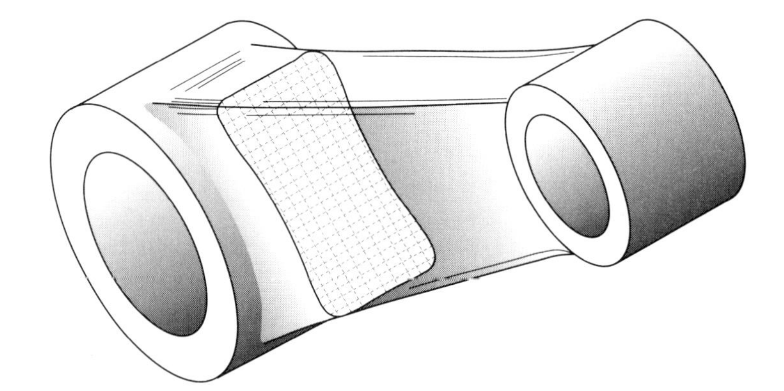

surface divides the body into two free bodies, shown separated in Fig. 2.26. Each part is subjected to interactions upon the intersection, actions and reactions, respectively, as well as the external loads on that portion. We recognize, however, that representation by a resultant on a cross section, say force and couple, is inadequate. To illustrate, suppose that each of the contiguous surfaces is subdivided into many small rectangular elements, as shown. Each elemental area is subject to a different force, $\Delta\mathbf{F}$, and the contiguous element of the separated body is subject to the reaction, $-\Delta\mathbf{F}$, as the elements at B and C. Each element is subject to a different force. Moreover, the magnitude and direction of the force $\Delta\mathbf{F}$ changes as the size of the element is changed. An elemental surface at a point P is shown in Fig. 2.27 with the force $\Delta\mathbf{F}$ that acts upon it. The area of the element is ΔA; the orientation of the element is established by

Fig. 2.26

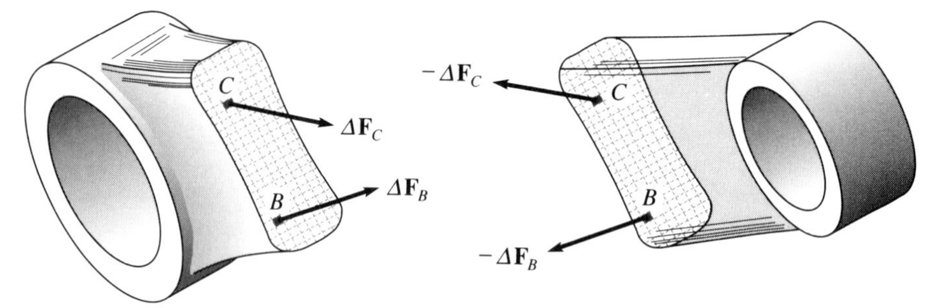

Fig. 2.27

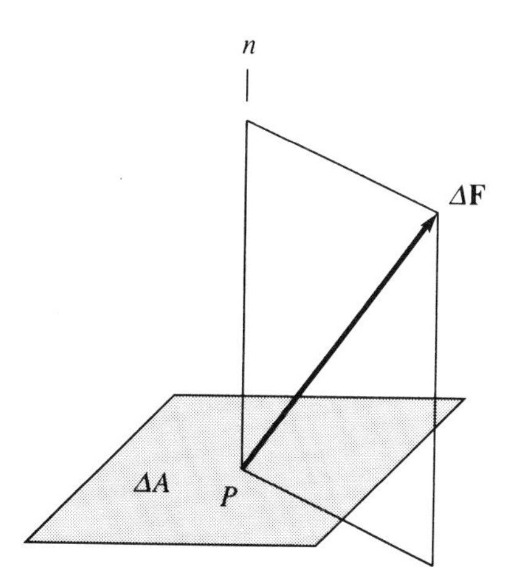

the direction of the normal n. Again, $\Delta\mathbf{F}$ depends on the size of ΔA. We are concerned with the nature and intensity at the specific point P. Therefore, we define the stress, a vector:

$$\boldsymbol{\sigma}_n \equiv \lim_{\Delta A \to 0} \frac{\Delta \mathbf{F}}{\Delta A} \tag{2.15}$$

Several features must be noted.

1. The stress $\boldsymbol{\sigma}_n$ depends on the orientation of the surface; hence, the subscript n is needed. At the same location the stress is generally different on surfaces of different orientation. A simple example illustrates that fact: Suppose that a thin bar is pulled by axial force. Upon a right section we expect a large stress; upon a section parallel to the axis we expect a very small (negligible) stress.

2. The stress presumes that $\Delta\mathbf{F}$ diminishes as ΔA diminishes. Mathematically speaking, we presume that the limit exists. Practically speaking, we assume that the quantity has physical significance. Of course, on a *microscopic* scale, real materials are not continuous. Metals are composed of crystals, which are, in turn, composed of atoms; impurities, even spaces, exist between crystals. Other materials, such as wood and concrete, are more heterogeneous. This means that the concept of stress, as defined here, must be used with some caution; it may not suffice to describe phenomena that originate within a microstructure. It is quite effective for the description of macroscopic phenomena and has proven most satisfactory for most analyses of structures and machines, particularly deformational behavior.

In practice, it is necessary to decompose the vector $\boldsymbol{\sigma}_n$ into components. Quite naturally, the vector can be decomposed into components normal to the surface and tangent to the surface; these rectangular components, denoted by σ_n and τ_n, are depicted in Fig. 2.28. The component σ_n is the *normal* component of stress. The component τ_n is tangent to the surface; it is called the *shear* component, because the action tends to shear the material (as scissors shear paper). If, at some point, we identify a direction n and give the normal component σ_n, the meaning is without ambiguity. If

Fig. 2.28

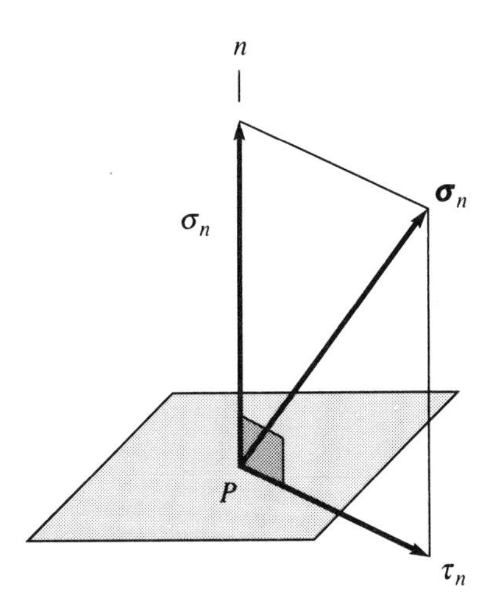

we give the shear component τ_n, the meaning is ambiguous, since that component could have any direction tangent to the surface. Physically, the consequences might be very different, depending upon the direction of the shear τ_n; the material might be more susceptible to shearing in certain directions, as wood shears more readily parallel to its grain. Accordingly, we select two perpendicular directions, such as s and t in Fig. 2.29, and decompose the shear τ_n into the two rectangular components τ_{ns} and τ_{nt}. Note that two subscripts are needed; the first identifies the orientation of the surface and the second identifies the direction.

Fig. 2.29

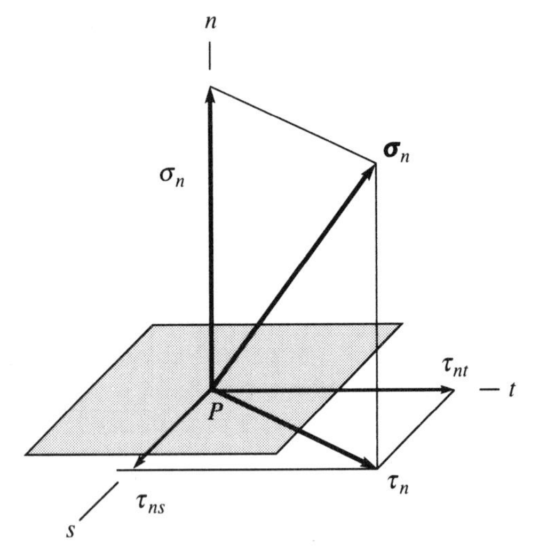

Notation: The symbols σ and τ are consistently used in this text to signify normal and shear components of stress. One subscript is needed to identify the surface on which the stress acts. The components of shear stress require a second subscript to indicate the direction of the action.

In any coordinate system, directions are indicated by the lines of the coordinates, a line along which only that coordinate changes. As examples, consider the rectangular coordinates (x, y, z) of Fig. 2.30(a) and the polar coordinates (r, θ, z) of Fig. 2.30(b). The circular line (r = constant, z = constant) in Fig. 2.30(b) is a θ line; the surface perpendicular to it is a θ surface, which is subject to the stress $\boldsymbol{\sigma}_\theta$ with components σ_θ, $\tau_{\theta z}$, and $\tau_{\theta r}$.

Fig. 2.30

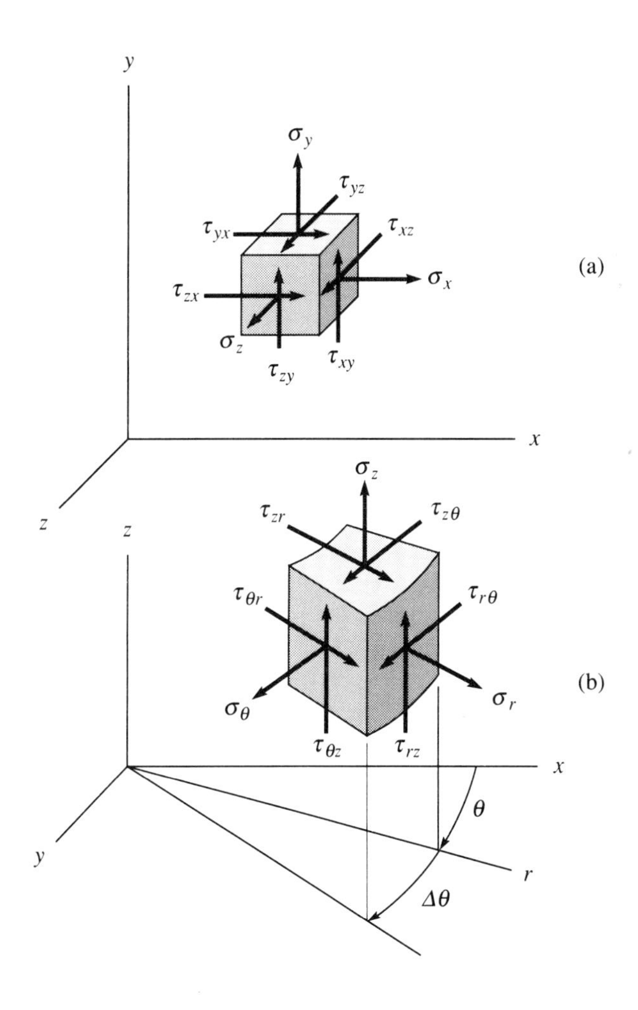

Now consider the two adjoining rectangular blocks of Fig. 2.31; the separated blocks are depicted at (a) and (b). Components of stress on the x face of block A are labeled in accordance with the stated convention; the normal is the $+x$-direction. The actions upon the contiguous face of block B are in opposite directions, in accordance

Fig. 2.31

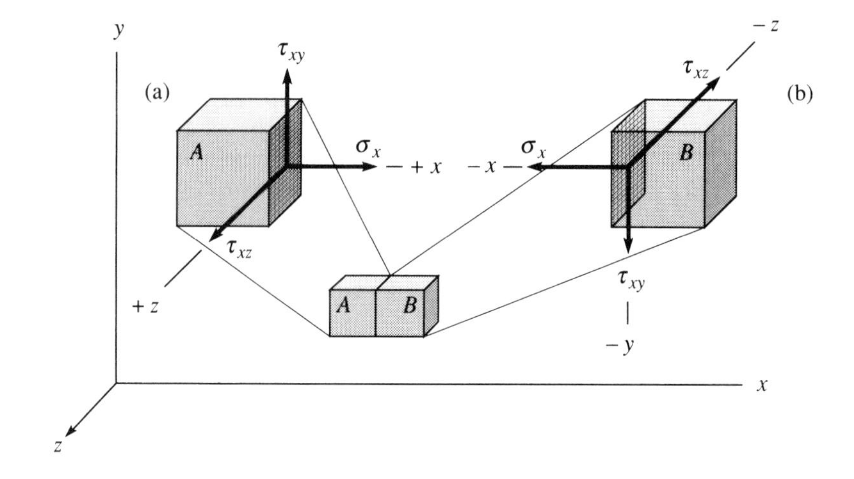

with Newton's law of action and opposing reaction. This opposite sense is signified, too, by the normal direction ($-x$) and tangent directions ($-y$ and $-z$).

2.5 The Relationship Between Stress and the Net Internal Force System

The actual distribution of internal forces on a sectioning surface is described by the stresses at each point of the surface. The net internal system is merely the resultant of the distributed stresses; it is fully determined by that distribution. We can express the resultant in terms of the stresses as follows.

Let ΔA be an element of area taken from a sectioning surface of a body, as shown in Fig. 2.32. The rectangular coordinates (n, s, t) are normal (n) and tangent (s, t) to the section. Normal and shear components of stress are indicated on the enlarged element ΔA. $\Delta\mathbf{F}$ is the internal force acting on ΔA, and ΔF, ΔV_s, and ΔV_t are the normal and two shear components of $\Delta\mathbf{F}$. Recall the definition

$$\boldsymbol{\sigma}_n(P) = \lim_{\substack{\Delta A \to 0 \\ P \text{ always in } \Delta A}} \left(\frac{\Delta\mathbf{F}}{\Delta A}\right)$$

If ΔA is small, we can write

$$\Delta\mathbf{F} = \boldsymbol{\sigma}_n \Delta A \tag{2.16}$$

where it is understood that the stress is evaluated for *some* point in ΔA. By a similar line of reasoning,

$$\Delta V_s = \tau_{ns} \Delta A \tag{2.17}$$

$$\Delta V_t = \tau_{nt} \Delta A \tag{2.18}$$

Fig. 2.32

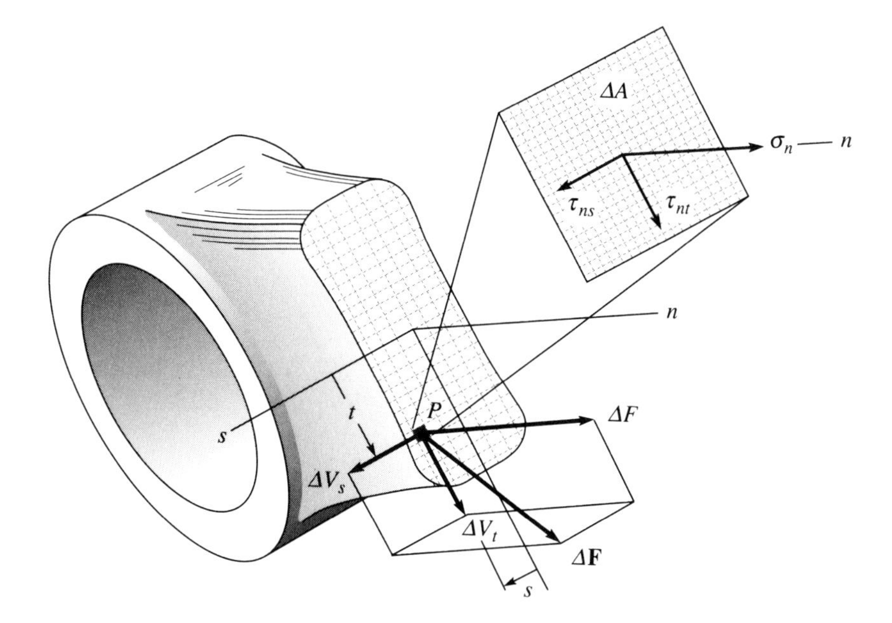

Furthermore, the moments [see Fig. 2.9(b)] of these forces about the s-, t-, and n-axes at O (the centroid of S) are given by

$$\Delta M_n \equiv \Delta T = s\,\Delta V_t - t\,\Delta V_s = [s\tau_{nt} - t\tau_{ns}]\,\Delta A \tag{2.19}$$

$$\Delta M_s = t\,\Delta F = t\sigma_n\,\Delta A \tag{2.20}$$

$$\Delta M_t = -s\,\Delta F = -s\sigma_n\,\Delta A \tag{2.21}$$

Now, ΔA is a typical element of the sectioning surface S. To find the components of the net internal force system, we add the contributions from each element of the surface. In the limit, these sums become integrals. Thus, the components of the net internal force system can be expressed as integrals of the stresses. Normal stresses give rise to normal forces and bending couples; shear stresses give rise to shear forces and twisting couples. This is illustrated in the following examples.

EXAMPLE 9

For the cylinder shown in Fig. 2.33, a cross section S is located by its distance x from one end. Suppose that the stresses on S are known to be constant. In Fig. 2.33, the stresses on the small element of area ΔA give rise to the forces

$$\Delta F = \sigma_x\,\Delta A, \qquad \Delta V_y = \tau_{xy}\,\Delta A, \qquad \Delta V_z = \tau_{xz}\,\Delta A$$

These contributions to the net internal force system on S are added for every element of area S. In the limit,

Fig. 2.33

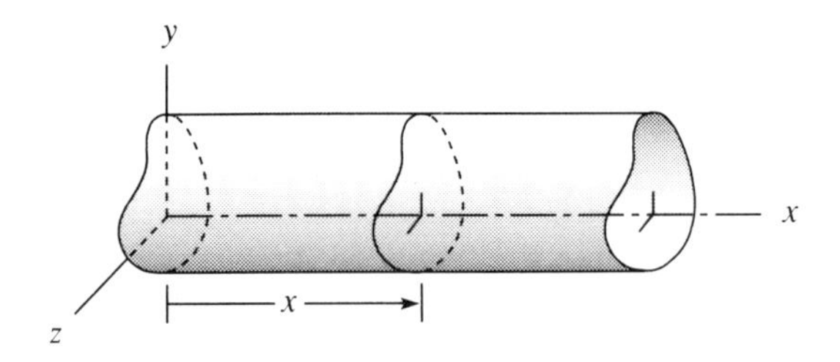

$$
\begin{aligned}
F &= \iint_S \sigma_x \, dA \\
V_y &= \iint_S \tau_{xy} \, dA \\
V_z &= \iint_S \tau_{xz} \, dA
\end{aligned}
\tag{2.22}
$$

where F is the normal force on S and V_y and V_z are the components of the shear force. In this particular example, the stresses σ_x, τ_{xy}, and τ_{xz} are constant on S; therefore, they can be taken outside the integral signs in Eq. (2.22), so that

$$
\begin{aligned}
F &= \sigma_x A \quad \text{or} \quad \sigma_x = \frac{F}{A} \\
V_y &= \tau_{xy} A \quad \text{or} \quad \tau_{xy} = \frac{V_y}{A} \\
V_z &= \tau_{xz} A \quad \text{or} \quad \tau_{xz} = \frac{V_z}{A}
\end{aligned}
$$

where A is the area of S. Note that the present example is an *exception*; usually, stresses are *variable*.

The twisting couple T and the components of the bending couple, M_y and M_z, are also easily computed. In Fig. 2.34, forces are shown acting on the element of area ΔA. The moments of these forces about the coordinate axes are

$$
\begin{aligned}
\Delta T &= y \, \Delta V_z - z \, \Delta V_y = y \tau_{xz} \, \Delta A - z \tau_{xy} \, \Delta A \\
\Delta M_y &= z \, \Delta F = z \sigma_x \, \Delta A \\
\Delta M_z &= -y \, \Delta F = -y \sigma_x \, \Delta A
\end{aligned}
$$

Summing these incremental moments over the entire area of S and again noting that the stresses are constant, we obtain

$$
T = \tau_{xz} \iint_S y \, dA - \tau_{xy} \iint_S z \, dA \tag{2.23}
$$

Fig. 2.34

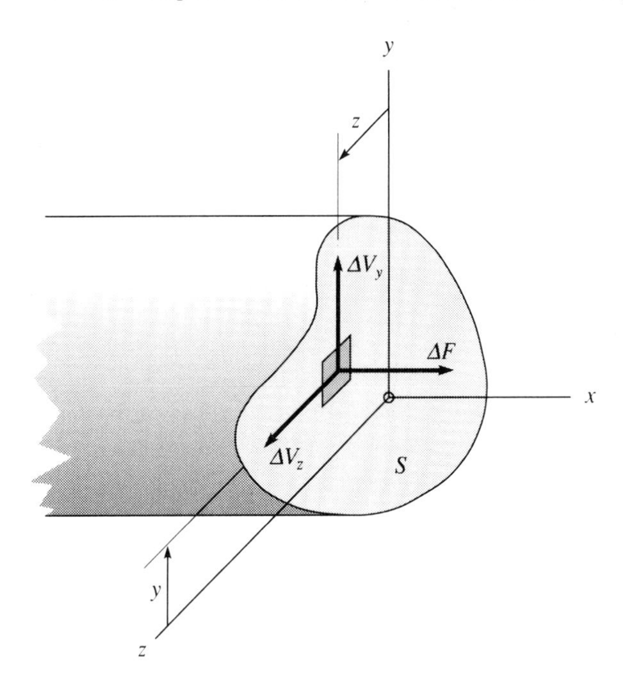

$$M_y = \sigma_x \iint_S z\,dA \tag{2.24}$$

$$M_z = -\sigma_x \iint_S y\,dA \tag{2.25}$$

If the origin of coordinates is at the centroid of S, *then*

$$\iint_S y\,dA = \iint_S z\,dA = 0$$

Consequently, all the integrals in Eqs. (2.23), (2.24), and (2.25) are zero; therefore,

$$T = M_y = M_z = 0$$

This means that a *constant* stress distribution on the section is represented by a force *at the centroid* with no bending or twisting couples. *Again, this is an exceptional circumstance.* ◆

EXAMPLE 10

Consider the same cylinder shown in Fig. 2.33, but in this case suppose that the shear stresses τ_{xy} and τ_{xz} on S are zero. Furthermore, suppose that the only nonzero stress, σ_x, is linearly distributed on S. In equation form,

$$\tau_{xy} = \tau_{xz} = 0 \tag{2.26}$$

$$\sigma_x = C_1 y + C_2 z + C_3 \tag{2.27}$$

where C_1, C_2, and C_3 are at most functions of x. In Fig. 2.34 only the normal force ΔF acts on the element of area ΔA, where

$$\Delta F = \sigma_x \Delta A = C_1 y \Delta A + C_2 z \Delta A + C_3 \Delta A$$

Adding these increments of normal force over the entire area of S, we obtain

$$F = C_1 \iint_S y\, dA + C_2 \iint_S z\, dA + C_3 \iint_S dA$$

The first two integrals are zero *if* y and z are measured from the centroid of S, and the last integral is simply the area A of S. Hence

$$F = C_3 A \quad \text{or} \quad C_3 = \frac{F}{A} \tag{2.28}$$

The contributions of ΔF to the components of the bending couple are

$$\Delta M_y = z\, \Delta F = z\sigma_x \Delta A = C_1 zy\, \Delta A + C_2 z^2\, \Delta A + C_3 z\, \Delta A$$

$$\Delta M_z = -y\, \Delta F = -y\sigma_x \Delta A = -C_1 y^2\, \Delta A - C_2 yz\, \Delta A - C_3 y\, \Delta A$$

(The third integral in each equation is again zero, because y and z are measured from the centroid of S). Then

$$M_y = C_1 \iint_S yz\, dA + C_2 \iint_S z^2\, dA \tag{2.29}$$

$$M_z = -C_1 \iint_S y^2\, dA - C_2 \iint_S yz\, dA \tag{2.30}$$

For convenience, set[2]

$$I_{yz} = \iint_S yz\, dA$$

$$I_z = \iint_S z^2\, dA \tag{2.31}$$

$$I_y = \iint_S y^2\, dA$$

Then Eqs. (2.29) and (2.30) become

$$M_y = C_1 I_{yz} + C_2 I_z \tag{2.32}$$

$$M_z = -C_1 I_y - C_2 I_{yz} \tag{2.33}$$

---

2 The reader will recognize these integrals as the product and the moments of inertia of the area of S with respect to the y- and z-axes in Fig. 2.34.

Solving Eqs. (2.32) and (2.33) for C_1 and C_2, we get

$$C_1 = -\frac{I_z M_z + I_{yz} M_y}{I_z I_y - I_{yz}^2} \tag{2.34}$$

$$C_2 = \frac{I_y M_y + I_{yz} M_z}{I_z I_y - I_{yz}^2} \tag{2.35}$$

Then, upon substituting Eqs. (2.28), (2.34), and (2.35) into Eq. (2.27), we obtain, finally,

$$\sigma_x = -\left[\frac{I_z M_z + I_{yz} M_y}{I_z I_y - I_{yz}^2}\right] y + \left[\frac{I_y M_y + I_{yz} M_z}{I_z I_y - I_{yz}^2}\right] z + \frac{F}{A}$$

From Eq. (2. 26) it follows that

$$V_y = V_z = T = 0$$

on S. ◆

EXAMPLE 11

Consider the circular cylinder shown in Fig. 2.35. Suppose that the normal stress σ_x is zero throughout the cylinder and that the shear stress (τ_x) is symmetric about the axis. Since $\sigma_x = 0$, the normal force F and bending couples, M_y and M_z, vanish.

Fig. 2.35

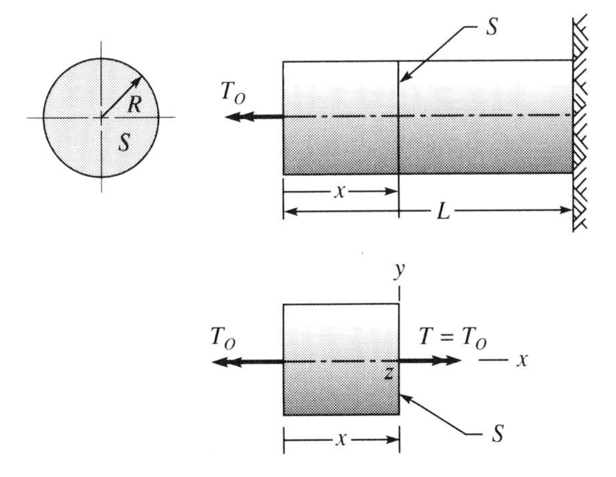

In view of the axial symmetry, it is convenient to use polar coordinates r and θ in the cross section S, as indicated in Fig. 2.36. This defines an r- and a θ-direction at each point of S. On the element of area ΔA shown in Fig. 2.37, the shear stresses τ_{xr} and $\tau_{x\theta}$ give rise to the force components

Fig. 2.36

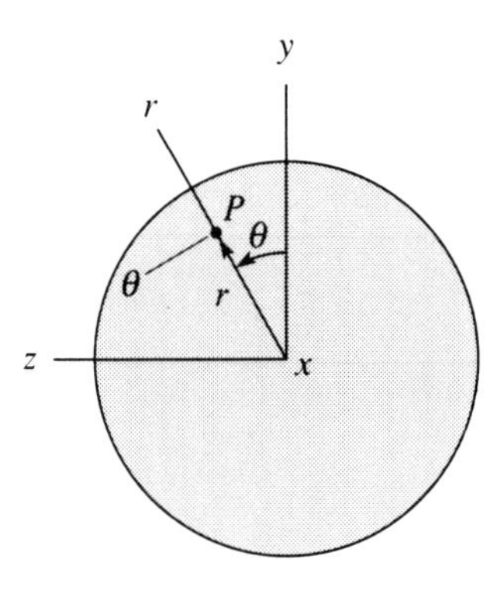

Fig. 2.37

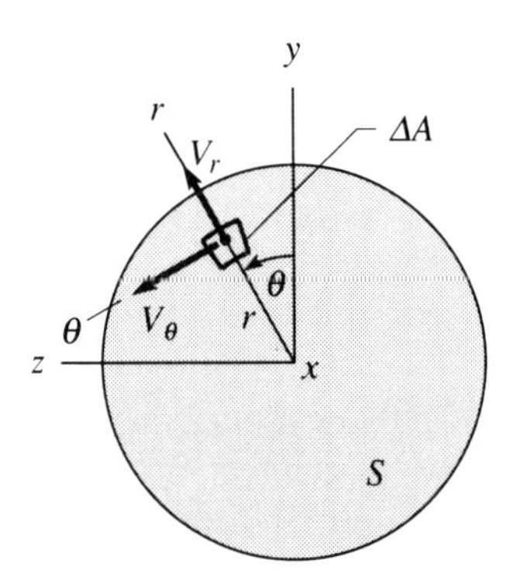

$$\Delta V_r = \tau_{xr}\,\Delta A$$

$$\Delta V_\theta = \tau_{x\theta}\,\Delta A$$

Since ΔV_r is always directed radially, it contributes nothing to the twisting couple on S. The contribution to T from ΔV_θ is

$$\Delta T = r\,\Delta V_\theta = r\tau_{x\theta}\,\Delta A$$

In polar coordinates, the element of area is

$$\Delta A = r\,\Delta r\,\Delta\theta, \tag{2.36}$$

so that

$$\Delta T = \tau_{x\theta} r^2\,\Delta r\,\Delta\theta$$

The twisting couple is the sum of all these increments from the entire area of S:

$$T\,(=T_0) = \int_0^{2\pi}\int_0^R \tau_{x\theta} r^2\,dr\,d\theta \tag{2.37}$$

The directions of ΔV_r and ΔV_θ change from point to point in S. To find the components of the shear force on S, we use the fixed y- and z-directions of Fig. 2.38.

Fig. 2.38

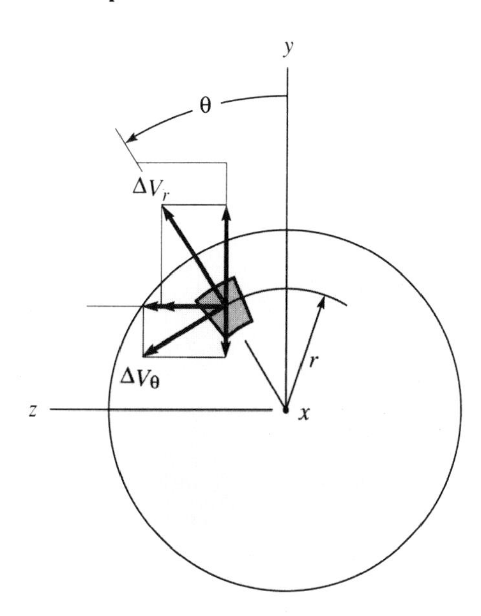

Then the increment of shear force on ΔA has components in the x- and y-directions given by

$$\Delta V_y = \Delta V_r \cos\theta - \Delta V_\theta \sin\theta = \tau_{xr}\cos\theta\,\Delta A - \tau_{x\theta}\sin\theta\,\Delta A$$

$$\Delta V_z = \Delta V_r \sin\theta + \Delta V_\theta \cos\theta = \tau_{xr}\sin\theta\,\Delta A + \tau_{x\theta}\cos\theta\,\Delta A$$

Over the entire area of S, the shear force components in the y- and z-directions are

$$V_y = \int_0^{2\pi}\int_0^R (\tau_{xr}\cos\theta) r\,dr\,d\theta - \int_0^{2\pi}\int_0^R (\tau_{x\theta}\sin\theta) r\,dr\,d\theta \tag{2.38}$$

$$V_z = \int_0^{2\pi}\int_0^R (\tau_{xr}\sin\theta) r\,dr\,d\theta + \int_0^{2\pi}\int_0^R (\tau_{x\theta}\cos\theta) r\,dr\,d\theta \tag{2.39}$$

In our example, we suppose that the shear stresses are radially symmetric; that is, τ_{xr} and $\tau_{x\theta}$ depend on r but not on θ. We can, therefore, complete the integrations with respect to θ in Eqs. (2.37), (2.38), and (2.39) to get

$$T_0 = 2\pi \int_0^R \tau_{x\theta} r^2\,dr \tag{2.40}$$

$$V_y = \int_0^{2\pi} \cos\theta\,d\theta \int_0^R \tau_{xr} r\,dr - \int_0^{2\pi} \sin\theta\,d\theta \int_0^R \tau_{x\theta} r\,dr = 0 \tag{2.41}$$

$$V_z = \int_0^{2\pi} \sin\theta\,d\theta \int_0^R \tau_{xr} r\,dr + \int_0^{2\pi} \cos\theta\,d\theta \int_0^R \tau_{x\theta} r\,dr = 0 \tag{2.42}$$

From Eqs. (2.41) and (2.42), we see that the cross section is free of shear force, whatever the distribution of τ_{xr} and $\tau_{x\theta}$ along a radius. But from Eq. (2.40), the value of T_0 can be determined only if we know the distribution of $\tau_{x\theta}$ along a radius, that is, $\tau_{x\theta}(r)$. For example, suppose that $\tau_{x\theta}$ varies along a diameter, as indicated in Fig. 2.39. It is zero at the axis of the cylinder ($r = 0$) and increases in proportion to the radius until it reaches a limiting value τ_0 (say at $r = b < R$). Over the remainder of the cross section it has value τ_0. This distribution of shear stress is given by the relation

$$\tau_{x\theta} = \begin{cases} \dfrac{r}{b}\tau_0, & 0 \le r \le b \\ \tau_0, & b \le r \le R \end{cases} \tag{2.43}$$

Fig. 2.39

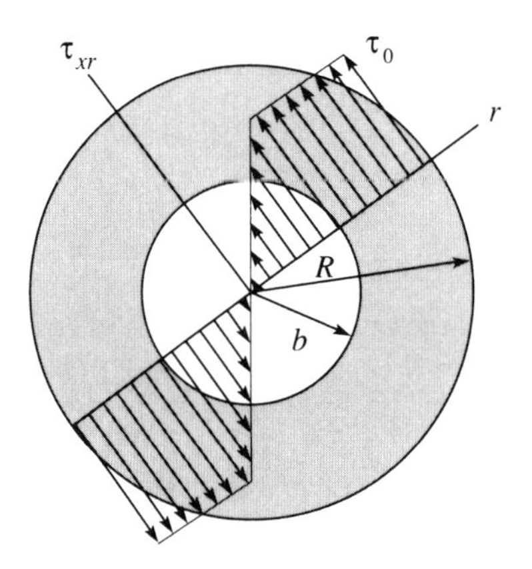

where b is the (unspecified) radius at which $\tau_{x\theta}$ assumes its limiting value. Equation (2.43) is substituted into Eq. (2.40). In order to evaluate the integral, we split it into two parts, one from 0 to b and the other from b to R:

$$\begin{aligned} T_0 &= 2\pi \int_0^b \tau_{x\theta} r^2\,dr + 2\pi \int_b^R \tau_{x\theta} r^2\,dr \\ &= 2\pi \int_0^b \frac{\tau_0}{b} r^3\,dr + 2\pi \int_b^R \tau_0 r^2\,dr \\ &= \frac{2\pi\tau_0}{3}\left[R^3 - \frac{b^3}{4}\right] \end{aligned} \tag{2.44}$$

If T_0 were known, Eq. (2.44) could be solved for b and the result put in Eq. (2.43) to determine completely $\tau_{x\theta}$. However, for this stress distribution to be possible, b must lie between 0 and R. From Eq. (2.44), T_0 must then have a value between $\frac{1}{2}\pi R^3\tau_0$ and $\frac{2}{3}\pi R^3\tau_0$. ◆

2.42 A straight bar with a 2-in. by 2-in. square cross section supports an axial load of 2000 lb. If the components of stress are constant on every cross section, compute their values at any point on a cross section.

2.43 A homogeneous circular cylinder of radius R and length L hangs under its own weight W, as shown in Fig. P2.43. If the stresses on any cross section are constant, determine $\sigma_x(x)$ at any point of the cylinder.

Fig. P2.43

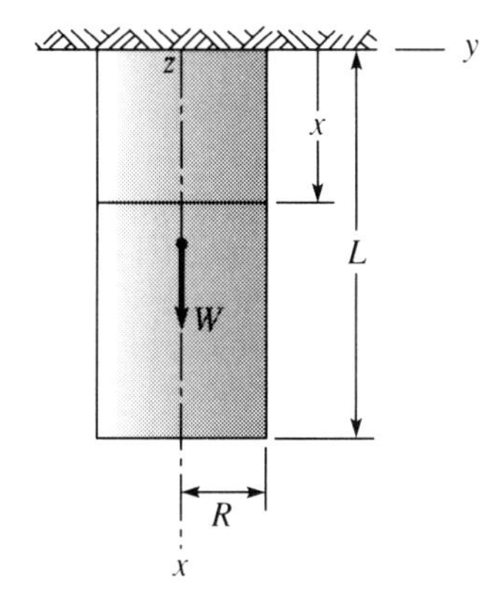

2.44 The normal stress that acts at a point on a cross section of the circular cylinder shown in Fig. P2.44 is proportional to the distance r of that point from the axis of the cylinder, that is, $\sigma_z = Kr$. Determine the constant K.

Fig. P2.44

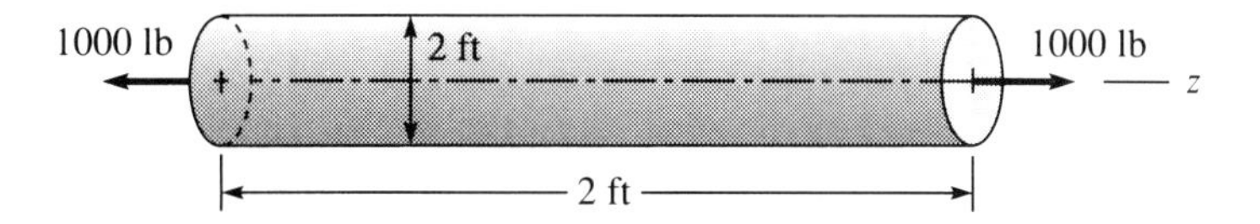

2.45 Repeat Prob. 2.44 for the hollow cylinder shown in Fig. P2.45.

Fig. P2.45

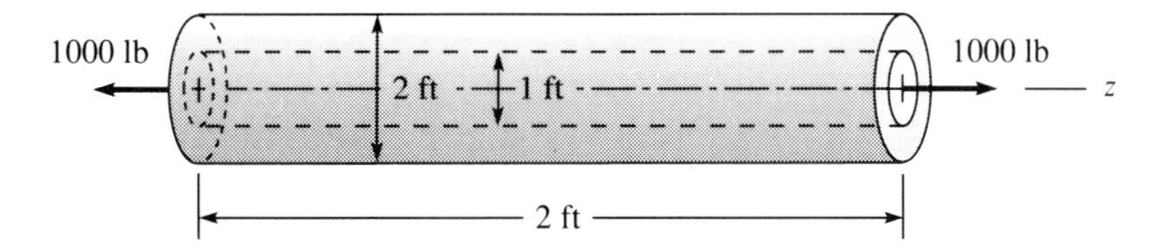

2.46 A straight bar with a 2-cm by 2-cm square cross section supports an axial load of 3000 kN. In terms of the coordinate system set up in a cross section (as shown in Fig. P2.46), $\sigma_x = K(1 - y^2)(1 - z^2)$, $\tau_{xy} = \tau_{zx} = 0$. Compute the value of σ_x at the point $y = \frac{1}{3}$, $z = \frac{1}{4}$ in a cross section.

Fig. P2.46

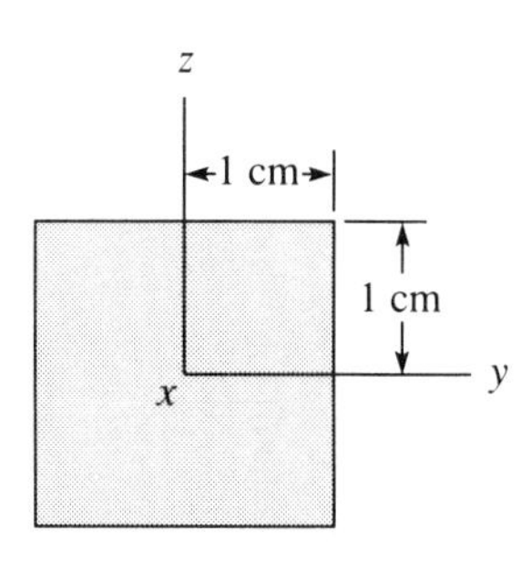

2.47 The circular cylinder shown in Fig. P2.47 is twisted by end couples. If the stresses on a cross section vary according to $\sigma_x = 0$, $\tau_{xr} = 0$, $\tau_{x\theta} = kr$, where k is an undetermined constant, express the value of k, and hence $\tau_{x\theta}$, in terms of T_0.

Fig. P2.47

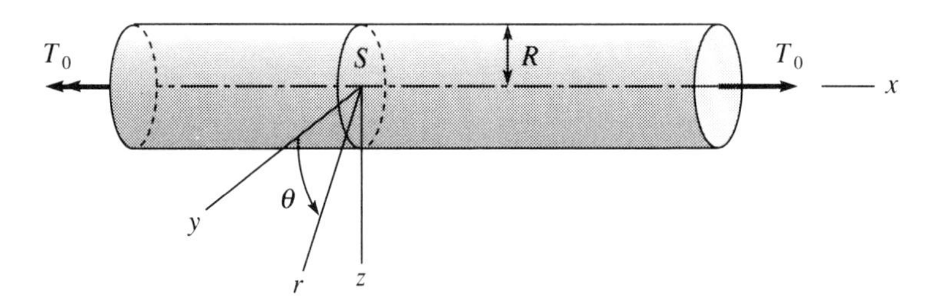

2.48 Suppose the stress distribution on a cross section of the circular cylinder of Prob. 2.47 is given by $\sigma_x = 0$, $\tau_{xr} = 0$, $\tau_{x\theta} = \tau_0$, where τ_0 is a constant. Express τ_0 in terms of T_0.

2.49 Suppose the stress distribution on a cross section of the circular cylinder of Prob. 2.47 is given by $\sigma_x = 0$, $\tau_{xr} = 0$, $\tau_{x\theta} = k\sqrt{r}$, where k is unknown. Express this constant k in terms of T_0.

2.50 Suppose the stress distribution on a cross section of the circular cylinder of Prob. 2.47 is given by $\sigma_x = 0$, $\tau_{xr} = 0$, $\tau_{x\theta} = k[1 - e^{-r/R}]$, where k is unknown. Express this constant k in terms of T_0.

2.51 Repeat Prob. 2.47 for the hollow cylinder shown in Fig. P2.51.

Fig. P2.51

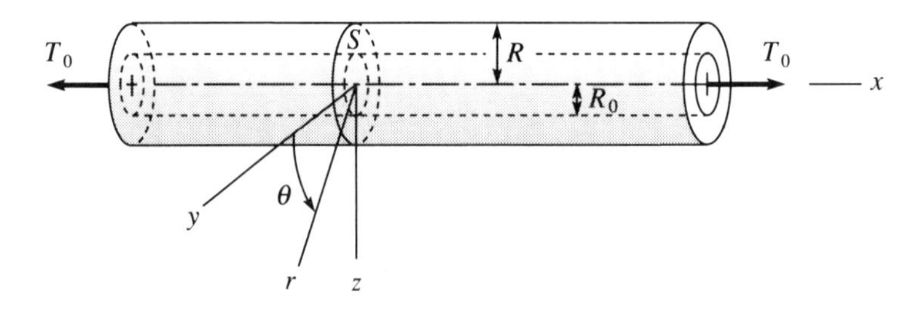

2.52 Repeat Prob. 2.48 for the hollow cylinder shown in Fig. P2.51.

2.53 Repeat Prob. 2.49 for the hollow cylinder shown in Fig. P2.51.

2.54 Repeat Prob. 2.50 for the hollow cylinder shown in Fig. P2.51.

2.55 The thin ring (0.4 in. thick) shown in Fig. P2.55 is subjected to a uniform pressure of 100 lb/in.^{2} on its inner surface. If the stresses on a cross section (such as S) are known to be constant, determine σ_θ. (*Note:* The width does not enter into the answer.)

Fig. P2.55

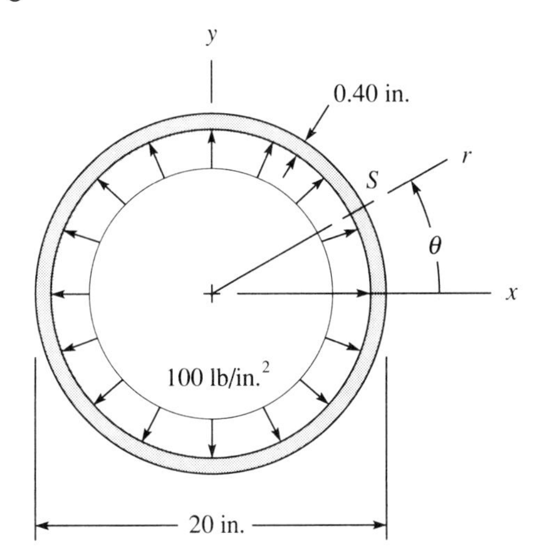

2.56 The stress distribution on the circular cross section shown in Fig. P2.56 is $\sigma_x = 0$, $\tau_{xy} = -20z$ MPa, $\tau_{xz} = +20y$ MPa. Determine the net internal force system on the cross section.

Fig. P2.56

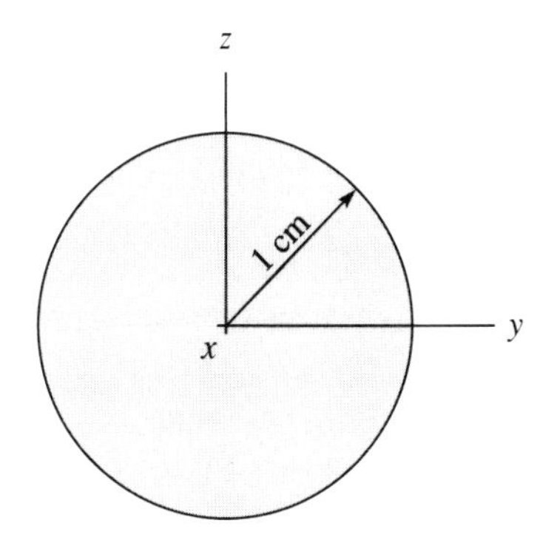

2.57 A straight bar with a thin rectangular cross section is twisted as shown in Fig. P2.57. The stresses on a cross section are given by $\sigma_x = 0$, $\tau_{xy} = 0$, $\tau_{xz} = ky$. Relate the constant k to the twisting couple T_0 and the dimensions h and w.

Fig. P2.57

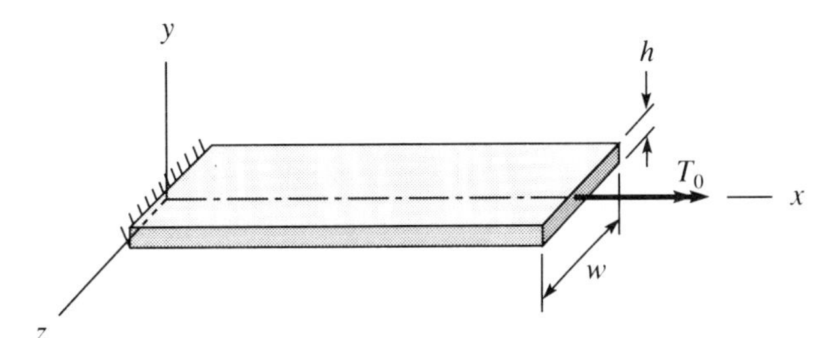

2.58 The stress distribution on the rectangular cross section shown in Fig. P2.58 is given by $\sigma_x = 1000y - 500z + 800$ psi, $\tau_{xy} = 200z$ psi, $\tau_{xz} = 0$ (y and z are measured in inches). What is the net internal force system on this cross section?

Fig. P2.58

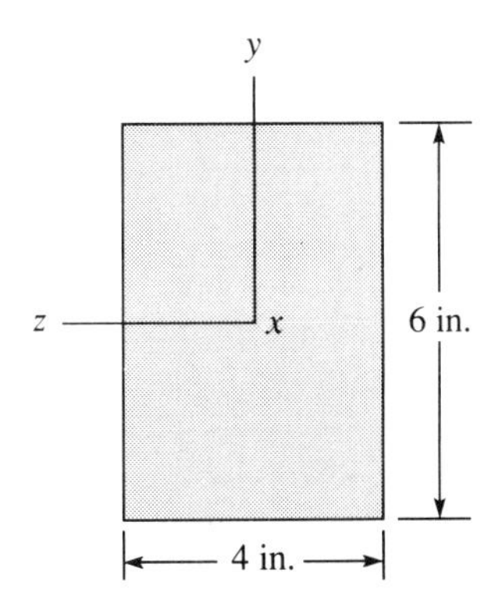

2.59 Repeat Prob. 2.58 with the following stress distribution: $\sigma_x = 4000y + 1000z$ lb/in.^{2}, $\tau_{xy} = 100(9 - y^2)$ lb/in.^{2}, $\tau_{xz} = 100(4 - z^2)$ lb/in^2.

2.60 On the T-shaped cross-section of Fig P2.60 the rectangular components of stress are $\sigma_x = 700(10 - y)$, $\tau_{xy}, \tau_{xz} = 0$ in kPa (kN/m^2). If the resultant is represented by a force at the origin ($y = z = 0$) and a couple, calculate *all* rectangular components of that force and couple.

Fig. P2.60

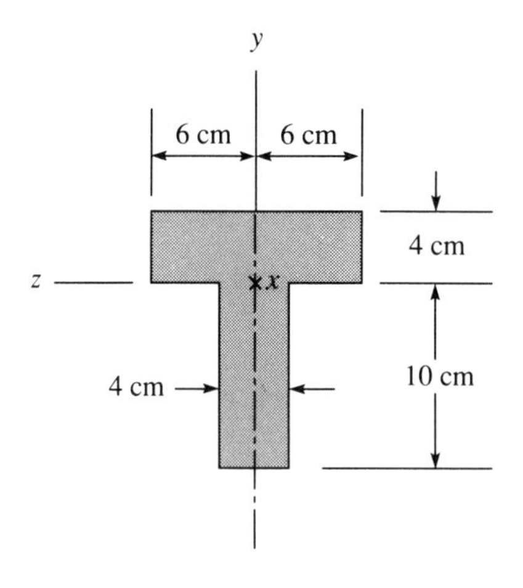

2.61 On the I-shaped cross section of Fig. P2.61, the components of stress are

$$\sigma_x = A + \frac{By}{5}, \qquad \tau_{xz} = 0 \qquad \text{(throughout)}$$

$$\tau_{xy} = \begin{cases} 0, & y > 4 \quad \text{(top flange)} \\ C, & 4 \geq y \geq -4 \\ 0, & -4 > y \quad \text{(bottom flange)} \end{cases}$$

If the resultant is represented by a force acting at the origin and a couple, express *all* components in terms of the constants A, B, and C.

Fig. P2.61

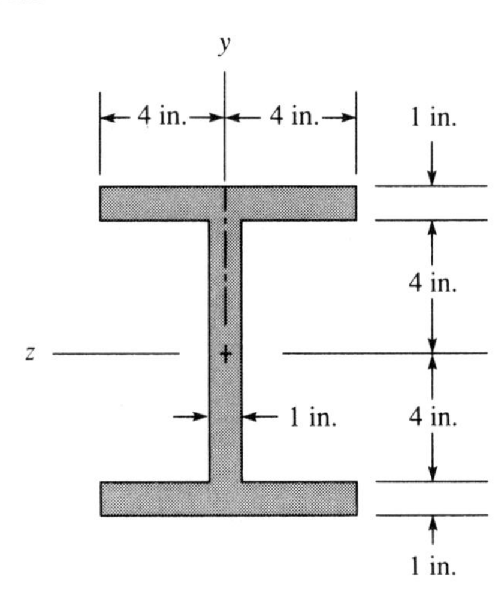

2.62 The beam in Fig. P2.62 has a circular cross section and is subjected to bending couples at its ends. The stress distribution on a cross section is given by $\sigma_x = Ay$, $\tau_{xy} = \tau_{xz} = 0$, where A is an undetermined constant. Find the value of A in terms of M_z.

Fig. P2.62

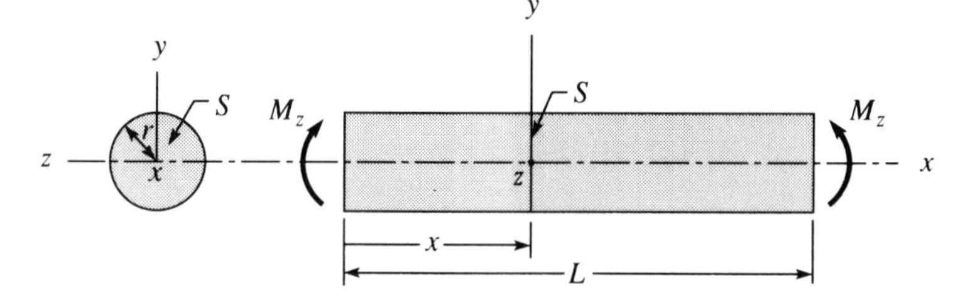

2.63 The resultant of the stress upon the rectangular cross section of Fig. P2.63 is a bending couple M_z *only*. The components of stress are

$$\sigma_x = \begin{cases} A(y+B), & y \geq -B \\ 2A(y+B), & y < -B \end{cases}$$

$$\tau_{xy} = \tau_{xz} = 0, \qquad A \text{ and } B \text{ constants}$$

Express the constants (A, B) in terms of the couple M_z and lengths h and b.

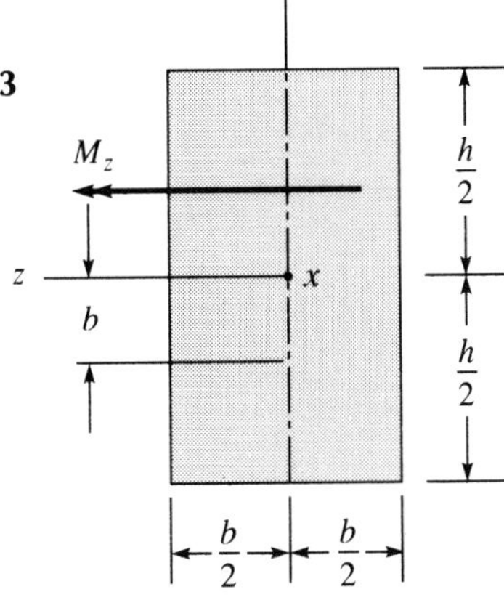

Fig. P2.63

2.64 Repeat Prob. 2.63 with the following stress distribution:

$$\sigma_x = \begin{cases} k\sqrt{y}, & y \geq 0 \\ -k\sqrt{-y}, & y \leq 0 \end{cases} \qquad \tau_{xy} = \tau_{xz} = 0$$

where k is a constant. Find its value in terms of M_z, h, and b.

2.65 Repeat Prob. 2.63 with the following stress distribution:

$$\sigma_x = \begin{cases} -\sigma_0, & y \geq c \\ \sigma_0 \dfrac{y}{c}, & -c \leq y \leq c \\ \sigma_0, & y \leq -c \end{cases} \qquad \tau_{xy} = \tau_{xz} = 0$$

where c is a constant between 0 and $h/2$ and σ_0 is a known number. Determine c and the maximum value (in terms of σ_0) that M_z can have. (Use the value of c between 0 and $h/2$ that maximizes M_z for a fixed σ_0).

2.66 A beam has the triangular cross section of Fig. P2.66. The distribution of stress upon the section is

$$\sigma_z = Cx + By, \qquad \tau_{zx} = \tau_{zy} = 0$$

If the resultant is represented by a force at the origin and a couple, express the components of each in terms of the constants a, B, and C.

Fig. P2.66

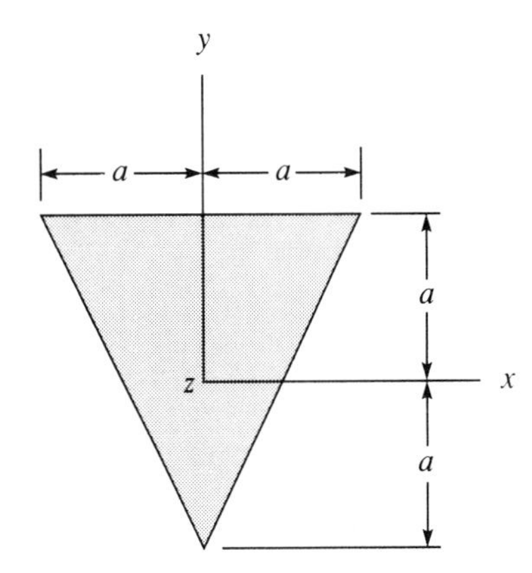

2.67 A member has the circular cross section of Fig. P2.67. The components of stress upon that section are

$$\sigma_z = Ay = Ar\sin\theta, \qquad \tau_{zr} = 0, \qquad \tau_{z\theta} = Br$$

The resultant can be represented by a force at the origin and a couple. Express the components of each in terms of the constants a, A, and B.

Fig. P2.67

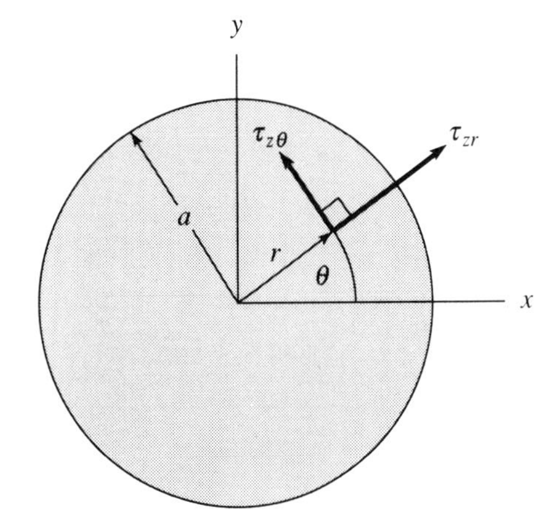

2.68 Repeat Prob. 2.67 if the diameter of the member in Fig. P2.67 is 10 cm and the components of stress are

$$\sigma_z = 10y, \qquad \tau_{zx} = -10y, \qquad \tau_{zy} = 10x$$

in MPa (y and x in cm).

2.69 A bent bar with a 2-in. by 2-in. square cross section is loaded as shown in Fig. P2.69. The stress distribution on a cross section such as S is assumed to be of the form $\sigma_x = Ay + B$, $\tau_{xy} = \tau_{xz} = 0$, where A and B are constants. Compute the values of A and B.

Fig. P2.69

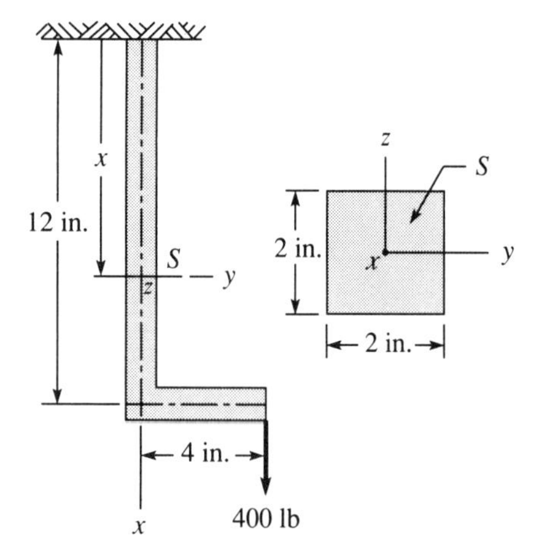

2.70 A beam with a rectangular cross section is loaded as shown in Fig. P2.70. The stress distribution on a cross section such as S is assumed to be of the form $\sigma_x = Ay + B$, $\tau_{xy} = C(25 - y^2)$ where A, B, and C may depend on x but not y or z. Determine A, B, and C.

Fig. P2.70

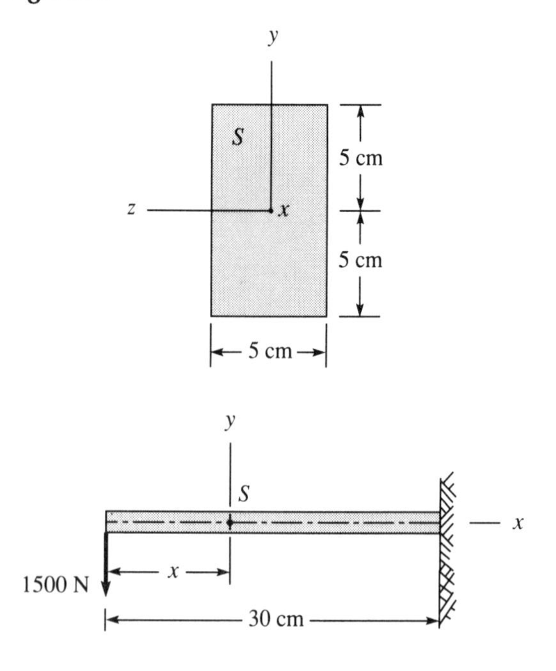

2.71 A triangular plate hangs under its own weight, as shown in Fig. P2.71. The plate weighs 0.2 lb/in³. The stresses on a cross section such as S are assumed to be of the form $\sigma_x = Ay + B$, $\tau_{xy} = \tau_{xz} = 0$, where A and B may depend on x but not y or z. Determine A and B.

Fig. P2.71

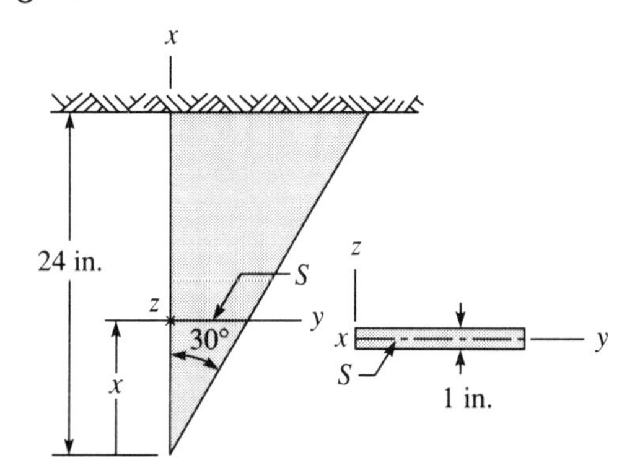

2.72 Opposing forces of ± 5 kN are applied axially to the inner and outer surface of the hollow cylinder in Fig. P2.72. On an intermediate cylindrical surface S

Fig. P2.72

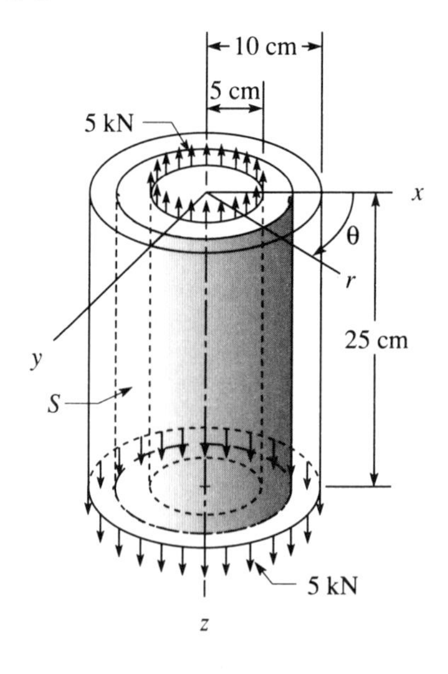

the stress distribution has the form $\sigma_r = 0, \tau_{r\theta} = 0$, $\tau_{rz} = K(r)$. Determine τ_{rz} and plot the variation of τ_{rz} from the inner to outer surfaces.

2.73 Opposing twisting couples of ± 400 in.-lb act upon the inner and outer surfaces of the hollow cylinder shown in Fig. P2.73. On an intermediate cylindrical surface S, the distribution of stress is $\sigma_r = 0$, $\tau_{rz} = 0$, and $\tau_{r\theta} = K(r)$. Determine $\tau_{r\theta}$ and plot the function between the inner and outer surfaces.

Fig. P2.73

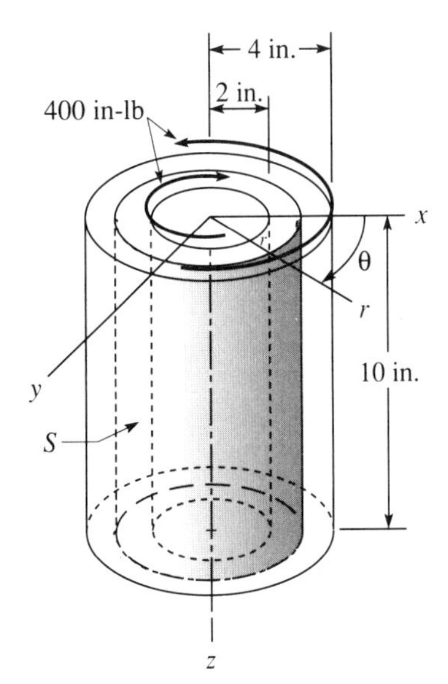

2.6 State of Stress

The concept of stress was introduced to characterize the intensity of the internal forces in the vicinity of an interior point of a body. In addition to specifying the point at which a stress acts, we must specify the plane on which it acts. The stress on the plane can then be defined by a normal stress and two shear stresses. Thus, at every point of the body a normal stress and two shear stresses act upon *every plane through the point*. The question now arises: How many stresses must be known at a point in order to determine the three stresses for every plane?

Stresses on Perpendicular Planes

A small rectangular block removed from the interior of a body is shown in Fig. 2.40. On each of its six faces the resultant of the internal forces is decomposed into three components. For clarity, this is shown only for the faces of the block visible in Fig. 2.40. In this way we see 18 distinct force components acting on the block. Each of these force components can be expressed in terms of an average stress. For example, the force components acting on the front face are

$$\Delta F_z' = \bar{\sigma}_z' \Delta x \Delta y, \qquad \Delta V_{zy}' = \bar{\tau}_{zy}' \Delta x \Delta y, \qquad \Delta V_{zx}' = \bar{\tau}_{zx}' \Delta x \Delta y$$

where bars are placed over average stresses to distinguish them from stresses at a point. Quantities with primes refer to the three faces visible in Fig. 2.40, whereas quantities without primes refer to the three faces passing through the corner P. In Fig. 2.41 the block is viewed from a point on the z-axis. For clarity, force components parallel to the z-axis are not shown.

Fig. 2.40

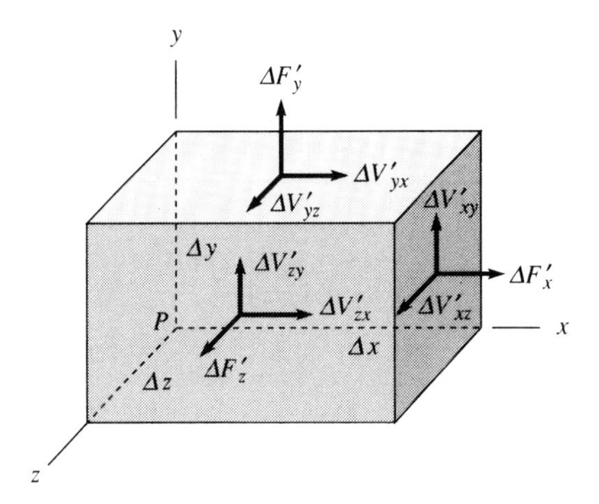

Fig. 2.41

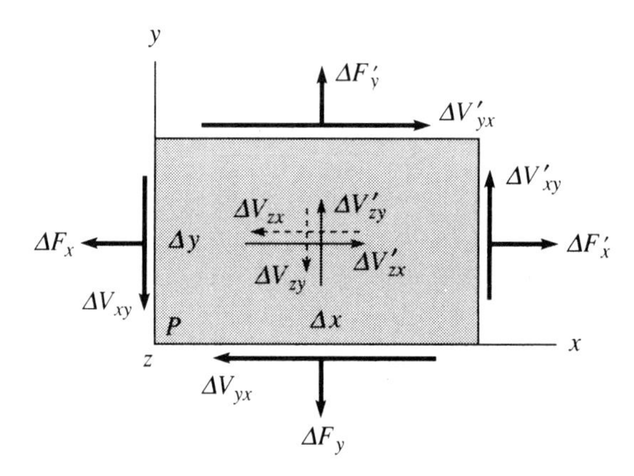

If the dimensions $\Delta x, \Delta y$, and Δz are made smaller, the opposite faces get closer together. The average stresses with primes approach the values of the corresponding average stresses without primes. Furthermore, these average stresses become (in the limit) the stresses associated with the point P. Because we wish to examine only the stresses at point P, we must eventually perform this limiting process. Sacrificing mathematical rigor for simplicity, we regard the values of corresponding average stresses on opposite faces as equal to the values of the stresses at P. For example, the forces on the front and back faces are taken as

$$\Delta F_z = \Delta F_z' = \sigma_z \Delta x \Delta y, \qquad \Delta V_{zy} = \Delta V_{zy}' = \tau_{zy} \Delta x \Delta y$$

$$\Delta V_{zx} = \Delta V_{zx}' = \tau_{zx} \Delta x \Delta y$$

where σ_z, τ_{zy}, and τ_{zx} are evaluated at the point P. The view of Fig. 2.41 is redrawn in Fig. 2.42. Again, for clarity, those forces perpendicular to the xy-plane are not shown.

Fig. 2.42

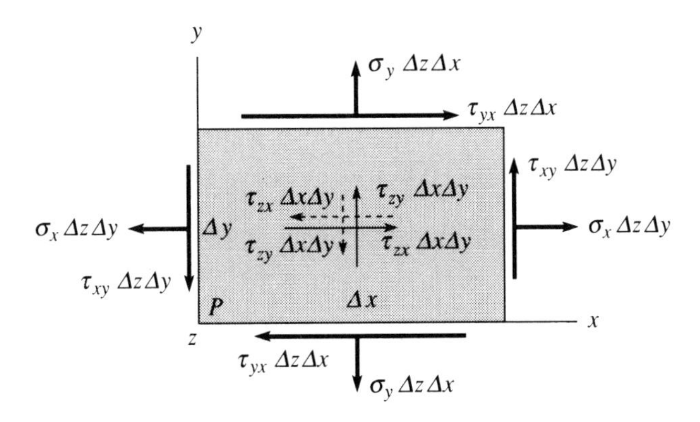

Equilibrium of an Element

Let us consider the equilibrium of the element. The three force conditions of equilibrium are identically satisfied,[3] because the force components on opposite faces cancel in pairs; that is,

$$\sum F_x = 0, \qquad \sum F_y = 0, \qquad \sum F_z = 0$$

However, the sum of the moments is not identically zero. For example, from Fig. 2.42, the sum of moments about the z-axis (through point P) is

$$\begin{aligned}\sum M_z &= (\tau_{xy}\,\Delta y\,\Delta z)\,\Delta x - (\tau_{yx}\,\Delta x\,\Delta z)\,\Delta y \\ &= (\tau_{xy} - \tau_{yx})\,\Delta x\,\Delta y\,\Delta z\end{aligned}$$

For equilibrium, $\sum M_z = 0$. Upon dividing out the factor $\Delta x\,\Delta y\,\Delta z$, we conclude that

$$\tau_{xy} = \tau_{yx} \tag{2.45}$$

Similarly, summing moments about the x- and y-axis, respectively, we conclude that

$$\tau_{zy} = \tau_{yz} \tag{2.46}$$

$$\tau_{zx} = \tau_{xz} \tag{2.47}$$

The number of distinct stresses needed to specify the internal forces acting on the elementary block is thereby reduced to six, namely,

$$\sigma_x, \sigma_y, \sigma_z, \tau_{xy}, \tau_{yz}, \tau_{xz} \tag{2.48}$$

where τ_{xy} is written for both τ_{xy} and τ_{yx}, and so on.

We show later that these six components evaluated at P are sufficient to determine the stresses on any plane through the point. In fact, the six components associated with any three mutually perpendicular planes at P suffice to determine the stresses on any other plane. The three normal stresses and the three (distinct) shear stresses

---

3 This concept is examined more fully in Sec. 2.13.

associated with any three mutually perpendicular directions at P are said to define the *state of stress* at P.

2.7 Stresses on Any Surface

Suppose that the six components in a rectangular system (x, y, z) are known at a point. Then the stresses on those surfaces are given by the vectors

$$\boldsymbol{\sigma}_x = \sigma_x \hat{\mathbf{i}} + \tau_{xy} \hat{\mathbf{j}} + \tau_{xz} \hat{\mathbf{k}} \tag{2.49a}$$

$$\boldsymbol{\sigma}_y = \tau_{yx} \hat{\mathbf{i}} + \sigma_y \hat{\mathbf{j}} + \tau_{yz} \hat{\mathbf{k}} \tag{2.49b}$$

$$\boldsymbol{\sigma}_z = \tau_{zx} \hat{\mathbf{i}} + \tau_{zy} \hat{\mathbf{j}} + \sigma_z \hat{\mathbf{k}} \tag{2.49c}$$

where $\hat{\mathbf{i}}$, $\hat{\mathbf{j}}$, and $\hat{\mathbf{k}}$ denote the unit vectors in the directions of the coordinates x, y, and z, respectively. Now suppose that we are interested in the stress upon a surface with some other orientation. The orientation of the surface in question can be identified by its unit normal $\hat{\mathbf{n}}$; this direction can be specified by the angles that the normal forms with axes x, y, z or, quite simply, by the direction cosines l_n, m_n, n_n. Then the unit normal has the form

$$\hat{\mathbf{n}} = l_n \hat{\mathbf{i}} + m_n \hat{\mathbf{j}} + n_n \hat{\mathbf{k}} \tag{2.50}$$

To determine the stress upon the surface, we suppose that a plane passes through the body with the normal $\hat{\mathbf{n}}$. The plane slices through the three rectangular planes (x, y, z) near the point, and together these planes enclose the small tetrahedron of Fig. 2.43. Our inclined face has normal $\hat{\mathbf{n}}$; the rectangular faces have normals $-\hat{\mathbf{i}}$, $-\hat{\mathbf{j}}$, and $-\hat{\mathbf{k}}$. This small tetrahedron can be viewed as a free body, acted upon by the stresses $\boldsymbol{\sigma}_n$ on the inclined face and $-\boldsymbol{\sigma}_x$, $-\boldsymbol{\sigma}_y$, and $-\boldsymbol{\sigma}_z$ on the rectangular faces. The free-body diagram of Fig. 2.43 is acted upon by forces; these are the mean stresses multiplied by the respective areas. Those areas are denoted by $\Delta A_x, \Delta A_y$, and ΔA_z, where

Fig. 2.43

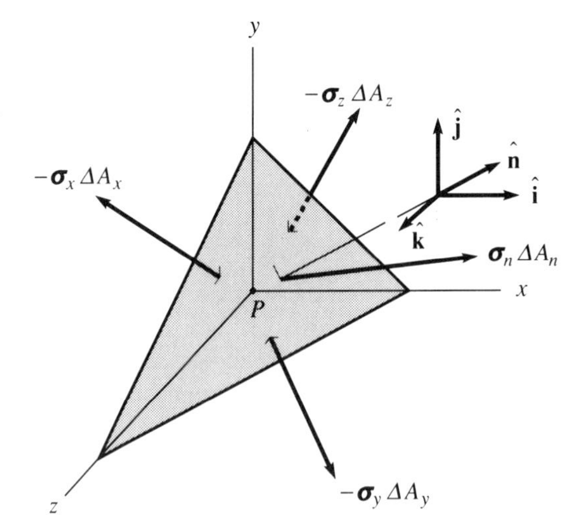

the subscripts signify the orientation of the face. Equilibrium requires that the sum of all forces vanish:

$$\sum \mathbf{F} = \boldsymbol{\sigma}_n \Delta A_n - \boldsymbol{\sigma}_x \Delta A_x - \boldsymbol{\sigma}_y \Delta A_y - \boldsymbol{\sigma}_z \Delta A_z = 0 \tag{2.51a}$$

or

$$\boldsymbol{\sigma}_n = \frac{\Delta A_x}{\Delta A_n}\boldsymbol{\sigma}_x + \frac{\Delta A_y}{\Delta A_n}\boldsymbol{\sigma}_y + \frac{\Delta A_z}{\Delta A_n}\boldsymbol{\sigma}_z \tag{2.51b}$$

Of course, we understand that the elemental tetrahedron is shrunk to the point; then the equation gives the stress at the point. The ratios of the areas depend on the orientation of the inclined face, or the direction of the normal $\hat{\mathbf{n}}$. In fact, the relations are simply

$$\frac{\Delta A_x}{\Delta A_n} = l_n, \qquad \frac{\Delta A_y}{\Delta A_n} = m_n, \qquad \frac{\Delta A_z}{\Delta A_n} = n_n \tag{2.52}$$

Then the equation of equilibrium takes the conventional form:

$$\boldsymbol{\sigma}_n = l_n \boldsymbol{\sigma}_x + m_n \boldsymbol{\sigma}_y + n_n \boldsymbol{\sigma}_z \tag{2.53}$$

This result is most important: If the stresses are known for the three directions (x, y, z), in other words, the six components

$$\begin{matrix} \sigma_x & \tau_{xy} & \tau_{xz} \\ & \sigma_y & \tau_{yz} \\ & & \sigma_z \end{matrix} \tag{2.54}$$

are known, then the stress is determined on *any other* surface in accordance with Eq. 2.53. We need to specify only the orientation, or direction. The direction might be prescribed by the angles between the normal n and the directions x, y, z or by the direction cosines (l_n, m_n, n_n).

We might be interested in the normal component of $\boldsymbol{\sigma}_n$, namely, σ_n. (The material might be brittle and, therefore, susceptible to fracture caused by a normal stress.) In accordance with the definition, the normal component is the projection of the vector $\boldsymbol{\sigma}_n$ on the normal. This component is the scalar (dot) product of $\hat{\mathbf{n}}$ (2.50) and $\boldsymbol{\sigma}_n$ (2.53), where $\boldsymbol{\sigma}_x, \boldsymbol{\sigma}_y, \boldsymbol{\sigma}_z$ are given by (2.49):

$$\begin{aligned} \sigma_n = \hat{\mathbf{n}} \cdot \boldsymbol{\sigma}_n = {} & l_n l_n \sigma_x + l_n m_n \tau_{xy} + l_n n_n \tau_{yz} \\ & + m_n l_n \tau_{yx} + m_n m_n \sigma_y + m_n n_n \tau_{yz} \\ & + n_n l_n \tau_{zx} + n_n m_n \tau_{zy} + n_n n_n \sigma_z \end{aligned} \tag{2.55}$$

The component is depicted in Fig. 2.44.

We might require a particular shear component of the stress $\boldsymbol{\sigma}_n$. (The material might be susceptible to shearing action in this particular direction along the surface.) If so, we must prescribe the direction in question. Most conveniently, we could give the direction cosines (l_t, m_t, n_t) of the tangential direction t depicted in Fig. 2.44. The unit tangent vector is then

Fig. 2.44

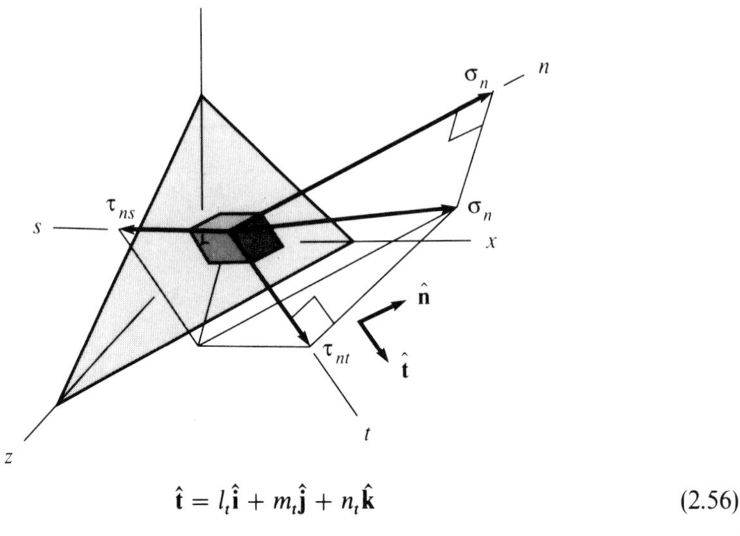

$$\hat{\mathbf{t}} = l_t\hat{\mathbf{i}} + m_t\hat{\mathbf{j}} + n_t\hat{\mathbf{k}} \tag{2.56}$$

The required component on the surface (n) in the direction (t) is τ_{nt}, the projection of $\boldsymbol{\sigma}_n$ on the line t. It is also obtained as the scalar (dot) product of (2.53) and (2.56):

$$\begin{aligned}\tau_{nt} = \hat{\mathbf{t}}\cdot\boldsymbol{\sigma}_n &= l_t l_n \sigma_x + l_t m_n \tau_{xy} + l_t n_n \tau_{xz} \\ &\quad + m_t l_n \tau_{yx} + m_t m_n \sigma_y + m_t n_n \tau_{yz} \\ &\quad + n_t l_n \tau_{zx} + n_t m_n \tau_{zy} + n_t n_n \sigma_z\end{aligned} \tag{2.57}$$

In this manner, we can compute any component of stress upon any surface at a point if the six components are known. We say that the *state of stress* is determined by the six components.

It is noteworthy that the foregoing results and equations also hold if the body is in motion. A simple argument follows: Suppose that the element of Fig. 2.43 were in motion with an acceleration **a**. The mass density is a number, say ρ, and the mass of the element is $\rho\,\Delta V$, where ΔV denotes the volume. Since the sum of the forces equals the product of the mass and acceleration, the right side of (2.51a) is not zero but is $\rho\mathbf{a}\,\Delta V$. Again, the equation is divided by the area ΔA_n and, upon passing to the limit $\Delta V/\Delta A_n \to 0$, we obtain the previous result (2.53).

The following example serves to illustrate the foregoing description of a state and the transformation from one rectangular system to another.

EXAMPLE 12

A state of stress is described by six components; these need not be given with reference to the same rectangular system. Suppose that the four components ($\tau_{xy}, \tau_{xz}, \tau_{yz}$, and σ_z) are given in one system, (x, y, z), and the two components (σ_n, τ_{nt}) in another,

(n, t, z), as shown here. From the information, the definitions, and equilibrium, we can determine *all* components in either system; that is, we can fill in the blanks:

$$\begin{matrix} \sigma_x & \tau_{xy} & \tau_{xz} \\ \tau_{yx} & \sigma_y & \tau_{yz} \\ \tau_{zx} & \tau_{zy} & \sigma_z \end{matrix} = \begin{matrix} --- & 0 & 0 \\ --- & --- & 50 \text{ ksi} \\ --- & --- & 80 \text{ ksi} \end{matrix}$$

$$\begin{matrix} \sigma_n & \tau_{nt} & \tau_{nz} \\ \tau_{tn} & \sigma_t & \tau_{tz} \\ \tau_{zn} & \tau_{zt} & \sigma_z \end{matrix} = \begin{matrix} 60 & 40 & --- \\ --- & --- & --- \text{ ksi} \\ --- & --- & --- \text{ ksi} \end{matrix}$$

Of course, the relative orientations of the systems must be known. Here, the orthogonal directions n, t are also orthogonal to z and inclined to x and y, as shown in Fig. 2.45.

Fig. 2.45

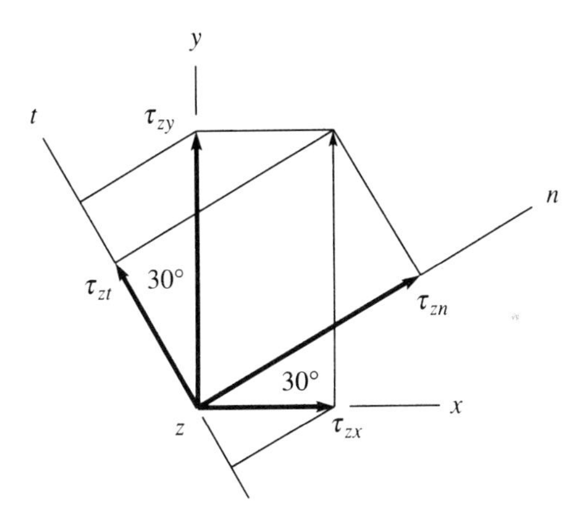

Coordinates (x, y, z) and (n, t, z)

We recall Eqs. (2.45), (2.46), and (2.47), which enforce equilibrium of moment and assert the symmetry of the shear components with respect to their subscripts. Accordingly,

$$\tau_{yx} = \tau_{xy} = 0, \qquad \tau_{zx} = \tau_{xz} = 0, \qquad \tau_{tn} = \tau_{nt} = 40 \text{ ksi} \quad [276 \text{ MPa}]$$

$$\tau_{zy} = \tau_{yz} = 50 \text{ ksi} \quad [345 \text{ MPa}]$$

From the definitions, the shear stress upon the z surface is given by the two orthogonal components, τ_{zx} and τ_{zy}. This shear stress can be expressed as well by the components τ_{zn} and τ_{zt}. In other words, it is a matter of projecting the vector $\boldsymbol{\tau}_z$ in the directions n or t. From the figure we see that

$$\tau_{zt} = -\tau_{zx} \sin 30^\circ + \tau_{zy} \cos 30^\circ$$
$$= 50 \cos 30^\circ = 43.3 \text{ ksi} \quad [298 \text{ MPa}]$$
$$\tau_{zn} = \tau_{zx} \cos 30^\circ + \tau_{zy} \sin 30^\circ$$
$$= 50 \sin 30^\circ = 25 \text{ ksi} \quad [172 \text{ MPa}]$$

Either result can be achieved by an appeal to equilibrium. The free body in Fig. 2.46 is a wedge acted upon by the forces on the $-x$, $-y$, and n faces. For simplicity, we show only the relevant forces (stresses × areas); A denotes the area of the inclined (n) face. Equilibrium of the forces gives the preceding result for τ_{nz}. The same result is obtained from Eq. (2.55) by forming the scalar product $\hat{\mathbf{k}} \cdot \boldsymbol{\sigma}_n$, as in Eq. (2.57); since $l_n = \cos\theta$, $m_n = \sin\theta$, and $n_n = 0$,

$$\hat{\mathbf{k}} \cdot \boldsymbol{\sigma}_n = \hat{\mathbf{k}} \cdot \boldsymbol{\sigma}_x \cos\theta + \hat{\mathbf{k}} \cdot \boldsymbol{\sigma}_y \sin\theta$$
$$= \tau_{xz} \cos\theta + \tau_{yz} \sin\theta$$

Fig. 2.46

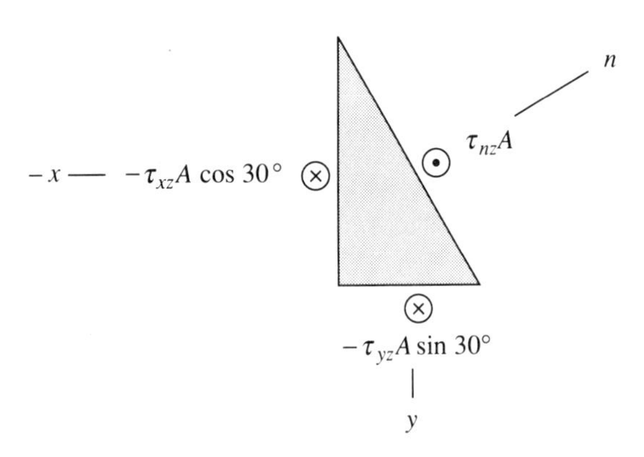

Free body

The remaining components, σ_y and σ_t, are also obtained by enforcing equilibrium. Each of the free bodies shown in Fig. 2.47 is acted upon by one or more unknown actions ($A\sigma_x, A\sigma_y, A\sigma_t$).

From the first free-body diagram (F.B.D.), $\sum F_y = 0$ gives

$$\sigma_y \sin 30^\circ = 60 \sin 30^\circ + 40 \cos 30^\circ, \qquad \sigma_y = 129 \text{ ksi} \quad [891 \text{ MPa}]$$

and $\sum F_x = 0$ gives

$$\sigma_x \cos 30^\circ = 60 \cos 30^\circ - 40 \sin 30^\circ, \qquad \sigma_x = 37 \text{ ksi} \quad [254 \text{ MPa}]$$

From the second F.B.D., $\sum F_t = 0$ gives

$$\sigma_t = \sigma_y \cos^2 30^\circ + \sigma_x \sin^2 30^\circ = 106 \text{ ksi} \quad [732 \text{ MPa}]$$

Fig. 2.47

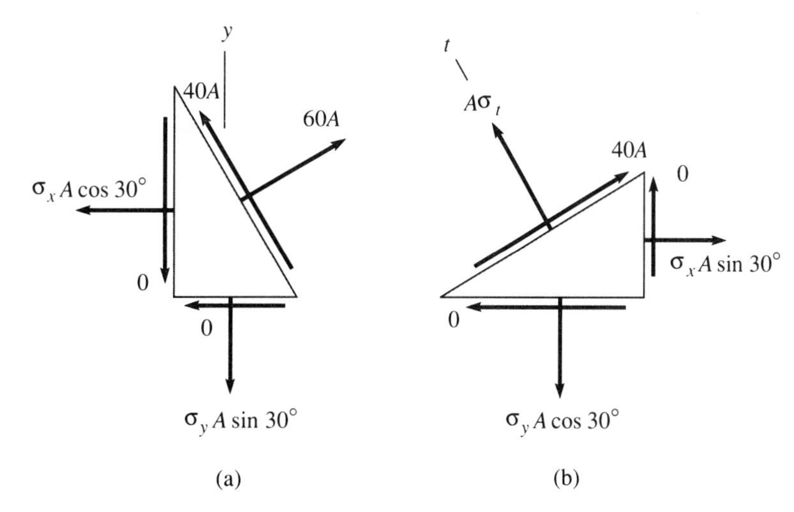

Free bodies

The reader can verify that the latter component, too, is the normal component of $\boldsymbol{\sigma}_t$. Indeed, the important message here is that transformations such as Eq. (2.55) and (2.57) embody no more than the *definitions* and equilibrium. ◆

PROBLEMS

2.74 On a plane through a point P of a body, two rectangular coordinates (x, y, z) and (n, t, z) are oriented as shown in Fig. P2.74. Solve the following problems by the method of the previous section; that is, sketch the free bodies (wedge-shaped bodies) by passing planes normal to lines n and t, write the equations of equilibrium, and solve for the required components of stress.

a. $\sigma_x = 2000$ psi, $\sigma_y = 4000$ psi, $\sigma_z = 0$
$\tau_{xy} = 4000$ psi, $\tau_{xz} = 2000$ psi, $\tau_{yz} = 0$
$\theta = 30°$
Obtain $\sigma_n, \sigma_t, \tau_{nt}, \tau_{nz}, \tau_{tz}$.

b. $\sigma_x = -30$ MPa, $\sigma_y = 0$, $\sigma_z = 20$ MPa
$\tau_{xy} = 30$ MPa, $\tau_{xz} = 20$ MPa, $\tau_{yz} = 10$ MPa
$\theta = 15°$
Obtain $\sigma_n, \sigma_t, \tau_{nt}, \tau_{nz}, \tau_{tz}$.

c. $\sigma_x = 0$, $\sigma_y = 0$, $\tau_{xy} = 6$ MPa, $\sigma_z = 0$
$\tau_{xz} = \tau_{yz} = 0$, $\theta = -45°$
Obtain $\sigma_n, \sigma_t, \tau_{nt}, \tau_{nz}, \tau_{tz}$.

d. $\sigma_x = -10$ ksi, $\sigma_y = 10$ ksi, $\sigma_z = 10$ ksi
$\tau_{xy} = \tau_{xz} = \tau_{yz} = 0$, $\theta = 45°$
Obtain $\sigma_n, \sigma_t, \tau_{nt}, \tau_{nz}, \tau_{tz}$.

Fig. P2.74

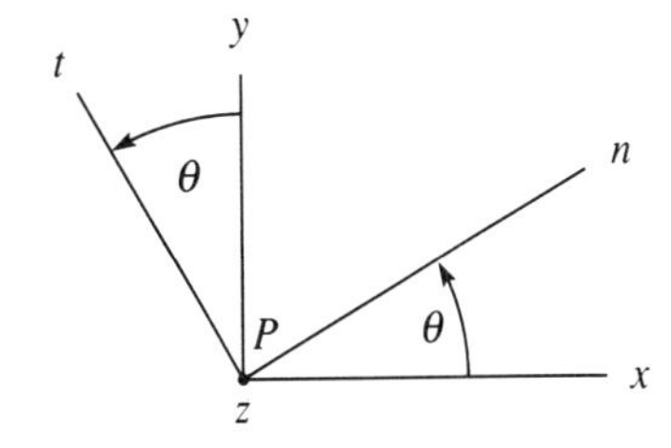

2.75 The wedge-shaped element of Fig. P2.75 is subjected to the uniform stresses indicated in the figure. Enforce equilibrium to determine the components

σ_x, σ_y, τ_{xy}. (*Hint:* Other free-body diagrams are always admissible.)

Fig. P2.75

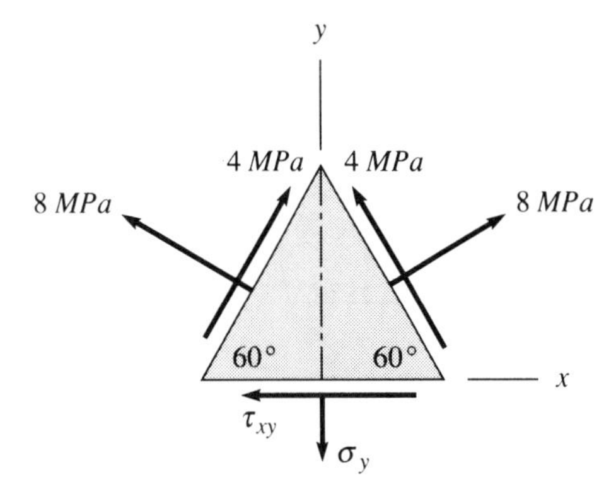

2.76 Some components of stress are given in the rectangular system (x, y, z) of Fig. P2.76:

$$\sigma_x = 20 \text{ ksi}, \qquad \sigma_z = -10 \text{ ksi}$$

$$\tau_{xy} = \tau_{yz} = 0$$

At the same point, two components are given in the rectangular system (x', y', z):

$$\sigma_{x'} = 19 \text{ ksi}, \qquad \tau_{y'z} = -10 \text{ ksi}$$

Employ the method of the preceding section, namely, statics, to determine the components τ_{xz} and σ_y.

Fig. P2.76

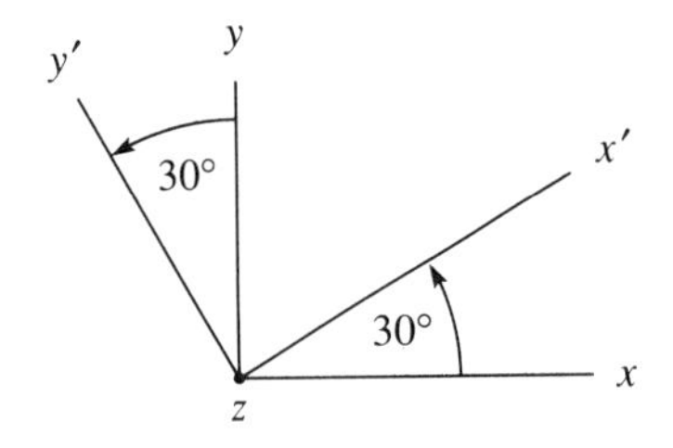

2.77 At a point of a body the components of stress are given in a rectangular system (x, y, z); the six components are presented in the array:

$$\begin{matrix} \sigma_x & \tau_{xy} & \tau_{xz} \\ & \sigma_y & \tau_{yz} \\ & & \sigma_z \end{matrix} = \begin{matrix} 0 & 100 & 0 \\ & 200 & 0 \\ & & -200 \end{matrix} \quad (\text{MPa})$$

Lines p and q of Fig. P2.77 lie in the yz-plane. Determine components τ_{px} and τ_{qx}. (*Hint:* These are two rectangular components of the stress acting on the x surface, much as τ_{zx} and τ_{zy} act on the z surface.

Fig. P2.77

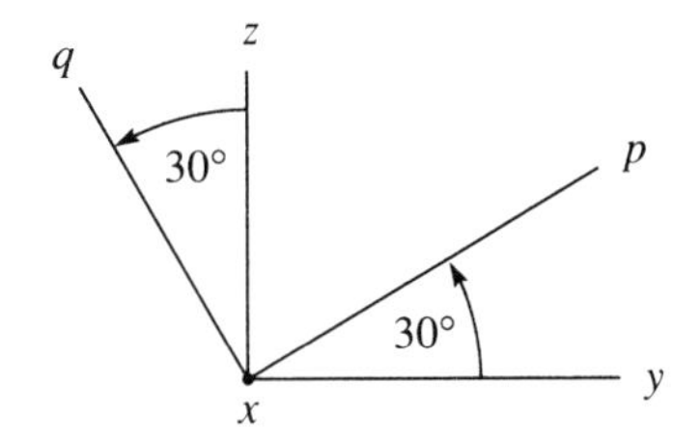

2.78 Five components of stress are given in terms of the rectangular system (x, y, z) at a point within a steel member:

$$\begin{matrix} \sigma_x & \tau_{xy} & \tau_{xz} \\ & \sigma_y & — \\ & & \sigma_z \end{matrix} = \begin{matrix} 20 & 0 & 0 \\ & -30 & — \\ & & 15 \end{matrix} \quad (\text{ksi})$$

One component is given in the rectangular system $(\bar{x}, \bar{y}, z)$:

$$\tau_{\bar{x}z} = 20 \text{ ksi}$$

Determine all components of stress in both systems (Fig. P2.78).

Fig. P2.78

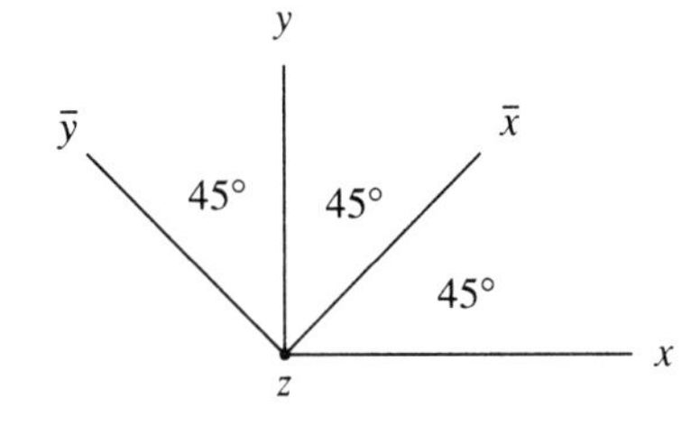

2.79 Components of stress at a point of a body are given in a rectangular system (x, y, z):

$$\begin{matrix} \sigma_x & \tau_{xy} & \tau_{xz} \\ & \sigma_y & \tau_{yz} \\ & & \sigma_z \end{matrix} = \begin{matrix} 100 & 0 & 0 \\ & -80 & 60 \\ & & 60 \end{matrix} \quad (\text{MPa})$$

Unit vectors in these directions (x, y, z) are designated $\hat{\mathbf{i}}, \hat{\mathbf{j}}, \hat{\mathbf{k}}$. The unit normal vector to a particular surface at this point is

$$\hat{\mathbf{n}} = \frac{1}{2}\hat{\mathbf{i}} + \frac{1}{2}\hat{\mathbf{j}} + \frac{1}{\sqrt{2}}\hat{\mathbf{k}}$$

Determine the stress (vector) on this surface at the point. Determine also the component of normal stress σ_n.

2.80 At a point of a body, components of stress are known in the rectangular systems (x, y, z) and (n, t, z). The known components are

$$\sigma_x = 20 \text{ ksi}, \qquad \sigma_y = -10 \text{ ksi}, \qquad \sigma_z = 10 \text{ ksi}$$

$$\tau_{xz} = 20 \text{ ksi}, \qquad \tau_{nt} = 30 \text{ ksi}, \qquad \tau_{nz} = 10 \text{ ksi}$$

The orientations are shown in Fig. P2.80. Determine the components τ_{xy}, τ_{zt}, and τ_{zy}.

Fig. P2.80

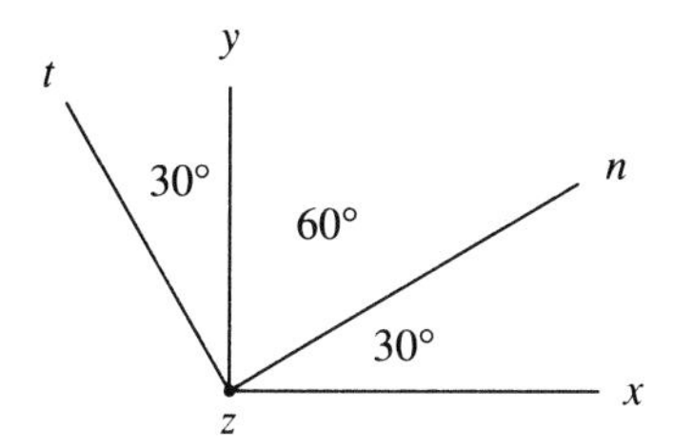

2.81 Four components of stress are known for the coplanar directions (x, y), (m, n), and (s, t) (shown in Fig. P2.81):

$$\sigma_n = 20 \text{ ksi}, \qquad \tau_{mn} = 12 \text{ ksi}$$

$$\sigma_s = 30 \text{ ksi}, \qquad \tau_{ts} = 6 \text{ ksi}$$

Determine the components σ_y and τ_{xy}. (*Hint:* These are *statically* determinate.)

Fig. P2.81

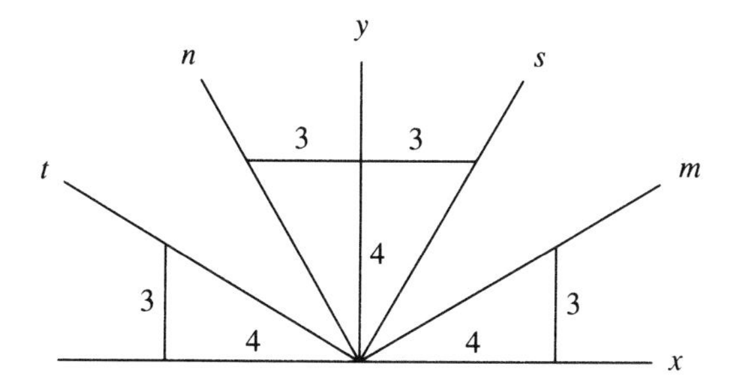

2.82 A state of stress is defined by $\sigma_x, \sigma_y, \sigma_z, \tau_{xy}, \tau_{xz}, \tau_{yz}$. What is the normal stress on a plane whose normal makes equal angles with the coordinate axes?

2.83 A state of stress is defined throughout a body by σ_x, σ_y, σ_z, τ_{xy}, τ_{xz}, τ_{yz}. What is the normal stress at a point on the plane that has intercepts $-1, 3, 5$ with the x-, y-, and z-axis, respectively?

2.84 A state of stress is defined by the three normal stresses and the three shear stresses associated with three mutually orthogonal directions. If each of the shear stresses is zero and each of the normal stresses is less than s in absolute value, show that the maximum absolute value of a normal stress on any place is less than s.

2.85 A small cube of material rests at the bottom of the sea, where the pressure is 60,000 lb/in^2. What are the normal and shear stresses on any plane in the cube?

***2.86** Let x, y, z represent a set of mutually perpendicular directions at a point, and let n, t, s represent any other set of mutually perpendicular directions at the point. Show that

$$\sigma_n + \sigma_t + \sigma_s = \sigma_x + \sigma_y + \sigma_z$$

***2.87** Under the conditions of Problem 2.86, show that

$$\begin{aligned} &\sigma_x\sigma_y - \tau_{xy}^2 + \sigma_y\sigma_z - \tau_{yz}^2 + \sigma_x\sigma_z - \tau_{xz}^2 \\ &= \sigma_n\sigma_t - \tau_{nt}^2 + \sigma_t\sigma_s - \tau_{ts}^2 + \sigma_n\sigma_s - \tau_{ns}^2 \end{aligned}$$

***2.88** Under the conditions of Problem 2.86, show that

$$\begin{vmatrix} \sigma_x & \tau_{xy} & \tau_{xz} \\ \tau_{yx} & \sigma_y & \tau_{yz} \\ \tau_{zx} & \tau_{zy} & \sigma_z \end{vmatrix} = \begin{vmatrix} \sigma_n & \tau_{nt} & \tau_{ns} \\ \tau_{nt} & \sigma_t & \tau_{ts} \\ \tau_{ns} & \tau_{ts} & \sigma_s \end{vmatrix}$$

*2.8 Properties of a State of Stress

In the preceding section we obtained the remarkable result expressed by Eq. (2.53) [and Eqs. (2.55) and (2.57) as well] that the stress upon any surface can be expressed as a *linear* combination of the stresses on three surfaces. More specifically, the components of stress on the surface (n) can be expressed as a linear combination of the six components on the three rectangular faces (x, y, z). Much can be learned by examining these equations of transformation, (2.55) and (2.57). They are quadratic forms in the direction cosines. Specifically, the normal component σ_n is a quadratic function $\sigma_n(l_n, m_n, n_n)$. We can ask: In what direction does the normal component attain the maximum value? This question might be one of practical importance, if a material is susceptible to failure under a tensile force. Any smooth function attains a maximum (or minimum) only if the derivatives vanish—that is, $\partial\sigma_n/\partial l_n = \partial\sigma_n/\partial m_n = \partial\sigma_n/\partial n_n = 0$; the function σ_n is said to be *stationary*. Here, the mathematical exercise is complicated by the fact that the three variables are not independent; specifically, by (2.50),

$$\hat{\mathbf{n}} \cdot \hat{\mathbf{n}} = l_n^2 + m_n^2 + n_n^2 = 1 \tag{2.58}$$

Omitting the details, we give the required *stationary* criteria. The cosines (l_n, m_n, n_n) are to satisfy three linear equations:

$$\begin{aligned} (\sigma_x - \lambda)l_n + \tau_{xy}m_n + \tau_{xz}n_n &= 0 \\ \tau_{yx}l_n + (\sigma_y - \lambda)m_n + \tau_{yz}n_n &= 0 \\ \tau_{zx}l_n + \tau_{zy}m_n + (\sigma_z - \lambda)n_n &= 0 \end{aligned} \tag{2.59}$$

Additionally, λ is to be determined by the condition that the determinant formed by the coefficients of the cosines (l_n, m_n, n_n) in Eq. (2.59) must vanish:

$$\begin{vmatrix} (\sigma_x - \lambda) & \tau_{xy} & \tau_{xz} \\ \tau_{yx} & (\sigma_y - \lambda) & \tau_{yz} \\ \tau_{zx} & \tau_{zy} & (\sigma_z - \lambda) \end{vmatrix} = 0 \tag{2.60}$$

The expansion of this determinant Eq. (2.60) is a cubic equation in the unknown λ:

$$\lambda^3 - I_1\lambda^2 + I_2\lambda - I_3 = 0 \tag{2.61}$$

Here, the coefficients are functions of the *known* components. For example, I_1 is the simple linear form:

$$I_1 = \sigma_x + \sigma_y + \sigma_z \tag{2.62}$$

I_2 and I_3 are, respectively, quadratic and cubic forms in the known components. To determine these special directions, one must first solve the cubic equation (2.61), which determines three real numbers $(\lambda_1, \lambda_2, \lambda_3)$. For each number, we return to Eqs. (2.58) and (2.59) (any *two* of the latter) to determine three directions. These three

---

Note: Asterisks indicate optional sections.

directions are the *principal directions* of stress. The three numbers $(\lambda_1, \lambda_2, \lambda_3)$ are the *normal stresses* for those directions. One is a maximum and one is a minimum.

We can also ask: Upon what surface is the shear stress a maximum? This too is a mathematical exercise. According to Eq. (2.57) a component of shear τ_{nt} is a function of six cosines $(l_n, m_n, n_n; l_t, m_t, n_t)$. One must find two directions that satisfy the stationary criteria, but the variables are not independent. Again, we bypass the mathematical exercise and give the remarkable results: The directions of maximum shear bisect the principal directions of maximum and minimum normal stress.

An important attribute of the quantities I_1, I_2, and I_3 in Eq. (2.61) is this: We note that the three roots of the cubic Eq. (2.61) are determined by the coefficients I_1, I_2, I_3; those roots are the principal values. Those principal stresses cannot depend upon the choice of coordinates, and, therefore, the coefficients must also be independent of such choice. We say that these quantities are *invariant* (with respect to coordinate transformations). For example, if (x, y, z) and (x', y', z') are two different rectangular coordinates,

$$I_1 = \sigma_x + \sigma_y + \sigma_z = \sigma_{x'} + \sigma_{y'} + \sigma_{z'}$$

The expansion of the determinant equation (2.60) gives the explicit forms of the invariants I_2 and I_3; that expansion is left as an exercise.

2.9 Principal Directions; Maximum Normal and Shear Stresses

Any state of stress is fully determined by six components, for example, $\sigma_x, \sigma_y, \sigma_z, \tau_{xy}, \tau_{xz}, \tau_{yz}$. In other words, we can determine, from the six components, the stress upon any other surface at the point. It is a mathematical exercise (sketched in the preceding section) to determine various other properties. Specifically, we can ask the following:

1. Are there directions (surfaces) of zero shear stress?
2. In what direction is the normal stress a maximum?
3. In what direction is the shear stress a maximum?

The questions are interesting but are also of extreme practical importance. Ultimately the engineer must be concerned with the failure of members, machines, and structures; such failure often hinges on the magnitudes and directions of a normal or shear stress, depending upon the properties of a material. The very important answers follow:

1. At each point of a stressed body, there exist *three mutually perpendicular directions of zero shear stress.* In other words, at every point we can imagine a small *rectangular* element, as depicted in Fig. 2.48; the faces of the element experience only normal stress $(\lambda_1, \lambda_2, \lambda_3)$. These three perpendicular directions are called the *principal directions* of stress. The normal stresses are called the *principal stresses.*

Fig. 2.48

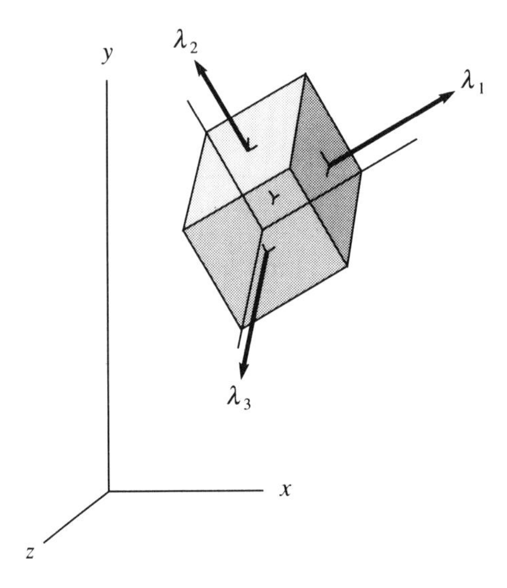

2. One of the principal stresses is the *maximum normal stress* at the point. One of the principal stresses is the *minimum normal stress.*

3. The *directions of maximum shear stress bisect* the principal *directions of maximum and minimum normal stress.* The directions of maximum shear stress are depicted in Fig. 2.49. Here, principal directions 1 and 3 are the directions of maximum and minimum normal stress (λ_1 and λ_3); principal direction 2 is normal to the page. The maximum shear stress is *always* given by the simple, but useful, formula

$$\tau_{\text{max}} = \tfrac{1}{2}|\sigma_{\text{max}} - \sigma_{\text{min}}| \tag{2.63}$$

Note that the normal stress on these surfaces of maximum shear is not, in general, zero. Coincidentally, the normal stress on these surfaces is $\sigma = \frac{1}{2}(\sigma_{\text{max}} + \sigma_{\text{min}})$.

Fig. 2.49

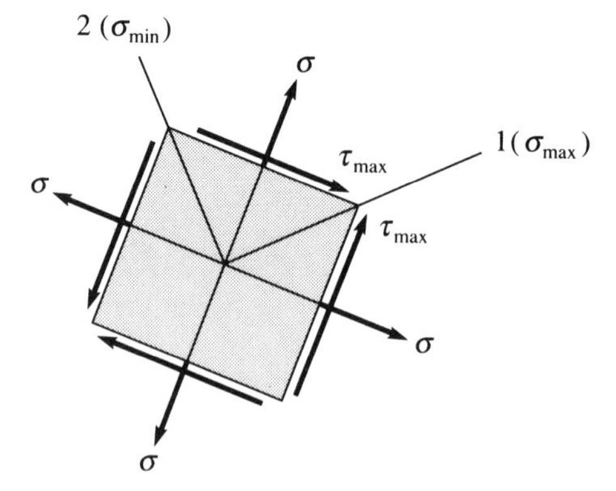

Before leaving this discussion of state of stress at a point, it is important to reemphasize certain aspects of it. Suppose that a small rectangular parallelopiped is removed from a body, as shown in Fig. 2.50. Whatever the orientation of this element, a knowledge of the six stresses acting on the faces of the element is sufficient to determine the stresses on any other plane. If we are able to pick an element whose edges are parallel to the principal directions, then we need only the three principal stresses to specify the state of stress. Thus, for the element shown in Fig. 2.50(a), there are—in general—six different stresses acting on the faces. However, in Fig. 2.50(b), the element is oriented so that the faces are principal planes, and there are, therefore, only three different stresses acting on the faces. *Either set of stresses can be used to specify the state of stress at the point.* Obviously, the use of principal stresses will lead to simpler formulas.

Fig. 2.50

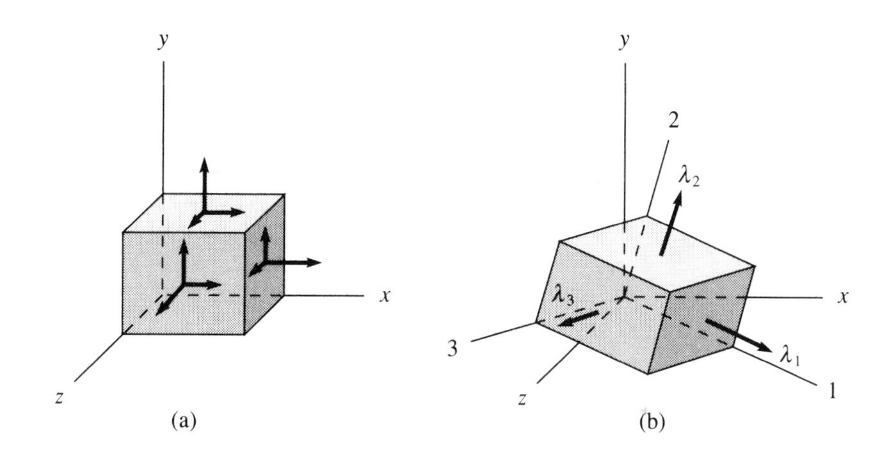

There is, however, one unfortunate feature connected with the use of principal stresses. Unless the principal directions can be determined easily, it is generally more of a problem to find the principal planes than it is to use the six stresses associated with a fixed set of axes. Furthermore, the principal directions can change from point to point. Therefore, if it is necessary to "orient" the state of stress, there is generally no advantage in using principal stresses.

In some instances it is desirable to relate some quantity to the state of stress at a point. Orientation will not be important. Under these circumstances, the use of principal stresses usually simplifies the relations involved.

2.10 | Special States of Stress

Several special states of stress deserve attention. The most special circumstance exists if all three principal stresses are equal [Eq. (2.61) has but one root]. Then, in accordance with Eq. (2.63), the maximum shear stress is *zero*; stated otherwise, *no* shear stress exists on *any* surface at such a point. Every direction may be called a principal direction. If one imagines a tiny sphere at the point, as in Fig. 2.51(a), then every elemental area of the surface experiences the same normal stress ($\sigma_n = T, \tau_n = 0$). This is called a *hydrostatic* state of stress. If we were submerged in water (hydro) at a great depth and if the water were quiescent (static), then every elemental surface of our bodies would experience the same pressure.

Fig. 2.51

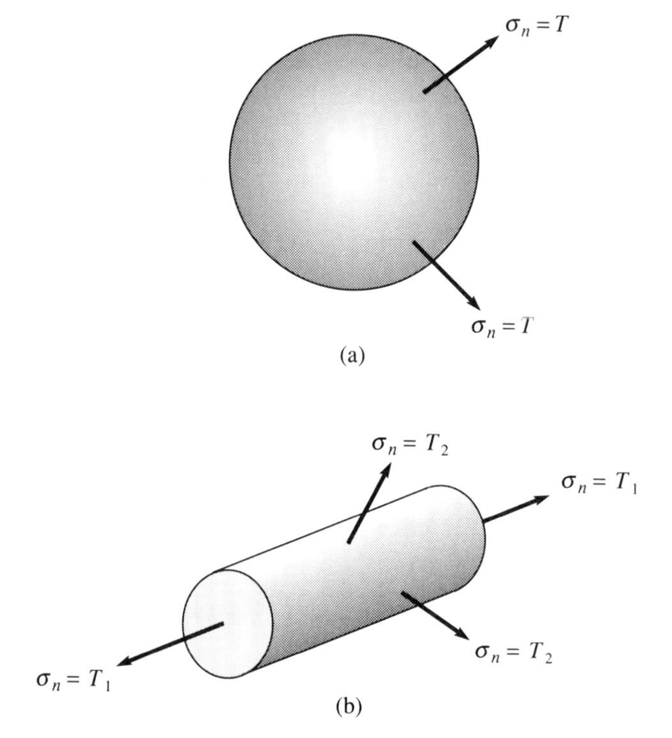

Another special circumstance exists if only one principal stress is distinct. [Two roots of (2.61) are equal.] There is then only one distinct principal direction. We can imagine a tiny circular cylinder at such point, as depicted in Fig. 2.51(b). The axis lies in the one distinct principal direction; a right cross section experiences that distinct principal stress ($\sigma_n = T_1$). Every element of the cylindrical surface experiences only normal stress [$\sigma_n = T_2$, the repeated root of Eq. (2.61)]; every radial direction may be called a principal direction.

Another special state exists if the three principal stresses are $\sigma_1 = T, \sigma_2 = -T$, $\sigma_3 = 0$. This state is depicted in Fig. 2.52(a). The surfaces that bisect the two principal directions (1 and 2) experience *only* the shear stress ($\tau_{rs} = \tau_{sr} = T$), as illustrated in Fig. 2.52(b). The transformation from the principal directions to the bisecting directions is given by Eqs. (2.55) and (2.57) or by the special forms (2.67) and (2.68).

Fig. 2.52

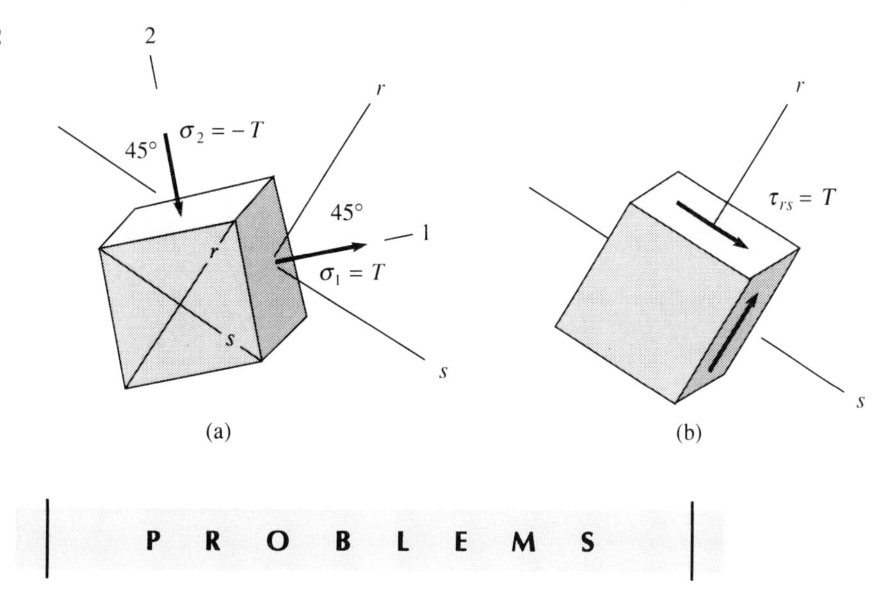

PROBLEMS

2.89 State all properties of the principal directions of stress.

2.90 A state of stress is given by the rectangular components:

$$\begin{matrix} \sigma_x & \tau_{xy} & \tau_{xz} \\ & \sigma_y & \tau_{yz} \\ & & \sigma_z \end{matrix} = \begin{matrix} 20 & 0 & 0 \\ & -10 & 0 \\ & & 40 \end{matrix} \text{ ksi}$$

Calculate the maximum shear stress and clearly identify the surface on which it acts.

2.91 On the cross section of a beam (normal to the axis z), the components of stress are

$$\sigma_x = \sigma_y = \tau_{xy} = \tau_{xz} = 0$$

$$\sigma_z = yz \text{ MPa}, \qquad \tau_{yz} = 10(y^2 - 9) \text{ MPa}$$

where the coordinates are given in centimeters.

a. Compute the maximum shear stress at the point ($y = 3$ cm, $z = 50$ cm).

b. Compute the maximum and minimum normal stresses at the points on the axis ($y = 0$). (*Hint:* Consider the equilibrium of a wedge-shaped element with two faces in principal directions.)

2.92 What is the maximum shear stress at a point where the three principal stresses are equal?

2.93 At a point of a body

$$\tau_{xy} = \tau_{\max} = 4 \text{ MPa}, \qquad \sigma_{\max} = 10 \text{ MPa}$$

$$\sigma_z = 4 \text{ MPa}$$

Show the principal directions relative to the rectangular directions (x, y, z) and give the principal stresses acting upon the surfaces.

2.94 By expanding the determinant of Eq. (2.60) verify Eq. (2.62) and also obtain the expressions for the invariants I_2 and I_3.

2.11 Special Case: Stress on a Plane Parallel to a Coordinate Line

Suppose that the direction n is normal to the z-axis. Then a perpendicular plane cuts the xz- and yz-planes to form the thin wedge shown in Fig. 2.53(a) with the point P at one corner. This is a free-body diagram. The resultant of the internal forces on the inclined face has components $\Delta F_n, \Delta V_{nt}$, and ΔV_{nz} in the n- (normal), t- (tangential), and z- (tangential) directions, respectively.

Fig. 2.53

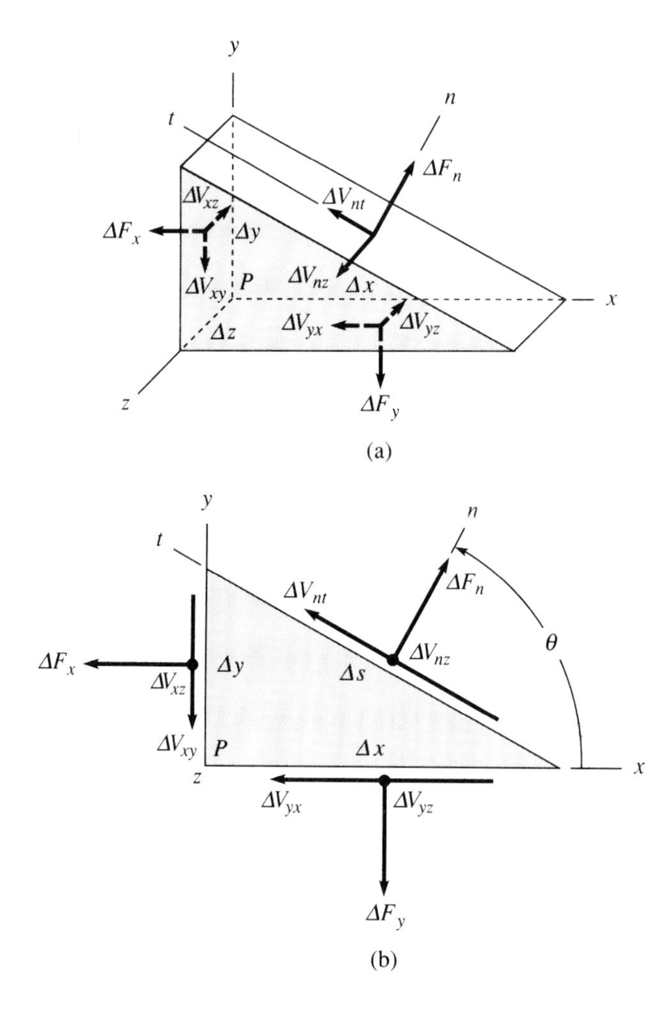

The wedge is redrawn in the planar view of Fig. 2.53(b). The force components on the front and rear faces of the wedge are not shown in either figure, because they cancel in the equations that equilibrate forces. The orientation of the inclined face is specified by θ, the angle between the x-axis and n-axis. From Fig. 2.53(b), the quantities Δx, Δy, Δs, and θ are related:

$$
\begin{aligned}
\Delta x &= \Delta s \sin\theta \\
\Delta y &= \Delta s \cos\theta
\end{aligned}
\tag{2.64}
$$

According to the usual notation, σ_n denotes the normal stress on the inclined face, τ_{nt}, the shear stress in the t-direction, and τ_{nz}, the shear stress in the z-direction. The components of force on each of the faces may be expressed in terms of stresses and areas. Thus, on the inclined face,

$$\Delta F_n = \sigma_n \,\Delta s\,\Delta z$$

$$\Delta V_{nt} = \tau_{nt}\,\Delta s\,\Delta z$$

$$\Delta V_{nz} = \tau_{nz}\,\Delta s\,\Delta z$$

Also, by using Eq. (2.64), the force components on the other two faces of the wedge are

$$\Delta F_x = \sigma_x\,\Delta y\,\Delta z = \sigma_x \cos\theta\,\Delta s\,\Delta z$$

$$\Delta V_{xy} = \tau_{xy}\,\Delta y\,\Delta z = \tau_{xy} \cos\theta\,\Delta s\,\Delta z$$

$$\Delta V_{xz} = \tau_{xz}\,\Delta y\,\Delta z = \tau_{xz} \cos\theta\,\Delta s\,\Delta z$$

$$\Delta F_y = \sigma_y\,\Delta x\,\Delta z = \sigma_y \sin\theta\,\Delta s\,\Delta z$$

$$\Delta V_{yx} = \tau_{yx}\,\Delta x\,\Delta z = \tau_{yx} \sin\theta\,\Delta s\,\Delta z$$

$$\Delta V_{yz} = \tau_{yz}\,\Delta x\,\Delta z = \tau_{yz} \sin\theta\,\Delta s\,\Delta z$$

Here we have taken the liberty of replacing average stress values by the values at the point P. It is understood that the dimensions of the wedge are "infinitesimally" small, so that (in the limit) the inclined face passes through point P.

Let us now write the equations for equilibrium of the forces. Moments were equilibrated previously to ascertain the equality $\tau_{xy} = \tau_{yx}$. For equilibrium, the sum of the forces in the n-direction must vanish:

$$\Delta F_n - \Delta F_x \cos\theta - \Delta V_{xy} \sin\theta - \Delta F_y \sin\theta - \Delta V_{yx} \cos\theta = 0$$

or, in terms of stress components,

$$[\sigma_n - \sigma_x \cos^2\theta - \sigma_y \sin^2\theta - 2\tau_{xy} \sin\theta\cos\theta]\,\Delta s\,\Delta z = 0 \tag{2.65}$$

Similarly, the sum of the forces in the t-direction must also be zero:

$$\Delta V_{nt} + \Delta F_x \sin\theta - \Delta V_{xy} \cos\theta - \Delta F_y \cos\theta + \Delta V_{yx} \sin\theta = 0$$

or

$$[\tau_{nt} + (\sigma_x - \sigma_y)\sin\theta\cos\theta - \tau_{xy}(\cos^2\theta - \sin^2\theta)]\,\Delta s\,\Delta z = 0 \tag{2.66}$$

The common factor $\Delta s\,\Delta z$ can be divided out of Eqs. (2.65) and (2.66), so that

$$\sigma_n = \sigma_x \cos^2\theta + \sigma_y \sin^2\theta + 2\tau_{xy} \sin\theta\cos\theta \tag{2.67}$$

$$\tau_{nt} = (\sigma_y - \sigma_x)\sin\theta\cos\theta + \tau_{xy}(\cos^2\theta - \sin^2\theta) \tag{2.68}$$

Note that Eqs. (2.67) and (2.68) are special forms of Eqs. (2.55) and (2.57), wherein

$$l_n = \cos\theta, \qquad m_n = \sin\theta, \qquad n_n = 0$$

$$l_t = -\sin\theta, \qquad m_t = \cos\theta, \qquad n_t = 0$$

These equations may be put in another form by substituting the trigonometric identities:

$$\cos^2\theta = \tfrac{1}{2} + \tfrac{1}{2}\cos 2\theta$$

$$\sin^2\theta = \tfrac{1}{2} - \tfrac{1}{2}\cos 2\theta$$

$$\sin\theta\cos\theta = \tfrac{1}{2}\sin 2\theta$$

Then Eqs. (2.67) and (2.68) become

$$\sigma_n = \frac{\sigma_x + \sigma_y}{2} + \frac{\sigma_x - \sigma_y}{2}\cos 2\theta + \tau_{xy}\sin 2\theta \tag{2.69}$$

$$\tau_{nt} = -\frac{\sigma_x - \sigma_y}{2}\sin 2\theta + \tau_{xy}\cos 2\theta \tag{2.70}$$

A third equation defining τ_{nz} is obtained by summing forces in the z-direction:

$$\Delta V_{nz} - \Delta V_{xz} - \Delta V_{yz} = 0$$

which reduces to

$$\tau_{nz} = \tau_{xz}\cos\theta + \tau_{yz}\sin\theta \tag{2.71}$$

If the values of $\sigma_x, \sigma_y, \tau_{xy}, \tau_{xz}$, and τ_{yz} are known at a point, the stresses at that point on any other plane (parallel to the z-axis) can be determined by this procedure. It is simply a matter of writing equilibrium equations for the forces acting on a properly chosen wedge. Note that only σ_x, σ_y, and τ_{xy} are required to compute σ_n and τ_{nt} for n and t, any perpendicular directions in the xy-plane.

Extreme care must be exercised in the use of Eqs. (2.69), (2.70), and (2.71). If the angle θ is measured differently (say between the y- and n-axes) or if the t-axis is opposite to that shown, the signs of the trigonometric functions appearing in these equations could be altered. For this reason it is usually preferable to work directly from a free-body diagram of a wedge.

We can pose the following question: What direction n gives a maximum normal stress σ_n? By the necessary criterion, applied to Eq. (2.69),

$$\frac{d\sigma_n}{d\theta} = -(\sigma_x - \sigma_y)\sin 2\theta + 2\tau_{xy}\cos 2\theta = 0 \tag{2.72a}$$

or

$$\tan 2\theta = \frac{2\tau_{xy}}{\sigma_x - \sigma_y} \tag{2.72b}$$

Any values $(\sigma_x, \sigma_y, \tau_{xy})$ determine two angles (θ_1, θ_2), which differ by 90°. By comparison of Eqs. (2.72a) and (2.70), we see that the component τ_{nt} vanishes on the plane. Note that this is not necessarily a principal direction, since the component τ_{nz} need not vanish. *If* we know also that the z-direction is a principal direction, *then* $\tau_{zn} = \tau_{nz} = 0$, and the directions (θ_1, θ_2) determined by Eq. (2.72) are the remaining principal directions.

Equations (2.69) and (2.70) apply only if direction n is normal to axis z. Their applicability is, therefore, limited. They are useful primarily if one principal direction (z) is known a priori. Then the other two principal directions are determined by Eq. (2.72); subsequently, the other two principal stresses are obtained from Eq. (2.69), with the appropriate directions (θ_1 and θ_2). The most common circumstance for such application occurs at a *free surface* of a member; free implies *free of stress*—in particular, free of shear stress. Such a free surface is a principal surface; the normal is a principal direction and the principal stress is zero. Then the other two principal directions are tangent to the surface and are determined in accordance with Eq. (2.72).

EXAMPLE 13

At the free surface of a body (direction z is normal to the surface, x and y are tangent) the components of stress are

$$\sigma_x = 100 \text{ ksi} \quad [689 \text{ MPa}]$$

$$\tau_{xy} = 30 \text{ ksi} \quad [207 \text{ MPa}]$$

$$\sigma_y = 20 \text{ ksi} \quad [138 \text{ MPa}]$$

Because the surface is free, that is, does not bear upon another body,

$$\sigma_z = \tau_{zx} = \tau_{zy} = 0$$

We wish to determine the maximum and minimum values of the normal stress and the maximum shear stress. Depending upon the material, these stresses could cause a failure, fracture, or yielding.

Since $\tau_{zx} = \tau_{zy} = 0$, the z-direction (normal to the surface) is one principal direction. It follows that the other two directions are tangent to the surface (in the xy-plane). One principal stress is $\sigma_1 = \sigma_z = 0$. The other directions of vanishing shear—that is, principal directions—are determined by Eq. (2.70):

$$\tau_{nt} = 0$$

$$\frac{\sin 2\theta}{\cos 2\theta} = \frac{2\tau_{xy}}{\sigma_x - \sigma_y} = \tan 2\theta$$

We note that this also determines the directions in the xy-plane that provide stationary values of normal stress in accordance with Eq. (2.72a, b). With the given numerical values,

$$\tan 2\theta = \frac{2 \times 30}{100 - 20} = \frac{3}{4}$$

Hence $2\theta_2 = 36.87°$ and $2\theta_3 = 216.87°$, so $\theta_2 = 18.43°$ and $\theta_3 = 108.43°$. These give the other two principal directions. The corresponding normal stresses (principal stresses) follow according to Eq. (2.69):

$$\sigma_{2,3} = \tfrac{1}{2}(120) + \tfrac{1}{2}(80)(\pm\tfrac{4}{5}) + 30(\pm\tfrac{3}{5})$$

$$= 60 \pm \tfrac{160}{5} \pm \tfrac{90}{5} = 110 \text{ ksi} \quad \text{or} \quad 10 \text{ ksi} \quad [758 \text{ MPa or } 69 \text{ MPa}]$$

The three principal stresses are

$$\sigma_1 = 0, \qquad \sigma_2 = 110 \text{ ksi} \quad [758 \text{ MPa}] \qquad \sigma_3 = 10 \text{ ksi} \quad [69 \text{ MPa}]$$

Therefore, the maximum and minimum normal stresses are

$$\sigma_{\text{max}} = \sigma_2 = 110 \text{ ksi} \quad [758 \text{ MPa}], \qquad \sigma_{\text{min}} = \sigma_1 = 0$$

In accordance with Eq. (2.63), the maximum shear stress is

$$\tau_{\text{max}} = \tfrac{1}{2}|\sigma_{\text{max}} - \sigma_{\text{min}}| = 55 \text{ ksi} \quad [379 \text{ MPa}]$$

Figure 2.54 shows the principal directions and the directions of maximum shear; the latter bisect the principal directions 1 and 2.

Fig. 2.54

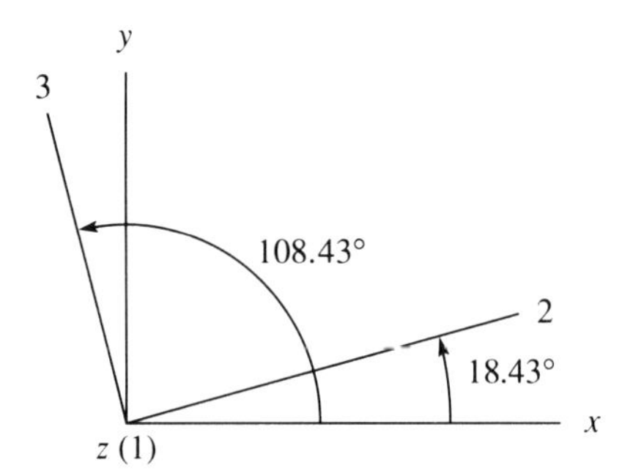

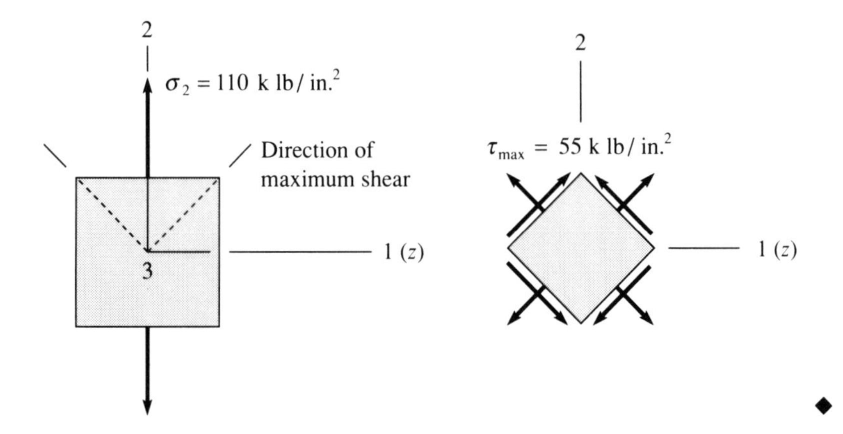

◆

EXAMPLE 14

Let us now examine the state of stress expressed by the following rectangular components:

$$\sigma_x = 200 \text{ MPa}, \qquad \tau_{xy} = -200 \text{ MPa}, \qquad \tau_{xz} = -100 \text{ MPa}$$
$$\sigma_y = -100 \text{ MPa}, \qquad \tau_{yz} = 100 \text{ MPa}, \qquad \sigma_z = 200 \text{ MPa}$$

Directions (n, t) perpendicular to z that bear no component τ_{nt} are obtained according to Eq. (2.72): $\tau_{nt} = 0$ or

$$\tan 2\theta = \frac{2\tau_{xy}}{\sigma_x - \sigma_y} = -\frac{4}{3}$$

Therefore, $2\theta_n = -53.13^\circ$ and $2\theta_t = +126.87^\circ$. Hence, $\theta_n = -26.57^\circ$ and $\theta_t = +63.43^\circ$.

These are *not* principal directions. To be a principal direction, the *net* shear stress must vanish. It is not enough that $\tau_{nt} = 0$ on the surface with normal n; the component $\tau_{nz} = \tau_{zn}$ must also vanish. Here, by projecting the components τ_{zx} and τ_{zy}, we have

$$\begin{aligned}
\tau_{zn} = \tau_{nz} &= \tau_{zx}\cos\theta_n + \tau_{zy}\cos(90 - \theta_n) \\
&= -100\cos(-26.57^\circ) + 100\cos(116.57^\circ) \\
&= -89.4 - 44.7 = -134.1 \text{ MPa}
\end{aligned}$$

Note: In the present case none of the given directions (x, y, z) is a principal direction. The principal directions can be determined only by reverting to the description of Sec. 2.7 "Stresses on Any Surface," solving the cubic equation (2.61) of Sec. 2.8, and obtaining the directions (cosines l_n, m_n, n_n) that satisfy Eqs. (2.59). ◆

PROBLEMS

2.95 A rectangular block of wood is loaded as shown in Fig. P2.95. Compute the average normal stress and shear stress on a plane parallel to the grain of the wooden block.

Fig. P2.95

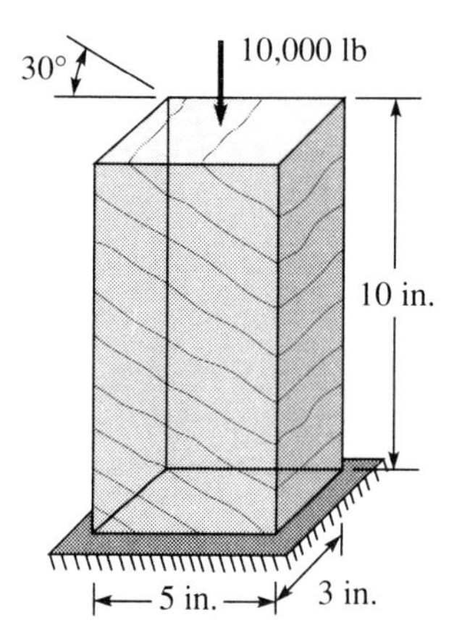

2.96 Determine the average normal stress and shear stress on a plane parallel to the grain in the block of wood shown in Fig. P2.96.

Fig. P2.96

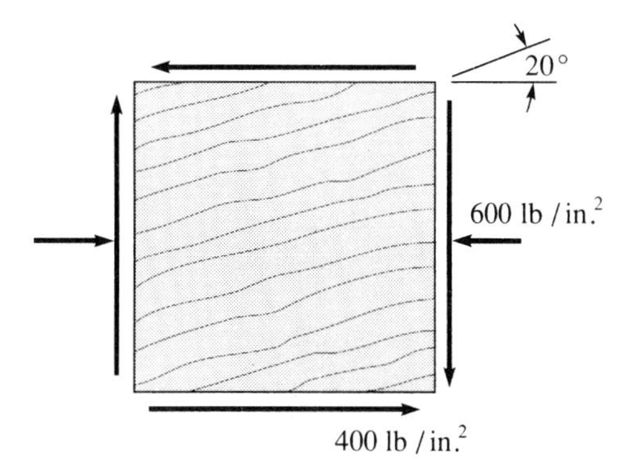

2.97 On a plane through a point P in a body, two rectangular coordinate systems are constructed, as shown in Fig. P2.97. Solve the following problems by sketching the appropriate wedge (or wedges), writing the proper equilibrium equations, and solving them for the required quantities.

a. $\sigma_x = 3000$ psi, $\sigma_y = 1200$ psi, $\tau_{xy} = -700$ psi, $\theta = 120°$; find $\sigma_n, \sigma_t, \sigma_{nt}$.

b. $\sigma_n = -40$ MPa, $\sigma_t = -35$ MPa, $\tau_{nt} = -17.5$ MPa, $\theta = 30°$; find $\sigma_x, \sigma_y, \tau_{xy}$.

c. $\sigma_x = 4000$ psi, $\sigma_y = 1000$ psi, $\sigma_n = 8000$ psi, $\theta = 45°$; find τ_{nt}.

d. $\sigma_x = 6$ MPa, $\sigma_y = 72$ MPa, $\sigma_n = 6.6$ MPa, $\tau_{xy} = 0.6$ MPa; find θ.

e. $\sigma_x = 4000$ psi, $\sigma_y = 8000$ psi, $\tau_{xy} = -2000$ psi, $\tau_{nt} = 0$; find σ_n, σ_t.

f. $\sigma_x = 1400$ psi, $\sigma_y = 600$ psi, $\sigma_n = 1800$ psi, $\tau_{xy} = 600$ psi; find θ.

g. $\sigma_n = -90$ MPa, $\sigma_t = -30$ MPa, $\tau_{xy} = 75$ MPa, $\tau_{nt} = 90$ MPa; find $\sigma_x, \sigma_y, \theta$.

Fig. P2.97

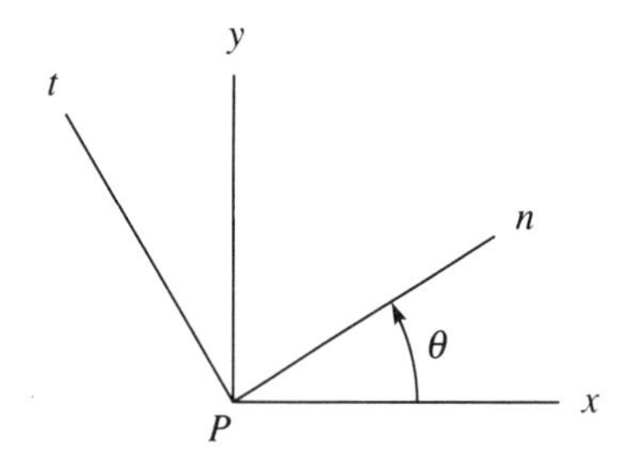

2.98 The block of wood shown in Fig. P2.98 is loaded so that $\sigma_y = -3.5$ MPa and $\sigma_z = \tau_{xy} = \tau_{zx} = \tau_{yz} = 0$.

Fig. P2.98

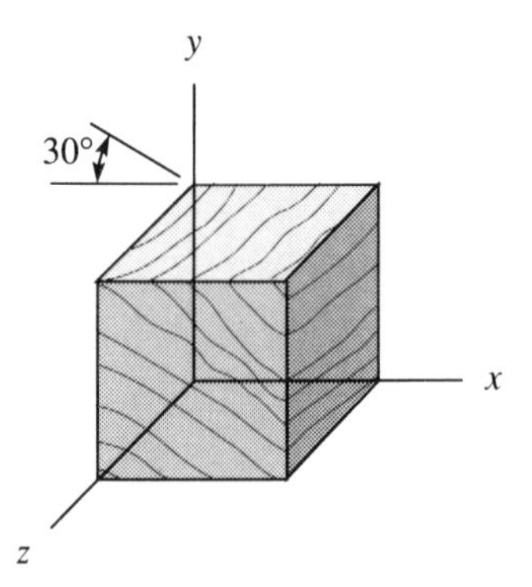

If the block fails when the shear stress along the grain exceeds +2.8 MPa, what range of values can σ_x assume without causing failure?

2.99 The rectangular block shown in Fig. P2.99 is made by gluing two wedges together. The joint fails if the shear stress on the interface exceeds 1000 psi. What range of values can σ_y assume if $\sigma_x = -500$ psi, $\tau_{xy} = 1000$ psi, and $\sigma_z = \tau_{zx} = \tau_{yz} = 0$?

Fig. P2.99

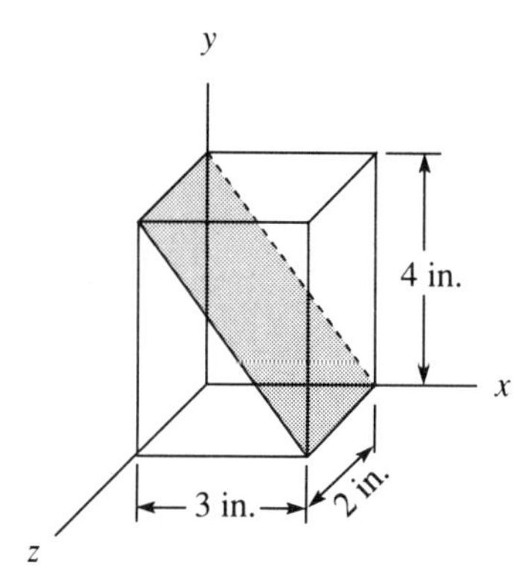

2.100 The rectangular block shown in Fig. P2.100 is made by gluing two wedges together. The joint fails if the normal stress on the interface exceeds 500 psi (tension). What is the maximum value σ_x may take if $\sigma_y = -400$ psi and $\sigma_z = \tau_{xy} = \tau_{zx} = \tau_{yz} = 0$?

Fig. P2.100

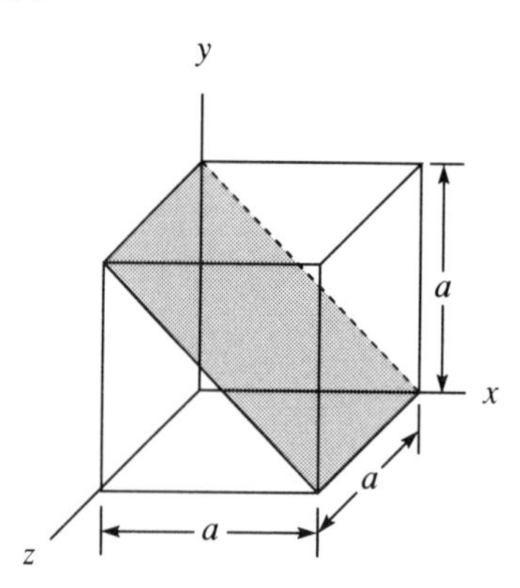

2.101 The rectangular block shown in Fig. P2.101 is composed of two wedges glued together. The joint fails if the normal stress on the interface becomes positive or if the shear stress there exceeds 500 psi. What range of values can σ_y assume if $\sigma_x = -1000$ psi, $\tau_{xy} = 500$ psi, and $\sigma_z = \tau_{yz} = \tau_{zx} = 0$?

Fig. P2.101

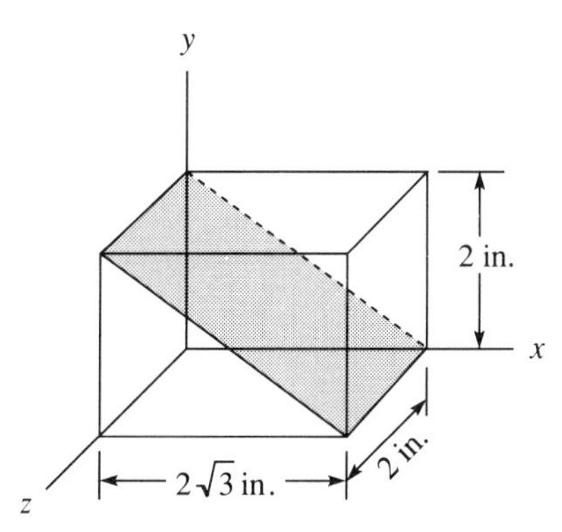

2.102 Given: $\sigma_x = -80$ MPa, $\sigma_y = 4.8$ MPa, $\sigma_z = 80$ MPa, $\tau_{xy} = 64$ MPa, $\tau_{zx} = 16$ MPa, and $\tau_{yz} = 0$. Calculate the stresses on a plane parallel to the z-axis with a normal bisecting the right angle between the x- and y-axes.

2.103 A rectangular block is in equilibrium subjected to the uniform "stress" S at an angle α to the normal on two opposite faces and the uniform shear stress τ on two other faces, as shown in Fig. P2.103. Compute σ_n and τ_{nt} in terms of S and α.

Fig. P2.103

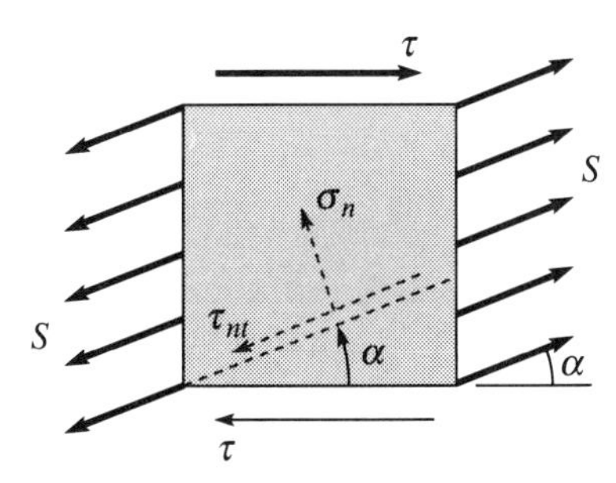

2.104 The only nonzero stress component at a point is τ_{xy}; that is, $\sigma_x = \sigma_y = \sigma_z = \tau_{xz} = \tau_{yz} = 0$. If θ is the angle from the positive x-axis to the n-axis (positive toward y), for what value of θ is σ_n a maximum or minimum?

2.105 The state of stress at a point on the outer surface of a cylindrical pressure vessel is given by $\sigma_\theta = 8000$ psi, $\sigma_z = 4000$ psi, and $\sigma_r = \tau_{r\theta} = \tau_{rz} = \tau_{\theta z} = 0$, where (r, θ, z) are cylindrical coordinates with z along the axis of the cylinder. What is the state of stress associated with the orthogonal planes, two of which have normals bisecting the right angles between the r- and z-coordinate directions? What is the state on the planes if their normals bisect the r- and θ-directions?

2.106 If n and t denote any set of perpendicular directions in the xy-plane, show that, at a given point, (a) $\sigma_n + \sigma_t = \sigma_x + \sigma_y$ and (b) $\sigma_n\sigma_t - \tau_{nt}^2 = \sigma_x\sigma_y - \tau_{xy}^2$.

2.107 What values of angle θ provide a minimum or maximum value τ_{nt} in Eq. (2.70)? Compare these with the values given by Eq. (2.72).

2.108 Let **n**, **t**, and **s** be three coplanar vectors, as shown in Fig. P2.108. Compute τ_{ns} in terms of σ_n, σ_s, and σ_t.

Fig. P2.108

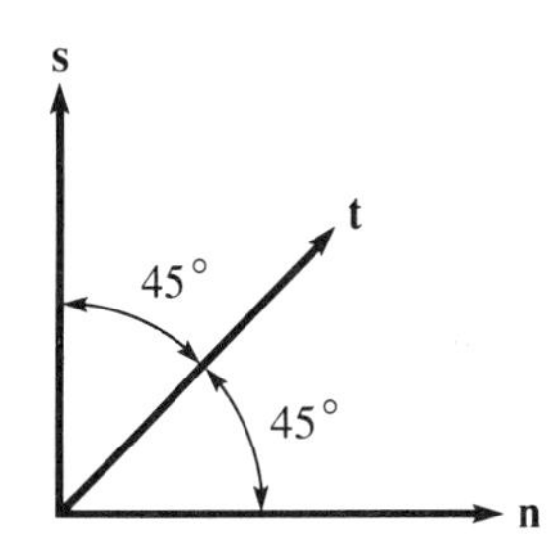

2.109 Let **n**, **t**, and **s** be three coplanar vectors, as shown in Fig. P2.109. Compute τ_{xy} in terms of σ_n, σ_s, and σ_t.

Fig. P2.109

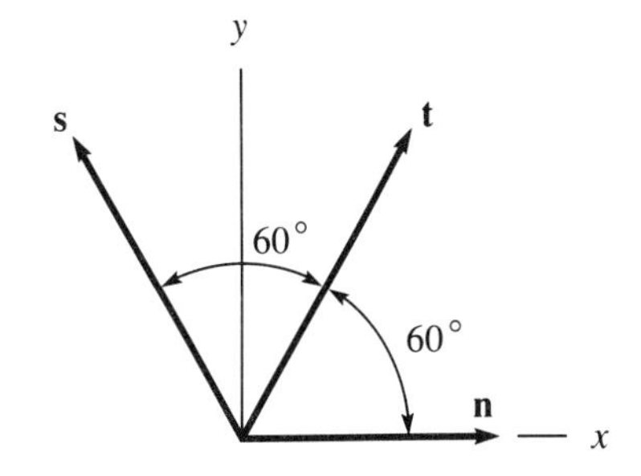

***2.110** Let **n**, **t**, and **s** be three coplanar vectors, as shown in Fig. P2.110. Compute σ_x, σ_y, and τ_{xy} in terms of $\sigma_n, \sigma_t, \sigma_s$ and $\theta_n, \theta_t, \theta_s$.

Fig. P2.110

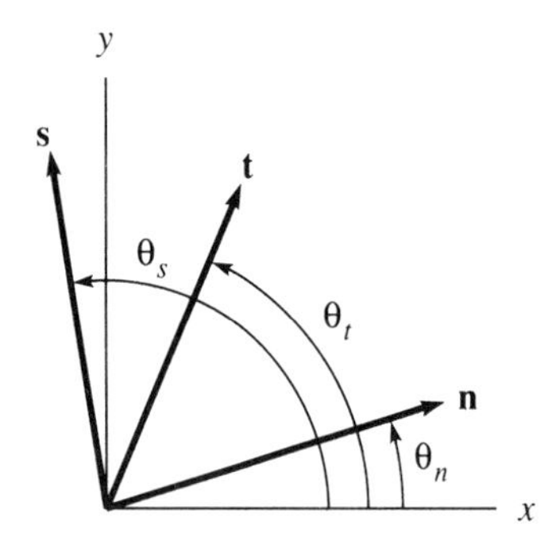

2.111 For each state of stress listed here, obtain *all* components in the rectangular system (n, t, z)—that is, $\sigma_n, \tau_{nt}, \tau_{nz}, \sigma_t, \tau_{tz}$, and σ_z. See Fig. P2.111.

a. $\sigma_x = 80$ ksi, $\tau_{xy} = \tau_{xz} = 0$;
$\sigma_y = -20$ ksi, $\tau_{yz} = 30$ ksi, $\sigma_z = 60$ ksi;
$\theta = 45°$

b. $\sigma_x = -60$ MPa, $\tau_{xy} = \tau_{xz} = 0$;
$\sigma_y = 60$ MPa, $\tau_{yz} = 30$ MPa, $\sigma_z = -20$ MPa;
$\theta = 30°$

c. $\sigma_x = 100$ MPa, $\tau_{xy} = 0$, $\tau_{xz} = 40$ MPa;
$\sigma_y = -20$ MPa, $\tau_{yz} = 0$, $\sigma_z = 80$ MPa;
$\theta = 30°$

d. $\sigma_x = 20$ ksi, $\tau_{xy} = 0$, $\tau_{xz} = 10$ ksi;
$\sigma_y = -12$ ksi, $\tau_{yz} = 0$, $\sigma_z = 10$ ksi;
$\theta = 30°$

e. $\sigma_x = 100$ MPa, $\tau_{xy} = \tau_{xz} = 0$;
$\sigma_y = 60$ MPa, $\tau_{yz} = 40$ MPa, $\sigma_z = 0$;
$\theta = 45°$

Fig. P2.111

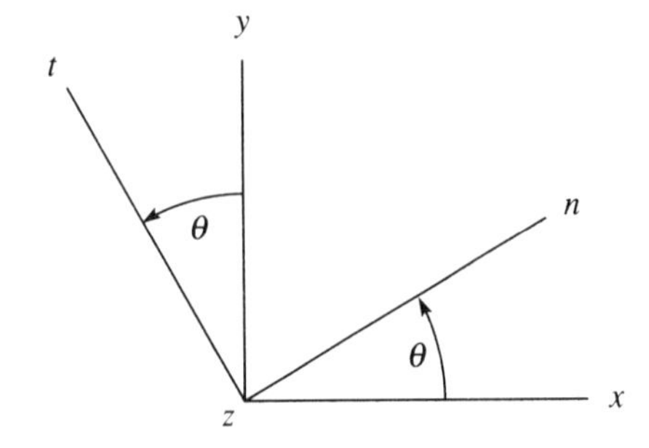

f. $\sigma_x = 20$ ksi, $\tau_{xy} = 20$ ksi, $\tau_{xz} = -10$ ksi;
$\sigma_y = -10$ ksi, $\tau_{yz} = 10$ ksi, $\sigma_z = 20$ ksi;
$\theta = 45°$

2.112 Locate the principal directions, determine the maximum normal stress and maximum shear stress, and locate the directions of the maximum shear stress for the state given in Prob. 2.111(b).

2.113 Repeat Prob. 2.112 for the state given in Prob. 2.111(c).

2.114 Repeat Prob. 2.112 for the state given in Prob. 2.111(d).

2.115 What can you say about the principal directions for the state of stress given in Prob. 2.111(f)?

2.116 Five components of stress are given in the rectangular system (x, y, z):

$$\tau_{xy} = 0, \qquad \tau_{xz} = \tau_{yz} = 40 \text{ ksi}$$
$$\sigma_y = 80 \text{ ksi}, \qquad \sigma_z = -20 \text{ ksi}$$

Rectangular coordinates (n, t) lie in the xy-plane of Fig. P2.116. The normal component σ_n equals 50 ksi. Compute the remaining components of stress in the system (n, t, z).

Fig. P2.116

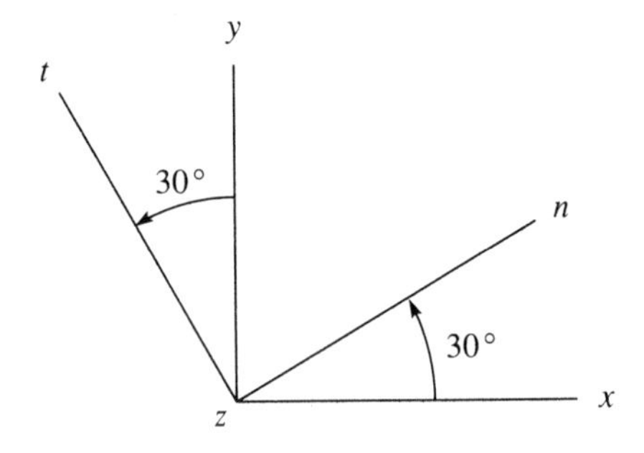

2.117 For each of the states of stress listed here, compute the principal stresses and give the direction numbers of the principal directions. Also compute the maximum shear stress and give the direction numbers of the plane on which it acts.

a. $\sigma_x = -36$ MPa, $\tau_{xz} = +24$ MPa, $\sigma_y = \sigma_z = \tau_{xy} = \tau_{yz} = 0$

b. $\tau_{yz} = 10{,}000$ psi, $\sigma_x = \sigma_y = \sigma_z = \tau_{xy} = \tau_{xz} = 0$

c. $\sigma_x = 2000$ psi, $\sigma_y = -2000$ psi, $\sigma_z = \tau_{xy} = \tau_{xz} = \tau_{yz} = 0$

d. $\sigma_x = 40$ MPa, $\sigma_y = -10$ MPa, $\tau_{xy} = -20$ MPa, $\sigma_z = \tau_{zx} = \tau_{yz} = 0$

e. $\sigma_x = 800$ psi, $\sigma_y = -800$ psi, $\tau_{xy} = 600$ psi, $\sigma_z = \tau_{zx} = \tau_{yz} = 0$

f. $\sigma_x = -2.0$ MPa, $\sigma_z = -8.0$ MPa, $\tau_{zx} = -5.2$ MPa, $\sigma_y = \tau_{xy} = \tau_{yz} = 0$

g. $\sigma_x = 800$ psi, $\sigma_y = -600$ psi, $\sigma_z = 1000$ psi, $\tau_{xy} = 400$ psi, $\tau_{zx} = \tau_{yz} = 0$

h. $\sigma_x = 9$ MPa, $\sigma_y = -6$ MPa, $\sigma_z = 12$ MPa, $\tau_{zx} = -1.5$ MPa, $\tau_{xy} = \tau_{yz} = 0$

i. $\sigma_x = 2000$ psi, $\sigma_y = -2000$ psi, $\sigma_z = -2000$ psi, $\tau_{yz} = 1000$ psi, $\tau_{xy} = \tau_{zx} = 0$

j. $\sigma_x = 3$ MPa, $\tau_{xy} = -3$ MPa, $\sigma_y = \sigma_z = \tau_{xy} = \tau_{xz} = 0$

k. $\sigma_x = 20{,}000$ psi, $\sigma_y = -5000$ psi, $\sigma_z = 15{,}000$ psi, $\tau_{xy} = -10{,}000$ psi, $\tau_{zx} = \tau_{yz} = 0$

2.118 At a point on the surface of a shaft, the state of stress is given by $\sigma_z = 3600$ psi, $\tau_{\theta z} = 2700$ psi, and $\sigma_r = \sigma_\theta = \tau_{r\theta} = \tau_{rz} = 0$, where (r, θ, z) are cylindrical coordinates with z along the axis of the shaft. Find the principal stresses at this point and indicate the principal directions in a sketch. What is the maximum shear stress at this point?

2.119 At a point at the surface of a spherical pressure vessel, the normal stress is σ_0 in every direction tangent to the surface and zero normal to the surface. The surface normal is known to be a principal direction. What is the maximum shear stress at the surface and on what plane(s) does it act?

2.120 Show that, if x, y, z are principal directions and $\sigma_x = \sigma_y = \sigma_z$, every direction is a principal direction.

2.121 Show that, if z is a principal direction and σ_x, σ_y, and σ_z are all increased by a fixed amount s, the principal directions are unchanged and the principal stresses are merely increased by the same amount s.

2.122 The xy-plane is a principal plane and the maximum principal stress in this plane is 10,000 lb/in.^{2}. If $\sigma_y = 4000$ lb/in.^{2} and $\tau_{xy} = 8000$ lb/in.^{2}, compute σ_x and show the principal directions in a sketch of the xy-plane.

2.123 The sum of two principal stresses is zero. If the maximum shear stress in planes parallel to the other principal directions is 30,000 lb/in.^{2}, compute the values of these two principal stresses.

2.124 Two principal stresses are $\sigma_1 = 20$ MPa and $\sigma_2 = -6$ MPa. Locate (in a sketch) the orthogonal directions perpendicular to the third principal direction for which the normal stresses are equal. What is this normal stress, and what is the shear stress associated with these directions?

2.125 At a point in a body, the state of stress is given by $\sigma_x = -10{,}000$ psi, $\tau_{xy} = 40{,}000$ psi, and $\sigma_z = \tau_{xz} = \tau_{yz} = 0$, with σ_y not given. What is the largest value σ_y may take if the maximum normal stress on planes parallel to the z-axis cannot exceed 90,000 psi?

2.126 The vectors **n**, **t**, and **s** lie in a principal plane, as shown in Fig. P2.126. Compute the principal stresses for directions in the plane and show the principal directions in a sketch if $\sigma_n = 100$ MPa, $\sigma_t = -20$ MPa, and $\sigma_s = 60$ MPa.

Fig. P2.126

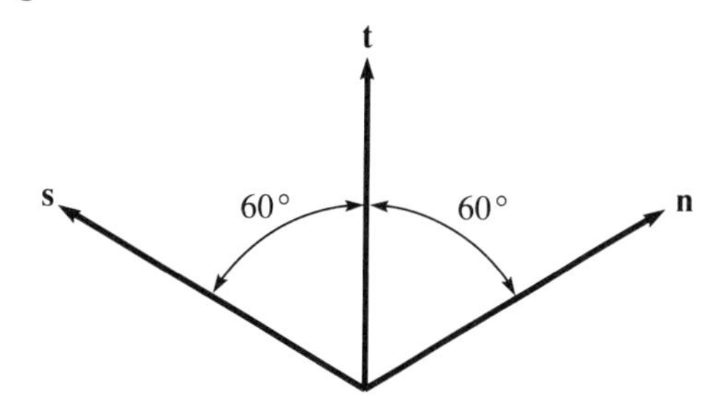

2.127 Repeat Prob. 2.126 for the directions **n**, **t**, and **s** shown in Fig. P2.127 with $\sigma_n = 10{,}000$ psi, $\sigma_t = 5000$ psi, and $\sigma_s = 2000$ psi.

Fig. P2.127

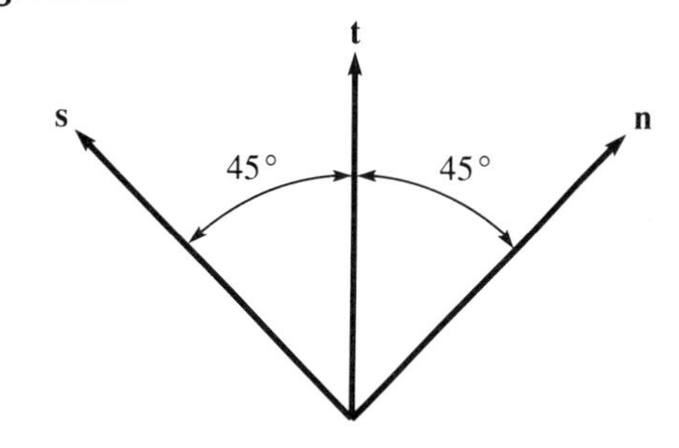

2.12 Mohr's Graphical Representation

Equations (2.69) and (2.70) give the normal stress σ_n and one component of the shear stress τ_{nt} on a surface with normal n. These hold if and only if directions n and t are perpendicular to the direction z. None of the components on the z surface $(\sigma_z, \tau_{zx}, \tau_{zy})$ enter. As noted, we can always locate a direction n such that $\tau_{nt} = 0$. The direction follows in accordance with Eqs. (2.70) and (2.72b). Now, let x' and y' denote these directions—that is, $\tau_{x'y'} = 0$—and let θ denote the angle between x' and n. Then, in accordance with Eqs. (2.69) and (2.70),

$$\sigma_n = \frac{\sigma_{x'} + \sigma_{y'}}{2} + \frac{\sigma_{x'} - \sigma_{y'}}{2} \cos 2\theta \tag{2.73}$$

$$\tau_{nt} = -\frac{\sigma_{x'} - \sigma_{y'}}{2} \sin 2\theta \tag{2.74}$$

We suppose that the components $(\sigma_{x'}, \sigma_{y'})$ are known. For brevity, let

$$\frac{\sigma_{x'} + \sigma_{y'}}{2} \equiv a, \qquad \frac{\sigma_{x'} - \sigma_{y'}}{2} \equiv b \tag{2.75a, b}$$

Then Eqs. (2.73) and (2.74) have the abbreviated forms

$$\sigma_n - a = b \cos 2\theta \tag{2.76}$$

$$-\tau_{nt} = b \sin 2\theta \tag{2.77}$$

These equations are the parametric equations of a circle, Mohr's circle; the center lies on the σ_n-axis in Fig. 2.55 at $\sigma_n = a$ and the radius is b. By squaring each of Eqs. (2.76) and (2.77) and adding the squares, we eliminate the parameter θ and obtain a familiar form:

$$(\sigma_n - a)^2 + \tau_{nt}^2 = b^2 \tag{2.78}$$

Note: To be entirely consistent with usual conventions, we plot $-\tau_{nt}$ versus σ_n. Then the radial line to the point $(\sigma_n, -\tau_{nt})$ on the circle forms the angle 2θ with the σ_n-axis. Moreover, rotation from $A(\sigma_{x'}, 0)$, which represents the stress in the given direction x', to the point $B(\sigma_n, -\tau_{nt})$, which represents the stress in direction n, has the sense of our original rotation (θ) from x' to n. Progressing around the circle in the counterclockwise sense through the angle $2\theta = \pi$ [through $\theta = \pi/2$ between n and n' in Fig. 2.55(b) and Fig. 2.55(c)], we find that $\tau_{n't'} = -\tau_{nt}$, again consistent with the earlier findings $(\tau_{xy} = \tau_{yx})$.

The graphical representation is an alternative to formulas (2.69) and (2.70). Consider the following example.

Fig. 2.55

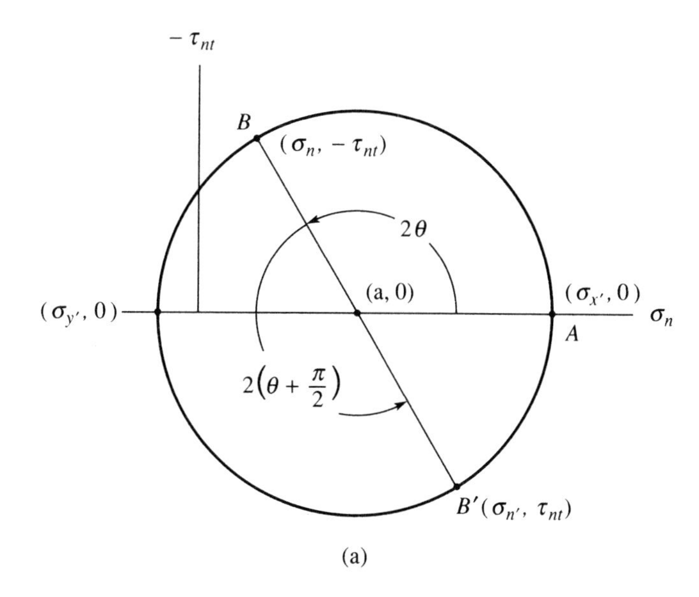

$-\tau_{nt}$
B
$(\sigma_n, -\tau_{nt})$
2θ
$(a, 0)$
$(\sigma_{y'}, 0)$
$(\sigma_{x'}, 0)$
σ_n
A
$2\left(\theta + \frac{\pi}{2}\right)$
$B'(\sigma_{n'}, \tau_{nt})$

(a)

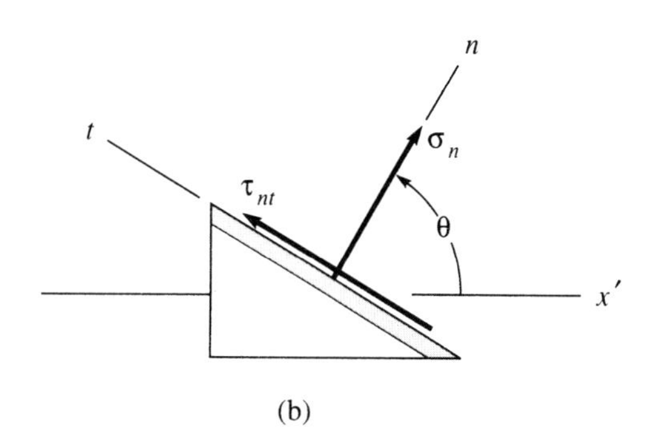

n
t
σ_n
τ_{nt}
θ
x'

(b)

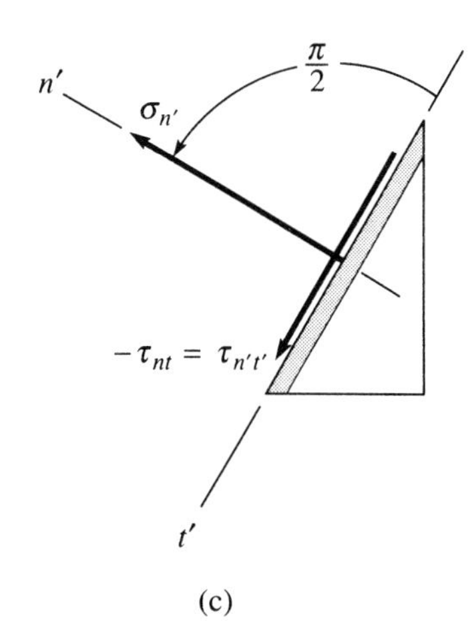

$\frac{\pi}{2}$
n'
$\sigma_{n'}$
$-\tau_{nt} = \tau_{n't'}$
t'

(c)

EXAMPLE 15

Suppose that we are given the following components:

$$\begin{matrix} \sigma_x & \tau_{xy} & \tau_{xz} \\ & \sigma_y & \tau_{yz} \\ & & \sigma_z \end{matrix} = \begin{matrix} 8 & -3 & 2 \\ & -2 & 4 \\ & & 4 \end{matrix} \text{ ksi}$$

We require the components on a surface with normal *n*, which is perpendicular to direction *z*, as shown in Fig. 2.56(a). The components of stress are shown on surfaces with normals *y* and *x* in Fig. 2.56(b) and (c), respectively, in accordance with the conventions used to derive Eq. (2.69), Eq. (2.70), and the plot of Mohr's circle. Since the rotation from *x* to *y* is an angle $\pi/2$, the corresponding points are diametrically

Fig. 2.56

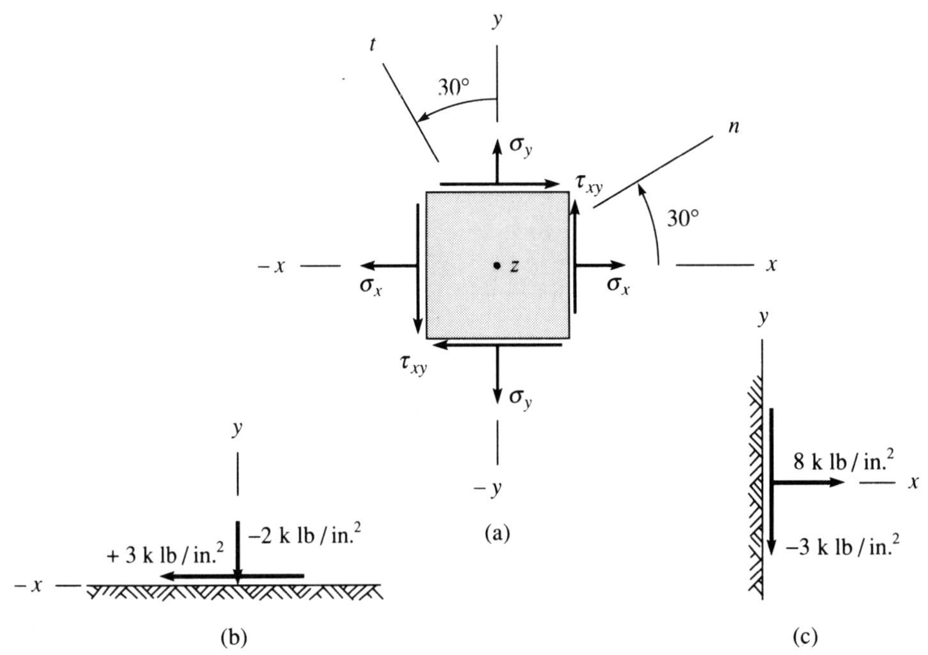

opposed on Mohr's circle; these are labeled (x, y) and $(y, -x)$ in Fig. 2.57. Note that $\tau_{xy} = -\tau_{y-x}$. These points are plotted, the diameter is drawn between them, and the circle is fully determined. The center is at $(3, 0)$ and the radius is

$$b = \sqrt{3^2 + 5^2} = \sqrt{34} \text{ ksi}$$

The angle 2α is

$$\tan^{-1} \tfrac{3}{5} = 30.96^\circ$$

Fig. 2.57

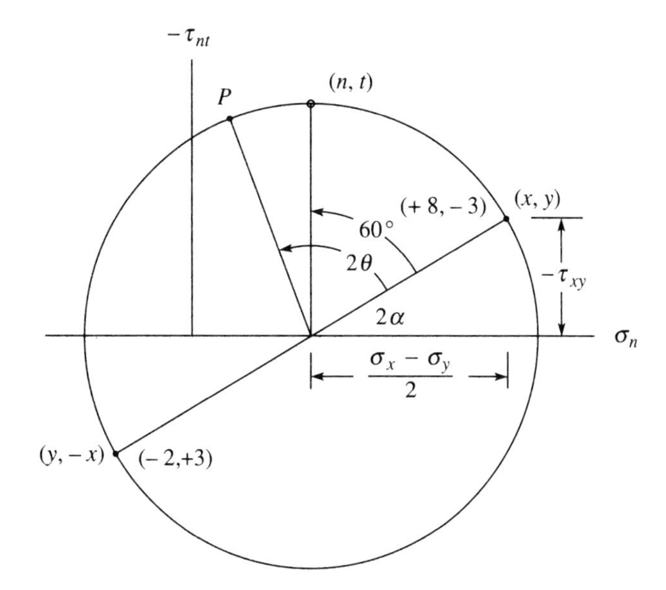

To reach the point corresponding to direction n, we must traverse the circle (counterclockwise) from the point labeled (x, y) through 60° (2 × 30°) to the point labeled (n, t). This lies near the top, where $\sigma_n = 3$ ksi and $\tau_{nt} = -5.8$ ksi. The component τ_{nz} cannot be obtained from the plot. We must revert to Eq. (2.71). Suppose that the angle between x and n were some other value θ. Then the components $(\sigma_n, -\tau_{nt})$ lie at point P of Fig. 2.57. Recall that the center lies at $(\sigma_x + \sigma_y)/2$; also, if the radius is denoted by b, then

$$\cos 2\alpha = \frac{\sigma_x - \sigma_y}{2b}, \qquad \sin 2\alpha = -\frac{\tau_{xy}}{b}$$

At point P,

$$\sigma_n = \frac{\sigma_x + \sigma_y}{2} + b\cos 2(\alpha + \theta) \tag{2.79}$$

$$-\tau_{nt} = b\sin 2(\alpha + \theta) \tag{2.80}$$

However,

$$\cos 2(\alpha + \theta) = \cos 2\alpha \cos 2\theta - \sin 2\alpha \sin 2\theta$$

$$= \frac{\sigma_x - \sigma_y}{2b}\cos 2\theta + \frac{\tau_{xy}}{b}\sin 2\theta$$

$$\sin 2(\alpha + \theta) = \cos 2\alpha \sin 2\theta + \sin 2\alpha \cos 2\theta$$

$$= \frac{\sigma_x - \sigma_y}{2b}\sin 2\theta - \frac{\tau_{xy}}{b}\cos 2\theta$$

When we substitute the latter into Eqs. (2.79) and (2.80), we recover Eqs. (2.69) and (2.70). Again, Mohr's circle is a graphical representation of the equations, which can be reconstructed from the circle. ◆

Principal Circles

We can plot one circle for all lines perpendicular to one principal direction, another for all lines perpendicular to the second principal direction, and another for all lines perpendicular to the third. The three circles might appear as shown in Fig. 2.58. The points labeled 1, 2, and 3 correspond to the principal directions; the corresponding values of normal stress σ_n are the principal stresses σ_1, σ_2, and σ_3. The minimum and maximum principal stresses are σ_1 and σ_3. The maximum shear stress is the radius of the largest principal circle; consistent with Eq. (2.63),

$$\tau_{max} = \left| \frac{\sigma_{max} - \sigma_{min}}{2} \right|$$

Fig. 2.58

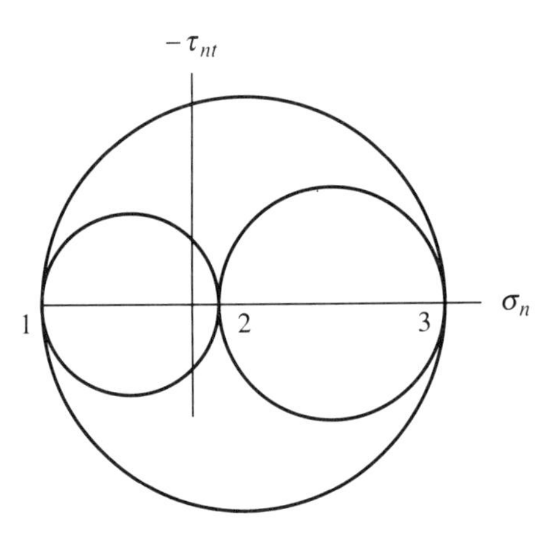

2.128 The xy-plane is known to be a principal plane. For each of the following cases, sketch Mohr's circle for stresses on planes perpendicular to the xy-plane. Find the other principal stresses and locate the principal directions in a sketch of the xy-plane.

a. $\sigma_x = 3000$ psi, $\sigma_y = 1000$ psi, $\tau_{xy} = 0$

b. $\sigma_x = 0$, $\sigma_y = 0$, $\tau_{xy} = -5000$ psi

c. $\sigma_x = 20$ MPa, $\sigma_y = 0$, $\tau_{xy} = 30$ MPa

d. $\sigma_x = 8000$ psi, $\sigma_y = 8000$ psi, $\tau_{xy} = 0$

e. $\sigma_x = 48$ MPa, $\sigma_y = 48$ MPa, $\tau_{xy} = 60$ MPa

f. $\sigma_x = -1200$ psi, $\sigma_y = 2000$ psi, $\tau_{xy} = 0$

2.129 The nt-rectangular coordinate system lies in the xy-coordinate plane, as shown in Fig. P2.129. Sketch Mohr's circle for stresses on planes perpendicular to this one in each case.

a. $\tau_{xy} = 2000$ psi, $\tau_{nt} = 0$, $\sigma_y = 1000$ psi

b. $\sigma_x = 7$ MPa, $\sigma_n = 7$ MPa, $\sigma_t = 7$ MPa

c. $\sigma_x = 12{,}000$ psi, $\sigma_y = 12{,}000$ psi, $\sigma_n = 6000$ psi

d. $\tau_{xy} = 707$ psi, $\tau_{nt} = 707$ psi, $\sigma_x + \sigma_y = 0$

Fig. P2.129

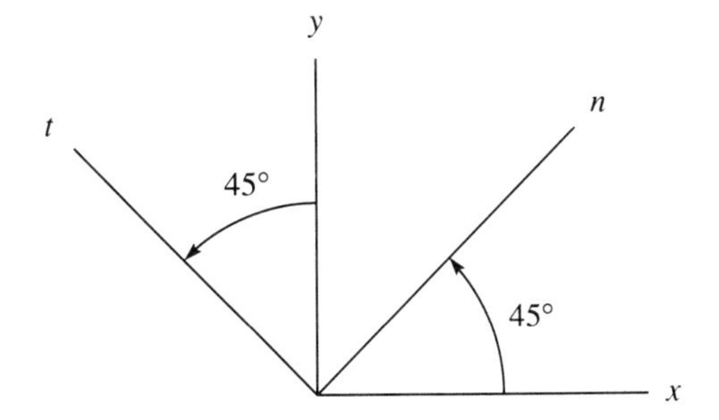

2.130 Solve the exercises in Prob. 2.97 by constructing Mohr's circle.

2.131 The principal stresses at a point are $\sigma_1 > \sigma_2 > \sigma_3$. Sketch, on the same set of axes, three Mohr's circles corresponding to stresses on planes perpendicular to each of the three principal planes.

2.132 From the geometrical construction of Mohr's circle, obtain the following formula for the maximum value of the shear stress τ_{nt}:

$$\tau_{\max} = \sqrt{\left(\frac{\sigma_x - \sigma_y}{2}\right)^2 + \tau_{xy}^2}$$

2.133 From the geometry of Mohr's circle, obtain the following formula for the maximum and minimum values of the normal stress σ_n:

$$\sigma_n(\max, \min) = \frac{\sigma_x + \sigma_y}{2} \pm \sqrt{\left(\frac{\sigma_x - \sigma_y}{2}\right)^2 + \tau_{xy}^2}$$

2.13 The Differential Equations of Equilibrium

Consider once again a small rectangular block removed from the interior of a body, as shown in Fig. 2.59. We had previously regarded the corresponding average stresses on opposite faces of the block as equal. This conclusion was justified on the basis that the block could be taken as small as we pleased and the faces, therefore, as close together as we pleased. Under these conditions the three force-equilibrium equations for the block are identically satisfied. In this section we consider the small differences in the stresses on opposite faces.

Fig. 2.59

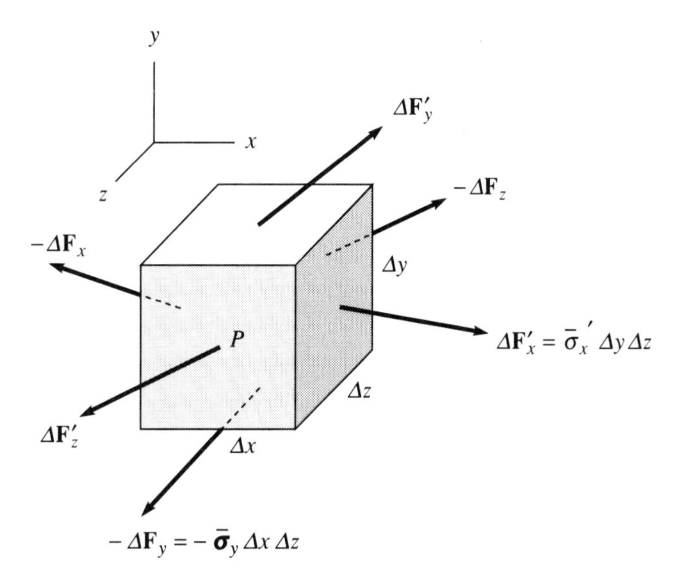

Once again primes are used for quantities on the three visible faces of the block in Fig. 2.59. Quantities associated with the hidden faces are written without primes. The bars above the stresses denote average values over the appropriate faces. With this notation, we can write

$$\Delta \mathbf{F}'_x = \bar{\boldsymbol{\sigma}}'_x \, \Delta y \, \Delta z \tag{2.81}$$

whereas

$$\Delta \mathbf{F}_x = \bar{\boldsymbol{\sigma}}_x \, \Delta y \, \Delta z \tag{2.82}$$

Similar equations hold for the forces upon faces normal to the y and z lines.

Let us assume that the body is in equilibrium. We further assume that no body forces, such as weight or electromagnetic forces, act. These assumptions are more stringent than necessary; they are employed here only to avoid complications. For equilibrium, the sum of the forces on all faces must vanish:

$$\Delta \mathbf{F}'_x - \Delta \mathbf{F}_x + \Delta \mathbf{F}'_y - \Delta \mathbf{F}_y + \Delta \mathbf{F}'_z - \Delta \mathbf{F}_z = 0$$

If we let the block get smaller and smaller ($\Delta x \to 0, \Delta y \to 0, \Delta z \to 0$), each term approaches zero, as anticipated previously; also $\boldsymbol{\sigma}'_x \to \bar{\boldsymbol{\sigma}}_x \to \boldsymbol{\sigma}_x$ in Eqs. (2.81) and (2.82). However, if we first divide by the volume of the block, we get

$$\frac{\bar{\boldsymbol{\sigma}}'_x - \bar{\boldsymbol{\sigma}}_x}{\Delta x} + \frac{\bar{\boldsymbol{\sigma}}'_y - \bar{\boldsymbol{\sigma}}_y}{\Delta y} + \frac{\bar{\boldsymbol{\sigma}}'_z - \bar{\boldsymbol{\sigma}}_z}{\Delta z} = 0 \tag{2.83}$$

The limit of each quotient in Eq. (2.83) is a partial derivative. Furthermore, these derivatives are evaluated at the point P. Equation (2.83) can thus be written as

$$\frac{\partial \boldsymbol{\sigma}_x}{\partial x} + \frac{\partial \boldsymbol{\sigma}_y}{\partial y} + \frac{\partial \boldsymbol{\sigma}_z}{\partial z} = 0 \tag{2.84}$$

In accordance with expression (2.49a, b, c) for the vectors $\boldsymbol{\sigma}_x, \boldsymbol{\sigma}_y$, and $\boldsymbol{\sigma}_z$, the one vectorial equation (2.84) represents three scalar equations. The three differential equations of equilibrium are then

$$\begin{aligned} \frac{\partial \sigma_x}{\partial x} + \frac{\partial \tau_{xy}}{\partial y} + \frac{\partial \tau_{xz}}{\partial z} &= 0 \\ \frac{\partial \tau_{xy}}{\partial x} + \frac{\partial \sigma_y}{\partial y} + \frac{\partial \tau_{yz}}{\partial z} &= 0 \\ \frac{\partial \tau_{xz}}{\partial x} + \frac{\partial \tau_{yz}}{\partial y} + \frac{\partial \sigma_z}{\partial z} &= 0 \end{aligned} \tag{2.85}$$

(Note that we have written τ_{xy} for τ_{yx}, and so on.)

These equations play a fundamental role in more thorough studies of deformable bodies but are not used explicitly in this text. To use them effectively, we need considerable facility with partial differential equations.

3 Deformation

Chapter 2 set forth the essential dynamic concepts. Here, we explore the kinematic concepts. These are entirely *geometric* and *independent* of any causes (forces, temperature, etc.) or the composition of the body (rubber, plastic, steel, etc.). The geometric analyses presuppose only that the medium is continuous and cohesive, as described in Sec. 3.1.

Because each elemental portion of a component (e.g., a connecting rod or a beam) can deform differently (e.g., one portion may stretch while another contracts), we must devise means to describe completely the deformation *at* each location. This is accomplished via the concept of *strain*, which is introduced in Sec. 3.3. The nature of *components*, the description of a *state* of strain, special properties, and special states are explored in the subsequent sections. Section 3.9 serves to illustrate the concepts via simple, yet important, problems.

The concept of strain is a very essential one. Every description of continuous media and all analyses of deformable bodies require this important tool. It is needed in our description of materials in Chapter 4 and in all subsequent analyses.

3.1 Preliminary Considerations

Deformation

The term *deformation* denotes any change in shape and size of a body or any of its parts. The deformation may be caused by the application of external loads or by changes of temperature; it may be visible to the naked eye or detectable only by delicate instrumentation. Deformations always occur when a body is subjected to forces. A rubber band will extend to several times its original length when placed around a bundle of papers. In contrast a person standing on a railroad rail will cause a slight, but unnoticeable, deformation of the rail. Another cause of deformation is a change in temperature. The operation of the bimetallic strip in a conventional thermostat is an example of temperature-induced deformation.

For the present, we disregard the causes of deformation and consider only the geometric problems associated with describing changes in size and shape of a body.

Cohesive, Continuous Media

To develop a reliable and workable theory of the behavior of deformable bodies, we assume that every body occupies all the space within its apparent surface boundaries. In other words, any arbitrarily small portion of volume within the body is a part of the body. Such an idealization of the matter is called a *continuous medium.*

Materials such as sand differ from those such as wood, concrete, or metal. Adjacent particles in a box of sand may be separated. On the other hand, a line of particles in a bar of steel remains intact throughout a deformation of the bar. An idealized continuous medium is said to be *cohesive* if the continuity of a line of particles must persist during deformation.

3.2 Displacement

A cylinder is shown in Fig. 3.1. Each particle of material within this cylinder occupies some point in space. One such particle occupies the point labeled P. We call point P the *initial position* of the particle. If a system of applied forces stretch the cylinder, it

Fig. 3.1

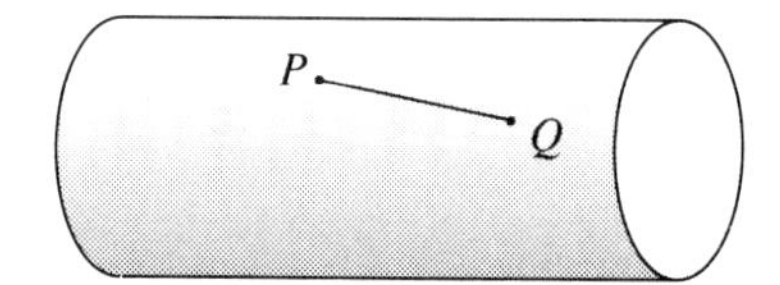

will become longer and thinner, as shown in Fig. 3.2. The particles of the cylinder change their positions in space. The particle of material that was originally at P now occupies a new position in space, denoted by P^*. The point P^* is called the *final position* of the particle. This particle, which was at P, has been displaced to P^*. Its *displacement* can be represented as a vector, $\mathbf{u}(P)$. The displacement vector $\mathbf{u}(P)$ depends, of course, on the initial position of the particle. Notice that our attention is focused on a particle and the positions it occupies in space.

Fig. 3.2

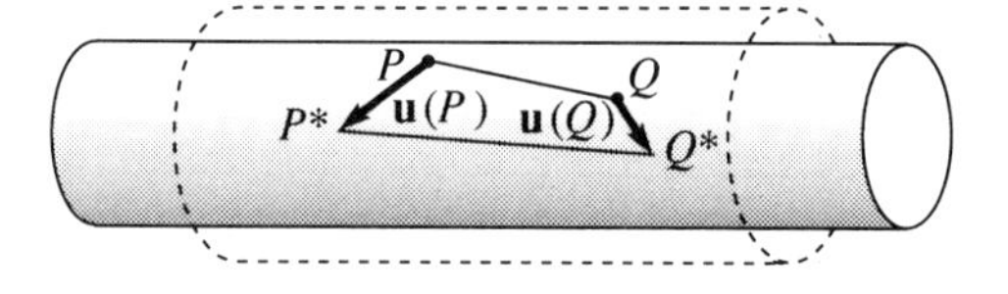

In Fig. 3.1, Q is the initial position of another particle of the cylinder. When the cylinder is stretched, this particle is displaced to Q^*. Since the displacement depends on the initial position of this particle, the displacement vector is denoted by $\mathbf{u}(Q)$.

In the deformed body, the length of P^*Q^* is generally different from the undeformed length, PQ. The particles undergo *relative* changes in position. It is possible to displace every particle of a body so that no change in distance occurs between any pair of particles. Such displacements are very special and are called *rigid body displacements.* In general, these displacements consist of a *translation* (all particles are moved the same distance in the same direction) and a *rotation* (all lines are rotated through the same angle about some fixed line). Imagine, for example, a piston moving in the cylinder of a stationary engine. As the piston moves from the top to the bottom of the cylinder, all the particles in the piston move parallel to the cylinder wall. The piston undergoes a translation. On the other hand, a flywheel that turns about a fixed axle undergoes a rotation, because, when the wheel turns, every radius turns through the same angle.

We are concerned only with displacements that change the *relative* positions of particles in a body, since deformation is a manifestation of such displacements. Therefore, no significance is attached to a rigid-body displacement in the remainder of this text.

3.3 Deformation and Strain

Deformation of a Body and Its Parts

Deformation is sensed by noting overall change in the shape and size of a body. When a balloon is inflated to a spherical form, the change in radius is sufficient to describe the balloon's alteration in size. If a solid circular cylinder is stretched so that it remains a circular cylinder, the changes in radius and length are enough to account for its final size. These observable effects are the results of changes in shape and size that occur in every portion of the body.

Strain

Two forms of deformation can be observed. They are illustrated by the body of Fig. 3.3(a), which is deformed into the shape of Fig. 3.3(b). The straight lines PQ and PR of Fig. 3.3(a) are cohesive, as chains of particles. They are, therefore, deformed to the continuous curves P^*Q^* and P^*R^* of Fig. 3.3(b). In general the lengths of the lines are changed, and also the angles between the lines have changed. This suggests that we devise means to measure changes in the lengths of line segments and changes of angles between line segments. These changes, which occur during deformation, are generally not uniform throughout a body. Some portions of a line segment will stretch; others will contract. Short segments of the same length originating at the same point but oriented in different directions will experience different extensions. Similarly, the changes in angle between segments will vary with position and orientation. To account for such variations, it is necessary to define quantities that characterize these changes in the neighborhood of a point. A quantity that measures the changes of *length* is an *extensional strain.* A quantity that measures changes of *angle* is a *shear strain.* Notice that strains measure only the relative positions of neighboring particles; consequently, a strain is independent of any rigid-body motion.

Fig. 3.3

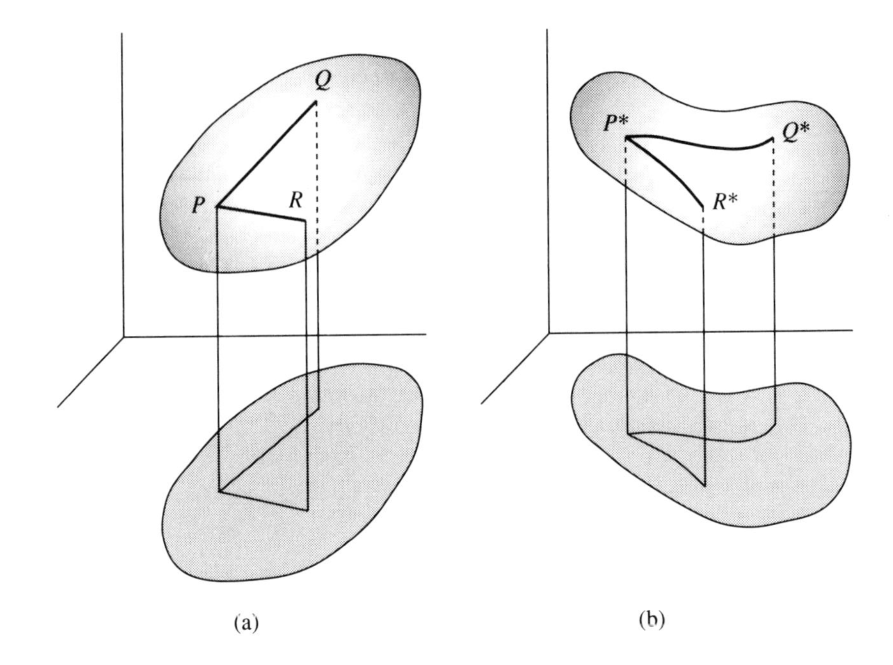

3.4 Extensional Strain

Figure 3.4(a) shows a cohesive body before deformation and a straight line of particles PQ in the interior. After a deformation, the line PQ is displaced and deformed into the continuous curve P^*Q^* of Fig. 3.4(b). The length of P^*Q^* is generally different from the length of PQ. The initial line PQ of Fig. 3.4(a) is shown subdivided into many

Fig. 3.4

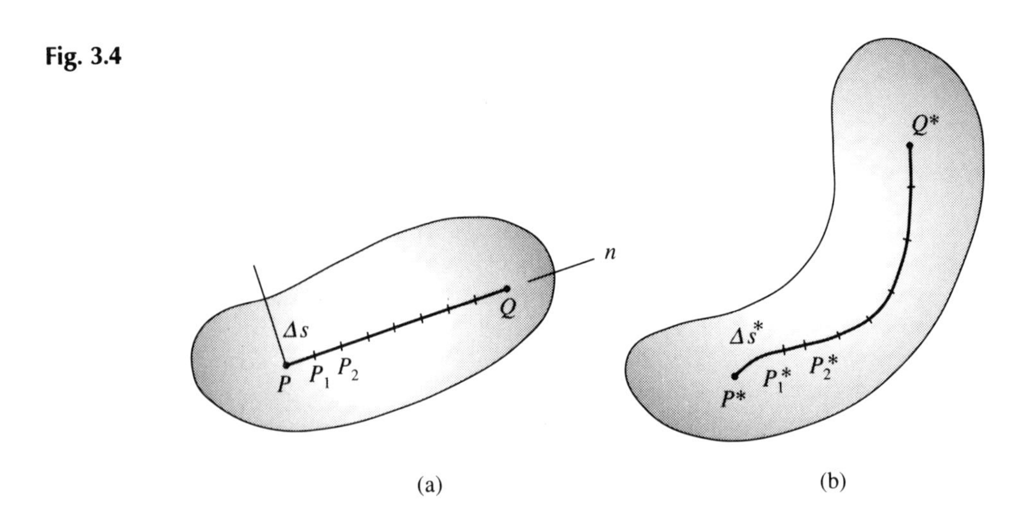

small, but *equal*, segments by labeling intermediate particles $P_1, P_2, \ldots$, which are displaced to $P_1^*, P_2^*, \ldots$, on P^*Q^* of Fig. 3.4(b). Note that, in general, each segment undergoes a different change; for example, $P^*P_1^*$ is longer than PP_1, whereas $P_1^*P_2^*$ is shorter than P_1P_2. Evidently the extension (or contraction) depends upon the location along the line and also the length of the segment. To describe the extensional deformation, we require a measure at each point; we must compare the original and deformed lengths, but we must do so as the lengths shrink to the point in question. Such a measure is called an *extensional strain.* Note, too, that a line with different orientation will, in general, experience a different extensional strain. The cylinder of Fig. 3.1 is a simple example: Lines in an axial direction are stretched, whereas lines in a radial direction are contracted. Any definition of extensional strain must characterize extension at the specific point and in the specific direction.

The length of one segment PP_1 in Fig. 3.4(a) is Δs. The direction of the segment is identified by n. After the deformation, the particles at P and P_1 are at P^* and P_1^* in Fig. 3.4(b), and the length has changed to Δs^*. We *define*[1] the *extensional strain* $\varepsilon_n(P)$ at the point P in the direction n by the limit

$$\varepsilon_n(P) = \lim_{\substack{\Delta s \to 0 \\ P_1 \to P}} \frac{\Delta s^* - \Delta s}{\Delta s} \tag{3.1}$$

Note that the lengths Δs and Δs^* diminish as P_1 approaches P, because the body is continuous and cohesive. The extensional strain is obtained as a ratio of two lengths, so it is a dimensionless quantity.

Let us examine the meaning of the notation for extensional strain. To specify $\varepsilon_n(P)$, we must (1) specify the point P, (2) specify the direction n, and (3) give the numerical value of $\varepsilon_n(P)$.

The quantity $\varepsilon_n(P)$ enables us to compute the approximate change in length of a short line segment in the body when the body is deformed. The particular segment originates at P in the undeformed body and is in the direction n. At P, in the direction n, we can choose a very short segment of length Δs. From Eq. (3.1), we have a good approximation for the final length of this segment:

$$\Delta s^* \doteq [1 + \varepsilon_n(P)]\,\Delta s \tag{3.2}$$

If $\varepsilon_n(P)$ is positive, the segment of length Δs has extended; if $\varepsilon_n(P)$ is negative, this segment has shortened. It is also possible that $\varepsilon_n(P)$ is zero. If so, there is no change in length of the segment.

To determine the change in length of a line of particles of finite length, we need to know the extensional strain at every point along the line in the direction of the line. In Fig. 3.4(a), the line PQ has an initial length L, which has been divided into m equal segments of length Δs. After deformation the line is deformed to P^*Q^*; each of the

1 Many alternative definitions are admissible. We adopt the most convenient. Here, we adopt definition (3.1), which is conventional in engineering, conceptually simple, and yet adequate for our purposes.

particles $P_1, P_2, \ldots$, has been displaced to $P_1^*, P_2^*, \ldots$. In general, each segment has a new length: $P^*P_1^*$ has length Δs_1^*; $P_1^*P_2^*$ has length Δs_2^*; and so on. Using Eq. (3.2) and denoting P by P_0, we have

$$\begin{aligned} \Delta s_1^* &= [1 + \varepsilon_n(P_0)]\,\Delta s \\ \Delta s_2^* &= [1 + \varepsilon_n(P_1)]\,\Delta s \\ &\vdots \end{aligned} \tag{3.3}$$

If L^* is the length of P^*Q^*, then

$$L^* \doteq \sum_{k=0}^{m-1} [1 + \varepsilon_n(P_k)]\,\Delta s$$

By passing to the limit, we obtain

$$L^* = \int_0^L [1 + \varepsilon_n(P)]\,ds = L + \int_0^L \varepsilon_n(P)\,ds \tag{3.4}$$

Thus, knowing $\varepsilon_n(P)$ at every point along PQ, Eq. (3.4) enables us to compute the exact value of the new length, P^*Q^*.

EXAMPLE 1

To illustrate this idea, consider the straight wire shown in Fig. 3.5(a). Suppose that a nonuniform heating of the wire causes an extensional strain, ε_x, given by

$$\varepsilon_x = k\left(\frac{x}{L}\right)^2$$

Fig. 3.5

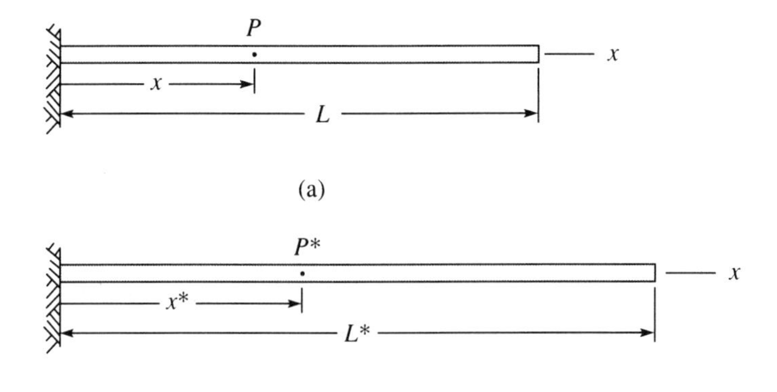

We can determine the change in length of the wire by computing the change in length of the line of particles along the x-axis. The new length L^* is obtained by placing the expression for ε_x in Eq. (3.4):

$$L^* = L + \int_0^L \varepsilon_x(x)\,dx = L + \int_0^L k\left(\frac{x}{L}\right)^2 dx = L + \frac{kL}{3}$$

Consequently, the change in length of the wire is

$$\Delta L = L^* - L = \frac{kL}{3} \quad \blacklozenge$$

PROBLEMS

3.1 A wire 300 cm long is subjected to nonuniform heating so that the extensional strain along the wire is proportional to the distance from one end of the wire. If the wire stretches 3 cm, what is the extensional strain at the center of the wire?

3.2 A cable is fixed at its upper end and supports a mine elevator car at the bottom of a 1000-ft shaft. The cable is strained uniformly by the weight of the elevator car by an amount equal to 0.001. At each point it is strained additionally (by its own weight) by an amount that is proportional to the length of cable below the point. If the extensional strain at the upper end of the cable is 0.002, what is the total elongation of the cable?

3.3 A wire is stretched in such a manner that the extensional strain along the wire varies with the square of the distance from one end. If the original length of the wire is 50 cm and the extensional strain at the center of the wire is 0.025 cm, compute the increase in length of the wire.

3.4 A wire is in the form of a circular hoop, as shown in Fig. P3.4. It is heated nonuniformly so that the extensional strain along the hoop is given by $\varepsilon_\theta = k\sin^2\theta$. If the initial radius is $R = 20$ cm and $k = 10^{-4}$, compute the radius when heated.

Fig. P3.4

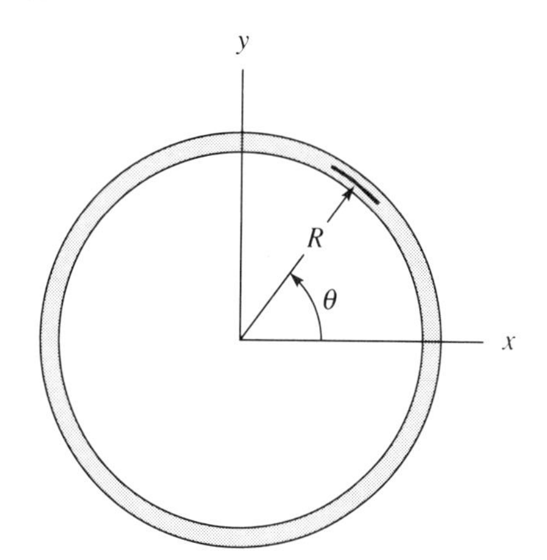

3.5 A wire lies initially along the x-axis, as shown in Fig. P3.5. It is displaced to the line $y = mx$ and then to $y = nx$ without any point of the wire displacing in the x-direction. Compute $\varepsilon_m(x)$, the strain in the intermediate case (along $y = mx$), and also compute $\varepsilon_n(x)$ the strain in the final state (along $y = nx$). If the wire had been initially unstrained in the position along $y = mx$, what would have been the extensional strain if it were then displaced to $y = nx$? Denoting this strain by $\varepsilon_{n-m}(x)$, show that if ε_m and ε_n are small, then to first-order terms,

$$\varepsilon_n = \varepsilon_m + \varepsilon_{n-m}$$

Fig. P3.5

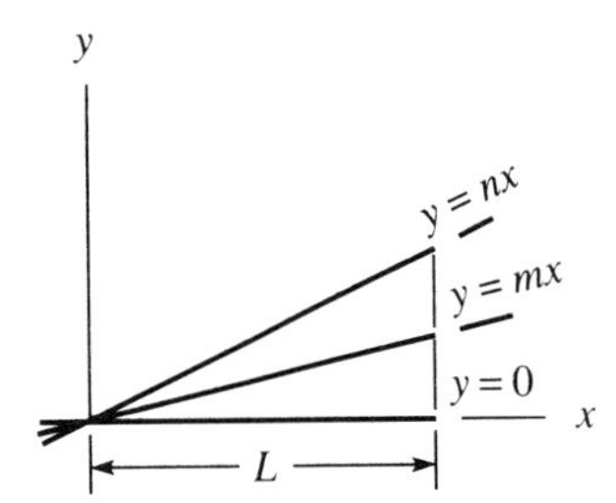

3.6 A straight piece of wire is stretched onto a parabola, as shown in Fig. P3.6, so that no point of the wire displaces in the x-direction. Compute the extensional strain at any point along the wire.

Fig. P3.6

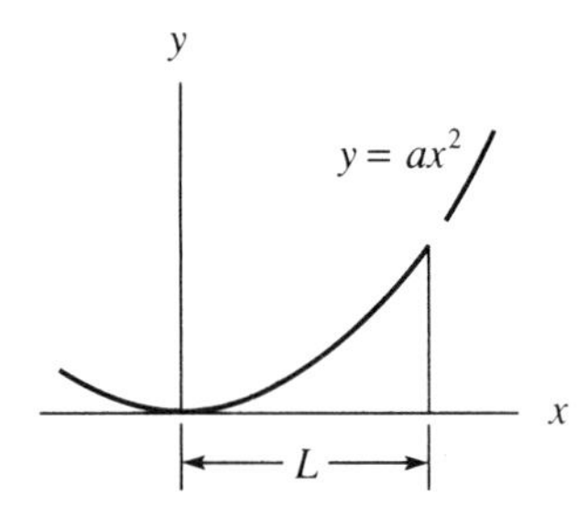

3.7 A straight piece of wire PQ lies along the line $y = mx$, as shown in Fig. P3.7. The wire is strained and displaced to the line $y = nx$ in such a way that a point originally at x is displaced to $x^2/2$. Show that the extensional strain at any point along the wire is given by

$$\varepsilon = \sqrt{\frac{1 + n^2}{1 + m^2}}x - 1$$

(where x is the original coordinate of the point).

Fig. P3.7

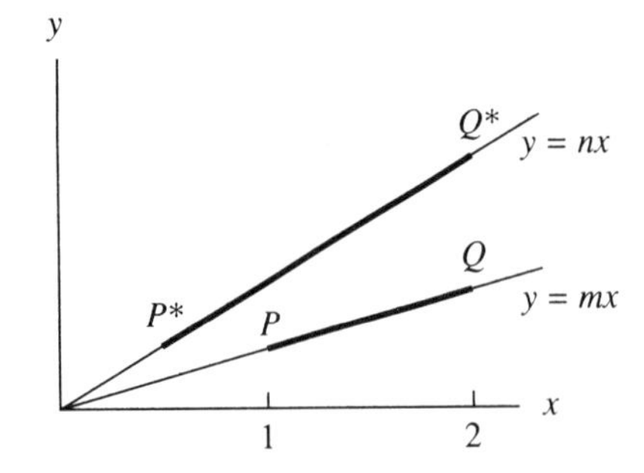

3.8 We might have defined extensional strain on the basis of final length; that is, instead of Eq. (3.1), we would write

$$\varepsilon_n'(P) = \lim_{\substack{P_1 \to P \\ \text{along } n}} \frac{\Delta s^* - \Delta s}{\Delta s^*}$$

Show that the difference in the values computed by this definition and by Eq. (3.1) is

$$\varepsilon_n(P) - \varepsilon_n'(P) = \varepsilon_n(P)\varepsilon_n'(P)$$

Hence, it is a second-order term (if the strain is small).

3.9 On the basis of the definition of extensional strain given by Eq. (3.1), is there any limit to how large (algebraically) an extensional strain can be? Is there a limit to how small (algebraically) an extensional strain can be? If the alternative definition of Prob. 3.8 is used, how does this affect the answers to these questions?

3.10 It is sometimes desirable to use another definition of strain. Instead of Eq. (3.1), we would write

$$\varepsilon_n''(P) = \lim_{\substack{P_1 \to P \\ \text{along } n}} \frac{1}{2}\left[\frac{(\Delta s^*)^2 - (\Delta s)^2}{(\Delta s)^2}\right]$$

Show that this differs from the extensional strain defined in Eq. (3.1) by a second-order term in the extensional strain.

3.11 Another definition of extensional strain is the so-called logarithmic strain $\bar{\varepsilon}_n$. An increment, or differential, of the logarithmic strain ($\delta\bar{\varepsilon}_n$) is defined as the increment of current length $\delta(\Delta S')$ divided by the current length $\Delta S'$:

$$\delta\bar{\varepsilon}_n \equiv \lim_{\Delta S' \to 0} \frac{\delta(\Delta S')}{\Delta S'}$$

The logarithmic strain is obtained by integrating from the initial (ΔS) to final (ΔS^*) length. If ε_n denotes the conventional strain defined by Eq. (3.1), show that $\bar{\varepsilon}_n = \ln(1 + \varepsilon_n)$.

3.5 Average Extensional Strain

Let the initial length of a segment be L and its final length be L^*. The ratio

$$\varepsilon = \frac{L^* - L}{L} \tag{3.5}$$

is called the *average extensional strain* of the line segment.

EXAMPLE 2

Let us use the concept of average extensional strain to describe the change in length of a wire. The rigid bar ABC in Fig. 3.6(a) rotates through an angle θ. Then ε for the wire can be computed by noting the displacement of B. B^* has as coordinates $(L\cos\theta, -L\sin\theta)$; therefore,

$$L^* = \sqrt{(L - L\cos\theta)^2 + (L + L\sin\theta)^2} = L\sqrt{3 + 2(\sin\theta - \cos\theta)}$$

so that

$$\varepsilon = \sqrt{3 + 2(\sin\theta - \cos\theta)} - 1$$

Fig. 3.6

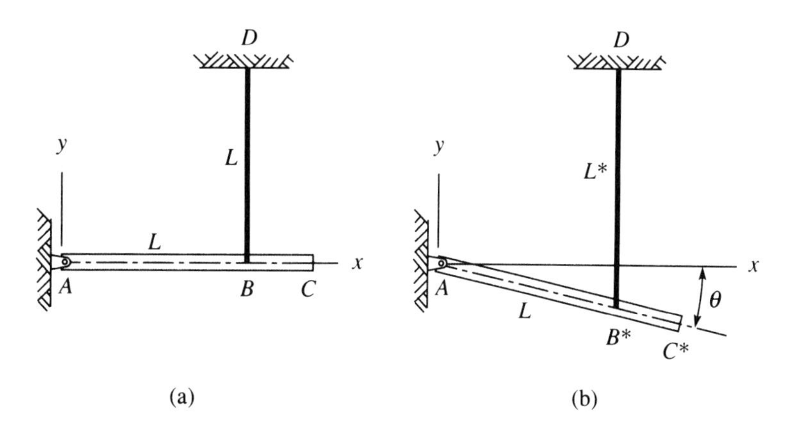

If θ is small, we can employ the small-number algebra of Sec. 1.3 to get

$$\sqrt{3 + 2(\sin\theta - \cos\theta)} \doteq 1 + \theta$$

Consequently,

$$\varepsilon \doteq \theta \quad \blacklozenge$$

EXAMPLE 3

The two wires shown in Fig. 3.7 are joined at P and to the fixed horizontal surface at B and C. The joint P is displaced to P^*; horizontal and vertical components of displacement are u and v, as shown. Let us obtain the exact expression for the average extensional strains of each wire in terms of u and v and then determine the linear approximation of each strain, valid if $u/L \ll 1$, $v/L \ll 1$. Finally, let us determine the error in the linear approximation when $u = v = 0.010L$; for instance, if $L = 2$ m (about 80 in.), $u = v = 2$ cm (about 0.8 in.).

Fig. 3.7

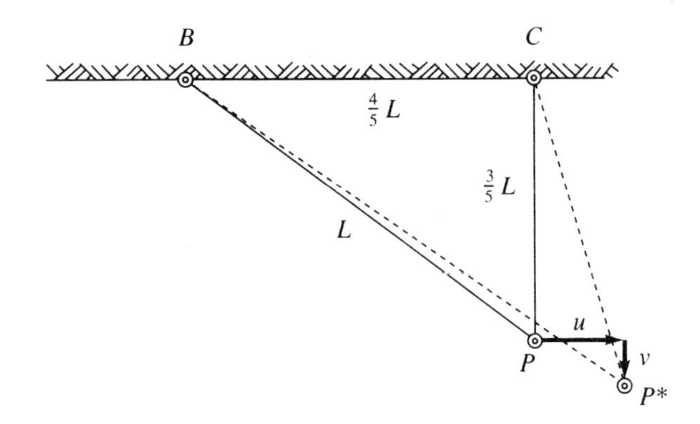

By the Pythagorean theorem the new lengths of the wires are

$$BP^* = \sqrt{\left(\frac{4}{5}L + u\right)^2 + \left(\frac{3}{5}L + v\right)^2}$$

$$= L\sqrt{\left(\frac{4}{5} + \frac{u}{L}\right)^2 + \left(\frac{3}{5} + \frac{v}{L}\right)^2}$$

$$CP^* = L\sqrt{\left(\frac{u}{L}\right)^2 + \left(\frac{3}{5} + \frac{v}{L}\right)^2}$$

The extensional strains, ε_B and ε_C, of wires BP and CP, respectively, follow from the definition:

$$\varepsilon_B = \frac{BP^*}{L} - 1$$

$$= \left[\left(\frac{4}{5} + \frac{u}{L}\right)^2 + \left(\frac{3}{5} + \frac{v}{L}\right)^2\right]^{1/2} - 1 = 0.01400197$$

$$\varepsilon_C = \frac{5}{3}\left[\left(\frac{u}{L}\right)^2 + \left(\frac{3}{5} + \frac{v}{L}\right)^2\right]^{1/2} - 1 = 0.01680327$$

By means of the bionomial theorem, we obtain the linear approximations

$$\varepsilon_B \doteq \frac{4}{5}\frac{u}{L} + \frac{3}{5}\frac{v}{L} = 0.01400000$$

$$\varepsilon \doteq \frac{5}{3}\frac{v}{L} = 0.01666667$$

The errors of these approximations are -0.014% and -0.81%. Note the relatively small errors, although the displacements are larger than we normally encounter in engineering practice. ◆

3.12 A wire segment, initially 600 cm long, is stretched to a length of 600.6 cm. Determine the average extensional strain of the wire.

3.13 A thin wire 1 ft long is stretched around a cylinder of radius 2 in. Determine the average extensional strain of the wire.

3.14 A balloon has an initial radius of 12 cm. More air is blown into it to increase the radius to 15 cm. What is the average extensional strain of a great circle on the balloon's surface (measured from the initial state)?

3.15 A circular hoop 25 in. in diameter is expanded uniformly by heating, so that the area it encloses is increased by 0.5π in^2. What is the approximate (first-order) average extensional strain in the hoop?

3.16 The linkage shown in Fig. P3.16 consists of four rigid bars, AB, BC, CD, and DA, and a wire AC fastened diagonally across the frame. If the square frame is deformed into the rhombus shown, compute the average extensional strain of AC.

3.17 A wire AB has one end attached at the origin (A) in a rectangular system (x, y). The other end (B) is initially at the point ($x = 600$ mm, $y = 800$ mm). End B is then displaced to the point ($x = 640$ mm, $y = 820$ mm). Compute the average extensional strain. What is the error (%) if you use a first-order approximation—that is, linear—in the displacement (40 mm and 20 mm in the directions x and y)?

Fig. P3.16

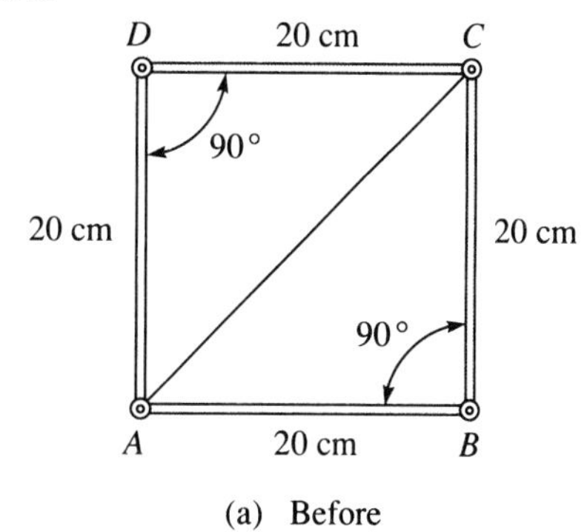

(a) Before

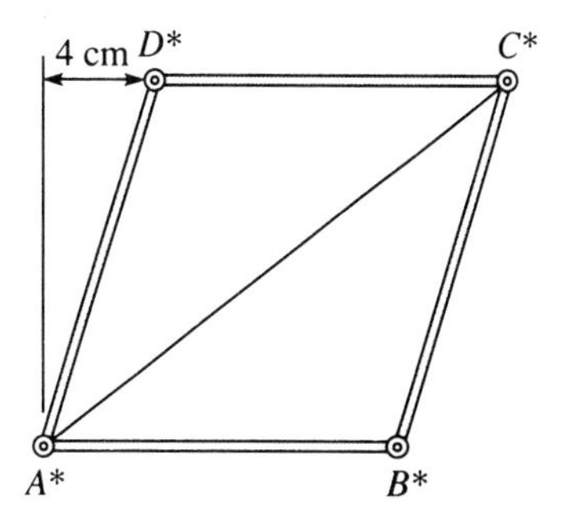

(b) After

3.18 The rods BC, CD, and DA are connected by pins at A, B, C, and D (Fig. P3.18). The assembly is held in the rectangular position by taut cables AC and BD. If the assembly is moved so that the horizontal displacement of joint D is 20 cm to the right, compute the average extensional strain of each wire to three decimal places. Assume that the rods are rigid.

Fig. P3.18

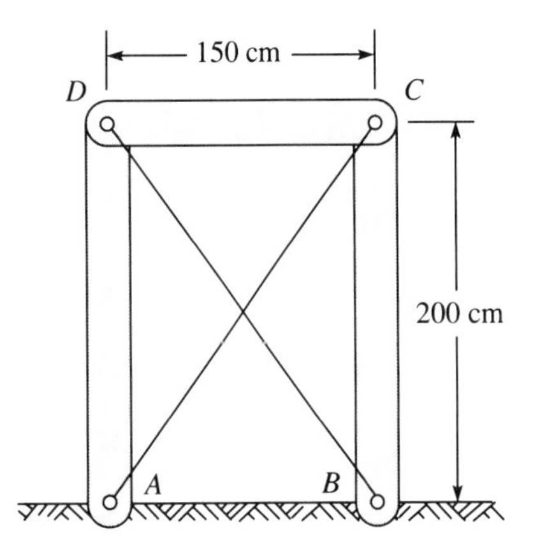

3.19 The point C is displaced to C^*, as shown in Fig. P3.19. The horizontal and vertical components of this displacement are u and v. Express the average extensional strains of AC and BC in terms of u, v, and L. If u and v are small, what are the approximate expressions of these average extensional strains?

Fig. P3.19

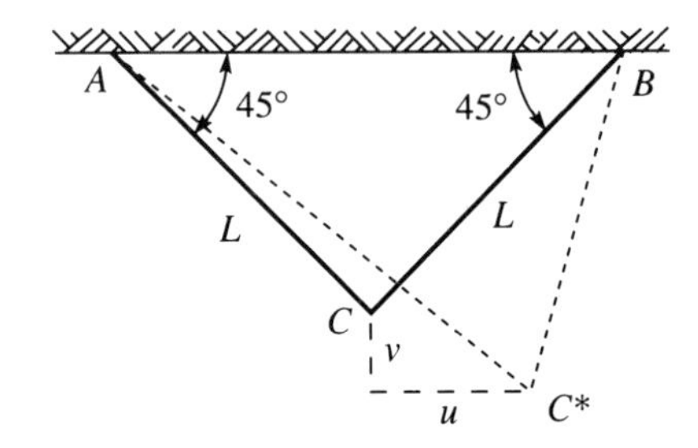

3.20 The point D in Fig. P3.20 is displaced vertically by an amount $0.01L$. Obtain the approximate (first-order) average extensional strains of AD, BD, and CD.

Fig. P3.20

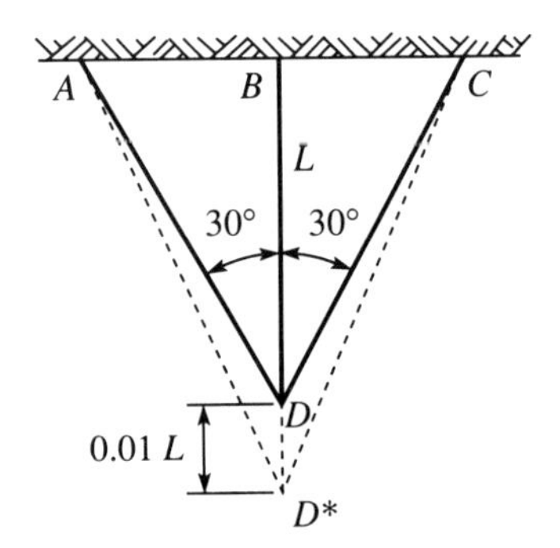

3.21 A straight wire PQ is strained and displaced so that it becomes the straight wire P^*Q^* shown in Fig. P3.21. The horizontal and vertical components of the small displacement of P to P^* are u_P and v_P; for the

Fig. P3.21

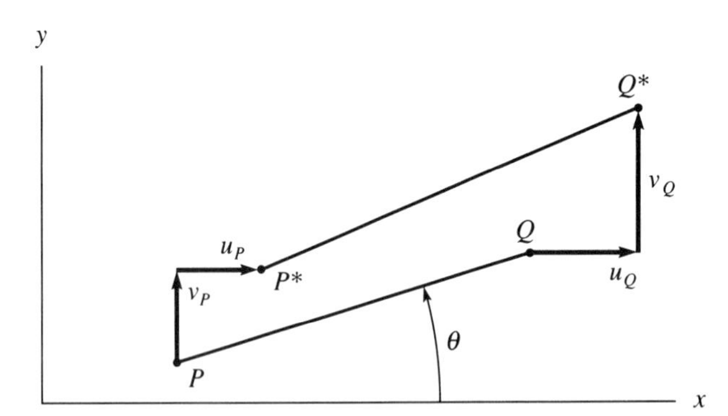

small displacement of Q to Q^*, they are u_Q and v_Q. Show that the average extensional strain of the wire is approximately

$$\varepsilon = \frac{u_Q - u_P}{L}\cos\theta + \frac{v_Q - v_P}{L}\sin\theta$$

where L is the length of PQ.

3.22 A pin-connected truss is shown in Fig. P3.22. Joint P displaces as point P in Fig. P3.21 ($u_P = 0.020$ in., $v_P = 0.010$ in.) and joint Q displaces as point Q of Fig. P3.21 ($u_Q = 0.030$ in., $v_p = 0.016$ in.). Obtain an approximation of the average extensional strain of the member PQ. Can you justify your approximation?

Fig. P3.22

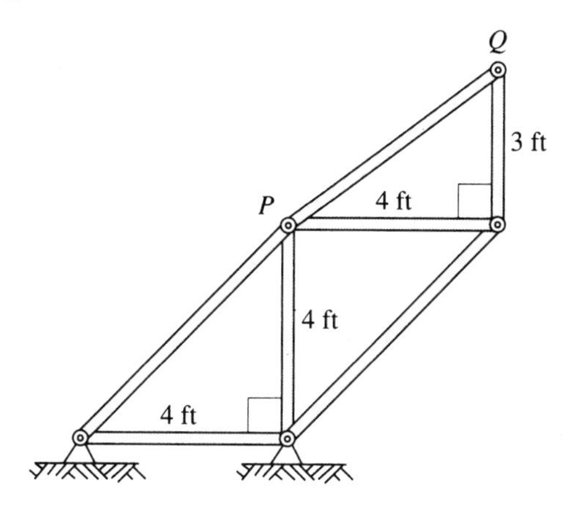

3.23 A rectangular frame is stretched into another rectangular frame, as shown in Fig. P3.23. The average extensional strain of AB is $\bar{\varepsilon}_1$, of AD is $\bar{\varepsilon}_2$, and of AC is $\bar{\varepsilon}_3$. If these strains are small, show that $\bar{\varepsilon}_3 = \bar{\varepsilon}_1 \cos^2\theta + \bar{\varepsilon}_2 \sin^2\theta$.

Fig. P3.23

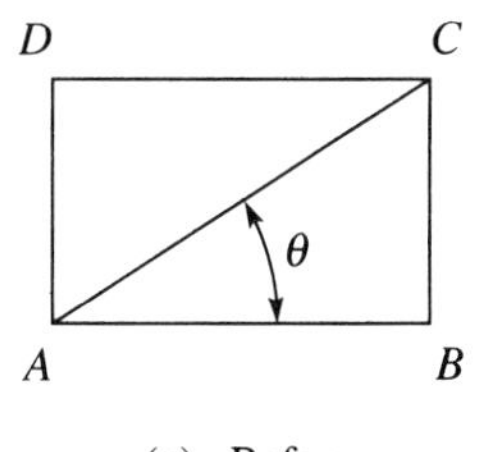

(a) Before

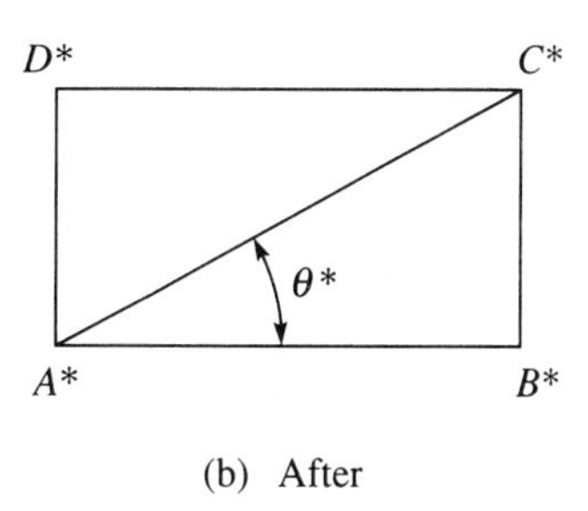

(b) After

3.24 The bar AB is rigid and is supported by two wires DB and CB, as shown in Fig. P3.24. If the bar rotates through a small angle θ, express the average extensional strain of DB and CB in terms of θ.

Fig. P3.24

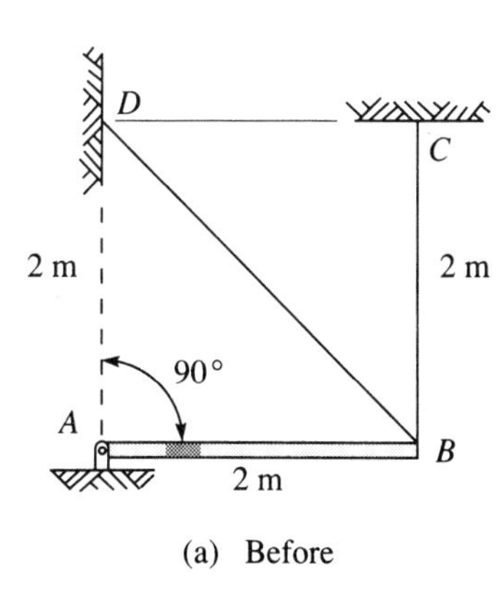

(a) Before

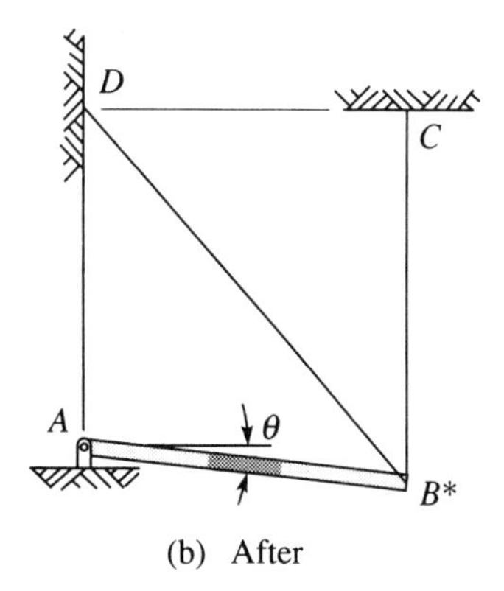

(b) After

3.25 The "rigid" bar DO in Fig. P3.25 is pinned at D and supported by wires OC and OB. Assume that points D, C, and B are fixed. If the bar DO is rotated (clockwise) 0.1000 rad about D, (a) compute the average extensional strain of each wire, *accurate* to four decimal places, and (b) give the first-order (linear) approximation of these strains, also accurate to four decimal places.

Fig. P3.25

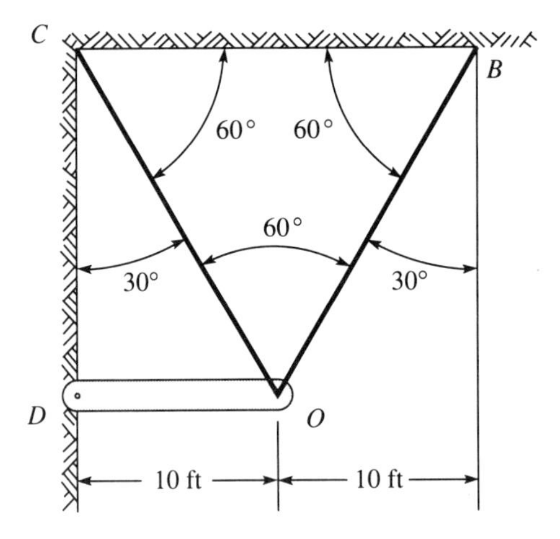

3.26 Three wires support a bar, as shown in Fig. P3.26. If the left end of the bar is raised an amount Δ and the bar is rotated through at an angle θ (both Δ and θ are small), what are the average extensional strains in the wires?

Fig. P3.26

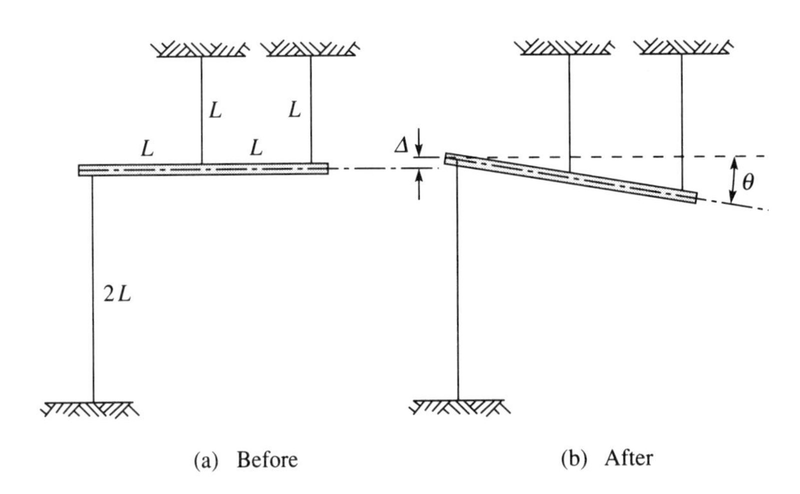

(a) Before (b) After

3.27 The sides a, b, and c of a rectangular block are increased by the small amounts $\Delta a, \Delta b$, and Δc, respectively, without changing the right angles at the edges. Compute the average extensional strain of each side, $\bar{\varepsilon}_a$, $\bar{\varepsilon}_b$, and $\bar{\varepsilon}_c$, and, in terms of these, the approximate average extensional strain of a diagonal. What is the approximate fractional increase in volume of the block?

3.28 Three wires support an object P at the point $(4, 4, 4)$, as shown in Fig. P3.28. If the object is displaced to the point $(4.1, 3.8, 4.2)$, compute the approximate average extensional strain in each of the three wires.

Fig. P3.28

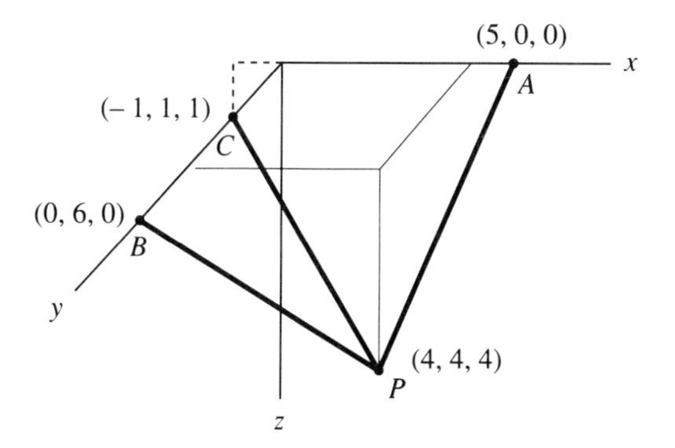

3.29 A straight wire connects the origin O to the point $P(x, y, z)$, as shown in Fig. P3.29. If P is displaced slightly to $P^*(x + \Delta x, y + \Delta y, z + \Delta z)$ while the wire remains straight and O does not move, show that the average extensional strain of the wire is given approximately by

$$\bar{\varepsilon} = \frac{x\,\Delta x + y\,\Delta y + z\,\Delta z}{x^2 + y^2 + z^2}$$

Fig. P3.29

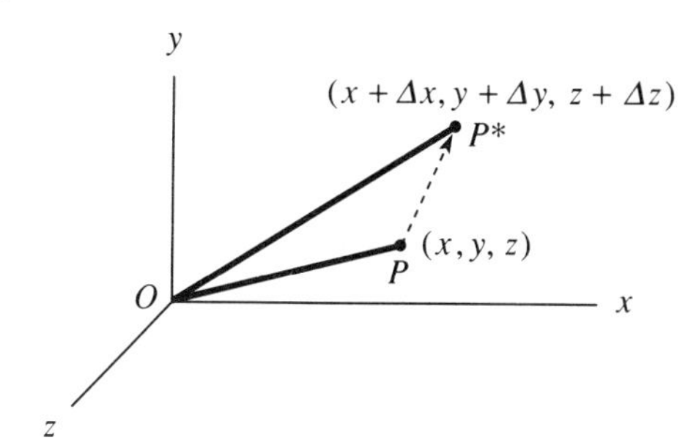

3.30 Three wires are connected to an object at $P(16, 0, 0)$, as shown in Fig. P3.30. If P is displaced slightly to $P^*(16 + u, v, w)$, where u, v, and w are small, determine the approximate average extensional strain in each of the wires (a linear combination of u, v, w).

Fig. P3.30

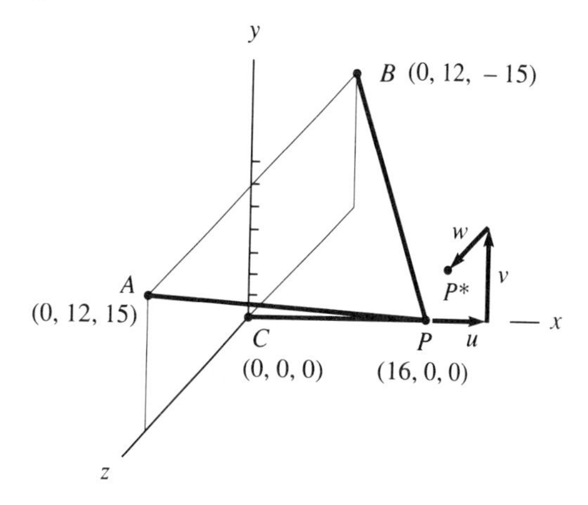

3.31 Compute the average extensional strain for the wire PQ of Prob. 3.7.

3.32 A straight wire of length L is strained so that the pointwise extensional strain along the wire is given by $\varepsilon_x(x) = k(x/L)^2$, where x is the distance from one end of the wire. What is the average extensional strain of the wire?

3.33 Compute the average extensional strain of the wire of Prob. 3.6.

3.6 | Shear Strain

In Fig. 3.8(a), two perpendicular directions n and t are shown at the point P. The particles along PQ deform into the curve P^*Q^*, and those along PR deform into P^*R^*. The angle between the curves P^*R^* and P^*Q^* at P^* is ϕ_{nt}^* (in radians). The shear strain at P associated with the perpendicular directions n and t is defined by[2]

---

2 Many alternative definitions are possible. The engineer adopts the one that serves most conveniently. Here, we give the definition that is conceptually simple, yet adequate.

Fig. 3.8

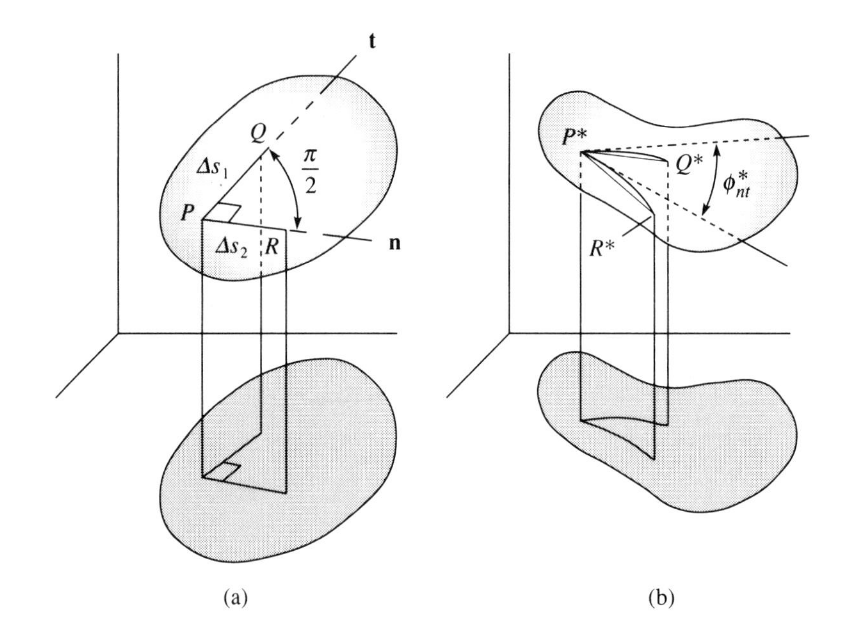

$$\gamma_{nt}(P) = \frac{\pi}{2} - \phi^*_{nt} \tag{3.6}$$

If ϕ^*_{nt} is less than $\pi/2$, the shear strain is positive. If ϕ^*_{nt} is greater than $\pi/2$, the shear strain is negative. In order to specify $\gamma_{nt}(P)$, we must (1) identify the point P, (2) specify the two perpendicular directions at P, and (3) give a numerical value for the shear strain γ_{nt}. We may also define $\gamma_{nt}(P)$ as

$$\gamma_{nt}(P) = \frac{\pi}{2} - \lim_{\substack{\Delta S_1 \to 0 \\ \Delta S_2 \to 0}} \angle R^*P^*Q^* \tag{3.7}$$

where $\angle R^*P^*Q^*$ is the angle between the chords P^*Q^* and P^*R^* in Fig. 3.8(b). Clearly, the limiting value of this angle is ϕ^*_{nt}. This alternative definition is sometimes useful for computing shear strains.

EXAMPLE 4

To illustrate shear strain, let us suppose that the rectangular plate in Fig. 3.9(a) with edges parallel to the rectangular x- and y-axes, deforms such that particles displace only in the y-direction and all lines parallel to the x-axis deform to similar parabolas, as the x-axis ($y = 0$) deforms to $y = ax^2$ in Fig. 3.9(b).

By definition the slope of the deformed (x) line is

$$\frac{dy}{dx} = \tan\theta = 2ax$$

Fig. 3.9

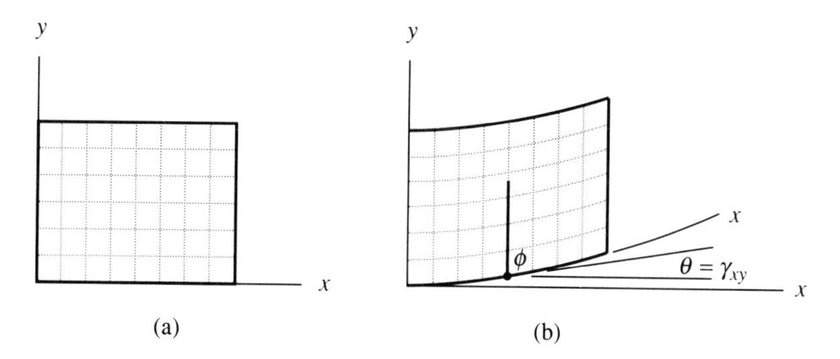

Because the vertical lines (y lines) remain vertical, the shear strain is

$$\gamma_{xy} = \frac{\pi}{2} - \phi = \tan^{-1}\theta = \tan^{-1} 2ax$$

Now, suppose that the dimensions of the plate are 8 cm by 5 cm and $a = 1/80$. Then

$$\gamma_{xy} = \tan^{-1}\frac{x}{40}$$

The maximum value is at the right edge, $x = 8$, where

$$\gamma_{xy} = \tan^{-1} 0.200 = 0.197 \text{ (rad)}$$

This is a large shear strain ($\theta = 11.3°$). In most structural problems, strains are of the order 10^{-3}. Still, for this large shear strain, the slope ($dy/dx = 0.200$) differs little from the angle ($\theta = 0.197$).

Note that the strains vary throughout most members; in this example the shear strain (γ_{xy}) varies from zero at the left edge to the maximum (0.200) at the right. ◆

PROBLEMS

3.34 A rectangular plate is deformed into a parallelogram, as shown in Fig. P3.34. The lengths of the edges are unchanged. Compute the approximate (first-order) shear strain $\gamma_{xy}(P)$.

Fig. P3.34

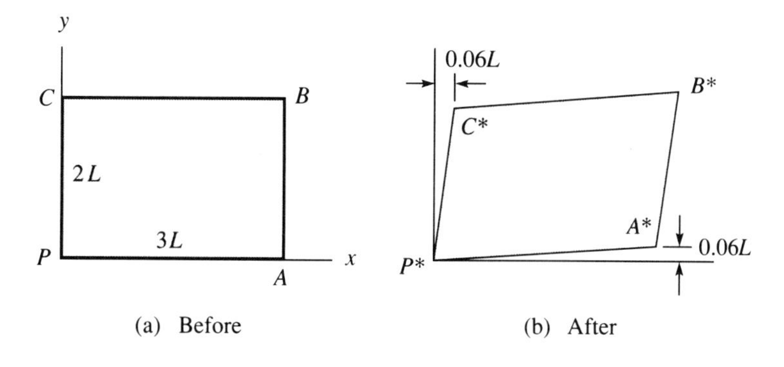

3.35 Obtain an exact expression for the extensional strain in the deformed plate of Fig. 3.9 (See Example 4). Obtain the approximation (linear in the small number a). What is the maximum error of the approximation if the plate has given dimensions of 8 cm by 5 cm and $a = 1/80$?

3.36 A thin triangular plate is deformed as indicated in Fig. P3.36. Compute an approximate value for the shear strain $\gamma_{st}(P)$.

Fig. P3.36

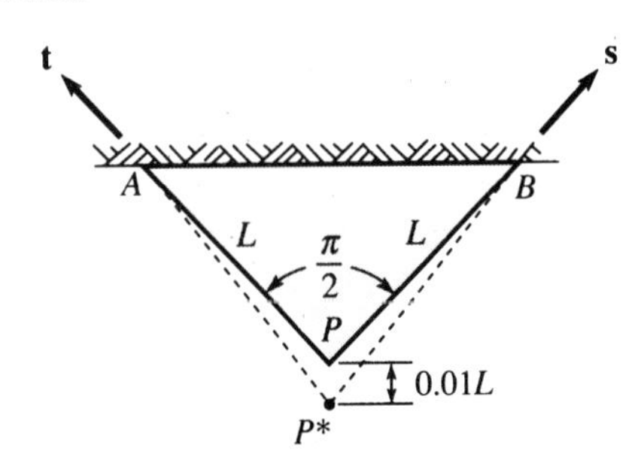

3.37 In the x, y-rectangular coordinate system, point A is at $(1, 0)$ and point B is at $(0, 1)$. If the origin remains fixed and the strains are uniform, compute the approximate value of γ_{xy} when B is displaced to $(-0.05, 0.95)$ and A, to $(1.10, 0.05)$.

3.38 A thin plate is in the form of a 30°-60°-90° triangle and is supported along its hypotenuse, as shown in Fig P3.38. It is subjected to the uniform strains $\varepsilon_s = 0.004$, $\varepsilon_t = 0.003$. Compute the approximate value of $\gamma_{st}(P)$.

Fig. P3.38

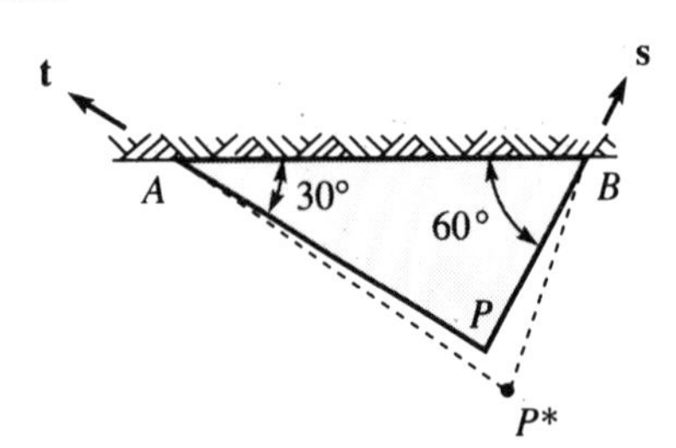

3.39 The rectangle shown in Fig. P3.39 is deformed into a parallelogram. The average extensional strains along PA and PB are zero; along the diagonal PC the extensional strain is ε, a small number. Compute the shear strain $\gamma_{xy}(P)$ in terms of a, b, and ε (assuming γ_{xy} is also small).

Fig. P3.39

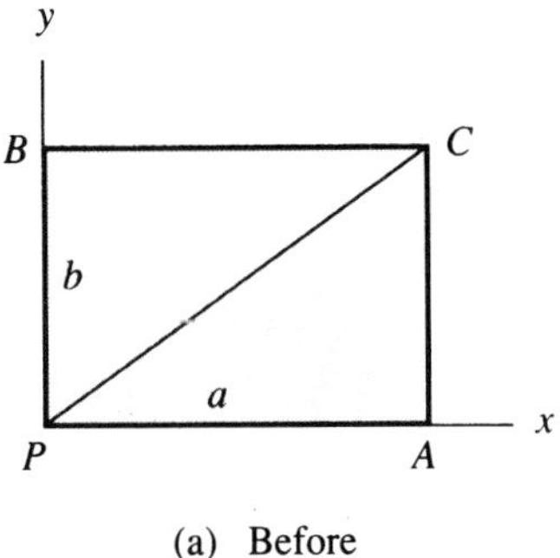

(a) Before

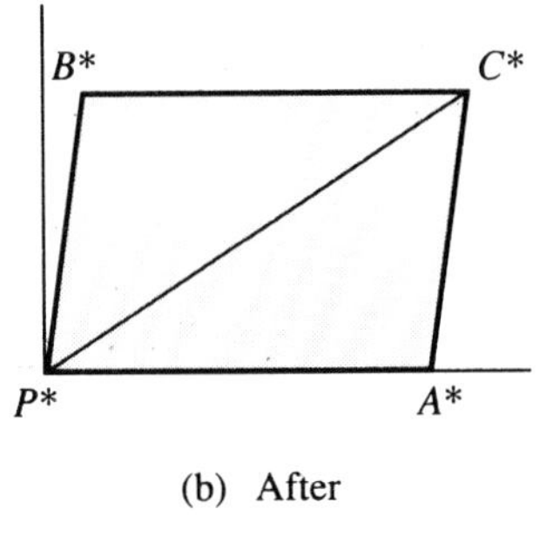

(b) After

3.40 For the rectangle in Fig. P3.40, the average extensional strains of PA, PB, and PC are small and have the values ε_x, ε_y, and ε_r, respectively. Determine γ_{xy} (assuming it is also small) in terms of $a, b, \varepsilon_x, \varepsilon_y$, and ε_r. Eliminate a and b from this relation by introducing the angle θ.

Fig. P3.40

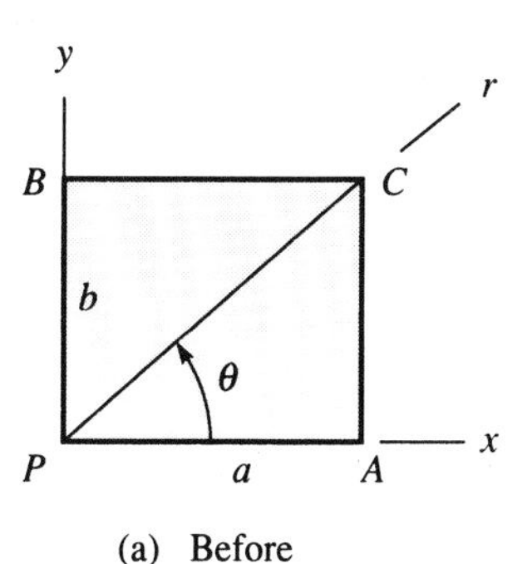

(a) Before

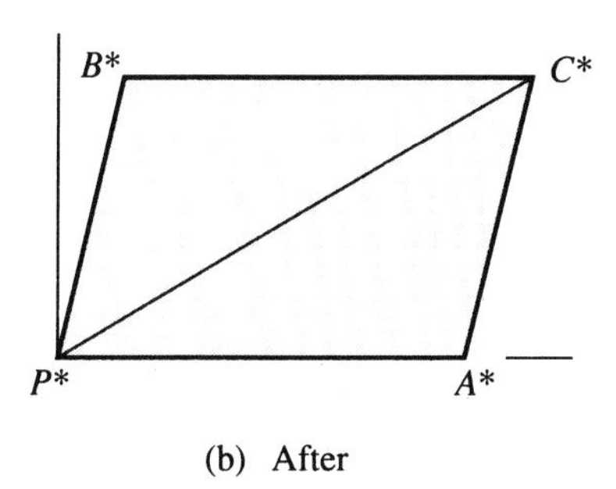

(b) After

3.41 A thin rectangular plate is deformed uniformly into another rectangle, as shown in Fig. P3.41. It elongates 10% of its original length and contracts 3% of its original width. Compute an approximate value of the shear strain γ_{st}.

Fig. P3.41

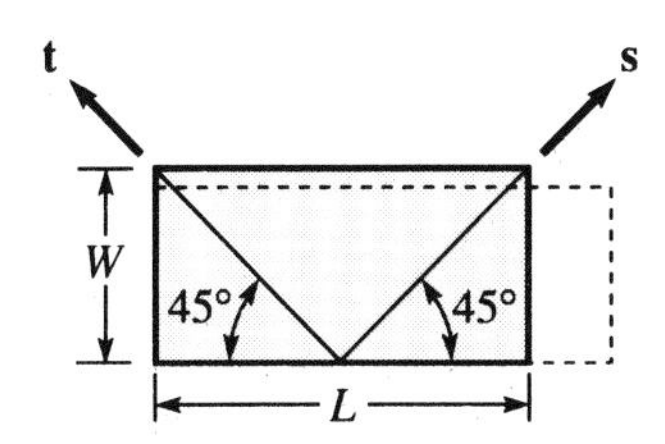

3.42 A thin rectangular plate is deformed uniformly into another rectangle, as shown in Fig. P3.42. The extensional strain along its length is 0.02, and the width remains unchanged. Compute an approximate value of the shear strain γ_{st}.

Fig. P3.42

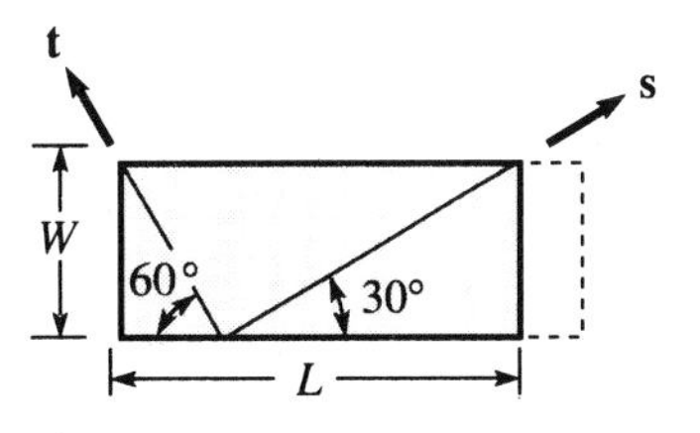

3.43 The adjacent rectangles shown in Fig. P3.43 are deformed into parallelograms by the small uniform shear strain γ_{xy}, with ε_x and ε_y both zero. Compute γ_{st} in terms of a, b, and γ_{xy}. Eliminate a and b from the resulting expression by introducing the angle θ.

Fig. P3.43

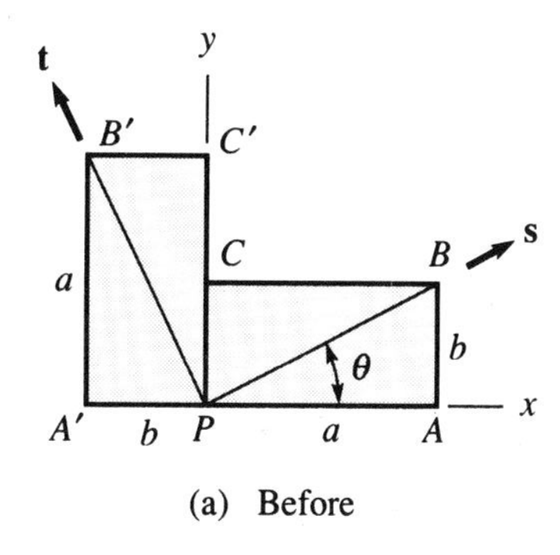

(a) Before

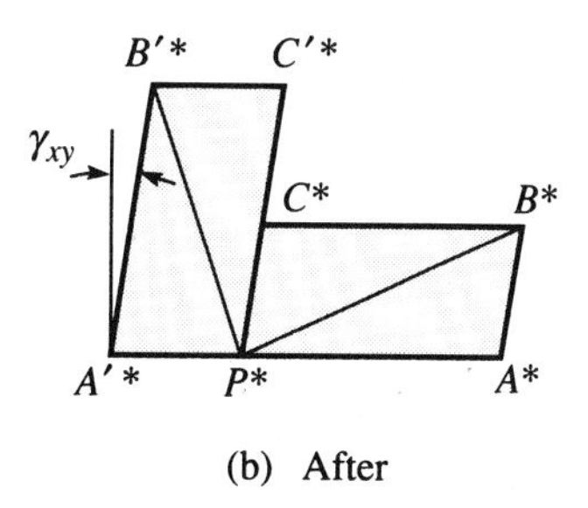

(b) After

3.44 The adjacent rectangles shown in Fig. P3.44 are deformed into different rectangles by the small uniform extensional strains ε_x and ε_y, with γ_{xy} zero. Compute γ_{st} in terms of a, b, ε_x, and ε_y. Eliminate a and b from the resulting expression by introducing the angle θ.

Fig. P3.44

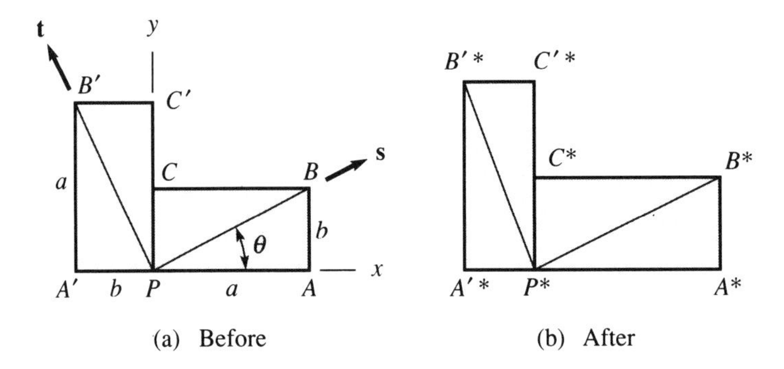

(a) Before (b) After

3.45 The rectangular plate in Fig. P3.45 is deformed homogeneously (each elemental rectangle deforms similarly) to the parallelogram. If the extensional strains ε_x and ε_y (along the edges) and the shear strain γ_{xy} (between the edges) are prescribed, (a) obtain a formula for the extensional strain ε_n (of the diagonal) in terms of $\varepsilon_x, \varepsilon_y, \gamma_{xy}$, and angle θ, and (b) obtain the approximation that expresses strain ε_n as a linear combination of strains $\varepsilon_x, \varepsilon_y$, and γ_{xy}, valid when $\varepsilon_x \ll 1$, $\varepsilon_y \ll 1$, and $\gamma_{xy} \ll 1$.

Fig. P3.45

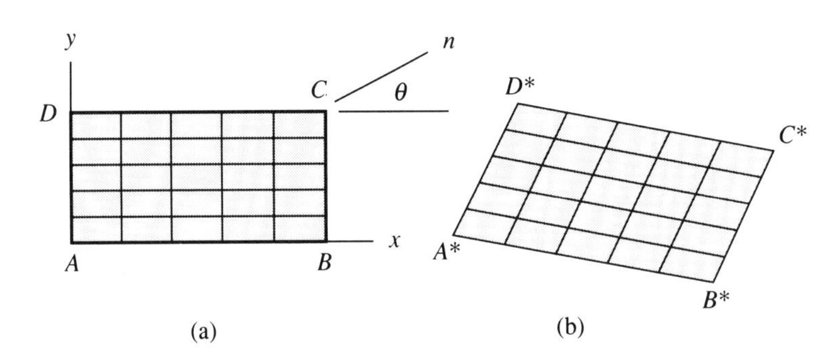

(a) (b)

3.46 A thin membrane is stretched between two concentric hoops, as shown in Fig. P3.46. The inner hoop is held fixed and the outer hoop is rotated through the (not necessarily small) angle ϕ. If the radial lines in the membrane remain straight, compute the shear strain γ_{rt} at the inner hoop. Simplify this result if the angle ϕ is, in fact, small.

Fig. P3.46

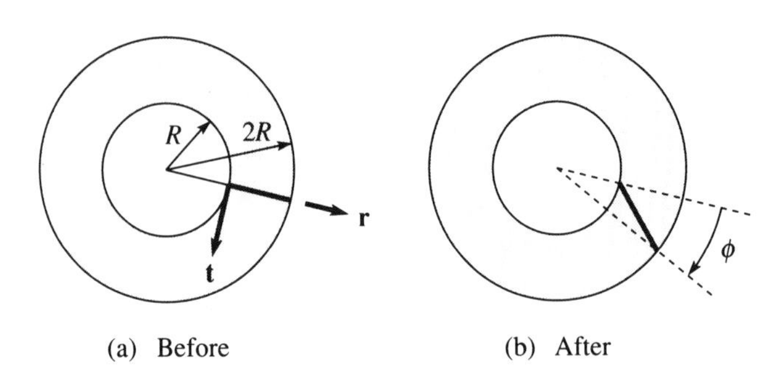

(a) Before (b) After

3.47 Let **s** and **t** define two perpendicular directions, and suppose that $\gamma_{st}(P)$ is the shear strain between these directions at a point P. If $\mathbf{p} = -\mathbf{s}$ and $\mathbf{q} = -\mathbf{t}$, what are the relations between the associated shear strains $\gamma_{pt}(P)$, $\gamma_{sq}(P)$, and $\gamma_{pq}(P)$ and the original $\gamma_{st}(P)$? Does a similar situation occur for the extensional strains $\varepsilon_p(P)$ and $\varepsilon_s(P)$, for example?

***3.48** OP and OQ are two straight lines in a body. In the xyz-rectangular coordinate system, the point O is at the origin, P is at $(\frac{1}{2}, \frac{1}{2}, 1/\sqrt{2})$ and Q is at $(-\frac{1}{2}, -\frac{1}{2}, 1/\sqrt{2})$; consequently, OP and OQ are perpendicular to each other. The point P displaces slightly to P^* and the point Q, to Q^*. The (small) components of these displacements are (u_P, v_P, w_P) and (u_Q, v_Q, w_Q), respectively. If point O remains fixed at the origin and the lines OP and OQ remain straight, show that the shear strain between these lines is approximately

$$\gamma_{PQ} = \frac{1}{2}(u_Q - u_P) + \frac{1}{2}(v_Q - v_P) + \frac{1}{\sqrt{2}}(w_Q + w_P)$$

3.49 The square plate shown in Fig. P3.49 is subjected to the shear strain $\gamma_{xy} = 0.001x + 0.002y$ (x and y in feet). If side AB remains straight and fixed on the x-axis and all lines parallel to the x-axis remain parallel to it, compute the displacement of point D in the x-direction. (*Hint:* Note the magnitude of γ_{xy}).

Fig. P3.49

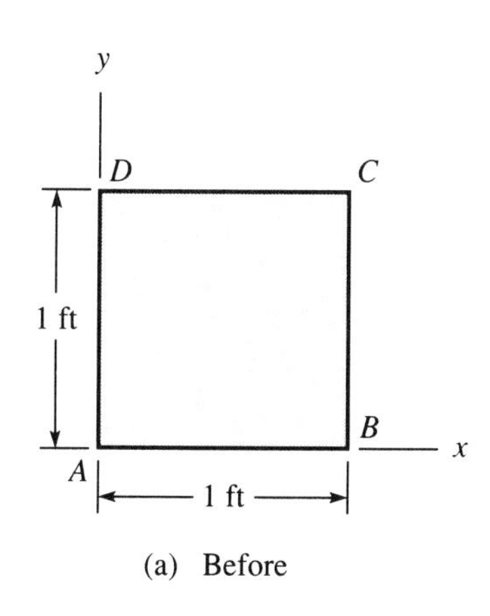

(a) Before

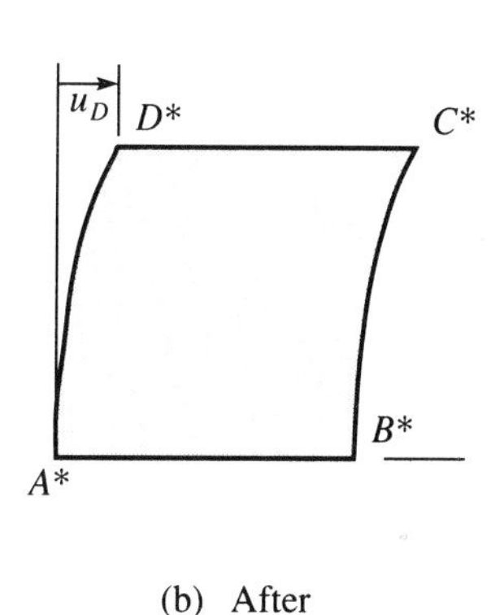

(b) After

3.50 The rectangular plate shown in Fig. P3.50 is subjected to the shear strain

$$\gamma_{xy} = a\frac{x}{L} + b\frac{y}{M} + c\left(\frac{x}{L}\right)\left(\frac{y}{M}\right)$$

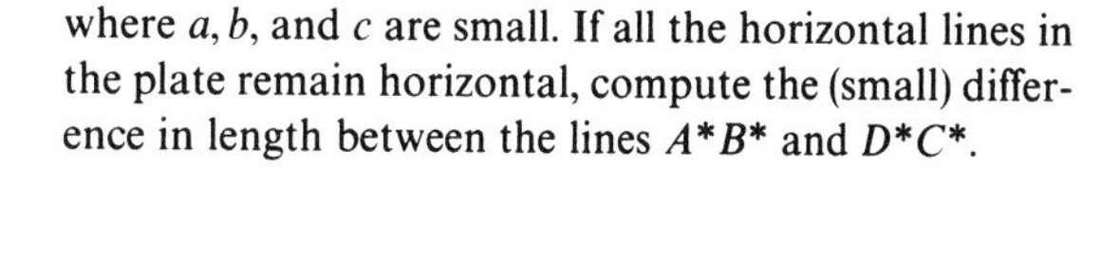

where a, b, and c are small. If all the horizontal lines in the plate remain horizontal, compute the (small) difference in length between the lines A^*B^* and D^*C^*.

Fig. P3.50

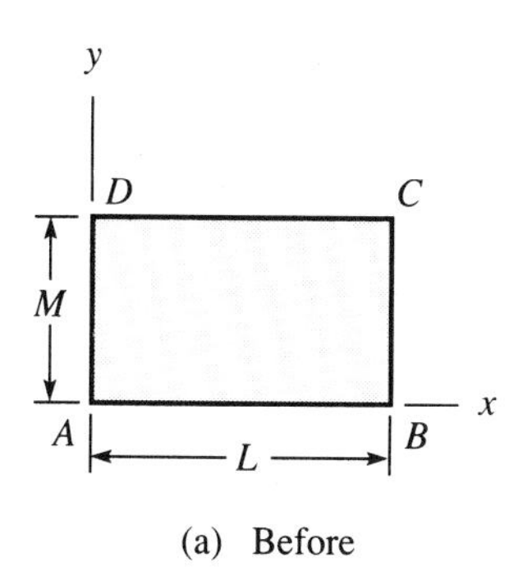

(a) Before

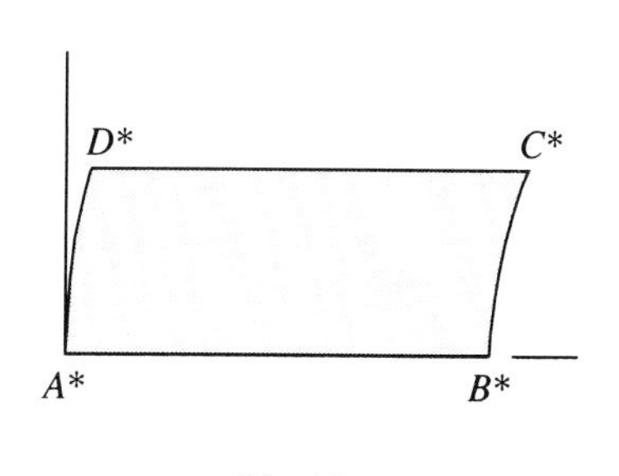

(b) After

***3.51** An xy rectangular coordinate system is defined in the surface of a thin plate, as shown in Fig. P3.51. The plate is deformed in such a way that the circle $x^2 + y^2 = R^2$ becomes the ellipse

$$x^2 + \frac{y^2}{(1+\eta)^2} = R^2$$

with all points of the plate displacing only radially from the origin. Treat η as a small number and compute the shear strain $\gamma_{r\theta}$ (between the radial and tangential directions) at any point on the circle.

Fig. P3.51

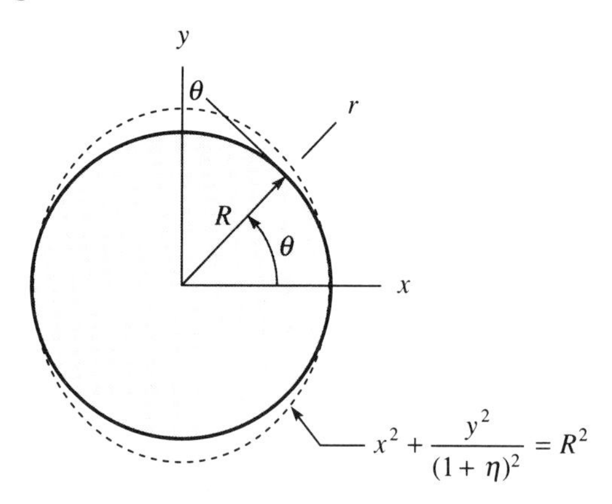

*3.7 Alternative Measures of Strain

Definitions of strain, such as the extensional strain of Eq. (3.1) or the shear strain of Eq. (3.6), are somewhat arbitrary. The practitioner uses the simplest one that meets his or her requirements. Another definition of extensional strain is sometimes more convenient; with the same basic quantities of Fig. 3.4, we can define an alternative extensional strain $\tilde{\varepsilon}_n$:

$$\tilde{\varepsilon}_n(P) = \lim_{\substack{\Delta s \to 0 \\ P_1 \to P}} \frac{1}{2}\left[\left(\frac{\Delta s^*}{\Delta s}\right)^2 - 1\right] \tag{3.8}$$

Instead of the approximation (3.2), we obtain

$$\Delta s^* = \sqrt{1 + 2\tilde{\varepsilon}_n}\,\Delta s \tag{3.9}$$

Also, Eqs. (3.2) and (3.8) give, in the limit at P,

$$\begin{aligned} \tilde{\varepsilon}_n &= \tfrac{1}{2}[(1+\varepsilon_n)^2 - 1] \\ &= \varepsilon_n + \tfrac{1}{2}\varepsilon_n^2 \end{aligned} \tag{3.10}$$

This alternative measure ($\tilde{\varepsilon}_n$) serves as well as the previous measure (ε_n). If one is known, then the other can be computed by means of Eq. (3.10). Both vanish if there is no stretching or contracting—that is, if $\Delta s^* = \Delta s$. Note, too, that extensional strains are usually very small; in a typical member of a machine or structure, $\varepsilon_n = O(10^{-3})$. In such a case, there is no practical difference between the values for ε_n and $\tilde{\varepsilon}_n$; the difference is usually too small to measure.

Likewise, we could adopt an alternative definition of a shear strain. With the angle ϕ_{nt}^* of Fig. 3.8, we can define a shear strain

$$\bar{\gamma}_{nt} = \cos\phi_{nt}^* = \sin\gamma_{nt} \tag{3.11}$$

This, too, serves as well as the previous measure. Both vanish if the angle is unchanged—that is, if $\phi_{nt}^* = \pi/2$. Again, the strain is often small $[O(10^{-3})]$; then the difference has no practical consequences.

The choice of definitions is a matter of practical and mathematical convenience. It is important, especially if the strains are large, to be consistent; any comparisons between theories and experiments must compare like quantities. Here, we introduce and use the earlier definitions, ε_n and γ_{nt} of Eqs. (3.1) and (3.6), because they are conceptually simple and yet are adequate for our needs.

PROBLEMS

3.52 If the extensional strain ε_n, according to the definition Eq. (3.1), has the following values, compute the extensional strain $\tilde{\varepsilon}_n$ according to the definition Eq. (3.8).

$\varepsilon_n = 10 \times 10^{-4}$ (a typical machine part)

$\varepsilon_n = 3 \times 10^{-4}$ (a structural component)

$\varepsilon_n = 10 \times 10^{-2}$ (a "soft" plastic member)

$\varepsilon_n = 10 \times 10^{-1}$ (a rubberlike part)

3.53 If the shear strain γ_{nt}, according to the definition Eq. (3.6), has the following values, compute the shear strain $\bar{\gamma}_{nt}$ according to the definition Eq. (3.11).

$\gamma_{nt} = 10 \times 10^{-4}$ (a typical steel or aluminum part)

$\gamma_{nt} = 10 \times 10^{-3}$ (certain composite members)

$\gamma_{nt} = 10 \times 10^{-1}$ (a rubber component)

3.54 A rectangular block of rubber is homogeneously deformed (all elements experience the same deformation) to another rectangle. In the directions of the edges, the extensional strains are

$$\varepsilon_x = 10 \times 10^{-2}, \qquad \varepsilon_y = 8 \times 10^{-2},$$
$$\varepsilon_z = -3 \times 10^{-2}$$

Compute the change (%) in volume if these are interpreted (a) according to the definition Eq. (3.1) and (b) according to the definition Eq. (3.8).

3.55 Repeat Prob. 3.54 if the values are

$$\varepsilon_x = 10 \times 10^{-4}, \qquad \varepsilon_y = 8 \times 10^{-4},$$
$$\varepsilon_z = -3 \times 10^{-4}$$

3.56 A bungee cable is 100 cm long. One end is attached at the origin (0, 0), and the other is initially at (80 cm, 60 cm). The latter is then displaced to the point (100 cm, 80 cm). Calculate the extensional strain ε_n and $\tilde{\varepsilon}_n$ according to the definitions Eq. (3.1) and (3.8), respectively. Which calculation requires less arithmetic?

3.57 Rework Prob. 3.17 using the definition Eq. (3.8) instead of Eq. (3.1) for the extensional strain. Compare the differences. Which computation requires more arithmetic?

3.58 Do Prob. 3.19 using the definition Eq. (3.8) instead of Eq. (3.1).

3.59 Do Prob. 3.22 with the extensional strain defined by Eq. (3.8). Do you think that the design of a steel truss is affected by the choice of definitions, Eq. (3.1) versus Eq. (3.8)?

3.60 Do Prob. 3.39 using the definition Eqs. (3.8) and (3.11) for the extensional strain and shear strain, respectively.

3.61 Do Prob. 3.40 if the extensional strains $(\varepsilon_x, \varepsilon_y, \varepsilon_r)$ and the shear strain (γ_{xy}) are interpreted in accordance with Eqs. (3.8) and (3.11), respectively.

3.62 Use the data of Prob. 3.41 to determine (exactly) the shear strain $\bar{\gamma}_{st}$ in accordance with Eq. (3.11). Obtain the first-order approximation.

3.8 State of Strain

The deformation of a body can be completely described in terms of extensional and shear strain. To understand how this can be accomplished, imagine cutting the original body into a large number of very small blocks. Any body can be subdivided into small blocks in numerous ways. The manner in which the body is subdivided determines the size and orientation of each of the small blocks. For example, the body could be subdivided into blocks whose edges are parallel to a set of rectangular coordinate axes, x, y, z. A few such blocks are shown for the body in Fig. 3.10(a), and a typical block is illustrated in Fig. 3.10(b). If these blocks are very small, their deformed counterparts are approximately parallelepipeds. This result follows from the observation that very short line segments remain approximately straight and very small plane surfaces remain approximately plane after deformation.

Fig. 3.10

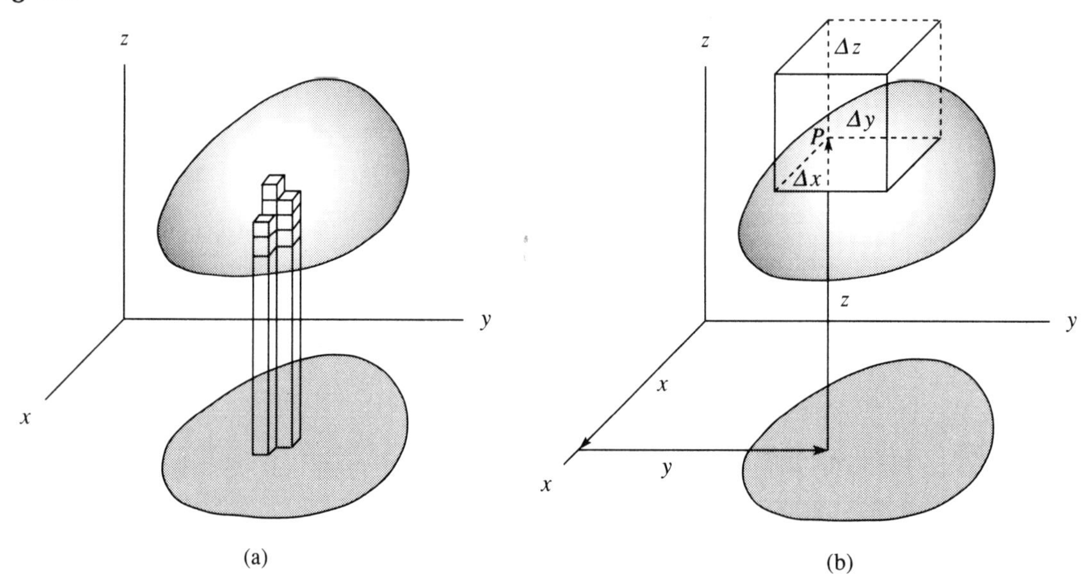

Thus, a small rectangular block becomes a parallelepiped, as illustrated by Fig. 3.11. Consider the process by which the rectangular block of Fig. 3.11(a) can become the parallelepiped shown in Fig. 3.11(d).

1. The sides PB, PD, and PH of the block in Fig. 3.11(a) can be stretched until their lengths are equal to those of P^*B^*, P^*D^*, and P^*H^*, respectively, as in Fig. 3.11(b).

2. The sides of the new block in Fig. 3.11(b) can be skewed until the angles between the edges agree with those of the final state, as shown by Fig. 3.11(c).

Fig. 3.11

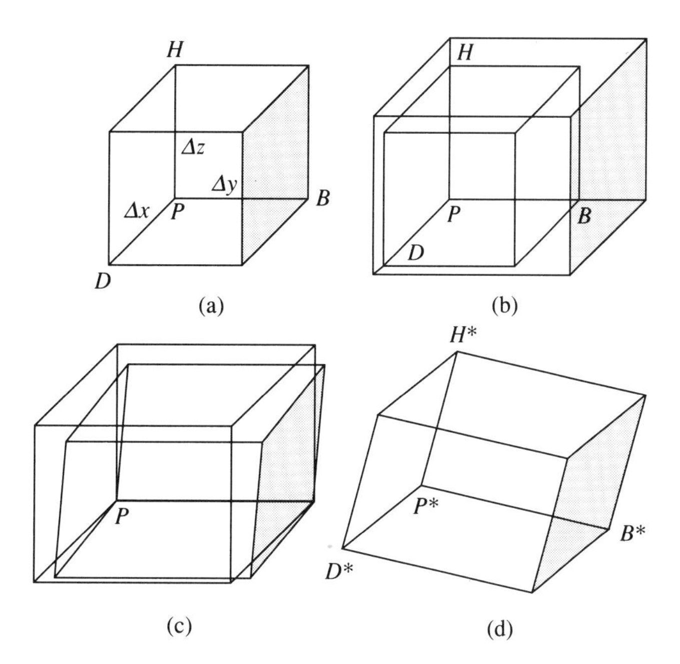

3. The parallelepiped can be displaced as a rigid body until it is coincident with the final position.

This sequence of events can be classified as follows:

1. A change in size (extension or contraction) } Deformation
2. A change in shape (shear or distortion) }

3. A rigid-body translation } Rigid Body Motion
4. A rigid-body rotation }

In any actual deformation, all these events would occur simultaneously. This breakdown is given only as an aid in understanding the process.

The deformation of the rectangular block is characterized by the changes in length of its edges and the changes in angle between the edges. Thus the method we have used to describe the change in length of a line segment can be applied in turn to the three concurrent edges of the block to determine their final lengths. If at P we know the extensional strains for the x-, y-, and z-directions, then the change in size of the block is determined by these extensional strains. According to the notation introduced in Sec. 3.4, these normal strains are written as $\varepsilon_x(P)$, $\varepsilon_y(P)$, and $\varepsilon_z(P)$. The original block has dimensions $\Delta x, \Delta y$, and Δz. Hence, the new lengths of the sides in the deformed state are approximately

$$(1 + \varepsilon_x)\Delta x, \qquad (1 + \varepsilon_y)\Delta y, \qquad (1 + \varepsilon_z)\Delta z$$

To describe the distortion or change in shape, we can introduce the shear strains between the pairs of directions x and y, y and z, and z and x. The new angles between these edges are

$$\frac{\pi}{2} - \gamma_{xy}(P), \qquad \frac{\pi}{2} - \gamma_{yz}(P), \qquad \frac{\pi}{2} - \gamma_{zx}(P)$$

The deformation of the block is now completely determined, because we have the lengths of three concurrent edges of the parallelepiped and the angles between the edges. If the six strains

$$\varepsilon_x, \varepsilon_y, \varepsilon_z, \gamma_{xy}, \gamma_{yz}, \gamma_{zx} \tag{3.12}$$

are known at every point throughout the body, then, by imagining the body to be composed of very small rectangular blocks with edges parallel to the x-, y-, and z-coordinate axes, we can determine the shape and size of the deformed blocks. Because the change in size and shape of a body is the net result of all changes that occur in its elementary portions, the deformed state of the entire body is defined.[3]

There are other ways of subdividing a body into small blocks. The manner of subdivision is dictated by the coordinate system used. One additional example will serve to illustrate this. Suppose that cylindrical coordinates are used instead of Cartesian coordinates. If the body is subdivided into small blocks whose edges are along the coordinate directions, then, as we move throughout the body, these small blocks change their orientation. Figure 3.12 shows a few typical blocks and how the orienta-

Fig. 3.12

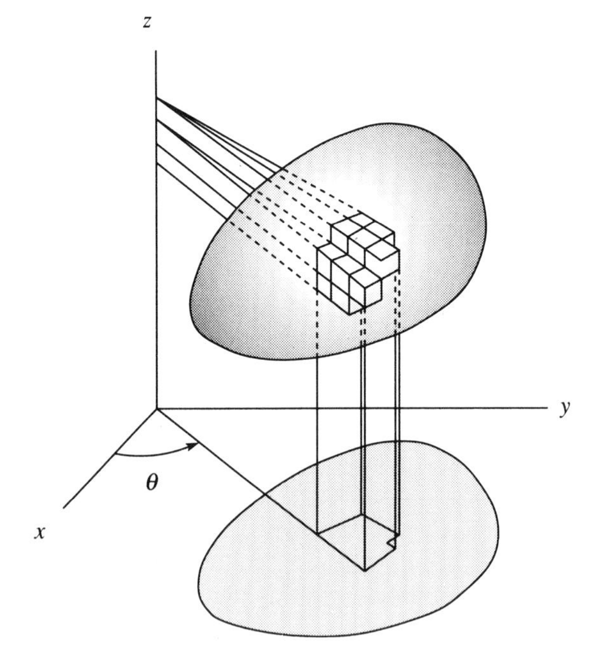

3 Upon further study (see Sec. 3.22) we find that these strains must satisfy certain differential equations to ensure the continuity of the body.

tion of these blocks changes as we change position within the body. A typical block, as in Fig. 3.13, is located by the point P, and its orientation is determined by the three edges that intersect at this point. In this case one of the edges is in the direction of the radius, which is denoted by r. A second edge is in the direction of the z-axis, and the third edge, perpendicular to the other two, is in a direction denoted by θ; it is the direction of increasing θ. The changes in lengths of these edges can be characterized by the extensional strains associated with these directions. Thus, the extensional strains are $\varepsilon_r(P)$, $\varepsilon_\theta(P)$, and $\varepsilon_z(P)$. Although the small blocks obtained in this manner are not exactly rectangular, they are approximately so, and, as the subdivision becomes finer, the approximation gets better. We may take the lengths of the edges of the original block as $\Delta r, r\,\Delta\theta$, and Δz. The lengths of these edges in the deformed block are

$$(1 + \varepsilon_r)\,\Delta r, \qquad (1 + \varepsilon_\theta) r\,\Delta\theta, \qquad (1 + \varepsilon_z)\,\Delta z$$

Fig. 3.13

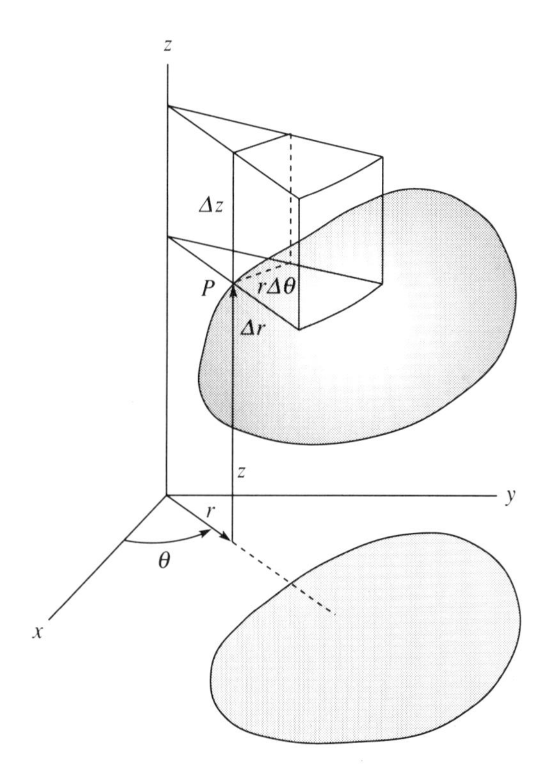

The new angles between these edges are

$$\frac{\pi}{2} - \gamma_{r\theta}(P), \qquad \frac{\pi}{2} - \gamma_{\theta z}(P), \qquad \frac{\pi}{2} - \gamma_{rz}(P)$$

In this case the six strains

$$\varepsilon_r, \varepsilon_\theta, \varepsilon_z, \gamma_{r\theta}, \gamma_{\theta z}, \gamma_{rz} \tag{3.13}$$

given throughout the body, will determine the change in size and shape of every small block and, therefore, of the entire body.

Clearly, either set of strains, Eqs. (3.12) or (3.13), can be utilized to describe the deformation of a given body. In fact, if three extensional strains and three shear strains associated with three perpendicular directions are known throughout the body, the deformation of the body is completely described. The set of directions used to specify the strain is a matter of convenience and will vary from problem to problem.

Any set of six strains that describe the deformation in the neighborhood of a particle is said to determine the *state of strain* at the point which the particle occupies.

3.9 Severity of Deformation

Bodies subjected to forces or changes in temperature exhibit deformations of varying degree. A rubber band can be stretched to several times its original length. Thin flat plates can be rolled into pipe. Boilers under internal pressure at elevated temperatures experience small changes in diameter and length. A rafter in a building bends slightly while supporting the roof above it. Circular shafts twist, but only slightly, when used to transmit torque.

In a great variety of situations, the deformation of a body is not too severe. Such small deformations are characterized by placing certain limits on the order of magnitude of strains. A strain, either extensional or shear, is considered small if it is small *compared to unity*—that is, $\varepsilon_n \ll 1$, $\gamma_{nt} \ll 1$. Most theories of deformable bodies, structures, and machines are founded on the assumption of small strains. This does not necessarily preclude large rotations and displacements.

Long, thin members, such as rods and thin sheets, can be rolled into a cylindrical form with very little stretching. A long, thin shaft can experience a large relative rotation of its opposite ends, even though the strains remain small. In general, *thin* bodies can experience large rotations and large deflections, though the strains are small throughout. On the other hand, large strains are usually accompanied by large rotations.

To illustrate a circumstance of small strain accompanied by a large rotation, consider a thin straight strip of steel, such as a conventional hacksaw blade, as shown in Fig. 3.14. The strip is clamped at one end; a transverse load is applied to the other. The load can cause a very large displacement of the loaded end and the rotation might easily exceed 45°, yet the *relative* rotation of any two neighboring lines remains small. Hence, the shear strains and also the extensional strains are small throughout (so small that no permanent deformations result).

To appreciate the occurrence of large rotations, which usually accompany large strains, we need only consider the definition of a shear strain, which can be large only if a large rotation occurs (see Fig. 3.8).

Most problems in this book involve only small strains. A few applications admit large rotations and, indeed, a later study of thin beams (Chapter 7) requires that we specifically include the relative rotations in our analysis.

Fig. 3.14

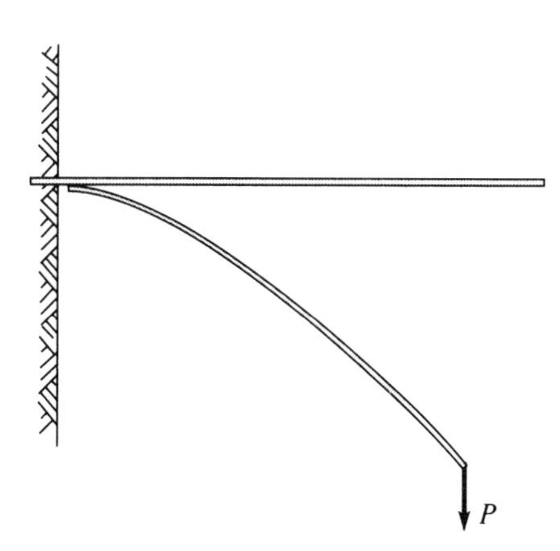

3.10 Superposition of Small Strains

An important consequence of the assumption of small strain is that successive strains are additive. For example, suppose a short line segment Δx is subjected to an extensional strain ε_x' so that its length becomes $\Delta x^* = (1 + \varepsilon_x')\,\Delta x$. If the element is then extended again, so that the additional strain is ε_x'', its new length is

$$\Delta x^{**} = (1 + \varepsilon_x'')\,\Delta x^* = (1 + \varepsilon_x'')(1 + \varepsilon_x')\,\Delta x$$

If the strains are small compared to unity, only first-order terms are significant, so

$$\Delta x^{**} \doteq (1 + \varepsilon_x' + \varepsilon_x'')\,\Delta x$$

Then the result of these successive strains is the strain

$$\varepsilon_x = \lim_{\Delta x \to 0} \frac{\Delta x^{**} - \Delta x}{\Delta x} \doteq \varepsilon_x' + \varepsilon_x''$$

We therefore say that the successive strains may be superimposed. Similarly, the result of two successive shear strains γ_{xy}' and γ_{xy}'', which are small, is the net shear strain

$$\gamma_{xy} \doteq \gamma_{xy}' + \gamma_{xy}''$$

In general, suppose that

$$\varepsilon_x', \varepsilon_y', \varepsilon_z', \gamma_{xy}', \gamma_{yz}', \gamma_{zx}'$$

describe a state of small strain throughout a body. If, in addition to this state, we impose a second state of small strain,

$$\varepsilon_x'', \varepsilon_y'', \varepsilon_z'', \gamma_{xy}'', \gamma_{yz}'', \gamma_{zx}''$$

then, to first-order terms, the net effect is a state of strain given by

$$\begin{aligned} \varepsilon_x &= \varepsilon_x' + \varepsilon_x'', & \varepsilon_y &= \varepsilon_y' + \varepsilon_y'', & \varepsilon_z &= \varepsilon_z' + \varepsilon_z'' \\ \gamma_{xy} &= \gamma_{xy}' + \gamma_{xy}'', & \gamma_{yz} &= \gamma_{yz}' + \gamma_{yz}'', & \gamma_{zx} &= \gamma_{zx}' + \gamma_{zx}'' \end{aligned}$$

This result will be used later to determine the response of a structure subjected to simultaneous stretching, twisting, and bending.

EXAMPLE 5

To illustrate the superposition of strain, let us reconsider Example 3 of Sec. 3.5, wherein the displacement of point P has components u and v. We have seen that the linear approximation introduces a small error when $u = v = 0.01L$. Suppose that the components u and then v were imposed in order. In the first displacement (u), the strains are

$$\varepsilon'_B = \left[\left(\frac{4}{5} + \frac{u}{L}\right)^2 + \left(\frac{3}{5}\right)^2\right]^{1/2} - 1 = 0.008017857$$

$$\varepsilon'_C = \frac{5}{3}\left[\left(\frac{u}{L}\right)^2 + \left(\frac{3}{5}\right)^2\right]^{1/2} - 1 = 0.000138879$$

In the second displacement (v), the strains are

$$\varepsilon''_B = \frac{\left[\left(\frac{4}{5} + \frac{u}{L}\right)^2 + \left(\frac{3}{5} + \frac{v}{L}\right)^2\right]^{1/2}}{\left[\left(\frac{4}{5} + \frac{u}{L}\right)^2 + \left(\frac{3}{5}\right)^2\right]^{1/2}} - 1 = 0.005936517$$

$$\varepsilon''_C = \frac{\left[\left(\frac{u}{L}\right)^2 + \left(\frac{3}{5} + \frac{v}{L}\right)^2\right]^{1/2}}{\left[\left(\frac{u}{L}\right)^2 + \left(\frac{3}{5}\right)^2\right]^{1/2}} - 1 = 0.016662075$$

If we add these successive strains, we obtain

$$\varepsilon'_B + \varepsilon''_B = 0.01395437, \qquad \varepsilon'_C + \varepsilon''_C = 0.01680095$$

Of course the latter is erroneous, but the errors are exceedingly small: -0.3% and -0.01%.

The linear approximations of the successive strains are

$$\varepsilon'_B = \frac{4}{5}\frac{u}{L}, \qquad \varepsilon'_C = 0$$

and

$$\varepsilon''_B = \frac{3}{5}\frac{v}{L}, \qquad \varepsilon''_C = \frac{5}{3}\frac{v}{L}$$

The sums are the same as the previous result. In the context of first-order (linear) approximations, the successive strains are additive. ◆

PROBLEMS

3.63 A rubber band is initially 20.0 cm long and is stretched uniformly to a length of 24.0 cm and then to a final length of 30.0 cm. Compute the average extensional strain incurred during deformation from the initial to final lengths and compare this to the sum of the successive strains. Use the definition of Eq. (3.1).

3.64 In an initial state (a) a wire is anchored at one end $A(x=0, y=0, z=0)$ and held at the other $B(400\,\text{cm}, 300\,\text{cm}, 0)$. In a second state (b), the end B has been moved to $(401\,\text{cm}, 301\,\text{cm}, 0)$. Subsequently, (c) the end B is moved to $(402\,\text{cm}, 300\,\text{cm}, 4\,\text{cm})$.

Jack observed only the states (a) and (b) and computed the average strain ε_b in state (b). Jennifer observed only state (b) and subsequent state (c) and, therefore, computed the average strain ε_{cb} associated with that transition.

Compute the total (average) strain ε_{ca} associated with the change from initial state (a) to final state (c) and compare this with the sum $\varepsilon_b + \varepsilon_{cb}$. Use the definition of Eq. (3.1).

3.65 A spherical balloon is initially inflated with a diameter 24 cm. It is subsequently inflated to a final diameter of 30 cm. Assuming that the deformation is homogeneous (every element experiences the same strain), compute the extensional strain ε_θ (compare lengths of any great circle) according to the definition of Eq. (3.1). Does superposition apply; that is, can we add subsequent strains of similar magnitudes to obtain the final strain?

3.65 If you were treating a problem in cylindrical coordinates (r, θ, z), for example, the shaft in a rotary engine, how would you label the six components of strain and how would you interpret them geometrically?

3.67 A rectangular block of rubber is deformed homogeneously (strains are constant throughout) to another block in two stages. The first deformation causes strains in the direction of the edges (x, y, z), as follows:

$$\varepsilon'_x = 0.100, \qquad \varepsilon'_y = 0.200, \qquad \varepsilon'_z = -0.100$$
$$\gamma'_{xy} = \gamma'_{xz} = \gamma'_{yz} = 0$$

A subsequent deformation causes the "additional" strains; that is, the previously deformed state is the reference state:

$$\varepsilon''_x = 0.200, \qquad \varepsilon''_y = 0.100, \qquad \varepsilon''_z = 0.100$$
$$\gamma''_{xy} = \gamma''_{xz} = \gamma''_{yz} = 0$$

Compute the actual strains incurred during the deformation from the initial to final states, and determine the errors if the successive strains were merely superposed. Use the definition of Eq. (3.1).

***3.68** Use the data of Prob. 3.67 to compute the extensional strain along the diagonal of a block that has edges of length $X = 80$ cm, $Y = 36$ cm, and $Z = 48$ cm. Compare the final extensional strain of that line with the approximation obtained by superposing (adding) the successive strains incurred during the subsequent deformations. Use the definition of Eq. (3.1).

***3.69** Three states of strain are given for the directions shown in Fig. P3.69. In each case $\varepsilon_z = \gamma_{xz} = \gamma_{yz} = 0$ and

1. $\varepsilon'_x = 0$, $\varepsilon'_y = 0.1000$, $\gamma'_{xy} = 0$
2. $\varepsilon''_x = 0.0500$, $\varepsilon''_y = 0.1000$, $\gamma''_{xy} = 0$
3. $\varepsilon_x = 0.0500$, $\varepsilon_y = 0.2100$, $\gamma_{xy} = 0$

Note that state (3) results from the successive strains (1) and (2); for example, $\varepsilon_y = (1+\varepsilon'_y)(1+\varepsilon''_y) - 1$, in accordance with Eq. (3.1).

Obtain the shear strains γ'_{nt}, γ''_{nt}, and γ_{nt}. (*Hint:* Consider the deformation that carries the square $(a \times a)$ into a rectangle and determine $\gamma_{nt} = \pi/2 - \phi_{nt}$).

Fig. P3.69

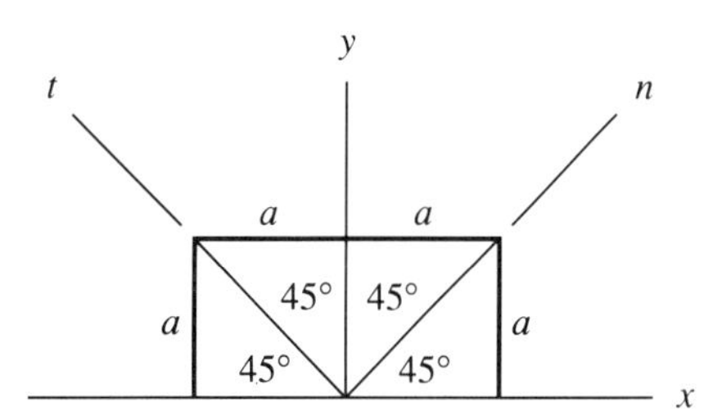

3.11 Some Elementary Strain-Displacement Analyses

Most fabricated structures consist of a number of interconnected members. The simplest types of members are all prismatic or cylindrical in shape. The distinguishing feature or characteristic of these simple members is the nature of the loads that they transmit or resist. Members of a pin-connected truss are subjected to loads that tend to stretch or compress them. A torque transmitted by means of a shaft tends to twist the shaft. The rafters of a building must support the load of the roof and thus tend to bend.

Individually, each of these members is a body. The complete deformational behavior of any of these bodies requires a knowledge of the displacements of all particles; however, some of this information is incidental. In a member of a pin-connected truss, for example, our main concern is the amount of its extension or contraction. The other deformational changes of the member are of secondary importance.

Analyses of elementary members generally begin with some assumptions or observations with regard to the general nature of the deformation of the member. These assumptions characterize the significant changes in the geometry of the member and are usually based upon the experience and observations of the engineer. The assumptions allow the investigator to relate certain strains to the significant displacements. To illustrate this let us consider a number of examples in detail.

EXAMPLE 6

A variety of structural elements are cylindrical or prismatic in shape and are subjected to loads that tend to stretch or compress the member. For example, the pin-connected truss in Fig. 3.15(a) consists of two members, AB and BC. When the load P is applied,

Fig. 3.15

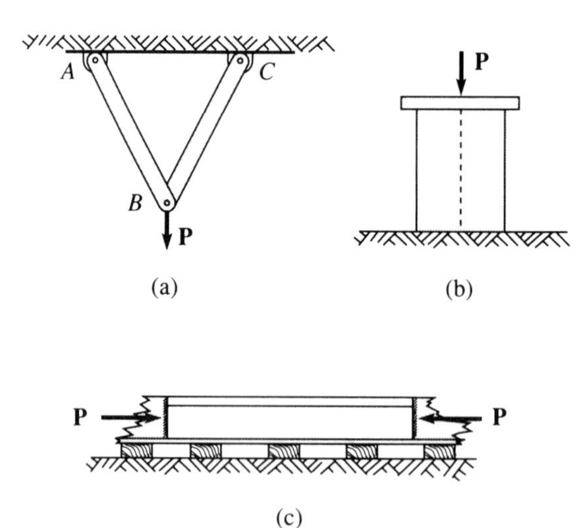

each of these members is stretched. Figure 3.15(b) shows a concrete cylinder supporting a bearing plate that carries a load P along the axis of the cylinder. Fig. 3.15(c) shows a section of rail subjected to a temperature change. As the rail tends to elongate, the adjacent portions restrain this expansion. Each of these members is a cylindrical body whose geometric behavior can be adequately described by the statements that (1) the axis of the cylinder remains straight after deformation and (2) any plane cross section of particles remains plane and perpendicular to the axis after deformation.

Suppose that the cylinder of Fig. 3.16(a) deforms according to these assumptions. If S represents a plane cross section of particles at a distance x from the left end of the cylinder, then S deforms into the plane S^*, as indicated in Fig. 3.16(b). We denote its displacement by $u(x)$. For convenience we take the left end of the cylinder to be fixed; that is, $u(0) = 0$. The state of strain throughout the cylinder describes the nature of

Fig. 3.16

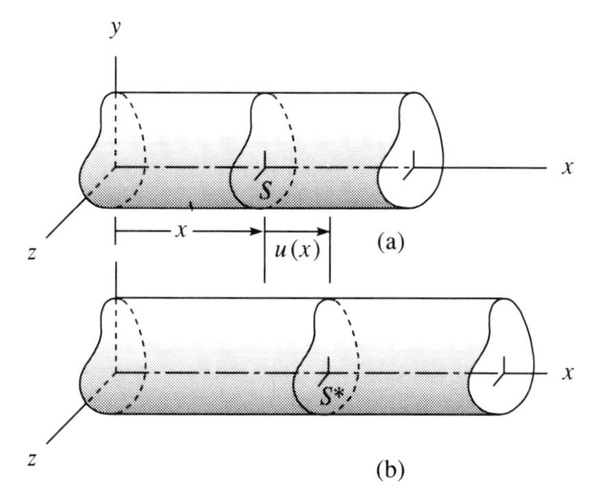

deformation of every small rectangular block in the interior of the cylinder. Figure 3.17 shows one typical block in position in the bar and also in a separate view. This block is contained between two plane cross sections S and S_1 that are a distance Δx

Fig. 3.17

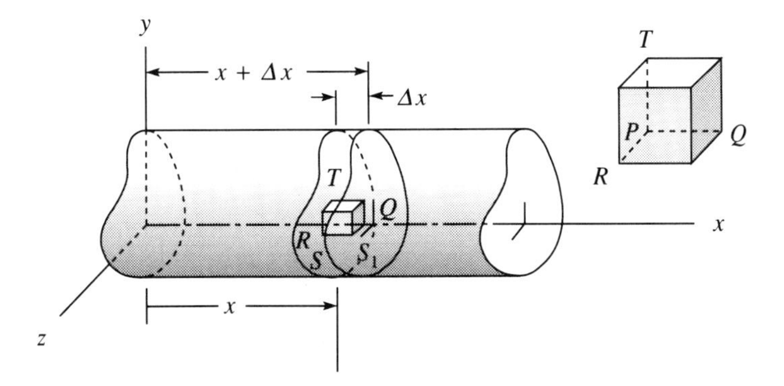

apart. When the cylinder is stretched, this block changes size. A description of the state of strain at P requires a complete knowledge of the displacements of all particles in the neighborhood of P. Our assumptions about the way in which the cylinder deforms do not provide this information; only the displacement of particles in the x-direction. Consequently, all that we can do is describe the stretching of the edge PQ of the rectangular block in Fig. 3.17, and hence relate ε_x to the axial displacement of the cross sections.

Figure 3.18 illustrates the two sections S and S_1 and the segment PQ between them. S is at a distance x and S_1 is at a distance $x + \Delta x$ from the left end of the cylinder. After deformation, S^* is at a distance $x + u(x)$ and S_1^* is at a distance $x + \Delta x + u(x + \Delta x)$ from the left end of the cylinder. The extensional strain in the x-direction at P is

$$\varepsilon_x(P) = \lim_{\substack{Q \to P \\ \text{along } x}} \frac{P^*Q^* - PQ}{PQ}$$

Fig. 3.18

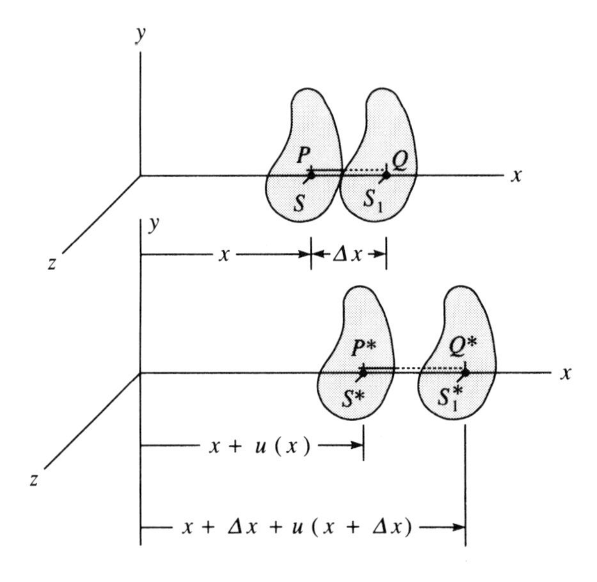

Since the strains are small,

$$PQ = \Delta x, \qquad P^*Q^* \doteq \Delta x + u(x + \Delta x) - u(x)$$

and we have

$$\varepsilon_x(P) = \lim_{\Delta x \to 0} \frac{u(x + \Delta x) - u(x)}{\Delta x}$$

Recall from calculus that this limit is the derivative of $u(x)$ with respect to x. Hence,

$$\varepsilon_x(x) = \frac{du}{dx} \tag{3.14}$$

(We replace P by x because every particle in the cross section at x experiences the same extensional strain in the x-direction in this particular problem.) The strains $\varepsilon_y, \varepsilon_z, \gamma_{xy}, \gamma_{yz}, \gamma_{zx}$ remain unknown. This should have been anticipated because our displacement assumptions describe only the axial component of displacement of particles. However, the significant deformation of the cylinder (axial extension or contraction) is described by ε_x. ◆

EXAMPLE 7

Let us consider a circular shaft that is twisted; that is, one end of the shaft is rotated about the axis relative to the other end. Such deformations result when a shaft is used to transmit torque. The drive shaft of an automobile, for example, behaves in this manner as it transmits power to the rear wheels.

When a circular shaft is twisted slightly, the deformation can be described by the statements that (1) the axis of the cylinder remains straight after deformation and (2) all radii in a cross section remain straight and rotate through the same angle about the axis of the cylinder.

Figure 3.19 illustrates a shaft that has been twisted according to these assumptions. A typical radius, OA, is shown in Fig. 3.19(a) at a distance z from the rear end of the cylinder. For convenience we shall assume that the back end of the cylinder is fixed. After deformation, the radius OA has rotated to OA^*, as shown in Fig. 3.19(b).

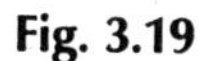
Fig. 3.19

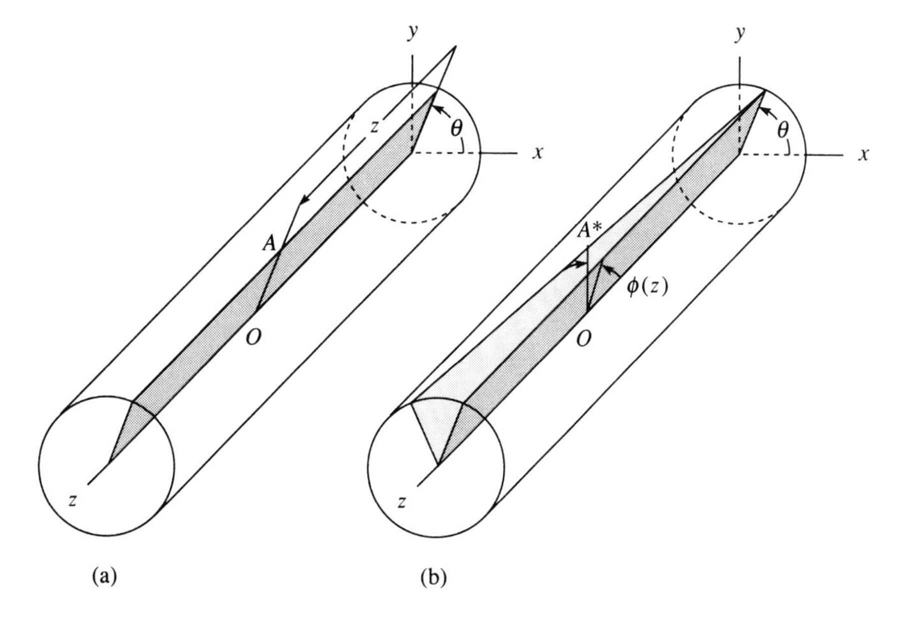

Generally, the amount of rotation will depend on the location of the cross section. We therefore denote it by $\phi(z)$, where z is the distance of the cross section from the rear end of the cylinder. Because we have assumed that the rear end of the cylinder is fixed, $\phi(0) = 0$.

Figure 3.20 depicts a typical small block in the cylinder before deformation. The block is contained between the cross sections S and S_1, which are a distance Δz apart.

Fig. 3.20

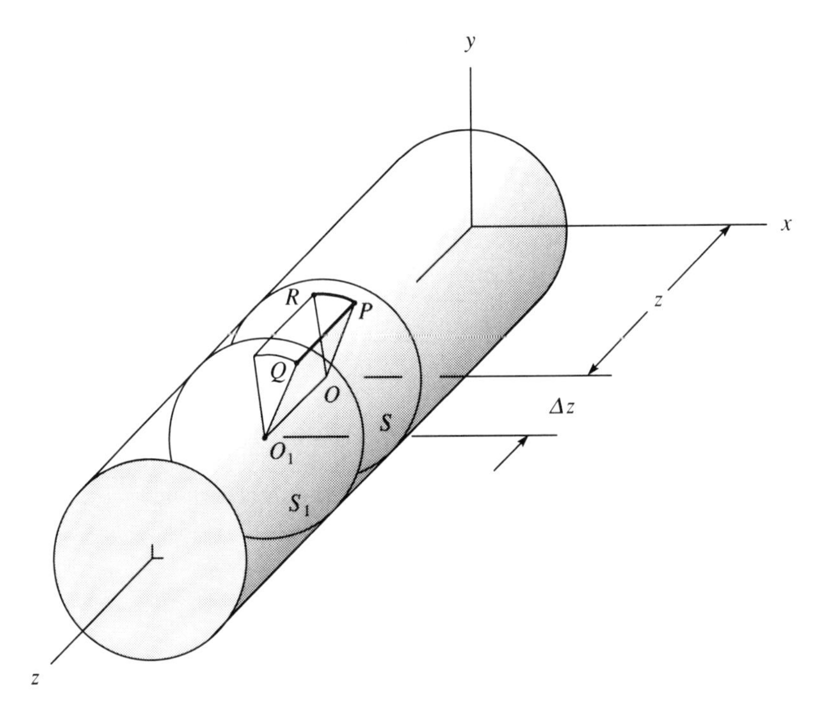

When the cylinder twists, S_1 rotates relative to S, which causes a change in the angle QPR. It is possible to relate the shear strain $\gamma_{z\theta}(P)$ to the rotation $\phi(z)$. To compute $\gamma_{z\theta}$, consider the edges PQ (in the axial direction) and PR (in the tangential direction) of the block of Fig. 3.20. These edges are shown in Fig. 3.21 before and after deformation.

When the cylinder twists, the radius OP, which initially makes an angle θ with the x-axis, rotates through an angle $\phi(z)$. Consequently, OP^* makes an angle $\theta + \phi(z)$ with the x-axis. On the other hand, the radius O_1Q, which initially makes an angle θ with the x-axis, rotates into a position O_1Q^*, which in turn makes an angle $\theta + \phi(z + \Delta z)$ with the x-axis. The angle is different because O_1Q is in a plane cross section at a distance $z + \Delta z$ from the rear end of the cylinder. After deformation, the segment P^*R^* remains in the cross section S^*. If, in Fig. 3.21(b), the line P^*M is drawn parallel

Fig. 3.21

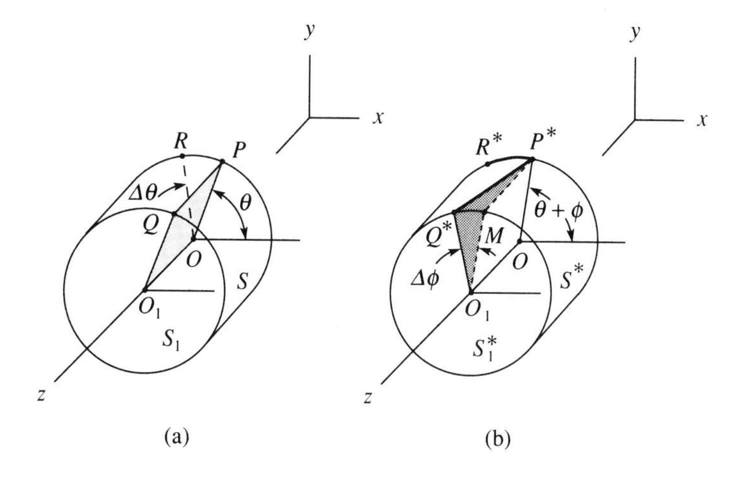

to the axis of the cylinder, the angle MP^*Q^* represents the change in the right angle QPR. Consequently, from the definition of shear strain, we have

$$\gamma_{z\theta}(P) = \lim_{\substack{Q \to P \\ \text{along } z}} \angle MP^*Q^*$$

Because all strains and relative rotations are assumed to be small,

$$MQ^* = r[\phi(z + \Delta z) - \phi(z)]$$

$$P^*M = \Delta z$$

and it follows that

$$\angle MP^*Q^* \doteq \frac{r[\phi(z + \Delta z) - \phi(z)]}{\Delta z}$$

Consequently

$$\gamma_{z\theta}(P) = \lim_{\substack{Q \to P \\ \text{along } z}} \angle MP^*Q^* = \lim_{\Delta z \to 0} \frac{r[\phi(z + \Delta z) - \phi(z)]}{\Delta z}$$

and

$$\gamma_{z\theta}(P) = r\frac{d\phi}{dz} \tag{3.15}$$

In summary, we have related the shear strain associated with the axial and tangential directions at P to the rate of rotation of cross sections along the axis of the cylinder. To obtain the remaining components of the state of strain at P, we need some information about how particles in the neighborhood of P displace in the axial and radial directions. Since nothing has been said about these components of displacement, no other strain components can be computed. ◆

EXAMPLE 8

As a final example, let us consider long, straight members of a structure or machine that carry loads transverse to their length. Such members are called *beams*. A floor joist in a building is a typical beam. The geometric behavior of a beam can be very complex, depending on the nature of the cross section of the member and the manner of application of the loads that cause deformation. In this example, we consider beams that deform according to the statements that (1) there is one line of particles parallel to the axis of the beam which remains in a plane after deformation and does not stretch or contract; this line is called the *neutral axis* of the beam, and (2) any cross section of particles perpendicular to the neutral axis remains perpendicular to it after deformation occurs. This type of behavior is called *bending*.

Suppose that the beam in Fig. 3.22(a) bends. For definiteness, we take the x-axis coincident with the initial position of the neutral axis and place the y-axis so that the xy-plane contains the neutral axis after deformation. Let N be the piercing point of the neutral axis in the cross section S before bending occurs. After deformation, the particle initially at N has displaced to N^*.

Fig. 3.22

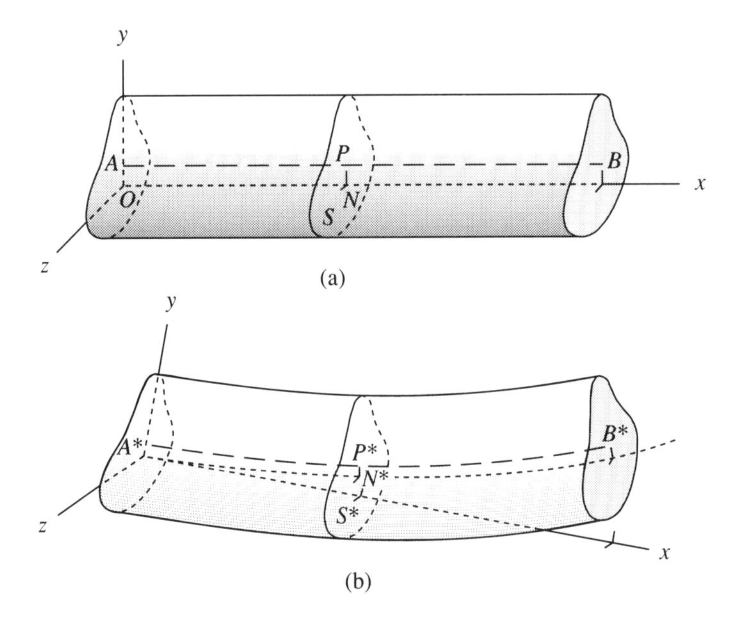

Figure 3.23 shows a typical small rectangular block contained between the cross sections S and S_1 a distance Δx apart. As the beam bends, the cross section S_1 rotates relative to S about the z-axis. When this occurs, the edges of the block parallel to the x-direction change length. Consider the profile view in Fig. 3.24 of the cross sections S and S_1, the neutral axis, and PQ before and after deformation. If P and Q are a distance above the neutral axis before deformation, then P^* and Q^* remain approximately the same distance from the neutral axis after deformation. Furthermore, since

Fig. 3.23

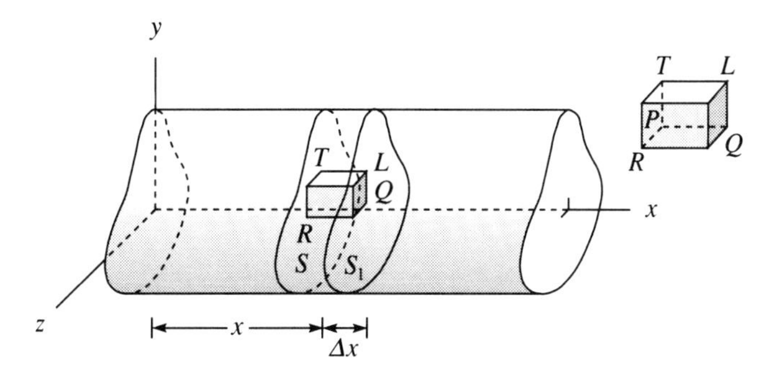

Fig. 3.24

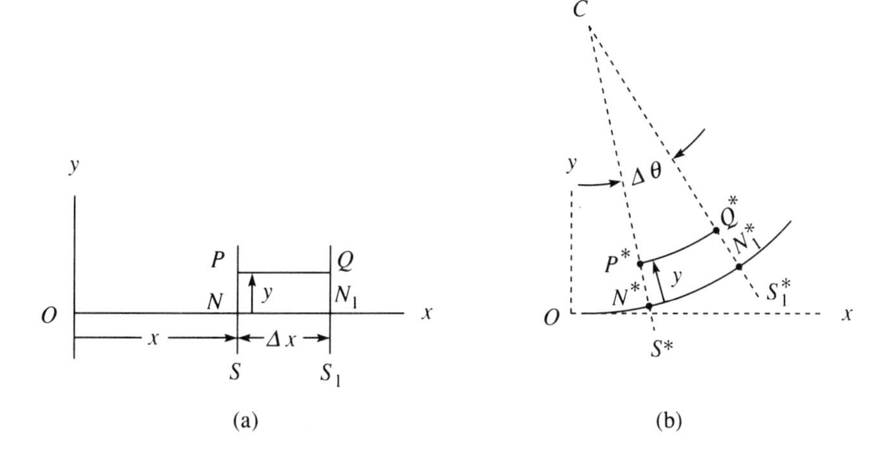

the nearby cross sections S^* and S_1^* are perpendicular to the neutral axis, their profiles, when extended, intersect near C, the center of curvature for the point N^* of the deformed neutral axis. The distance CN^* is the radius of curvature of the neutral axis at N^*; it is denoted by ρ. Consequently, if $\Delta\theta$ is the dihedral angle between the planes of the cross sections S^* and S_1^*, then

$$P^*Q^* = CP^*\,\Delta\theta \doteq (\rho - y)\,\Delta\theta \tag{3.16}$$

Since the neutral axis does not stretch or contract,

$$N^*N_1^* = \Delta x$$

But, from Fig. 3.24, we also have

$$N^*N_1^* = \rho\,\Delta\theta$$

Consequently,

$$\Delta\theta = \frac{\Delta x}{\rho}$$

When this is substituted in Eq. (3.16), P^*Q^* can be written as

$$P^*Q^* \doteq (\rho - y)\frac{\Delta x}{\rho}$$

From the definition of extensional strain in the x-direction at P, we have

$$\varepsilon_x(P) = \lim_{\substack{Q\to P \\ \text{along } x}} \frac{P^*Q^* - PQ}{PQ}$$

Consequently,

$$\varepsilon_x(P) = \lim_{\Delta x\to 0} \frac{(\rho - y)\dfrac{\Delta x}{\rho} - \Delta x}{\Delta x} = -\frac{y}{\rho} \qquad (3.17)$$

This relates the extensional strain in the axial direction at P to the distance y above (or below) the neutral axis, and the radius of curvature ρ of the neutral axis at the cross section containing P.

Nothing has been said about the displacement of P in the cross section itself, so the remaining components of the state of strain cannot be considered here. ◆

PROBLEMS

3.70 The plate shown in Fig. P3.70 is sheared in such a way that all horizontal lines remain parallel and all vertical lines remain vertical. Compute γ_{xy} anywhere in the plate.

Fig. P3.70

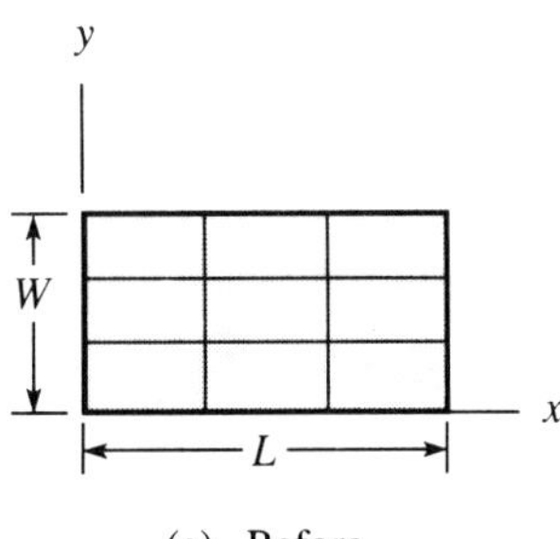

(a) Before

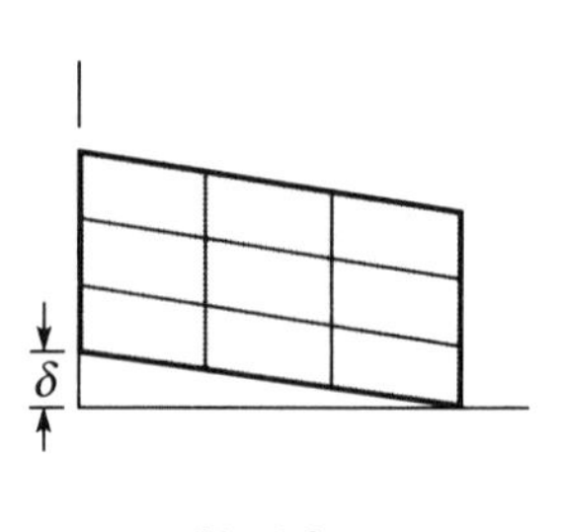

(b) After

3.71 A square plate is deformed as shown in Fig. P3.71. The edges AB and AD are held rigid; DC is stretched uniformly in such a way that no point of the plate displaces in the y-direction and vertical lines on the plate remain straight. Obtain ε_x and γ_{xy} at any point of the plate in terms of e, L, and the original coordinates of the point.

Fig. P3.71

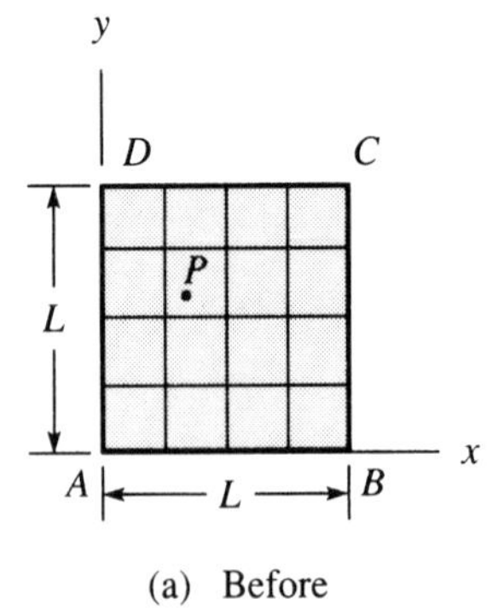

(a) Before

Fig. P3.71
(*cont.*)

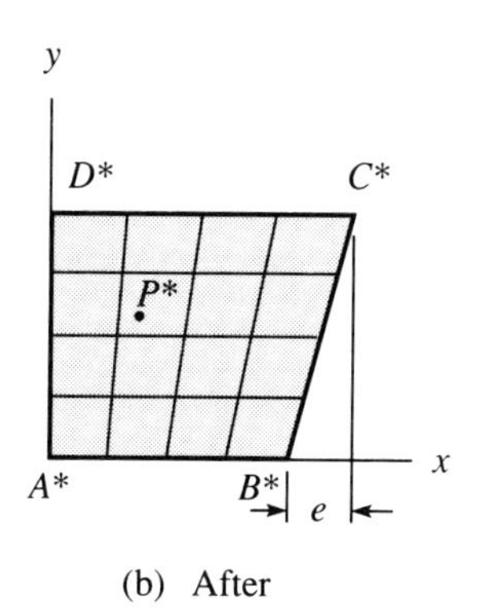

(b) After

3.72 The rectangular plate shown in Fig. P3.72 is deformed in such a way that no point of the plate displaces in the x-direction, and lines in the plate that were originally horizontal remain straight. If the vertical displacement along edges AD and BC is given by $V_1(y)$ and $V_2(y)$, respectively, and $V_1(y)$ and $V_2(y)$ are *small*, then show that (approximately)

$$\varepsilon_x(x, y) = 0$$

$$\varepsilon_y(x, y) = \frac{x}{L}\frac{d}{dy}V_2(y) + \left(1 - \frac{x}{L}\right)\frac{d}{dy}V_1(y)$$

$$\gamma_{xy} = \frac{V_2(y) - V_1(y)}{L}$$

Fig. P3.72

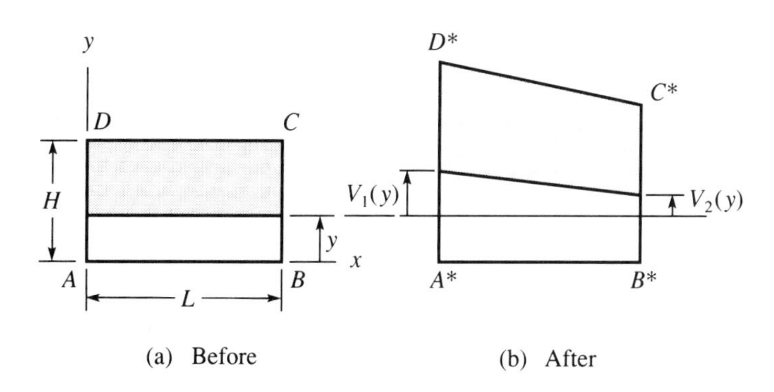

(a) Before (b) After

3.73 The triangular plate shown in Fig. P3.73 is fixed along the edge AB, and the vertex C is displaced downward a small amount δ. This deformation occurs in such a way that edges AC and BC remain straight, and all vertical and horizontal lines in the plate remain straight and parallel to their original positions. Compute $\varepsilon_s(P)$ and $\varepsilon_t(P)$ for any point $P(s, t)$ in the plate.

Fig. P3.73

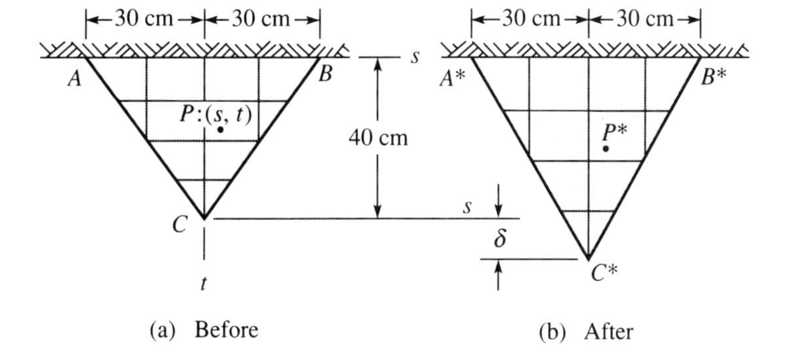

(a) Before (b) After

3.74 The hollow circular cylinder shown in Fig. P3.74 expands in such a way that each particle of material is displaced outward (along a radius) a small amount $u(r)$, which could depend on the magnitude of the radius to the point in question. Express $\varepsilon_r(P)$ and $\varepsilon_\theta(P)$ for the point $P(r, \theta, z)$ in the cylindrical shell in terms of $u(r)$.

Fig. P3.74

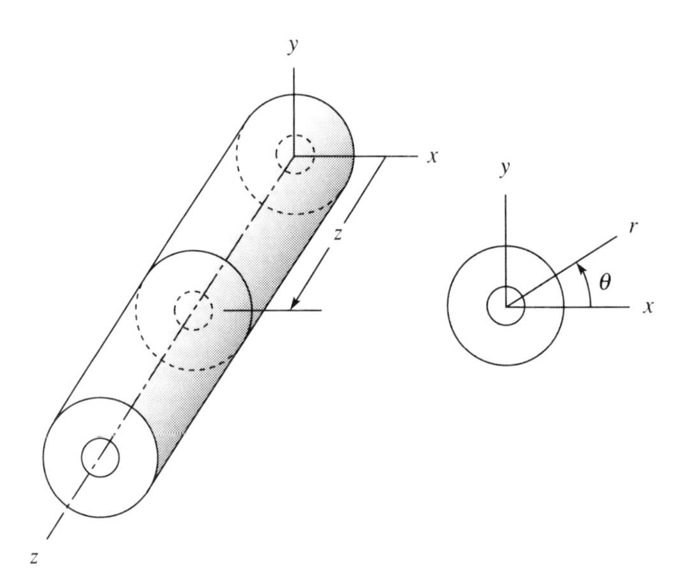

3.75 A hollow spherical shell has an inner radius $r = a$ and an outer radius $r = b$. If concentric spherical surfaces in the shell remain spherical and concentric [increasing their radius by the amount $u(r)$, which could depend on the magnitude of the original radius], obtain an expression for the extensional strain $\varepsilon_t(r)$ in any tangential direction (perpendicular to a radius) in terms of r and $u(r)$. Also obtain an expression for the extensional strain $\varepsilon_r(r)$ along a radius. If the shell is thin [$(b - a)/a$ small] and if $u(r)$ is constant through the thickness and also small compared to a, show that $\varepsilon_t(r)$ is approximately constant throughout the shell.

3.76 In Prob. 3.74, suppose that each cross section of the cylinder displaces a small amount $w(z)$ parallel to the z-axis in addition to the radial displacement $u(r)$. The cross sections remain plane and perpendicular to the axis. Determine completely the state of strain at $P(r, \theta, z)$; that is, $\varepsilon_r, \varepsilon_\theta, \varepsilon_z, \gamma_{r\theta}, \gamma_{\theta z}, \gamma_{rz}$.

3.77 A circular cylinder is stretched and twisted simultaneously. By combining the assumptions of Examples 6 and 7, obtain the significant (small) strains.

3.78 A circular cylinder is bent and twisted simultaneously. Combine the assumptions of Examples 7 and 8 (modifying them as necessary to avoid contradiction) and obtain the significant strains.

3.79 A prismatic bar is stretched and bent simultaneously. Combine the assumptions of Examples 6 and 8 (modifying them as necessary to avoid contradiction) and obtain the significant strains.

3.80 A thin bar with a square cross section is $12\frac{3}{8}$ in. long and 0.001 in. thick. It is wrapped around a circle 4 in. in diameter so that the ends of the strip just meet (in a plane surface). Assume that the cross sections of the bar remain plane and end up perpendicular to the arc of the circle. Compute the average extensional strain along the center of the bar.

3.81 A rectangular strip of material 60 cm long and 0.1 cm thick is bent into a circular arc with a radius of 60 cm. If the cross sections remain plane and perpendicular to the arc and the outer edge of the strip is unstrained after bending, compute the extensional strains (tangent to the arc) throughout the strip.

3.82 The circular ring segment shown in Fig. P3.82 is deformed into a rectangular strip in such a way that the circumferential lines become straight and the radial lines remain perpendicular to them. Furthermore, there is a circumferential line NN that experiences no extensional strain. Determine the extensional strain $\varepsilon_\theta(P)$ at the point P shown in the figure in terms of y and R. Simplify this relation if y is small compared to R.

Fig. P3.82

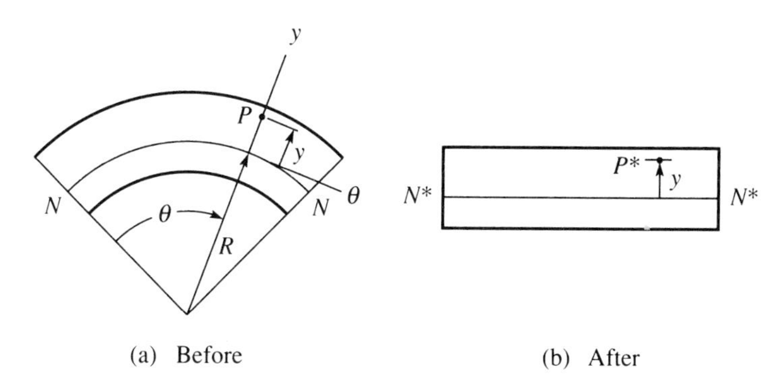

(a) Before (b) After

3.83 The hollow cylinder shown in Fig. P3.83 is deformed so that every cross section such as S deforms into a shallow conical surface. Neither the inner nor the outer cylinder is deformed. The inner cylinder is displayed axially a distance Δ, relative to the outer cylinders so that every cross section deforms to a similar cone. Express the shear strain $\gamma_{zr}(P)$ in terms of Δ, a, and b.

Fig. P3.83

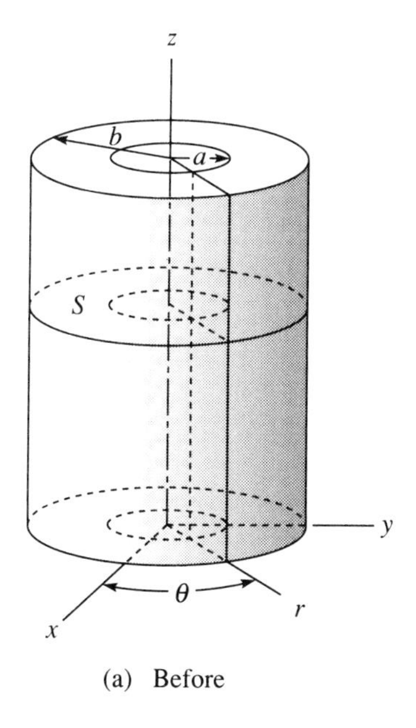

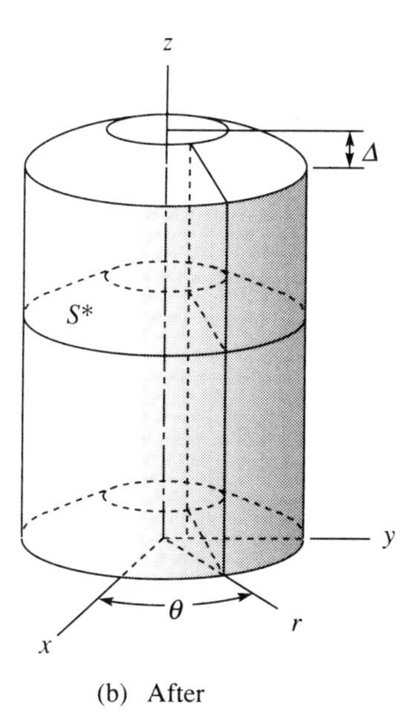

(a) Before (b) After

3.84 A hollow cylinder is deformed so that the inner surface remains fixed and the outer surface rotates through a small angle ϕ, as shown in Fig. P3.84. If radial planes remain plane and cylindrical surfaces remain circular, determine $\gamma_{r\theta}(P)$.

Fig. P3.84

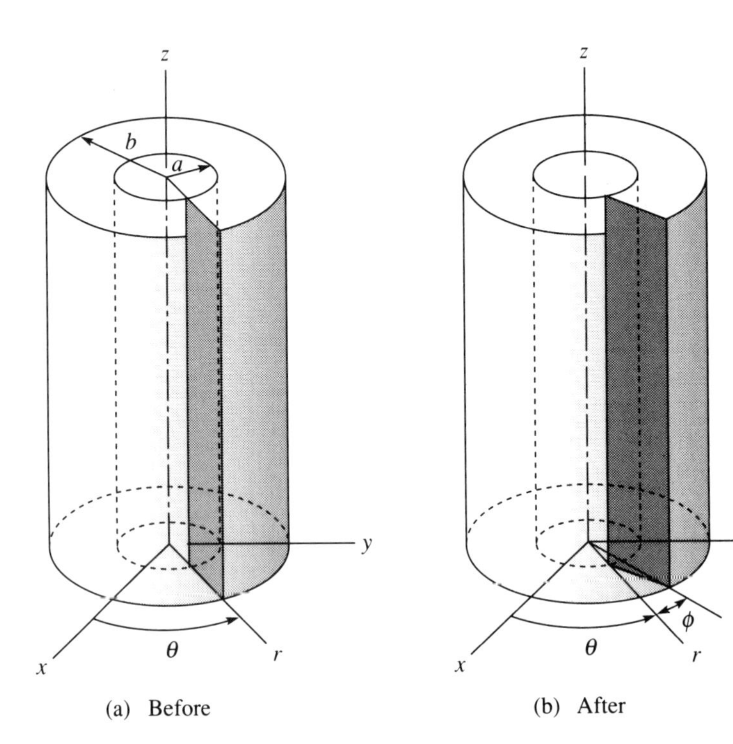

(a) Before (b) After

3.85 A segment of a circular ring, as shown in Fig. P3.85, is deformed in such a way that (a) there is an arc of particles that remains in the plane of the ring and deforms without stretching into another circular arc, called the neutral axis, and (b) any cross section of particles perpendicular to this neutral axis remains plane and perpendicular to it. If the initial radius of the neutral axis is R and its final radius is ρ, show that the extensional strain at the point P a distance y inside the neutral axis is given by

$$\varepsilon_\theta(P) = -\frac{y}{\rho}\left[\frac{1-\dfrac{\rho}{R}}{1-\dfrac{y}{R}}\right]$$

Fig. P3.85

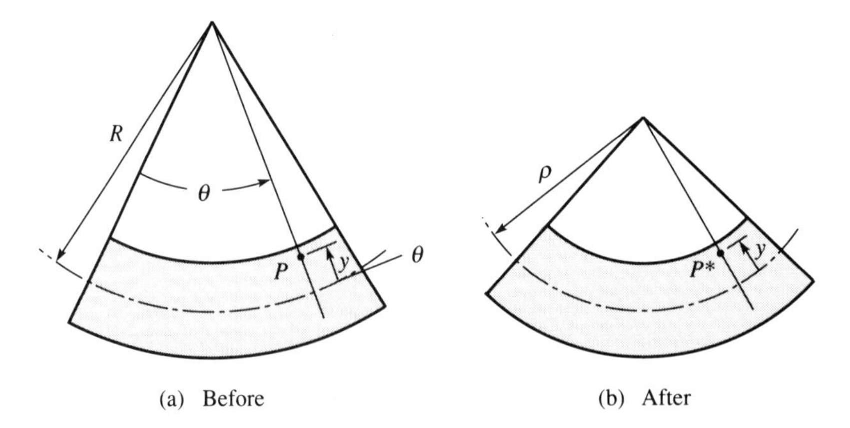

(a) Before (b) After

Hence, show that this reduces to the result obtained for bending of straight bars in Example 8, if the bar is initially straight.

3.86 A thin, slightly tapered shaft (Fig. P3.86) is twisted in accordance with the assumptions for twisting of cylinders in Example 7. The shear strain between a meridional line and a circumferential line on the surface of the *slightly* tapered shaft is $\gamma_{m\theta}(P)$. Express this strain as a function $\gamma_{m\theta}(z)$.

Fig. P3.86

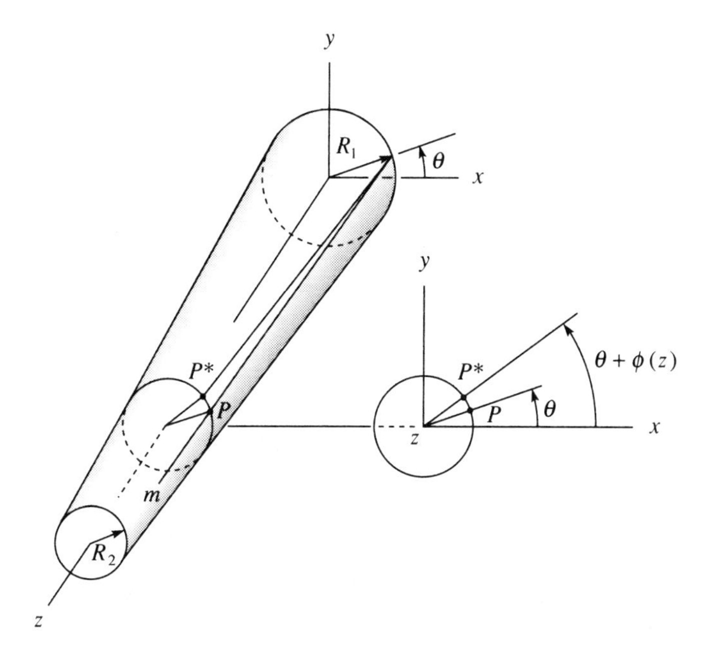

3.12 | Completeness of the State of Strain

As indicated in Sec. 3.8, the deformation of a body is completely described by a knowledge of six strains, three extensional strains and three shear strains, at each point throughout the body. These strains are associated with three mutually perpendicular directions at each point. Occasionally, it is necessary to know an extensional strain or a shear strain at a point for directions other than those used to define the state of strain at the point. For example, in experimental work, extensional strains are measured in several directions at a point on the surface of a body. From these the investigator must be able to obtain the shear strain between two perpendicular directions in the surface at the point of measurement.

Because six strains at a point define the deformation in the neighborhood of the point, the strains associated with any other directions at that point are determined. It should be possible to express these other strains in terms of the six given strains and the directions for which these other strains are required.

In Fig. 3.25, P represents a point in a body. The state of strain at P is determined by the six quantities

Fig. 3.25

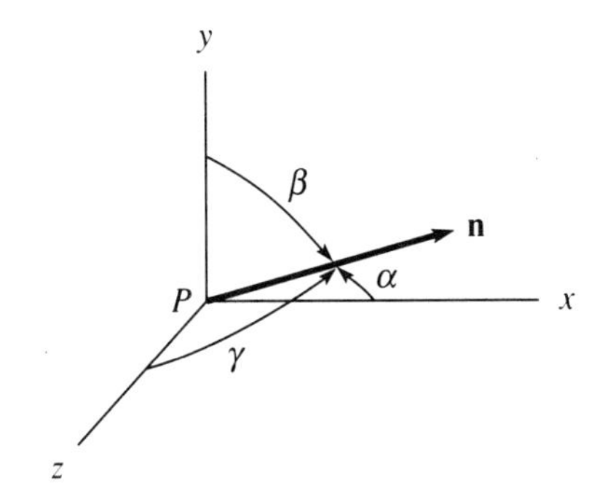

$$\varepsilon_x, \varepsilon_y, \varepsilon_z, \gamma_{xy}, \gamma_{yz}, \gamma_{zx}$$

where x, y, and z represent any three mutually perpendicular directions at P. Suppose that we require the extensional strain at P in a direction n that is specified by direction angles α, β, γ, or, equivalently, by direction cosines l_n, m_n, n_n.

To obtain the required strain $\varepsilon_n(P)$, we might proceed as follows: We can select a short segment PQ of length Δs in the direction **n**, as shown in Fig. 3.26(a). The segment PQ forms the diagonal of the rectangular block with edges PA, PB, and PC in the directions x, y, and z, respectively. The lengths of the edges are given by the direction cosines:

$$PA = l_n \Delta s, \qquad PB = m_n \Delta s, \qquad PC = n_n \Delta s$$

Fig. 3.26

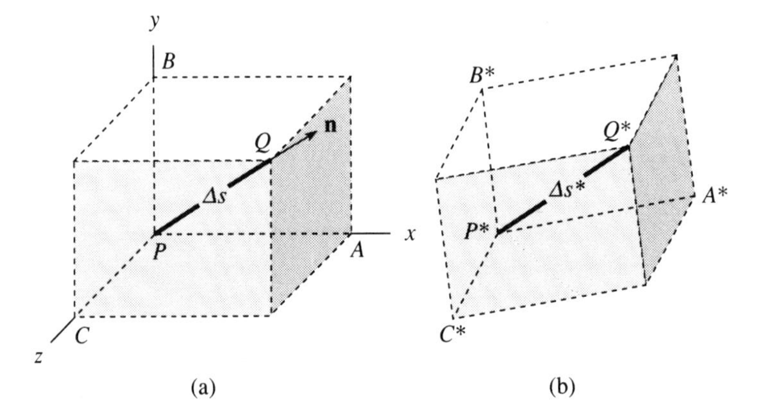

When the body deforms, this block becomes a parallelepiped, as shown in Fig. 3.26(b). PQ becomes P^*Q^*, a diagonal of the parallelepiped. The lengths of the sides of the parallelepiped are

$$P^*A^* = l_n(1 + \varepsilon_x)\Delta s, \qquad P^*B^* = m_n(1 + \varepsilon_y)\Delta s, \qquad P^*C^* = n_n(1 + \varepsilon_z)\Delta s$$

Also, the angles between adjacent edges are

$$\angle A^*P^*B^* = \frac{\pi}{2} - \gamma_{xy}, \qquad \angle A^*P^*C^* = \frac{\pi}{2} - \gamma_{xz}, \qquad \angle B^*P^*C^* = \frac{\pi}{2} - \gamma_{yz}$$

The parallelepiped is completely determined by the lengths of its edges and the angles between the edges. In particular, the length of the diagonal P^*Q^* can be expressed in terms of these six quantities, which are already expressed in terms of $\varepsilon_x, \varepsilon_y, \varepsilon_z, \gamma_{xy}, \gamma_{xz}, \gamma_{yz}, l_n, m_n, n_n,$ *and* Δs. The extensional strain $\varepsilon_n(P)$ is given by the definition

$$\varepsilon_n(P) = \lim \frac{P^*Q^* - PQ}{PQ} \tag{3.18}$$

Note here that the length $PQ = \Delta s$ cancels from the numerator and denominator. The result, an expression for $\varepsilon_n(P)$, is a function of the six strains $(\varepsilon_x, \varepsilon_y, \varepsilon_z, \gamma_{xy}, \gamma_{xz}, \gamma_{yz})$ and the direction cosines (l_n, m_n, n_n). If the strains are small, then the strain, ε_n is given by the linear approximation

$$\varepsilon_n(P) \doteq l_n^2\varepsilon_x + l_n m_n \gamma_{xy} + l_n n_n \gamma_{xz} + m_n^2 \varepsilon_y + m_n n_n \gamma_{yz} + n_n^2 \varepsilon_z \tag{3.19}$$

Now, suppose that we require the shear strain for two directions **n** and **t**, which are initially perpendicular at P in Fig. 3.27. The directions can be specified in various ways, but, in any case, we know the direction cosines l_n, m_n, n_n, and l_t, m_t, n_t. By the preceding arguments, we can always express the extensional strains $\varepsilon_n(P)$ and $\varepsilon_t(P)$ in terms of the six components $(\varepsilon_x, \varepsilon_y, \varepsilon_z, \gamma_{xy}, \gamma_{xz}, \gamma_{yz})$; $\varepsilon_n(P)$ is a function of l_n, m_n, n_n; $\varepsilon_t(P)$ is a function of l_t, m_t, n_t. We can now select a direction s that bisects **n** and **t**. The direction cosines of that line are

$$\frac{1}{\sqrt{2}}(l_n + l_t), \qquad \frac{1}{\sqrt{2}}(m_n + m_t), \qquad \frac{1}{\sqrt{2}}(n_n + n_t)$$

Fig. 3.27

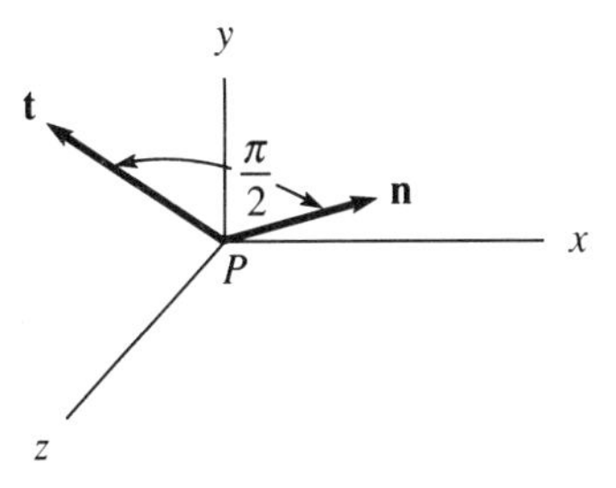

Again, the extensional strain of that line can be expressed in terms of the six components $\varepsilon_x, \varepsilon_y, \varepsilon_z, \gamma_{xy}, \gamma_{xz}, \gamma_{yz}$ and the direction cosines (given previously). Now, consider the square with edges PA and PB along **n** and **t** and diagonal PC along **s**, as depicted in Fig. 3.28. This deforms to a parallelogram with edges P^*A^* and P^*B^* and diagonal P^*C^*. In Fig. 3.28 we view *planar* views of both the original square $PACB$ and the deformed parallelogram $P^*A^*C^*B^*$. If the original length of the diagonal is Δs, then the deformed lengths follow:

Fig. 3.28

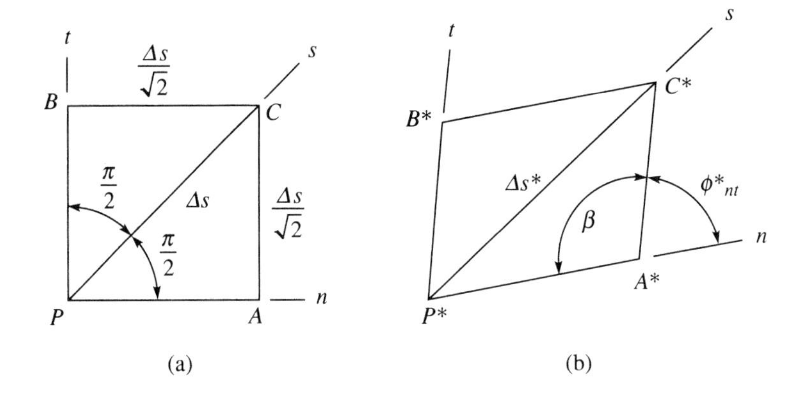

$$P^*C^* = (1 + \varepsilon_s)\,\Delta s, \qquad P^*A^* = \frac{(1 + \varepsilon_n)\,\Delta s}{\sqrt{2}}$$

$$P^*B^* = \frac{(1 + \varepsilon_t)\,\Delta s}{\sqrt{2}}$$

The length of each segment is known in terms of $\varepsilon_n, \varepsilon_s, \varepsilon_t$, and Δs, which, in turn, can be expressed in terms of $\varepsilon_x, \varepsilon_y, \varepsilon_z, \gamma_{xy}, \gamma_{yz}, \gamma_{xz}, l_n, m_n, n_n, l_t, m_t, n_t$, and Δs, as previously described. It is a geometric exercise (by the law of cosines) to express the angle β in terms of these quantities; length Δs cancels. Note that $\gamma_{nt} = \beta - \pi/2$. It is thereby established that the shear strain for *any* directions $(\mathbf{n}, \mathbf{t})$ can be expressed in terms of the six given components, $\varepsilon_x, \varepsilon_y, \varepsilon_z, \gamma_{xy}, \gamma_{xz}, \gamma_{yz}$, the direction cosines of those lines. If the strains are small, then the strain γ_{nt} is given by the approximation

$$\begin{aligned}\tfrac{1}{2}\gamma_{nt}(P) = {} & l_n l_t \varepsilon_x + \tfrac{1}{2} l_n m_t \gamma_{xy} + \tfrac{1}{2} l_n n_t \gamma_{xz} \\ & + \tfrac{1}{2} m_n l_t \gamma_{xy} + m_n m_t \varepsilon_y + \tfrac{1}{2} m_n n_t \gamma_{yz} \\ & + \tfrac{1}{2} n_n l_t \gamma_{xz} + \tfrac{1}{2} n_n m_t \gamma_{yz} + n_n n_t \varepsilon_z \end{aligned} \tag{3.20}$$

The reader will note the remarkable similarities between the linear approximations in (3.19) and (3.20). Both are quadratic forms in the direction cosines; of course, ε_n depends only on the direction n, whereas γ_{nt} depends on both directions $\mathbf{n}$ and $\mathbf{t}$. Actual derivations of (3.19) and (3.20) are given in the following section.

Remark on Notations and Transformations

The reader may recall a similar transformation of the components of stress, specifically Eqs. (2.55) and (2.57). Indeed, we need only replace the σ and τ in Eqs. (2.55) and (2.57) by ε and $\gamma/2$, respectively, to obtain Eqs. (3.19) and (3.20). From a mathematical viewpoint, the six components of stress and the *corresponding* six components of strain transform as the components of a symmetric second-order Cartesian tensor.

The mathematical jargon need not impress the engineer, but the recurrence of this linear transformation is noteworthy. This linear transformation of the six symmetrical components calls for an alternative system of notation. To this end, let

$$\begin{matrix} \sigma_x & \tau_{xy} & \tau_{xz} \\ & \sigma_y & \tau_{yz} \\ & & \sigma_z \end{matrix} \quad \equiv \quad \begin{matrix} \sigma_{xx} & \sigma_{xy} & \sigma_{xz} \\ & \sigma_{yy} & \sigma_{yz} \\ & & \sigma_z \end{matrix}$$

Similarly, let

$$\begin{matrix} \varepsilon_x & \dfrac{\gamma_{xy}}{2} & \dfrac{\gamma_{xz}}{2} \\ & \varepsilon_y & \dfrac{\gamma_{yz}}{2} \\ & & \varepsilon_z \end{matrix} \quad \equiv \quad \begin{matrix} \varepsilon_{xx} & \varepsilon_{xy} & \varepsilon_{xz} \\ & \varepsilon_{yy} & \varepsilon_{yz} \\ & & \varepsilon_{zz} \end{matrix}$$

Again, these are merely alternative notations. The repeated subscripts signal a normal stress (e.g., σ_{xx}) or extensional strain (e.g., ε_{xx}), whereas the dissimilar subscripts signal a shear stress (e.g., $\sigma_{xy} = \tau_{xy}$) or a shear strain (e.g., $\varepsilon_{xy} = \gamma_{xy}/2$). Then, we can write the equations of the linear transformation in the same form; to illustrate, Eqs. (3.19) and (3.20) assume the following forms:

$$\begin{aligned} \varepsilon_{nn} = {} & l_n l_n \varepsilon_{xx} + l_n m_n \varepsilon_{xy} + l_n n_n \varepsilon_{xz} \\ & + m_n l_n \varepsilon_{yx} + m_n m_n \varepsilon_{yy} + m_n n_n \varepsilon_{yz} \\ & + n_n l_n \varepsilon_{zx} + n_n m_n \varepsilon_{zy} + n_n n_n \varepsilon_{zz} \\ \varepsilon_{nt} = {} & l_n l_t \varepsilon_{xx} + l_n m_t \varepsilon_{xy} + l_n n_t \varepsilon_{xz} \\ & + m_n l_t \varepsilon_{yx} + m_n m_t \varepsilon_{yy} + m_n n_t \varepsilon_{yz} \\ & + n_n l_t \varepsilon_{zx} + n_n m_t \varepsilon_{zy} + n_n n_n \varepsilon_{zz} \end{aligned}$$

The same equations give the transformation of stress components; we need only replace ε by σ.

If n and t are orthogonal to z, then $n_n = n_t = 0$, and we have the two-dimensional versions:

$$\begin{aligned} \varepsilon_{nn} = {} & l_n l_n \varepsilon_{xx} + l_n m_n \varepsilon_{xy} \\ & + m_n l_n \varepsilon_{yx} + m_n m_n \varepsilon_{yy} \\ \varepsilon_{nt} = {} & l_n l_t \varepsilon_{xx} + l_n m_t \varepsilon_{xy} \\ & + m_n l_t \varepsilon_{yx} + m_n m_t \varepsilon_{yy} \end{aligned}$$

In Appendix B, we show that the area integrals ($I_y = I_{yy}, I_z = I_{zz}, I_{yz}$) in Eq. (2.31) also transform via these equations.

EXAMPLE 9

To illustrate the transformation of strains, consider the specific orthogonal directions n and t shown in Fig. 3.29. Axes n, t, and z lie in a common plane, which is inclined 45° to the x- and z-axes. From the geometry

$$l_n = n_n = \frac{1}{2}, \qquad m_n = \frac{1}{\sqrt{2}}$$

$$l_t = n_t = -\frac{1}{2}, \qquad m_t = \frac{1}{\sqrt{2}}$$

Fig. 3.29

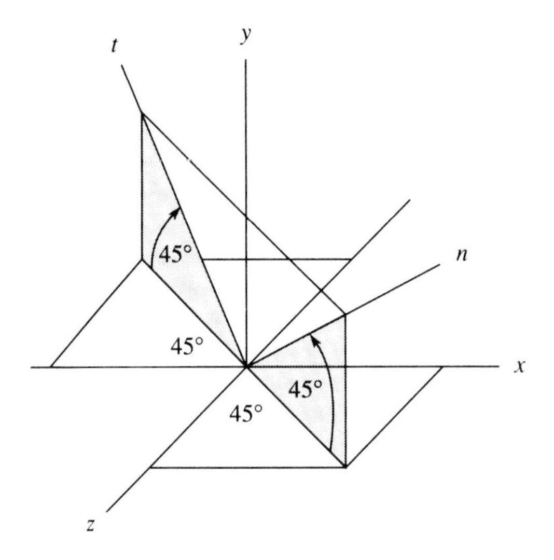

the reader can verify the orthogonality—that is, $l_n l_t + m_n m_t + n_n n_t = 0$. Now, suppose that the state of strain at the point is given by the six components ($\mu = 10^{-6}$)

$$\begin{bmatrix} \varepsilon_x & \gamma_{xy} & \gamma_{xz} \\ & \varepsilon_y & \gamma_{yz} \\ & & \varepsilon_z \end{bmatrix} = \begin{bmatrix} 100 & 60 & -80 \\ & -60 & 100 \\ & & 80 \end{bmatrix} \mu$$

By Eq. (3.19)

$$\varepsilon_n = \left[\frac{1}{4}(100) + \frac{1}{2\sqrt{2}}(60) + \frac{1}{4}(-80) + \frac{1}{2}(-60) + \frac{1}{2\sqrt{2}}(100) + \frac{1}{4}(80)\right]\mu$$
$$= 52\mu$$

By Eq. (3.20)

$$\gamma_{nt} = 2\left[-\frac{1}{4}(100) + \frac{1}{2\sqrt{2}}(60) - \frac{1}{4}(-80) + \frac{1}{2}(-60) - \frac{1}{2\sqrt{2}}(100) - \frac{1}{4}(80)\right]\mu$$
$$= -40.9\mu \; \blacklozenge$$

PROBLEMS

3.87 A state of strain at a point is given by

$$\varepsilon_x = \varepsilon_y = \varepsilon_z = \varepsilon, \qquad \gamma_{xy} = \gamma_{yz} = \gamma_{xz} = 0$$

Show that, if n, t, and s denote any three mutually perpendicular directions at the point,

$$\varepsilon_n = \varepsilon_t = \varepsilon_s = \varepsilon, \qquad \gamma_{nt} = \gamma_{ts} = \gamma_{ns} = 0$$

3.88 The state of strain at a point is given by the six strains associated with three mutually perpendicular directions. If each of these is less than e (in absolute value), show that the absolute value of the extensional strain in any direction at the point must be less than $4e$.

3.89 At a point on the surface of a circular cylinder, the circumferential strain is 4×10^{-4}. The longitudinal strain is 2×10^{-4}, and the radial strain is zero. The shear strains associated with these directions are also zero. Compute the extensional strain in the direction that makes equal angles with the preceding three directions.

3.90 Let x, y, z represent a set of mutually perpendicular directions at a point, and suppose n, t, s represent any other set of mutually perpendicular directions at the point. Show that

$$\varepsilon_n + \varepsilon_t + \varepsilon_s = \varepsilon_x + \varepsilon_y + \varepsilon_z$$

***3.91** Under the conditions of Prob. 3.90, show that

$$\varepsilon_n\varepsilon_t - \left(\frac{\gamma_{nt}}{2}\right)^2 + \varepsilon_t\varepsilon_s - \left(\frac{\gamma_{ts}}{2}\right)^2 + \varepsilon_s\varepsilon_n - \left(\frac{\gamma_{sn}}{2}\right)^2$$
$$= \varepsilon_x\varepsilon_y - \left(\frac{\gamma_{xy}}{2}\right)^2 + \varepsilon_y\varepsilon_z - \left(\frac{\gamma_{yz}}{2}\right)^2 + \varepsilon_z\varepsilon_x - \left(\frac{\gamma_{zx}}{2}\right)^2$$

***3.92** Under the conditions of Prob. 3.90, show that

$$\begin{vmatrix} \varepsilon_n & \frac{\gamma_{nt}}{2} & \frac{\gamma_{ns}}{2} \\ \frac{\gamma_{nt}}{2} & \varepsilon_t & \frac{\gamma_{ts}}{2} \\ \frac{\gamma_{ns}}{2} & \frac{\gamma_{ts}}{2} & \varepsilon_s \end{vmatrix} = \begin{vmatrix} \varepsilon_x & \frac{\gamma_{xy}}{2} & \frac{\gamma_{xz}}{2} \\ \frac{\gamma_{xy}}{2} & \varepsilon_y & \frac{\gamma_{yz}}{2} \\ \frac{\gamma_{xz}}{2} & \frac{\gamma_{yz}}{2} & \varepsilon_z \end{vmatrix}$$

*3.13 Transformation of Strain Components

In the preceding section we argued that six components completely describe the state of strain at a point. Again, this means that the six components determine any other component at the point, an extensional strain of any line, or the shear strain between any two lines. Here, we derive the equations for that transformation.

A particle of an undeformed body is at point P in Fig. 3.30 and is located by the vector $\mathbf{r}$. Small changes of position accompany a change in each of the coordinates; for example, the change Δx carries the vector from P to A along the x line as $\mathbf{r}$ undergoes the change

Fig. 3.30

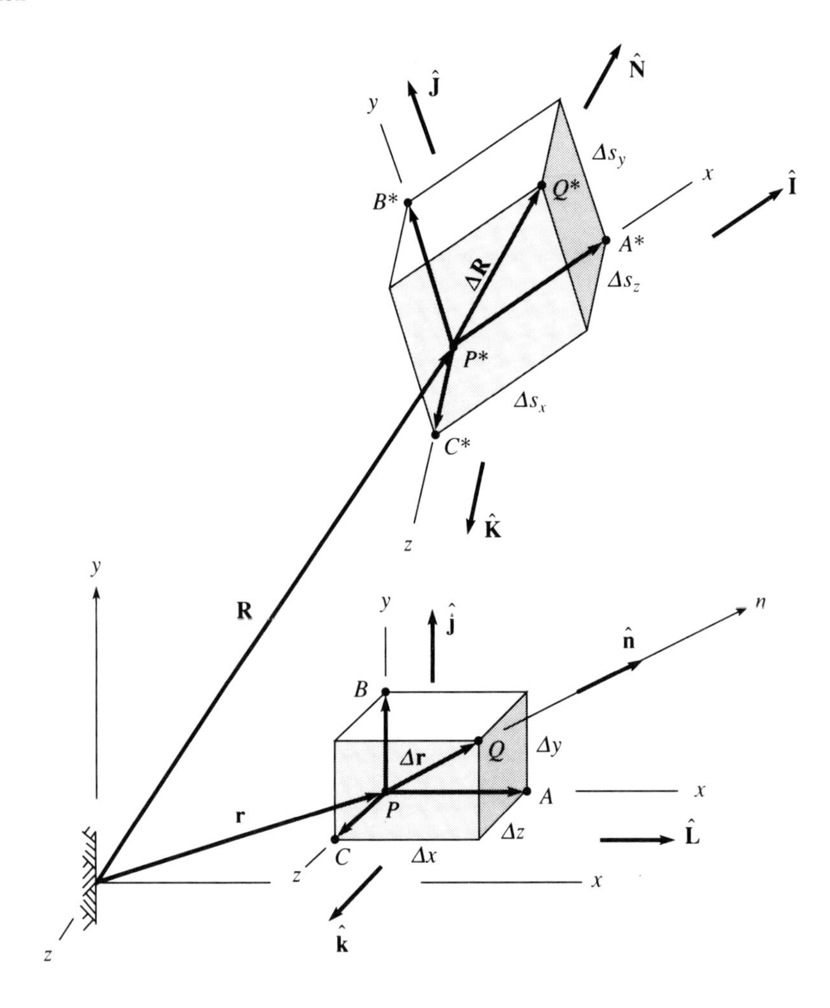

$$\Delta\mathbf{r}_x = \frac{\partial \mathbf{r}}{\partial x}\Delta x = \hat{\mathbf{i}}\,\Delta x \tag{3.21}$$

Here, $\hat{\mathbf{i}}, \hat{\mathbf{j}}$, and $\hat{\mathbf{k}}$ denote unit vectors in the directions of the rectangular lines x, y, and z, respectively. An arbitrary change in position carries the vector from P to Q; the change of position is given by the vector

$$\begin{aligned}\Delta\mathbf{r} &= \frac{\partial \mathbf{r}}{\partial x}\Delta x + \frac{\partial \mathbf{r}}{\partial y}\Delta y + \frac{\partial \mathbf{r}}{\partial z}\Delta z \\ &= \hat{\mathbf{i}}\,\Delta x + \hat{\mathbf{j}}\,\Delta y + \hat{\mathbf{k}}\,\Delta z\end{aligned}$$

Let Δs_n denote the length of this initial line segment PQ; $\Delta s_n = |\Delta\mathbf{r}|$. The subscript n signifies the initial orientation of the line in the direction of the unit vector $\hat{\mathbf{n}}$:

$$\hat{\mathbf{n}} = \lim_{\Delta s_n \to 0} \frac{\Delta \mathbf{r}}{\Delta s_n} = \frac{\partial \mathbf{r}}{\partial s_n}$$

$$= \frac{\partial \mathbf{r}}{\partial x}\frac{\partial x}{\partial s_n} + \frac{\partial \mathbf{r}}{\partial y}\frac{\partial y}{\partial s_n} + \frac{\partial \mathbf{r}}{\partial z}\frac{\partial z}{\partial s_n}$$

$$= \hat{\mathbf{i}}\frac{\partial x}{\partial s_n} + \hat{\mathbf{j}}\frac{\partial y}{\partial s_n} + \hat{\mathbf{k}}\frac{\partial z}{\partial s_n}$$

where

$$\frac{\partial x}{\partial s_n}, \frac{\partial y}{\partial s_n}, \frac{\partial z}{\partial s_n} = \lim_{\Delta s_n \to 0} \frac{\Delta x}{\Delta s_n}, \frac{\Delta y}{\Delta s_n}, \frac{\Delta z}{\Delta s_n}$$

These are the direction cosines (l_n, m_n, n_n):

$$\hat{\mathbf{n}} = \frac{\partial \mathbf{r}}{\partial s_n} = l_n\hat{\mathbf{i}} + m_n\hat{\mathbf{j}} + n_n\hat{\mathbf{k}} \tag{3.22}$$

Upon deformation, the particle at P is transported to P^*, located now by the new vector $\mathbf{R}$. Particles at A, B, and C are moved to new locations A^*, B^* and C^*. A very important feature is the deformation, which has *extended* (or contracted) each line and *sheared* the angles between the lines. The particle at A^* is now located relative to P^* by the vector

$$\Delta \mathbf{R}_x = \frac{\partial \mathbf{R}}{\partial x}\Delta x$$

Unlike Eq. (3.21), the vector $\partial \mathbf{R}/\partial x$ is *not* a unit vector. Here the change ΔR_x that accompanies the change Δx has the extended length P^*A^*. Indeed, by our earlier definition of an extensional strain,

$$\lim_{\Delta x \to 0}\left(\frac{P^*A^*}{PA} - 1\right) \equiv \varepsilon_x$$

Stated otherwise, if $\hat{\mathbf{I}}$ is the unit vector in the new direction of the deformed x line, then

$$\lim_{\Delta x \to 0} \frac{\Delta \mathbf{R}_x}{\Delta x} \equiv \frac{\partial \mathbf{R}}{\partial x} = (1 + \varepsilon_x)\hat{\mathbf{I}}$$

Likewise,

$$\frac{\partial \mathbf{R}}{\partial y} = (1 + \varepsilon_y)\hat{\mathbf{J}}$$

$$\frac{\partial \mathbf{R}}{\partial z} = (1 + \varepsilon_z)\hat{\mathbf{K}}$$

where $\hat{\mathbf{I}}, \hat{\mathbf{J}}$, and $\hat{\mathbf{K}}$ are unit vectors along the displaced lines x, y, and z. Then, for the line P^*Q^*

$$\lim_{\Delta s_n \to 0} \frac{\Delta \mathbf{R}}{\Delta s_n} = \frac{\partial \mathbf{R}}{\partial x}\frac{\partial x}{\partial s_n} + \frac{\partial \mathbf{R}}{\partial y}\frac{\partial y}{\partial s_n} + \frac{\partial \mathbf{R}}{\partial z}\frac{\partial z}{\partial s_n}$$
$$\frac{\partial \mathbf{R}}{\partial s_n} = l_n(1 + \varepsilon_x)\hat{\mathbf{I}} + m_n(1 + \varepsilon_y)\hat{\mathbf{J}} + n_n(1 + \varepsilon_z)\hat{\mathbf{K}} \tag{3.23}$$

Just as the line PA is extended to P^*A^* by extensional strain ε_x, so the line PQ is extended to P^*Q^* by the extensional strain ε_n. If $\hat{\mathbf{N}}$ is the unit vector in the direction P^*Q^*, then

$$\frac{\partial \mathbf{R}}{\partial s_n} = (1 + \varepsilon_n)\hat{\mathbf{N}} = l_n(1 + \varepsilon_x)\hat{\mathbf{I}} + m_n(1 + \varepsilon_y)\hat{\mathbf{J}} + n_n(1 + \varepsilon_z)\hat{\mathbf{K}} \tag{3.24}$$

Now, we must note that the edges of the block are no longer perpendicular. Indeed, if ϕ_{xy} denotes the angle between the deformed x and y lines—that is, between P^*A^* and P^*B^*—then

$$\cos \phi_{xy} = \hat{\mathbf{I}} \cdot \hat{\mathbf{J}}$$

By our earlier definition of the shear strain $\gamma_{xy}, \gamma_{xz}, \gamma_{yz}$,

$$\sin \gamma_{xy} = \cos \phi_{xy} = \hat{\mathbf{I}} \cdot \hat{\mathbf{J}} \tag{3.25a}$$

$$\sin \gamma_{xz} = \hat{\mathbf{I}} \cdot \hat{\mathbf{K}}, \qquad \sin \gamma_{yz} = \hat{\mathbf{J}} \cdot \hat{\mathbf{K}} \tag{3.25b,c}$$

The extensional strain ε_n can now be evaluated by means of Eqs. (3.24) and (3.25).

$$\begin{aligned}\frac{\partial \mathbf{R}}{\partial s_n} \cdot \frac{\partial \mathbf{R}}{\partial s_n} &= (1 + \varepsilon_n)^2 \\ &= l_n(1 + \varepsilon_x)[l_n(1 + \varepsilon_x) + m_n(1 + \varepsilon_y)\sin \gamma_{xy} + n_n(1 + \varepsilon_z)\sin \gamma_{xz}] \\ &\quad + m_n(1 + \varepsilon_y)[l_n(1 + \varepsilon_x)\sin \gamma_{yx} + m_n(1 + \varepsilon_y) + n_n(1 + \varepsilon_z)\sin \gamma_{yz}] \\ &\quad + n_n(1 + \varepsilon_z)[l_n(1 + \varepsilon_x)\sin \gamma_{zx} + m_n(1 + \varepsilon_y)\sin \gamma_{zy} + n_n(1 + \varepsilon_z)]\end{aligned} \tag{3.26}$$

The result seems messy, but it does give the extensional strain ε_n *exactly* in terms of the six given components $(\varepsilon_x, \varepsilon_y, \varepsilon_z, \gamma_{xy}, \gamma_{xz}, \gamma_{yz})$. Two simplifications are possible. The first possibility is to adopt the pragmatic view that we intend to deal primarily with small strains; that is, the ε's and γ's are very small compared to unity. This is certainly true in the members of most machines and structures. Accordingly, we can expand both sides of (3.26) in powers of these small quantities. We note that $\hat{\mathbf{n}} \cdot \hat{\mathbf{n}} = l_n^2 + m_n^2 + n_n^2 = 1$ and also that $\sin \gamma \doteq \gamma$. When we retain only the linear terms of Eq. (3.26), we obtain the approximation

$$\begin{aligned}\varepsilon_n = {} & l_n^2 \varepsilon_x + l_n m_n \tfrac{1}{2}\gamma_{xy} + l_n n_n \tfrac{1}{2}\gamma_{xz} \\ & + m_n l_n \tfrac{1}{2}\gamma_{yx} + m_n^2 \varepsilon_y + m_n n_n \tfrac{1}{2}\gamma_{yz} \\ & + n_n l_n \tfrac{1}{2}\gamma_{zx} + n_n m_n \tfrac{1}{2}\gamma_{zy} + n_n^2 \varepsilon_z\end{aligned} \tag{3.27}$$

A second simplification of the messy equation (3.26) is possible. This is accomplished without approximations by introducing alternative definitions of extensional and shear strains. Let us define an extensional strain $\tilde{\varepsilon}_n$ in the n-direction (similarly in any direction):

$$\tilde{\varepsilon}_n = \lim_{\Delta s_n \to 0} \frac{1}{2}\left[\left(\frac{P^*Q^*}{PQ}\right)^2 - 1\right]$$

$$= \frac{1}{2}\left[\frac{\partial \mathbf{R}}{\partial S_n} \cdot \frac{\partial \mathbf{R}}{\partial S_n} - 1\right]$$

$$= \varepsilon_n + \frac{1}{2}\varepsilon_n^2 \tag{3.28}$$

The right side of Eq. (3.28) follows immediately from Eq. (3.24). This measure of strain ($\tilde{\varepsilon}_n$) serves as well as the other (ε_n); given ε_n, we can compute $\tilde{\varepsilon}_n$, and vice versa. Both strains vanish if the line is not extended or contracted. Let us define a shear strain for lines x and y (similarly for any initially orthogonal pair):

$$\tilde{\gamma}_{xy} \equiv \tfrac{1}{2}(1 + \varepsilon_x)(1 + \varepsilon_y)\cos\phi_{xy} \tag{3.29a}$$

$$= \tfrac{1}{2}(1 + \varepsilon_x)(1 + \varepsilon_y)\sin\gamma_{xy} \tag{3.29b}$$

This measure of shear ($\tilde{\gamma}_{xy}$) also serves as well as the other (γ_{xy}). Both vanish if the lines remain orthogonal. With these alternative definitions for the extensional and shear strains and the identity $l^2 + m^2 + n^2 = 1$, expression (3.26) assumes the form

$$\begin{aligned}\tilde{\varepsilon}_n = {} & l_n^2\tilde{\varepsilon}_x + l_n m_n \tilde{\gamma}_{xy} + l_n n_n \tilde{\gamma}_{xz} \\ & + m_n l_n \tilde{\gamma}_{yx} + m_n^2 \tilde{\varepsilon}_y + m_n n_n \tilde{\gamma}_{yz} \\ & + n_n l_n \tilde{\gamma}_{zx} + n_n m_n \tilde{\gamma}_{zy} + n_n^2 \tilde{\varepsilon}_z \end{aligned} \tag{3.30}$$

The form of Eq. (3.30) differs from (3.27) only in the factor $\frac{1}{2}$ in each of the shear terms. This is to be expected: The previous result, (3.27), is valid only for small strains, but then, according to Eqs. (3.28) and (3.29b), we have the valid approximations

$$\tilde{\varepsilon}_n \doteq \varepsilon_n, \qquad \tilde{\gamma}_{xy} \doteq \tfrac{1}{2}\gamma_{xy}, \ldots$$

The preceding results provide the means to determine an extensional strain in any direction at a point if we have six components ($\varepsilon_x, \varepsilon_y, \varepsilon_z, \gamma_{xy}, \gamma_{xz}, \gamma_{yz}$). Now, let us derive the equation(s) to obtain the shear strain for two initially orthogonal lines.

At the point P of Fig. 3.31 the directions n and t are orthogonal; unit vectors $\hat{\mathbf{n}}$ and $\hat{\mathbf{t}}$ are given in the forms

$$\hat{\mathbf{n}} = l_n\hat{\mathbf{i}} + m_n\hat{\mathbf{j}} + n_n\hat{\mathbf{k}}$$

$$\hat{\mathbf{t}} = l_t\hat{\mathbf{i}} + n_t\hat{\mathbf{j}} + n_t\hat{\mathbf{k}}$$

They satisfy the following conditions:

$$\hat{\mathbf{n}}\cdot\hat{\mathbf{n}} = l_n^2 + m_n^2 + n_n^2 = 1 \tag{3.31a}$$

$$\hat{\mathbf{t}}\cdot\hat{\mathbf{t}} = l_t^2 + m_t^2 + n_t^2 = 1 \tag{3.31b}$$

$$\hat{\mathbf{n}}\cdot\hat{\mathbf{t}} = l_n l_t + m_n m_t + n_t n_t = 0 \tag{3.31c}$$

Fig. 3.31

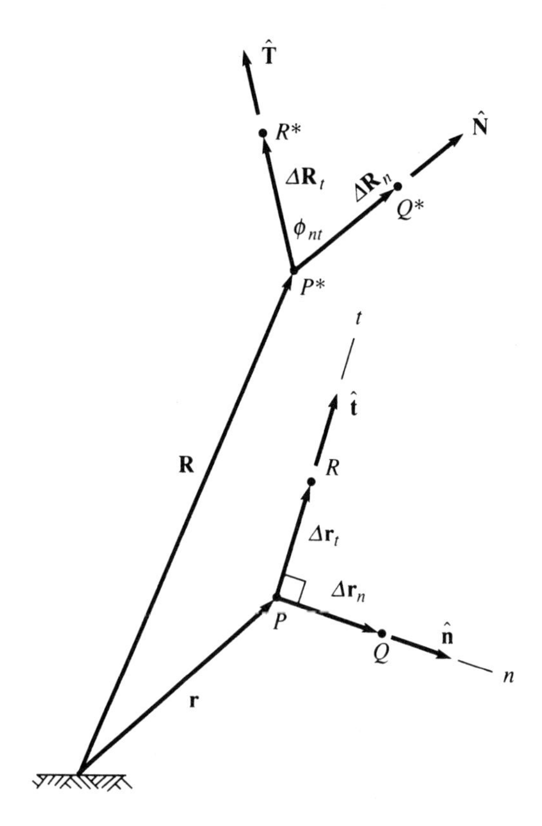

Our attention focuses on the two orthogonal line elements PQ and PR. Each is defined by a relative change in the vector $\mathbf{r}$, namely,

$$\Delta\mathbf{r}_n = (l_n\hat{\mathbf{i}} + m_n\hat{\mathbf{j}} + n_n\hat{\mathbf{k}})\,\Delta s_n \tag{3.32a}$$

$$\Delta\mathbf{r}_t = (l_t\hat{\mathbf{i}} + m_t\hat{\mathbf{j}} + n_t\hat{\mathbf{k}})\,\Delta s_t \tag{3.32b}$$

Here, we call upon the earlier arguments, for example,

$$l_n = \frac{\partial x}{\partial s_n}, \qquad \Delta x = l_n\,\Delta s_n, \ldots$$

Deformation transports the particle at P to the position P^* defined by the vector $\mathbf{R}$. The neighboring particles at Q and R are transported to Q^* and R^*. It is always the relative positions of particles that determine deformation and strain. The positions of Q^* and R^* *relative* to P^* are given by the changes in the vector $\mathbf{R}$, namely,

$$\Delta\mathbf{R}_n = \left(\frac{\partial\mathbf{R}}{\partial x}l_n + \frac{\partial\mathbf{R}}{\partial y}m_n + \frac{\partial\mathbf{R}}{\partial z}n_n\right)\Delta s_n \tag{3.33a}$$

$$\Delta\mathbf{R}_t = \left(\frac{\partial\mathbf{R}}{\partial x}l_t + \frac{\partial\mathbf{R}}{\partial y}m_t + \frac{\partial\mathbf{R}}{\partial z}n_t\right)\Delta s_t \tag{3.33b}$$

The reader will note the similarities between Eqs. (3.32) and (3.33). More important are the differences; specifically, the vectors $\Delta\mathbf{R}_n$ and $\Delta\mathbf{R}_t$ are extended (or contracted) and are not, in general, orthogonal. Following the previous notation, we have

$$\frac{\partial \mathbf{R}}{\partial x} = (1 + \varepsilon_x)\hat{\mathbf{I}} \tag{3.34a}$$

$$\frac{\partial \mathbf{R}}{\partial y} = (1 + \varepsilon_y)\hat{\mathbf{J}} \tag{3.34b}$$

$$\frac{\partial \mathbf{R}}{\partial z} = (1 + \varepsilon_z)\hat{\mathbf{K}} \tag{3.34c}$$

$$\frac{\partial \mathbf{R}}{\partial s_n} = (1 + \varepsilon_n)\hat{\mathbf{N}} \tag{3.34d}$$

$$\frac{\partial \mathbf{R}}{\partial s_t} = (1 + \varepsilon_t)\hat{\mathbf{T}} \tag{3.34e}$$

Here, as before, $\hat{\mathbf{I}}, \hat{\mathbf{J}}$, and $\hat{\mathbf{K}}$ are unit vectors in directions of the *deformed* x, y, and z lines (not shown in Fig. 3.31), and $\hat{\mathbf{N}}$, and $\hat{\mathbf{T}}$ are unit vectors in the directions of the *deformed* n and t lines. From Eqs. (3.33a) and (3.34a, b, c, d) we obtain, as before, Eq. (3.24); from Eq. (3.34e) we also obtain a similar equation for that vector tangent to the deformed t line:

$$(1 + \varepsilon_n)\hat{\mathbf{N}} = l_n(1 + \varepsilon_x)\hat{\mathbf{I}} + m_n(1 + \varepsilon_y)\hat{\mathbf{J}} + n_n(1 + \varepsilon_z)\hat{\mathbf{K}} \tag{3.35a}$$

$$(1 + \varepsilon_t)\hat{\mathbf{T}} = l_t(1 + \varepsilon_x)\hat{\mathbf{I}} + m_t(1 + \varepsilon_y)\hat{\mathbf{J}} + n_t(1 + \varepsilon_z)\hat{\mathbf{K}} \tag{3.35b}$$

Now, the shear strain γ_{nt} can be obtained from the earlier definition:

$$\sin \gamma_{nt} = \cos \phi_{nt} = \hat{\mathbf{N}} \cdot \hat{\mathbf{T}}$$

Clearly, we can obtain a precise, albeit messy, equation for γ_{nt}. Note that the extensional strains, ε_n and ε_t, can be obtained from previous equations. Here is the product of the vectors of Eq. (3.35a, b):

$$(1 + \varepsilon_n)(1 + \varepsilon_t)\hat{\mathbf{N}} \cdot \hat{\mathbf{T}} = (1 + \varepsilon_n)(1 + \varepsilon_t)\sin \gamma_{nt} \tag{3.36a}$$

$$\begin{aligned} &= l_n(1 + \varepsilon_x)[l_t(1 + \varepsilon_x) + m_t(1 + \varepsilon_y)\sin \gamma_{xy} + n_t(1 + \varepsilon_z)\sin \gamma_{xz}] \\ &\quad + m_n(1 + \varepsilon_y)[l_t(1 + \varepsilon_x)\sin \gamma_{yx} + m_t(1 + \varepsilon_y) + n_t(1 + \varepsilon_z)\sin \gamma_{yz}] \\ &\quad + n_n(1 + \varepsilon_z)[l_t(1 + \varepsilon_x)\sin \gamma_{zx} + m_t(1 + \varepsilon_y)\sin \gamma_{zy} + n_t(1 + \varepsilon_z)] \end{aligned} \tag{3.36b}$$

Again, two simplifications are possible. We can adopt the pragmatic view that we expect to treat only small strains. Accordingly, the two sides of Eq. (3.36) can be expanded in powers of the *small* strains (ε's and γ's). Constant terms cancel by the conditions (3.31a, b, c). When we retain only the linear terms of Eq. (3.36), we obtain the approximation

$$\tfrac{1}{2}\gamma_{nt} = l_n l_t \varepsilon_x + l_n m_t \tfrac{1}{2}\gamma_{xy} + l_n n_t \tfrac{1}{2}\gamma_{xz}$$
$$+ m_n l_t \tfrac{1}{2}\gamma_{yx} + m_n m_t \varepsilon_y + m_n n_t \tfrac{1}{2}\gamma_{yz}$$
$$+ n_n l_t \tfrac{1}{2}\gamma_{zx} + n_n m_t \tfrac{1}{2}\gamma_{zy} + n_n n_t \varepsilon_z \qquad (3.37)$$

The reader will note the remarkable similarities between Eqs. (3.37) and (3.27), the equations for ε_n and γ_{nt}. Both are linear in the components of strain; both are quadratic forms in the direction cosines of the initial lines n and t.

Again, an alternative means of simplifying Eq. (3.36) is possible and requires no approximation. This is achieved by adopting the alternative definitions of strain. An extensional strain ε_n is defined by Eq. (3.28); a shear strain is defined by Eq. (3.29). Accordingly, the expression (3.36) assumes the following simpler form:

$$\tilde{\gamma}_{nt} \equiv \tfrac{1}{2}(1 + \varepsilon_n)(1 + \varepsilon_t)\cos\phi_{nt}$$
$$= l_n l_t \tilde{\varepsilon}_x + l_n m_t \tilde{\gamma}_{xy} + l_n n_t \tilde{\gamma}_{xz}$$
$$+ m_n l_t \tilde{\gamma}_{yx} + m_n m_t \tilde{\varepsilon}_y + m_n n_t \tilde{\gamma}_{yz}$$
$$+ n_n l_t \tilde{\gamma}_{zx} + n_n m_t \tilde{\gamma}_{zy} + n_n n_n \tilde{\varepsilon}_z \qquad (3.38)$$

This version, like Eq. (3.30), is exact. Again, when the strains are small, $\tilde{\varepsilon}_n \doteq \varepsilon_n$, $\tilde{\gamma}_{nt} \doteq \tfrac{1}{2}\gamma_{nt}$, and so on, and then the approximation (3.37) differs from (3.38) only in the factor $\tfrac{1}{2}$, which appears in the shear terms of the latter.

Finally, we emphasize that Eqs. (3.30) and (3.38) are exact linear equations that transform components $\tilde{\varepsilon}_x, \tilde{\gamma}_{xy}, \tilde{\gamma}_{xz}, \tilde{\varepsilon}_y, \tilde{\gamma}_{yz}, \tilde{\varepsilon}_z$ in one Cartesian system, (x, y, z), to components $\tilde{\varepsilon}_n, \tilde{\gamma}_{nt}, \tilde{\gamma}_{ns}, \tilde{\varepsilon}_t, \tilde{\gamma}_{ts}, \tilde{\varepsilon}_s$ in another Cartesian system, (n, t, s). Compare these with Eqs. (3.19) and (3.20); they are the same except for the notation ($\tilde{\varepsilon}$ replaces ε and $\tilde{\gamma}$ replaces $\gamma/2$). Equations (3.19) and (3.20) entail only the approximation of small strain ($\varepsilon \ll 1$, $\gamma \ll 1$); rotations and displacements may be large. Equations (3.30) and (3.38) are exact; strains also may be large.

PROBLEMS

3.93 Verify the approximation of Eq. (3.26) by Eq. (3.27) when all strains are small compared to unity; for example, $\varepsilon_x \ll 1$, $\gamma_{xy} \ll 1, \ldots$.

3.94 Verify the approximation of Eq. (3.36) by Eq. (3.37) when all strains are small compared to unity; for example, $\varepsilon_x \ll 1$, $\gamma_{xy} \ll 1, \ldots$.

3.95 The rectangular plate of Fig. P3.95(a) is homogeneously deformed to the parallelepiped of Fig. 3.95(b). From the geometry of the latter and the definitions of the strains $\tilde{\varepsilon}_n$ [Eq. (3.28)] and $\tilde{\gamma}_{nt}$ [Eq. (3.29a)], obtain the exact equation

$$\tilde{\varepsilon}_n = \tilde{\varepsilon}_x \cos^2\theta + 2\tilde{\gamma}_{xy}\cos\theta\sin\theta + \tilde{\varepsilon}_y \sin^2\theta$$

Note: a, b, and l are eliminated via the relations $a/l = \cos\theta$, $b/l = \sin\theta$.

Fig. P3.95

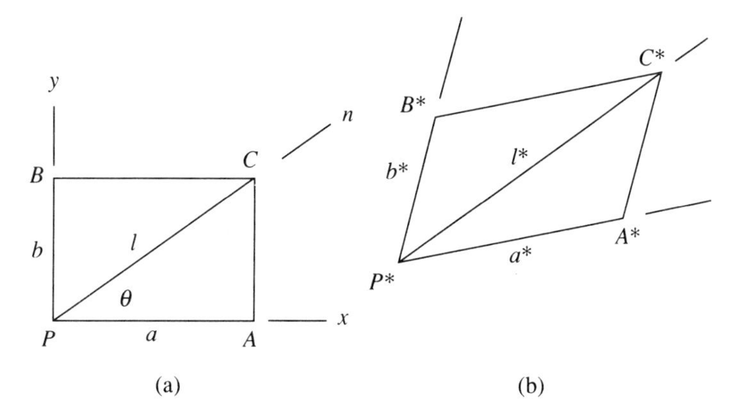

(a) (b)

3.96 Consider the rectangular plate of Fig. P3.95(a), which is homogeneously deformed to the parallelepiped of Fig. 3.95(b). With the definition of Eq. (3.29a), show that the shear strain $\tilde{\gamma}_{xy}$ is given exactly by the formula

$$\begin{aligned}\tilde{\gamma}_{xy} &= \sec\theta \csc\theta(\varepsilon_n^2 + \tfrac{1}{2}\varepsilon_n^2)\\ &\quad - \tan\theta(\varepsilon_y + \tfrac{1}{2}\varepsilon_y^2)\\ &\quad - \cot\theta(\varepsilon_x + \tfrac{1}{2}\varepsilon_x^2)\end{aligned}$$

3.97 If x, y, and z represent any set of mutually orthogonal directions at a point, and n, t, and s represent *any* other set of mutually orthogonal directions at the point, show that

$$\tilde{\varepsilon}_n + \tilde{\varepsilon}_t + \tilde{\varepsilon}_s = \tilde{\varepsilon}_x + \tilde{\varepsilon}_y + \tilde{\varepsilon}_z$$

***3.98** Show that the result given in Prob. 3.91 holds *exactly* if the strains $(\varepsilon, \gamma/2)$ are replaced by those defined by Eq. (3.28) and Eq. (3.29); that is, replace ε by $\tilde{\varepsilon}$ and $\gamma/2$ by $\tilde{\gamma}$.

***3.99** Show that the result given in Prob. 3.92 holds exactly if the strains $(\varepsilon, \gamma/2)$ are replaced by those defined by Eq. (3.28) and Eq. (3.29); that is, replace ε by $\tilde{\varepsilon}$ and $\gamma/2$ by $\tilde{\gamma}$.

3.14 Principal Directions; Maximum and Minimum Values of Strains

The geometric relations, specifically Eqs. (3.19) and (3.20) [or the equivalent (3.27), (3.30) and (3.37), (3.38)], which enable us to obtain extensional and shear strains for arbitrarily oriented lines, also enable us to examine the state of strain. In particular, we can vary the orientation (l_n, m_n, n_n) of the direction n in Eq. (3.19) and ascertain the direction that provides the maximum or minimum extensional strain. We may vary the orientations $(l_n, m_n, n_n; l_t, m_t, n_t)$ of directions n and t in Eq. (3.20) and determine directions of zero shear strain or the directions of maximum shear strain. Such investigation provides important information from a theoretical viewpoint and also from a practical viewpoint. The magnitude of an extensional strain or shear strain might limit the use of a material.

The mathematics is not unlike the search for the maximum of a function of one variable. To find the maximum (or minimum) of a smooth function $y(x)$, we must impose the condition of stationarity, $dy/dx = 0$. Here, the strain ε_n is a function of three variables (l_n, m_n, n_n), which are not independent $(l_n^2 + m_n^2 + n_n^2 = 1)$. The strain γ_{nt} is a function of six variables $(l_n, m_n, n_n; l_t, m_t, n_t)$, which are also interrelated. In view of the complexities, we omit the mathematical derivations and state the important results:

1. There exist, at every point of the continuous body, three *mutually perpendicular* directions of *zero shear strain.* These directions are called the *principal directions* of strain. The *extensional* strains in these three directions are called the *principal strains.* Fig. 3.32 provides a graphic illustration. The edges of the small cube at P lie in the principal directions. After deformation the particle at P has been transported to P^*. The cube is rectangular, since the shear strains of the edges are zero.

Fig. 3.32

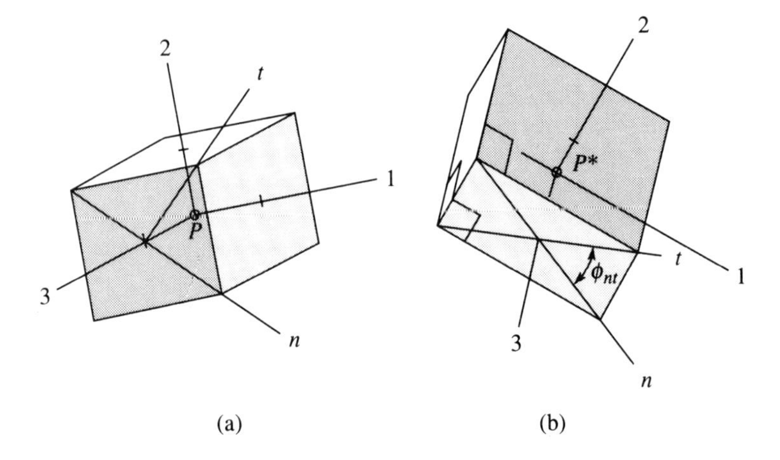

2. One principal strain is the *maximum extensional* strain, and one is the *minimum extensional* strain.

3. The directions of *maximum shear* strain *bisect* the principal directions of maximum and minimum extensional strain. In Fig. 3.32, 1 and 2 are the directions of maximum and minimum extensional strain; the values are λ_1 and λ_2. The directions n and t, which *initially* bisect 1 and 2 at Point P_1, are the directions of maximum shear strain. The latter is given by the simple, but general, formula[4]

$$\gamma_{\max}(=\gamma_{nt}) = |\lambda_1 - \lambda_2| \tag{3.39}$$

*3.15 Determination of Principal Strains and Directions

The actual determination of principal directions (cosines l, m, n) requires the solution of the homogeneous equations:

4 This formula is always valid for small strain ($\gamma_{nt} \ll 1$) but does not precisely give the angle ($\pi/2 - \phi_{nt}$) when that angle is large.

$$(\varepsilon_x - \lambda)l + \tfrac{1}{2}\gamma_{xy}m + \tfrac{1}{2}\gamma_{xz}n = 0 \tag{3.40a}$$

$$\tfrac{1}{2}\gamma_{yz}l + (\varepsilon_y - \lambda)m + \tfrac{1}{2}\gamma_{yz}n = 0 \tag{3.40b}$$

$$\tfrac{1}{2}\gamma_{zx}l + \tfrac{1}{2}\gamma_{zy}m + (\varepsilon_z - \lambda)n = 0 \tag{3.40c}$$

These have a solution if and only if the determinant of the coefficients vanishes —that is, if

$$\begin{vmatrix} (\varepsilon_x - \lambda) & \tfrac{1}{2}\gamma_{xy} & \tfrac{1}{2}\gamma_{xz} \\ \tfrac{1}{2}\gamma_{yx} & (\varepsilon_y - \lambda) & \tfrac{1}{2}\gamma_{yz} \\ \tfrac{1}{2}\gamma_{zx} & \tfrac{1}{2}\gamma_{zy} & (\varepsilon_z - \lambda) \end{vmatrix} = 0 \tag{3.41a}$$

The latter is a cubic equation of the form:

$$\lambda^3 - I_1\lambda^2 + I_2\lambda - I_3 = 0 \tag{3.41b}$$

The labeling of the coefficients (I_1, I_2, I_3) is quite arbitrary. These are polynomials in the components $\varepsilon_x, \varepsilon_y, \varepsilon_z, \gamma_{xy}, \gamma_{xz}, \gamma_{yz}$, which follow from the expansion of the determinant; I_1 is linear, I_2 is quadratic, and I_3 is cubic. The roots of (3.41) are the principal strains. When these have been computed, the direction for each is obtained by solving any two of the three linear equations (3.40) *and* the equation

$$l^2 + m^2 + n^2 = 1 \tag{3.42}$$

3.16 Dilatation

Dilatation measures the change of volume at a point; it might be called volumetric strain. Let ΔV^* denote the initial volume of an elemental region enclosing a point P and ΔV^* be the volume of the *same* material deformed and displaced to P^*. We require a measure of the change in the limit as the region shrinks to the point. We can adopt a definition analogous to the extensional strain (3.1) and define a dilatation e:

$$e \equiv \lim_{\Delta V \to 0} \frac{\Delta V^*}{\Delta V} - 1 \tag{3.43}$$

If we consider a small cube with faces in the principal directions, then the deformed counterpart is a rectangular parallelepiped. If the initial cube has edges of length $\Delta x = \Delta y = \Delta z$, then the deformed rectangle has edges of length $(1 + \varepsilon_x)\Delta x$, $(1 + \varepsilon_y)\Delta x$, and $(1 + \varepsilon_z)\Delta x$. In this case, the foregoing definition gives

$$\begin{aligned} e &= (1 + \varepsilon_x)(1 + \varepsilon_y)(1 + \varepsilon_z) - 1 \\ e &= \varepsilon_x + \varepsilon_y + \varepsilon_z + \varepsilon_x\varepsilon_y + \varepsilon_x\varepsilon_z + \varepsilon_y\varepsilon_z + \varepsilon_x\varepsilon_y\varepsilon_z \end{aligned} \tag{3.44}$$

Of course, if the directions were not principal directions, then the cube would be deformed to a parallelepiped and the dilatation would also depend on the components of shear strain. As previously noted, the geometry of such deformed parallelepipeds is fully determined by the six components $\varepsilon_x, \varepsilon_y, \varepsilon_z, \gamma_{xy}, \gamma_{xz}, \gamma_{yz}$ and the initial length Δx. It is, therefore, a geometric exercise to obtain the expression for the dilatation e. Suffice it to note that the components of shear strain appear only in terms that are quadratic and higher degree. For small strains, the dilatation is given by the linear approximation

$$e \doteq \varepsilon_x + \varepsilon_y + \varepsilon_z \tag{3.45}$$

3.17 | Special States of Strain

Several special states deserve our attention. The most special circumstance occurs if all three principal strains [the three roots of Eq. (3.41)] at a point are equal. It follows, in accordance with Eq. (3.39), that no shear strain occurs for *any* directions at that point. This situation is illustrated graphically by Fig. 3.33. The small sphere at the point P is transported during the deformation to P^*. Lines may be rotated, but every radial line experiences the same extensional strain and the angle between any two lines remains unchanged.

Fig. 3.33

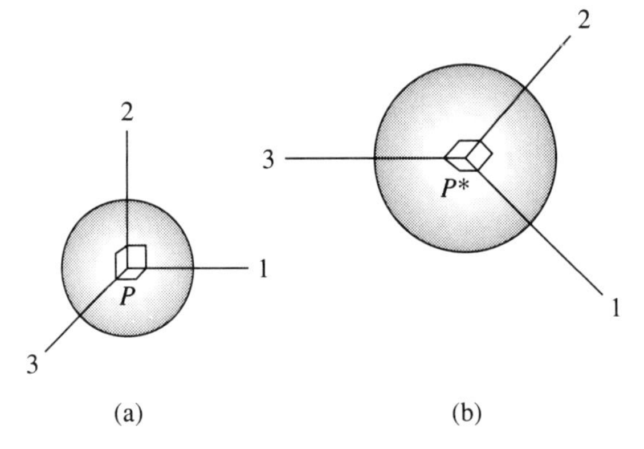

Another special circumstance arises if there is but one distinct principal direction of strain [Eq. (3.41) has two equal roots]. This situation is graphically illustrated in Fig. 3.34, wherein a small circular cylinder at the point P is deformed to another circular cylinder at P^*. The axis lies in the direction of the one distinct principal strain (λ_1). Every radial line experiences the same extensional strain ($\lambda_2 = \lambda_3 \neq \lambda_1$), and the

Fig. 3.34

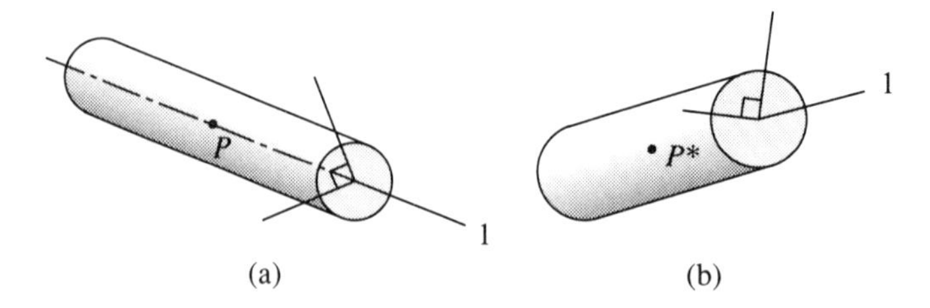

angle between any two radial lines is unchanged. Note that the maximum shear strain occurs between lines that are inclined 45° to the axis.

Another special state occurs if two principal strains are equal in magnitude but opposite in sign and the third is zero ($\lambda_1 = -\lambda_2, \lambda_3 = 0$). By the preceding equations, the strain components for the directions that bisect the principal directions, 1 and 2, are *all* zero except the one shear strain. The situation is depicted in Fig. 3.35; for clarity it is viewed in the 1-2-plane. The one nonzero shear strain, which is also the maximum shear strain,[5] is

$$\gamma_{nt} = \gamma_{\max} = 2\lambda_1$$

The state is *simple shear*. Relative to the rectangular directions n, t, and 3, γ_{nt} is the only nonzero strain. Note the simple shearing of the shaded element ($\gamma_{nt} = \pi/2 - \phi_{nt}$).

Fig. 3.35

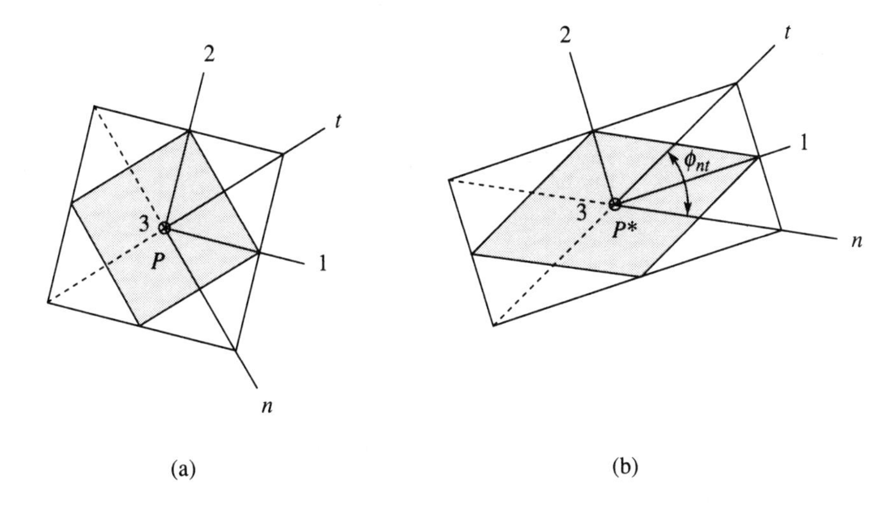

PROBLEMS

3.100 At a point of a body the principal strains are

$$\varepsilon_1 = 0.200, \qquad \varepsilon_2 = -0.100, \qquad \varepsilon_3 = 0.100$$

Determine (a) the dilatation e and (b) the maximum change in a right angle (see Fig. 3.35). *Hint:* A cube with edges in the principal directions deforms to a rectangle. What are the errors of the approximations, Eq. (3.45) and Eq. (3.39)?

3.101 At a point the principal directions are aligned with the rectangular system, $(1, 2, 3)$, of Fig. 3.35. Another rectangular system, $(n, t, 3)$, is oriented as shown; lines n and t lie on the diagonals of the square. If the principal strains are

$$\varepsilon_1 = -\varepsilon_2 = \varepsilon, \qquad \varepsilon_z = 0$$

use the transformations of Sec. 3.12, namely, Eqs.

5 This formula is always valid for small strain ($\gamma_{nt} \ll 1$) but does not precisely give the angle ($\pi/2 - \phi_{nt}$) when that angle is large.

(3.19) and (3.20), to show that

$$\gamma_{nt} = 2\varepsilon, \qquad \gamma_{n3} = \gamma_{t3} = \varepsilon_n = \varepsilon_t = \varepsilon_3 = 0$$

In words, the state is simple shear.

3.102 A state of small strain (all components are small compared to unity) has the following properties: The dilatation is $e = 16 \times 10^{-4}$. The maximum shear strain is $\gamma_M = 12 \times 10^{-4}$, and the maximum extensional strain is $\varepsilon_M = 20 \times 10^{-4}$. Determine all principal strains in accordance with the approximations for small strain.

3.103 A block of rubber is uniformly strained such that two principal strains are $\varepsilon_1 = 1.000$ and $\varepsilon_2 = -0.500$, in accordance with the engineering definition. If the rubber is incompressible (cannot change volume), compute the third principal strain.

3.104 At a point a body experiences a state of simple shear strain $\gamma_{nt} = 60 \times 10^{-5}$. All other components in the orthogonal directions (n, t, s) are zero. Compute all principal strains and the dilatation.

3.105 Directions 1 and 2 in Fig. P3.105 are principal directions, which experience extensional strains according to the engineering definition:

Fig. P3.105

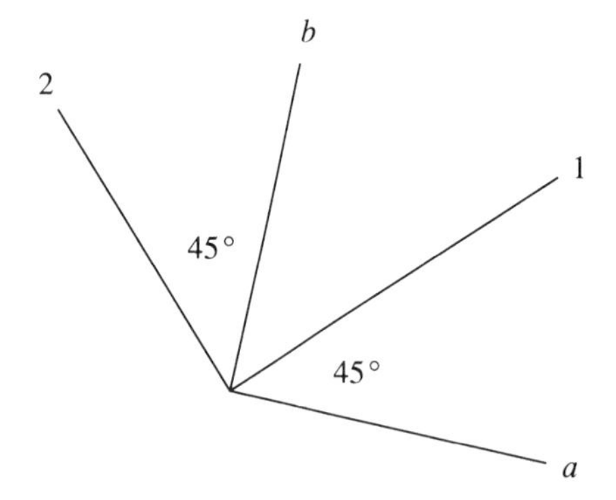

$$\varepsilon_1 = 0.100, \qquad \varepsilon_2 = -0.060$$

Prior to the deformation, lines a and b are orthogonal and lie in the plane of 1 and 2. Determine (a) the change in the right angle between lines a and b and (b) the extensional strains of lines a and b. *Hint:* Examine the deformation of a square with edges parallel to 1 and 2 and diagonals along a and b. Compare your results to the approximations given by Eqs. (3.20) and (3.19).

3.106 Principal strains are

$$\varepsilon_x = 40 \times 10^{-2} \quad \text{and} \quad \varepsilon_y = -20 \times 10^{-2}$$

If direction x' bisects the x- and y-directions and y' bisects the $-x$- and y-directions, then compute *exactly* the extensional strains $\varepsilon_{x'}$ and $\varepsilon_{y'}$. Consider the deformation of a square with edges in the x- and y-directions and diagonals in the x'- and y'-directions. What is the error of the approximation obtained by the linear transformation, Eq. (3.19)?

3.107 With the data of Prob. 3.106, compute the change of the angle between lines in the directions x' and y'. Compare your result with the shear strain given by Eq. (3.39) (assume $\varepsilon_z = 0$).

3.108 Principal strains at a point are

$$\varepsilon_1 = 0.0400, \qquad \varepsilon_2 = -0.0200, \qquad \varepsilon_3 = 0.0100$$

Calculate the dilatation and the error of the approximation obtained by Eq. (3.45).

3.109 Principal strains at a point are

$$\varepsilon_1 = 0.100, \qquad \varepsilon_2 = -0.1000, \qquad \varepsilon_3 = 0$$

These are not small. The maximum shear strain is the change of angle between the perpendicular lines that bisect the directions 1 and 2. Examine the deformation of a square with edges in the directions 1 and 2; the diagonals experience the maximum shear strain. Compute the change of that angle and compare it with the value given by the formula $\gamma_{max} \doteq \varepsilon_1 - \varepsilon_2$.

3.18 Strains in a Surface

Sometimes the engineer is concerned only with the deformations of a surface. Suppose that P in Fig. 3.36(a) lies in a surface; x and y are orthogonal lines in the surface and z is normal to the surface. An arbitrary direction n in the surface is completely specified

Fig. 3.36

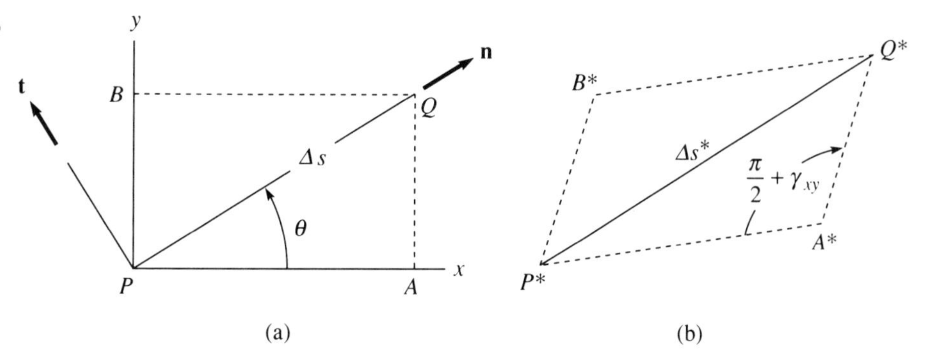

by one angle θ between the directions n and x. Likewise, direction t, also in the surface but perpendicular to n, is specified by the angle θ. We suppose that the components of strain $\varepsilon_x, \varepsilon_y, \gamma_{xy}$ are known, and we require the extensional strain ε_n and shear strain γ_{nt}. The reader, who has assimilated the general transformation of Sec. 3.13 or the results of Sec. 3.12, will recognize this special case. Here, the direction cosines are

$$l_n = \cos\theta, \qquad m_n = \sin\theta, \qquad n_n = 0$$

$$l_t = -\sin\theta, \qquad m_t = \cos\theta, \qquad n_t = 0$$

The reader can substitute these values into Eqs. (3.19) and (3.20), immediately obtain Eqs. (3.46) and (3.48), and then proceed to the alternative forms (3.49) and (3.50). Similarly, Eqs. (3.51) and (3.52) are special forms of (3.30) and (3.38). Alternatively, the reader may choose to follow the present derivation for this special case.

To derive the equation for ε_n we focus on a small segment PQ of length Δs along direction n. This defines the rectangle $PAQB$. During a deformation, the particle at P is transported to P^*; x and y lines are generally skewed but are still in the surface. The small rectangle $PAQB$ is deformed to the parallelogram $P^*A^*Q^*B^*$ in Fig. 3.36(b).

From the definition of shear strains it follows that

$$\angle P^*A^*Q^* \doteq \frac{\pi}{2} + \gamma_{xy}(P)$$

and from the definition of extensional strain

$$P^*A^* \doteq (1 + \varepsilon_x)PA = (1 + \varepsilon_x)\cos\theta\,\Delta s$$

$$P^*B^* \doteq (1 + \varepsilon_y)PB = (1 + \varepsilon_y)\sin\theta\,\Delta s$$

Using the law of cosines to compute $(\Delta s^*)^2$ gives us

$$\begin{aligned}(\Delta s^*)^2 &= (P^*A^*)^2 + (P^*B^*)^2 - 2(P^*A^*)(P^*B^*)\cos\angle P^*A^*Q^* \\ &= (1 + \varepsilon_x)^2\,\Delta s^2\cos^2\theta + (1 + \varepsilon_y)^2\,\Delta s^2\sin^2\theta \\ &\quad - 2(1 + \varepsilon_x)(1 + \varepsilon_y)\,\Delta s^2\sin\theta\cos\theta\cos\left(\frac{\pi}{2} + \gamma_{xy}\right)\end{aligned}$$

If the strains are small, we can apply the small-number algebra of Sec. 1.5 to reduce this expression to

$$\Delta s^* \doteq \Delta s(1 + \varepsilon_x \cos^2\theta + \varepsilon_y \sin^2\theta + \gamma_{xy} \sin\theta\cos\theta)$$

Hence the extensional strain at P in the direction n is given by

$$\begin{aligned}\varepsilon_n(P) &= \lim_{\substack{Q\to P\\ \text{along } n}} \frac{P^*Q^* - PQ}{PQ} = \lim_{\Delta s\to 0} \frac{\Delta s^* - \Delta s}{\Delta s} \\ &= \varepsilon_x \cos^2\theta + \varepsilon_y \sin^2\theta + \gamma_{xy} \sin\theta\cos\theta \end{aligned} \tag{3.46}$$

To obtain the shear strain $\gamma_{nt}(P)$ in the plane case, we take a second direction t in the surface perpendicular to n. This direction is illustrated in Fig. 3.37(a). Along n and t we choose segments of lengths Δs_1 and Δs_2, respectively. These segments, labeled PQ and PR in Fig. 3.37(a), determine two rectangles, as shown. After the body deforms, these adjacent rectangles become the adjacent parallelograms of Fig. 3.37(b). (They remain adjacent because the body does not tear or fold.) Now, from Fig. 3.37(b),

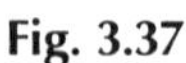
Fig. 3.37

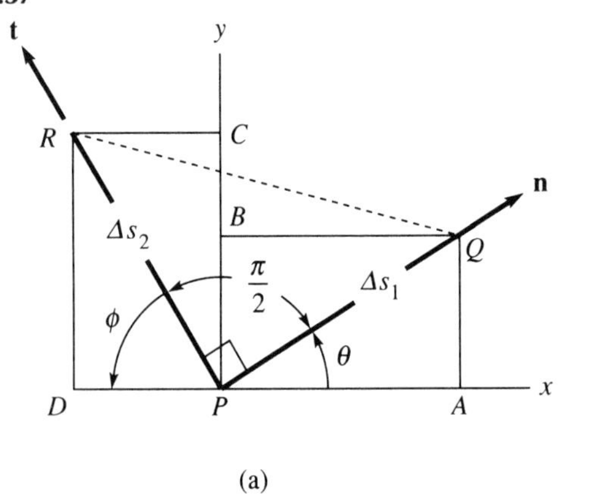

(a)

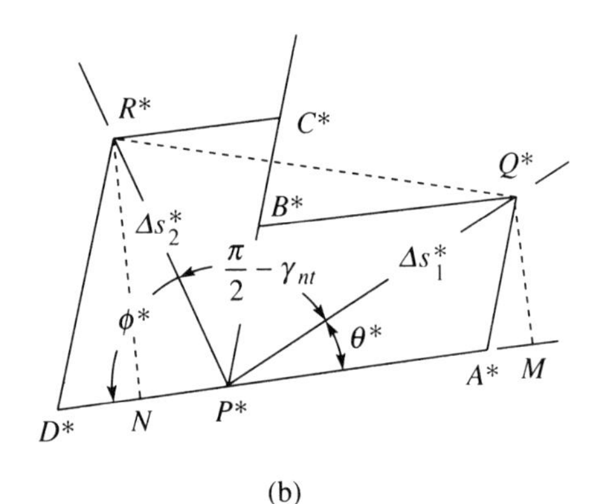

(b)

$$\theta^* + \phi^* = \frac{\pi}{2} + \gamma_{nt}(P)$$

or

$$\gamma_{nt}(P) = (\theta^* + \phi^*) - \frac{\pi}{2}$$

Since γ_{nt} is small, we can write

$$\begin{aligned}\gamma_{nt}(P) \doteq \sin\gamma_{nt}(P) &= \sin\left[(\theta^* + \phi^*) - \frac{\pi}{2}\right] \\ &= \cos(\theta^* + \phi^*) = \cos\theta^*\cos\phi^* + \sin\theta^*\sin\phi^*. \end{aligned} \tag{3.47}$$

From Fig. 3.37(b), it follows that

$$\sin\theta^* = \frac{MQ^*}{P^*Q^*} = \frac{A^*Q^*\sin\left(\frac{\pi}{2} - \gamma_{xy}\right)}{P^*Q^*} = \frac{(1+\varepsilon_y)\,\Delta s\sin\theta\cos\gamma_{xy}}{(1+\varepsilon_n)\,\Delta s}$$

$$\doteq (\sin\theta)(1 + \varepsilon_y - \varepsilon_n)$$

$$\sin\phi^* = \frac{NR^*}{P^*R^*} = \frac{D^*R^*\sin\left(\frac{\pi}{2} - \gamma_{xy}\right)}{P^*R^*} = \frac{(1+\varepsilon_y)\,\Delta s\cos\theta\cos\gamma_{xy}}{(1+\varepsilon_t)\,\Delta s}$$

$$\doteq (\cos\theta)(1 + \varepsilon_y - \varepsilon_t)$$

$$\cos\theta^* = \frac{P^*A^* + A^*M}{P^*Q^*} = \frac{P^*A^* + A^*Q^*\cos\left(\frac{\pi}{2} - \gamma_{xy}\right)}{P^*Q^*}$$

$$= \frac{(1+\varepsilon_x)\,\Delta s\cos\theta + (1+\varepsilon_y)\,\Delta s\sin\theta\sin\gamma_{xy}}{(1+\varepsilon_n)\,\Delta s}$$

$$\doteq (\cos\theta)(1 + \varepsilon_x - \varepsilon_n) + \gamma_{xy}\sin\theta,$$

$$\cos\phi^* = \frac{P^*D^* - ND^*}{P^*R^*} = \frac{P^*D^* - D^*R^*\cos\left(\frac{\pi}{2} - \gamma_{xy}\right)}{P^*R^*}$$

$$= \frac{(1+\varepsilon_x)\,\Delta s\sin\theta - (1+\varepsilon_y)\,\Delta s\cos\theta\sin\gamma_{xy}}{(1+\varepsilon_t)\,\Delta s}$$

$$\doteq (\sin\theta)(1 + \varepsilon_x - \varepsilon_t) + \gamma_{xy}\cos\theta$$

These expressions have been simplified on the basis that all the strains are small. Substituting these expressions in Eq. (3.47) and discarding the higher-order terms, we obtain

$$\gamma_{nt}(P) = 2(\varepsilon_y - \varepsilon_x)\sin\theta\cos\theta + \gamma_{xy}(\cos^2\theta - \sin^2\theta) \tag{3.48}$$

Equations (3.46) and (3.48) relate the strains $\varepsilon_n, \gamma_{nt}$ associated with any two perpendicular directions in the surface to the strains associated with the x and y lines—that is, $\varepsilon_x, \varepsilon_y$, and γ_{xy}.

It is often convenient to simplify Eqs. (3.46) and (3.48) by introducing the double-angle formulas for the sine and cosine [see Eqs. 1.9(a) and 1.9(b)]. When this is done, we obtain the following as alternative expressions for Eqs. (3.46) and (3.48):

$$\varepsilon_n(P) = \tfrac{1}{2}(\varepsilon_x + \varepsilon_y) + \tfrac{1}{2}(\varepsilon_x - \varepsilon_y)\cos 2\theta + \tfrac{1}{2}\gamma_{xy}\sin 2\theta \tag{3.49}$$

$$\gamma_{nt}(P) = -(\varepsilon_x - \varepsilon_y)\sin 2\theta + \gamma_{xy}\cos 2\theta \tag{3.50}$$

Again, we can introduce alternative definitions of the strains: Instead of ε_n and γ_{nt}, let us employ

$$\tilde{\varepsilon}_n \equiv \varepsilon_n + \tfrac{1}{2}\varepsilon_n^2$$

$$\tilde{\gamma}_{nt} = \tfrac{1}{2}(1 + \varepsilon_n)(1 + \varepsilon_t)\sin\gamma_{nt}$$

Also, instead of $\varepsilon_x, \varepsilon_y$, and γ_{xy}, let us employ $\tilde{\varepsilon}_x$, $\tilde{\varepsilon}_y$, and $\tilde{\gamma}_{xy}$, defined in the same way. Then, without any approximation, we obtain

$$\tilde{\varepsilon}_n(P) = \tilde{\varepsilon}_x \cos^2\theta + \tilde{\varepsilon}_y \sin^2\theta + 2\tilde{\gamma}_{xy}\sin\theta\cos\theta \tag{3.51}$$

$$\tilde{\gamma}_{nt}(P) = (\tilde{\varepsilon}_y - \tilde{\varepsilon}_x)\sin\theta\cos\theta + \tilde{\gamma}_{xy}(\cos^2\theta - \sin^2\theta) \tag{3.52}$$

Within a surface, we may seek the direction of maximum or minimum extension strain. The necessary condition follows from Eq. (3.46) or (3.49), namely,

$$\frac{d\varepsilon_n}{d\theta} = -(\varepsilon_x - \varepsilon_y)\sin 2\theta + \gamma_{xy}\cos 2\theta = 0 \tag{3.53a}$$

or, in another form,

$$\tan 2\theta = \frac{\gamma_{xy}}{(\varepsilon_x - \varepsilon_y)} \tag{3.53b}$$

For any given real number the function $\tan 2\theta$ assumes that value in the interval $0 \leq 2\theta \leq \pi$ and repeats the value at intervals of π; roots θ_1 and θ_2 differ by $\pi/2$. Stated otherwise, there are two orthogonal directions at each point of the surface that satisfy Eq. (3.53). One is the direction of maximum extensional strain and the other is the direction of minimum extensional strain. According to Eqs. (3.50) and (3.53a) the shear strain vanishes in these directions. This means that at every point of the surface, we can find two orthogonal lines that *remain* orthogonal after the deformation. This circumstance is depicted in Fig. 3.38. Here, a small square at a point P is transported and deformed to the rectangle at P^*. Lines 1 and 2 are the directions of maximum and minimum extensional strain.

Fig. 3.38

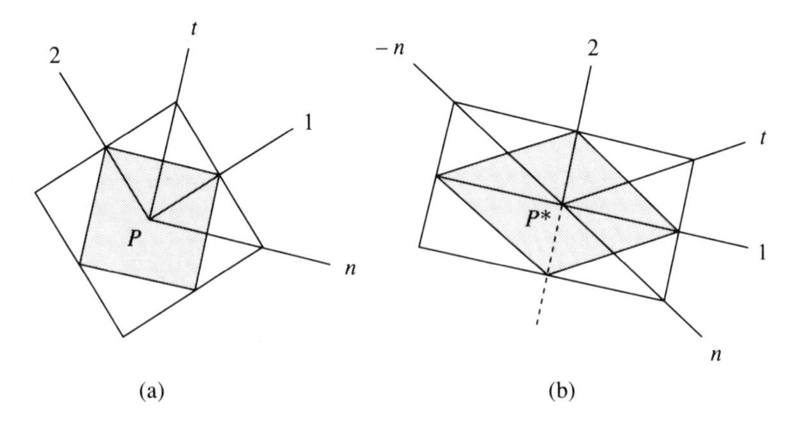

The directions of maximum (or minimum) shear strain follow from Eq. (3.50) and the condition

$$\frac{d\gamma_{nt}}{d\theta} = -2(\varepsilon_x - \varepsilon_y)\cos 2\theta - 2\gamma_{xy}\sin 2\theta = 0 \tag{3.54a}$$

or

$$\tan 2\theta = -\frac{(\varepsilon_x - \varepsilon_y)}{\gamma_{xy}} \tag{3.54b}$$

It follows from Eqs. (3.53b) and (3.54b) that the directions (n and t) of maximum (or minimum) shear strain bisect the directions of maximum (or minimum) extensional strain. This is depicted in Fig. 3.38. If ε_{max} and ε_{min} denote these extremal values of extensional strain, then the maximum shear strain follows:

$$\gamma_{max} = |\varepsilon_{max} - \varepsilon_{min}| \tag{3.55}$$

Note that these directions in the surface may *not* be the principal directions. They are the principal directions if and only if the normal direction z is a principal direction. Note, too, that the foregoing equations apply only to small strain, with the exception of Eqs. (3.51) and (3.52), which embody alternative definitions of strain.

EXAMPLE 10

Suppose that the strains in orthogonal (x, y) directions at a point on a surface are as follows:

$$\varepsilon_x = 700\mu, \qquad \gamma_{xy} = +600\mu$$

$$\varepsilon_y = -100\mu$$

when $\mu = 10^{-6}$. According to Eq. (3.50) the directions of zero shear strain follow:

$$\tan 2\theta = \frac{\gamma_{xy}}{\varepsilon_x - \varepsilon_y} = \frac{600}{800} = +\frac{3}{4}$$

$$\sin 2\theta = \pm\frac{3}{5}, \qquad \cos 2\theta = \pm\frac{4}{5}$$

$$\theta = 18.4^\circ,\ 108.4^\circ$$

According to Eq. (3.53), these are the directions of extremal extensional strains. The latter are given by Eq. (3.49):

$$\varepsilon(18.4^\circ) = \tfrac{1}{2}(600) + \tfrac{1}{2}(800)\tfrac{4}{5} + \tfrac{1}{2}(600)\tfrac{3}{5} = 800\mu$$

$$\varepsilon(108.4^\circ) = \tfrac{1}{2}(600) - \tfrac{1}{2}(800)\tfrac{4}{5} - \tfrac{1}{2}(600)\tfrac{3}{5} = -200\mu$$

In accordance with Eq. (3.54) the extremal values of the shear bisect the directions of the maximum and minimum extensional strain; that is,

$$\tan 2\theta = -\tfrac{4}{3}$$

$$\theta = 63.4^\circ,\ 153.4^\circ$$

The maximum shear is given by Eq. (3.50) [the absolute value, by Eq. (3.55)]:

$$\gamma_{\max} = -(800)(-\tfrac{4}{5}) + 600(\tfrac{3}{5}) = 1000\mu$$

PROBLEMS

3.110 Carry out the details of establishing Eqs. (3.49) and (3.50) from Eqs. (3.46) and (3.48).

3.111 At point P in a surface, two rectangular coordinate systems are constructed as shown in Fig. P3.111. All strains given here have been multiplied by 10^6, indicated by the symbol μ, which stands for micro-, or 10^{-6}. Thus a strain of $\varepsilon = 300\mu$ is actually $\varepsilon = 300 \times 10^{-6} = 0.0003$.

a. $\varepsilon_x = 200\mu$, $\varepsilon_y = 400\mu$, $\gamma_{xy} = 800\mu$, $\theta = 30°$; find ε_n, ε_t, and γ_{nt}.

b. $\varepsilon_x = -300\mu$, $\varepsilon_y = 0$, $\gamma_{xy} = 600\mu$, $\theta = 15°$; find ε_n, ε_t, and γ_{nt}.

c. $\varepsilon_x = 0$, $\varepsilon_y = 0$, $\gamma_{xy} = 1200\mu$, $\theta = -45°$; find ε_n, ε_t, and γ_{nt}.

d. $\varepsilon_x = 400\mu$, $\varepsilon_y = 400\mu$, $\gamma_{xy} = 0$, $\theta = 45°$; find ε_n, ε_t, and γ_{nt}.

e. $\varepsilon_n = -1200\mu$, $\varepsilon_t = -600\mu$, $\gamma_{nt} = -600\mu$, $\theta = -60°$; find ε_x, ε_y, and γ_{xy}.

f. $\varepsilon_x = 400\mu$, $\varepsilon_y = 100\mu$, $\varepsilon_n = 800\mu$, $\theta = 45°$; find γ_{nt}.

g. $\varepsilon_x = 1000\mu$, $\varepsilon_y = 1200\mu$, $\varepsilon_n = 1100\mu$, $\gamma_{xy} = 200\mu$; find θ.

Fig. P3.111

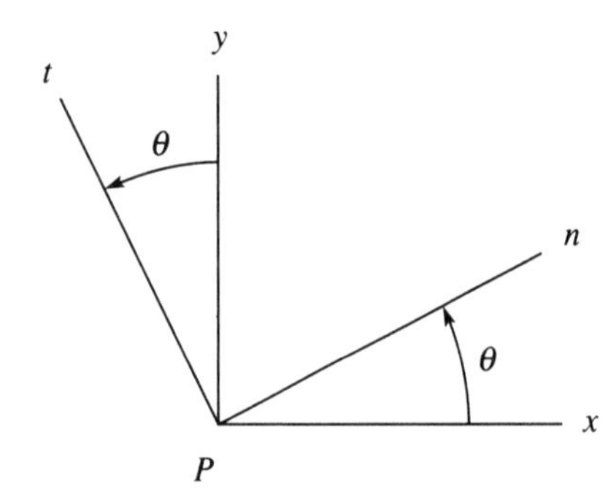

h. $\varepsilon_x = 400\mu$, $\varepsilon_y = 800\mu$, $\gamma_{xy} = -400\mu$, $\gamma_{nt} = 0$; find ε_n and ε_t.

i. $\varepsilon_x = 700\mu$, $\varepsilon_y = 300\mu$, $\varepsilon_n = 900\mu$, $\gamma_{xy} = 600\mu$; find θ

j. $\varepsilon_n = 600\mu$, $\varepsilon_t = -200\mu$, $\gamma_{xy} = 1000\mu$, $\gamma_{nt} = 1200\mu$; find ε_x, ε_y, and θ.

3.112 Let **n** and **t** be perpendicular vectors and let **p** = −**n** and **q** = −**t**. Use Eqs. (3.49) and (3.50) to find how $\varepsilon_p(P)$, $\varepsilon_q(P)$, $\gamma_{pt}(P)$, $\gamma_{nq}(P)$, and $\gamma_{pq}(P)$ are related to $\varepsilon_n(P)$, $\varepsilon_t(P)$, and $\gamma_{nt}(P)$.

3.113 A point P has rectangular coordinates (x, y) and polar coordinates (r, θ). If $x = 4$, $y = 3$, $\varepsilon_x(P) = 0.001$, $\varepsilon_y(P) = 0.002$, and $\gamma_{xy}(P) = 0$, determine $\varepsilon_r(P)$, $\varepsilon_\theta(P)$, and $\gamma_{r\theta}(P)$.

3.114 If x and y denote two perpendicular directions and n and t denote any other set of perpendicular directions in the same plane, show that, at a given point,

a. $\varepsilon_n + \varepsilon_t = \varepsilon_x + \varepsilon_y$

b. $\varepsilon_n\varepsilon_t - \left(\dfrac{\gamma_{nt}}{2}\right)^2 = \varepsilon_x\varepsilon_y - \left(\dfrac{\gamma_{xy}}{2}\right)^2$

3.115 What value of θ in Eq. (3.49) gives a maximum or minimum value for $\varepsilon_n(P)$?

3.116 What value of θ in Eq. (3.50) gives a maximum or minimum value for $\gamma_{nt}(P)$?

3.117 Let **n**, **t**, and **s** be three coplanar vectors separated 120° from each other, as shown in Fig. P3.117. If x and y are perpendicular directions (in the same plane), show that at a point ε_x, ε_y, and γ_{xy} can be determined from a knowledge of the three extensional strains $\varepsilon_n, \varepsilon_t, \varepsilon_s$ (and θ).

Fig. P3.117

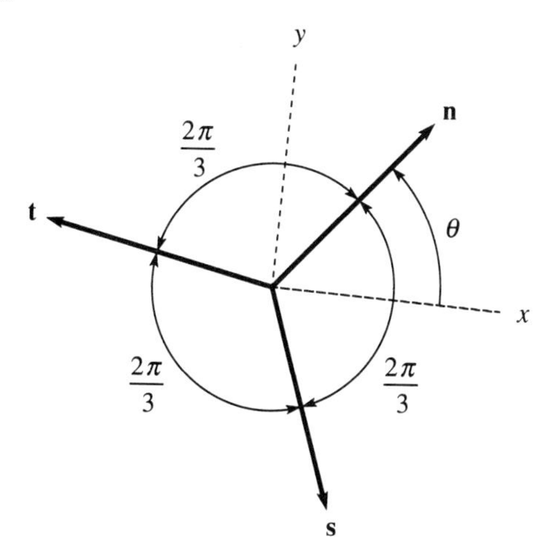

3.118 Let $\theta = 0$ in Fig. P3.117 and derive the formula that expresses (a) ε_y in terms of ε_n, ε_t, and ε_s and (b) γ_{xy} in terms of ε_n, ε_t and ε_s.

3.119 Determine the principal strains and locate the associated principal directions with respect to an xyz rectangular coordinate system for each of the following strain states. The z-axis is known to be a principal direction in all cases, and μ stands for 10^{-6}.

a. $\varepsilon_x = 200\mu$, $\varepsilon_y = -400\mu$, $\gamma_{xy} = 800\mu$

b. $\varepsilon_x = -300\mu$, $\varepsilon_y = 0$, $\gamma_{xy} = 600\mu$

c. $\varepsilon_x = 0$, $\varepsilon_y = 0$, $\gamma_{xy} = 1200\mu$

d. $\varepsilon_x = 400\mu$, $\varepsilon_y = 400\mu$, $\gamma_{xy} = 0$

e. $\varepsilon_x = -1200\mu$, $\varepsilon_y = -600\mu$, $\gamma_{xy} = -600\mu$

3.120 If z is a principal direction and if $\varepsilon_x = \varepsilon_y$ and $\gamma_{xy} = 0$ at a point, show that every direction in the xy-plane is a principal direction.

3.121 In Fig. P3.121, x, y, and n are in a plane perpendicular to a principal direction. Compute the principal strains and sketch the location of the principal directions if $\varepsilon_x = -0.0008$, $\varepsilon_y = 0.0004$, and $\varepsilon_n = 0.0006$. Also compute γ_{xy}.

Fig. P3.121

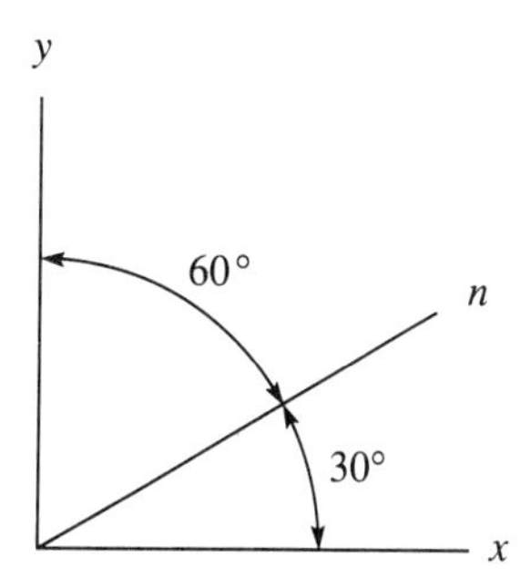

3.122 In Fig. P3.122, x, y, and n are in a plane perpendicular to a principal direction. Compute the principal strains and sketch the location of the principal directions if $\varepsilon_x = 0.0002$, $\varepsilon_y = -0.0002$, and $\varepsilon_n = 0$. Also compute γ_{xy}.

Fig. P3.122

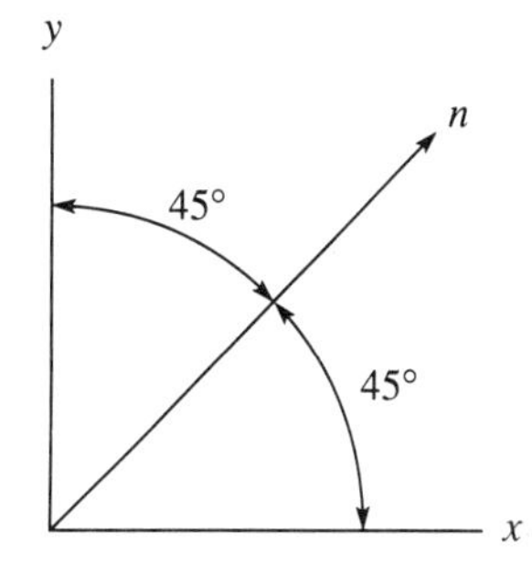

3.123 In Fig. P3.123, x, y, n, and t are in a plane perpendicular to a principal direction. Compute the principal strains and sketch the location of the principal directions if $\varepsilon_x = 0$, $\varepsilon_n = 0.0004$, $\varepsilon_t = -0.0004$. Also compute ε_y and γ_{xy}.

Fig. P3.123

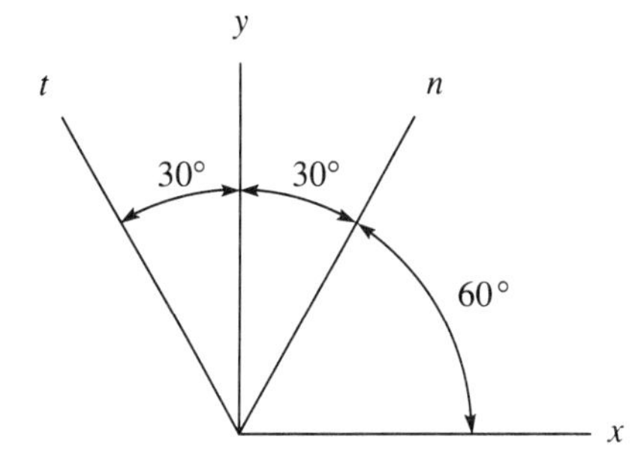

3.124 A thin strip is deformed so that the longitudinal extension strain (along the strip) is 0.0008 and the lateral extension strain is -0.00024. The shearing strain associated with these directions is zero. Find the orthogonal directions (in the plane of the strip) along which the extensional strains are equal. What is the shear strain associated with these directions?

3.125 The xy-plane is perpendicular to a principal direction. If $\varepsilon_y = 0.0003$, $\gamma_{xy} = -0.0008$, and the minimum extensional strain in the xy-plane is -0.0001, compute the maximum extensional strain in the xy-plane.

3.126 For the states of strain given in Prob. 3.119, determine the maximum shear strain and locate the associated directions in the xy-plane.

3.127 Strain components in a rectangular system (x, y, z) are

$$\varepsilon_x = 600\mu, \quad \gamma_{xy} = 600\mu, \quad \gamma_{xz} = 0$$
$$\varepsilon_y = -200\mu, \quad \gamma_{yz} = 0, \quad \varepsilon_z = 900\mu$$

Locate the principal directions relative to the rectangular system (x, y, z) and determine the maximum extensional strain, the maximum shear strain, and the directions of the maximum shear.

3.128 The extensional strains in the three directions (x, n, y) in a surface, as shown in Fig. P3.128, are $\varepsilon_x = 600\mu$, $\varepsilon_n = 1000\mu$, and $\varepsilon_y = -200\mu$. Compute the shear strain γ_{xy}.

Fig. P3.128

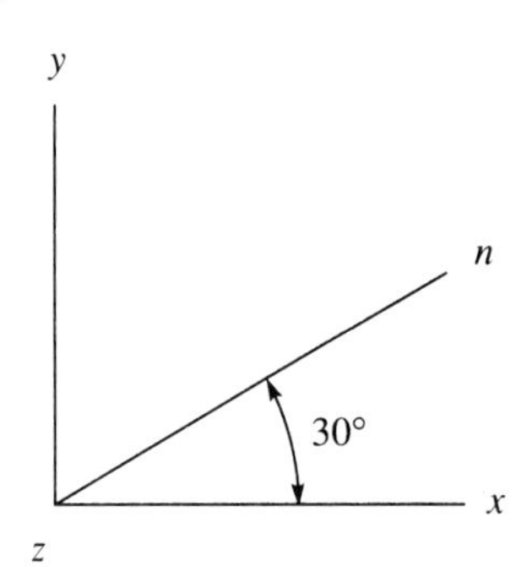

3.129 Use the data of Prob. 3.128. Additionally, z is a principal direction and $\varepsilon_z = 0$. Determine the other principal directions, the other principal strains, and the maximum shear strain.

3.130 Extensional strains are known for three directions, as shown in Fig. P3.130 on the smooth surface of a body:

$$\varepsilon_x = 20 \times 10^{-4}, \quad \varepsilon_n = 6 \times 10^{-4}, \quad \varepsilon_t = 18 \times 10^{-4}$$

Calculate (a) the shear strain γ_{nt} and (b) the extensional strain ε_y.

Fig. P3.130

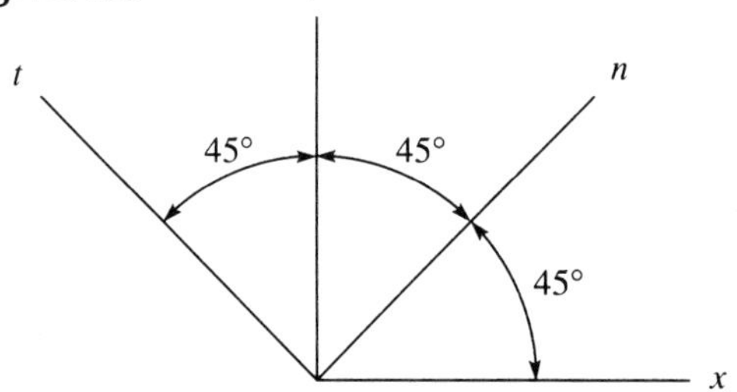

3.19 Mohr's Graphical Representation

According to the preceding equations, the components of strain $(\varepsilon_n, \gamma_{nt})$ for any orthogonal directions in a surface are given by Eqs. (3.49) and (3.50) [or, equivalently, (3.46) and (3.48)]. Also, according to the preceding arguments, there exist always two orthogonal directions (1 and 2) in which the shear strain vanishes. If the angle θ is measured from the direction 1, then $\gamma_{12} = 0$ and Eqs. (3.49) and (3.50) assume the simpler forms

$$\varepsilon_n = +\frac{\varepsilon_1 + \varepsilon_2}{2} + \frac{\varepsilon_1 - \varepsilon_2}{2}\cos 2\theta \tag{3.56}$$

$$-\tfrac{1}{2}\gamma_{nt} = \frac{\varepsilon_1 - \varepsilon_2}{2}\sin 2\theta \tag{3.57}$$

Let us assume that the extremal values, ε_1 and ε_2, are known; for brevity, let

$$a \equiv \frac{\varepsilon_1 + \varepsilon_2}{2}, \qquad b = \frac{\varepsilon_1 - \varepsilon_2}{2} \tag{3.58a,b}$$

Then Eqs. (3.56) and (3.57) take the abbreviated forms

$$\varepsilon_n - a = b\cos 2\theta \tag{3.59}$$

$$-\tfrac{1}{2}\gamma_{nt} = b\sin 2\theta \tag{3.60}$$

These are the parametric equation of a circle. Directions, 1, 2, n, and t are shown in Fig. 3.39(a), and the plot of $-\gamma_{nt}/2$ versus ε_n is shown in Fig. 3.39(b). The labeled points correspond to the states 1, 2, n, and t. Observe that the shear strains corresponding to the points n and t have opposite signs. This is a consequence of the conventions: At any point, say n, the value is the shear strain between that line and the perpendicular line in the counterclockwise sense (n rotates toward t). Thus, at t the value is the shear strain between lines t and $-n$. If during deformation the angle (shear strain) between n and t decreases (positive shear), then the angle (shear strain) between t and $-n$ increases (negative shear) in the same amount. (See Fig. 3.39).

Fig. 3.39

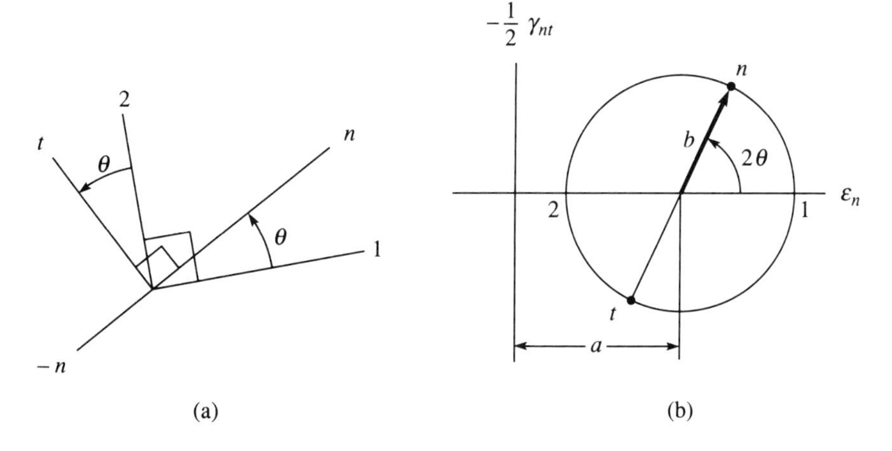

PROBLEMS

3.131 States of strain in a surface are given by three components $(\varepsilon_x, \gamma_{xy}, \varepsilon_y)$ in an x, y rectangular system. Here the normal direction (z) is a principal direction.

Sketch Mohr's circle for each of the following states of strain. From this sketch, find the principal strains and locate the directions of the maximum and minimum extensional strains in the x, y rectangular coordinate system. μ signifies 10^{-6}.

a. $\varepsilon_x = 100\mu$, $\varepsilon_y = 400\mu$, $\gamma_{xy} = 200\mu$

b. $\varepsilon_x = -1200\mu$, $\varepsilon_y = 200\mu$, $\gamma_{xy} = -1600\mu$

c. $\varepsilon_x = 0$, $\varepsilon_y = 0$, $\gamma_{xy} = 2000\mu$

d. $\varepsilon_x = 0$, $\varepsilon_y = 2000\mu$, $\gamma_{xy} = 0$

e. $\varepsilon_x = 800\mu$, $\varepsilon_y = 800\mu$, $\gamma_{xy} = 1600\mu$

f. $\varepsilon_x = 800\mu$, $\varepsilon_y = 800\mu$, $\gamma_{xy} = 0$

3.132 The nt-coordinate system is related to the xy-coordinate system as shown in Fig. P3.132. If $\gamma_{xy} = 0.0004$, $\gamma_{nt} = 0$, and $\varepsilon_y = 0.0001$, draw Mohr's circle.

Fig. P3.132

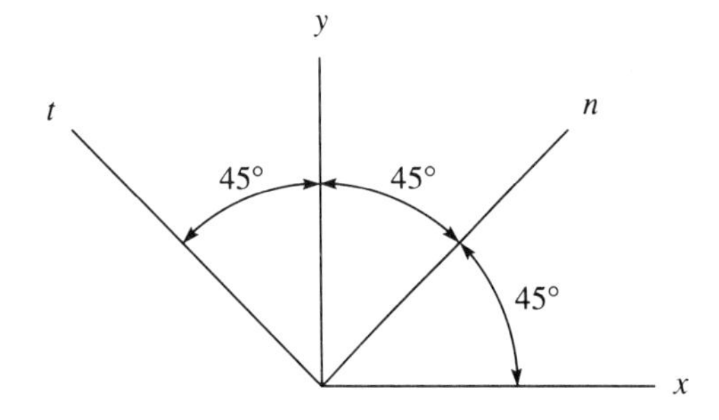

3.133 Draw Mohr's circle for Prob. 3.121.

3.134 Draw Mohr's circle for Prob. 3.122.

3.135 Draw Mohr's circle for Prob. 3.123.

3.136 Starting with the rules for constructing Mohr's circle, establish Eqs. (3.49) and (3.50).

3.137 The principal strains at a point are ε_1, ε_2, and ε_3. Assume $\varepsilon_1 > \varepsilon_2 > \varepsilon_3$ and sketch, on the same set of axes, the three Mohr's circles corresponding to the three planes perpendicular to the principal directions.

3.138 Draw three Mohr's circles for the data of Prob. 3.131(d)—that is, one for each of the directions (x-y), (x-z), and (y-z)—if $\varepsilon_z = -600\mu$.

3.139 Draw three Mohr's circles for the data of Prob. 3.131(f)—that is, one for each of the directions (x-y), (x-z), and (y-z)—if $\varepsilon_z = -200\mu$.

3.140 Show that, if the extensional strains at a point in any three coplanar but nonparallel directions are small, the extensional strain in any other direction in the plane and also the shear strain between any pair of perpendicular directions in the plane must be small.

3.141 Show that, at a point in a surface, two shear strains and one extensional strain in the surface are sufficient to determine the state of strain in the surface at that point.

3.142 Show that three—or, indeed, any number of—shear strains in a surface cannot be used to determine the state of strain in the surface.

3.20 Measurement of Strain

Average extensional strains in the surface of a body can be determined in different ways. A widely used method is to bond a thin wire to the surface of the body. When the body deforms, the wire deforms with it; the electrical resistance of the wire then increases or decreases as the wire elongates or contracts. This change in resistance is proportional to the change in length of the wire. Consequently the average extensional strain can be measured by the changes in resistance. When the length of the wire is very short, the average extensional strain of the wire will approach the pointwise extensional strain. These measurements, in conjunction with Eq. (3.49), can be used to determine the state of strain in the surface of a body.

Suppose that the point P in Fig. 3.40 is in the surface of a body and that x and y are any two perpendicular directions in the surface. If the extensional strain ε_a has been measured, it follows from Eq. (3.46) that

$$\varepsilon_a = \varepsilon_x \cos^2 \theta_a + \varepsilon_y \sin^2 \theta_a + \gamma_{xy} \sin \theta_a \cos \theta_a$$

Fig. 3.40

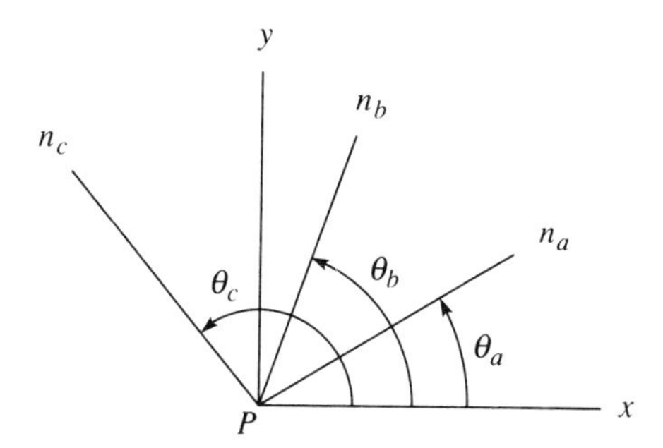

This is one equation in three unknowns, $\varepsilon_x, \varepsilon_y$, and γ_{xy}. Measurements of the extensional strains, ε_b and ε_c, in two different directions, b and c, provide two similar equations:

$$\varepsilon_b = \varepsilon_x \cos^2 \theta_b + \varepsilon_y \sin^2 \theta_b + \gamma_{xy} \sin \theta_b \cos \theta_b$$

$$\varepsilon_c = \varepsilon_x \cos^2 \theta_c + \varepsilon_y \sin^2 \theta_c + \gamma_{xy} \sin \theta_c \cos \theta_c$$

The three equations provide a unique solution for the unknowns $\varepsilon_x, \varepsilon_y, \gamma_{xy}$; it is essential that the directions, a, b, and c, are distinct. Of course, we can use Eq. (3.49) instead of (3.46). Much simplification results if measurements are made in orthogonal directions, x and y, and a third direction a. Then the solution of one equation determines the strain γ_{xy}. In any case, three measurements are enough to determine completely the state of strain *in the surface*, but, of course, no information about strains in the normal direction z is available.

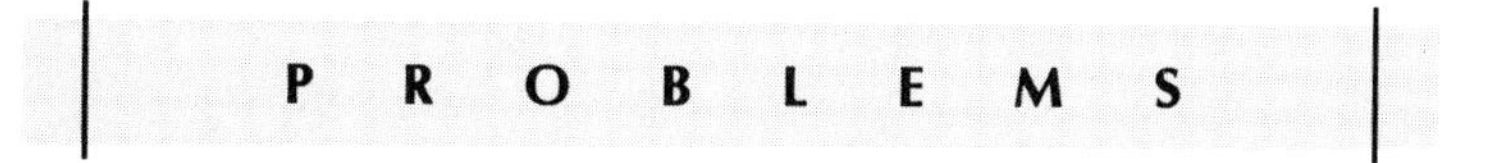

PROBLEMS

3.143 Obtain $\varepsilon_x, \varepsilon_y$, and γ_{xy} from the measurements given here, which are taken with the 45° strain rosette of Fig. P3.143. The symbol μ signifies 10^{-6}.

a. $\varepsilon_1 = 200\mu, \varepsilon_2 = 600\mu, \varepsilon_3 = -100\mu$

b. $\varepsilon_1 = -1200\mu, \varepsilon_2 = 400\mu, \varepsilon_3 = 0$

c. $\varepsilon_1 = 2000\mu$, $\varepsilon_2 = -1200\mu$, $\varepsilon_3 = -800\mu$

d. $\varepsilon_1 = 100\mu$, $\varepsilon_2 = -400\mu$, $\varepsilon_3 = 100\mu$

e. $\varepsilon_1 = 1000\mu$, $\varepsilon_2 = 0$, $\varepsilon_3 = -400\mu$

Fig. P3.143

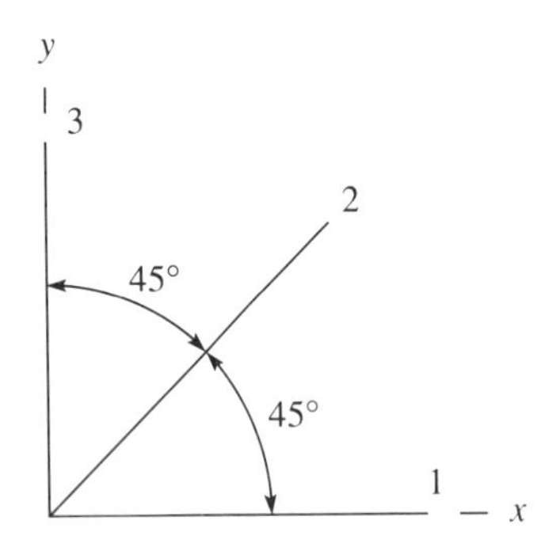

3.144 Obtain $\varepsilon_x, \varepsilon_y$, and γ_{xy} from the measurements given here, which are taken with the 60° strain rosette shown in Fig. P3.144. The symbol μ signifies 10^{-6}.

a. $\varepsilon_1 = 100\mu$, $\varepsilon_2 = 100\mu$, $\varepsilon_3 = 100\mu$

b. $\varepsilon_1 = 1000\mu$, $\varepsilon_2 = -800\mu$, $\varepsilon_3 = 200\mu$

c. $\varepsilon_1 = -600\mu$, $\varepsilon_2 = 0$, $\varepsilon_3 = 600\mu$

d. $\varepsilon_1 = 0$, $\varepsilon_2 = 400\mu$, $\varepsilon_3 = 400\mu$

e. $\varepsilon_1 = -1000\mu$, $\varepsilon_2 = -2000\mu$, $\varepsilon_3 = -3000\mu$

Fig. P3.144

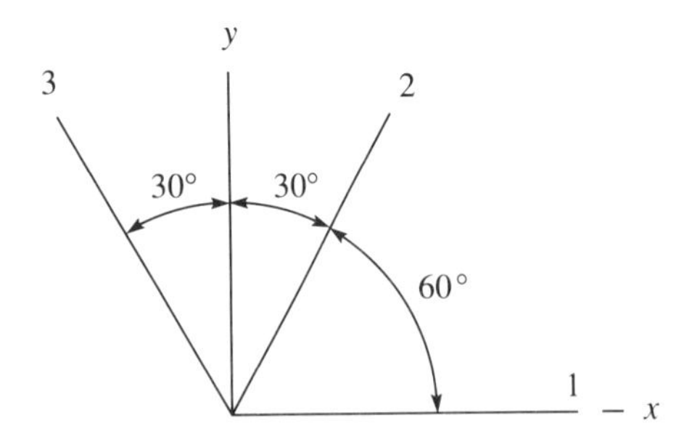

3.145 Strain measurements in the *a*-, *b*-, and *c*-directions shown in Fig. P3.145 are ε_a, ε_b, and ε_c. If *t* is in the direction of the maximum extensional strain and *s* is in the direction of the minimum extensional strain (*a*, *b*, *c*, *t*, and *s* in the same plane) determine ε_t, ε_s, and ϕ.

Fig. P3.145

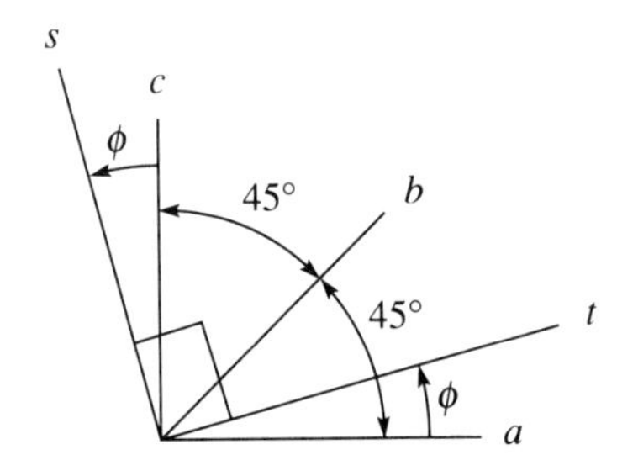

3.146 Repeat Prob. 3.145 for the 60° rosette of Fig. P3.146.

Fig. P3.146

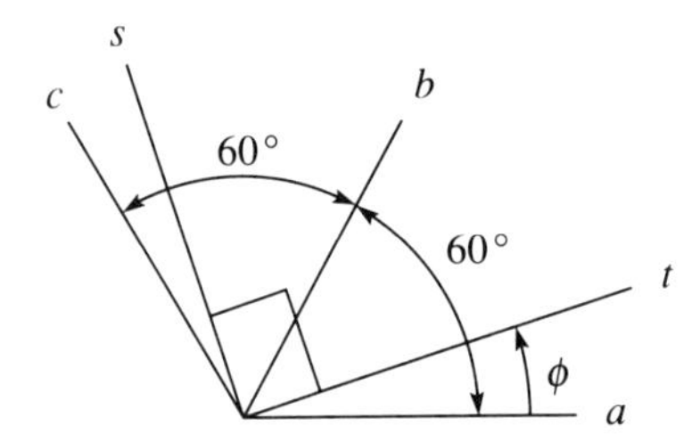

3.147 Repeat Prob. 3.145 for a rosette with the general configuration of Fig. P3.147. Angles α, β, and γ are arbitrary but unequal; that is, directions *a*, *b*, and *c* are not parallel.

Fig. P3.147

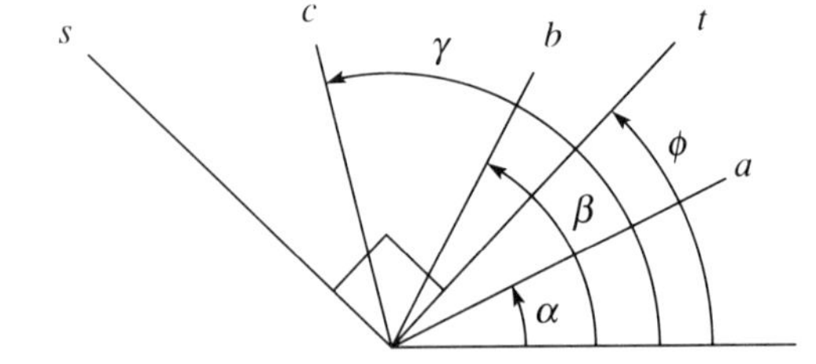

*3.21 Strain-Displacement Relations

As noted previously, the deformed configuration of a body is fully determined if the position (vector **R**) is known for every particle, or, stated otherwise, if the displacement (vector **V**) is known. The displacement **V** carries the particle from its original position (vector **r**) to its current position (vector **R**), as depicted in Fig. 3.41, wherein particle P is transported to P^* by the displacement **V**; mathematically,

$$\mathbf{R} = \mathbf{r} + \mathbf{V} \tag{3.61}$$

Fig. 3.41

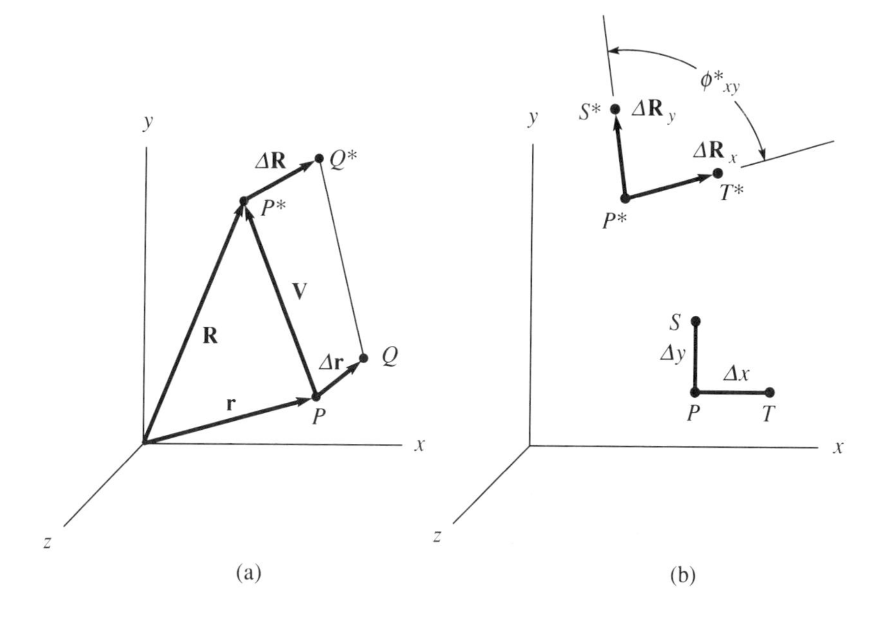

The displacement, per se, need not cause deformation and strain, because the displacement might constitute only a rigid-body motion. For example, if the displacement of all particles were the same, then the body would be rigidly translated; then, in Fig. 3.41, the displacement of Q to Q^* would be same as the displacement of P to P^* and line P^*Q^* would have the same length as PQ. It is only the *relative* displacement of neighboring particles that can produce strains; we must compare the lengths of vectors $\Delta\mathbf{R}$ with $\Delta\mathbf{r}$. (These define the *relative* positions.)

To determine the extensional strain ε_x, we consider a segment PT, initially parallel to the x-axis at P, as shown in Fig. 3.41(b). The segment has initial length Δx and the initial position of T relative to P is

$$\Delta\mathbf{r} = \Delta x\hat{\mathbf{i}}$$

Following a motion, P is displaced to P^* and T to T^*, and the position of T^* relative to P^* is (approximately)

$$\Delta \mathbf{R} \doteq \frac{\partial \mathbf{R}}{\partial x} \Delta x$$

If ΔS_x denotes the length of P^*T^*, then (approximately)

$$\Delta S_x^2 \doteq \Delta \mathbf{R}_x \cdot \Delta \mathbf{R}_x$$

$$\doteq \frac{\partial \mathbf{R}}{\partial x} \cdot \frac{\partial \mathbf{R}}{\partial x} \Delta x^2$$

The extensional strain is obtained in accordance with definition (3.1):

$$\varepsilon_x \equiv \lim_{\Delta x \to 0} \frac{\Delta S_x - \Delta x}{\Delta x}$$

$$= \sqrt{\frac{\partial \mathbf{R}}{\partial x} \cdot \frac{\partial \mathbf{R}}{\partial x}} - 1 \tag{3.62}$$

Alternatively, with definition (3.8)

$$\tilde{\varepsilon}_x \equiv \lim_{\Delta x \to 0} \frac{1}{2}\left[\frac{\partial \mathbf{R}}{\partial x} \cdot \frac{\partial \mathbf{R}}{\partial x} - 1\right] \tag{3.63}$$

From Eq. (3.61) we have

$$\frac{\partial \mathbf{R}}{\partial x} = \frac{\partial}{\partial x}(\mathbf{r} + \mathbf{v}) = \hat{\mathbf{i}} + \frac{\partial \mathbf{V}}{\partial x} \tag{3.64a,b}$$

The latter can be substituted into Eqs. (3.62) or (3.63) to obtain the desired strain in terms of the displacement. The strain $\tilde{\varepsilon}_x$ of Eq. (3.63) provides a simpler result, which follows:

$$\tilde{\varepsilon}_x = \frac{\partial \mathbf{V}}{\partial x} \cdot \hat{\mathbf{i}} + \frac{1}{2} \frac{\partial \mathbf{V}}{\partial x} \cdot \frac{\partial \mathbf{V}}{\partial x} \tag{3.65}$$

Previously, we showed [see Eq. (3.10)] that the difference between the strains of Eqs. (3.62) and (3.63), between ε_x and $\tilde{\varepsilon}_x$, is inconsequential *if* strains are *small.* Then, Eq. (3.65) also provides a satisfactory approximation of the former (ε_x).

To determine the shear strain γ_{xy}, we must consider two lines, PT and PS, which are *initially* parallel to the x- and y-axes, respectively, as shown in Fig. 3.41(b). Following a motion, P is displaced to P^*; T and S are displaced to T^* and S^*. The positions of these particles, *relative* to P, are (approximately)

$$\Delta \mathbf{R}_x \doteq \frac{\partial \mathbf{R}}{\partial x} \Delta x, \qquad \Delta \mathbf{R}_y = \frac{\partial \mathbf{R}}{\partial y} \Delta y$$

The shear strain γ_{xy} is given by definition (3.6); in the limit, $\Delta x \to 0$, $\Delta y \to 0$,

$$\gamma_{nt} = \frac{\pi}{2} - \phi_{xy}^*$$

If ΔS_x and ΔS_y denote the lengths of $\Delta \mathbf{R}_x$ and $\Delta \mathbf{R}_y$, then

$$\cos \phi_{xy}^* = \frac{\Delta \mathbf{R}_x \cdot \Delta \mathbf{R}_y}{\Delta S_x \Delta S_y} = \sin \gamma_{xy}$$

But $\Delta S_x = (1 + \varepsilon_x)\Delta x$, $\Delta S_y = (1 + \varepsilon_y)\Delta y$. Therefore,

$$\sin \gamma_{xy} = \frac{\dfrac{\Delta \mathbf{R}_x}{\Delta x} \cdot \dfrac{\Delta \mathbf{R}_y}{\Delta y}}{(1 + \varepsilon_x)(1 + \varepsilon_y)}$$

In the limit, $\Delta x \to 0$, $\Delta y \to 0$,

$$\sin \gamma_{xy} = \frac{\dfrac{\partial \mathbf{R}}{\partial x} \cdot \dfrac{\partial \mathbf{R}}{\partial y}}{(1 + \varepsilon_x)(1 + \varepsilon_y)}$$

By definition (3.29a),

$$\tilde{\gamma}_{xy} = \frac{1}{2} \frac{\partial \mathbf{R}}{\partial x} \cdot \frac{\partial \mathbf{R}}{\partial y} \tag{3.66}$$

Just as $\partial \mathbf{R}/\partial x$ is given by Eq. (3.64b),

$$\frac{\partial \mathbf{R}}{\partial y} = \hat{\mathbf{j}} + \frac{\partial \mathbf{V}}{\partial y} \tag{3.67}$$

Then, by substituting Eqs. (3.64) and (3.67) into Eq. (3.66), we obtain

$$\tilde{\gamma}_{xy} = \frac{1}{2}\left(\frac{\partial \mathbf{V}}{\partial x} \cdot \tilde{\mathbf{j}} + \frac{\partial \mathbf{V}}{\partial y} \cdot \tilde{\mathbf{i}}\right) + \frac{1}{2} \frac{\partial \mathbf{V}}{\partial x} \cdot \frac{\partial \mathbf{V}}{\partial y} \tag{3.68}$$

Previously we expressed the strain $\tilde{\gamma}_{xy}$ in terms of the strain γ_{xy} [see Eq. (3.29b)]. When the strains are small, $\gamma_{xy} \doteq 2\tilde{\gamma}_{xy}$. Then, Eq. (3.68) also provides a satisfactory approximation of the former ($\gamma_{xy} = 2\tilde{\gamma}_{xy}$).

The expression for $\tilde{\varepsilon}_y$ or $\tilde{\varepsilon}_z$ follows immediately from Eq. (3.65); we need only replace x by y or z. The expression for $\tilde{\gamma}_{yz}$ (or $\tilde{\gamma}_{xz}$) follows immediately from Eq. (3.68); we need only replace x by y and y by z (or y by z).

The expressions for strains in terms of displacement can be elaborated in terms of components. The displacement can be expressed in terms of rectangular components:

$$\mathbf{V} = u\hat{\mathbf{i}} + v\hat{\mathbf{j}} + w\hat{\mathbf{k}}$$

The corresponding versions of Eqs. (3.65) and (3.68) follow:

$$\tilde{\varepsilon}_x = \frac{\partial u}{\partial x} + \frac{1}{2}\left(\frac{\partial u}{\partial x}\right)^2 + \frac{1}{2}\left(\frac{\partial v}{\partial x}\right)^2 + \frac{1}{2}\left(\frac{\partial w}{\partial x}\right)^2 \tag{3.69}$$

$$\tilde{\gamma}_{xy} = \frac{1}{2}\left(\frac{\partial v}{\partial x} + \frac{\partial u}{\partial y}\right) + \frac{1}{2}\frac{\partial u}{\partial x}\frac{\partial u}{\partial y} + \frac{1}{2}\left(\frac{\partial v}{\partial x}\frac{\partial v}{\partial y}\right) + \frac{1}{2}\left(\frac{\partial w}{\partial x}\frac{\partial w}{\partial y}\right) \tag{3.70}$$

Again, if the strains are small, then $\varepsilon_x \doteq \tilde{\varepsilon}_x$ and $\gamma_{xy} \doteq 2\tilde{\gamma}_{xy}$. The strains are usually small in the members of structures and machines. Then these relations are valid irrespective of the magnitudes of displacements and rotations. If the rotations as well as the strains are everywhere small, we can delete the nonlinear terms of Eqs. (3.65), (3.68), (3.69), and (3.70)—that is, the products. The complete set of the linear approximations follows:

$$\varepsilon_x = \frac{\partial u}{\partial x}, \qquad \varepsilon_y = \frac{\partial v}{\partial y}, \qquad \varepsilon_z = \frac{\partial w}{\partial z}$$

$$\gamma_{xy} = \frac{\partial u}{\partial y} + \frac{\partial v}{\partial x}$$

$$\gamma_{xz} = \frac{\partial u}{\partial z} + \frac{\partial w}{\partial x} \tag{3.71}$$

$$\gamma_{yz} = \frac{\partial v}{\partial z} + \frac{\partial w}{\partial y}$$

Here, $\tilde{\varepsilon}_x = \varepsilon_x$, $2\tilde{\gamma}_{xy} = \gamma_{xy}$, and so on. Note that the foregoing strain-displacement equations are valid *only* for Cartesian coordinates.

PROBLEMS

3.148 According to the foregoing equations, exact equations such as Eqs. (3.69) and (3.70), or the approximations of Eq. (3.71), if the displacements (u, v, w) are linear in the coordinates (x, y, z), then all strains are constant (so-called homogeneous states). Determine the states caused by the following displacements, interpret them, and examine the errors caused by the linear approximations, Eq. (3.71):

a. $u = Kx$, $v = -my$, $w = -mz$, where K and m are positive constants

b. $u = -\alpha y$, $v = \beta x$, $w = 0$, where α and β are constant. What occurs if $\alpha = \beta$? What occurs if $\alpha = -\beta$?

3.149 Use exact equations such as Eq. (3.69) and (3.70) and also the approximations of Eq. (3.71) to determine the states of strain given by the following displacements, interpret them, and note the errors of the approximations:

a. $u = -\phi zy$, $v = +\phi zx$, $w = 0$, where ϕ is a constant

b. $u = -\kappa yx$, $v = \frac{1}{2}\kappa x^2$, $w = 0$

3.150 Use the approximations of Eq. (3.71) to obtain the state of strain given by the following displacements:

$$u = Kx - \alpha y, \qquad v = -my + \alpha x, \qquad w = -mz$$

Compare this with the data and results of Prob. 3.148(a) and (b).

3.151 Use the approximations of Eq. (3.71) to obtain the state of strain given by the following displacement:

$$u = Kx - \kappa yx, \qquad v = -my + \tfrac{1}{2}\kappa x^2, \qquad w = -mx$$

Note that these displacements are the sums of those given in Prob. 3.148(a) and 3.149(b). How do the results compare?

3.152 Use the approximations of Eq. (3.71) to obtain the state of strain given by the following displacements:

$$u = Kx - \phi zy, \qquad v = -my + \phi zx, \qquad w = -mz$$

Note that this deformation is obtained by adding the displacements of Prob. 3.148(a) and Prob. 3.149(a). What is the effect of such addition?

3.153 Determine the strains, according to the linear approximations of Eqs. (3.71), caused by arbitrary linear displacements (u, v, w), for example,

$$u = U + U_x x + U_y y + U_z z$$

where the coefficients are constants (U, U_x, U_y, U_z). Altogether the displacements incorporate 12 such constants. Because there are only 6 strain components, what is the physical (kinematic) significance of the six additional constants?

3.154 In the approximation of a deformed body by finite elements, we could approximate a rectangular element by a "trilinear" displacement such as

$$u = U + U_x x + U_y y + U_z z + U_{xy} xy + U_{yz} yz + U_{xz} xz + U_{xyz} xyz$$

The displacement (components u, v, w) incorporate 24 coefficients (constants). Since the displacements of the eight corners involve 24 displacements (8×3), the coefficients can be expressed in terms of the motion of those corners ("nodes" in the body). Examine components ε_x and γ_{xy} and note the roles of the coefficients. What can you say about the roles of constants U and U_{xy}?

*3.22 Compatibility Conditions

In most practical cases of structures and machines, the strains and rotations are small enough to use the linear strain-displacement equations (3.71). Then these give the state of strain throughout the body in terms of the displacement. In other words, the *six* components of strain are determined by the *three* functions u, v, and w. We might, therefore, suspect that not any state of strain can exist in a body, since we have six relations among three quantities. In particular, it has always been assumed that lines of particles within the body remain continuous and do not overlap. (See Sec. 3.2). Consequently, the only possible states of strain in a cohesive media are those resulting from displacements that do not destroy the continuity of line segments. Mathematically, we need to establish sufficient conditions upon the six strains in order to ensure the existence of the continuous displacements. It is a simpler task to establish necessary conditions. To illustrate the latter, consider the three relations

$$\varepsilon_x = \frac{\partial u}{\partial x}, \qquad \varepsilon_y = \frac{\partial v}{\partial y}, \qquad \gamma_{xy} = \frac{\partial u}{\partial y} + \frac{\partial v}{\partial x}$$

Differentiating the first of these with respect to y twice, the second with respect to x twice, and the third with respect to y and x successively, we obtain

$$\frac{\partial^2 \varepsilon_x}{\partial y^2} = \frac{\partial^3 u}{\partial y^2 \partial x}, \qquad \frac{\partial^2 \varepsilon_y}{\partial x^2} = \frac{\partial^3 v}{\partial x^2 \partial y}$$

$$\frac{\partial^2 \gamma_{xy}}{\partial x \partial y} = \frac{\partial^3 u}{\partial x \partial y^2} + \frac{\partial^3 v}{\partial x \partial y \partial x}$$

Now, u and v must be such that the order of differentiation is interchangeable. Hence, it follows that

$$\frac{\partial^2 \varepsilon_x}{\partial y^2} + \frac{\partial^2 \varepsilon_y}{\partial x^2} = \frac{\partial^2 \gamma_{xy}}{\partial x \partial y}$$

Five other independent relations of this nature may also be derived. The entire set is as follows:

$$\begin{aligned}
\frac{\partial^2 \varepsilon_x}{\partial y^2} + \frac{\partial^2 \varepsilon_y}{\partial x^2} &= \frac{\partial^2 \gamma_{xy}}{\partial x \partial y}, & 2\frac{\partial^2 \varepsilon_x}{\partial y \partial z} &= \frac{\partial}{\partial x}\left(-\frac{\partial \gamma_{yz}}{\partial x} + \frac{\partial \gamma_{xz}}{\partial y} + \frac{\partial \gamma_{xy}}{\partial z}\right) \\
\frac{\partial^2 \varepsilon_y}{\partial z^2} + \frac{\partial^2 \varepsilon_z}{\partial y^2} &= \frac{\partial^2 \gamma_{yz}}{\partial y \partial z}, & 2\frac{\partial^2 \varepsilon_y}{\partial z \partial x} &= \frac{\partial}{\partial y}\left(-\frac{\partial \gamma_{zx}}{\partial y} + \frac{\partial \gamma_{yx}}{\partial z} + \frac{\partial \gamma_{yz}}{\partial x}\right) \qquad (3.72) \\
\frac{\partial^2 \varepsilon_z}{\partial x^2} + \frac{\partial^2 \varepsilon_x}{\partial z^2} &= \frac{\partial^2 \gamma_{zx}}{\partial z \partial x}, & 2\frac{\partial^2 \varepsilon_z}{\partial x \partial y} &= \frac{\partial}{\partial z}\left(-\frac{\partial \gamma_{xy}}{\partial z} + \frac{\partial \gamma_{zy}}{\partial x} + \frac{\partial \gamma_{zx}}{\partial y}\right)
\end{aligned}$$

These six relations, called *compatibility conditions*, are necessary conditions imposed on the state of strain in order that no tears or folds occur in the body. It can also be demonstrated that these conditions are sufficient. Therefore, only those states of strain that satisfy these six partial differential equations are permissible states of strain in a cohesive media.

These compatibility relations complete our discussion of small strain and small rotation.

PROBLEMS

3.155 Verify that the deformation described in Prob. 3.49 satisfies linear equations (3.72) for compatibility.

3.156 Use the first of the compatibility equations (3.72) to obtain the general form of the function $\varepsilon_x(x, y)$ in Prob. 3.50.

3.157 If the strains in a body are constant and strains and rotations are small [Eqs. (3.71) are applicable], what are the general forms of the displacements $u(x, y, z)$, $v(x, y, z)$, and $w(x, y, z)$? Interpret any unspecified constants.

4 Mechanical Behavior of Materials

In the preceding chapters we have developed specific tools: Stress serves to describe fully the nature and intensity of internal force *at* a point of a continuous medium. Strain serves to describe completely the nature and severity of deformation *at* a point of a continuous medium. Both *stress* and *strain* are employed to describe the behavior of materials, but other variables, notably *temperature*, also play a role.

Any description of materials in terms of stress, strain, and temperature must be accompanied by one important caveat: Such mathematical models usually presume *continuity*, as described in Sec. 4.1. Certain phenomena have their origins in the microstructure of the material, which contains discrete irregularities and *discontinuities*; then the concept of a continuum may be inadequate.

The material, its properties, and, then, a mathematical model of the material provide the links between the deformation and external loads (and temperature), between the kinematic variables (e.g., strains), the dynamic variables (e.g., stresses), and the temperature. To appreciate the role of the material, the reader need only imagine axial forces upon a steel bar, the same forces upon a geometrically similar rubber bar, and the consequent elongations.

With the dynamic and kinematic tools of Chapters 2 and 3, we are in a position to explore the behavior of materials, some practical theories, and useful mathematical models. Then, in subsequent chapters, we utilize those mathematical models of the materials, the dynamics (equations of equilibrium), and the kinematics to effect the solution of basic problems that arise in actual structures and machines.

4.1 Preliminary Considerations

Behavior and Properties of Materials

All material bodies are affected in some way by their environment, by changes of environment, and by the application of external forces. The nature of these effects depends upon the size and shape of the body, the nature of the environment and the applied forces, and the material composition of the body. A rubber band will stretch to several times its original length but recovers its size and shape when the forces stretching it are removed. It sustains no permanent deformation. A paper clip is readily bent but does not return to its original shape when the forces bending it are removed. The paper clip can experience severe permanent deformation without fracture. In contrast, a piece of chalk bends only slightly before it fractures.

Often the deformation of a material is related to the duration of loading. For example, a piece of taffy deforms steadily under a constant force. Such gradual and continuous deformation is known as *creep*.

The response of a member to applied loads or temperature changes depends to some extent upon the manner in which these are applied. If very hot water is suddenly poured into a glass jar, the jar may fracture. However, if the jar is slowly heated to the same temperature, it probably will not fracture. Often suddenly applied forces (impact loads) cause fracture when slowly applied forces do not. Many repetitions of a load may cause fracture, even though the same load is too small to cause any damage when applied only a few times. The axle of a passenger car may fracture in this way after many hours of use.

Under applied loads, the same material may exhibit different behavior in different environments. In particular, heat tends to soften many materials and reduce their strength. Boiler tubes, which operate at high temperatures, creep and would eventually fracture if they were not replaced. Very low temperatures have an entirely different effect on materials. Many materials become brittle at low temperatures; their resistance to impact loads is greatly reduced. Other environmental factors, such as radiation, corrosion, weathering, wear, and moisture, influence the behavior of materials.

The purpose of analyzing deformable bodies is to relate causes and effects. For example, the load on a beam is a cause; its effect is a deformation. The cause-effect relation depends on the nature of the cause (the type of loading and environmental conditions) and on the properties of the body. Size and shape enter into the relation of cause and effect; these are properties of the particular body. However, certain properties are characteristic of the material of which the body is composed.

In the present chapter we wish to describe the mechanical behavior of the material and those properties that characterize it. To do this we seek mathematical relations between causes and effects, which are independent of the size and shape of any particular body. Relationships that utilize the variables of stress and strain accomplish this. A relation between a cause (stress) and an effect (strain) involves only the material properties. Such variables as temperature, time, or humidity may also enter into the cause-effect relationship. In any case, these relations describe only the behavior of the material, and not the behavior of a particular body. The quantities which serve to characterize the mechanical behavior of a material are known as *mechanical properties*.

Structure of Materials

The response of any structure or machine is related to the physical attributes of the material of which the structure is composed. Every small portion of a material body is itself a structure composed of elementary constituents arranged in quite irregular fashion. A metal, for example, may appear quite uniform to the unaided eye; microscopically, it consists of many small crystals of varied shapes and random orientations. Each such crystal is composed of aggregates of atoms (which are often arranged in imperfect order) and voids. These irregularities and imperfections influence the resistance of materials to deformation and fracture. A full understanding of the behavior of materials hinges upon knowledge of the smallest structural units and the manner in which they interact. We will return to this idea after a brief examination of some concepts that are useful in describing the nature of a material.

Homogeneity The term homogeneous is used to describe something that possesses uniformity. *A material body is homogeneous if every similar portion possesses the same physical attributes and properties.* In view of previous statements it is unlikely that the common materials of engineering are truly homogeneous. However, if we examine large enough portions of the material, small irregularities and differences in the various portions become unimportant because of the random distribution and orientation of the small constituents. The lack of homogeneity is significant only on a very small or microscopic scale. Where this is true, it is permissible to regard the material as homogeneous or uniform, meaning that any portion (of a certain minimum size) has properties that are essentially the same as every similar portion. The extent to which a material may be regarded as homogeneous depends upon the size of the body being analyzed and also upon the effects being studied. A very small filament consisting of a few metal crystals cannot be treated as a homogeneous body. Also, certain phenomena of material behavior have their origins in the small-scale irregularities of material structure. This is particularly true of the phenomenon of fracture as the result of repeated loading. A study of such effects must take into consideration the true non-homogeneous structure of the material.

Isotropy The structure of materials is often oriented within the material body. The structure of wood, for example, consists of parallel layers that form with each year's growth. Its appearance, properties, and behavior are dependent on direction: From experience we know that it is far more difficult to cut through these layers than it is to split the wood along the layers. Our interest in the orientation of material structure is limited to its influence on properties and behavior. *A material that exhibits neither structural orientation nor properties dependent on directions within the material is said to be isotropic. Materials that possess directional properties are called anisotropic.* On a very small scale, nearly all materials are anisotropic. For example, the crystals within a metal possess certain directional properties. A single crystal usually has one or more lines of symmetry. Its appearance and properties then depend upon the point from which it is viewed. Noncrystalline materials also display an anisotropic character when they are examined on a small enough scale. In the process of formation, the molecules in rubberlike materials and plastics are often elongated in one or more directions. However, large enough portions of materials often appear and act as though they were isotropic. This results from a random distribution and orientation of the various constituents.

The reader should take care not to confuse isotropy and homogeneity. Homogeneity means that each similar and similarly oriented portion behaves just as every other such portion; hence, a material may be homogeneous and anisotropic. Most engineering materials may be regarded as both homogeneous and isotropic. Notable exceptions are timber and certain ductile metals that become anisotropic when subjected to severe permanent deformation. An important example of the latter is common structural steel, which is cold-rolled (formed by rolling, forging, or drawing at *subcritical temperatures*. A flat bar of cold-rolled steel may exhibit properties in directions parallel to its axis that are quite different than those properties it exhibits perpendicular to its axis.

Continuous Media In the analyses of deformable bodies it is impossible to take full account of the complex structure of materials. We must, therefore, introduce a simple conceptual model of physical materials: the concept of a *continuous medium*. This concept has been mentioned previously and is, in fact, used in our analyses of stress and strain. The continuous medium contains no voids; every portion of the body is completely filled with matter. The properties of the medium may vary from point to point, but such variations are usually so gradual that they admit to a mathematical description. A continuous medium may be both nonhomogeneous and anisotropic.

Cohesion The present study is primarily concerned with solid materials, so that adjacent portions of material adhere to one another except when fracture occurs. It is not possible for adjacent portions of the material to be separated or to pass through one another. Furthermore, if adjoining portions must adhere, lines and surfaces embedded within the body can never be destroyed, except when fracture occurs. A material of this kind is termed *cohesive*. Throughout this study of deformable bodies we shall regard materials as *cohesive continuous media*.

Determination of Mechanical Behavior

The behavior and properties of materials are dependent on the interrelationships of elementary constituents. For example, the behavior of metals depends upon the arrangement of atoms within crystals, the bonds between these atoms, the size, shape, distribution, and orientation of the various crystals, and the ever-present impurities. The behavior and properties of nonmetallic materials are related to their elementary structure in a similar fashion. A complete knowledge of the behavior of a material requires a study of the smallest structural units and the manner in which they are bound together. Ideally, such studies would lead to mathematical statements of the mechanical behavior of materials and a quantitative knowledge of their properties. Because of the extreme complexity of the microstructure of most materials, this goal has been attained only in a few special cases (studies of a single crystal). These studies are important, however, because they reveal the origins of various phenomena. Certain of these phenomena are discussed later in the light of more recent observations of the microstructure of materials.

It is not essential at this time that we explain why materials behave as they do. Our purpose is to present the essential features of material behavior and to formulate suitable relationships of cause and effect. The various types of behavior of materials may be observed in certain simple experiments, in which geometrically simple bodies are subjected to various kinds of external loads. Such observations often suggest possible relationships of cause and effect. However, experimentation is limited to relatively few situations, making it necessary to infer rather general relations from the observations of particular experiments. In any case, a statement of the behavior of a certain material is simply a postulate. The engineer adopts the most convenient mathematical expression that provides satisfactory agreement with experiment. The validity of these relations, when applied to diverse problems, are therefore open to question. Although we may postulate innumerable cause-effect relations, the validity of a relation is firmly established only through repeated successful use to predict the behavior of a variety of structures and machines.

4.2 Simple Tension, Compression, and Shear Tests

The essential features of material behavior are revealed by means of simple experiments. The most common of these involves the application of static loads at room temperature. Normally, atmospheric conditions have little effect, so that only loads and deformations need be observed and measured. This is the nature of the simple tension, compression and shear tests.

Simple Tension Test

The simple tension test consists of the gradual application of an axial force to extend a prismatic bar. The cross section of the test specimen is uniform and usually of circular or rectangular shape. The important data to be recorded are the measurements of axial force and the corresponding length and width of the bar. The test is performed by gripping the specimen at both ends in the two heads of a testing device. The bar is extended by moving one of the heads and keeping the other fixed. The applied force is measured by a balancing system similar to that in a conventional weighing scale. A test setup is shown in Fig. 4.1. A preselected portion of the specimen is marked by points Q and R. This portion is termed the "gage length," since it is marked for observation and measurement. The initial gage length is denoted by L. Points Q and R are located far enough from the clamped ends that the portion QR is unaffected by local disturbances at the grips. Care must be taken to align the grips so that the bar experiences axial extension without bending.

As an illustration, consider the test of a typical structural metal (an aluminum or steel alloy). For simplicity, suppose that the specimen has a uniform circular cross section of diameter d. At regular intervals, during continuous stretching of the bar, recordings are made of the length of portion QR of the bar, the diameter of the bar,

Fig. 4.1

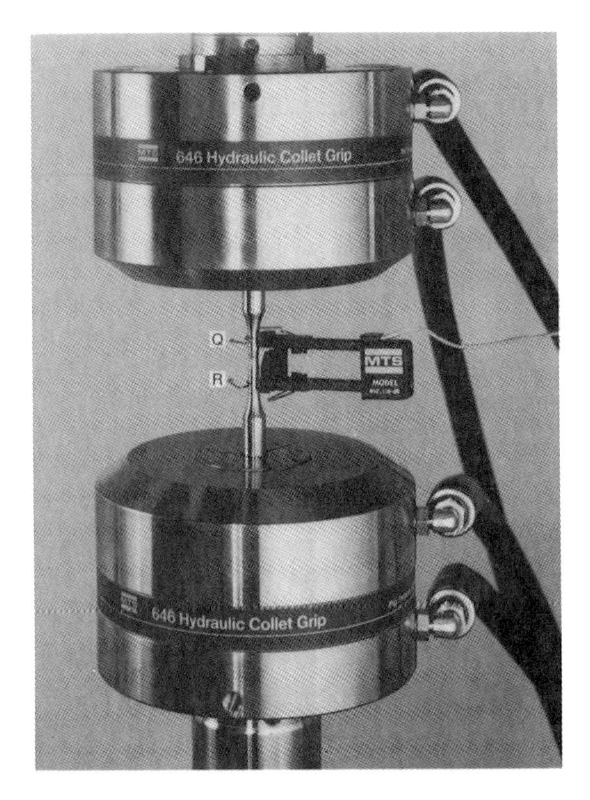

Tensile specimen with attached extensiometer, shown in the collet of a hydraulic testing machine. (Courtesy MTS Systems Corporation)

and the axial force. As the bar is stretched, the distance between Q and R increases. Let us denote the new length by L^* and the axial force at the same time by P. The increase in the gage length is $e = L^* - L$. From the data taken during the test, a plot of load P versus elongation e may be made. From sufficient data the continuous curve $OABCD$ of Fig. 4.2 may be drawn.

Fig. 4.2

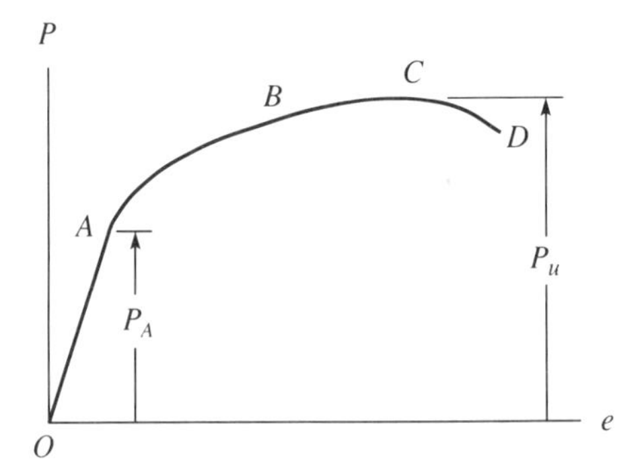

Several important features of this diagram should be noted. The initial portion OA is approximately a straight line; that is, for loads less than P_A, the elongation e is proportional to the load P. Additional extension of the bar requires a greater load until a maximum force P_u (at point C) is attained. This maximum force P_u is termed the ultimate load; it is the greatest force that the specimen can resist. Subsequently the bar elongates under the action of less and less force until it fractures (point D). The decrease in load prior to fracture is attributed to a phenomenon known as *necking*. Necking is a localized deformation that reduces the cross-sectional area near the location of the fracture, as shown in Fig. 4.3. Necking begins to manifest itself when the load approaches the ultimate load P_u. At the time of fracture there is a rather obvious and very localized contraction in diameter. Both necking and the resulting drop in load are pronounced only in ductile materials.

Fig. 4.3

However, even a slight extension of the bar is accompanied by a reduction in diameter. If d^* denotes the reduced diameter and e' is the change of diameter, then $e' = d^* - d$. Note that because d^* is smaller than d, the change e' is negative. Prior to necking, the contraction e' is nearly constant along the gage length. Furthermore, if the loads are less than P_A, the contraction e' is proportional to the extension e, and hence also proportional to the applied load P.

The relationships among load, elongation, and transverse contraction depend not only on the material, but also on the size and shape of the specimen. The greater the cross section of the specimen, the greater are the forces required to effect any changes. Also, a longer gage length will result in a greater elongation for any given force. Our aim is to describe the material behavior without reference to a particular body. With this in mind, the axial force P is divided by the original cross-sectional area, a constant A; the elongation e is divided by the original gage length, a constant L; and the contraction e' is divided by the original diameter, a constant d. This results in new variables: the average normal stress,

$$\bar{\sigma} = \frac{P}{A} \tag{4.1}$$

the average axial strain,

$$\bar{\varepsilon} = \frac{e}{L} = \frac{L^*}{L} - 1 \tag{4.2}$$

and the average transverse strain,

$$\bar{\varepsilon}' = \frac{e'}{d} = \frac{d^*}{d} - 1 \tag{4.3}$$

A plot of $\bar{\sigma}$ versus $\bar{\varepsilon}$ yields the curve $OABCD$ of Fig. 4.4. This curve is similar in every way to the P-e curve of Fig. 4.2, because it is obtained from the P-e diagram by dividing each of the ordinates by the constant A and the corresponding abscissas by the constant L. Indeed, the only difference between the $\bar{\sigma}$-$\bar{\varepsilon}$ and P-e diagrams is a change of scale. Hence all the previous remarks about the P-e diagram apply also to the $\bar{\sigma}$-$\bar{\varepsilon}$ diagram. However, the relationship of average stress $\bar{\sigma}$ to average strain $\bar{\varepsilon}$ is virtually independent of the size of the specimen. Only in the event of severe deformation (necking) is the $\bar{\sigma}$-$\bar{\varepsilon}$ relation greatly influenced by the dimensions of the specimen. For brevity in the following discussion, the quantities $\bar{\sigma}$, $\bar{\varepsilon}$, and $\bar{\varepsilon}'$ are referred to as the stress, strain, and transverse strain, respectively. The reader should bear in mind, however, that these quantities are certain average values as defined by Eqs. (4.1), (4.2), and (4.3).

Fig. 4.4

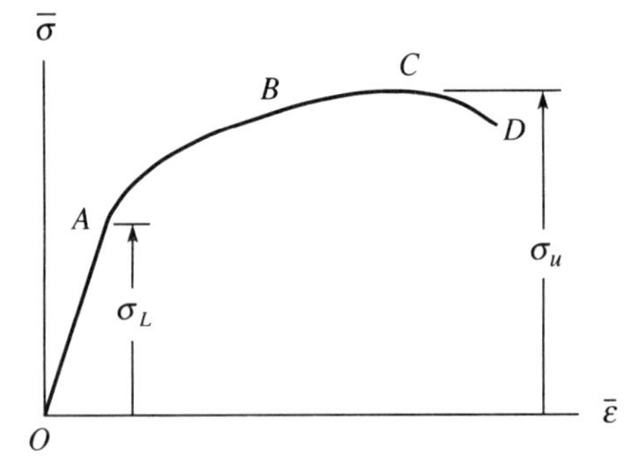

Modulus of Elasticity and Poisson's Ratio Let us examine the essential features of the $\bar{\sigma}$-$\bar{\varepsilon}$ diagram of Fig. 4.4. The initial portion OA is approximately a straight line. For this initial portion, the $\bar{\sigma}$-$\bar{\varepsilon}$ relation is expressed by the linear equation

$$\bar{\sigma} = E\bar{\varepsilon} \tag{4.4}$$

where E is a constant of proportionality, the slope of the line OA. The constant E is a property of the material, called the *modulus of elasticity*.

For the initial portion OA, the transverse contraction e' is proportional to the elongation e but has the opposite sign. Therefore, the transverse strain $\bar{\varepsilon}'$ is related to the axial strain $\bar{\varepsilon}$ by the linear equation

$$\bar{\varepsilon}' = -\nu\bar{\varepsilon} \tag{4.5}$$

where ν is a positive constant of proportionality, called *Poisson's ratio*. It follows from Eq. (4.4) that $\bar{\varepsilon}'$ is also linearly related to the normal stress:

$$\bar{\varepsilon}' = -\frac{\nu}{E}\bar{\sigma} \tag{4.6}$$

Eqs. (4.4), (4.5), and (4.6) apply only to the initial linear portion of the $\bar{\sigma}$-$\bar{\varepsilon}$ curve.

Proportional Limit When the stress $\bar{\sigma}$ attains a certain value, denoted by σ_L, the $\bar{\sigma}$-$\bar{\varepsilon}$ curve departs noticeably from the straight line OA. This deviation from the linear relation of Eq. (4.4) is marked by the point A of Fig. 4.4. The stress σ_L is termed the *proportional limit stress* (often simply the *proportional limit*).

Elastic Limit So long as the applied force is sufficiently small, the specimen recovers its original size and shape upon removal of the force. This recovery of size and shape characterizes elastic behavior. The deformation which disappears when the force is removed is termed an *elastic deformation*. However, if the stress exceeds a certain value, σ_E (not shown), the specimen does not completely recover its size and shape when the force is removed. A permanent deformation is sustained. The stress σ_E is termed the *elastic limit*. The action which results in a permanent deformation is termed *yielding*.

The values σ_L and σ_E are not easily established. It is extremely difficult to determine accurately the stress at which the curve deviates from a straight line, and it is equally difficult to detect the earliest occurrence of permanent deformation. The precision with which these quantities are obtained is limited by the accuracy of the measuring devices. There is, in fact, good reason to doubt that any material exhibits a truly linear $\bar{\sigma}$-$\bar{\varepsilon}$ relation. Also, careful studies of deformation indicate that some permanent changes of structure accompany even the least application of force.

Yield Stress For most metals the proportional limit and elastic limit appear to be so close that, for practical purposes, a single value $\sigma_0(\doteq \sigma_L \doteq \sigma_E)$ characterizes the transition from linear as well as elastic behavior. Henceforth we will refer to this stress σ_0 as the *yield stress*. The term *yielding* signifies the occurrence of permanent deformation. If the stress is less than the yield stress, then, for practical purposes, the observed behavior is elastic, and the stress-strain relation is linear in accordance with Eq. (4.4). We say that this is *linearly elastic behavior*.

Elastic and Plastic Behavior Stress exceeding the yield stress σ_0 will cause permanent deformation. Suppose, for example, that the bar is continuously loaded to the point B of Fig. 4.2 or 4.4 and then continuously unloaded. The P-e and $\bar{\sigma}$-$\bar{\varepsilon}$ diagrams have been redrawn in Figs. 4.5 and 4.6, respectively. In either figure the unloading curve BO_P is practically a straight line, parallel to OA. (The deviation from linearity is exaggerated in the diagrams.) After the load is entirely removed, a permanent deformation remains, which is termed a *plastic* deformation. However, not all of the deformation caused by the loading is permanent. A part of the deformation is elastic (recoverable). In Fig. 4.6, the plastic (permanent) strain is represented by OO_P, and the elastic strain, which is recovered during unloading, is represented by O_PO_B.

Fig. 4.5

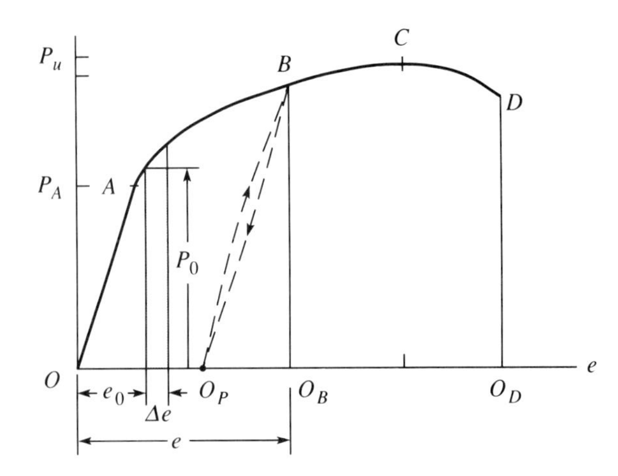

Fig. 4.6

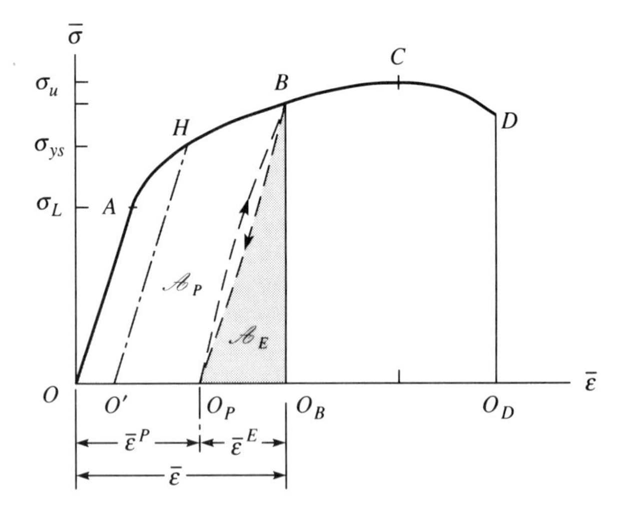

Strain Hardening Suppose that the specimen has been loaded and unloaded according to the path OBO_P. When the specimen is reloaded, the $\bar{\sigma}$-$\bar{\varepsilon}$ plot traces the indicated curve from O_P to B. This curve is again almost straight and parallel to OA. In other words, the $\bar{\sigma}$-$\bar{\varepsilon}$ relation is essentially linear. Furthermore, the behavior is essentially elastic, since unloading practically retraces the path to O_P. So long as the stress does not exceed the value at B, the behavior is approximately linear and elastic. Because of the previous plastic deformation (OO_P), a stress greater than that at B is required to cause *additional* plastic deformation. The yield stress for the plastically deformed material is greater than that for the virgin material. We say that the material has *strain hardened.*

Yield Strength As mentioned previously, the stress that initiates yielding is not well defined. For this reason, engineers sometimes use another quantity to characterize resistance to plastic deformation. This is the *yield strength,* defined as follows:

From a preselected point O' of the $\bar{\varepsilon}$-axis, a line $O'H$ is drawn parallel to the line OA. The stress determined by the intersection of this line and the $\bar{\sigma}$-$\bar{\varepsilon}$ curve is termed the *yield strength,* denoted by $\sigma_{y.s.}$. This value depends on the amount by which the point O' is "offset" from the origin O; hence, the yield strength has a definite value only when the offset OO' is specified. According to the previous discussion, unloading from point H follows very nearly the straight line HO'. This means that the preselected offset OO' represents, approximately, the permanent strain caused by loading to the yield strength. We may, therefore, interpret the yield strength as that average stress which causes a prescribed amount of plastic strain in simple tension.

Ultimate Strength Quite often the engineer requires some measure of the maximum resistance of a material to a simple tensile force. This is given by the *ultimate strength,* σ_u, indicated in Fig. 4.4. It is simply the ultimate force P_u divided by the original cross-sectional area of the specimen.

Work and Energy Throughout the previous discussion, the basic concepts of force and displacement have been employed to describe observed behavior. Another concept of great importance in the analyses of deformable bodies is that of *work* and *energy.* Let us therefore consider the work done in stretching the portion QR of the bar. Suppose that the portion QR, acted upon by a force P_0, has elongated the amount e_0. For simplicity, imagine that the point Q is held fixed, as shown in Fig. 4.7. During an

Fig. 4.7

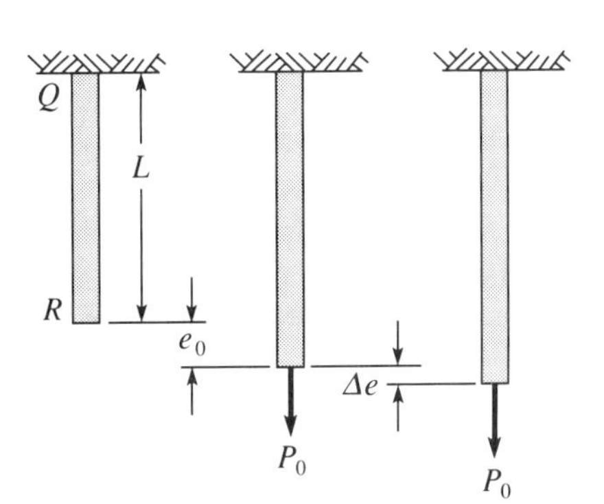

infinitesimally small additional elongation Δe, the work done by the force P_0 is

$$\Delta W = P_0 \Delta e$$

This small amount of work is represented by the shaded area in Fig. 4.5. The total work done to effect an elongation e is then given by

$$W = \int_0^e P\,de \tag{4.7}$$

This integral is represented by the area $OABO_B$ of Fig. 4.5. Work of amount W is required to cause an elongation of amount e ($= OO_B$). Similarly, the total work done to fracture the bar is represented by the area $OABDO_D$ under the entire P-e curve of Fig. 4.5.

Suppose now that Eq. (4.7) is divided by the original area A and the original gage length L. This yields

$$\frac{W}{AL} = \int_0^e \frac{P}{A}\frac{de}{L} = \int_0^{\bar{\varepsilon}} \bar{\sigma}\,d\bar{\varepsilon} \tag{4.8}$$

The product AL is simply the original volume; hence, the left side of Eq. (4.8) is work per unit volume. The right side is represented by the shaded area of Fig. 4.6, the area $OABO_B$. For brevity, let us denote the work per unit volume by U, so that

$$U = \frac{W}{AL} = \int_0^{\bar{\varepsilon}} \bar{\sigma}\,d\bar{\varepsilon} \tag{4.9}$$

The quantity U is called the *work density.*

Toughness The total work density required to fracture the specimen is represented by the area $OABDO_D$ under the entire $\bar{\sigma}$-$\bar{\varepsilon}$ curve of Fig. 4.6. This provides some measure of the ability of the material to resist fracture. It is termed the *toughness.*

Strain-Energy Density The work density required to cause the strain of amount $\bar{\varepsilon}$ is represented by the shaded area $OABO_B$ of Fig. 4.6. Removing the load recovers a part of this strain (O_PO_B), and the other part (OO_P) represents a plastic deformation. The work density expended during the cycle of application and removal of load is represented by the area $OABO_P$. This much of the work density represents the energy dissipated during the process of plastically deforming the material, which causes the permanent strain OO_P. The area O_PBO_B represents the work density required to cause the elastic deformation represented by the elastic strain O_PO_B. The total work is represented graphically by the sum of these areas, $A_P + A_E$. Area A_P represents the dissipated energy; area A_E represents the energy recoverable during unloading. The latter is termed *potential*-energy, *elastic*-energy, or *strain*-energy density. In particular, the strain-energy density when the stress $\bar{\sigma}$ reaches the proportional limit σ_L is

$$U_L = \frac{\sigma_L^2}{2E}$$

Since the proportional limit σ_L is generally very near the yield stress σ_0, the preceding energy density approximates the energy density required to initiate yielding in simple tension. It is called the *modulus of resilience* and is represented by the area under the linear portion (OA) of the $\bar{\sigma}$-$\bar{\varepsilon}$ curve of Fig. 4.6.

Ductility Certain engineering applications demand material that can undergo severe permanent (plastic) deformation. The ability of materials to withstand plastic deformation is termed *ductility*. The processes of forging, stamping, and extrusion depend on the ductility of materials. Furthermore, ductility is essential if a structure is to absorb energy prior to fracture. Hence the use of a ductile material provides a measure of safety in certain structural applications.

The plastic deformation incurred in the simple tension test is indicative of ductility. Measurements of the specimen before and after a test to fracture can be used to measure ductility. The *percent reduction* of the cross-sectional area and *percent elongation* of the gage length are commonly accepted measures of ductility. Because of necking, both measures are dependent on the gage length and, to some extent, on the size and shape of the material specimen. For this reason, test specifications require the use of a standard specimen.

Structural Steel The preceding discussion of simple tension has revealed the essential characteristics of a typical ductile metal. Although the observed behavior is representative of many commonly used structural metals, certain other materials may exhibit a quite different behavior. The behavior of common structural steel (hot-rolled low-carbon steel) is not entirely unlike the ductile metal just described. Its behavior under simple tension is illustrated by the diagram of Fig. 4.8. Among the common engineering materials, it must be regarded as extremely ductile. The total plastic strain at fracture is more than one hundred times the elastic strain. The distinctive feature of structural steel is the portion BC of the $\bar{\sigma}$-$\bar{\varepsilon}$ diagram, which is very nearly a straight

Fig. 4.8

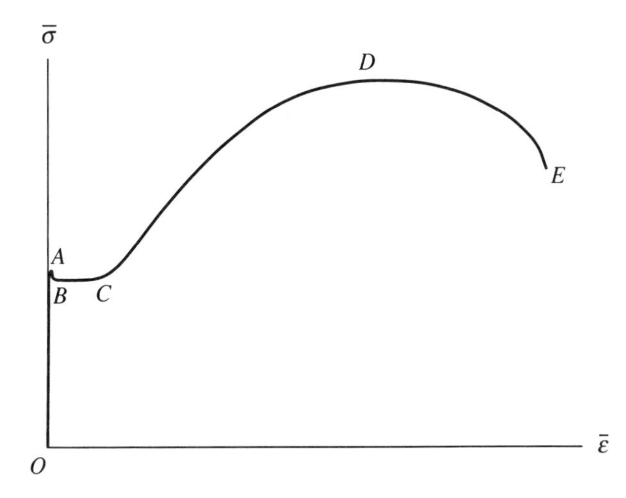

horizontal line. At point A the $\bar{\sigma}$-$\bar{\varepsilon}$ curve has a horizontal tangent. Additional plastic strain is accompanied by a decrease in stress. At point B the curve again has a horizontal tangent. The stress corresponding to point A is known as the *upper yield point*, and the stress corresponding to point B is known as the *lower yield point*. The "flat" portion BC represents a plastic strain approximately twenty times the elastic strain. In most other respects the behavior of structural steel is similar to the behavior of the ductile metal described in the earlier parts of this section.

Brittle Material The behavior of a typical brittle material is illustrated by the $\bar{\sigma}$-$\bar{\varepsilon}$ diagram of Fig. 4.9. Little plastic deformation occurs prior to fracture (point B). Furthermore, most brittle materials exhibit a linear $\bar{\sigma}$-$\bar{\varepsilon}$ relation almost to the point of fracture; that is, the entire curve is essentially a straight line.

Fig. 4.9

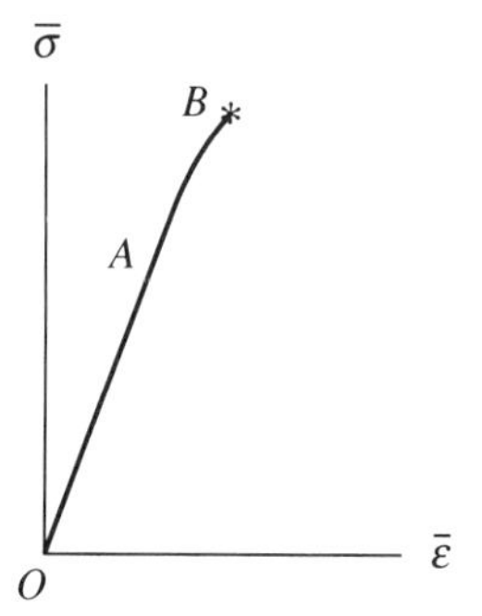

Rubberlike Material Not all materials display a linear relationship between the stress $\bar{\sigma}$ and the strain $\bar{\varepsilon}$. Many rubberlike materials exhibit a nonlinear $\bar{\sigma}$-$\bar{\varepsilon}$ relation, as illustrated by Fig. 4.10. Although the $\bar{\sigma}$-$\bar{\varepsilon}$ curve is nonlinear, the body may return to its original size and shape when the applied forces are removed. Indeed, such materials exhibit very little tendency toward plastic deformation. If a specimen is loaded to point B and the load is removed, the curve for the unloading may not retrace the loading curve OAB. However, if the load is very slowly removed, the

Fig. 4.10

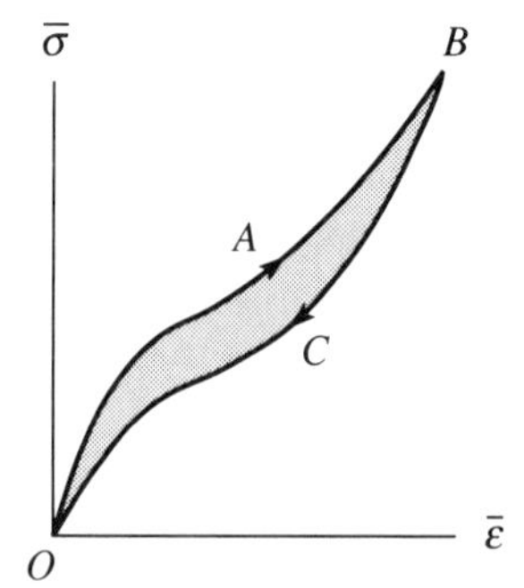

unloading curve BCO terminates at the origin O. Thus, there is no permanent deformation. If the unloading curve does not retrace the loading curve, the (shaded) area enclosed by the loading and unloading curves in Fig. 4.10 represents dissipated energy. The cycle $OABCO$ is frequently termed a *hysteresis loop.*

True Stress In the definition of the average stress $\bar{\sigma}$ in Eq. (4.1), the force P acting upon a transverse section is divided by the original cross-sectional area A. Because the cross-sectional area is reduced by deformation, the quantity $\bar{\sigma}$ is not a "true" average stress. This is particularly significant where necking occurs, because the axial force P then acts upon a reduced area. The specimen deforms under less force, and the average stress (based on original area) actually decreases as necking progresses. A more meaningful quantity is the so-called true average stress obtained by dividing the axial force P by the corresponding true (reduced) area. The plot of true average stress for a typical tension test of a ductile material is compared to the conventional plot in Fig. 4.11. Because, in the analyses of deformable bodies, we are always concerned with forces acting upon and within a deformed body, the discussions of Chapter 2 refer to true stresses. However, the deformations are not usually so severe that the distinction between original and deformed areas is important. For example, the difference between the curves of Fig. 4.11 is significant only for very large strains. No distinction between true and conventional stress is required in analyses of the members of machines and structures, but it is essential in analyses of severe inelastic deformations in forging and extrusion or the large elastic deformation of rubberlike materials.

Fig. 4.11

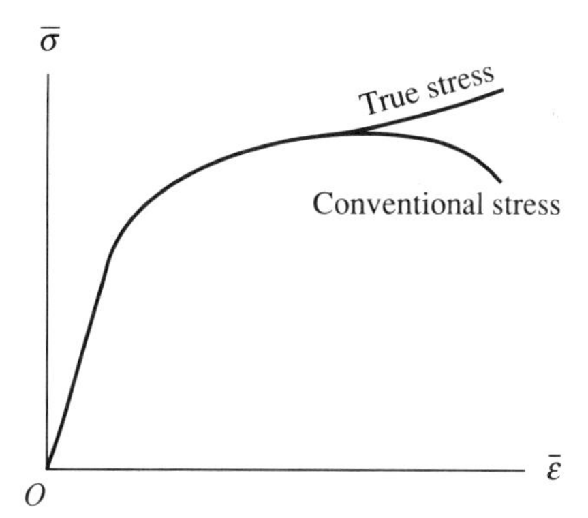

Simple Compression Test

Basically, the simple compression test differs from the simple tension test only in the signs of the quantities that describe the causes and effects. An axial compressive force results in axial contraction of the test specimen. In accordance with previous sign conventions, the average stress $\bar{\sigma}$ and the average axial strain $\bar{\varepsilon}$ are negative. As a result of the applied compressive force, the test bar experiences a positive transverse strain $\bar{\varepsilon}'$. The relationships between the quantities $\bar{\sigma}$, $\bar{\varepsilon}$, and $\bar{\varepsilon}'$ are much like those observed in the tension test.

It is particularly informative to show the $\bar{\sigma}$-$\bar{\varepsilon}$ curves for tension and compression in a single diagram. Thus, in Fig. 4.12 the curves $OABC$ and $OA'B'C'$ are obtained from a simple tension test and a simple compression test, respectively. For many common materials, these two curves are quite similar if the strain magnitudes are not much greater than the strain at initial yielding. A portion OA' of the curve $OA'B'C'$ is very nearly a straight line extension of AO. Accordingly, the linear $\bar{\sigma}$-$\bar{\varepsilon}$ relation of Eq. (4.4) is, in most cases, a valid approximation for average stress magnitudes which do not exceed the proportional limits in compression or in tension. In most cases, the proportional limit and the yield stress in compression have approximately the same magnitudes as those in tension.

Fig. 4.12

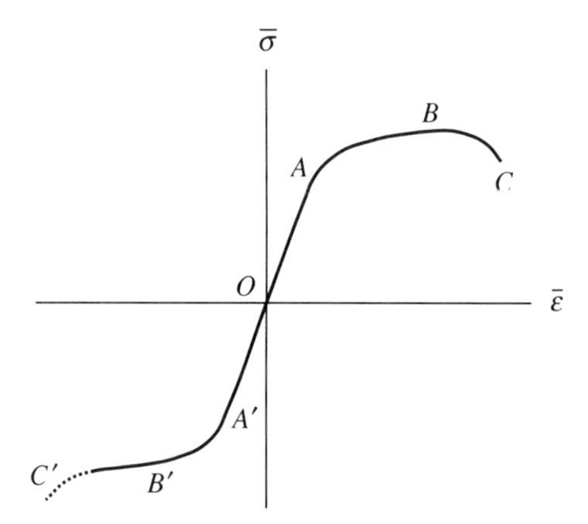

For large magnitudes of average strain, the conventional tension and compression curves may be quite different. This is truer of some materials than of others. A very ductile material subjected to axial compression flattens into a thin disk. The applied load must, therefore, increase almost without limit as the deformation progresses. At the same time, the average strain is limited, since the change in length of the disk cannot exceed the original length.

Simple Shear Test

The effects of shear forces are most readily studied by tests of thin circular cylindrical tubes subjected to twisting couples. Suitable fixtures at the ends maintain symmetry with respect to the tube axis. Under these conditions it is reasonable to assume that the forces on any transverse section consist of shear forces distributed symmetrically about the axis. Any small element of the tube, which is oriented as shown in Fig. 4.13(a), is subject to the action of shear forces as shown in Fig. 4.13(b). The average shear stress $\bar{\tau}$ is given by

Fig. 4.13

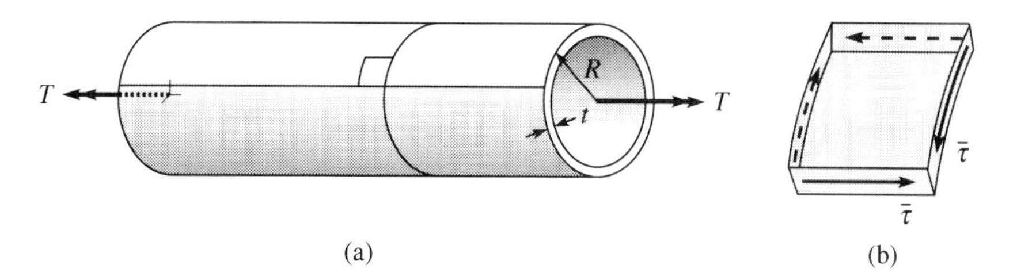

$$\bar{\tau} = \frac{T}{2\pi\bar{R}t}$$

where T is the applied twisting couple, $\bar{R}$ is the mean radius of the tube, and t is the thickness of the wall. Since the tube is thin, we suppose that the shear stress on any element is approximately $\bar{\tau}$.

Let us consider the nature of the deformations caused by the twisting couple. It is unlikely that the radius or length would change significantly under the action of such external loading. That this is true is verified by experimental observations. The predominant effect is a rotation of transverse sections (described in Chapter 3). The force system is identical on every transverse section. Consequently, if the tube is uniform throughout its length, the angle of twist θ (per unit of length) is constant along the tube. If ϕ denotes the rotation of one end relative to the other, the angle of twist per unit length is given by

$$\theta = \frac{\phi}{L}$$

where L is the length of the tube. The deformation of a small element is then predominantly a shear strain, as illustrated in Fig. 4.14. The average shear strain, according to Fig. 4.14(b), is

$$\bar{\gamma} = \arctan \bar{R}\frac{\Delta\phi}{\Delta L} = \arctan\frac{\bar{R}\phi}{L} = \arctan \bar{R}\theta$$

Fig. 4.14

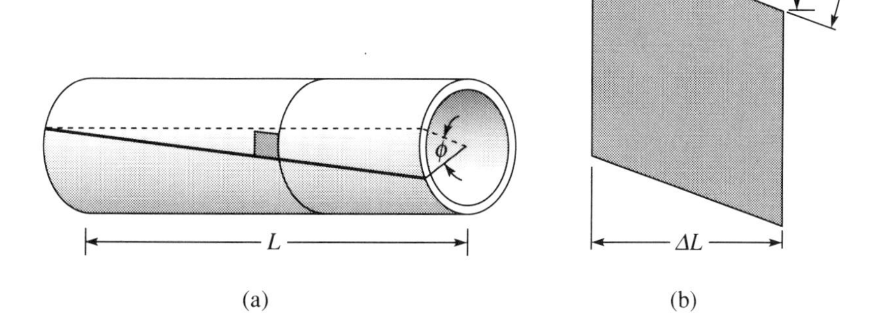

It is essentially the same for each similar element. Since the tube is thin, we suppose that the shear strain at any point is approximately $\bar{\gamma}$. In any event, the computations of average stresses and strains must suffice, as it is never possible to measure stresses and strains directly.

In the torsion test, measurements of the rotation ϕ and the corresponding torque T are required. The additional measurements of L, R, and t will suggest that these dimensions do not change significantly.

Let us consider a typical ductile material, the same material that was examined in the tension and compression tests. During the gradual and continuous twisting of the cylinder, a plot of average shear stress $\bar{\tau}$ versus the average shear strain $\bar{\gamma}$ traces the curve $OABC$ shown in Fig. 4.15. Reversing the sense of the applied torque would give a similar curve, $OA'B'C'$. If the material is isotropic, the two curves are identical in shape. The character of the shear stress-strain relation is very much like that of the tension-compression stress-strain relation. For all practical purposes, the portion $A'OA$ of the diagram is a straight line. Thus, for stress magnitudes within the shear proportional limits, $\bar{\tau} = G\bar{\gamma}$, where G is a constant of proportionality. This constant G, a property of the material, is called the *shear modulus* of elasticity. A shear yield stress τ_0 can be defined as the least average shear stress which causes noticeable plastic strain. Unloading from point B to O_P traccs a curvc BO_P which is essentially linear and parallel to the initial portion of the loading curve OA. Thus the shear strain incurred by loading to point B may be regarded as composed of an elastic part and a plastic part, represented in the diagram by the line segments O_PO_B and OO_P, respectively.

Fig. 4.15

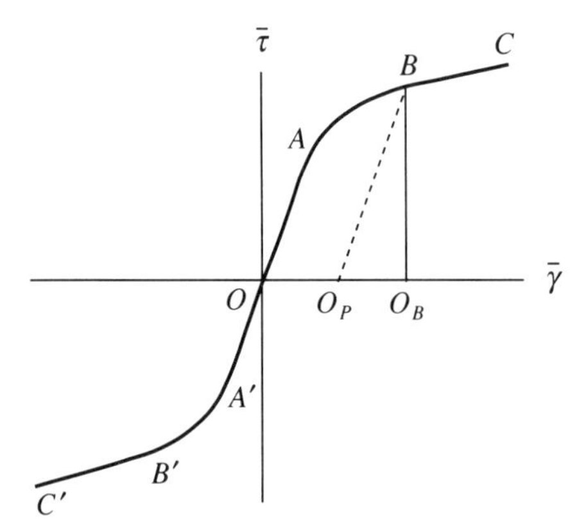

The previous statements about work and energy densities in tension and compression are also applicable here. The work done (per unit volume) in causing a strain of amount OO_B is represented by the area OBO_B. The energy dissipated as a result of plastic deformation is represented by the area OBO_P, and the elastic (recoverable)

strain energy by the area $O_P B O_B$. Plastic deformation alters the material properties, changes the subsequent yield stresses, and decreases ductility.

It should be noted that there is, in general, no irregularity in the deformation, such as the necking phenomenon observed in tension. However, if the tube is very thin, there is a possibility of localized buckling or collapse, which destroys the geometrical symmetry of the tube. Data obtained during such irregular deformations cannot serve to establish the relation between $\bar{\tau}$ and $\bar{\gamma}$.

Most ductile metals have a shear yield stress τ_0 that is approximately one-half of the tensile yield stress σ_0 obtained in tension or compression tests. In a simple tension test the greatest average shear stress occurs on a plane inclined 45° to the plane of maximum normal stress. This maximum shear stress is one-half the average tensile stress. Hence, these tests suggest that yielding occurs in *each* case when the maximum shear stress for that case attains the value τ_0.

The average shear stress-strain diagrams for other materials may differ markedly from that of Fig. 4.15. Brittle materials may be expected to exhibit an almost linear shear stress-strain relation to fracture; other materials may be quite nonlinear in their stress-strain relation. For common structural steel, the shear stress-strain relation is, in general, much like its tension or compression relations, although the magnitudes in the two types of relations differ.

PROBLEMS

4.1 A uniform bar 50 cm long with a cross-sectional area of 3.20 cm^2 is made from a soft aluminum alloy. During a tensile test of the specimen, the extension of a 20-cm gage length and corresponding loads were recorded as in Table P4.1.

Table P4.1

Load (N)	Extension (mm)
0	0.0590
4,000	0.0936
8,000	0.1316
16,000	0.1992
24,000	0.2663
32,000	0.3508
40,000	0.4422
48,000	0.5701
52,000	0.6797
56,000	0.8624
60,000	1.5787
62,000	2.1269
62,400	Fracture

Plot the stress-strain curve and determine (a) the modulus of elasticity, (b) yield strength (offset = 0.002), and (c) ultimate strength.

4.2 A structural steel specimen was machined to a diameter of 0.700 in. over an 8-in. gage length. It was tested to fracture in tension, yielding the data in Table P4.2. At fracture, the gage length was 10.23 in. and the minimum diameter was 0.475 in. Draw the stress-strain diagram and determine as many of the mechanical properties as possible from the test data.

Table P4.2

Load (lb)	Extension (in.)
1,400	0.001
2,900	0.002
4,300	0.003
5,550	0.004
7,000	0.005
8,300	0.006
9,700	0.007
11,000	0.008
12,400	0.009
13,650	0.010
15,000	0.011
16,150	0.012
16,000	0.013
15,600	0.014
15,400	0.015
15,400	0.020
15,400	0.1
19,350	0.2
22,650	0.4
24,150	0.6
24,900	0.8
25,300	1.0
25,600	1.2
25,700	1.4
25,700	1.6
25,500	1.8
23,600	2.0
18,300	2.2
17,800	Fracture

4.3 A concrete cylinder with a 6-in. diameter and a 12-in. gage length is tested in compression. The results are given in Table P4.3. Plot the stress-strain diagram and determine E.

Table P4.3

Compressive Load (lb)	Contraction (in.)
2,900	0.0003
5,000	0.0006
7,300	0.0009
9,600	0.0012
12,600	0.0015
16,400	0.0021
18,700	0.0024
20,800	0.0027
23,100	0.0030
25,400	0.0033
27,000	0.0036
30,000	0.0039
32,000	0.0042
34,300	0.0045
36,400	0.0048
38,500	0.0051
40,700	0.0054
42,700	0.0057
44,600	0.0060
46,400	0.0063
48,100	0.0066
49,700	0.0069
51,400	0.0072
52,800	0.0075
54,500	0.0078

4.4 A hollow cast-iron cylinder with a 10-cm gage length, 8.26-cm inside diameter, and 8.77-cm outside diameter was tested in torsion. From the data in Table P4.4, plot a stress-strain diagram and determine G.

Table P4.4

Torque (mN)	Twist (rad)
554	0.00051
893	0.00092
1288	0.00139
1695	0.00185
2085	0.00231
2333	0.00275
2797	0.00324
3124	0.00370
3407	0.00414
3769	0.00464
3927	0.00508
4187	0.00556
4384	0.00600
4554	0.00651
4752	0.00700
4910	0.00741
5006	0.00797
5198	0.00859
5384	0.00924
5531	0.01000
5707	0.01092
5763	0.01179
5933	0.01283

4.3 Linearly Elastic (Hookean) Behavior

The analysis of any engineering problem requires a suitable mathematical description of the physical situation. The problem must be sufficiently simple mathematically to permit an analysis, yet realistic enough so that the solution will predict the response of the physical system with reasonable accuracy. As such, the mathematical description is an idealization. Our immediate attention focuses on the simplest idealization of solid materials, those that exhibit attributes of *elasticity*, *isotropy*, and *linearity* in the relations between strain, stress, and temperature. Linearly elastic materials are often termed hookean after Robert Hooke, who proposed the idealization. Since most of our applications also presume isotropy, we employ the acronym HI for hookean isotropic. When the HI material is also homogeneous—that is, the properties are constant throughout the body—we use the elaboration HI-HO, which represents hookean, isotropic, and homogeneous. The reader will be burdened with no other acronyms; HI and HI-HO materials are so prevalent in our introductory applications that such abbreviation seems justified.

Experiments indicate that most engineering materials exhibit a linear relation between the applied forces and displacements—and, therefore, between the average stress and the corresponding average strains—so long as such forces or stresses do not exceed certain limits. Furthermore, the observed behavior is also elastic within this limited range. These elastic deformations are usually quite small, so that the strain components are small compared to unity ($\varepsilon, \gamma \ll 1$). In view of these observations, it is natural to postulate linear relations between the components of stress and strain in order to describe the small elastic deformations of such materials. Furthermore, many of the common materials are isotropic on a macroscopic scale, so that the relations between components of stress and strain do not depend upon the directions associated with the components.

Although much of our study employs the HI-HO idealization, the reader must be alert to limitations. The engineer often employs materials that exhibit nonlinear or inelastic behavior and intentionally designs with anisotropic materials such as composites, wherein filaments, or fibers of high strengths, are oriented to give directional properties of strength or stiffness.

4.4 Consequences of Linearity —Superposition of Strains

In the preceding study of strain (Sec. 3.10), we observed that small strains can be superposed. If two states of small strain are imposed, whether successively or simultaneously, they are additive; for example, if a body experiences two small strains, ε_x' and ε_x'', the practical result is a small strain,

$$\varepsilon_x \doteq \varepsilon_x' + \varepsilon_x'' \tag{4.10}$$

"Small" means small *compared* to unity; that is, $\varepsilon_x' \ll 1$, $\varepsilon_x'' \ll 1$, which justifies the neglect of higher-order terms.

Our study of stress preceded any thorough discussion of deformation, but clearly the imposition of stress (or temperature) causes deformation, and, in particular, the area of a surface is generally altered. For example, a simple stress $\bar{\sigma}_x' = F_x'/A'$ causes a change in area ($A' \to A''$) such that an additional force F_x'' results in a stress $\bar{\sigma}_x = (F_x' + F_x'')/A''$. Then, strictly speaking, $\bar{\sigma}_x \neq \bar{\sigma}_x' + \bar{\sigma}_x''$. However, if the strains are small, then the dilation of the area is also small; specifically, if x, y, and z are principal directions, then

$$A'' = (1 + \varepsilon_y')(1 + \varepsilon_z')A'$$

Again, if the strains are small compared to unity, then $A'' \doteq A'$ and

$$\sigma_x \doteq \sigma_x' + \sigma_x'' \tag{4.11}$$

The superposition of strains and stresses applies to arbitrary states provided that the former are small compared to unity. We now use superposition as a simple and direct way to synthesize the linear strain-stress-temperature equations for HI material.

4.5 Hookean-Isotropic (HI) Behavior

Consider the simple state of stress depicted in Fig. 4.16. If the material is hookean (a typical metal within the proportional limit), then the stress causes a proportional extensional strain

$$\varepsilon_x' = \frac{1}{E}\sigma_x \tag{4.12}$$

Fig. 4.16

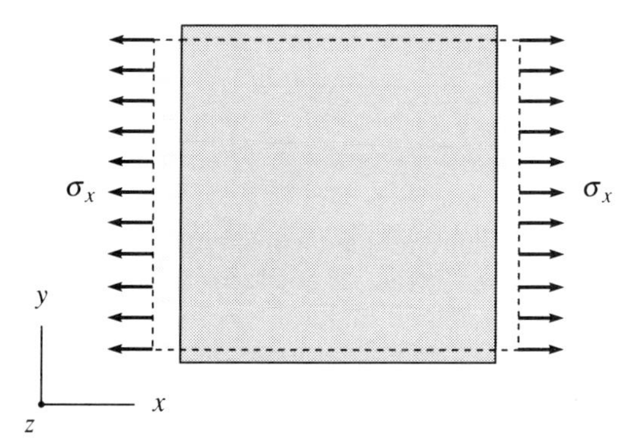

The coefficient $(1/E)$ can only be inferred from the tension and/or compression tests; see Eq. (4.4). Also, from such tests we observe proportional strains (contractions) in the transverse directions (y and z):

$$\varepsilon_y' = \varepsilon_z' = -\frac{\nu}{E}\sigma_x \tag{4.13}$$

These are identical, since the material is *isotropic*. The coefficient $(-\nu/E)$ must be inferred from tests; see Eqs. (4.5) and (4.6). The constant E is the modulus of elasticity (also known as Young's modulus). It has the dimensions of stress [force/length^2]. The constant ν, known as Poisson's ratio, is dimensionless. The modulus of elasticity can be determined by a simple tension or compression test. However, Poisson's ratio is not normally obtained in this way.

Now suppose that the same block of material is acted upon by the uniform stress σ_y only. Because the material is isotropic, it shows no preference for any direction. Consequently, the resulting extensional strains are

$$\varepsilon_y'' = \frac{1}{E}\sigma_y, \qquad \varepsilon_x'' = \varepsilon_z'' = -\frac{\nu}{E}\sigma_y$$

and if the normal stress σ_z acts alone, the resulting extensional strains are

$$\varepsilon_z''' = \frac{1}{E}\sigma_z, \qquad \varepsilon_x''' = \varepsilon_y''' = -\frac{\nu}{E}\sigma_z$$

In view of the linearity of these relations, the small strains caused by the combined application of the normal stresses, σ_x, σ_y, and σ_z, are obtained by superposition. For example, the resulting strain ε_x is the sum of the strains ε_x', ε_x'', and ε_x''' caused by each of the stress components. The effects of this combination of stress components are the strain components

$$\varepsilon_x = \frac{1}{E}[\sigma_x - \nu(\sigma_y + \sigma_z)] \tag{4.14}$$

$$\varepsilon_y = \frac{1}{E}[\sigma_y - \nu(\sigma_z + \sigma_x)] \tag{4.15}$$

$$\varepsilon_z = \frac{1}{E}[\sigma_z - \nu(\sigma_x + \sigma_y)] \tag{4.16}$$

To this point we have considered only the relations between the normal stresses and extensional strains associated with the perpendicular directions x, y, and z. In the case of an isotropic material, the block of Fig. 4.16 remains rectangular under the action of the normal stresses σ_x, σ_y, σ_z. That is, these normal stresses cause no shear strains γ_{xy}, γ_{yz}, γ_{zx}. This is a consequence of *isotropy* and is *not limited to hookean behavior.* It implies that principal directions of stress ($\tau_{xy} = \tau_{xz} = \tau_{yz} = 0$) are also principal directions of strain ($\gamma_{xy} = \gamma_{xz} = \gamma_{yz} = 0$) and vice versa.

Consider now the state of stress depicted in Fig. 4.17(a); the only nonzero components are $\sigma'_x = \tau = -\sigma'_y$. By the equations of transformation (see Sec. 2.10, Fig. 2.52, and/or Sec. 2.11), we obtain the components of stress in the rectangular system x, y, z, wherein x', y' are rotated 45° about z:

$$\begin{matrix} \sigma_x & \tau_{xy} & \tau_{xz} \\ & \sigma_y & \tau_{yz} \\ & & \sigma_z \end{matrix} = \begin{matrix} 0 & \tau & 0 \\ & 0 & 0 \\ & & 0 \end{matrix}$$

In other words, this is a state of simple shear, $\tau_{xy} = \tau_{yx} = \tau$, as depicted in Fig. 4.17(b).

Fig. 4.17

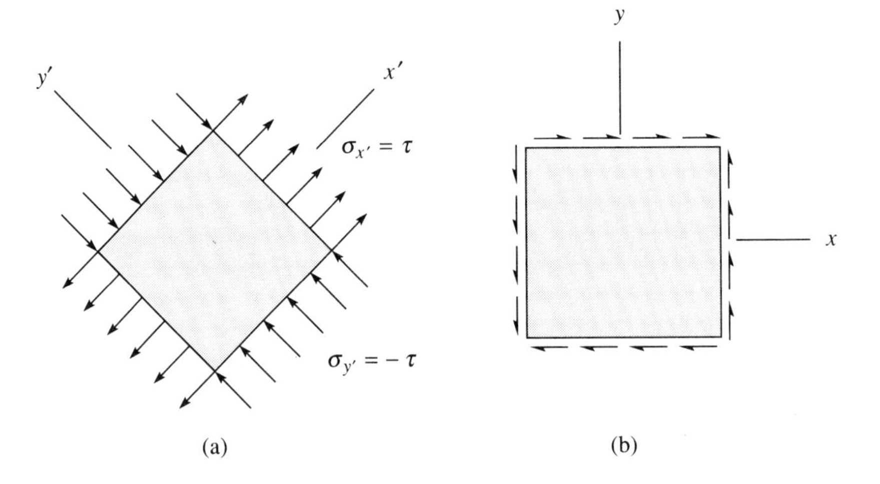

By Eqs. (4.14), (4.15), and (4.16), the state of strain for this HI material has the following components in the system (x', y', z'):

$$\varepsilon_{x'} = +\frac{(1+\nu)}{E}\tau, \qquad \varepsilon_{y'} = -\frac{(1+\nu)}{E}\tau$$

$$\varepsilon_{z'} = \gamma_{x'y'} = \gamma_{x'z'} = \gamma_{y'z'} = 0$$

The components in the system (x, y, z) are obtained by transformations (3.19) and (3.20) (see Sec. 3.12 and/or Sec. 3.13):

$$\varepsilon_x = \varepsilon_y = \varepsilon_z = \gamma_{yz} = \gamma_{xz} = 0$$

$$\gamma_{xy} = \frac{2(1+\nu)}{E}\tau = \frac{2(1+\nu)}{E}\tau_{xy} \tag{4.17}$$

One might infer that a simple state of shear stress τ_{xy}, as shown in Fig. 4.17(b), causes only the shear strain γ_{xy}. That conclusion is supported by experiments on HI materials, such as the torsional test of Fig. 4.13. However, such experiments alone do not show that the coefficient is expressed in terms of the modulus E and Poisson ratio ν in the form $2(1 + \nu)/E$. Indeed, another property, the shear modulus, can be defined as follows:

$$G \equiv \frac{E}{2(1+\nu)} \tag{4.18}$$

The modulus G might be obtained by the test of Fig. 4.13 and the modulus E, by the tension test; then the Poisson ratio can be determined by Eq. (4.18). In any case, we note that the complete linear system of strain-stress equations for the HI material incorporates *only two* independent coefficients (mechanical properties).

Since the behavior is isotropic, two equations like (4.17) apply to the other rectangular components, τ_{xz} and τ_{yz}. These and Eqs. (4.14), (4.15), (4.16), and (4.17) constitute the six strain-stress relations for the HI material. For convenience these are collected here:

$$\varepsilon_x = \frac{1}{E}[\sigma_x - \nu(\sigma_y + \sigma_z)] \tag{4.19}$$

$$\varepsilon_y = \frac{1}{E}[\sigma_y - \nu(\sigma_z + \sigma_x)] \tag{4.20}$$

$$\varepsilon_z = \frac{1}{E}[\sigma_z - \nu(\sigma_x + \sigma_y)] \tag{4.21}$$

$$\gamma_{xy} = \frac{1}{G}\tau_{xy} \tag{4.22}$$

$$\gamma_{yz} = \frac{1}{G}\tau_{yz} \tag{4.23}$$

$$\gamma_{zx} = \frac{1}{G}\tau_{zx} \tag{4.24}$$

Often it is necessary to compute the stress components from a knowledge of the strain components. For example, the strains may be determined by experimental means, using electric resistance strain gages. If the material is linearly elastic, the stresses may be found by solving Eqs. (4.19) through (4.24) for the stresses in terms of strains. The solution follows:

$$\sigma_x = \frac{E}{(1 - 2\nu)(1 + \nu)}[(1 - \nu)\varepsilon_x + \nu(\varepsilon_y + \varepsilon_z)] \tag{4.25}$$

$$\sigma_y = \frac{E}{(1 - 2\nu)(1 + \nu)}[(1 - \nu)\varepsilon_y + \nu(\varepsilon_z + \varepsilon_x)] \tag{4.26}$$

$$\sigma_z = \frac{E}{(1 - 2\nu)(1 + \nu)}[(1 - \nu)\varepsilon_z + \nu(\varepsilon_x + \varepsilon_y)] \tag{4.27}$$

$$\tau_{xy} = G\gamma_{xy} \tag{4.28}$$

$$\tau_{yz} = G\gamma_{yz} \tag{4.29}$$

$$\tau_{zx} = G\gamma_{zx} \tag{4.30}$$

One additional property is occasionally employed in descriptions of HI materials; it is the coefficient of proportionality between a hydrostatic state of stress (see Sec. 2.10) and the dilatation (see Sec. 3.16). In the present case, strains are small so that the dilatation (or volumetric strain) is given by the linear equation

$$e = \varepsilon_x + \varepsilon_y + \varepsilon_z$$

It follows from (4.19), (4.20), and (4.21) that

$$e = \frac{3(1 - 2\nu)}{E} \cdot \frac{(\sigma_x + \sigma_y + \sigma_z)}{3} \tag{4.31}$$

The factor $(\sigma_x + \sigma_y + \sigma_z)/3$ is the *mean* normal stress, also termed the hydrostatic part of the state of stress. Then the coefficient of proportionality characterizes the resistance to dilatation (i.e., volumetric change). The reciprocal, like E and G, has the dimensions of stress; like E and G, it is called a modulus, the *bulk modulus*:

$$K = \frac{E}{3(1 - 2\nu)} \tag{4.32}$$

We recall that the modulus E and the Poisson ratio are positive; also, the bulk modulus must be positive. Therefore, the denominator of (4.32) must remain positive; it follows that

$$0 < \nu \leq \tfrac{1}{2}$$

As the Poisson ratio approaches $\frac{1}{2}$, the bulk modulus approaches infinity; stated otherwise, no amount of pressure can cause dilatation. When $\nu = \frac{1}{2}$, the HI material is said to be incompressible.

One final word about HI behavior: Materials can be hookean and elastic, yet nonhomogeneous. For example, a composite might be formed with particles disbursed more densely in some portions; then the properties (E, G, K, v) would be variable. However, most common materials (e.g., metals) are nearly homogeneous; then the properties are constant throughout. To avoid ambiguity, we identify these with HI-HO.

PROBLEMS

4.5 Hooke's law is frequently written in the form

$$\sigma_x = \lambda e + 2\mu\varepsilon_x, \quad \sigma_y = \lambda e + 2\mu\varepsilon_y, \quad \sigma_z = \lambda e + 2\mu\varepsilon_z$$

$$\tau_{xy} = \mu\gamma_{xy}, \quad \tau_{yz} = \mu\gamma_{yz}, \quad \tau_{xz} = \mu\gamma_{xz}$$

The constants λ and μ are called *Lame's constants*, and $e = \varepsilon_x + \varepsilon_y + \varepsilon_z$ is called the *dilatation*. What are the relationships (a) between E, v and λ, μ and (b) between E, G and λ, μ?

4.6 A body is in a plane state of strain if $\varepsilon_z = \gamma_{xz} = \gamma_{yz} = 0$ throughout the body. Show that Hooke's law, using $E^* = E/(1 - v^2)$ and $v^* = v/(1 - v)$, may be written in the two-dimensional form

$$\varepsilon_x = \frac{1}{E^*}(\sigma_x - v^*\sigma_y), \qquad \varepsilon_y = \frac{1}{E^*}(\sigma_y - v^*\sigma_x),$$

$$\gamma_{xy} = \frac{2(1 + v^*)}{E^*}\tau_{xy}$$

4.7 A body is in a plane state of stress if $\sigma_z = \tau_{xz} = \tau_{yz} = 0$ throughout the body. Show that Hooke's law can be written in the form

$$\sigma_x = \lambda^* e^* + 2\mu\varepsilon_x, \quad \sigma_y = \lambda^* e^* + 2\mu\varepsilon_y, \quad \tau_{xy} = \mu\gamma_{xy}$$

where $e^* = \varepsilon_x + \varepsilon_y$, $\lambda^* = 2\mu\lambda/(\lambda + 2\mu)$ and λ, μ are Lame's constants, defined in Prob. 4.5.

4.8 Show that the dilatation $e = \varepsilon_x + \varepsilon_y + \varepsilon_z$ is related to the mean normal stress $\sigma_m = 1/3(\sigma_x + \sigma_y + \sigma_z)$ by

$$e = \frac{3(1 - 2v)}{E}\sigma_m$$

4.9 An incompressible material can experience no volume change. If the material obeys Hooke's law and the strains are small, what must Poisson's ratio be? (See Prob. 4.8).

4.10 A rectangular xyz-coordinate system is constructed at a point P on the surface of a linearly elastic body, with the xy-plane tangent to the surface. The n-direction in the xy-plane is defined as shown in Fig. P4.10. If $\sigma_z = \tau_{xz} = \tau_{yz} = \tau_{xy} = 0$, show that

$$\sigma_x = \frac{E}{1 + v}\varepsilon_n$$

Fig. P4.10

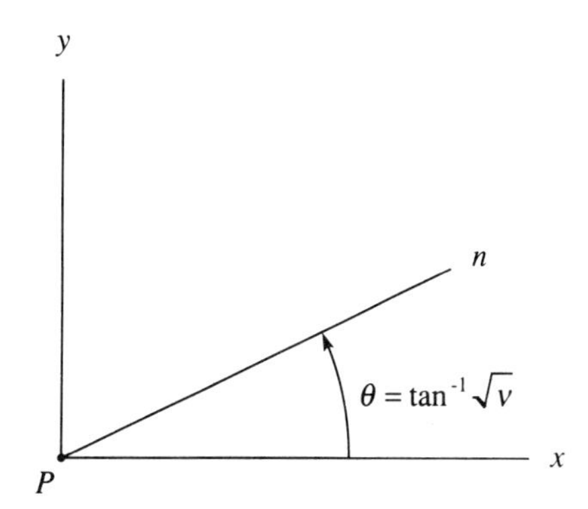

4.11 Show that, for a linearly elastic isotropic material, the principal directions of strain coincide with the principal directions of stress.

4.12 A steel member ($E = 30 \times 10^6$ lb/in.^{2}, $v = 0.3$) is subjected to the stresses

$$\sigma_x = 15{,}000 \text{ lb/in.}^2, \quad \sigma_y = -5000 \text{ lb/in.}^2, \quad \sigma_z = 0$$

$$\tau_{xy} = -8000 \text{ lb/in.}^2, \quad \tau_{xz} = 0, \quad \tau_{yz} = 0$$

Determine the principal strains and the principal directions.

4.13 For a steel, $E = 200 \times 10^3$ MPa, and $v = \frac{1}{3}$. Determine the state of stress that corresponds to the following state of strain if the material obeys Hooke's law:

$$\varepsilon_x = 0.001, \quad \varepsilon_y = -0.005, \quad \varepsilon_z = 0$$
$$\gamma_{xy} = -0.0025, \quad \gamma_{yz} = -0.0025, \quad \gamma_{zx} = 0$$

4.14 For aluminum ($E = 10^7$ lb/in.^{2}, $\nu = 0.3$), compute the state of strain that corresponds to the following state of stress if the material obeys Hooke's law:

$$\sigma_x = 30{,}000 \text{ lb/in.}^2, \quad \sigma_y = -15{,}000 \text{ lb/in.}^2,$$
$$\sigma_z = -15{,}000 \text{ lb/in.}^2$$
$$\tau_{xy} = 5{,}000 \text{ lb/in.}^2, \quad \tau_{yz} = 7{,}500 \text{ lb/in.}^2,$$
$$\tau_{xz} = 10{,}000 \text{ lb/in.}^2$$

4.15 The state of stress at a point P in a body is given by

$$\sigma_x = 70 \text{ MPa}, \quad \sigma_y = 105 \text{ MPa}, \quad \sigma_z = 0$$
$$\tau_{xy} = 70 \text{ MPa}, \quad \tau_{yz} = 0, \quad \tau_{xz} = 0$$

If the material obeys Hooke's law ($E = 84 \times 10^3$ MPa, $\nu = 0.25$) compute the extensional strain in the direction $(1/\sqrt{3}, 1/\sqrt{3}, 1/\sqrt{3})$ at P.

4.16 The state of strain at a point P in a body is given by

$$\varepsilon_x = 0.0006 \quad \varepsilon_y = -0.0003 \quad \varepsilon_z = 0.003$$
$$\gamma_{yz} = -0.001, \quad \gamma_{xz} = 0, \quad \gamma_{xy} = 0.001$$

If the material obeys Hooke's law with $E = 30 \times 10^6$ psi and $\nu = \frac{1}{3}$, what is the component of stress σ_n where the direction n is specified by $(1/2, 1/2, 1/\sqrt{2})$?

4.17 The principal strains at a point of a hookean isotropic body are $\varepsilon_1 = 100 \times 10^{-5}$, $\varepsilon_2 = -200 \times 10^{-5}$, and $\varepsilon_3 = 120 \times 10^{-5}$. The modulus and Poisson ratio are $E = 200 \times 10^3$ MPa, $\nu = 0.25$. Determine the mean normal stress, maximum normal stress, and maximum shear stress.

4.18 At a point in a steel member ($E = 30 \times 10^3$ ksi, $\nu = \frac{1}{3}$, six components of stress are known in a rectangular system (x, y, z):

$$\sigma_x = 80 \text{ ksi}, \quad \tau_{xy} = 40 \text{ ksi}, \quad \tau_{xz} = 60 \text{ ksi}$$
$$\sigma_y = -40 \text{ ksi}, \quad \tau_{yz} = 20 \text{ ksi}$$
$$\sigma_z = 100 \text{ ksi}$$

Determine the extensional strain in a direction n; direction cosines are $(\sqrt{3}/2, 1/2, 0)$ relative to the rectangular system (x, y, z).

4.19 Strain components in a rectangular system are

$$\begin{matrix} \varepsilon_x, & \gamma_{xy}, & \gamma_{xz} \\ & \varepsilon_y, & \gamma_{yz} \\ & & \varepsilon_z \end{matrix} = \begin{matrix} 120, & 0, & 0 \\ & -80, & 0 \\ & & 150 \end{matrix} \times 10^{-6}$$

If this material is steel ($E = 200 \times 10^3$ MPa, $\nu = 0.25$), compute the maximum normal stress and maximum shear stress.

4.20 A state of stress is given by the components in a rectangular system (x, y, z):

$$\begin{matrix} \sigma_x, & \tau_{xy}, & \tau_{xz} \\ & \sigma_y, & \tau_{yz} \\ & & \sigma_z \end{matrix} = \begin{matrix} 10 & 0 & 0 \\ & 50 & -30 \\ & & -30 \end{matrix} \text{ ksi}$$

The material is steel with HI properties $E = 30 \times 10^3$ ksi, $\nu = 0.25$. Determine the principal directions of stress and strain, the maximum extensional strain, and the maximum shear stress.

4.6 Plane Stress of a Hookean Isotropic Body

No stress acts upon a free surface (i.e., a surface not contacting another body). Therefore, the normal to a free surface identifies a principal direction of stress; moreover, that principal stress is zero. If z denotes the normal direction, then $\tau_{xz} = \tau_{yz} = \sigma_z = 0$. According to Eqs. (4.19) through (4.24),

$$\varepsilon_x = \frac{1}{E}(\sigma_x - \nu\sigma_y) \tag{4.33a}$$

$$\varepsilon_y = \frac{1}{E}(-\nu\sigma_x + \sigma_y) \tag{4.33b}$$

$$\varepsilon_z = -\frac{\nu}{E}(\sigma_x + \sigma_y) \tag{4.33c}$$

$$\gamma_{xy} = \frac{1}{G}\tau_{xy}, \qquad \gamma_{xz} = \gamma_{yz} = 0 \tag{4.34a,b,c}$$

As previously noted, the normal direction is also a principal direction of strain. Note, too, that the extensional strain in the normal direction (ε_z) is *not* generally zero.

The inverse of the preceding equations gives the nonvanishing stresses:

$$\sigma_x = \frac{E}{1-\nu^2}(\varepsilon_x + \nu\varepsilon_y) \tag{4.35a}$$

$$\sigma_y = \frac{E}{1-\nu^2}(\nu\varepsilon_x + \varepsilon_y) \tag{4.35b}$$

$$\tau_{xy} = G\gamma_{xy} \tag{4.36}$$

Since the extensional strains can be measured on the free surface (see Sec. 3.20), the strains for any directions in the surface (e.g., ε_x, ε_y, γ_{xy}) can be determined. Then, by means of the preceding equations, the nonzero stresses ($\sigma_x, \sigma_y, \tau_{xy}$) can be determined and, by means of (4.33c), the extensional strain (ε_z) can be found.

Because the normal to the free surface is a principal direction, the remaining two principal directions lie in the surface. Consequently, they can be determined by the special forms of Sec. 3.18 or 2.11.

Since the strains in a *free* surface are readily measured by electrical-resistance gages (see Sec. 3.20) and since three components of stress are zero, one can determine the states of strain *and* stress *if* the body is known to be composed of the HI-HO material. An example serves to illustrate such computation.

EXAMPLE 1

States of Strain and Stress at the Surface of a HI-HO Body The strain gage depicted in Fig. 4.18 measures the extensional strains in the directions *a*, *b*, and *c* on the free surface of a steel member. The behavior is HI within the elastic limit; the modulus and Poisson ratio are $E = 30 \times 10^3$ ksi (approximately 210 MPa) and $\nu = 0.25$. The measured values are

$$\varepsilon_a = 100 \times 10^{-5}, \qquad \varepsilon_c = 40 \times 10^{-5}, \qquad \varepsilon_b = -20 \times 10^{-5}$$

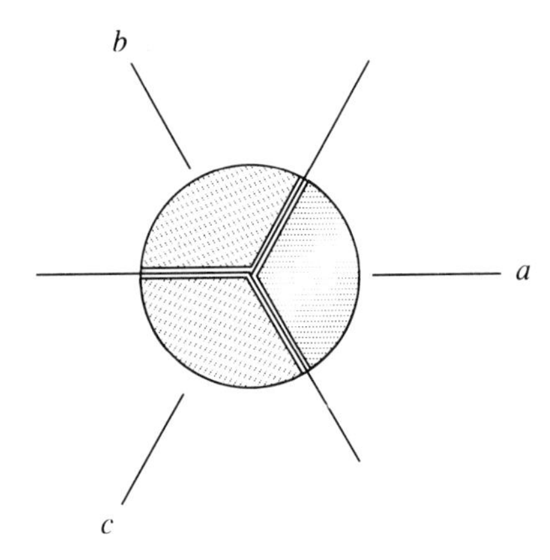

Fig. 4.18
Strain gage—120° rosette

We refer to the equations of transformations as given in Sec. 3.20 and let x denote the direction of a, y denote the orthogonal direction in the surface, and z denote the normal direction. Then

$$\varepsilon_c = \varepsilon_a \cos^2 120^\circ + \varepsilon_y \sin^2 120^\circ + \gamma_{xy} \sin 120^\circ \cos 120^\circ$$

$$\varepsilon_b = \varepsilon_a \cos^2 240^\circ + \varepsilon_y \sin^2 240^\circ + \gamma_{xy} \sin 240^\circ \cos 240^\circ$$

Substituting the numerical values, we have the two equations:

$$40 \times 10^{-5} = 100 \times 10^{-5} \times \frac{1}{4} + \varepsilon_y \frac{3}{4} - \gamma_{xy} \frac{\sqrt{3}}{4}$$

$$-20 \times 10^{-5} = 100 \times 10^{-5} \times \frac{1}{4} + \varepsilon_y \frac{3}{4} + \gamma_{xy} \frac{\sqrt{3}}{4}$$

The solution follows:

$$\varepsilon_y = -20 \times 10^{-5}, \qquad \gamma_{xy} = -40\sqrt{3} \times 10^{-5}$$

Since the normal (z) is a principal direction of stress ($\sigma_z = \tau_{zx} = \tau_{zy} = 0$) and the material is isotropic, the normal (z) is also a principal direction of strain. The remaining principal directions lie in the surface; they are determined by Eq. (3.53):

$$\tan 2\theta = \frac{\gamma_{xy}}{(\varepsilon_x - \varepsilon_y)}$$

$$= \frac{-40\sqrt{3}}{100 + 20} = -\frac{1}{\sqrt{3}}$$

or

$$2\theta = -30^\circ, \quad +150^\circ$$

$$\theta = -15^\circ, \quad +75^\circ$$

The principal strains are then obtained by Eq. (3.49):

$$\varepsilon_1(-15^\circ) = (40 + 40\sqrt{3}) \times 10^{-5} = 109 \times 10^{-5}$$

$$\varepsilon_2(+75^\circ) = (40 - 40\sqrt{3}) \times 10^{-5} = -29 \times 10^{-5}$$

In the foregoing computations we employed only the isotropic property. The principal stresses now follow from Eqs. (4.35a,b):

$$\sigma_1 = \frac{E}{1-\nu^2}(\varepsilon_1 + \nu\varepsilon_2) = 32.6 \text{ ksi}$$

$$\varepsilon_2 = \frac{E}{1-\nu^2}(\varepsilon_2 + \nu\varepsilon_1) = -0.6 \text{ ksi}$$

The extensional strain in the third principal direction is given by Eq. (4.33c):

$$\varepsilon_z = -\frac{\nu}{E}(\sigma_1 + \sigma_2) = -26 \times 10^{-5}$$

Often the maximum shear stress is important; by Eq. (2.63),

$$\tau_{\max} = \tfrac{1}{2}|\sigma_{\max} - \sigma_{\min}|$$
$$= \tfrac{1}{2}(32.6 + 0.6) = 16.6 \text{ ksi}$$

Alternatively, the maximum shear strain is given by Eq. (3.55)

$$\gamma_{\max} = |\varepsilon_{\max} - \varepsilon_{\min}|$$
$$= (109 + 29) \times 10^{-5}$$
$$= 138 \times 10^{-5}$$

Then, by Eq. (4.36),

$$\tau_{\max} = G\gamma_{\max} = \frac{E}{2(1+\nu)}\gamma_{\max} = 16.6 \text{ ksi}$$

The latter computation serves only as an additional exercise and confirmation of our arithmetic. ◆

PROBLEMS

4.21 On the *free* surface of a HI-HO body, extensional strains are measured in perpendicular directions (x, y):

$$\varepsilon_x = 10 \times 10^{-5}, \qquad \varepsilon_y = 16 \times 10^{-5}$$

For this material

$$E = 12 \times 10^3 \text{ ksi}, \qquad \nu = 0.30$$

a. Compute the normal stress components, σ_x and σ_y.

b. If $\gamma_{xy} = 0$, compute the maximum shear stress at this point of the body.

4.22 Extensional strains along the principal lines in a stress-free surface of a HI-HO body are

$$\varepsilon_1 = 20 \times 10^{-5}, \quad \varepsilon_2 = 100 \times 10^{-5}$$

The modulus and Poisson ratio are

$$E = 5 \times 10^6 \text{ lb/in.}^2 \qquad \nu = 0.25$$

Determine (a) the principal stresses, (b) the maximum shear stress, and (c) the third principal strain.

4.23 At a point on the free surface of a HI-HO member, two principal stresses are

$$\sigma_1 = 200 \text{ MPa}, \qquad \sigma_2 = 70 \text{ MPa}$$

The modulus and Poisson ratio are

$$E = 200 \times 10^3 \text{ MPa}, \qquad \nu = 0.25$$

Determine the three principal strains and the maximum shear strain.

4.24 On the free surface of a steel member ($E = 200 \times 10^3$ MPa, $\nu = 0.25$) strains are measured in the three directions (a, b, c) (see Fig. P4.24):

$$\varepsilon_a = 60 \times 10^{-5}, \quad \varepsilon_b = 20 \times 10^{-5}, \quad \varepsilon_c = -20 \times 10^{-5}$$

Compute (a) the components of stress, σ_a, σ_c, and τ_{ac}, and (b) the extensional strain, ε_z (z is normal to the surface).

Fig. P4.24

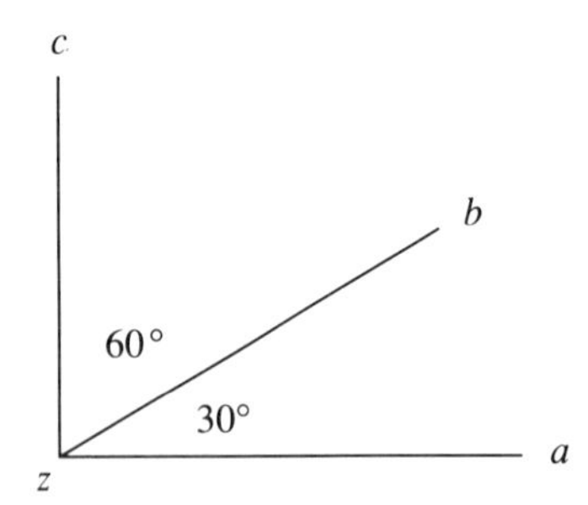

4.25 Components of stress at a point P (see Fig. P4.25) on the free surface of an aluminum member ($E = 12 \times 10^6$ psi, $\nu = 0.30$) are

$$\sigma_x = 40 \text{ ksi}, \qquad \sigma_n = 20 \text{ ksi}, \qquad \sigma_y = -60 \text{ ksi}$$

Determine the following components of stress and strain:

$$\tau_{xy}, \qquad \gamma_{xy}, \quad \text{and} \quad \varepsilon_x$$

Fig. P4.25

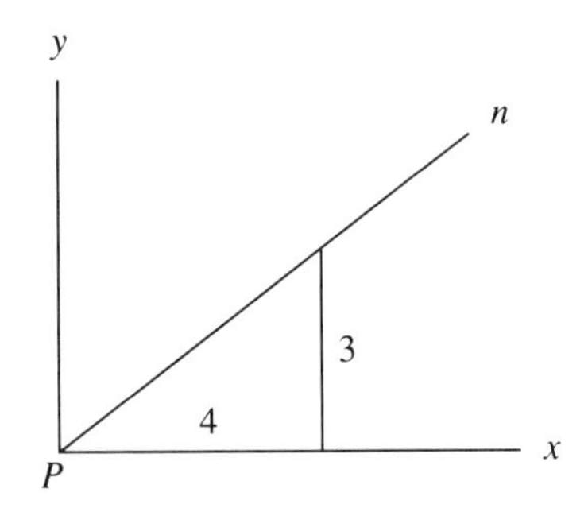

4.26 Within the free surface of a steel member (see Tables 4.1 and 4.2 on pp. 251–252) two principal strains are

$$\varepsilon_1 = -30 \times 10^{-5}, \qquad \varepsilon_2 = 90 \times 10^{-5}$$

The yield stress of this steel is 180 MPa in tension and in compression, but yielding is initiated by the maximum shear stress of 90 MPa. If the behavior is HI-HO to the onset of yielding, compute the maximum normal stress and maximum shear stress in this state of strain and justify the HI-HO assumption.

4.27 A state of plane stress exists at the free surface of a HI-HO body. Strain measurements are taken at a point P with a 45° strain rosette. The extensional strains in the directions 1, 2, and 3 on the surface (Fig. P4.27) and the elastic properties, E and ν, are given here. (i) Locate the principal directions relative to direction 1, (ii) obtain the principal stresses, and (iii) determine the maximum shear stress.

a. $\varepsilon_1 = 20 \times 10^{-5}$, $\varepsilon_2 = 30 \times 10^{-5}$, $\varepsilon_3 = 40 \times 10^{-5}$, $E = 30 \times 10^3$ ksi, $\nu = 0.25$

b. $\varepsilon_1 = 100 \times 10^{-4}$, $\varepsilon_2 = 30 \times 10^{-4}$, $\varepsilon_3 = 20 \times 10^{-4}$, $E = 200 \times 10^3$ MPa, $\nu = 0.25$

c. $\varepsilon_1 = 90 \times 10^{-4}$, $\varepsilon_2 = 40 \times 10^{-4}$, $\varepsilon_3 = -10 \times 10^{-4}$, $E = 10 \times 10^3$ ksi, $\nu = 0.30$

Fig. P4.27

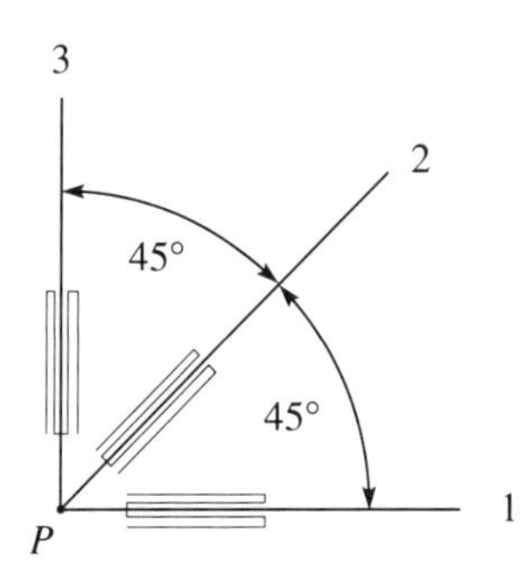

4.28 Strain measurements are recorded at a "point" P on the free surface of a HI-HO member with a 60° strain rosette as depicted in Fig. P4.28. From the values given here determine (i) the principal directions, (ii) the maximum normal stress, and (iii) the maximum shear stress:

a. $\varepsilon_1 = 10 \times 10^{-4}$, $\varepsilon_2 = 4 \times 10^{-4}$, $\varepsilon_3 = 4 \times 10^{-4}$, $E = 30 \times 10^3$ ksi, $\nu = 0.25$

b. $\varepsilon_1 = 40 \times 10^{-5}$, $\varepsilon_2 = 40 \times 10^{-5}$, $\varepsilon_3 = -60 \times 10^{-5}$, $E = 70 \times 10^3$ MPa, $\nu = 0.25$

Fig. P4.28

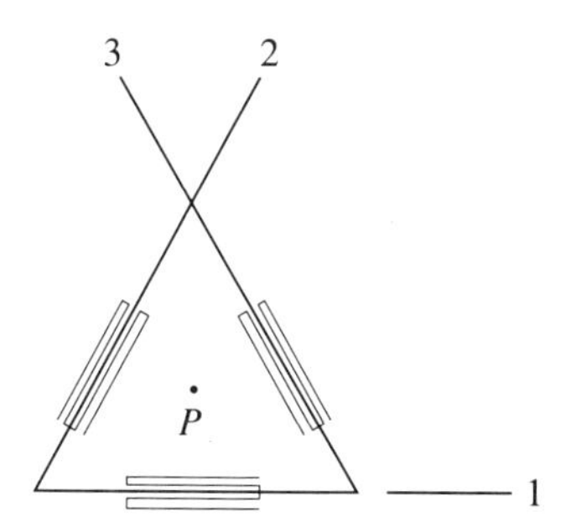

4.29 Three extensional strains are measured at a point on the free surface of a HI-HO member; see Fig. P4.29. With the following strains and properties, compute all components of stress and strain in the rectangular system (n, t, z).

$\varepsilon_x = 20 \times 10^{-5}$, $\varepsilon_n = 6 \times 10^{-5}$, $\varepsilon_t = 18 \times 10^{-5}$

$E = 30 \times 10^3$ ksi, $\nu = 0.30$

Fig. P4.29

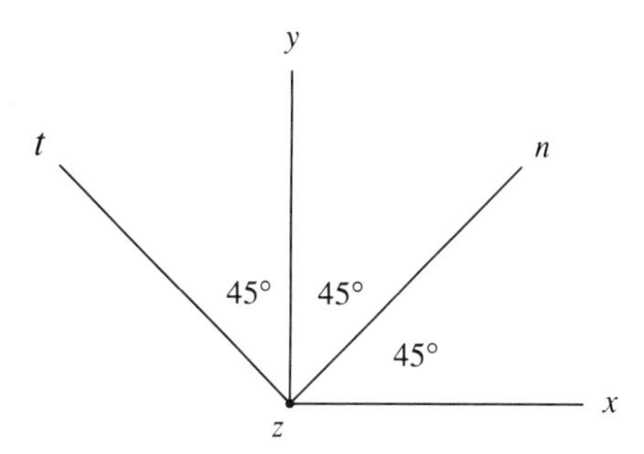

4.30 Extensional strains are measured at a point P on the free surface of steel part; $E = 200 \times 10^3$ MPa and $\nu = 0.25$. See Fig. P4.30.

$\varepsilon_1 = 20 \times 10^{-5}$, $\varepsilon_2 = 10 \times 10^{-5}$, $\varepsilon_3 = -30 \times 10^{-5}$

Determine (a) the normal stress σ_1, (b) the principal directions, and (c) the maximum shear stress.

Fig. P4.30

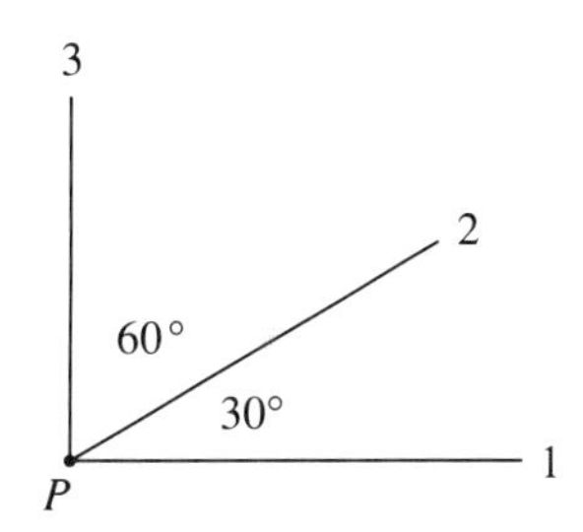

4.31 Extensional strains are measured at a point of a HI-HO member with the 60° rosette of Fig. P4.28:

$$(\varepsilon_1, \varepsilon_2, \varepsilon_3) = (20, -20, 10) \times 10^{-4}$$

The moduli for this material are $E = 30 \times 10^3$ ksi and $G = 12 \times 10^3$ ksi. Determine the normal stress σ_1.

***4.32** Measurements of extensional strain are made upon a free surface with the 45° rosette of Fig. P4.27. The material is HI-HO. Express the principal stresses in terms of the quantities ε_1, ε_2, ε_3, E, and ν.

***4.33** Repeat Prob. 4.32 for the 60° rosette of Fig. P4.28.

4.7 Effects of Temperature

Thermal Expansion

Almost all materials tend to expand when heated. The increase in volume that accompanies a rise in temperature is termed *thermal expansion*. Sometimes a temperature change has no other significant effect. Let us examine this tendency of materials to expand when heated.

If a material is isotropic and free to expand, every line experiences the same extensional strain when the temperature is increased. The freely expanding material experiences no shear deformation. The simplest relation between these extensional strains and the temperature changes that cause them are the linear relations

$$\varepsilon_x = \varepsilon_y = \varepsilon_z = \alpha\,\Delta Q \tag{4.37a,b,c}$$

where ΔQ denotes the temperature rise and α is a constant, the coefficient of thermal expansion. Experiments show that the linear relations of Eq. (4.37) are a satisfactory approximation for the solution of many engineering problems. The coefficient of thermal expansion α is a property of the particular material; some typical values are given in Tables 4.1 and 4.2 on pages 251 and 252.

A uniform rise in the temperature of a free homogeneous body tends to cause a uniform expansion of every portion. Consequently, strains result, although there may be no internal forces (or stresses) present. Nonuniform temperature distributions tend to cause greater expansion in some portions than in others, which may cause high stress intensities in some parts of the body. Sudden localized heating, called *thermal shock*, causes severe nonuniform temperature distributions that may fracture brittle materials.

The linear relations of Eq. (4.37) determine the strains that result from a temperature change ΔQ, *provided* that the material is free to expand. If the material is not free to expand, stresses arise. Consider, for example, the rectangular bar of Fig. 4.19. At some initial temperature, Q_1, there are no stresses or strains present. The lateral surfaces are free, but the end faces are constrained by rigid surfaces. Suppose that the bar is uniformly heated to the temperature Q_2. The bar tends to expand but is con-

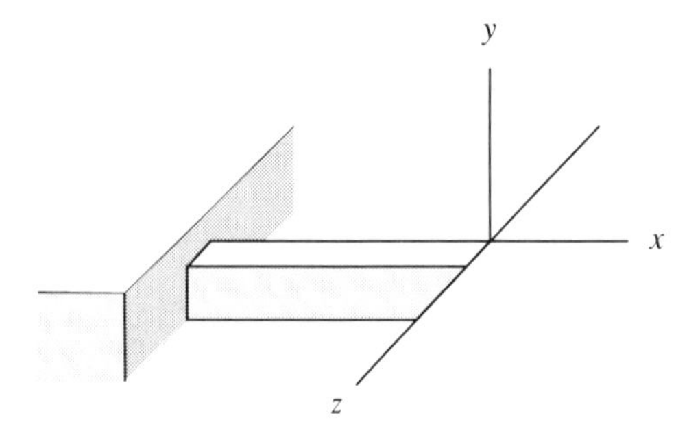

Fig. 4.19
Bar between rigid walls

strained by the rigid walls; hence, normal stresses are developed on the end faces. For simplicity we suppose that the conditions of stress and strain are uniform throughout the bar. If the bar were free, the effects of the temperature rise would be the extensional strains

$$\varepsilon_x' = \varepsilon_y' = \varepsilon_z' = \alpha\,\Delta Q = \alpha(Q_2 - Q_1)$$

The presence of an axial stress $\sigma_x = -p$, a pressure exerted on the end faces by the constraining walls, would also cause deformation. For linearly elastic behavior, the extensional strains caused by this stress are

$$e_x'' = -\frac{p}{E}, \qquad \varepsilon_z'' = \varepsilon_y'' = \left(\frac{\nu}{E}\right)p$$

When the bar is heated, both the effects of the rise in temperature, ΔQ, and the axial pressure, p, are present. Because the strains are small, these effects (strains) may be superimposed. Because of the constraining walls, the combined effect is zero net axial strain:

$$\varepsilon_x = \varepsilon_x' + \varepsilon_x'' = \alpha\,\Delta Q - \frac{p}{E} = 0 \tag{4.38}$$

Consequently, the axial stress induced by heating is

$$\sigma_x = -p = -E\alpha\,\Delta Q \tag{4.39}$$

The preceding is a simple example of a thermoelastic problem, in which strain-temperature and stress-strain relations are linear. Since, in such cases, the significant feature of the solution is the superposition of effects, the most general problem of this type may be treated by adding the thermal strain $\alpha\,\Delta Q$ to each one of Eqs. (4.19), (4.20), and (4.21). Often these linear relations are adequate for the solution of thermal stress problems of HI bodies. For future reference we collect complete strain-stress-temperature relations for the HI material:

$$\varepsilon_x = \frac{1}{E}(\sigma_x - \nu\sigma_y - \nu\sigma_z) + \alpha\,\Delta Q \tag{4.40a}$$

$$\varepsilon_y = \frac{1}{E}(-\nu\sigma_x + \sigma_y - \nu\sigma_z) + \alpha\,\Delta Q \tag{4.40b}$$

$$\varepsilon_z = \frac{1}{E}(-\nu\sigma_x - \nu\sigma_y + \sigma_z) + \alpha\,\Delta Q \tag{4.40c}$$

$$\gamma_{xy} = \frac{1}{G}\tau_{xy}, \qquad \gamma_{xz} = \frac{1}{G}\tau_{xz}, \qquad \gamma_{yz} = \frac{1}{G}\tau_{yz} \tag{4.41a,b,c}$$

Effects of Temperature on Mechanical Behavior

Temperature also influences the mechanical behavior of materials. At high temperatures, most materials soften and lose strength, and metals corrode faster. At low temperatures some ductile materials become brittle.

PROBLEMS

4.34 A steel spacer is subjected to a uniform temperature increase: $\Delta Q = 400°\text{F}$. It is constrained against any expansion in one direction z ($\varepsilon_z = 0$) but is otherwise free. Determine all components of stress. Refer to Tables 4.1 and 4.2 (pages 251–252) for typical properties of steel and assume HI-HO behavior.

4.35 If a HI-HO body were constrained against any deformation, what is the state of stress caused by a uniform temperature change?

4.36 If a HI-HO body remains stress-free—that is, unconstrained—during a temperature change, what is the state of strain caused by such change?

*4.8 Work and Energy

From our earlier analyses of deformation (Chapter 3) we recall that a sufficiently small rectangular element deforms to a parallelepiped, as depicted in Fig. 4.20. This is an adequate perception, because we eventually pass to the limit ($\Delta x, \Delta y, \Delta z \to 0$).

Fig. 4.20

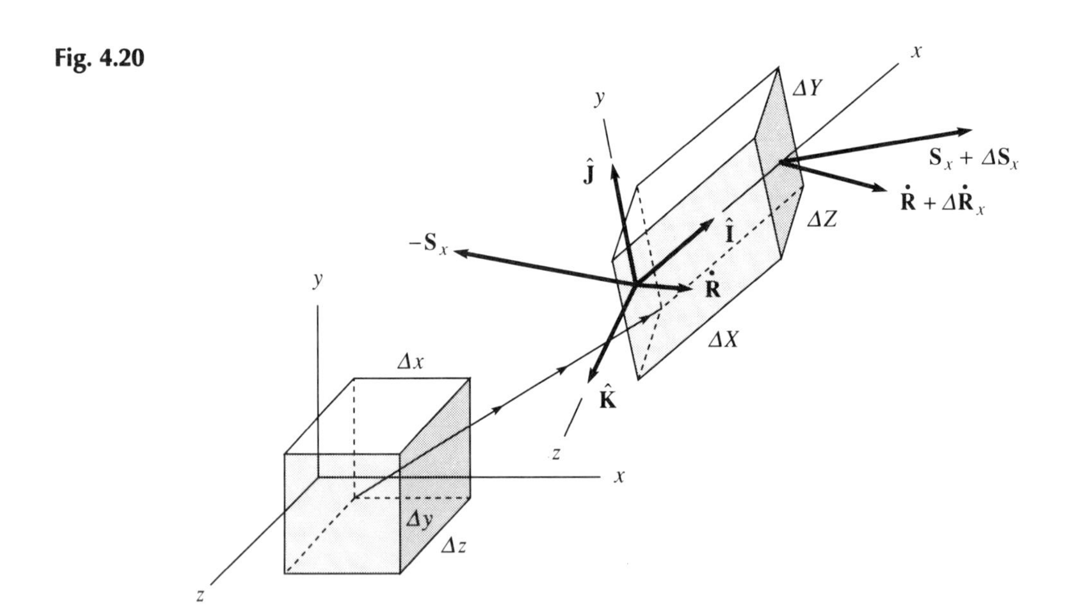

Unit vectors along the deformed x, y, z lines are designated $\hat{\mathbf{I}}$, $\hat{\mathbf{J}}$, and $\hat{\mathbf{K}}$, respectively. Deformed lengths along the edges are ΔX, ΔY, and ΔZ, where by the conventional definitions of the extensional strains, $\Delta X = (1 + \varepsilon_x)\,\Delta x$, and so on; therefore,

$$\hat{\mathbf{I}} = \frac{\partial \mathbf{R}}{\partial X} = \frac{1}{(1 + \varepsilon_x)} \frac{\partial \mathbf{R}}{\partial x}, \ldots$$

Otherwise stated,

$$\frac{\partial \mathbf{R}}{\partial x} = (1 + \varepsilon_x)\hat{\mathbf{I}}, \qquad \frac{\partial \mathbf{R}}{\partial y} = (1 + \varepsilon_y)\hat{\mathbf{J}}, \qquad \frac{\partial \mathbf{R}}{\partial z} = (1 + \varepsilon_z)\hat{\mathbf{K}} \tag{4.42}$$

To avoid confusion the stresses are shown only on the left (x) and right ($x + \Delta x$) faces. These stresses are $\mathbf{S}_x$ and $\mathbf{S}_x + \Delta\mathbf{S}_x$, respectively. Since we anticipate the limit ($\Delta x \to 0$), the first-order approximation suffices:

$$\Delta \mathbf{S}_x = \frac{\partial \mathbf{S}_x}{\partial x} \Delta x, \ldots$$

Now, consider a small incremental movement wherein the position, vector $\mathbf{R}$, is changed by an increment $\dot{\mathbf{R}}$. Also, the vector $\mathbf{R}$ changes with location, so that the movement on the right face is approximately

$$\Delta \dot{\mathbf{R}}_x = \frac{\partial \dot{\mathbf{R}}}{\partial x} \Delta x \tag{4.43}$$

The net work done by the force $-\mathbf{S}_x \Delta y \, \Delta z$ on the left face and by $(\mathbf{S}_x + \Delta\mathbf{S}_x) \Delta y \, \Delta z$ on the right face is

$$-\mathbf{S}_x \cdot \dot{\mathbf{R}} \, \Delta y \, \Delta z + (\mathbf{S} + \Delta \mathbf{S}_x) \cdot (\dot{\mathbf{R}} + \Delta \dot{\mathbf{R}}_x) \, \Delta y \, \Delta z$$

Similar expressions provide the work done by the forces upon the remaining (y and z) faces. Anticipating the limit, wherein the higher-order terms vanish, we give the essential terms in the incremental work:

$$\begin{aligned} \dot{W} = &\left(\mathbf{S}_x \cdot \frac{\partial \dot{\mathbf{R}}}{\partial x} + \mathbf{S}_y \cdot \frac{\partial \dot{\mathbf{R}}}{\partial y} + \mathbf{S}_z \cdot \frac{\partial \dot{\mathbf{R}}}{\partial z} \right) \Delta x \, \Delta y \, \Delta z \\ &+ \left(\frac{\partial \mathbf{S}_x}{\partial x} + \frac{\partial \mathbf{S}_y}{\partial y} + \frac{\partial \mathbf{S}_z}{\partial z} \right) \cdot \dot{\mathbf{R}} \, \Delta x \, \Delta y \, \Delta z \end{aligned} \tag{4.44}$$

The final parenthetical factor in Eq. (4.44) vanishes for an element in equilibrium; this follows from the equilibrium equation (2.84), as applied to the deformed element. The work (4.44) can be divided by the initial volume ($\Delta x \, \Delta y \, \Delta z$) to give the work per unit volume.[1]

$$\dot{w} = \left(\mathbf{S}_x \cdot \frac{\partial \dot{\mathbf{R}}}{\partial x} + \mathbf{S}_y \cdot \frac{\partial \dot{\mathbf{R}}}{\partial y} + \mathbf{S}_z \cdot \frac{\partial \dot{\mathbf{R}}}{\partial z} \right) \tag{4.45}$$

The last form can be reformulated by expressing the stresses in terms of components. To be meaningful, the stresses must be expressed in terms of components that have physical relevance to the lines and surfaces of the deformed element. Accordingly, we

1 Terms previously omitted would vanish in the limit ($\Delta x, \Delta y, \Delta x \to 0$) after division.

can refer them to the system $(\hat{\mathbf{I}}, \hat{\mathbf{J}}, \hat{\mathbf{K}})$:

$$\begin{aligned} \mathbf{S}_x &= S_x\hat{\mathbf{I}} + S_{xy}\hat{\mathbf{J}} + S_{xz}\hat{\mathbf{K}} \\ \mathbf{S}_y &= S_{yx}\hat{\mathbf{I}} + S_y\hat{\mathbf{J}} + S_{yz}\hat{\mathbf{K}} \\ \mathbf{S}_z &= S_{zx}\hat{\mathbf{I}} + S_{zy}\hat{\mathbf{J}} + S_z\hat{\mathbf{K}} \end{aligned} \tag{4.46}$$

The reader will notice that the system $(\hat{\mathbf{I}}, \hat{\mathbf{J}}, \hat{\mathbf{K}})$ differs little from the system $(\hat{\mathbf{i}}, \hat{\mathbf{j}}, \hat{\mathbf{k}})$ when the motions, rotations, and deformations are small. *Only then* can we identify the preceding components with their counterparts $(\sigma_x, \tau_{xy}, \tau_{xz}, \sigma_y, \tau_{yz}, \sigma_z)$ that act upon the surfaces in the fixed rectangular system (x, y, z). Substituting the forms (4.46) into Eq. (4.45), we obtain

$$\begin{aligned} \dot{w} = {} & S_x\hat{\mathbf{I}}\cdot\frac{\partial\dot{\mathbf{R}}}{\partial x} + S_{xy}\hat{\mathbf{J}}\cdot\frac{\partial\dot{\mathbf{R}}}{\partial x} + S_{xz}\hat{\mathbf{K}}\cdot\frac{\partial\dot{\mathbf{R}}}{\partial x} \\ & + S_{yx}\hat{\mathbf{I}}\cdot\frac{\partial\dot{\mathbf{R}}}{\partial y} + S_y\hat{\mathbf{J}}\cdot\frac{\partial\dot{\mathbf{R}}}{\partial y} + S_{yz}\hat{\mathbf{K}}\cdot\frac{\partial\mathbf{R}}{\partial y} \\ & + S_{zx}\hat{\mathbf{I}}\cdot\frac{\partial\dot{\mathbf{R}}}{\partial z} + S_{zy}\hat{\mathbf{J}}\cdot\frac{\partial\dot{\mathbf{R}}}{\partial z} + S_z\hat{\mathbf{K}}\cdot\frac{\partial\mathbf{R}}{\partial z} \end{aligned} \tag{4.47}$$

Let us define incremental components of strain[2] as follows:

$$\dot{\varepsilon}_x = \hat{\mathbf{I}}\cdot\frac{\partial\dot{\mathbf{R}}}{\partial x}, \qquad \dot{\varepsilon}_y = \hat{\mathbf{J}}\cdot\frac{\partial\dot{\mathbf{R}}}{\partial y}, \qquad \dot{\varepsilon}_z = \hat{\mathbf{K}}\cdot\frac{\partial\dot{\mathbf{R}}}{\partial z} \tag{4.48}$$

Additionally,[3]

$$\begin{aligned} \dot{\varepsilon}_{xy} &\doteq \frac{1}{2}\left(\hat{\mathbf{I}}\cdot\frac{\partial\dot{\mathbf{R}}}{\partial y} + \hat{\mathbf{J}}\cdot\frac{\partial\dot{\mathbf{R}}}{\partial x}\right) \doteq \dot{\varepsilon}_{yx} \\ \dot{\varepsilon}_{xz} &\doteq \frac{1}{2}\left(\hat{\mathbf{I}}\cdot\frac{\partial\dot{\mathbf{R}}}{\partial z} + \hat{\mathbf{K}}\cdot\frac{\partial\dot{\mathbf{R}}}{\partial x}\right) \doteq \dot{\varepsilon}_{zx} \\ \dot{\varepsilon}_{yz} &\doteq \frac{1}{2}\left(\hat{\mathbf{J}}\cdot\frac{\partial\dot{\mathbf{R}}}{\partial z} + \hat{\mathbf{K}}\cdot\frac{\partial\dot{\mathbf{R}}}{\partial y}\right) \doteq \dot{\varepsilon}_{zy} \end{aligned} \tag{4.49}$$

In contradistinction, we define also incremental "rotations":

$$\begin{aligned} \dot{\omega}_z &= \frac{1}{2}\left(\hat{\mathbf{J}}\cdot\frac{\partial\mathbf{R}}{\partial x} - \hat{\mathbf{I}}\cdot\frac{\partial\dot{\mathbf{R}}}{\partial y}\right) \\ \dot{\omega}_y &= \frac{1}{2}\left(\hat{\mathbf{I}}\cdot\frac{\partial\dot{\mathbf{R}}}{\partial z} - \hat{\mathbf{K}}\cdot\frac{\partial\dot{\mathbf{R}}}{\partial x}\right) \\ \dot{\omega}_x &= \frac{1}{2}\left(\hat{\mathbf{K}}\cdot\frac{\partial\dot{\mathbf{R}}}{\partial y} - \hat{\mathbf{J}}\cdot\frac{\partial\mathbf{R}}{\partial z}\right) \end{aligned} \tag{4.50}$$

---

2 Since $\hat{\mathbf{I}}\cdot\hat{\mathbf{I}} = 1$, $\hat{\mathbf{I}}\cdot\hat{\mathbf{I}} = 0$, Eqs. (4.48) are consistent with Eqs. (4.43).

3 These increments differ little from the increments of ε_{xy}, ε_{xz}, and ε_{yz} and only to the extent that the deformed lines are stretched, as ΔX differs from Δx.

From Eqs. (4.49) and (4.50), it follows that

$$
\begin{aligned}
\hat{\mathbf{I}}\cdot\frac{\partial\dot{\mathbf{R}}}{\partial y} &= \dot{\varepsilon}_{xy} - \dot{\omega}_z, &\quad \hat{\mathbf{J}}\cdot\frac{\partial\dot{\mathbf{R}}}{\partial x} &= \dot{\varepsilon}_{xy} + \dot{\omega}_z \\
\hat{\mathbf{I}}\cdot\frac{\partial\dot{\mathbf{R}}}{\partial z} &= \dot{\varepsilon}_{xz} + \dot{\omega}_y, &\quad \hat{\mathbf{K}}\cdot\frac{\partial\dot{\mathbf{R}}}{\partial x} &= \dot{\varepsilon}_{xz} - \dot{\omega}_y \\
\hat{\mathbf{J}}\cdot\frac{\partial\dot{\mathbf{R}}}{\partial z} &= \dot{\varepsilon}_{yz} - \dot{\omega}_x, &\quad \hat{\mathbf{K}}\cdot\frac{\partial\dot{\mathbf{R}}}{\partial y} &= \dot{\varepsilon}_{yz} + \dot{\omega}_x
\end{aligned}
\tag{4.51}
$$

When Eqs. (4.48) and (4.51) are substituted into Eq. (4.47), the latter assumes the form

$$
\begin{aligned}
\dot{w} = {} & S_x\dot{\varepsilon}_x + S_{xy}\dot{\varepsilon}_{xy} + S_{xz}\dot{\varepsilon}_{xz} \\
& + S_{yx}\dot{\varepsilon}_{yx} + S_y\dot{\varepsilon}_y + S_{yz}\dot{\varepsilon}_{yz} \\
& + S_{zx}\dot{\varepsilon}_{zx} + S_{zy}\dot{\varepsilon}_{zy} + S_z\dot{\varepsilon}_z \\
& + (S_{xy} - S_{yx})\dot{\omega}_z + (S_{zx} - S_{xz})\dot{\omega}_y + (S_{yz} - S_{zy})\dot{\omega}_x
\end{aligned}
\tag{4.52}
$$

A close examination of Eqs. (4.50) reveals that these do represent rigid rotations of the element. We might also argue that the work expended in the deformation is manifested entirely by those terms that contain strain increments, so that those containing the "rotations" can represent only work upon the rigid motion. The latter must vanish for arbitrary rotations if the element is in equilibrium.[4] Therefore, we must have

$$
S_{xy} = S_{yx}, \qquad S_{xz} = S_{zx}, \qquad S_{yz} = S_{zy} \tag{4.53}
$$

Recalling also the symmetry of the strain increments—that is, Eqs. (4.49), a final form of the work (per unit of initial volume) follows:

$$
\begin{aligned}
\dot{w} = S_x\dot{\varepsilon}_x + 2S_{xy}\dot{\varepsilon}_{xy} &+ 2S_{xz}\dot{\varepsilon}_{xz} \\
+ S_y\dot{\varepsilon}_y &+ 2S_{yz}\dot{\varepsilon}_{yz} \\
&+ S_z\dot{\varepsilon}_z
\end{aligned}
\tag{4.54}
$$

This result is a cornerstone in all subsequent analyses, which are founded upon principles of work and energy. Because we admitted the arbitrary rotations of the lines and surfaces, Eq. (4.54) applies to large as well as small deformations. If the strains are small, they are related to the increments of engineering strains as follows: Increments of the extensional strains are, as the notations imply, $\dot{\varepsilon}_x$, $\dot{\varepsilon}_y$, $\dot{\varepsilon}_z$; increments of the shear strains differ by the factor $\frac{1}{2}$; that is,

$$
\dot{\varepsilon}_{xy} = \tfrac{1}{2}\dot{\gamma}_{xy}, \qquad \dot{\varepsilon}_{xz} = \tfrac{1}{2}\dot{\gamma}_{xz}, \qquad \dot{\varepsilon}_{yz} = \tfrac{1}{2}\dot{\gamma}_{yz}
$$

If the strains are small, then the deformed element is practically rectangular and the unit vectors $(\hat{\mathbf{I}}, \hat{\mathbf{J}}, \hat{\mathbf{K}})$ differ from the fixed vectors $(\hat{\mathbf{i}}, \hat{\mathbf{j}}, \hat{\mathbf{k}})$ by a rigid rotation. Relative to

---

4 The result holds also for a body in motion, because the rotational inertia is a higher-order quantity—that is, it vanishes in the limit.

the material and its deformation, the components of stress in (4.54) are the physical counterparts of those given previously $(\sigma_x, \tau_{xy}, \tau_{xz}, \sigma_y, \tau_{yz}, \sigma_z)$. Indeed, *if* it is understood that any rigid motion is precluded, then the work expended on small strain can be taken in the form

$$\begin{aligned} \dot{w} = \sigma_x \dot{\varepsilon}_x + \tau_{xy}\dot{\gamma}_{xy} + \tau_{xz}\dot{\gamma}_{xz} \\ + \sigma_y \dot{\varepsilon}_y + \tau_{yz}\dot{\gamma}_{yz} \\ + \sigma_z \dot{\varepsilon}_z \end{aligned} \tag{4.55}$$

The latter form is entirely adequate for analyses of hookean materials wherein strains remain very small.

From our previous observation of material responses to simple loadings, we know that the work expended ($\dot{w}$) can be dissipated in permanent strain or recoverable as elastic energy. In the latter case the increment of work ($\dot{w}$) is the differential of an elastic potential, a function (u) of the strain components. That differential (du) accompanies the changes of strain, which are then also recoverable; in other words, the strain increments are differentials of the strains (e.g., $\dot{\varepsilon} = d\varepsilon_x$). It therefore follows that

$$\begin{aligned} \dot{w} = du = \frac{\partial u}{\partial \varepsilon_x} d\varepsilon_x + \frac{\partial u}{\partial \varepsilon_{xy}} d\varepsilon_{xy} + \frac{\partial u}{\partial \varepsilon_{xz}} d\varepsilon_{xz} \\ + \frac{\partial u}{\partial \varepsilon_{yx}} d\varepsilon_{yx} + \frac{\partial u}{\partial \varepsilon_y} d\varepsilon_y + \frac{\partial u}{\partial \varepsilon_{yz}} d\varepsilon_{yz} \\ + \frac{\partial u}{\partial \varepsilon_{zx}} d\varepsilon_{zx} + \frac{\partial u}{\partial \varepsilon_{zy}} d\varepsilon_{zy} + \frac{\partial u}{\partial \varepsilon_z} d\varepsilon_z \end{aligned} \tag{4.56}$$

In the case of elastic deformation, the two forms (4.54) and (4.56) are equivalent for *arbitrary* differentials of the strains. Therefore, we must have

$$\begin{aligned} S_x = \frac{\partial u}{\partial \varepsilon_x}, \quad S_{xy} = \frac{\partial u}{\partial \varepsilon_{xy}}, \quad S_{xz} = \frac{\partial u}{\partial \varepsilon_{xz}} \\ S_y = \frac{\partial u}{\partial \varepsilon_y}, \quad S_{yz} = \frac{\partial u}{\partial \varepsilon_{yz}} \\ S_z = \frac{\partial u}{\partial \varepsilon_z} \end{aligned} \tag{4.57}$$

Equations (4.57) provide a general basis for the stress-strain relations of any *elastic* material under isothermal deformations.

4.9 Strain Energy of the Hookean Isotropic Material

From our previous observations of hookean isotropic behavior in simple tension (see Fig. 4.7), we note that the work done (per unit volume) has the form

$$u = \frac{\sigma_x^2}{2E} \tag{4.58}$$

where σ_x is the only nonzero component of stress. Since $\sigma_x = E\varepsilon_x$ in simple tension, the Eq. (4.58) has the alternative form

$$u = \frac{E}{2}\varepsilon_x^2 \tag{4.59}$$

In like manner the work done under simple shear of the HI material has similar form:

$$u = \frac{G}{2}\gamma_{xy}^2 \tag{4.60}$$

where γ_{xy} is the only nonzero component of strain. Forms (4.59) and (4.60) represent the strain energy (per unit volume) of the HI material in simple extension and in simple shear, respectively. These represent potentials of stored energy.

In general, the material of a deformed body experiences simultaneously the six components of strain. Then an increment of work (per unit volume) of the material has the form

$$\begin{aligned} dw = {} & \sigma_x d\varepsilon_x + \sigma_y d\varepsilon_y + \sigma_z d\varepsilon_z \\ & + \tau_{xy} d\gamma_{xy} + \tau_{xz} d\gamma_{xz} + \tau_{yz} d\gamma_{yz} \end{aligned} \tag{4.61}$$

The deformed state is fully defined by the six components of strain. Consequently, the strain energy, in general, is a function of the six strain components:[5]

$$u(\varepsilon_x, \varepsilon_y, \varepsilon_z, \gamma_{xy}, \gamma_{xz}, \gamma_{yz})$$

The work done upon an elastic body is a conservative process; that work is manifested in the change of the function u. It is a "potential," and the incremental change is a differential:

$$\begin{aligned} du = {} & \frac{\partial u}{\partial \varepsilon_x} d\varepsilon_x + \frac{\partial u}{\partial \varepsilon_y} d\varepsilon_y + \frac{\partial u}{\partial \varepsilon_z} d\varepsilon_z \\ & + \frac{\partial u}{\partial \gamma_{xy}} d\gamma_{xy} + \frac{\partial u}{\partial \gamma_{xz}} d\gamma_{xz} + \frac{\partial u}{\partial \gamma_{yz}} d\gamma_{yz} \end{aligned} \tag{4.62}$$

Now, $dw = du$ for arbitrary changes; for instance, the equality must hold if any one of the strains is changed. It follows that

$$\begin{aligned} \sigma_x &= \frac{\partial u}{\partial \varepsilon_x}, & \sigma_y &= \frac{\partial u}{\partial \varepsilon_y}, & \sigma_z &= \frac{\partial u}{\partial \varepsilon_z} \\ \tau_{xy} &= \frac{\partial u}{\partial \gamma_{xy}}, & \tau_{xz} &= \frac{\partial u}{\partial \gamma_{xz}}, & \tau_{yz} &= \frac{\partial u}{\partial \gamma_{yz}} \end{aligned} \tag{4.63}$$

---

5 Thermal effects are excluded here. Otherwise, the function would also depend on thermal variables.

If the material is *hookean*, then each of Eqs. (4.63) is a *linear* homogeneous form. Such form can follow only if the potential u is a homogeneous *quadratic* form in the strain components. If the material is *isotropic*, then the shear and extensional strains are uncoupled, according to Eqs. (4.25) through (4.30); also, isotropy demands that the three extensional strains enter in like manner and that the three shear strains enter in like manner. The form of u for the HI material must be[6]

$$u = \tfrac{1}{2}A(\varepsilon_x^2 + \varepsilon_y^2 + \varepsilon_z^2) + B(\varepsilon_x\varepsilon_y + \varepsilon_x\varepsilon_z + \varepsilon_y\varepsilon_z) + \tfrac{1}{2}C(\gamma_{xy}^2 + \gamma_{xz}^2 + \gamma_{yz}^2) \quad (4.64)$$

According to Eqs. (4.63) and (4.64),

$$\sigma_x = A\varepsilon_x + B(\varepsilon_y + \varepsilon_z)$$

$$\tau_{xy} = C\gamma_{xy}$$

Comparing these with Eqs. (4.25) and (4.28), we conclude that

$$A = \frac{(1-\nu)E}{(1-2\nu)(1+\nu)}, \qquad B = \frac{\nu E}{(1-2\nu)(1+\nu)}, \qquad C = G = \frac{E}{2(1+\nu)} \quad (4.65)$$

and

$$u = \frac{E}{2(1-2\nu)(1+\nu)}[(1-\nu)(\varepsilon_x^2 + \varepsilon_y^2 + \varepsilon_z^2) + 2\nu(\varepsilon_x\varepsilon_y + \varepsilon_x\varepsilon_z + \varepsilon_y\varepsilon_z)] + \frac{G}{2}[\gamma_{xy}^2 + \gamma_{xz}^2 + \gamma_{yz}^2] \quad (4.66)$$

This is the general form of the potential for the HI material.

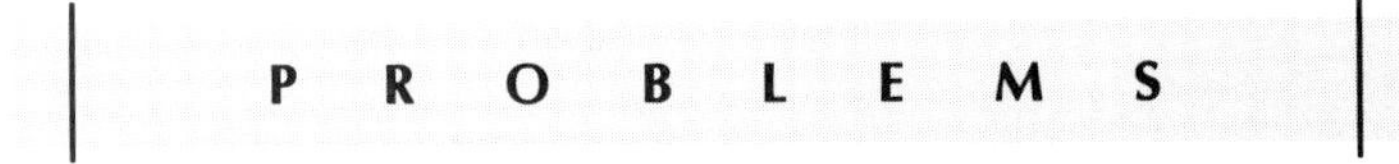

***4.37** The potential energy (per unit volume) of a HI material is a homogeneous quadratic function of the components of strain. One form in terms of principal strains is

$$u = \frac{C}{2}(\varepsilon_1 + \varepsilon_2 + \varepsilon_3)^2 + \frac{D}{9}[(2\varepsilon_1 - \varepsilon_2 - \varepsilon_3)^2 + (2\varepsilon_2 - \varepsilon_1 - \varepsilon_3)^2 + (2\varepsilon_3 - \varepsilon_1 - \varepsilon_2)^2]$$

Note: The second term vanishes in simple dilatation ($\varepsilon_1 = \varepsilon_2 = \varepsilon_3$). The first term vanishes in the absence of dilatation ($\varepsilon_1 + \varepsilon_2 + \varepsilon_3 = 0$). Express the coefficients C and D in terms of the moduli K and G. See Eqs. (4.31) and (4.32).

***4.38** A hookean material exhibits similar behavior with respect to all directions in a plane (xy) but acts

6 Terms of the form $\gamma_{xy}\gamma_{xz}$ are precluded by the isotropy and Eqs. (4.63), such as $\tau_{xy} = \partial u/\partial \gamma_{xy}$.

differently in the transverse direction (z). The strain energy function has the form

$$u = C_1(\varepsilon_x^2 + \varepsilon_y^2) + 2C_2\varepsilon_x\varepsilon_y + \tfrac{1}{2}(C_1 - C_2)\gamma_{xy}^2 + C_3(\varepsilon_x + \varepsilon_y)\varepsilon_z + C_4(\gamma_{xz}^2 + \gamma_{yz}^2) + C_5\varepsilon_z^2$$

A state of simple shear stress results if only σ_x and σ_y are nonzero and $\sigma_x = -\sigma_y$. Show that $\sigma_x - \sigma_y = 2(C_1 - C_2)(\varepsilon_x - \varepsilon_y)$. Show also that a simple state of shear gives $\tau_{xy} = (C_1 - C_2)\gamma_{xy}$.

*4.10 Hookean Anisotropic and Orthotropic Behavior

Many common materials, notably wood, exhibit directional properties. Modern *composites* are often *designed* to possess specific anisotropy: Glass, graphite, or kevlar fibers are oriented in matrices of various plastics (e.g., epoxy) to provide greater strength and stiffness in the direction(s) of their orientation(s). Usually the fibers are relatively small and are so uniformly dispersed that approximation as a continuum is acceptable. The mechanisms of failure are varied and complex; debonding of the constituents, crack propagation, and fracture all play roles. Subsequent to such failure, we can often describe the deformational behavior as linear and elastic, *but* anisotropic—that is, hookean anisotropic (HA).

In Sec. 4.8 we showed that elastic behavior is characterized by the existence of a strain-energy density (energy per unit volume), which (in the absence of thermal effects) is a function (u) of the six components of strain. Moreover, the stress-strain relations are derivable from that function, as follows:

$$S_x = \frac{\partial u}{\partial \varepsilon_x}, \quad S_{xy} = \frac{\partial u}{\partial \varepsilon_{xy}}, \quad S_{xz} = \frac{\partial u}{\partial \varepsilon_{xz}}$$

$$S_y = \frac{\partial u}{\partial \varepsilon_y}, \quad S_{yz} = \frac{\partial u}{\partial \varepsilon_{yz}}$$

$$S_z = \frac{\partial u}{\partial \varepsilon_z}$$

In a theory of *small* (e.g., $\varepsilon \ll 1$) strain *and* small rotation, the components of stress (S) and strain (ε) are approximately those of the linear theory ($\sigma_x, \tau_{xy}, \tau_{xz}, \sigma_y, \tau_{yz}, \sigma_z$; $\varepsilon_x, \frac{1}{2}\gamma_{xy}, \frac{1}{2}\gamma_{xz}, \varepsilon_y, \frac{1}{2}\gamma_{yz}, \varepsilon_z$). See Sec. 4.9.

If a material is hookean (the stress-strain equations are homogeneous linear forms), then the strain-energy function must be a homogeneous quadratic form (in the six components of strain). The *most general* homogeneous quadratic form incorporates 21 coefficients, as follows:

$$\begin{aligned} u = E_{11}\varepsilon_x^2 &+ E_{12}\varepsilon_x\varepsilon_y + E_{13}\varepsilon_x\varepsilon_z + \underline{E_{14}\varepsilon_x\varepsilon_{yz}} + \underline{E_{15}\varepsilon_x\varepsilon_{xz}} + \underline{E_{16}\varepsilon_x\varepsilon_{xy}} \\ &+ E_{22}\varepsilon_y^2 + E_{23}\varepsilon_y\varepsilon_z + \underline{E_{24}\varepsilon_y\varepsilon_{yz}} + \underline{E_{25}\varepsilon_y\varepsilon_{xz}} + \underline{E_{26}\varepsilon_y\varepsilon_{xy}} \\ &+ E_{33}\varepsilon_z^2 + \underline{E_{34}\varepsilon_z\varepsilon_{yz}} + \underline{E_{35}\varepsilon_z\varepsilon_{xz}} + \underline{E_{36}\varepsilon_z\varepsilon_{xy}} \end{aligned}$$

$$+ E_{44}\varepsilon_{yz}^2 + \underline{E_{45}\varepsilon_{yz}\varepsilon_{xz}} + \underline{E_{46}\varepsilon_{yz}\varepsilon_{xy}}$$
$$+ E_{55}\varepsilon_{xz}^2 + \underline{E_{56}\varepsilon_{xz}\varepsilon_{xy}} + E_{66}\varepsilon_{xy}^2$$

The reader can verify that the 21 terms include all products; hence, the most general hookean material is characterized by 21 elastic properties (E_{nm}). *If* the hookean anisotropic (HA) material is also homogeneous (HA-HO), *then* these properties are constants.

Special circumstances occur frequently in practice. Often a thin shell-like portion of an airframe, or a pressure vessel, is formed of laminations such that each layer, or laminate, contains parallel fibers. The fibers provide the strength and stiffness in the directions tangent to the shell; properties in the normal direction (z) are distinctly different. If the fibers of successive laminations are orthogonal, then the material acquires the properties of *orthotropy*. Specifically, if the fibers lie in the orthogonal directions (x, y), then the material is said to be orthotropic with respect to the three directions (x, y, z). The physical properties of such a material are such that it appears (mechanically or structurally) the same when viewed from positive or negative positions on a line of orthotropy (x, y, or z). In other words, the mathematical description of the material (HA-HO) must be unchanged if x and $-x$ (y or $-y$, z or $-z$) are interchanged. We note that changing x to $-x$ means that the shear strains γ_{xy} and γ_{xz} change sign. The quadratic u would be changed if terms containing a multiplier γ_{xy} or γ_{xz} were present; such terms must be absent. The same argument applies to the other directions of orthotropy (y and z). It follows that every term that contains *one* such multiplier ($\gamma_{xy}, \gamma_{xz}, \gamma_{yz}$) must vanish if the material is orthotropic with respect to those directions (x, y, z). The underlined terms in the quadratic u do not appear if the behavior is orthotropic. The *orthotropic* material is characterized by nine *elastic* properties ($E_{11}, E_{12}, E_{13}, E_{22}, E_{23}, E_{33}, E_{44}, E_{55}, E_{66}$).

The interested reader can explore other special forms of anisotropy in numerous books on elasticity.[7]

4.11 Plastic Behavior and Nonlinear Elasticity

In general, nonlinear stress-strain relationships cause mathematical difficulties in the solution of engineering problems. The occurrence of plastic deformation introduces further difficulty in all but the simplest situations. In either case, the aforementioned principle of superposition is no longer applicable.

It is not our purpose here to formulate general stress-strain relations for nonlinearly elastic or plastic behaviors. However, it is well to understand the essential difference between nonlinear elasticity and plasticity. A simple example serves to illustrate this distinction. Suppose the stress-strain relation for a certain material in

7 See, for instance, A. E. H. Love, *A Treatise on the Mathematical Theory of Elasticity* (New York: Cambridge University Press, 1927); (Mineola, N.Y.: Dover Publications, 1944); G. Wempner, *Mechanics of Solids* (New York: McGraw-Hill, 1973); (Alphen aan den Rijn, The Netherlands: Sijthoff & Noordhoff, 1981).

simple tension is as shown in Fig. 4.21. If the material is perfectly elastic, its behavior is described by the curve $OABC$. In this case the stress is uniquely determined by the strain (and vice versa), irrespective of the previous deformation (unloading retraces the loading curve). On the other hand, suppose that the material is not elastic. Then plastic deformation occurs during loading to B, and the unloading curve, BO', does not retrace the original loading curve. Consequently, the strain is not uniquely determined by the stress. For example, when the material is subjected to the tensile stress σ, as shown in Fig. 4.21, the extensional strain is ε if the loading has followed the curve OA, and ε' if the loading has followed the curve $OABA'$. Thus, a plastic deformation depends upon the entire loading history. Of course, for combined states of stress, the loading history may be much more complex, and the stress-strain relations are further complicated.

Fig. 4.21

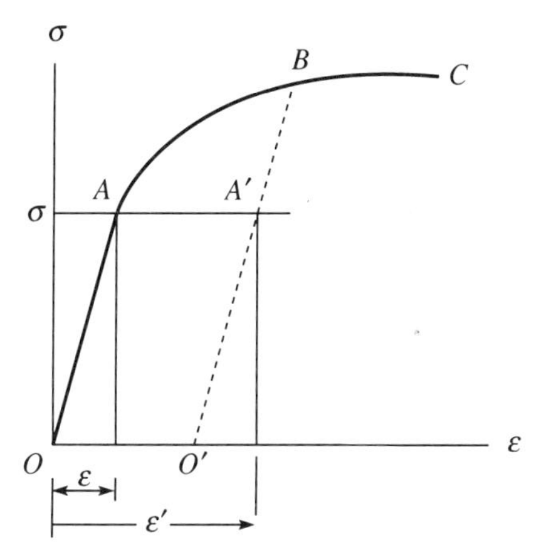

In the special case of a simple stress, either tension, compression, or shear, there is no essential difference between elastic and inelastic behavior, *so long as there is no unloading*. For example, in Fig. 4.21, if the material experiences no unloading, the stress-strain relation is represented in its entirety by the curve $OABC$. In this special circumstance only, the plastic stress-strain relation is expressed by the equation of a curve.

Many important problems are adequately described by simple states of stress. We will restrict our discussion to these, namely, simple tension (or compression) and simple shear. Also, conditions of static loading seldom cause the unloading of any material element. Under these conditions the mathematical treatment of plastic behavior is the same as nonlinearly elastic behavior, and the material behavior is adequately described by a single valued relation between the two variables, σ_x and ε_x (or τ_{xy} and γ_{xy}). A relation between the simple tensile (or compressive) stress σ_x and the transverse extensional strains ε_y and ε_z is seldom required. In the case of a simple shear stress τ_{xy}, it is usually safe to assume that the only significant effect is the associated shear strain γ_{xy}.

A satisfactory stress-strain relation must provide reasonable agreement with experimental data. At the same time the relationship must be as simple, mathematically, as possible. The choice of a relation in any particular problem depends not only upon the resistive character of the material, but also upon the mathematical complexity of the problem. Very elementary problems may permit refinements not possible in more complex situations. In the following discussion, we will consider but a few of the many possible stress-strain relations for the simple state of stress.

Linearly Elastic–Ideally Plastic Material

Figure 4.22(a) illustrates a stress-strain relation that is often used as an idealization for certain metals, notably low-carbon (structural) steel. The relation describes a linearly elastic behavior (line OA) to the yield stress σ_0, followed by plastic deformation at the constant stress σ_0. We refer to the behavior illustrated in Fig. 4.22(a) as *ideally plastic* or *perfectly plastic*. It is a reasonable approximation for mild steel so long as the deformations are not too severe. For relatively large plastic strains (say 0.02), mild steel experiences strain hardening, and the actual σ-ε curve rises.

Fig. 4.22

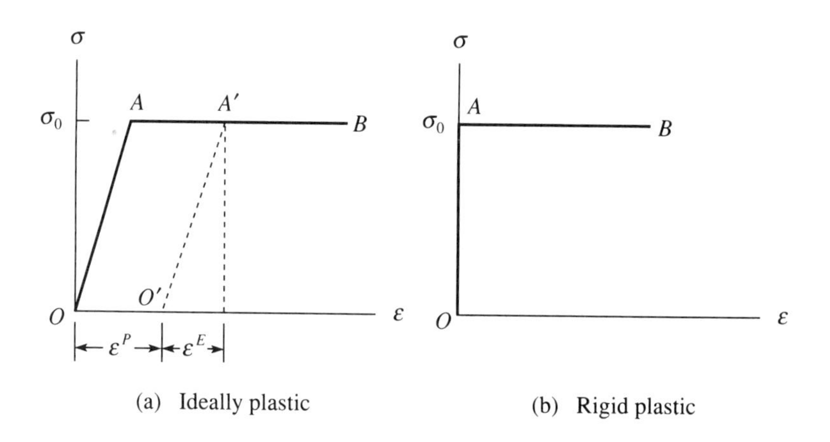

(a) Ideally plastic (b) Rigid plastic

Upon unloading, the behavior is linear and elastic; that is, unloading from point A' follows the straight line $A'O'$, which is parallel to OA. At the yield stress, the elastic strain is always

$$\varepsilon^E = \frac{\sigma_0}{E}$$

whereas the plastic strain ε^P can have any value whatever.

Rigid-Plastic Material

Sometimes plastic deformations are so great that elastic deformations are negligible. An idealization of this kind is referred to as a *rigid-plastic behavior*. It is applicable to processes of forming ductile materials, extrusion and drawing of wires and seamless tubes, or stamping parts for the bodies of automobiles. The rigid-plastic stress-strain relation is illustrated in Fig. 4.22(b). Such a material experiences no deformation until the stress reaches the yield point. If the material is rigid plastic, it can then experience unrestricted plastic deformation. However, a material element *within* a body does not necessarily experience unlimited strain when the yield stress is reached, because it may well be that the element is adjacent to material that has not yet yielded. If so, the deformation of the plastic element is constrained by the adjacent rigid material.

Linearly Elastic–Linear Hardening Material

Many common metals experience strain hardening during plastic deformation. That is, the stress-strain curve has a positive (nonzero) slope. A simple approximation of this is illustrated in Fig. 4.23(a). This stress-strain relation is linearly elastic (line OA) to the yield stress σ_0. Subsequent deformations are approximated by another linear relation, the straight line AB.

Fig. 4.23

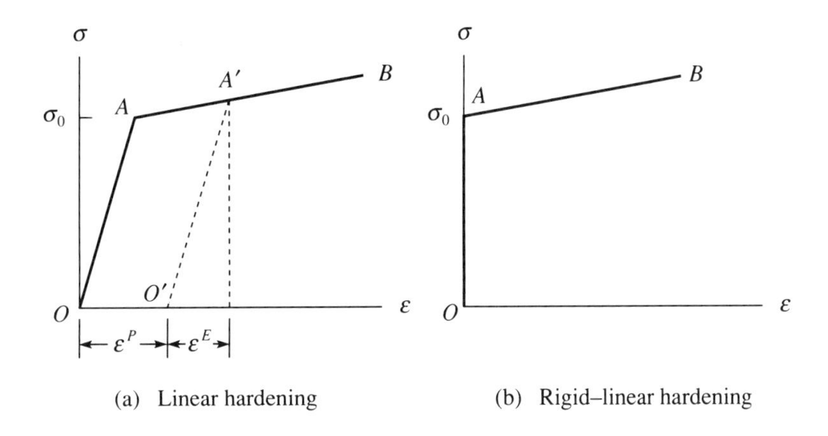

(a) Linear hardening (b) Rigid–linear hardening

Now suppose that unloading is also linear, so that the unloading curve from point A' traces the straight line $A'O'$, which is parallel to OA. Then the elastic strain is always

$$\varepsilon^E = \frac{\sigma}{E}$$

In addition there is a plastic strain when the stress exceeds σ_0. This, too, is a linear function of the stress, so that

$$\varepsilon^P = \frac{\sigma - \sigma_0}{kE} \qquad (\sigma > \sigma_0)$$

where k is a positive number. Hence the actual strain (when $\sigma > \sigma_0$) is

$$\varepsilon = \varepsilon^E + \varepsilon^P = \frac{(1 + k)\sigma}{kE} - \frac{\sigma_0}{kE} \qquad (\sigma > \sigma_0)$$

This equation is the equation of the straight line AB of Fig. 4.23(a). We refer to this idealization as a *linear-hardening* material.

The ideally plastic behavior is a special case of the linear-hardening stress-strain relation. It corresponds to $k = 0$.

Rigid Linear-Hardening Material

As before, if the elastic strain is negligible, the *rigid linear-hardening* idealization of Fig. 4.23(b) may be justified. This stress-strain relation is expressed by

$$\varepsilon = 0 \qquad (\sigma < \sigma_0)$$

$$\varepsilon = \varepsilon^P = \frac{\sigma - \sigma_0}{kE} \qquad (\sigma > \sigma_0)$$

Other Idealizations

In certain instances a smooth curve provides a more useful description of a nonlinear stress-strain relation because a single mathematical relation can then be used to describe the entire curve. For example, two relations are shown in Fig. 4.24. Both curves are closely related to the ideally plastic material. In both cases the slope at the origin is the elastic modulus E, and both curves approach the horizontal asymptote $\sigma = \sigma_0$.

Fig. 4.24

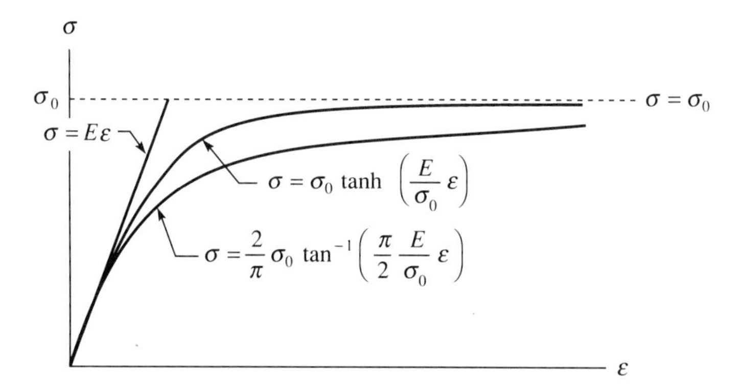

Another useful strain-stress relation is the Ramberg-Osgood equation:

$$\varepsilon = \frac{\sigma}{E}\left[1 + K\left(\frac{\sigma}{\sigma_0}\right)^n\right], \qquad n > 0$$

It follows that

$$\frac{d\sigma}{d\varepsilon} = E_T = \frac{E}{1 + (n+1)K(\sigma/\sigma_0)^n}$$

For all $n > 0$, the "tangent modulus" $E_T = E$ at the onset; that is, $\sigma = \varepsilon = 0$, as shown in Fig. 4.25. Also, in the limit as $n \to \infty$, the relation approaches the ideally plastic approximation depicted in Fig. 4.22(a).

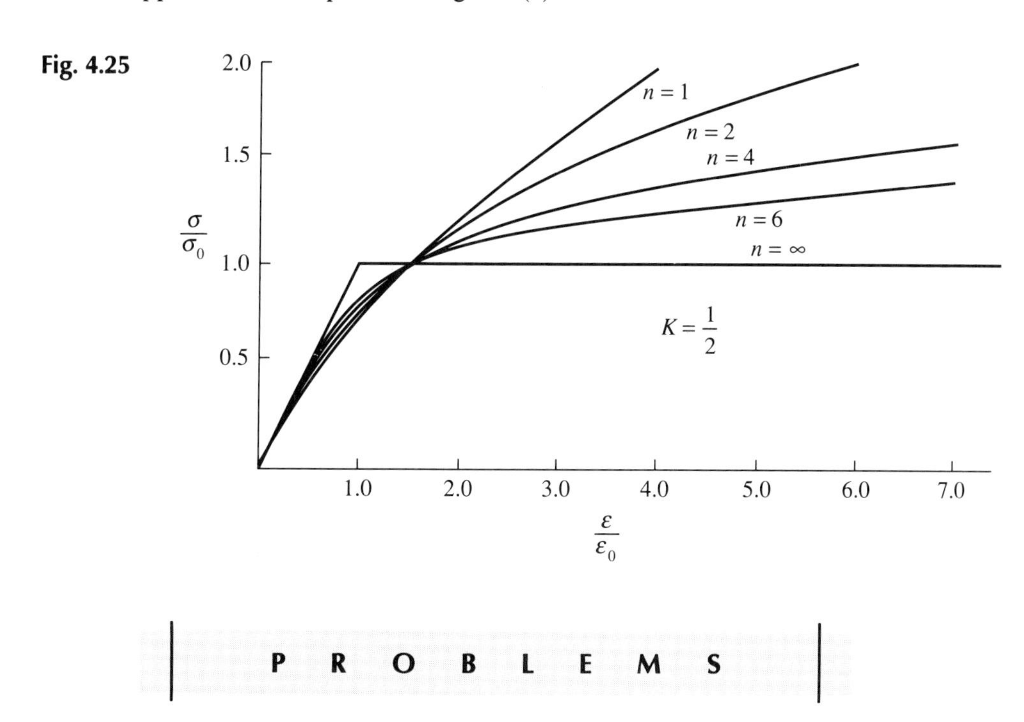

Fig. 4.25

PROBLEMS

4.39 A material is hookean ($E = 30 \times 10^3$ ksi) to the yield stress $\sigma_0 = 60$ ksi in tension and then exhibits linear hardening to fracture at a stress $\sigma_U = 120$ ksi and strain $\varepsilon_U = 60 \times 10^{-4}$. Obtain the equation governing the stress-strain relation in the inelastic range ($\sigma > \sigma_0$).

4.40 A very ductile material has an initial modulus $E = 10 \times 10^3$ ksi. At the strain $\varepsilon_0 = 10 \times 10^{-3}$, the stress is $\sigma_0 = 40$ ksi and the "tangent modulus" is $E_t = E/10$. Determine the parameters K and n in the Ramberg-Osgood equation that fits the given data.

4.41 A material exhibits hookean behavior ($E = 90 \times 10^3$ MPa) to a yield stress ($\sigma = 180$ MPa) and then strain-hardens. After a strain of 6×10^{-3} and stress of 300 MPa, the material exhibits nearly linear hardening to fracture at a strain of 16×10^{-3} with "tangent modulus" $E_t = E/9$. See Fig. P4.41. (a) Choose constants k and σ_0 for the model of ideal linear hardening and (b) choose constants K, σ_0, and n for the Ramberg-Osgood model. Plot your models versus the behavior.

Fig. P4.41

400
300
200
100
σ (MPa)
(16, 400)
(6, 300)
(2, 180)
5
10
15
20
$\varepsilon \times 10^3$

4.12 Time-Dependent Effects

Creep

The term *creep* denotes a continuing deformation under constant load. Some materials (lead, pure aluminum, wood, plastics) exhibit considerable creep at ordinary temperatures. Most metals creep significantly only at elevated temperatures. For this reason creep is a serious problem in high-temperature applications, such as boiler tubes or turbine blades. Often creep deformations proceed very slowly, so that their effects become significant only after long periods of time.

In a standard creep test the specimen is maintained under constant tensile load and at constant temperature while the strain and the corresponding time are recorded. The results of a creep test are shown in a plot of strain versus time. Typical curves are shown in Fig. 4.26(a) and 4.26(b). The slope of a curve represents the creep rate (strain per unit time). The creep rate and time to fracture are important quantities derived from a test. In general, an increase of temperature or stress causes a greater creep rate and a reduction in the time to fracture. These effects are illustrated in Fig. 4.26(a) and 4.26(b).

Fig. 4.26

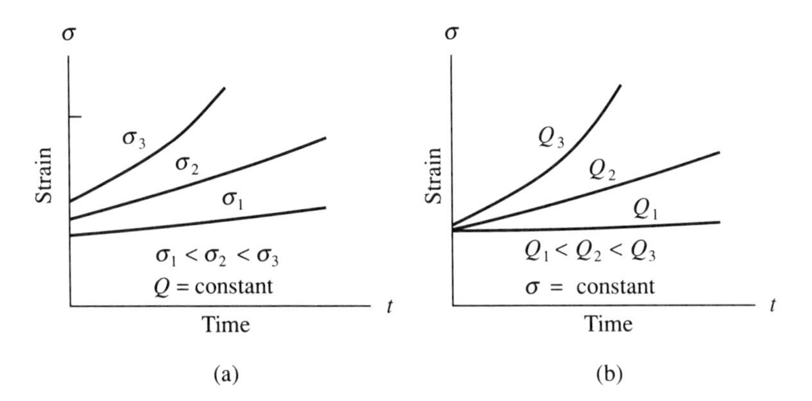

In tests of long duration the strain-time curve is linear during most of the test; that is, the deformation proceeds at a steady rate. Immediately before fracture the rate of deformation increases rapidly.

In practice, creep may occur very slowly over long periods of time, often many years. The designer seeks to avoid excessive creep deformation and, above all, a high creep rate which may lead to fracture. To do this the engineer must rely on extensive experimental data from creep tests, tests that are often of long duration. An alternative to overly conservative design is the common practice of periodically replacing parts to prevent significant creep damage.

Relaxation

In certain instances the load required to maintain a given deformation decreases with time. For example, if a bar were extended and then the length held fixed, the tensile stress might diminish as shown in Fig. 4.27. This behavior is known as *relaxation*. Basically, creep and relaxation are manifestations of the same phenomena; in one case (creep) the stress is maintained constant, and in the other (relaxation) the stress is held constant. A material that exhibits creep also exhibits relaxation. In most situations both effects are present and both are dependent on temperature.

Fig. 4.27

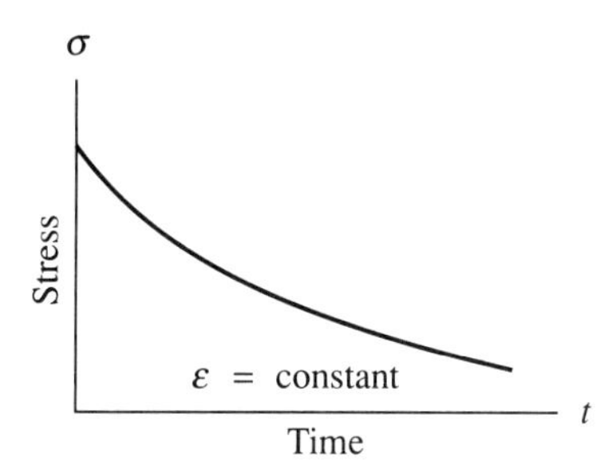

Impact Loads

In many cases materials respond to rapidly applied loads differently than they do to slowly applied loads. The complete explanation for this is not yet available but is thought to have its origins in certain microscopic irregularities of the material. However, the effects of rapid loading may be assessed, at least qualitatively, by testing materials under impact loads. Since it is virtually impossible to determine the magnitudes of very rapidly applied forces, measurements are made of the energy imparted to the specimen. The energy required to fracture a specimen in a standardized test is then used as a measure of resistance to impact loading.

Embrittlement of materials at low temperatures is vividly illustrated by impact tests of this kind. Tests are conducted at various temperatures and energy to fracture versus temperature is plotted. The curve shown in Fig. 4.28 is typical of some metals.

Fig. 4.28

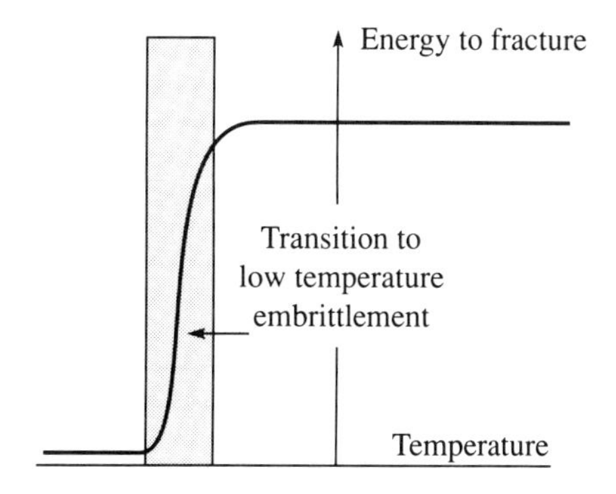

In this case a drastic reduction in impact resistance occurs over a very narrow temperature range.

4.13 Repeated Loading—Fatigue

Many repetitions of a load can cause fracture, although the stresses are far below the static ultimate strength of the material. Such fractures are termed *fatigue failures.* They are especially troublesome because they occur without warning. A minute crack starts at a point of high stress. This crack enlarges with each repetition of load until the material remaining is not sufficient to support the load. Often the fracture originates at a small surface scratch, a tool mark, a notch, a keyway, a small corroded spot, or an imperfection within the material.

In the usual fatigue test a specimen is subjected to alternating tensile and compressive stresses (of equal magnitude). The specimen is either subjected to this stress reversal until fracture, or the test is terminated after a preselected number of cycles. Many tests at varying stress levels are required for complete data. The result of each test is plotted on a diagram, with the magnitude of the alternating stress as ordinate and the number of cycles to fracture as abscissa. Usually the plotted points are widely scattered, as illustrated by Fig. 4.29. A curve drawn through these points represents only mean values of all experimental data. Such a curve is termed an *S-N* curve (stress versus number of cycles).

Fig. 4.29

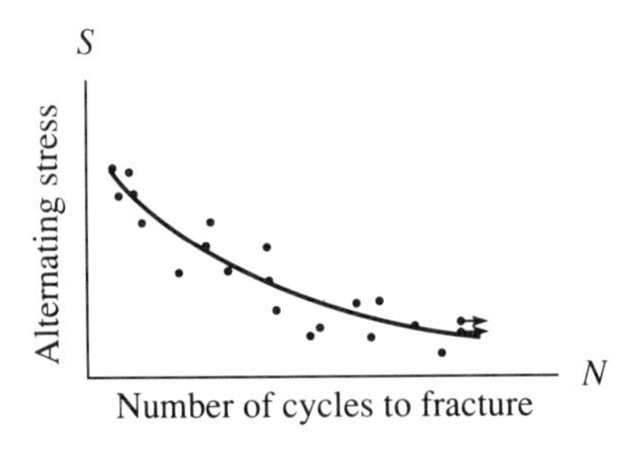

Fatigue Limit

The *S-N* curve for some materials approaches a horizontal line for a very large number of cycles. In this event the material is said to have a fatigue limit, *a stress magnitude below which most specimens will withstand an indefinite number of repetitions.* Certain materials, notably steels, have a fatigue limit. Most nonferrous metals and plastics have *S-N* curves similar to the aluminum alloy of Fig. 4.30. These materials may fracture at very low stresses if the stress is repeated often enough.

Fig. 4.30

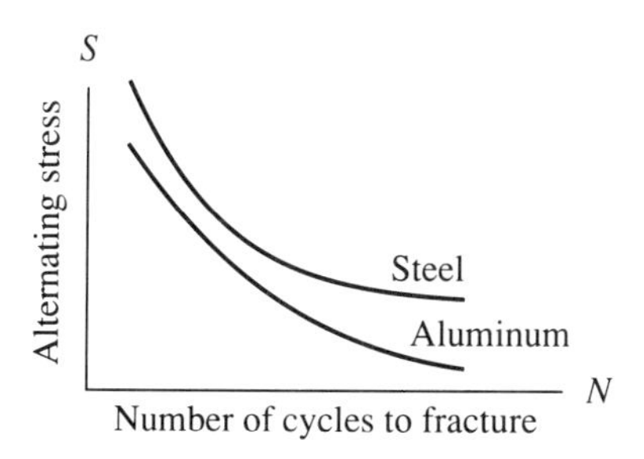

Fatigue Strength

The number of repetitions of stress before fatigue fracture is termed the *fatigue life*. The *fatigue strength* is defined in terms of a specified fatigue life. For any preselected fatigue life, the *S*-*N* diagram determines a stress magnitude, the fatigue strength of the material (for the prescribed life). Because of the unpredictability of fatigue failure, we can only say that the material will *probably* withstand this alternating stress for its prescribed fatigue life.

Effect of Mean Stress

The presence of a static load, in addition to a repeated load, accelerates the growth of fatigue cracks. This effect is evaluated by tests in which alternating stress is superimposed on a static (mean) stress. The effect of various mean stress levels is illustrated by the *S*-*N* curves of Fig. 4.31. In general the fatigue strength or the fatigue life is reduced.

Fig. 4.31

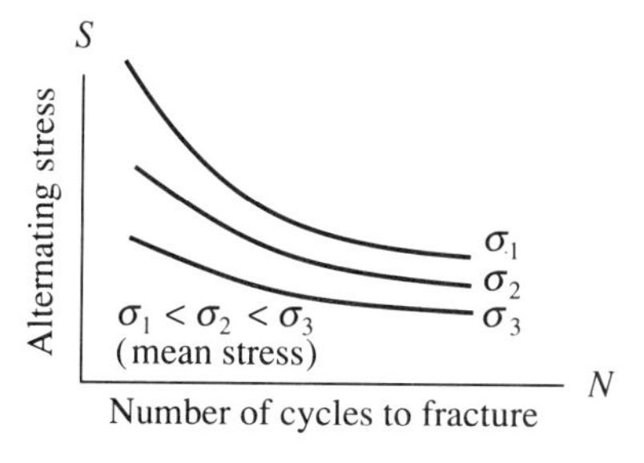

For a given fatigue life the influence of mean stress can be shown graphically by plotting the magnitude of the alternating stress versus the mean stress as in Fig. 4.32. This graphical representation is known as a *Soderberg diagram*. As the mean stress level approaches the ultimate strength, the magnitude of the alternating stress diminishes to zero. At the opposite extreme, as the mean stress level approaches zero, the magnitude of the alternating stress approaches the fatigue strength for alternating stress alone. In the absence of sufficient data, a reasonable approximation (usually conservative) is obtained by constructing a straight line through these two points.

Fig. 4.32

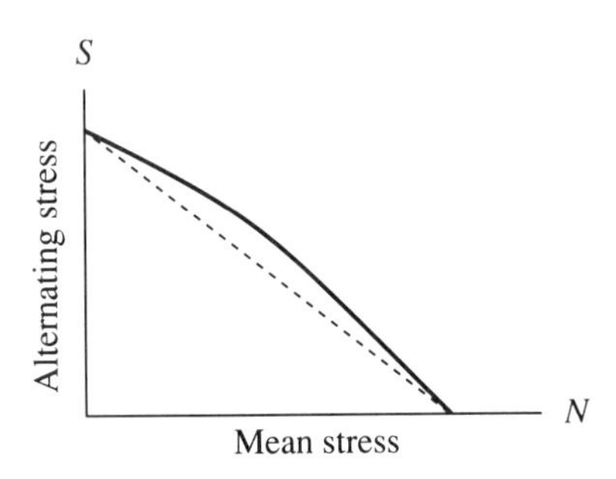

In view of the wide scatter in experimental data, it is not possible to predict fatigue strength accurately, nor is a stress below the fatigue limit a guarantee against fatigue failure. We can, at best, assess the *probability* of failure at a given stress. Furthermore, in practical applications it is difficult and often impossible to predict the magnitudes and frequencies of repeated loads. For example, the loads on the axle of an automobile depend on the road conditions, as well as the speed and the dynamic response of the loaded vehicle. The prediction of fatigue failure is a statistical matter. The designer can only utilize experimental data to reduce the probability of failure to an acceptable limit.

4.14 Yielding and Fracture Criteria

The strength of a material is characterized by its resistance to plastic deformation and its resistance to fracture. Unfortunately, the mechanisms of yielding and fracture are very complex. For this reason we content ourselves with a phenomenological approach; that is, we restrict ourselves to the matter of describing these phenomena rather than explaining them.

Previously we noted the resistance of a material to simple tension. Its behavior is predominantly elastic as long as the stress does not exceed the yield stress. Greater stress is accompanied by plastic deformation. Thus, for simple tension, a mathematical statement of the condition for incipient plastic flow is

$$\sigma = \sigma_0 \tag{4.67}$$

If the tensile stress reaches the ultimate strength σ_u, fracture is imminent. The condition for fracture is thus

$$\sigma = \sigma_u \tag{4.68}$$

Equations (4.67) and (4.68) are yield and fracture criteria for this simple state of stress. Neither criterion need be expressed in terms of stress but may equally well be written in terms of strain or even in terms of the work density required to effect yielding or fracture. For example, Eq. (4.67) may be replaced by

$$\varepsilon = \varepsilon_0, \quad \text{or} \quad U = U_0$$

where ε_0 and U_0 denote, respectively, the extensional strain and work density at the onset of plastic flow (in simple tension). Each also expresses a criterion for initial yielding.

In some machine parts or precision instruments any yielding of the material constitutes failure. If plastic deformation does not destroy the usefulness of a structure, fracture will certainly constitute failure. In the case of brittle materials, failure implies fracture. In any event, Eqs. (4.67) and (4.68) can serve as criteria to predict failure only for simple tension. In the most general case the material is subjected to the combined action of six stresses ($\sigma_z, \sigma_y, \sigma_z, \tau_{xy}, \tau_{yz}, \tau_{zx}$). Since the simultaneous action of all stresses will affect yielding and fracture in some manner, more general failure criteria are needed. They must take account of all possible combinations of the stresses. But it is virtually impossible to establish the criteria by testing materials under the many possible states of stress. For example, there is no means available for subjecting a material to a prescribed triaxial tension. Consequently, a *theory of failure* is needed.

A theory of failure sets forth the quantity (or quantities) that determines failure. The simplest theories ascribe failure to a single quantity. Failure is then supposed to occur when the specified quantity attains a certain critical value, which may be obtained from a simple test such as the tension test. The mathematical expression of the conditions for failure is termed the *failure criterion.* Clearly, we must distinguish between theories pertaining to yielding and those relating to fracture. If the occurrence of plastic deformation constitutes failure, an appropriate theory establishes a yield criterion. If failure is defined by fracture, the fracture criterion is established by a theory of fracture. In either case the criterion applies to an arbitrary state of stress. In particular, Eqs. (4.67) and (4.68) must be, respectively, special forms (the forms which apply to simple tension) of the general criteria for yielding and fracture.

Possibly the simplest point of view is to regard these criteria as limiting conditions imposed on the stresses. In other words, we may regard the general yield and fracture criteria as generalizations of Eqs. (4.67) and (4.68). Then, from the mathematical point of view, the concepts of yield and fracture criteria are quite similar. This does not mean that the physical events associated with the two phenomena are alike, but rather that the matter of formulating the criteria is similar. For this reason, it will be sufficient here to illustrate the foregoing ideas by discussing yield criteria.

4.15 Yield Criteria

Previous references to yielding pertained to simple states of stress. Then the criterion for yielding was the stipulation of a yield stress, either σ_0 in tension or τ_0 in shear. In general, a material is subject to six components of stress; all might affect the yielding. To appreciate the problem, suppose that the material were subject only to the two components σ_x and σ_y. Imagine a test wherein different combinations of these stresses are imposed; in each case, the values at the onset of yielding are noted. Then, we might plot these values, σ_x^0 and σ_y^0, as shown in Fig. 4.33. If the material is isotropic, then the curve intersects the respective axes at $\sigma_x = \sigma_y = \sigma_T$, the tensile yield stress, and at $\sigma_x = \sigma_y = -\sigma_C$, the compressive yield stress. In this relatively simple case, many

Fig. 4.33

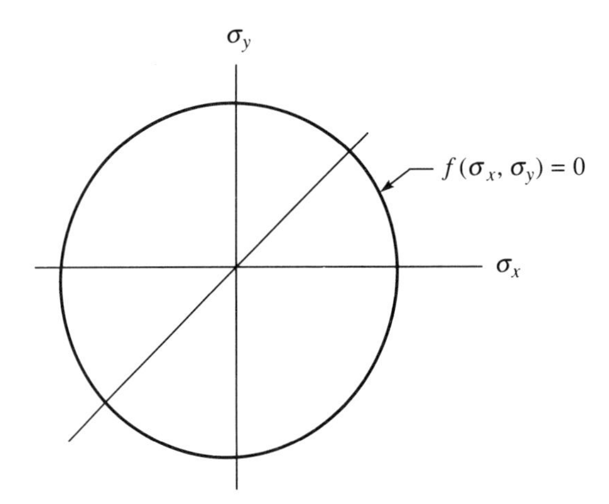

tedious tests are needed. The resulting curve of Fig. 4.33 can be approximated by an equation $f(\sigma_x, \sigma_y) = 0$, which constitutes the yield criterion. In general, a yield criterion must take the form

$$f(\sigma_x, \sigma_y, \sigma_z, \tau_{xy}, \tau_{xz}, \tau_{yz}) = 0$$

Clearly, this is a formidable experimental and theoretical task. Fortunately, the criterion is simplified by certain attributes of common metals. Specifically, most are essentially *isotropic*.

We first note that any state of stress is determined by the principal stresses $(\sigma_1, \sigma_2, \sigma_3)$ *and* the orientation of the principal directions (three angles would suffice). If the material is isotropic, then the orientation is irrelevant and the essential attributes of the state are the three principal stresses $(\sigma_1, \sigma_2, \sigma_3)$. Now, we can imagine tests that impose various combinations of the principal stresses and note the values that initiate yielding. Such experiments would provide the data to plot a surface in the three-dimensional space $(\sigma_1, \sigma_2, \sigma_3)$. That *yield* surface must enclose the origin. Progressing from the origin and following a path of loading—that is, changing the stresses $(\sigma_1, \sigma_2, \sigma_3)$—we would observe elastic behavior until our path strikes the predetermined yield surface. That surface can be approximated by a function

$$f(\sigma_1, \sigma_2, \sigma_3) = 0$$

Still, this is a hypothetical experiment that remains a formidable physical reality. Further observations and insights are needed to obtain a useful criterion.

One such observation follows from the extensive experiments of P. W. Bridgeman, who conducted tests of materials under hydrostatic states. These led to the conclusion that hydrostatic stress has little effect on yielding. Recall that a simple hydrostatic stress implies $\sigma_1 = \sigma_2 = \sigma_3$. That equation is represented by the line of Fig. 4.34, which forms equal angles with the axes, σ_1, σ_2 and σ_3. If we adopt the hypothesis that a hydrostatic state does not influence yielding, then the yield surface must enclose the line $\sigma_1 = \sigma_2 = \sigma_3$.

Fig. 4.34

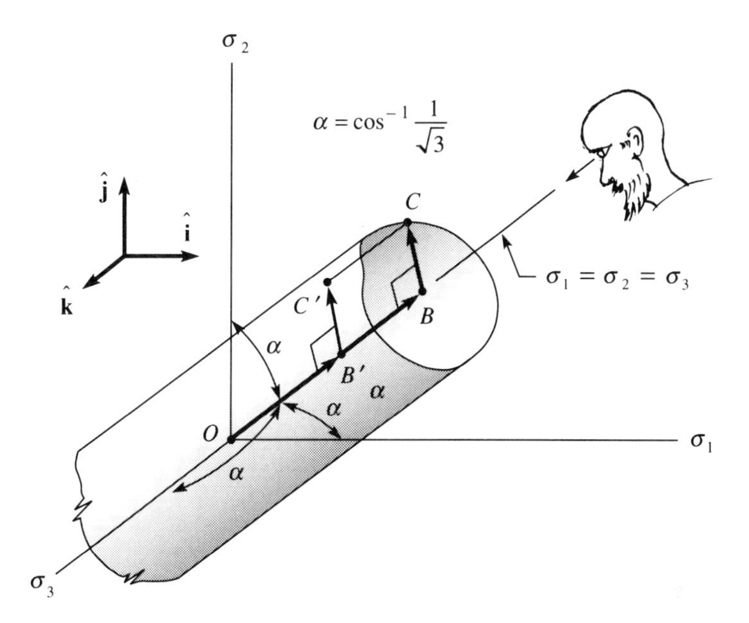

In the three-dimensional space of Fig. 4.34 $(\sigma_1, \sigma_2, \sigma_3)$, any state of stress is represented by a vector:

$$\boldsymbol{\sigma} = \sigma_1\hat{\mathbf{i}} + \sigma_2\hat{\mathbf{j}} + \sigma_3\hat{\mathbf{k}} \tag{4.69}$$

If that stress reaches the surface—for example, at point C'—yielding is initiated. Now the vector to C', namely, OC', can be decomposed into a hydrostatic part OB' along the line $\sigma_1 = \sigma_2 = \sigma_3$ and an orthogonal component $B'C'$. The latter is called the deviatoric part. The components of the hydrostatic part are equal; they are the mean of the normal components, in this case, the mean of the principal stresses:

$$\sigma_m = \frac{\sigma_1 + \sigma_2 + \sigma_3}{3}$$

The deviatoric components (the components of vector $B'C'$) are obtained by subtracting these mean values (the components of the hydrostatic part) from the components of stress. Accordingly, the deviator $B'C'$ is the vector

$$\boldsymbol{\sigma}' = (\sigma_1 - \sigma_m)\hat{\mathbf{i}} + (\sigma_2 - \sigma_m)\hat{\mathbf{j}} + (\sigma_3 - \sigma_m)\hat{\mathbf{k}} \tag{4.70}$$

By the hypothesis, supported by experiments, only the deviator effects the yielding. Then any additional hydrostatic component is inconsequential. Graphically, if point C' lies on the yield surface in Fig. 4.34, then the addition of $B'B$ takes the stress to point C *also* on the yield surface. The straight line $C'C$, parallel and equal to $B'B$, lies on the yield surface. In other words, the yield surface is cylindrical; generators, such as $C'C$, are parallel to the axis $OB'B$. The yield surface, and hence the function $f(\sigma_1, \sigma_2, \sigma_3)$, is determined by the shape of the cross section. A right cross section of the cylindrical surface is a curve, the intersection of the surface and the plane (the

π-plane) perpendicular to the line $\sigma_1 = \sigma_2 = \sigma_3$. The viewpoint, as depicted in Fig. 4.34, reveals the intersection in the π-plane, shown in Fig. 4.35; axes σ_1, σ_2, and σ_3 appear similarly foreshortened in this view. Since the material is presumed *isotropic*, the curve crosses each axis at σ^T, the tensile yield stress, and at $-\sigma^C$, the compressive yield stress. *If* the material yields at the same magnitudes of stress in simple tension and compression, then $\sigma^T = \sigma^C$.

Fig. 4.35

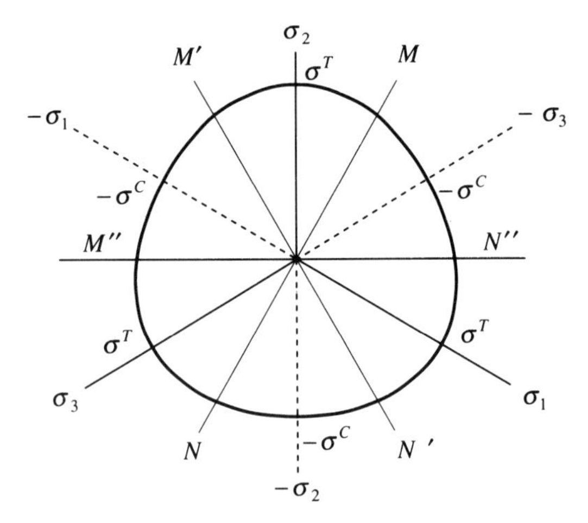

If we presume that the material is isotropic *and* also has the same magnitudes of yield stress in tension and compression, then the yield curve has certain symmetries, which are shown in Fig. 4.36. If a point A at (a, b, c) is on the curve, then the point A' at (a, c, b) is also, since this constitutes only an interchange in the roles of σ_2 and σ_3, which has no effect on the isotropic materials. In other words, the curve is symmetrical about the σ_1 line and likewise about the σ_2 and σ_3 lines. If

Fig. 4.36

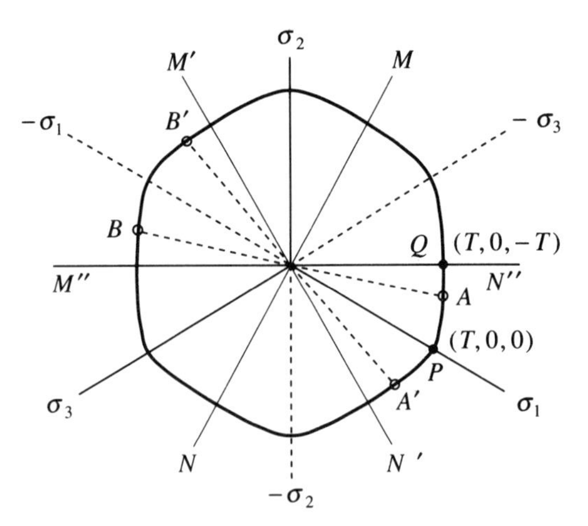

the points A at (a, b, c) and A' at (a, c, b) are on the curve, then the points B at $(-a, -b, -c)$ and B' at $(-a, -c, -b)$ are also, since the yielding is affected the same if the sign of the stress is reversed. We see that the curve is also symmetrical about line MN and likewise about the lines $M'N'$ and $M''N''$. Indeed, the entire curve is therefore determined by one 30° segment, for example, the segment PQ. Note that a simple tensile test ($\sigma_1 = T, \sigma_2 = \sigma_3 = 0$) determines point P; a simple shear test ($\sigma_1 = T, \sigma_2 = 0, \sigma_3 = -T$) determines point Q. Combinations of tension and shear are needed to ascertain intermediate points. The most prevalent and simple criteria presume isotropy and similar response upon reversal of sign; for example, tension and compression are equally effective. These are the Tresca and Mises criteria.

Tresca Theory

In a paper published in 1864, H. Tresca[8] proposed that metals yield when the maximum shear stress attains a certain critical value; that is, if the shear stress at a point on any plane reaches a specified value, the material experiences plastic deformation at that point. Because the theory takes no account of the orientation of the plane, it is strictly applicable only to isotropic materials.

Let us consider a mathematical expression of the Tresca theory. We recall from Chapter 2 that the maximum shear stress at a point is simply half the difference between the maximum and minimum principal stresses [see Eq. (2.63)]. Then, the criterion is expressed by one of six equations, depending on the relative magnitudes of the principal stresses:

$$\tfrac{1}{2}(\sigma_1 - \sigma_3) = \pm\tau_m \tag{4.71a,b}$$

$$\tfrac{1}{2}(\sigma_1 - \sigma_2) = \pm\tau_m \tag{4.71c,d}$$

$$\tfrac{1}{2}(\sigma_2 - \sigma_3) = \pm\tau_m \tag{4.71e,f}$$

where τ_m is the limiting value of the shear stress. This might be determined by a test in simple shear; then $\tau_m = \tau_0$, the yield stress in shear. However, in practice the value τ_m is more readily determined by a test in simple tension, say $\sigma = \sigma_1$ ($\sigma_2 = \sigma_3 = 0$). If $\sigma = \sigma_0$ denotes the tensile stress at the onset of yielding, then by Eqs. (4.71),

$$\tau_m = \frac{\sigma_0}{2} \tag{4.72}$$

The Tresca criterion of maximum shear stress is expressed by Eqs. (4.71). These are the equations of six planes, parallel to the line $\sigma_1 = \sigma_2 = \sigma_3$, which form a hexagonal cylinder. They intersect the π-plane to form the regular hexagon depicted in Fig. 4.37.

The criterion of Tresca presents difficulties, because the six equations in (4.71) are needed to express it generally. It is conveniently employed only when the relative magnitudes of the principal stresses are known. Then one equation of the six

8 H. Tresca, Compt. Rend., Acad. Sci., Paris, vol. 59 (1864): 754.

Fig. 4.37

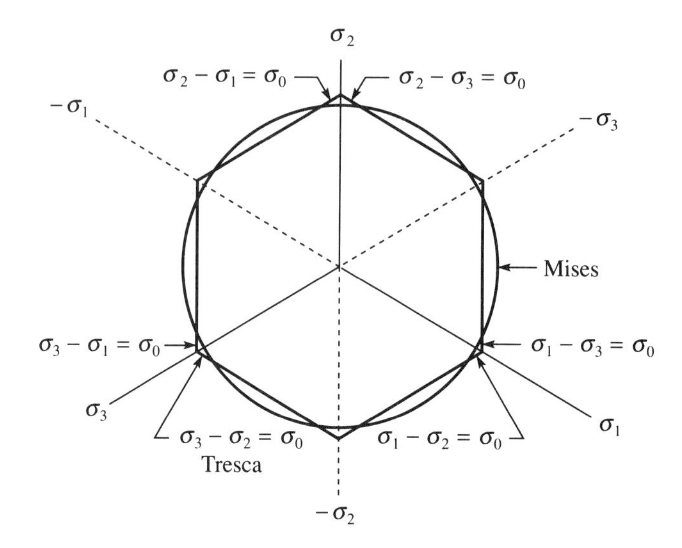

equations suffices. Geometrically speaking, only one plane (one line in Fig. 4.37) of the yield surface is then needed. A yield criterion that avoids the difficulty was proposed by R. von Mises in 1913.

Von Mises Theory

The von Mises[9] yield criterion replaces the six equations of the Tresca criterion by a single equation. In terms of principal stresses, the von Mises criterion is

$$(\sigma_1 - \sigma_3)^2 + (\sigma_2 - \sigma_3)^2 + (\sigma_1 - \sigma_2)^2 = C^2 \tag{4.73}$$

where C is an experimentally determined constant. In terms of the geometric representation, the von Mises criterion replaces the hexagonal cylinder of the Tresca criterion by a circular cylinder. Its axis is also the line $\sigma_1 = \sigma_2 = \sigma_3$.

Differences between the Tresca and von Mises criteria are evident from the curves of Fig. 4.37. However, these curves may be brought into reasonable agreement by proper choice of the constants C and τ_m. From a practical point of view, the important difference lies in the mathematical expression of the criteria. Quite often the choice of a yield criterion is a matter of mathematical convenience or necessity in the solution of a given problem. Either the Tresca or von Mises criterion will provide a suitable approximation for most metals. Yielding of some materials conforms more nearly to the Tresca criterion; in others, yielding is more accurately described by the von Mises criterion.

The von Mises criterion also has a physical interpretation. If the strain energy of an isotropic hookean material [see Eq. (4.66)] is expressed in terms of the components of stress, it is a quadratic form; specifically,

---

9 R. von Mises, Mechanik des festen Körper im plastishdeformablen Zustand, Nachr. Ges. Wiss., Göttingen, Math.-physik Klasse, (1913).

$$u = \frac{(1+\nu)}{2E}[(\sigma_1 - \sigma_m)^2 + (\sigma_2 - \sigma_m)^2 + (\sigma_3 - \sigma_m)^2] + \frac{1}{2K}\sigma_m^2 \tag{4.74}$$

In this form, we can interpret the final term as energy associated with the dilatation and the hydrostatic part (σ_m) of the stress. The initial term is associated with distortion; it depends only on the deviatoric part of the stress [see Eq. (4.70)]. From the definition $3\sigma_m = \sigma_1 + \sigma_2 + \sigma_3$, the von Mises criterion (4.73) has the alternative form

$$(\sigma_1 - \sigma_m)^2 + (\sigma_2 - \sigma_m)^2 + (\sigma_3 - \sigma_m)^2 = C^2/3 \tag{4.75}$$

Accordingly, the criterion has been interpreted[10] to mean that yielding is initiated when the distortional energy attains a specific limit $[(1+\nu)C^2/6E]$. In view of Eqs. (4.70) and (4.75), the constant C has also a geometric interpretation: The radius of the Mises circle is $R = \sqrt{\boldsymbol{\sigma}' \cdot \boldsymbol{\sigma}'} = C/\sqrt{3}$. Furthermore, if the constant is determined by a simple tensile test ($\sigma_1 = \sigma, \sigma_2 = \sigma_3 = 0$) and the yield stress is $\sigma = \sigma_0$, then $C = \sqrt{2}\sigma_0$. If the constant is determined by a simple shear test ($\sigma_1 = \tau, \sigma_2 = 0, \sigma_3 = -\tau$) and the yield stress is $\tau = \tau_0$, then $C = \sqrt{6}\tau_0$. If the constant τ_m in the Tresca criterion (4.71) and the constant C in the Mises criterion (4.73) are determined by a tensile test ($\tau_m = \sigma_0/2 = C/2\sqrt{2}$), the Mises circle circumscribes the Tresca hexagon. If the constants are determined by a shear test ($\tau_m = \tau_0 = C/\sqrt{6}$), the Mises circle inscribes the Tresca hexagon. In the first case, the Mises criterion predicts a higher yield stress in shear ($2/\sqrt{3}$). In the second case, the Tresca criterion predicts greater yield stress in tension ($2/\sqrt{3}$). In such worst cases, the error is 15%.

PROBLEMS

4.42 When tested in simple tension, a material exhibits yielding at a stress $\sigma_0 = 300$ MPa. This result is used to determine the parameters τ_m and C in the Tresca and Mises conditions, respectively. This same material is subsequently employed in circumstances of "radial loading"; that is, all components of stress increase in proportion.

$$\sigma_x = \sigma, \quad \tau_{xy} = \sigma R_{xy}, \quad \tau_{xz} = \sigma R_{xz}$$
$$\sigma_y = \sigma R_y, \quad \tau_{yz} = \sigma R_{yz}$$
$$\sigma_z = \sigma R_z$$

Determine the value σ that initiates yielding according to the Tresca criterion and the Mises criterion under the following circumstances:

a. $R_y = -1, R_{xy} = R_{xz} = R_{yz} = R_z = 0$

b. $R_y = 1, R_z = -1, R_{xy} = R_{xz} = R_{yz} = 0$

c. $R_y = R_z = 1, R_{xy} = R_{xz} = R_{yz} = 0$

d. $R_{xy} = 1, R_y = R_{xz} = R_{yz} = R_z = 0$

4.43 The material of Prob. 4.42 is subject to a state of simple shear $\tau_{xy} = \tau, \sigma_x = \tau_{xz} = \tau_{yz} = \sigma_y = \sigma_z = 0$. Determine the value of the stress τ that initiates yielding according to the Tresca criterion and according to the von Mises criterion.

4.44 A material yields in a state of simple tension at stress $\sigma = 50$ ksi. Determine the value T that initiates yielding according to the Tresca and von Mises criteria if

a. $\sigma_x = -T, \tau_{xy} = \tau_{xz} = \sigma_y = \tau_{yz} = \sigma_z = 0$

10 H. Hencky, "Zur Theorie plastischer Deformationen," *Z. angew. Math. Mechanik*, vol. 4 (1924).

b. $\tau_{xy} = \tau_{xz} = T$, $\sigma_x = \sigma_y = \tau_{yz} = \sigma_z = 0$

Note: Locate the principal directions in order to employ Eqs. (4.71) and (4.73).

***4.45** Determine the shear stress on the surface that makes equal angles with the principal directions and compare your result with Eq. (4.73).

4.16 Fracture

The phenomenon of fracture is quite different than yielding. We should, therefore, expect the mathematical form of a fracture criterion to differ markedly from a yield criterion. One outstanding difference is the influence of hydrostatic pressure, which experiments have shown to have marked effect on the resistance of materials to fracture. Very great pressures can even result in plastic flow of otherwise brittle materials; the material becomes more ductile and more resistant to fracture. On the other hand, fracture is readily effected by tensile stresses. Noting this, early investigators proposed a maximum normal stress theory of fracture. According to this theory, fracture occurs whenever the maximum (principal) normal stress attains a critical value, usually taken as the ultimate strength. The fracture criterion here is simply

$$\sigma_1, \sigma_2, \quad \text{or} \quad \sigma_3 = \sigma_u \tag{4.76}$$

Since the theory takes no account of the orientation of the maximum stress, it is strictly applicable only to isotropic materials.

In the case of brittle materials, Eq. (4.76) often provides a satisfactory fracture criterion. Fracture of ductile materials is seldom important, because the utility of a ductile structure is usually destroyed by the large plastic deformations that occur prior to fracture.

The reader must realize that the criterion (4.76) is merely the simplest. The phenomenon is very complex, and more elaborate theory is often needed.

4.17 Mechanism of Deformation in Metals

To understand the mechanical behavior of materials, it is necessary to examine the behavior of basic constituents. In metals the individual crystal is a basic structural element. Each crystal is composed of atoms (or ions), which are bound together by interacting forces of an electrodynamic character. In a perfectly formed crystal, the atoms are arranged in a regular pattern termed the *crystal lattice.* The simplified (two-dimensional) arrangement of Fig. 4.38(a) will serve for purposes of illustration.

During the application of force, we expect the crystal to deform as shown, for example, in Fig. 4.38(b). In this event the pattern is slightly altered, but the original configuration is restored upon removal of the forces. This is an elastic deformation of the crystal. On the other hand, sufficient force may cause a permanent disarrangement of the crystal lattice. Such a permanent disfiguration is illustrated by Fig. 4.38(c). In this case one layer of atoms *slips* over the adjacent layer. Thus, the crystal suffers a plastic deformation via the mechanism of slip. The observed deformations of metals are manifestations of such alterations of the many crystals that compose them.

Fig. 4.38

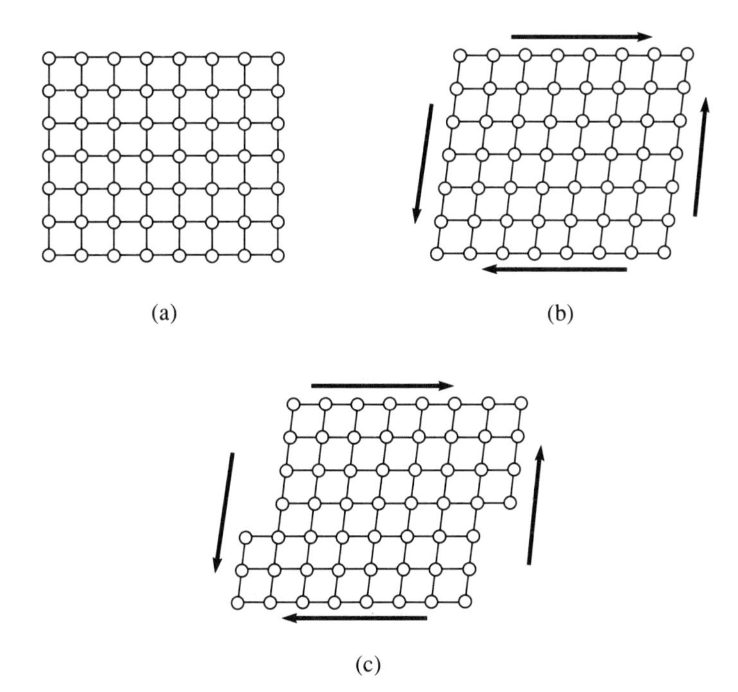

It is reasonable to suppose that the mechanical properties of a metal might be obtained by accurate analyses of the forces and displacements in individual crystals. To a limited extent, this is true. The elastic moduli of metals have been predicted with a fair degree of accuracy by analyses of interatomic forces that resist elastic deformations. However, estimates of the yield strength, which are based on slip in perfect crystals, are far in excess of observed values. Apparently, yielding does not occur in quite this manner.

The strength of metals is very much dependent on imperfections in crystals. Of particular importance are disarrangements of the crystal lattice, known as *dislocations.* One type of dislocation (edge dislocation) is illustrated by Fig. 4.39(a). The forces necessary to cause a movement of the dislocation are much less than those required to effect slip in the perfect crystal. It is necessary only to propagate the dislocation as shown in Fig. 4.39. By contrast, the shear force in Fig. 4.38 must simultaneously overcome the bonds between many atoms in order to cause slip in the perfect crystal. This mechanism has been likened to the movement of a wrinkle in a rug, shown in Fig. 4.40. The rug is easily displaced along the floor by moving a wrinkle from one end to the other, whereas much greater force is needed to slide the entire rug.

Each metal crystal has a certain inherent resistance to plastic deformation, depending on the type of crystal lattice and the strength of interatomic bonds. In actual metals many dislocations and impurities are present in each crystal. These impurities may take the form of small particles or foreign atoms within crystals. During deformation, dislocations move, join together, and even generate new dislocations. As the

Fig. 4.39

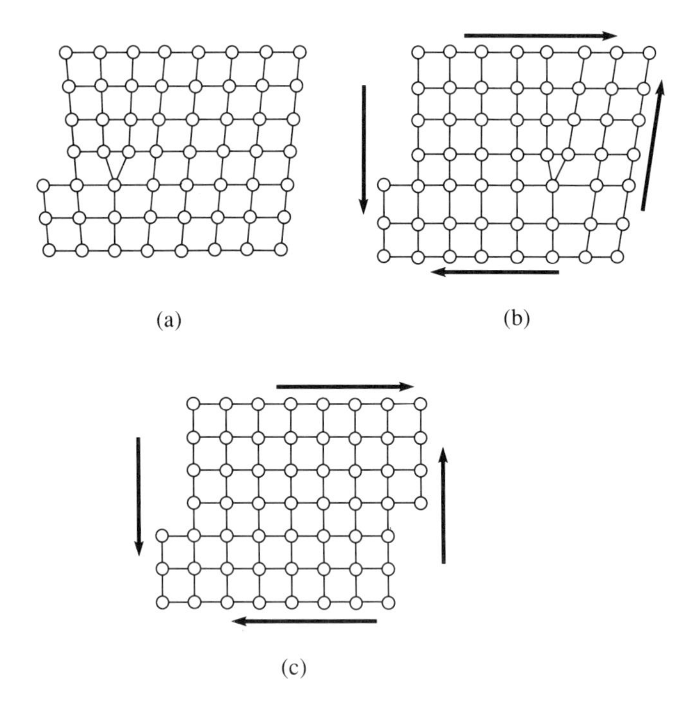

Fig. 4.40

plastic deformation progresses, the dislocations may interfere with one another and so impede further movement. Foreign particles (or atoms) and crystal boundaries also block the movement of dislocations. As the motions of the dislocations are arrested, the resistance to plastic flow increases. The metal is said to *strain harden*.

As the temperature increases the mobility of atoms increases. Foreign atoms tend to diffuse and dislocations tend to migrate around barriers. The obstacles which impede dislocation movement are dissipated by heating. Thus, at elevated temperatures, the metal may continue to deform plastically under constant force. This is the phenomenon called creep. If, on the other hand, a metal is cooled, the activity of atoms is decreased. The movement of dislocations becomes more difficult. Thus, we observe that the behavior of normally ductile metals is brittle in very cold environments.

Dislocations, as well as other crystal defects, play a major role in the mechanism of fracture. Since the movement of dislocations is effected by relatively little force, repeated loading can cause the movement of dislocations, even though stresses are well below the apparent yield stress. It would therefore appear that fatigue fractures originate with the movement of dislocations. Indeed, it appears that the elastic constants are the only mechanical properties which are not greatly influenced by these minute defects in the crystallographic structure.

4.18 Mechanical Properties of Some Common Engineering Materials

The values listed in Tables 4.1 and 4.2 should be regarded as typical values for purposes of comparison. They are not intended as design values, nor is the list of materials exhaustive. Other sources should be consulted for more extensive data.

Table 4.1

Properties of Some Common Engineering Materials
(Customary Units)

	Specific Weight 10^{-2} lb/in.^{3}	Modulus of Elasticity 10^6 lb/in.^{2}	Shear Modulus 10^6 lb/in.^{2}	Ultimate Tensile Strength 10^3 lb/in.^{2}	Tensile Yield Strength (0.2%) 10^3 lb/in.^{2}	Coefficient of Thermal Expansion 10^{-6}/°F
Structural steel	28	29	11	60	35	6.5
Stainless steel	28	28	11	120	80	9.6
High-carbon steel alloy	28	30	12	200	150	8.0
Aluminum (pure)	9.8	10	3.8	10	3	13.0
Aluminum (2014-T6)	10	10.6	4.0	70	60	13.0
Copper (hard-drawn)	32	17	6.4	55	45	9.3
Brass (cast)	30	14	5.2	45	20	10.4
Cast iron	27	14	5.5	25	—[b]	6.3
Titanium alloy	16	16	6.0	130	110	5.0
Magnesium alloy	6.5	6.5	2.4	35	23	14.5
Glass fibers	9.3	10	4.0	2000	—[b]	5.0
Concrete (typical)	8.5	3.5[a]	—	3.5[a]	—[c]	6.0
Timber (with grain)	2.0	1–2[a]	—	4–9[a]	—[c]	—[c]

[a] Compression. [b] Brittle material. [c] Not applicable.

Table 4.2

Properties of Some Common Engineering Materials
(SI Units)

	Specific Weight kN/m^3	Modulus of Elasticity 10^3 MPa	Shear Modulus 10^3 MPa	Ultimate Tensile Strength MPa	Tensile Yield Strength MPa	Coefficient of Thermal Expansion $10^{-6}/°C$
Structural steel	77	200	76	410	240	12
Stainless steel	77	190	76	820	550	17
High-carbon steel alloy	77	210	83	1400	1000	14
Aluminum (pure)	26.6	69	26	69	20	23
Aluminum (2014-T6)	28	73	28	480	410	23
Copper (hard-drawn)	87	120	44	380	310	17
Brass (cast)	81	97	36	310	140	19
Cast iron	72	97	38	170	—[b]	11
Titanium alloy	44	110	41	900	760	9.0
Magnesium alloy	17	45	17	240	160	26
Glass fibers	25	70	28	14,000	—[b]	9.0
Concrete (typical)	23	24[a]	—	24[a]	—[c]	11
Timber (with grain)	5.4	7–14[a]	—	30–60[a]	—[c]	—[c]

[a] Compression. [b] Brittle material. [c] Not applicable.

5 Fundamental Applications

Our basis for the mechanics and the analyses of solid bodies is analogous to a three-legged stool. The reader will recall that each of the three legs must support a part of the load. Here, the three legs are constructed in Chapters 2, 3, and 4:

1. *Force, stress, and equilibrium:* The basic mathematical tools are the equations of equilibrium. The *new* tool is stress, which is needed to describe behavior at a point.
2. *Deformation:* The basic mathematical tools are entirely geometrical. The *new* tool is strain, which is needed to describe behavior at a point.
3. *Behavior of materials:* The mathematical models of *continuous* media relate the dynamic variables (e.g., stresses), kinematic variables (e.g., strains), and temperature. Additional theories and concepts of failure delimit the models of continua.

The purpose of the present chapter is twofold: Most importantly, we must develop skill in utilizing our tools to obtain solutions. Secondarily, we obtain solutions to some fundamental problems of engineering importance; most notably, we obtain the classical solutions to the *extension* of prismatic bars, the *torsion* of circular cylindrical shafts, and the simple *bending* of symmetrical beams.

5.1 Fundamental Problems

The reader might ask what distinguishes our *fundamental* problems from others, as contained in subsequent chapters. Each of these *fundamental* problems is distinguished by *simplicity* of geometrical form, physical attributes, *and* loading. For example, we examine the torsion of a shaft, which is geometrically and physically symmetrical about an axis; it is subjected only to opposing couples upon the ends. The practical significance of such simplicity is a simple description of the deformation, state of strain, and stress, which are quite plausible (and in some instances, logical). Moreover, in all instances our results are in agreement with so-called exact solutions that derive from advanced theories of elasticity and plasticity. Of course, the ultimate justification for an engineer's solution rests on the agreement between predictions and actual observations or the results of more elaborate and precise theories.

Although the fundamental problems, as cited, admit simple analyses, they also have practical importance. By invoking the Saint-Venant principle (Sec. 5.2), the solution of simple extension can be applied to assemblies of axially loaded members (e.g., the trusses of Sec. 5.7); likewise, the solution of simple torsion can be applied to the assemblies of shafts (Sec. 5.10). In such applications the student is forewarned of limitations that are attributed to the local effects of the actual loadings. The extent of such effects is described in the following section.

Some exercises are intended to provoke the student's senses of engineering practice and to accept a degree of approximation by modifying the simple solutions to accommodate minor variations in form, properties, or loading (e.g., a bar or shaft may be *slightly* tapered). These provide a bridge between theoretical analyses and engineering practices with their attendant approximations and limitations.

5.2 Saint-Venant's Principle

More often than not, the loads applied to structural members are contact forces; that is, the external forces are applied to the member by physical contact with other bodies. Often these forces are transmitted to a member through a connecting device, such as the bolts or rivets that fasten members together. In such cases the actual distribution of the applied forces is so complicated that an accurate description of the load distribution is difficult. Fortunately, it is sometimes possible to assess the net effect of the loading without regard for the precise manner of load distribution. In most practical situations, the resultant of the distributed forces may be all that is needed.

As an illustration, consider the situations shown in Figs. 5.1 and 5.2. Identical rectangular bars have a square gridwork of lines drawn on their exteriors. These bars are subjected to axial loads applied near the ends. The manner of application of the applied forces is quite different in the two cases shown, but the resultant load is the same: an axial force acting along the centroid of the cross sections. Near the ends (in

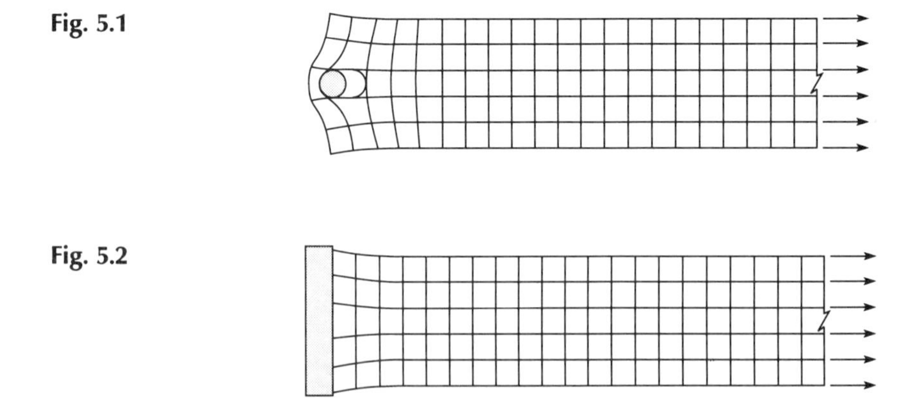

Fig. 5.1

Fig. 5.2

the vicinity of the applied loads) the deformations are quite different; severe distortions of the small squares are evident in both cases. However, the observed deformations are essentially the same throughout the central portion of these two bars; here the squares are deformed into rectangles. Apparently, the only significant differences in behavior are localized effects near the ends, where the loads are applied. If these members are long and slender, the overall stretching of the two bars is approximately the same.

The essential idea demonstrated by this example was elucidated by Barre de Saint-Venant in 1855.[1] In essence, Saint-Venant's principle states that *two different distributions of force acting on the same portion of a body have essentially the same effects on parts of the body that are sufficiently far from the region of application, provided that these force distributions have the same resultant.* This principle applies not only to axially loaded members, but to all types of loading.

According to Saint-Venant's principle, it is often possible to assess the net effect of a loading without regard for the manner of load application. In the preceding example the extension of the bar is essentially independent of the manner in which the load is applied. This statement is true because the load is applied over a relatively small portion of the member. A change in the load distribution has a localized effect. When these localized effects are of no consequence, the precise manner of load application is unimportant.

The principle of Saint-Venant is employed in a great variety of problems in which loads act upon relatively small portions of a body (essentially concentrated loads). These problems include many practical situations, especially those involving thin flexible bodies, such as bars, shafts, beams, springs, plates, and shells. Throughout the remainder of this text, the principle is tacitly employed unless specific details of loading are given. However, the reader should bear in mind that there are instances in which localized deformations and stresses are of the greatest importance. Fracture and fatigue failures are manifestations of local disturbances. Clearly, we cannot overlook local stress distributions when examining phenomena of this kind.

5.3 Axial Extension

The Problem

Often structures are composed of slender members that are subjected to axial loads, for example, the widely used bridge and building truss. The members of a truss can be treated as prismatic bars or rods subjected to axial forces (usually applied at the ends of the members). As an example, we consider a situation of this kind. The rod shown in Fig. 5.3 is subjected to an axial force P applied at the right end. For simplicity, consider the rod to be held fixed at the left end.

---

1 Barre de Saint-Venant, "Mémoire sur la Torsion des Prismes ..." *Mémoire des savants étrangers*, T. XIV (Paris: 1855), 297–99.

Fig. 5.3

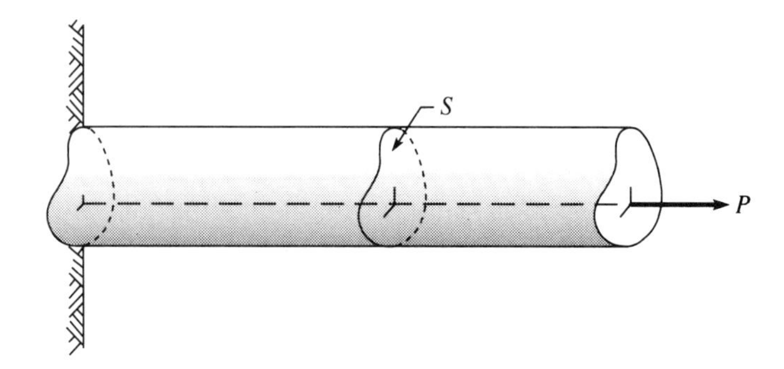

Let us consider a free-body diagram of a portion of this rod. The portion to the right of a cross section S is shown in Fig. 5.4. For equilibrium the net internal force system on the cross section is an axial force F, collinear and equal to the applied load P.

Fig. 5.4

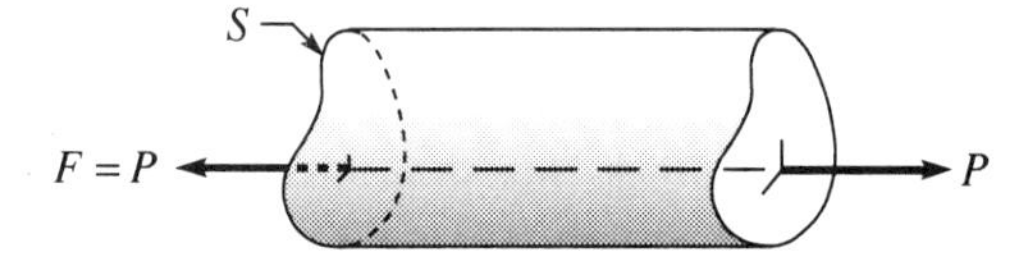

Description of the Deformation

Because of the nature of the loading, we expect the rod to stretch (or contract) in the axial direction. We will assume that the main features of the deformation are as follows: (1) The axis of the rod remains straight after deformation, and (2) a plane cross section remains plane and perpendicular to the axis, displacing only in the axial direction.

The rod is redrawn in Fig. 5.5 as it appears both before and after deformation. The x-axis is taken along the centroidal axis of the rod, and the origin of coordinates is at the fixed end. For clarity, the load is not shown in this figure.

In the undeformed rod [Fig. 5.5(a)] a cross section is located by the coordinate x, and a nearby cross section is located at $x + \Delta x$. After the axial load is applied, these planes are displaced as shown in Fig. 5.5(b). Since these cross sections remain plane and parallel, their deformed positions can be located by the coordinates x^* and $x^* + \Delta x^*$.

The slice that lies between these nearby cross sections is redrawn in Fig. 5.6(a) and 5.6(b), as it appears before and after deformation. If AB is any line segment between these cross sections and parallel to the axis, its length must be Δx. When the rod is deformed, the new length A^*B^* is Δx^*. The extensional strain is

$$\varepsilon_x = \lim_{\Delta x \to 0} \frac{\Delta x^* - \Delta x}{\Delta x} \tag{5.1a}$$

Fig. 5.5

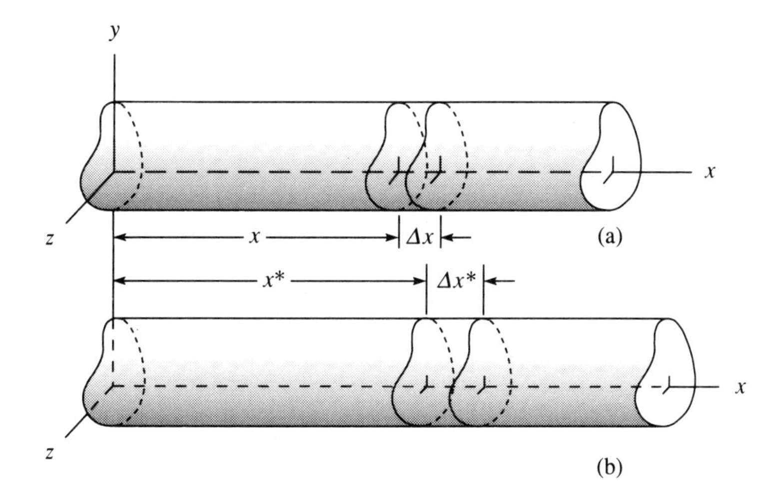

Fig. 5.6

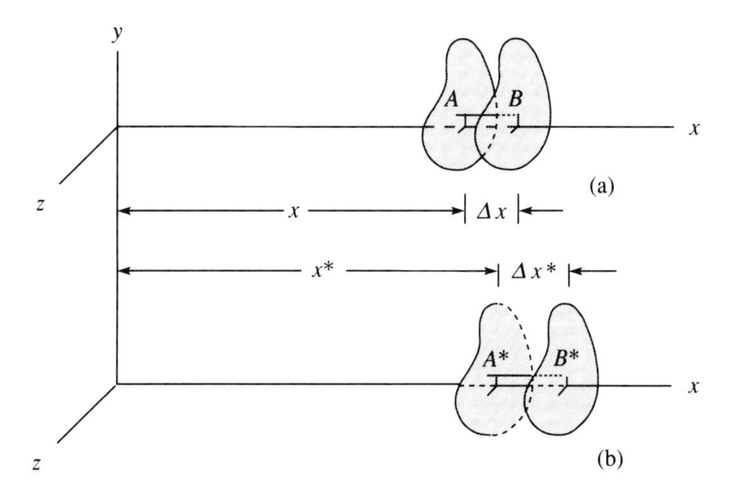

On any cross section the extension strain ε_x is constant; that is, it is independent of y and z. It might, of course, depend on the location of the cross section; that is, ε_x might depend on x but not on y or z.

The extension (or contraction) of the rod can also be obtained, according to Eq. (5.1a): For the slice shown in Fig. 5.6, the elongation is

$$\Delta e = \Delta x^* - \Delta x = \varepsilon_x \Delta x \tag{5.1b}$$

The total elongation of a member of length L is obtained as a summation and, in the limit ($\Delta x \to 0$), the integral

$$e = \int_0^L \varepsilon_x \, dx \tag{5.1c}$$

From the foregoing description we can say nothing about the extensional strains ε_y and ε_z; some additional information or approximation is needed. With this in mind let us examine the nature of the internal forces.

Statics

Consider a portion of the rod as drawn in Fig. 5.7. Because of the normal force F, we anticipate a normal force ΔF on the elemental area ΔA. This, of course, will have to come from a normal stress σ_x. On the other hand, if we look at the narrow rectangular prism, we might suppose that the normal stresses on its lateral faces are inconsequential, since the rod is itself a slender prism whose lateral surfaces are free of applied stress. Assuming that this is the case, we take $\sigma_y = \sigma_z = 0$ throughout the rod.

Fig. 5.7

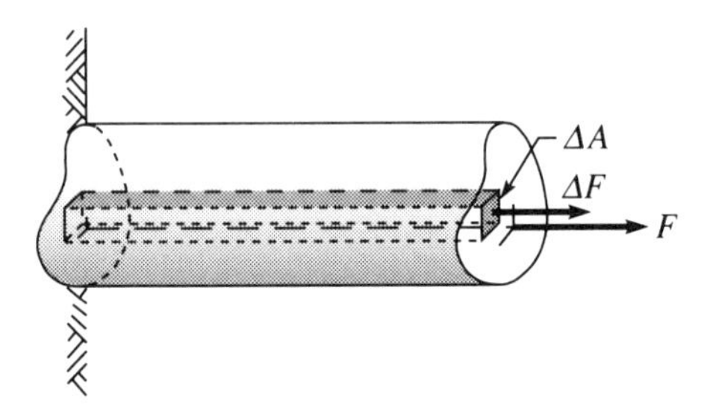

On the element of area ΔA in the cross section S shown in Fig. 5.8, the normal stress σ_x produces the increment of normal force

$$\Delta F = \sigma_x \Delta A$$

Fig. 5.8

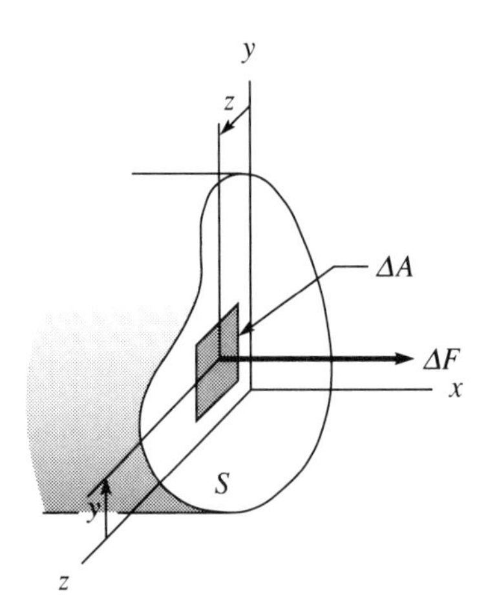

Then over the entire cross section, the total normal force is

$$F = \iint_S \sigma_x \, dA \tag{5.2}$$

The loading is an *axial* force; that is, the resultant is a force $P(= F)$ acting along the x-axis. Then, with respect to the y- and z-coordinates, $M_y = M_z = 0$. The moment of ΔF about the y-axis is

$$\Delta M_y = z \, \Delta F = \sigma_x z \, \Delta A$$

and, therefore,

$$M_y = \iint_S \sigma_x z \, dA = 0 \tag{5.3a}$$

Similarly,

$$M_z = -\iint_S \sigma_x y \, dA = 0 \tag{5.3b}$$

Equilibrium of the member is insured by Eqs. (5.2) and (5.3). To complete a solution, we require a description of the material.

Linearly Elastic Material

Suppose that the bar is composed of a linearly elastic and isotropic (HI = hookean isotropic) material. Then,

$$\begin{aligned} \varepsilon_x &= \frac{1}{E}(\sigma_x - \nu\sigma_y - \nu\sigma_z) \\ \varepsilon_y &= \frac{1}{E}(\sigma_y - \nu\sigma_z - \nu\sigma_x) \\ \varepsilon_z &= \frac{1}{E}(\sigma_z - \nu\sigma_x - \nu\sigma_y) \end{aligned} \tag{5.4a}$$

As previously noted, the lateral surfaces of the slender rod are stress-free. Therefore, we assume that σ_y and σ_z are negligible and Eqs. (5.4a) are simplified accordingly:

$$\varepsilon_x = \frac{\sigma_x}{E}, \qquad \varepsilon_y = \varepsilon_z = -\nu\frac{\sigma_x}{E} \tag{5.4b}$$

Let us now limit our analysis to *homogeneous* (HI-HO = hookean, isotropic, *and* homogeneous) bars. Then E is constant and σ_x, as ε_x, is independent of y and z.

With the *kinematic* equations (5.1), the *equilibrium* conditions (5.2) and (5.3), and the *hookean relations* (5.4), we can implement the solution.

Since σ_x is constant, the couples M_y and M_z of Eq. (5.3) vanish *if*

$$\iint_S z \, dA = \iint_S y \, dA = 0$$

In other words, the origin must be at the centroid of the cross section; the force $F = P$ acts at the centroid. Then, by Eq. (5.2)

$$\sigma_x = \frac{P}{A} \tag{5.5}$$

By Eqs. (5.4) and (5.5)

$$\varepsilon_x = \frac{P}{EA}, \qquad \varepsilon_y = \varepsilon_z = -\nu\frac{P}{EA} \tag{5.6}$$

By means of (5.1c) and (5.6), the elongation of the member follows:

$$e = \frac{PL}{EA} \tag{5.7}$$

Equations (5.5), (5.6), and (5.7) relate the most significant quantities: the applied load P, the normal stress σ_x, extensional strain ε_x, and total elongation e. Every cross section is acted upon by the same constant normal stress, and every cross section is subjected to the same constant extensional strain. Presumably this applies to the end faces as well as to all the intermediate cross sections, which implies that a very special support is required in order that the rod behave exactly as described.[2] Fortunately, Saint-Venant's principle enables us to apply this solution to a wide variety of problems. The manner in which the loads and reactions are applied does not significantly affect the overall deformation if the rod is slender. A similar statement can be made about the stresses in portions of the rod not near the ends.

5.4 Axial Extension of Nonhookean Rods

The description of the deformation of a slender bar subjected to axial loading was given earlier. This description is not limited to a hookean material. In other circumstances the axis of the bar remains straight, and cross sections remain perpendicular to it. For any homogeneous material, the extensional strain ε_x is constant on a cross section. Further analysis requires a description of that material.

Linear-Hardening Material

Suppose the rod is made of a linear-hardening material. The normal stress σ_x is related to the normal strain ε_x, as in Fig. 5.9. (This stress-strain curve presupposes that $\sigma_y = \sigma_z = 0$, which is again assumed to be true.) The behavior is linear and elastic as long as σ_x is less than the yield stress σ_0. Increases in σ_x beyond σ_0 are accompanied by plastic deformation (see Sec. 4.11). If we assume that the bar is not unloaded, the

---

2 If a prismatic bar is subjected to uniform normal stresses on its end faces, then the preceding solution is in accord with that obtained from more elaborate theories. Furthermore, it can be shown that $\sigma_y = \sigma_z = \tau_{xy} = \tau_{xz} = \tau_{yz} = 0$, $\varepsilon_y = \varepsilon_z = -\nu\varepsilon_x$, and $\gamma_{xy} = \gamma_{xz} = \gamma_{yz} = 0$.

Fig. 5.9

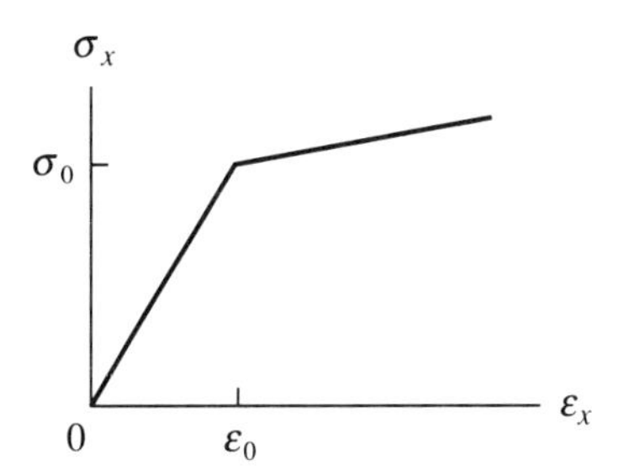

material behavior can be expressed by

$$\varepsilon_x = \begin{cases} \sigma_x, & \sigma_x \le \sigma_0 \qquad (5.8) \\ \dfrac{1+k}{kE}\sigma_x - \dfrac{\sigma_0}{kE}, & \sigma_x \ge \sigma_0 \qquad (5.9) \end{cases}$$

where k is a positive number.

Because ε_x is constant on any cross section, σ_x must also be constant on any cross section. It follows that the net internal force system consists of the normal force

$$F = \sigma_x A \tag{5.10}$$

where A is the cross-sectional area.

Two situations can arise. If F is less than $P_0 = \sigma_0 A$, the behavior is linearly elastic, and the substitution of Eq. (5.10) into Eq. (5.8) gives

$$\varepsilon_x = \frac{F}{EA}, \qquad F \le P_0 \tag{5.11a}$$

On the other hand, if F exceeds P_0, some plastic deformation occurs, and the substitution of Eq. (5.10) into Eq. (5.9) gives

$$\varepsilon_x = \frac{1+k}{kEA}F - \frac{\sigma_0}{kE}, \qquad F \ge P_0 \tag{5.11b}$$

We observed earlier that the normal force F on every cross section is equal to the load P, so Eqs. (5.10) and (5.11) become

$$\sigma_x = \frac{P}{A}, \tag{5.12}$$

$$\varepsilon_x = \begin{cases} \dfrac{P}{EA}, & P \le P_0 \qquad (5.13a) \\ \dfrac{1+k}{kEA}P - \dfrac{\sigma_0}{kE}, & P \ge P_0 \qquad (5.13b) \end{cases}$$

As before, the load P must lie along the centroidal axis of the bar, since a constant normal stress cannot give rise to bending couples.

The total elongation of the bar is given by

$$e = \int_0^L \varepsilon_x \, dx$$

so that

$$e = \begin{cases} \dfrac{PL}{EA}, & P \leq P_0 \quad (5.14a) \\[2ex] \dfrac{1+k}{k}\dfrac{PL}{EA} - \dfrac{\sigma_0 L}{kE}, & P \geq P_0 \quad (5.14b) \end{cases}$$

Consider the extension of the bar as the load P is gradually increased from zero. This is defined by Eq. (5.14), which is plotted in Fig. 5.10. Initially the behavior is linear and elastic; this is the same as the earlier example of this section. When the load P exceeds the value $P_0 = \sigma_0 A$, the material yields, and a permanent deformation is sustained by the bar. If the load were subsequently removed, unloading would not retrace the loading path.

Fig. 5.10

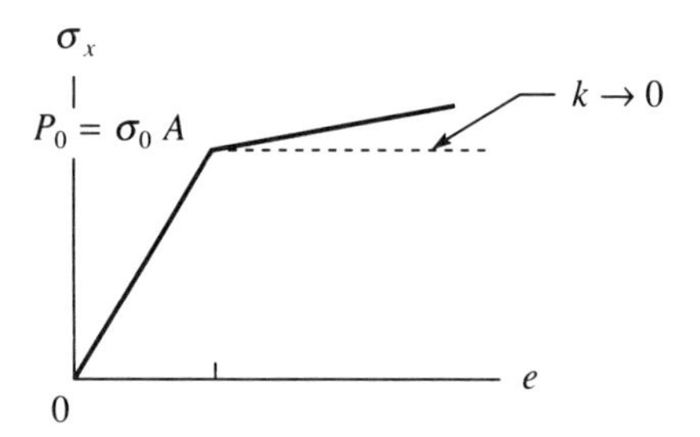

Ideally Plastic Material

A very important special case occurs when the constant k approaches zero. (The material in this case is ideally plastic and the load-elongation curve becomes flat, as indicated by the dotted line in Fig. 5.10). When the applied load attains the value $P_0 = A\sigma_0$, the bar undergoes unrestricted plastic deformation. This constitutes plastic collapse; because the bar cannot sustain a greater load, P_0 is termed the *limit load.*

5.5 Nature of the Analysis

A rigorous analysis of a deformable body requires a study of the deformation and equilibrium of each infinitesimal element of the body. This involves a study of the stress, strain, and displacement differentials from point to point throughout the body. Mathematically, such a treatment leads to the consideration of partial differential equations. To avoid these complexities, we resort to an approximate analysis based on assumptions concerning the significant features of the deformation. The analysis combines kinematic concepts (strain), material behavior (stress-strain laws), and dynamic relations (equilibrium equations) to obtain load-stress and load-deformation relations. The procedure follows the same general pattern for all the examples in this chapter. Let us review this procedure as it was applied to the axial extension of prismatic bars.

The assumptions concerning the significant features of the deformation are used to deduce how the strains vary in the member. This information will generally be incomplete. For example, in the axial extension problem the assumptions that the axis remains straight and cross sections remain perpendicular to it lead to the conclusion that the extensional strain on a cross section is constant. These assumptions contain no information about the other strains in the rod.

The material is described by relations between the strain and the stress distribution. Often the incompleteness of the strain description necessitates the introduction of other assumptions or simplifications before these stress-strain relations can be used. In the example of axial extension it was necessary to introduce the further assumption that the transverse normal stresses (σ_y and σ_z) vanish in order to use the stress-strain relations considered. It then followed that the normal stress on a cross section is constant.

Equilibrium conditions enable us to express the net internal force system on a sectioning surface in terms of the load. Since the stress distribution on certain sectioning surfaces is known, the stresses there can be related to the load. Thus, we found that the constant normal stress σ_x gave rise to a normal force equal to the load P, from which it followed that $\sigma_x = P/A$.

We return to the stress-strain relations to express the strains in terms of the external load. The significant features of the deformation are then related to the load, since they can be determined from a knowledge of the strains. In the example of axial extension we found, for the elastic material, $\varepsilon_x = \sigma_x/E = P/EA$. The significant feature of the deformation is the total elongation of the rod, which is related to the extension strain ε_x by $e = \int_0^L \varepsilon_x \, dx$. Hence, for the linearly elastic material, $e = PL/EA$.

EXAMPLE 1

The predominant features of the simple extension (or contraction) occur under many circumstances besides the direct application of a central force. To illustrate this, consider the mechanical assembly of the bolt B, which extends through the cylindrical tube T, as shown in the section of Fig. 5.11. Suppose for simplicity that the washers beneath the head and nut are stiff enough so that we can assume that they remain plane as the nut is advanced, the bolt is extended, and the tube is compressed. To be specific, suppose that the bolt is $\frac{1}{4}$ in. (nominal diameter) with 24 threads per inch and that it is made of steel ($E_B = 30 \times 10^6$ lb/in.^{2}, $\alpha_B = 6.5 \times 10^{-6}$/°F). The tube has inner and outer diameters $d = 0.600$ in. and $D = 0.800$ in. and is made of aluminum alloy ($E_T = 10 \times 10^6$ lb/in.^{2}, $\alpha_T = 13.5 \times 10^{-6}$/°F). At room temperature (70°F) the nut is

Fig. 5.11

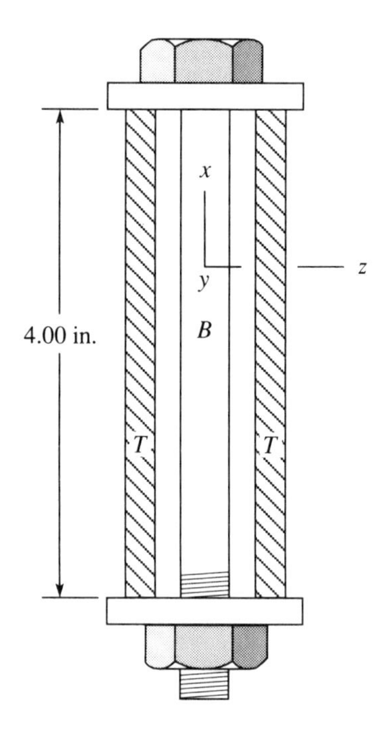

snug and is then advanced one-quarter turn ($\frac{1}{96}$ in.). The assembly is then also heated to 470°F. Our problem is to determine the mean normal stresses on cross sections of the bolt and the tube. The problem is a simple problem from an engineer's viewpoint, but one that serves to illustrate the essential features in the solution of a problem of deformable solids. Again, we must recognize that local effects occur near the ends and especially at the threads; yet simple extension provides a valid overall description in accordance with Saint-Venant's principle.

Firstly, let us consider the *kinematics*: Since the nut is advanced one turn, the effective length of the bolt (between the "rigid" washers) is reduced by $\frac{1}{96}$ in. The essential geometric constraint can be stated simply: The final effective length of the bolt and the tube must be the same (distance between the "rigid" washers). If L denotes the initial length of both, L_B^* and L_T^* are the final lengths, and e_B^* and e_T^* are the elongations of bolt and tube, respectively, then

$$L - \tfrac{1}{96} + e_B = L_B^* = L_T^* = L + e_T$$

or

$$e_B - e_T = \tfrac{1}{96} \tag{5.15}$$

Secondly, the *description* of both steel and aluminum under the circumstances is *hookean, isotropic*, and *homogeneous* (HI-HO). Moreover, our previous description of the axial extension suggests that the transverse normal stresses can be neglected in both components:

$$\sigma_y = \sigma_z = 0$$

Then the strain-stress-temperature relations for each are similar. If superscripts B and T signify steel bolt and aluminum tube, respectively, then

$$\varepsilon_x^B = \frac{\sigma_x^B}{E_B} + \alpha_B \Delta T \tag{5.16a}$$

$$\varepsilon_x^T = \frac{\sigma_x^T}{E_T} + \alpha_T \Delta T \tag{5.16b}$$

Finally, *equilibrium* must be enforced. The free-body diagram of Fig. 5.12 shows the axial force upon the bolt P_B and tube P_T. Notice that the axial forces, normal stresses, and "elongations" are assumed positive. Equilibrium dictates that

$$P_B + P_T = 0 \tag{5.17}$$

Fig. 5.12

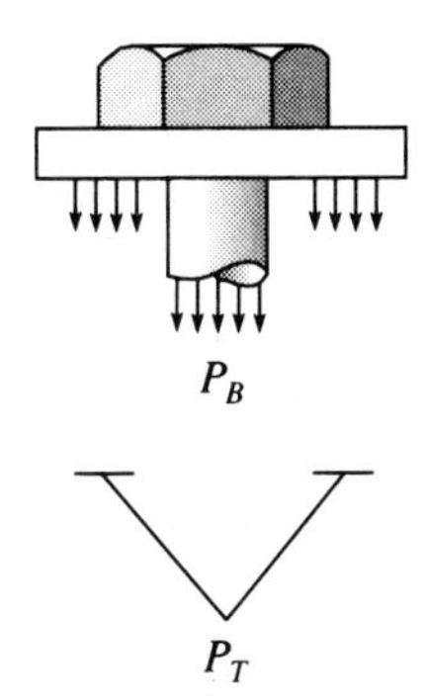

In accordance with previous descriptions,

$$P_B = \sigma_x^B A_B, \qquad P_T = \sigma_x^T A_T \tag{5.18}$$

Each of the three facets (*kinematics*, *material behavior*, and *equilibrium*) has now been addressed. It remains to utilize these equations to effect a solution. By Eqs. (5.17) and (5.18),

$$\sigma_x^B = -\frac{A_T}{A_B}\sigma_x^T \tag{5.19}$$

Then, σ_x^B can be eliminated from (5.16a,b) and the elongations determined, as follows

$$\begin{aligned} e_B &= \int_0^L \varepsilon_x^B \, dx = L\left(\frac{\sigma_x^B}{E_B} + \alpha_B \Delta T\right) \\ &= \left(-L\frac{A_T}{A_B}\frac{\sigma_x^T}{E_B} + \alpha_B \Delta T\right) \end{aligned}$$

Similarly,

$$e_T = L\left(\frac{\sigma_x^T}{E_T} + \alpha_T\,\Delta T\right)$$

Finally, the latter are substituted into the geometric constraint (5.15):

$$-L\left(\frac{1}{E_T} + \frac{A_T}{A_B E_B}\right)\sigma_x^T = +\frac{1}{96} + L(\alpha_T - \alpha_B)\,\Delta T$$

With the foregoing values ($L = 4.00$ in.), we obtain

$$\sigma_x^T = -21{,}700\ \text{lb/in.}^2 \qquad \sigma_x^B = +97{,}000\ \text{lb/in.}^2$$

The axial force upon the bolt and tube is 4770 lb. All numerical values are given to three decimal places of accuracy, because dimensions and properties are seldom more accurate in most practical circumstances. ◆

PROBLEMS

5.1 A structural steel channel, initially 3 m long, is stretched 0.80 cm. Calculate the permanent extension, regarding the steel as an ideally plastic material with $E = 200 \times 10^3$ MPa and $\sigma_0 = 300$ MPa.

5.2 An aluminum alloy bar is initially 48.000 in. long with a cross-sectional area of 1.500 in^2. The aluminum can be treated as a linear-hardening material with $E = 10^7$ psi, $\sigma_0 = 60{,}000$ psi, and $k = 0.10$, where $\varepsilon = \sigma/E + (\sigma - \sigma_0)/kE$ for $\sigma > \sigma_0$. The bar is stretched uniformly to an extensional strain of 0.017 by loads distributed uniformly on the ends. What load is needed to cause this strain, and how long is the bar after the load is removed?

5.3 The cylindrical bar in Fig. 5.3 is made of a material with a stress-strain law given by

$$\sigma_x = \sigma_0 \frac{E\varepsilon_x}{\sigma_0 + E\varepsilon_x}$$

Show that the load-extension relation is

$$e = \frac{\dfrac{PL}{EA}}{1 - \dfrac{P}{\sigma_0 A}}$$

Is there a limit load for this bar?

5.4 Suppose that the load on the bar in Problem 5.3 is $P = \frac{1}{4}\sigma_0 A$. If the material behavior upon unloading is linear, with the same slope as the initial portion of the loading curve, compute the permanent extension of the bar inflicted by the load. If the load had been doubled to $\frac{1}{2}\sigma_0 A$ before unloading, by what factor is the permanent extension of the bar increased? Repeat for the tripled load $\frac{3}{4}\sigma_0 A$.

5.5 A circular shaft has a diameter d. A collar is pressed onto the shaft and the shaft is then subjected to a tensile load P. Assume that the shaft is of a linearly elastic material (E, ν) and behaves according to the assumptions stated for axial extension. How much smaller than d must the original inside diameter of the collar be (before it is pressed on the shaft) so that it will not loosen when the load is applied to the shaft?

5.6 Two circular cylinders, one of steel ($E = 210 \times 10^3$ MPa) and one of aluminum ($E = 70 \times 10^3$ MPa) support a load applied by means of a rigid bearing plate, as shown in Fig. P5.6. If both cylinders are linearly elastic, determine the location x of the resultant load P so that cross sections of each bar remain plane and perpendicular to the axis of each bar.

Fig. P5.6

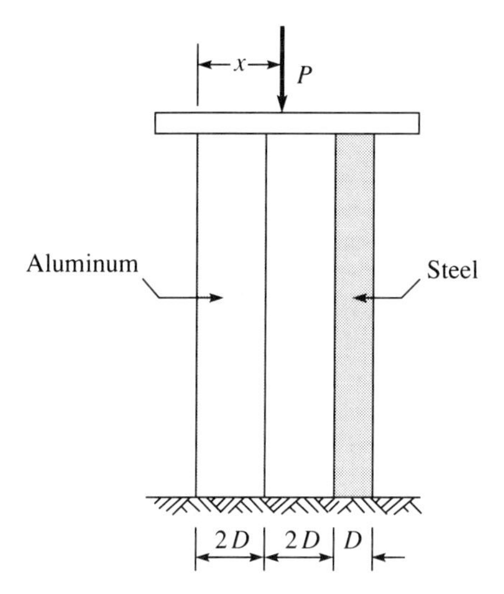

5.7 A member is composed of a series of thin strips of material bonded together so that the final member is a laminated bar with a rectangular cross section. Each of the strips is itself a linearly elastic material, but the modulus of elasticity varies from strip to strip across the width of the member according to $E = E_0[1 + (kx/D)]$. (Refer to Fig. P5.7). A rigid plate is attached to the free end of the laminated bar and a load P is applied as shown. Compute x so that plane cross sections remain plane and perpendicular to the axis of the bar as it stretches.

Fig. P5.7

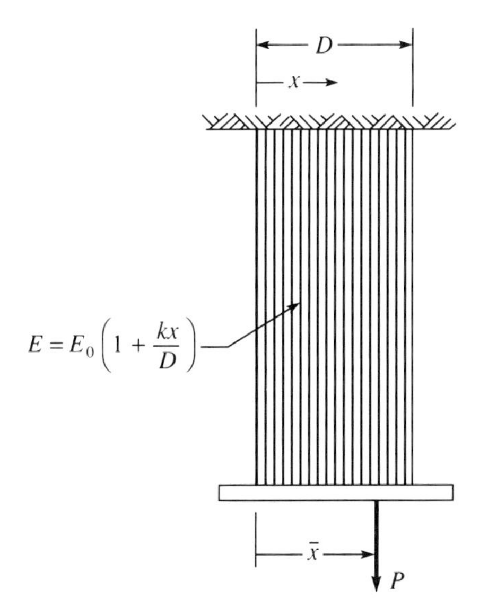

5.8 A wooden pile 12 in. in diameter is driven to a depth of 10 ft in a soil that supports the pile mainly by friction along the lateral surface. The pile supports a load of 9 tons (T). The intensity of the friction forces on the lateral surface of the pile varies linearly, doubling in value from the bottom to the top. Determine the normal stress on any cross section and the total contraction in the 10 ft of submerged pile. Assume that the wood is linearly elastic ($E = 2 \times 10^6$ lb/in.^{2}), and make the same assumptions for axial extension as were made in this section (ignoring the transverse normal stresses).

5.9 A cylindrical bar hangs under its own weight W, as shown in Fig. P5.9. The length of the bar is L and the cross-sectional area is A. Make the usual assumptions for axial extension: The axis of the bar remains straight; the cross sections remain plane and perpendicular to the axis. If necessary, also assume that the transverse normal stresses are zero. For linearly elastic material, show that

$$\sigma_x(x) = \frac{W}{A}\left(1 - \frac{x}{L}\right)$$

$$\varepsilon_x(x) = \frac{W}{EA}\left(1 - \frac{x}{L}\right)$$

Fig. P5.9

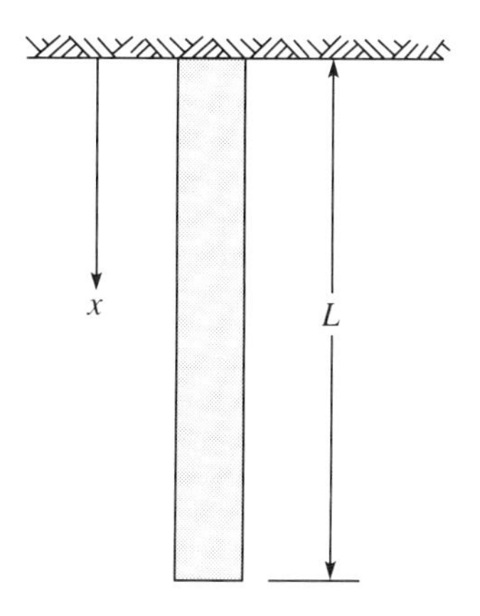

$$e = \frac{WL}{2EA}$$

where e is the total extension of the bar.

5.10 Repeat Problem 5.9 for a linear-hardening material:

$$\varepsilon_x = \begin{cases} \dfrac{\sigma_x}{E}, & \sigma_x \leq \sigma_0 \\ \dfrac{1+k}{kE}\sigma_x - \dfrac{\sigma_0}{kE}, & \sigma_x \geq \sigma_0 \end{cases}$$

Obtain results for values of W greater than $\sigma_0 A$.

5.11 Repeat Problem 5.9 for a material with a stress-strain law defined by

$$\sigma_x = \sigma_0 \frac{E\varepsilon_x}{\sigma_0 + E\varepsilon_x}$$

Obtain results for values less than $\sigma_0 A$.

5.12 A cylindrical bar is fixed at one end and subjected to a body force, as shown in Fig. P5.12. The point O attracts the slice of width Δl with the force

$$\Delta F = \frac{K}{l^2} A\,\Delta l$$

where A is the cross-sectional area of the bar, l is the distance from O to the slice, and K is a constant. Make the usual assumptions for axial extension (straight axis; plane cross sections remain perpendicular to the axis) and thereby show that for a linearly elastic material, the total extension of the bar is

$$e = \int_b^{b+L} \frac{K}{E}\left(\frac{1}{b} - \frac{1}{l}\right) dl$$

$$= \frac{K}{E}\left[\frac{L}{b} - \log\left(1 + \frac{L}{b}\right)\right]$$

Fig. P5.12

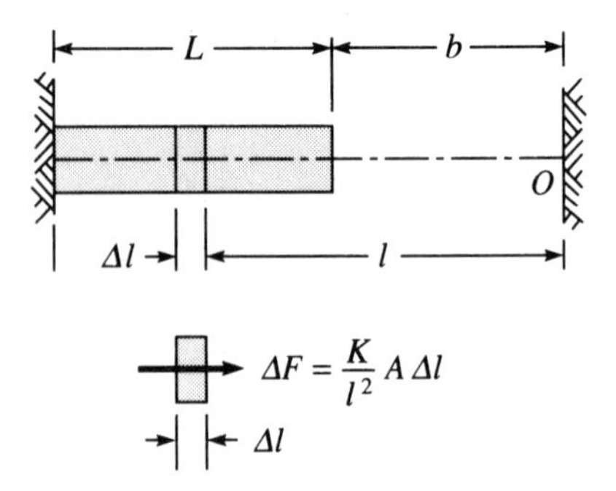

5.13 A conical bar with circular cross sections is loaded axially, as shown in Fig. P5.13. The radius of the cross section at x is given by

$$R(x) = R_0\left(1 - c\frac{x}{L}\right)$$

Fig. P5.13

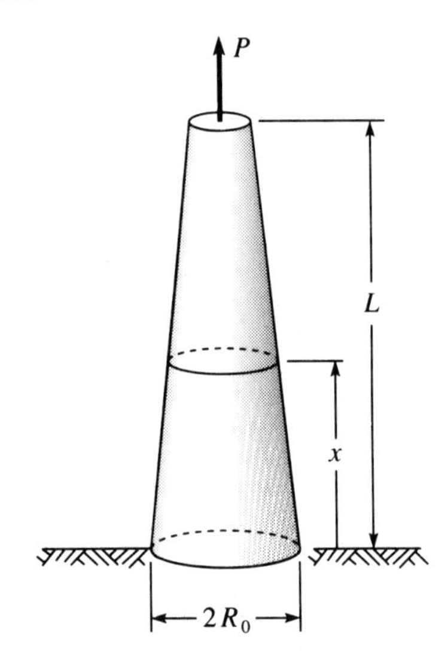

If the taper of the bar is small ($c \ll 1$), then the usual assumptions for axial extension are adequate; the axis of the bar remains straight and the cross sections remain plane and perpendicular to the axis. If necessary, also assume that the transverse normal stresses are zero. Then, for a linearly elastic bar satisfying these conditions, show that the load-stress and load-extension relations are

$$\sigma_x(x) = \frac{P}{A_0} \frac{1}{\left(1 - c\dfrac{x}{L}\right)^2}$$

$$e = \frac{PL}{EA_0} \frac{1}{(1-c)}$$

where we have set

$$A_0 = 2\pi R_0$$

5.14 Repeat Problem 5.13 for a linear-hardening material:

$$\sigma_x = \begin{cases} \dfrac{\sigma_x}{E}, & \sigma_x \le \sigma_0 \\ \dfrac{1+k}{kE}\sigma_x - \dfrac{\sigma_0}{kE}, & \sigma_x \ge \sigma_0 \end{cases}$$

5.15 The slightly tapered conical bar shown in Fig. P5.15 hangs under its own weight. The weight density of the bar is γ. Make the usual assumptions for axial extension, and thereby show that, for a linearly elastic bar, the total extension is given by $e = \gamma L^2/6E$.

Fig. P5.15

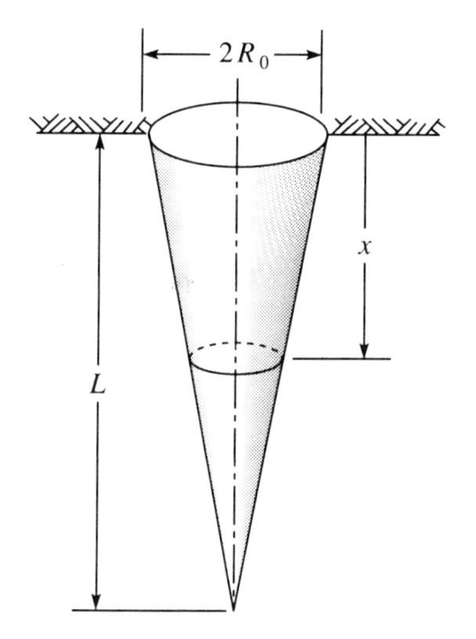

***5.16** A cylindrical rod is fixed at one end and free of all other externally applied loads. The material of the rod is set in motion in the axial direction only. Assume that at all times the axis of the bar is straight and that the particles of a cross section at any time always remain plane and perpendicular to the axis. The bar is shown in the unstrained state and also at some subsequent time t in Fig. P5.16.

a. Let $u(x, t)$ represent the axial displacement (from its unstrained location) of the cross section at x and at time t. Show that

$$\varepsilon_x(x, t) = \frac{\partial}{\partial x} u(x, t) \tag{P5.16a}$$

b. If the material is linearly elastic and we neglect the stresses σ_y and σ_z, show that

$$\sigma_x(x, t) = E \frac{\partial}{\partial x} u(x, t)$$

Hence, the normal force on any cross section at any time is

$$F(x, t) = EA \frac{\partial}{\partial x} u(x, t) \tag{P5.16b}$$

where A is the cross-sectional area.

c. Denote the lineal mass density of the bar by ρ so that the mass of the slice between S^* and S_1^* is $\Delta m = \rho\,\Delta x$. Because the slice is moving with an acceleration of $(\partial^2/\partial t^2)u(x, t)$, deduce from the equations of motion that

$$F(x + \Delta x, t) - F(x, t) = \rho\,\Delta x \frac{\partial^2}{\partial t^2} u(x, t)$$

Hence

$$\frac{\partial F}{\partial x} = \rho \frac{\partial^2}{\partial t^2} u(x, t) \tag{P5.16c}$$

d. Show that it follows from Eqs. (P5.16b) and (P5.16c) that the free axial vibration of a linearly elastic rod (assuming that plane cross sections remain plane

Fig. P5.16

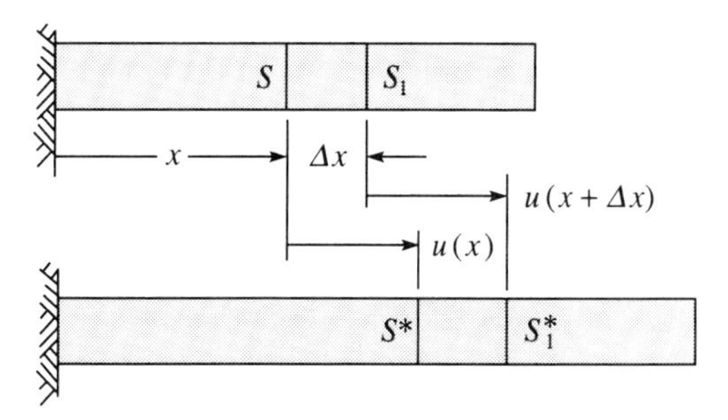

cross sections) satisfies the wave equation

$$\frac{\partial^2}{\partial x^2} u(x,t) - \frac{\rho}{EA} \frac{\partial^2}{\partial t^2} u(x,t) = 0 \quad \text{(P5.16d)}$$

5.17 The cylindrical rod in Fig. P5.17 is composed of bronze and steel; both have the same cross sections. Assume that these materials are HI-HO with the following properties (S denotes steel and B denotes bronze):

$$E_S = 200 \times 10^3 \text{ MPa}, \qquad \alpha_S = 3.6 \times 10^{-6}/^\circ\text{C}$$

$$E_B = 100 \times 10^3 \text{ MPa}, \qquad \alpha_B = 5.4 \times 10^{-6}/^\circ\text{C}$$

The member slips between the "rigid" walls at 20°C. Determine the normal stress upon a cross section when the temperature has been raised to 200°C. Employ the assumptions of simple axial deformation; for example, cross sections remain plane.

Fig. P5.17

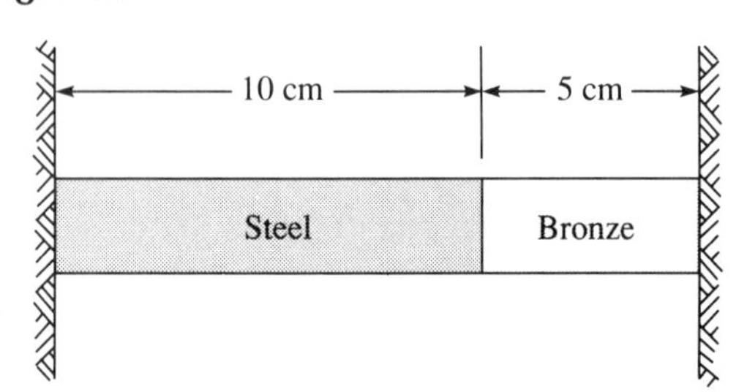

5.18 The steel bolt in Fig. P5.18 is inserted in the brass sleeve with their axes aligned. The nut is advanced until the assembly is intact but there is no significant stress upon either component. The bolt has a diameter of 0.375 in.; inner and outer diameters of the sleeve are 0.400 in. and 0.650 in. The temperature is then raised by 500°F. Compute the mean normal stress upon the cross section of each component. Employ the description of simple axial extension and use the properties given in Tables 4.1 and 4.2.

Fig. P5.18

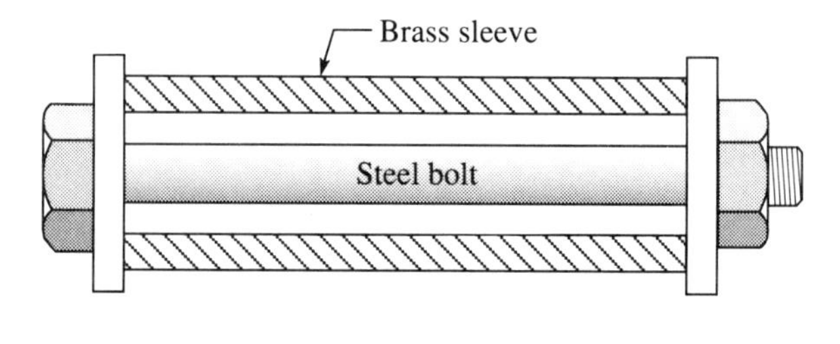

5.6 Axially Loaded Members: Applications

Most members that carry axial loads do not conform exactly to the description of simple extension given in Sec. 5.3. For example, loads may be applied at places other than the ends, or perhaps the cross section of the member is not constant along its length. Nevertheless, we can frequently apply the earlier results to those portions that are similar to the prototype.

Two results, derived for simple extension of HI-HO (hookean-homogeneous) bars, are used repeatedly. They are the load-stress relation,

$$\sigma = \frac{P}{A} \tag{5.20}$$

and the load-extension relation,

$$e = \frac{PL}{EA} \tag{5.21}$$

These are Eqs. (5.5) and (5.7), respectively; they are restated here for convenient reference. To use these intelligently requires a full awareness of their limitations.

Recall that the description of simple extension is applicable to a cylindrical or prismatic bar when it is subjected to end loads acting along its centroidal axis. Furthermore, Saint-Venant's principle allows us to apply this description to a slender bar, even though the exact distribution of the end loads is not known. The following examples illustrate the application of the simple extension solution.

EXAMPLE 2

Stepped Bar The member shown in Fig. 5.13(a) consists of two cylindrical portions with circular cross sections and a common centroidal axis. It is composed of a high-carbon steel ($E = 30 \times 10^6$ lb/in.^{2} = 200 GPa), which has a yield stress of $\sigma_0 = 70 \times 10^3$ lb/in.^{2} (480 MPa). Let us calculate the overall extension caused by the axial loads applied at locations B and D.

Fig. 5.13

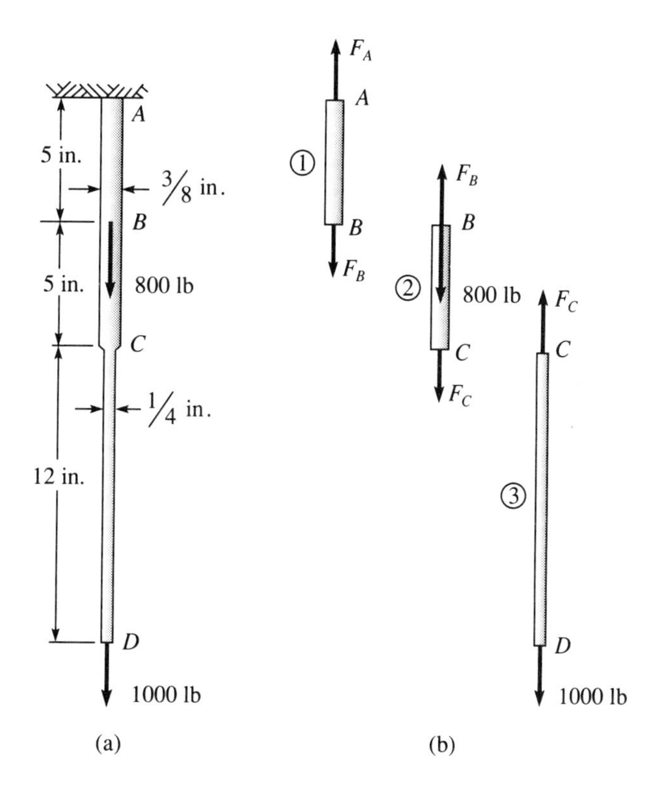

As a first step we consider the free-body diagrams of portions AB, BC, and CD, as shown in Fig. 5.13(b). We have separated parts AB and BC by a section immediately above the point of load application. This was an arbitrary choice, as we might equally well have cut the member just below B. Had we done so, the 800-lb (3560-N) load would have been shown acting at the lower end of the free body AB. The final result would be no different.

We denote the bodies AB, BC, and CD by the numbers 1, 2, and 3, respectively. The equilibrium conditions for the respective bodies are

$$F_A - F_B = 0, \qquad F_B - 800 - F_C = 0, \qquad F_C - 1000 = 0$$

Consequently,

$$F_A = F_B = 1800 \text{ lb} \quad (8000 \text{ N}), \qquad F_C = 1000 \text{ lb} \quad (4450 \text{ N})$$

Because these bodies are relatively slender and axially loaded at their ends, we can regard each as a case of simple extension. Let us assume, tentatively, that the behavior is linearly elastic. Then the elongations of each body, according to Eq. (5.21), are

$$e_1 = \frac{(1800)(5)}{(30 \times 10^6)\pi(\frac{3}{16})^2} = 0.00272 \text{ in.} \quad (0.00691 \text{ cm})$$

$$e_2 = \frac{(1000)(5)}{(30 \times 10^6)\pi(\frac{3}{16})^2} = 0.00151 \text{ in.} \quad (0.00384 \text{ cm})$$

$$e_3 = \frac{(1000)(12)}{(30 \times 10^6)\pi(\frac{1}{8})^2} = 0.00815 \text{ in.} \quad (0.02070 \text{ cm})$$

The total elongation of the bar is the sum of the elongations of its parts:

$$e = e_1 + e_2 + e_3 = 0.0124 \text{ in.} \quad (0.03145 \text{ cm})$$

This result is valid if the assumption of linear elasticity holds. To check this assumption, we can estimate the axial normal stress using Eq. (5.20). On a cross section of body 1,

$$\sigma_1 = \frac{1800}{\pi(\frac{3}{16})^2} = 16{,}300 \text{ lb/in.}^2 \quad (112 \text{ MPa})$$

on a cross section of body 2,

$$\sigma_2 = \frac{1000}{\pi(\frac{3}{16})^2} = 9050 \text{ lb/in.}^2 \quad (62 \text{ MPa})$$

and on a cross section of body 3,

$$\sigma_3 = \frac{1000}{\pi(\frac{1}{8})^2} = 20{,}400 \text{ lb/in.}^2 \quad (140 \text{ MPa})$$

Since these stresses do not exceed the yield stress σ_0, gross yielding in any portion is not possible. A word of caution: The preceding values are no indication of maximum stresses near locations A, B, C, or D. Localized stresses may be far greater. In some cases (repeated loading, brittle material), an accurate estimate of local stresses is needed. Such estimates are based upon methods not treated in this book.

The student should note that our complete solution of the problem requires (1) the conditions of *equilibrium*, (2) *kinematics* (e.g., relations between extensional strains and elongations), and (3) the *material behavior* [HI-HO behavior is implicit in Eq. (5.21)]. These three aspects are essential in a full solution of all problems of deformable bodies in equilibrium. ◆

EXAMPLE 3

A Statically Indeterminate Problem The stepped bar of Fig. 5.14(a) is composed of a linearly elastic material. Portions AB and BC are both cylindrical with a common centroidal axis. Ends A and B are rigidly clamped before the axial load P is applied via a rigid collar to the shoulder B. (The load is assumed to act along the centroidal axis.) We calculate the normal stress on cross sections of the stepped bar as follows.

Fig. 5.14

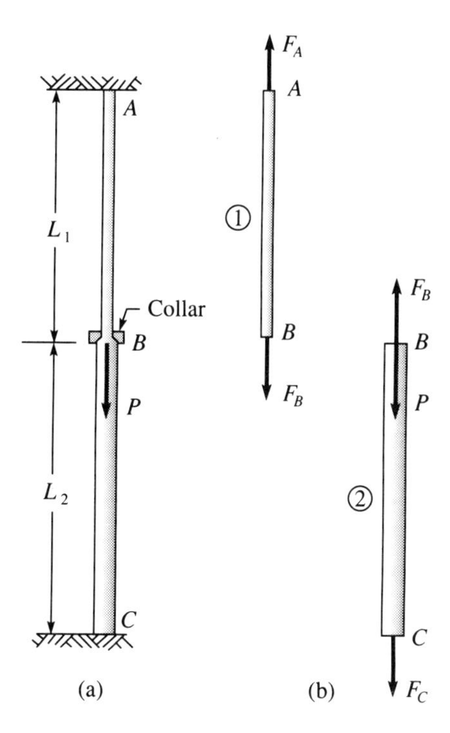

Again, free-body diagrams are drawn as in Fig. 5.14(b). *Equilibrium* conditions for the bodies, labeled 1 and 2, respectively, are

$$F_A - F_B = 0, \qquad F_B - P - F_C = 0 \tag{5.22}$$

In this case there are three unknown forces (F_A, F_B, F_C) but only two equilibrium equations. Hence the equilibrium conditions do not uniquely determine these forces, and the problem is said to be *statically indeterminate*. An additional equation is needed to compute these forces. To this end we consider the deformation of the member.

Assuming that each of the free bodies in Fig. 5.14(b) is a slender HI-HO bar, we can apply the results of simple extension. From Eq. (5.21),

$$e_1 = \frac{F_B L_1}{EA_1}, \tag{5.23}$$

$$e_2 = \frac{F_C L_2}{EA_2}, \tag{5.24}$$

where A_1 and A_2 are the cross-sectional areas of bars 1 and 2. Because of the end conditions, point A and C cannot displace. Furthermore, the two portions are not to be torn apart at B. This implies that the total elongation vanishes; that is,

$$e = e_1 + e_2 = 0 \tag{5.25}$$

This equation is a kinematic (geometric) constraint. It insures the continuity of the body and its connections. By substituting Eqs. (5.23) and (5.24) into Eq. (5.25), we have, in terms of forces, the additional equation

$$\frac{F_B L_1}{EA_1} + \frac{F_C L_2}{EA_2} = 0 \tag{5.26}$$

The simultaneous solution of Eqs. (5.22) and (5.26) (equilibrium and geometric conditions) gives

$$F_A = F_B = \frac{A_1 L_2 P}{A_1 L_2 + A_2 L_1}, \qquad F_C = -\frac{A_2 L_1 P}{A_1 L_2 + A_2 L_1} \tag{5.27}$$

Finally, the axial stress on a cross section of either portion is given by Eq. (5.20):

$$\sigma_1 = \frac{F_B}{A_1} = \frac{L_2 P}{A_1 L_2 + A_2 L_1}, \qquad \sigma_2 = \frac{F_C}{A_2} = -\frac{L_1 P}{A_1 L_2 + A_2 L_1} \tag{5.28}$$ ◆

EXAMPLE 4

Ideally Elastic-Plastic Material Suppose that the member of Example 2, Fig. 5.14(a), is composed of a steel that has the following properties: $E = 30 \times 10^6$ lb/in.^{2} (207×10^3 MPa) and $\sigma_0 = \pm 40{,}000$ lb/in.^{2} (± 276 MPa), where σ_0 is the tensile and compressive yield stress. To facilitate an analysis, let us assume hookean behavior to the yield stress and subsequent deformations in accordance with the ideally plastic behavior of Fig. 4.22a. The following dimensions are prescribed: $L_1 = 8$ in. (20 cm), $A_1 = 0.12$ in.^{2} (0.77 cm^2), $L_2 = 10$ in. (25 cm), $A_2 = 0.20$ in.^{2} (1.30 cm^2). We investigate the axial displacement of the rigid collar at B as the load P (measured in pounds) is slowly increased. The axial displacement of the collar is the total extension e_1 of part 1.

For sufficiently small load P, the behavior is linearly elastic as discussed in Example 2, so the axial displacement is obtained from Eqs. (5.23) and (5.27):

$$e_1 = \frac{L_1 L_2 P}{E(A_1 L_2 + A_2 L_1)} = \frac{P}{1.05 \times 10^6}\left(\frac{\text{in.}}{\text{lb}}\right) \quad \left[\frac{P}{1.84 \times 10^6}\left(\frac{\text{cm}}{\text{N}}\right)\right] \tag{5.29}$$

The stresses are given by Eq. (5.28):

$$\sigma_1 = \frac{L_2 P}{A_1 L_2 + A_2 L_1} = 3.57P/\text{in.}^2 \ (5.53 \times 10^3 P/\text{m}^2) \tag{5.30a}$$

$$\sigma_2 = \frac{L_1 P}{A_1 L_2 + A_2 L_1} = -2.86P/\text{in.}^2 \ (-4.43 \times 10^3 P/\text{m}^2) \tag{5.30b}$$

These relations are valid so long as neither stress exceeds the yield stress σ_0. According to Eq. (5.30), the material yields first throughout part 1. When $\sigma_1 = \sigma_0$,

$$P = P_0 = \frac{\sigma_0}{3.57} = 11{,}200 \text{ lb} \quad (49{,}800 \text{ N})$$

At this time the force F_B of Fig. 5.14(b) is

$$F_B = A_1 \sigma_0 = 4800 \text{ lb}$$

which is the limit load for bar 1. As the load P is subsequently increased, the force F_B remains constant. The force F_B is set equal to the limit load for bar 1, and the equilibrium of body 2 in Fig. 5.14(b) requires that

$$F_C = F_B - P = 4800 - P \tag{5.31}$$

Since portion 2 remains elastic, we can use Eqs. (5.24) and (5.25) to compute

$$e_1 = -e_2 = -\frac{F_C L_2}{E A_2} = \frac{P - 4800}{0.6 \times 10^6} \tag{5.32}$$

This relation holds until part 2 yields in compression, that is, when $\sigma_2 = -\sigma_0$. At that point

$$F_C = -A_2 \sigma_0 = -8000 \text{ lb}$$

and, from Eq. (5.31),

$$P = P_L = 12{,}800 \text{ lb} \quad (56{,}900 \text{ N}) \tag{5.33}$$

This is the limit load for the entire member, since neither portion can support further load.

The complete sequence of events is illustrated in Fig. 5.15. The initial elastic behavior follows Eq. (5.29). At the load P_0, part 1 yields. Additional loading causes contained plastic extension of bar 1; the behavior is described by Eq. (5.32). At the limit load P_L, portion 2 also yields; then the extension is unrestricted.

Now, suppose that the load is removed *after* some inelastic deformation of bar 1 but *before* yielding of bar 2. For simplicity, let us remove the load just as bar 2 is about to yield, i.e., at point B of Fig. 5.15. Bar 1 is inelastically deformed but responds

Fig. 5.15

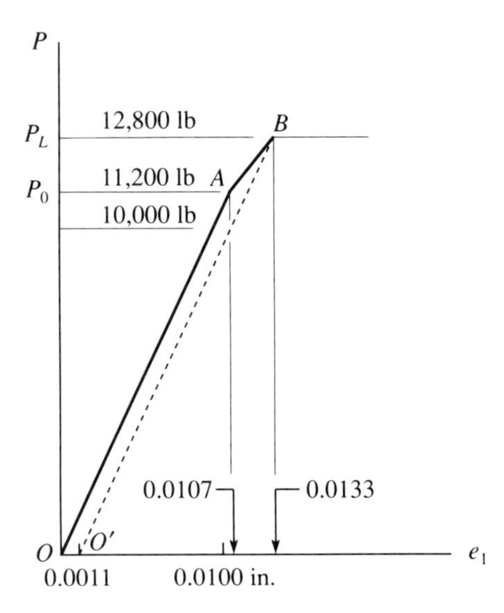

elastically to the unloading in accordance with the line $A'O'$ of Fig. 4.22. If $\Delta\sigma$, $\Delta\varepsilon$, ΔF, and Δe denote *changes* in mean normal stress, extensional strain, load, and elongation, respectively, then

$$\Delta\sigma_1 = E\,\Delta\varepsilon_1$$
$$\Delta F_B = \frac{EA_1}{L_1}\Delta e_1 \tag{5.34a}$$

Also, since bar 2 has not yielded,

$$\Delta F_C = \frac{EA_2}{L_2}\Delta e_2 \tag{5.34b}$$

The *residual* forces in bars 1 and 2, respectively, are

$$R_B = F_B + \Delta F_B = 4800 + \Delta F_B \tag{5.35a}$$

$$R_C = F_C + \Delta F_C = -8000 + \Delta F_C \tag{5.35b}$$

As before, *equilibrium* must be satisfied; now $P = 12{,}800 + \Delta P$ [see Eq. (5.22)], and

$$R_B - R_C = P + \Delta P$$

or

$$\Delta F_B - \Delta F_C = \Delta P \tag{5.36}$$

As before, the *kinematic* constraint must be satisfied [see Eq. (5.25)]:

$$\Delta e_1 + \Delta e_2 = 0 \tag{5.37}$$

By means of Eqs. (5.34), which incorporate the *material behavior*, Eq. (5.37) takes the form

$$\frac{L_1}{A_1}\Delta F_B + \frac{L_2}{A_2}\Delta F_C = 0 \tag{5.38}$$

The simultaneous solution of the equilibrium equation (5.36) and the kinematic constraint (5.38) gives

$$\Delta F_B = \frac{A_1 L_2}{A_1 L_2 + A_2 L_1}\Delta P, \qquad \Delta F_C = -\frac{A_2 L_1}{A_1 L_2 + A_2 L_1}\Delta P \tag{5.39a,b}$$

Then the changes of lengths Δe_1 and Δe_2 are obtained from Eq. (5.34):

$$\Delta e_1 = -\Delta e_2 = \frac{L_1 L_2}{E(A_1 L_2 + A_2 L_1)}\Delta P \tag{5.40}$$

Note that the change in length is proportional to the change in load; moreover, the proportionality is the same as the initial proportionality of Eq. (5.29). This unloading traces the dotted line BO' of Fig. 5.15. After complete removal of load, $\Delta P = -P_L = -12{,}800$ lb.; then by Eq. (5.40), $\Delta e_1 = -0.0122$ in., and by Eqs. (5.35) and (5.39), the residual forces are

$$R_B = R_C = -686 \text{ lb} \quad (-3050 \text{ N})$$

This simple example illustrates a common occurrence whenever a part of a member or structure suffers inelastic deformation while adjoining parts remain elastic. Here, both bars 1 and 2 remain in a state of residual compression. The state of each is illustrated by the stress-strain plot of Fig. 5.16 (similar to Fig. 4.22). During loading, bar 1 has been deformed elastically along path OA and then inelastically to point 1 at the yield stress σ_0; bar 2 has been deformed elastically from 0 to 2 at the yield stress $-\sigma_0$. Upon unloading, bar 1 traces path 1 to 1_R, whereas bar 2 retraces path 2 to 2_R. Both remain in states of residual compression. If the clamp at A were abruptly

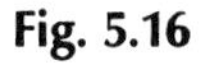
Fig. 5.16

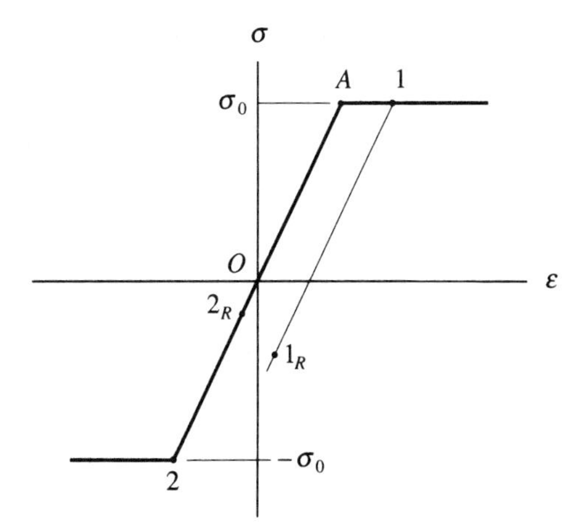

removed, the collar B would return to its initial position; then the stored energy in both bars would be released. ◆

EXAMPLE 5

A Reinforced Pier The pier shown in Fig. 5.17 is composed of concrete reinforced by eight steel rods. It is to support a compressive load $P = 328{,}000$ lb applied centrally by means of a rigid bearing plate. Here primes (′) refer to the concrete and double primes (″) refer to steel. The concrete is assumed hookean in simple compression if the stress is well within the ultimate strength, say,

$$-\frac{\sigma'_u}{2} < \sigma' < 0$$

Fig. 5.17

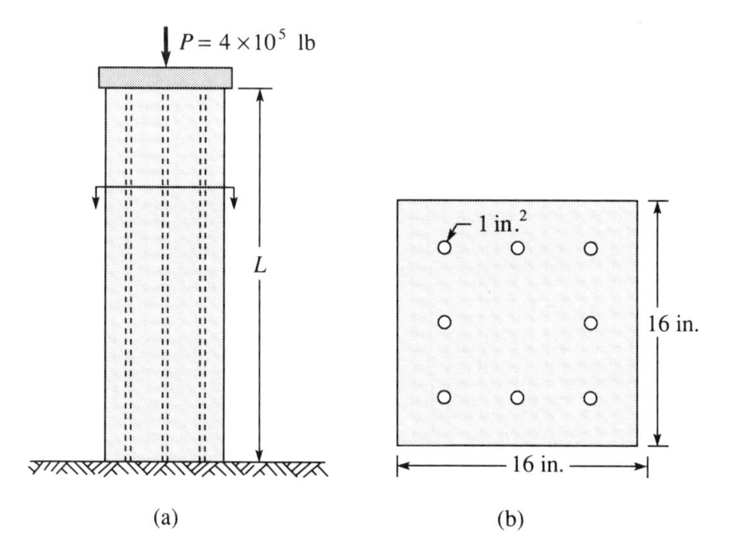

The steel is ideally elastic-plastic. Representative properties of the materials are $E' = 3 \times 10^6$ lb/in.² (20 GPa), $\sigma'_u = 3000$ lb/in.² (20 MPa), $E'' = 30 \times 10^6$ lb/in.² (200 GPa), and $\sigma''_0 = 35{,}000$ lb/in.² (241 MPa). The proposed design calls for each reinforcing rod to have a cross-sectional area of $A'' = 1$ in.² (6.45 cm²). The net cross-sectional area of the concrete is, therefore, $A' = 248$ in.² (1600 cm²). To check this design we will estimate the axial compressive stresses in the concrete and also in the steel.

To compute these stresses we tentatively assume HI-HO behavior of both materials. Also, in view of the symmetric arrangement of the reinforcing bars, we assume a simple contraction. Then the bearing plate remains horizontal, and all axial lines contract equally. Let e denote this axial contraction. According to Eq. (5.21), the compressive axial load on each reinforcing bar is

$$F'' = E''A''\frac{e}{L} \tag{5.41}$$

and the compressive axial load on the concrete is

$$F' = E'A'\frac{e}{L} \tag{5.42}$$

The total normal force on the cross section shown in Fig. 5.17(b) is carried partly by the steel bars and partly by the concrete. For *equilibrium*,

$$F' + 8F'' = P \tag{5.43}$$

Substituting Eqs. (5.41) and (5.42) into Eq. (5.43), we can solve for e/L:

$$\frac{e}{L} = \frac{P}{(E'A' + 8E''A'')}$$

Putting this into Eqs. (5.41) and (5.42), we obtain

$$F' = \frac{E'A'P}{E'A' + 8E''A''}, \qquad F'' = \frac{E''A''P}{E'A' + 8E''A''}$$

From Eq. (5.20) (recalling that both F' and F'' are compressive forces), we obtain, finally,

$$\sigma' = -\frac{F'}{A'} = \frac{-(3 \times 10^6)(328{,}000)}{(3 \times 10^6)(248) + (8)(30 \times 10^6)(1)} = -1000 \text{ lb/in.}^2 \quad (-6.9 \text{ MPa})$$

$$\sigma'' = -10{,}000 \text{ lb/in.}^2 \quad (-69 \text{ MPa})$$

The stress in the concrete is one-third the ultimate strength; the stress in the steel reinforcing rods is about one-third the yield stress. These stresses are well within the bounds of elastic behavior, validating our original assumption. ◆

Assume linearly elastic behavior unless otherwise stated.

5.19 A steel tie rod, 15 ft long, is stretched by axial end loads of 30,000 lb. Specifications require that the normal stress in the rod should not exceed 18,000 lb/in.^{2} and that the total extension must not be greater than 0.15 in. What is the smallest-diameter rod that will satisfy both requirements?

5.20 The cross section of an aluminum bar is a rectangle 2 cm by 3 cm. The bar is 80 cm long and is subjected to axial forces applied at its ends. It is specified that the total elongation must not be greater than 0.120 cm and that the normal stress should not exceed 140 MPa. Determine the largest possible load that satisfies both requirements.

5.21 A brass bar has a semicircular cross section. It is 150 cm long and carries an axial load of 220 kN applied at each end. Compute the radius of the cross section so that the total extension of the bar is 0.200 cm.

5.22 A casting weighing 20,000 lb is to be stored in a warehouse by placing each of its four corners on wooden blocks. One such block is shown in Fig. P5.22. The maximum compressive stress that the wood can carry perpendicular to the grain is 1000 lb/in.^{2}, and the maximum possible shear stress parallel to the grain is 200 lb/in^2. if the total weight of the casting is equally divided between the blocks, will they support the casting?

Fig. P5.22

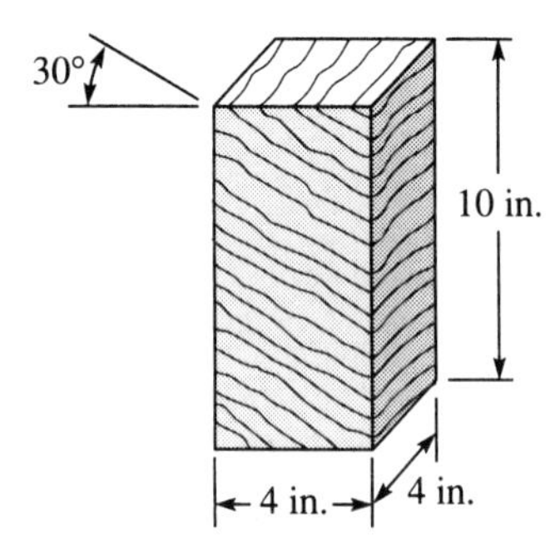

5.23 Member BC, as shown in Fig. P5.23, supports the end of a uniformly loaded platform AB. If BC is a rod of ideally plastic structural steel, $\sigma_0 = 331$ MPa, what is the smallest diameter BC can have without yielding?

Fig. P5.23

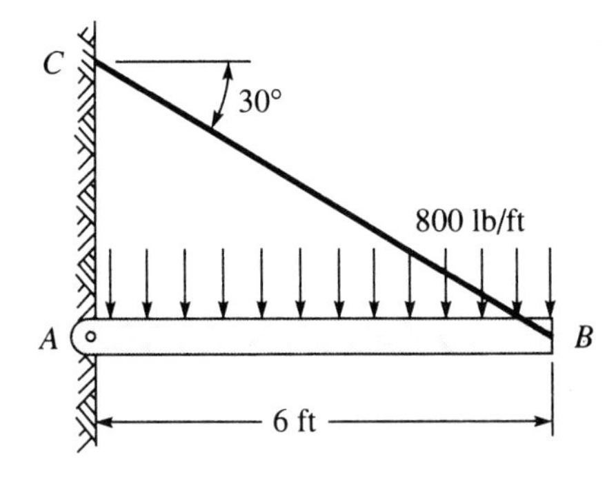

5.24 The central portion of a $\frac{1}{8}$-in.-diameter steel rod has a square cross section as shown in Fig. P5.24; that is, the diagonal of the cross section is $\frac{1}{8}$ in. If the steel rod is assumed to be ideally plastic, $\sigma_0 = 64{,}000$ lb/in.^{2}, determine the elongation of the rod when P is equal to one-half of the limit load P_0.

Fig. P5.24

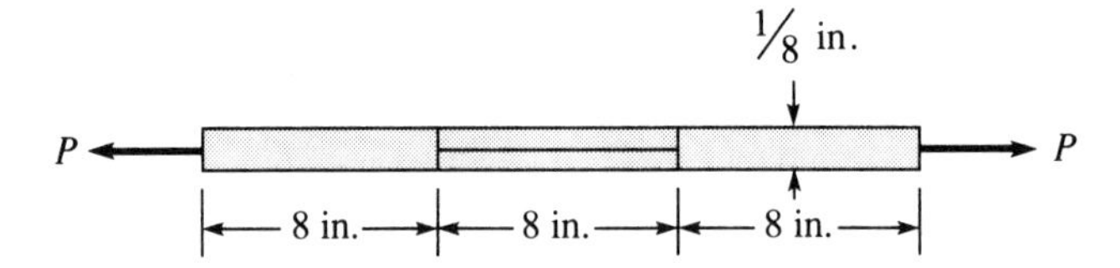

5.25 For the bar of Prob. 5.24, assume that the end portions are wrought iron, $\sigma_0' = 30{,}000$ lb/in.^{2}, and that the central portion is hot-rolled steel, $\sigma_0'' = 60{,}000$ lb/in^2. What is the limit load P_0 for this composite member?

5.26 A stepped aluminum rod is loaded as shown in Fig. P5.26; the cross-sections are circular. Compute the normal stress in each portion and the total elongation.

Fig. P5.26

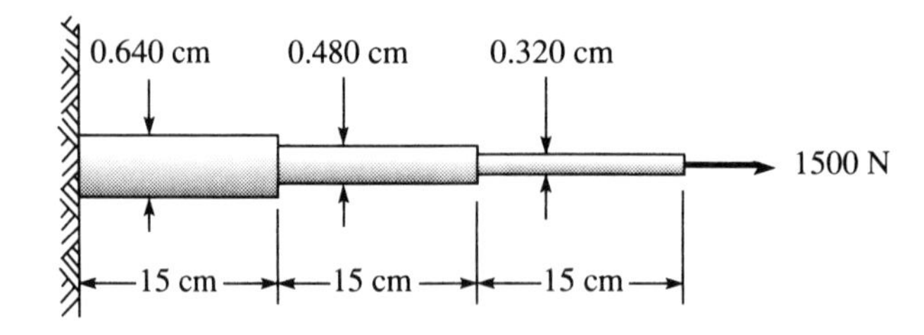

5.27 A steel and brass rod are welded end to end. Each is prismatic and has the cross-sectional area indicated in Fig. P5.27. Compute the force P needed to stretch the composite bar 0.0075 in.

Fig. P5.27

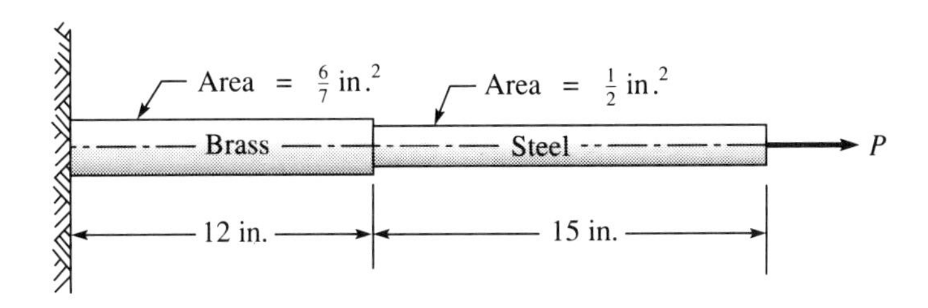

5.28 If each member of the bar in Prob. 5.27 is ideally plastic, determine the relation between P and the extension of the bar. The yield stress for the brass is $\sigma_0' = 40{,}000$ lb/in^2 and that for the steel is $\sigma_0'' = 60{,}000$ lb/in^2.

5.29 A steel bar 0.640 cm in diameter and 15 cm long is coated with a thin layer of porcelain ($E = 52 \times 10^3$ MPa), which fails at a tensile strain of 0.001. What is the maximum axial load to which the bar can be subjected before the porcelain cracks?

5.30 A steel rod is fixed between two walls, as shown in Fig. P5.30. Determine the load P that displaces the cross section C to the right 0.005 in.

Fig. P5.30

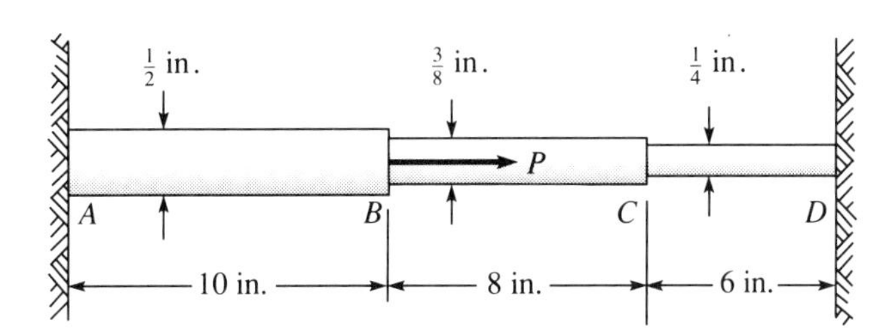

5.31 A steel and a brass rod are welded end to end, forming the composite member shown in Fig. P5.31. The member is unstressed when fixed to the rigid walls. Compute the normal stress in each part when load $P = 500$ kN.

5.32 Three aluminum bars are connected to a rigid plate, as shown in Fig. P5.32. The cross section of each is a rectangle $\frac{1}{2}$ in. by $\frac{3}{2}$ in., the members are fixed at A and B, and the assembly is symmetric about the centerline. Determine the displacement of the rigid plate caused by the axial load $P = 40{,}000$ lb.

Fig. P5.31

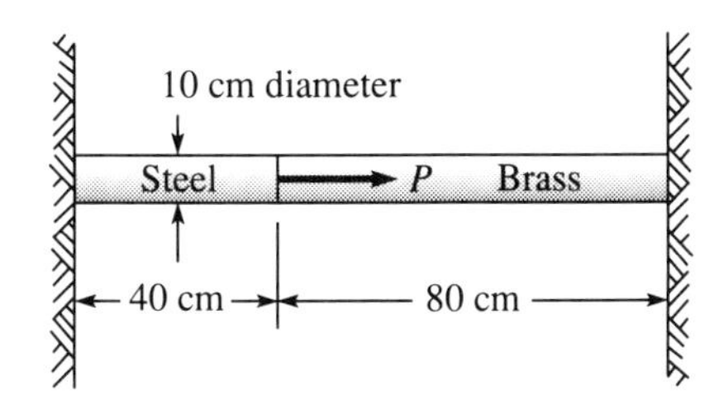

Fig. P5.32

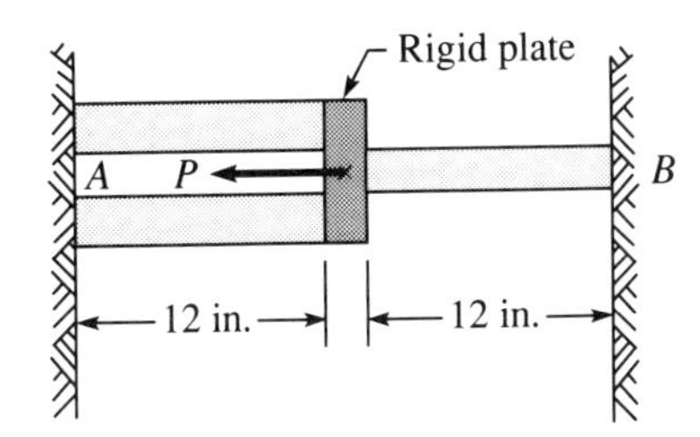

5.33 A mailing tube is constructed by wrapping the thin strip of cardboard shown in Fig. P5.33 about a cylinder of diameter D and gluing the seams along AB and CD. The cardboard fails in tension when the normal stress reaches a value σ_0, and the glue fails in shear at a value $\tau_0 = k\sigma_0$, where $0 < k < \frac{1}{2}$. Compute the relation between α and k so that the cardboard and glue fail simultaneously when the tube is axially loaded in tension.

Fig. P5.33

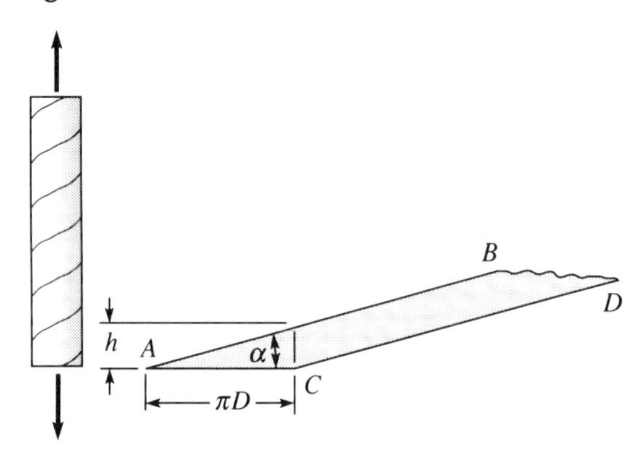

5.34 A load cell is constructed of two concentric steel cylinders, as shown in Fig. P5.34. If the load P is transmitted through a rigid plate, determine the load-extension relation and the maximum permissible load if the steel is to behave elastically. (Assume the steel has an elastic limit of 600 MPa.)

Fig. P5.34

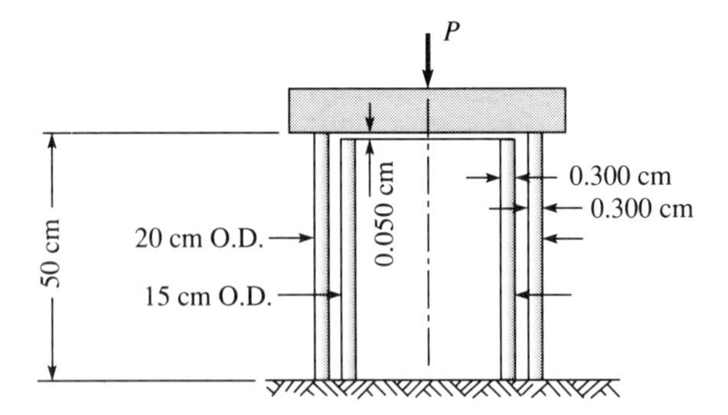

5.35 Three hollow aluminum cylinders have a common axis, as shown in Fig. P5.35. Compute the stress in each cylinder when the innermost cylinder is compressed 0.020 in.

Fig. P5.35

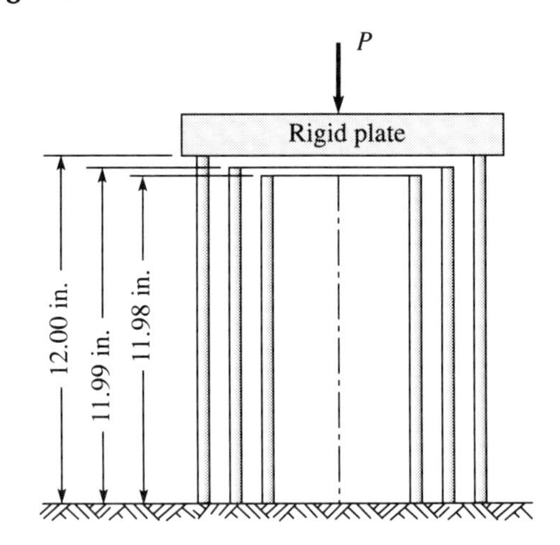

5.36 A hollow steel cylinder is 36 cm long with an inside diameter of 27 cm and a wall thickness of 3 cm. It is filled with concrete and loaded axially by means of a rigid bearing plate. If the axial stress in the concrete is 10 MPa, compute the axial load and the axial stress in the steel ($E_c = 21 \times 10^3$ MPa).

5.37 A prismatic concrete column contains a uniform steel reinforcing rod that lies on the axis of centroids of the cross sections. The cross-sectional area of the concrete is 140 in.^{2} and that of the reinforcing bar is 2 in^2. If an axial compressive load is applied through rigid end plates, calculate the percentage of this load carried by the concrete ($E_c = 3 \times 10^6$ lb/in^2).

5.38 A concrete pier has a square cross section 25 cm by 25 cm. It is reinforced by six 2.5-cm-diameter steel bars that are parallel to the axis of the pier and symmetrically located about this axis. If the pier supports an axial compressive load of 900 kN, what are the stresses in the steel and concrete ($E_c = 14 \times 10^3$ MPa)?

5.39 A reinforced concrete block has the cross section shown in Fig. P5.39. It supports a normal load P by means of a rigid bearing plate. Locate the line of action of P so that each reinforcing rod carries the same load ($E_c = 3 \times 10^6$ lb/in^2).

Fig. P5.39

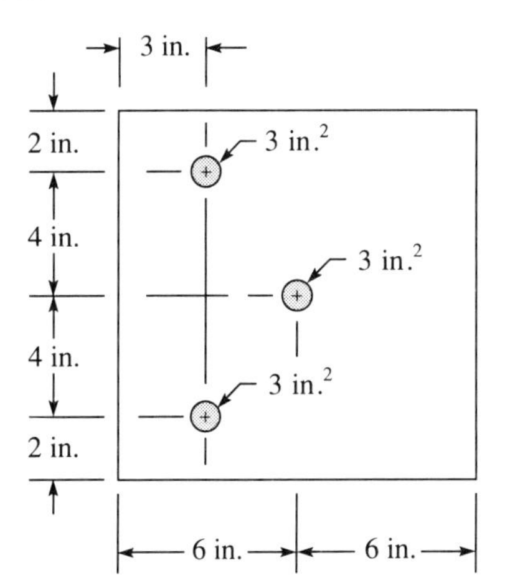

5.40 The fixture shown in Fig. P5.40 is attached to the plate by two $\frac{1}{4}$-in. steel ($\sigma_0 = 60{,}000$ lb/in^2) bolts that have 20 threads per inch. When mounting, the nuts are made snug and then given an additional 10° turn. Neglect the deformation of the plate and fixture and compute (a) the stresses in the bolts caused by tightening, and (b) the stresses in the bolts when the load $P = 2000$ lb is applied.

Fig. P5.40

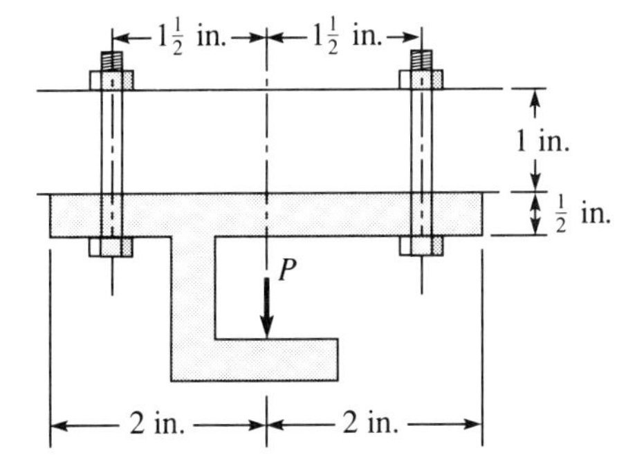

5.41 Assume that the bar ACD is rigid, as shown in Fig. P5.41. Member BC consists of a work-hardening material such that $\sigma = 210 \times 10^3\varepsilon$ for $\sigma \leq 210$ Mpa and $\sigma = 210 + 700(\varepsilon - 10^{-3})$ for $\sigma \geq 210$ MPa. Determine the relation between P and the angle of rotation, θ, of the bar ACD.

Fig. P5.41

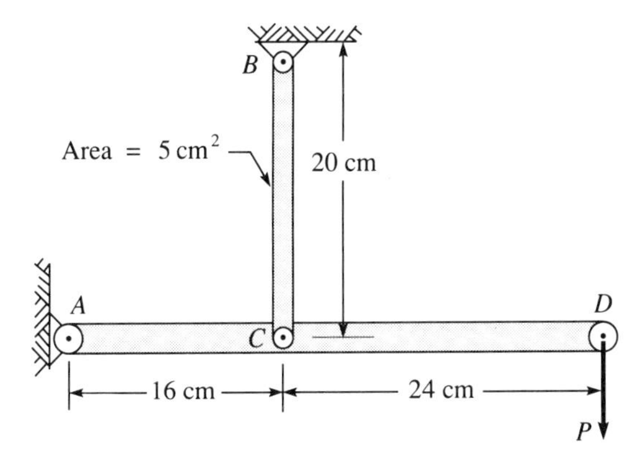

5.42 Two rigid members are joined together as shown in Fig. P5.42 by a $\frac{1}{2}$-in. steel bolt that has 20 threads per inch. The nut is turned down and tightened $\frac{1}{4}$ turn. Calculate the normal stress in the bolt when $P = 60{,}000$ lb.

Fig. P5.42

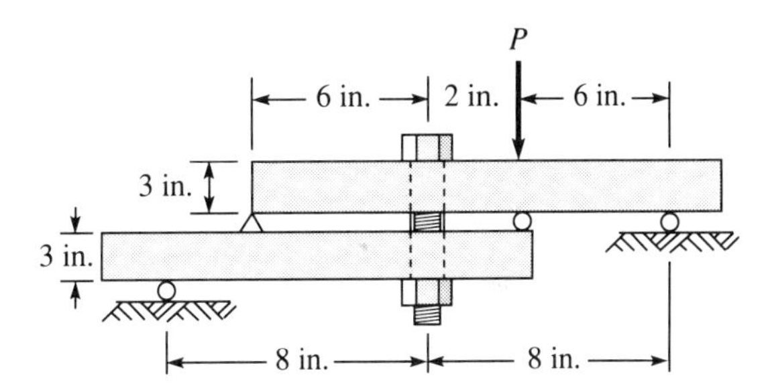

5.43 The brass sleeve shown in Fig. P5.43 is compressed by tightening the nut on the steel bolt. Each

Fig. P5.43

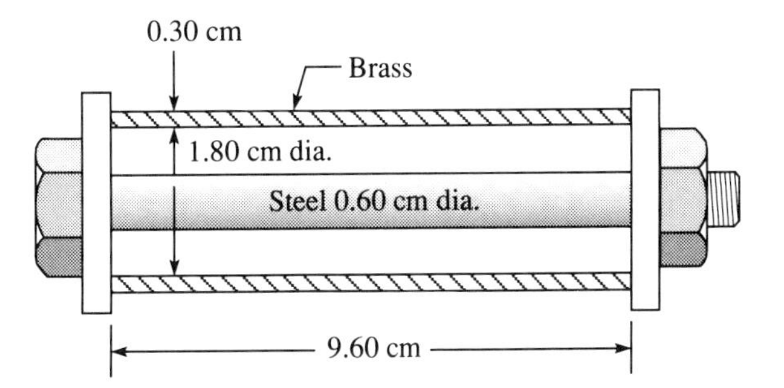

turn of the nut advances it 0.075 cm. What is the axial stress in the brass when the nut is advanced one-quarter turn?

5.44 If the materials in Prob. 5.43 are ideally plastic (for steel $\sigma_0 = 280$ MPa, for brass $\sigma_0 = 280$ MPa), determine the number of turns of the nut that will initiate yielding in either part.

5.45 In Fig. P5.45 *AB* is a steel bar with cross-sectional area 0.25 in.^{2}, and *ED* is a brass bar with cross-sectional area 0.5 in^2. If bar *BCD* is assumed rigid, compute *x* so that *BCD* remains horizontal.

Fig. P5.45

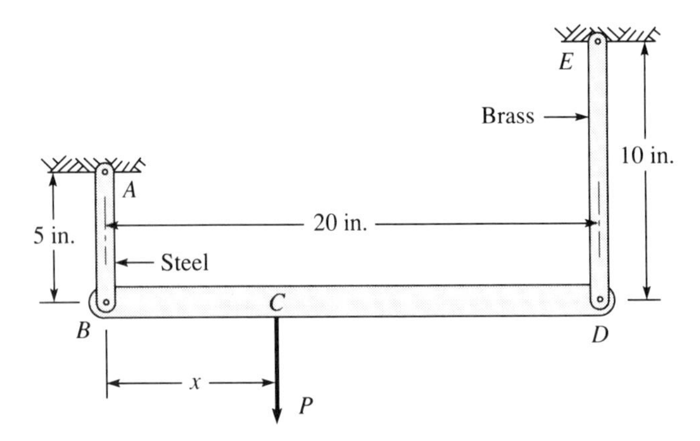

5.46 The rigid bar *AB* is pin-connected to the steel posts *AC* and *BD* and is subjected to a couple, as shown in Fig. P5.46. If the cross-sectional area of *BD* is 6 cm^2 and that of *AC* is 12 cm^2, what is the vertical displacement of the center of *AB*?

Fig. P5.46

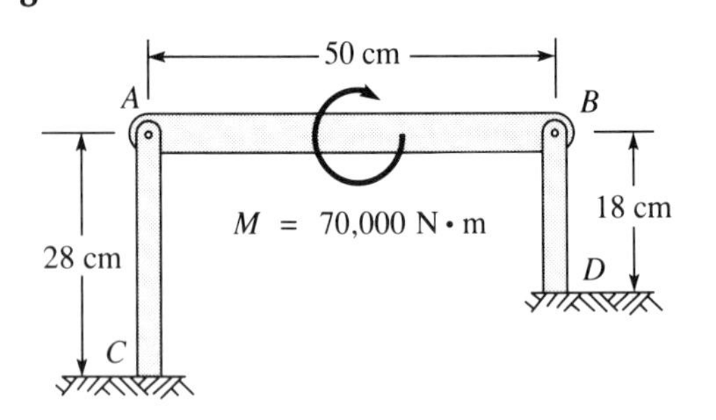

5.47 The rigid bar *AB* is supported by two wires of different materials, as shown in Fig. P5.47. The yield stress of wire 1 is twice that of wire 2, and their cross-sectional areas are the same. If both wires yield simultaneously under the load *P*, what is the ratio of the moduli E_1/E_2?

Fig. P5.47

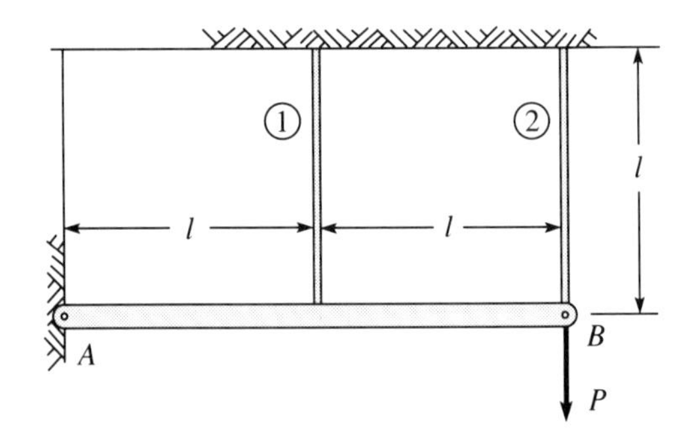

5.48 The rigid bar carries the load *W* and is supported by three wires, as shown in Fig. P5.48. Plot the load *W* versus the vertical displacement of the bar if each wire is ideally plastic with $\sigma_0 = 245$ MPa for steel and $\sigma_0 = 210$ MPa for aluminum. The diameter of each wire is 0.400 cm.

Fig. P5.48

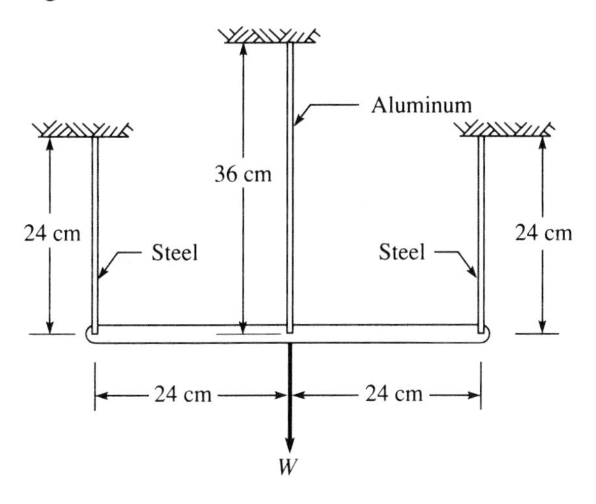

5.49 A bar $\frac{3}{8}$ in. square is composed of different materials, whose behavior is given in the stress-strain diagrams shown in Fig. P5.49. Plot the relation between P and the displacement of section AA if the member is unstressed when fixed in position.

Fig. P5.49

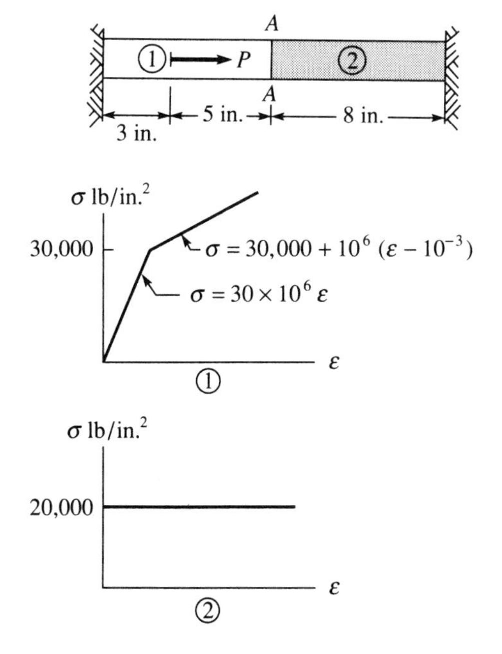

5.50 Two ideally plastic steel rods, $\sigma_0 = 50{,}000$ lb/in.^{2}, support a rigid bar AD as shown in Fig. P5.50. If each rod has a cross-sectional area of 0.40 in.^{2}, compute the displacement of point D (a) when $P = 16{,}000$ lb and (b) when $P = 22{,}000$ lb. Determine the residual stresses in each bar when the 22,000-lb load is removed.

Fig. P5.50

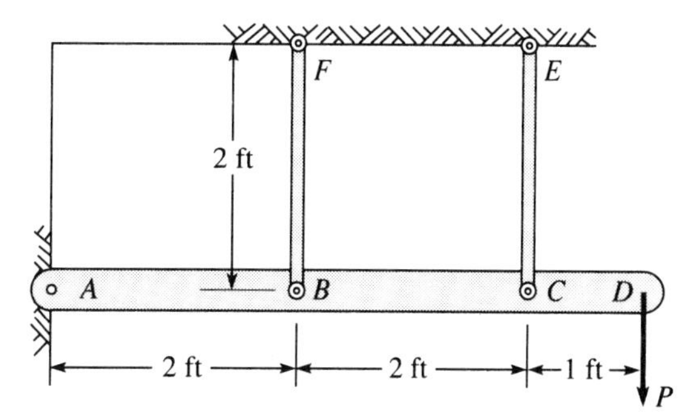

5.51 The rigid bar AB is supported by n equally spaced wires, as shown in Fig. P5.51. Each is of length L, cross-sectional area A, and ideally elastic-plastic with yield stress σ_0. Determine the relation between P and the rotation θ of the bar.

Fig. P5.51

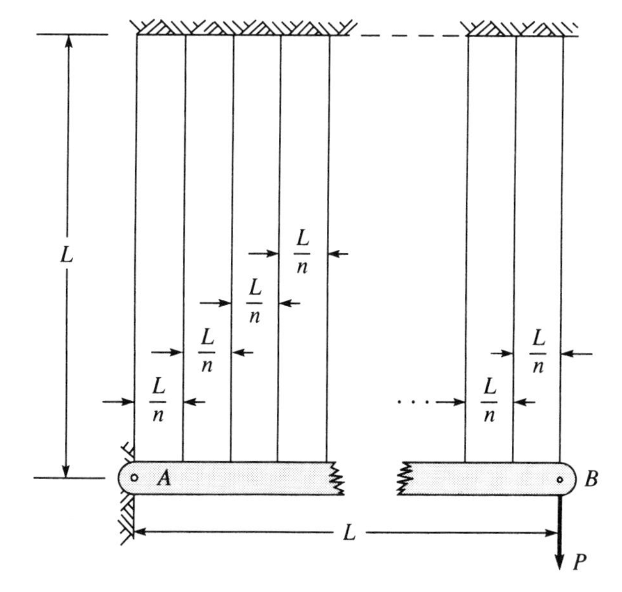

5.52 Assume that the rod of Problem 5.30 consists of a work-hardening material such that $\sigma = 10^7\varepsilon$ for $\varepsilon \leq 0.005$ and $\sigma = 5 \times 10^4 + 10^6(\varepsilon - 0.005)$ for $\varepsilon \geq 0.005$, where stress σ is given in pounds per square inch. Compute the stress in each portion when $P =$ 20,000 lb.

5.53 Three wires, one of aluminum, another of titanium, and a third of steel, hang under their own weights. If each wire is 150 cm long, compare the elongations of the wires.

5.54 What is the displacement of the cross section at C of the stepped-steel rod hanging under its own weight, as shown in Fig. P5.54?

Fig. P5.54

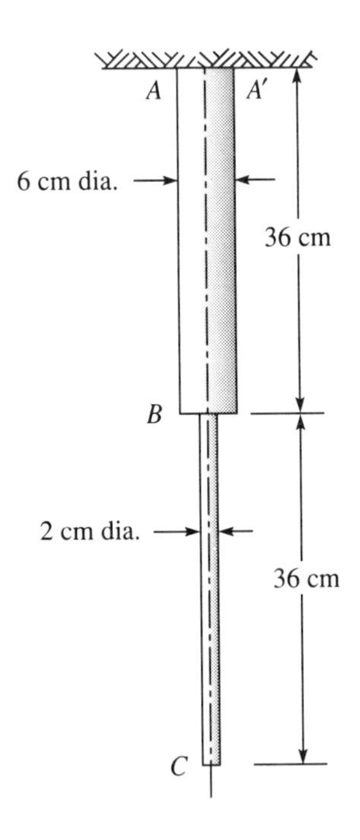

5.55 An aluminum rod hangs under its own weight. Additionally, an axial load equal to three times the weight of the rod acts at the lower end. The rod is 40 cm long and 2 cm in diameter. Compute the total extension.

5.56 A uniform bar hangs under its own weight. If $\sigma = E\varepsilon^{1/2}$, compute the total extension of the bar.

5.57 A tapered rod of square cross section and length L hangs under its own weight. Determine the total elongation if the area of a cross section is given by

$$A = A_0\left[1 - \frac{1}{2}\frac{x}{L}\right]^2$$

where x is the distance from the upper end.

5.58 A slightly tapered steel column in the form of a frustum of a cone rests on its base and supports its own weight. The length of the member is 120 cm, the radius of its base is 1 cm, and the radius of its upper end is $\frac{1}{2}$ cm. Compute the total contraction of the member.

5.59 A cylindrical rod of length L, radius r, modulus of elasticity E, and mass density ρ has a mass M attached to its end, as shown in Fig. P5.59. What is the total extension of the rod if it rotates with a constant angular velocity ω about an axle through O?

Fig. P5.59

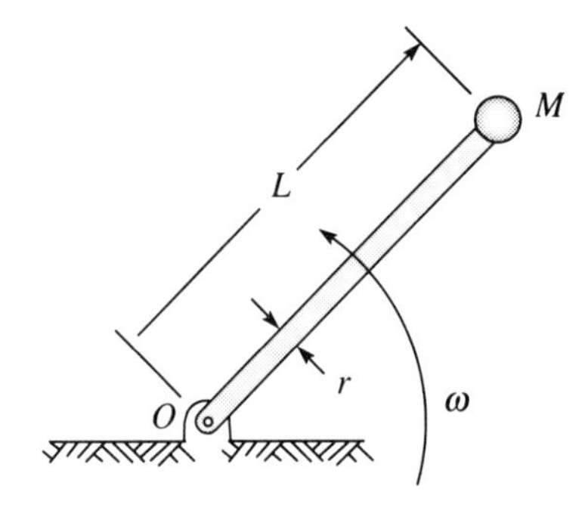

5.60 A steel turbine blade has a cross-sectional area of 0.280 in^2. The rotor has a 6-in. radius, and the blade is 3 in. long. Compute the extension of the blade when the rotor has an angular velocity of 600 rad/s.

5.61 A turbine blade is tapered as illustrated in Fig. P5.61. The cross-sectional area is given by the relation

$$A = A_0\left(1 - \frac{r - R_0}{3L}\right)$$

Compute the total extension of the blade if it rotates at n revolutions per minute.

Fig. P5.61

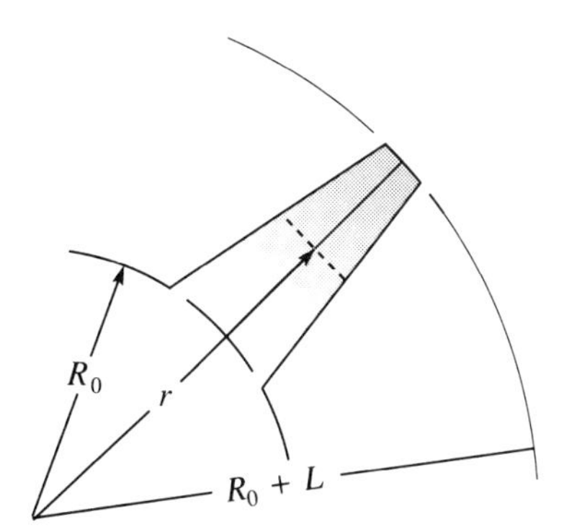

5.62 The shaft of Prob. 5.54 rotates about the diameter AA' of the section at A. At what angular velocity will the axial stress on section AA' reach a value of 200 MPa?

*5.7 Trusses

General Description A truss is a structure composed of slender members joined together at their ends. Portions of bridges, buildings, and airframes are often structures of this type. The members of a truss are relatively slender, because they are not intended to resist transverse (bending) loads. Hence, it is common practice to neglect the rotational resistance of the end connections; the effects of bending couples transmitted through these connections are neglected. Accordingly, we assume that the members are joined together by ideal pin connections. Hence, an ideal truss is composed of two force members; each member is acted upon by axial forces applied at the ends.

According to Saint-Venant's principle, the manner of load distribution at the end of a member has only a localized effect. Since the members of a truss are relatively slender, such effects are usually of secondary importance. The possibility of a stability failure (buckling) in the compressed members must be kept in mind. This problem is treated in Chapter 9. For the present we assume that all loads are well below their critical values. Then the results of the previous analyses of axially loaded members (Sec. 5.3) are applicable to the individual members of the ideal truss.

EXAMPLE 6

Statically Determinate Truss Let us determine the response of the simple coplanar truss of Fig. 5.18(a) to the vertical load W. For simplicity we suppose that both members have the same cross-sectional area A.

If u and v denote the horizontal and vertical displacements of joint A, the total extensions of members AB and AC are

Fig. 5.18

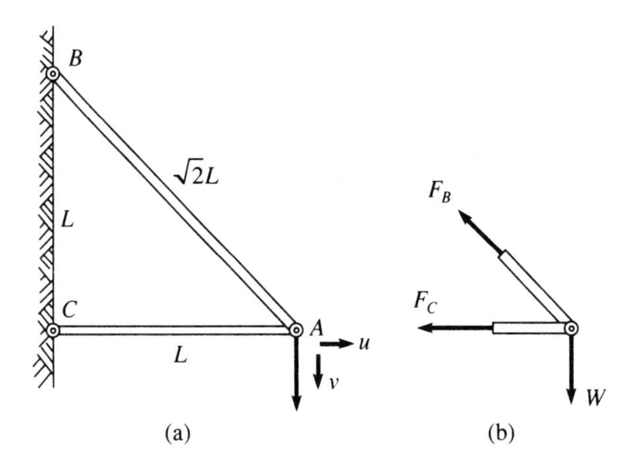

$$e_B = \frac{u}{\sqrt{2}} + \frac{v}{\sqrt{2}}, \qquad e_c = u \tag{5.44}$$

Here, as before, the strains and relative rotations are assumed small.

From equilibrium of forces at joint A, Fig. 5.18(b), the axial forces acting upon members AB and AC are

$$F_B = \sqrt{2}W, \qquad F_C = -W \tag{5.45}$$

According to Eq. (5.20), the normal stresses acting on cross sections of these members are

$$\sigma_B = \sqrt{2}\frac{W}{A}, \qquad \sigma_C = -\frac{W}{A} \tag{5.46}$$

Notice that the forces on each member have been determined by considering only equilibrium. It is this feature of the present problem that characterizes it as a statically determinate problem. However, to determine the deformation of the truss, we must introduce the material behavior.

Suppose that the members AB and AC are composed of the same ideally elastic-plastic material. From Eq. (5.46) it is apparent that no yielding will occur until the load W attains the value $W_0 = A\sigma_0/\sqrt{2}$. For $W < W_0$, the behavior of the truss members is hookean. Then, from Eqs. (5.21) and (5.45), the total extension of each member is

$$e_B = \frac{F_B(\sqrt{2}L)}{EA} = \frac{2WL}{EA} \tag{5.47a}$$

$$e_C = \frac{F_C L}{EA} = -\frac{WL}{EA}, \qquad \text{for } W < W_0 \tag{5.47b}$$

Substituting these results into Eq. (5.44), we solve for the displacements:

$$u = -\frac{WL}{EA}, \qquad v = (2\sqrt{2} + 1)\frac{WL}{EA}, \qquad \text{for } W < W_0 \tag{5.48}$$

Now suppose that the load attains the value $W_0 = A\sigma_0/\sqrt{2}$. The normal stress in member AB becomes $\sigma_B = \sigma_0$ [by Eq. (5.46)], and the member experiences unrestricted plastic deformation. The truss of Fig. 5.18 becomes, in effect, a mechanism, since the member AB can no longer prevent the rotation of bar AC about the point C. Consequently, the load $W = W_0$ is the limit load for this truss.

A plot of load W versus vertical displacement v is shown in Fig. 5.19. The relation is linear according to Eq. (5.48) for $W < W_0$. When the applied load reaches the limit load W_0, plastic collapse occurs. To a first approximation, the load-deflection relation follows the horizontal line $W = W_0$. However, as the deformation increases, the rotation of the members becomes appreciable. Then the equilibrium conditions, Eqs. (5.45), and the geometric relations, Eqs. (5.44), are no longer adequate. More accurate expressions are needed to account for the changes of geometry. The more precise analysis results in the dotted curve of Fig. 5.19. Evidently, the structure can withstand a load in excess of W_0—but only when the geometry of the structure is severely (and permanently) altered. Consequently, the limit load is a practical measure of the load-carrying capacity of the truss.

Fig. 5.19

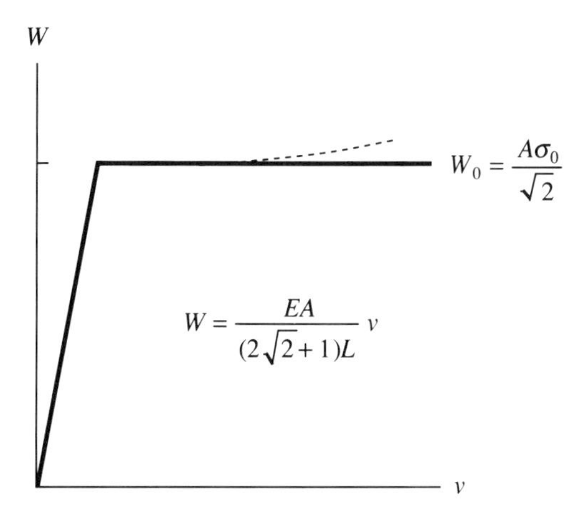

Note that a complete solution requires the *kinematic* relations (5.44), the *equilibrium* conditions (5.45), and the *material* descriptions (5.47). These are essential elements of every solution. Only in the simplest cases, such as the present, are stresses determined merely by equilibrium conditions. ◆

EXAMPLE 7

Statically Indeterminate Truss Now let us apply the same procedures to predict the behavior of the truss shown in Fig. 5.20(a). Each of the three members has the same cross-sectional area A.

Let u and v denote the horizontal and vertical displacements of joint A, as shown in Fig. 5.20(a). The total extensions of the members are

Fig. 5.20

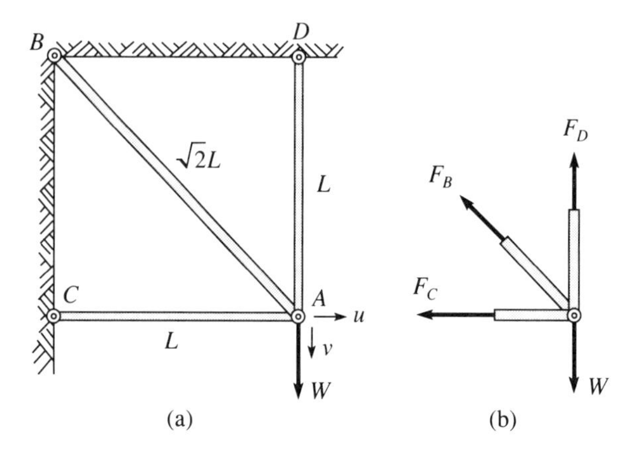

$$e_D = v, \qquad e_C = u, \qquad e_B = \frac{u}{\sqrt{2}} + \frac{v}{\sqrt{2}} \tag{5.49}$$

The geometric relations of Eq. (5.49) are essentially conditions of geometric continuity. They define the three extensions in terms of the two displacement components of point A so that the connection at A is not destroyed. Again, we suppose that the strains and rotations of members are small. From the conditions for equilibrium at joint A [Fig. 5.20(b)] we have

$$F_D + \frac{F_B}{\sqrt{2}} = W, \qquad F_C + \frac{F_B}{\sqrt{2}} = 0 \tag{5.50}$$

Since there are three unknown forces, F_D, F_B, and F_C, and only two independent equilibrium conditions [Eq. (5.50)], the problem is statically indeterminate. The solution of this problem differs slightly from that of the previous example because we must consider, *simultaneously*, the geometric relations, Eqs. (5.49), the equilibrium conditions, Eqs. (5.50), and the material behavior of the truss members.

Let us assume tentatively that the axial stress in each member is less than the yield stress. Then the behavior of each member is linearly elastic, and using Eq. (5.21), we write

$$e_D = \frac{F_D L}{EA}, \qquad e_C = \frac{F_C L}{EA}, \qquad e_B = \frac{F_B(\sqrt{2}L)}{EA} \tag{5.51}$$

Substituting Eq. (5.51) into the geometric relations of Eqs. (5.49), we express the three forces in terms of the two displacements:

$$F_D = \frac{EA}{L}v, \qquad F_C = \frac{EA}{L}u, \qquad F_B = \frac{EA}{2L}(u + v) \tag{5.52}$$

When Eqs. (5.52) are then substituted into the equilibrium conditions, Eqs. (5.50), the result is two equations in the two unknown displacements:

$$\left(1 + \frac{1}{2\sqrt{2}}\right)\frac{EA}{L}v + \frac{1}{2\sqrt{2}}\frac{EA}{L}u = W$$

$$\frac{1}{2\sqrt{2}}\frac{EA}{L}v + \left(1 + \frac{1}{2\sqrt{2}}\right)\frac{EA}{L}u = 0$$

The solution of these equations yields

$$v = \left(\frac{4+\sqrt{2}}{4+2\sqrt{2}}\right)\frac{WL}{EA} = 0.793\frac{WL}{EA}$$
$$u = -\left(\frac{\sqrt{2}}{4+2\sqrt{2}}\right)\frac{WL}{EA} = -0.207\frac{WL}{EA} \tag{5.53}$$

If the forces in Eq. (5.52) are divided by the area and the results of Eq. (5.53) are substituted for the displacements, we obtain the stresses in each member:

$$\sigma_D = \left(\frac{4+\sqrt{2}}{4+2\sqrt{2}}\right)\frac{W}{A} = 0.793\frac{W}{A}$$
$$\sigma_C = -\left(\frac{\sqrt{2}}{4+2\sqrt{2}}\right)\frac{W}{A} = -0.207\frac{W}{A} \tag{5.54}$$
$$\sigma_B = \left(\frac{2}{4+2\sqrt{2}}\right)\frac{W}{A} = 0.293\frac{W}{A}$$

These results are valid only if the entire truss behaves in a hookean manner. Because the largest stress occurs in member *AD*, no general yielding can occur until $\sigma_D = \sigma_0$. Then Eqs. (5.53) and (5.54) are valid so long as $\sigma_D < \sigma_0$, or, from the first of Eqs. (5.54), for

$$W < W_0 = \left(\frac{4+2\sqrt{2}}{4+\sqrt{2}}\right)\sigma_0 A = 1.26\sigma_0 A \tag{5.55}$$

When the load exceeds W_0, the member *AD* experiences general yielding. The hookean behavior of Eq. (5.51) no longer governs. Instead, member *AD* is assumed ideally plastic; the stress remains constant:

$$F_D = \sigma_0 A \qquad \text{for } W \geq W_0 \tag{5.56}$$

The truss is no longer statically indeterminate, since Eq. (5.56), along with the equilibrium conditions of Eqs. (5.50), determine the axial forces in the members:

$$F_B = \sqrt{2}(W - F_D) = \sqrt{2}(W - \sigma_0 A)$$
$$F_C = -\frac{F_B}{\sqrt{2}} = -W + \sigma_0 A \tag{5.57}$$

The relations in Eq. (5.57) are valid so long as members *AB* and *AC* do not yield. As the load W is increased from W_0, the stress in member *AB* increases until it reaches the yield stress. At this point, neither member AD nor member AB can resist the rotation of member AC, and the truss is reduced to a mechanism. Unrestricted plastic deformation occurs in both AD and AB. This limit load for the truss is found when $F_B = \sigma_0 A$ or from Eq. (5.57) when

$$W = W_L = \left(\frac{1+\sqrt{2}}{\sqrt{2}}\right)\sigma_0 A = 1.707\sigma_0 A$$

For $W_0 \leq W < W_L$, members AB and AC are linearly elastic; hence, the last two equations of (5.52) are still valid. In conjunction with Eq. (5.57), they may be solved for the displacements u and v:

$$\begin{aligned} u &= -\frac{WL}{EA} + \sigma_0 \frac{L}{E} \\ v &= (2\sqrt{2}+1)\left[\frac{WL}{EA} - \sigma_0 \frac{L}{E}\right], \qquad \text{for } W_0 \leq W < W_L \end{aligned} \tag{5.58}$$

A plot of the load W versus the vertical displacement v is shown in Fig. 5.21. For loads less than W_0 the relation is linear according to Eq. (5.53); subsequent increases in load are in accordance with Eq. (5.58) until the load reaches the limit load W_L. Plastic collapse is indicated by the horizontal line $W = W_L$ in Fig. 5.21.

Fig. 5.21

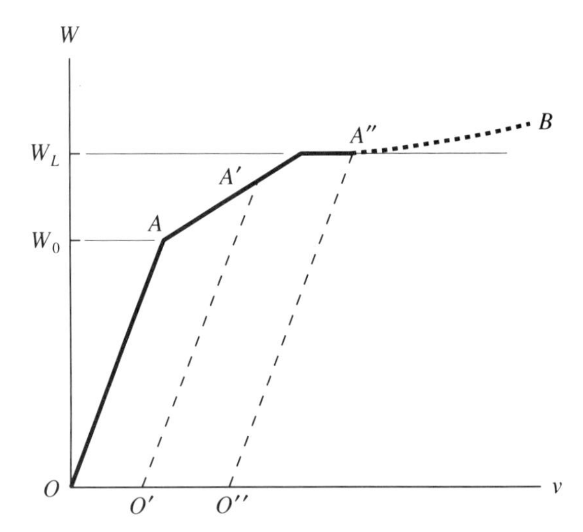

If the load is decreased at any stage, then the changes of stress and strain are linearly related (see Fig. 4.22); that is, $\Delta\sigma = E\,\Delta\varepsilon$ and $\Delta F = (EA)\Delta e/L$. This is true whether the member has been inelastically deformed or not. Accordingly, the incremental elongations are similar to Eq. (5.51):

$$\Delta e_D = \frac{L}{EA}\Delta F_D, \qquad \Delta e_C = \frac{L}{EA}\Delta F_C, \qquad \Delta e_B = \frac{\sqrt{2}L}{EA}\Delta F_B \tag{5.59}$$

Also, so long as geometric changes are small, the forces satisfy linear equilibrium conditions, such as Eq. (5.50):

$$\Delta F_D + \frac{\Delta F_B}{\sqrt{2}} = \Delta W, \qquad \Delta F_C + \frac{\Delta F_B}{\sqrt{2}} = 0 \tag{5.60}$$

Then, too, the incremental elongations and displacements are linearly related, as in Eq. (5.49):

$$\Delta e_D = \Delta v, \qquad \Delta e_C = \Delta u, \qquad \Delta e_B = \frac{\Delta u}{\sqrt{2}} + \frac{\Delta v}{\sqrt{2}} \tag{5.61}$$

The hookean relations (5.59), equilibrium conditions (5.60), and kinematical relations (5.61) govern the changes of forces, elongations, and displacements, just as Eqs. (5.51), (5.50), and (5.49) govern the initial responses. It follows that their solution also has the form of Eq. (5.53); specifically,

$$\Delta v = 0.793\frac{L\,\Delta W}{EA} \tag{5.62a}$$

$$\Delta u = -0.207\frac{L\,\Delta W}{EA} \tag{5.62b}$$

The plot of Eq. (5.62a) is a straight line, parallel to the initial portion OA of Fig. 5.21; the unloading follows either of the dotted lines, which depend on the extent of the prior deformation. Upon removal of the load from state A'', $\Delta W = -W_L$; then the changes of displacement are given by Eq. (5.62) and elongations and forces are given by Eq. (5.59) and (5.60), respectively. Again, a consequence of the inelastic deformations is a state of residual stress, as depicted in the free-body diagram of Fig. 5.22.

Fig. 5.22

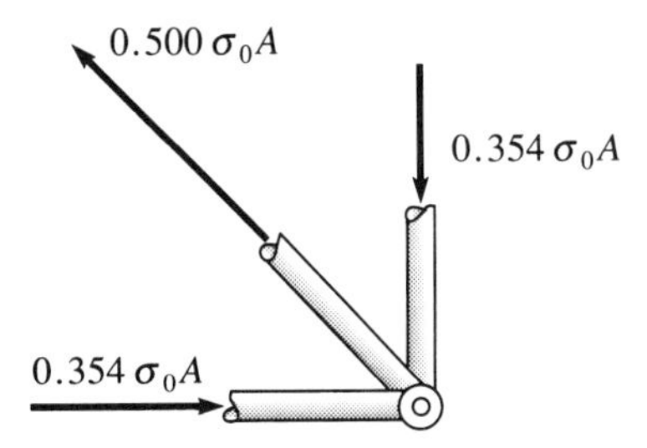

If the steel is very ductile, then large elongations, displacements, and rotations occur that invalidate the linear equations governing the equilibrium and kinematics. Also, most steels would exhibit some strain hardening. Such effects would cause the actual response to deviate from the foregoing description, as indicated by the dotted path OB of Fig. 5.21.

As noted in our introductory remarks, slender members under axial compression are liable to buckle. Unless lateral deflection is inhibited by adjoining supports, such buckling usually occurs at stresses, which are less than the compressive yield stress. Then buckling might actually delimit loading. Buckling and its consequences are examined in Chapter 9. For the present we presume that buckling is inhibited and that simple extensional deformations prevail. ◆

PROBLEMS

Assume linearly elastic behavior unless otherwise stated.

5.63 The members OA and OB are pin-connected at O, A, and B, as shown in Fig. P5.63. Bar OA is steel with a cross-sectional area of $\frac{1}{4}$ in.^{2}, and bar OB is aluminum with an area of $\frac{1}{5}$ in^2. Determine (a) the normal stress in each bar, and (b) the displacement of point O.

Fig. P5.63

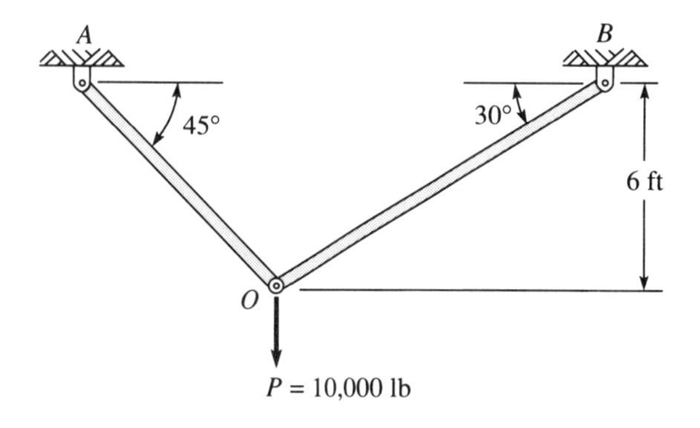

5.64 Two steel rods are pinned at C and pinned to the fixed surface at A and B (see Fig. P5.64). Both rods have the same constant cross section with area 5 cm^2. Joint C displaces vertically (direction BC) 0.15 cm under the load **P**. Determine the magnitude and direction (θ) of force **P**.

Fig. P5.64

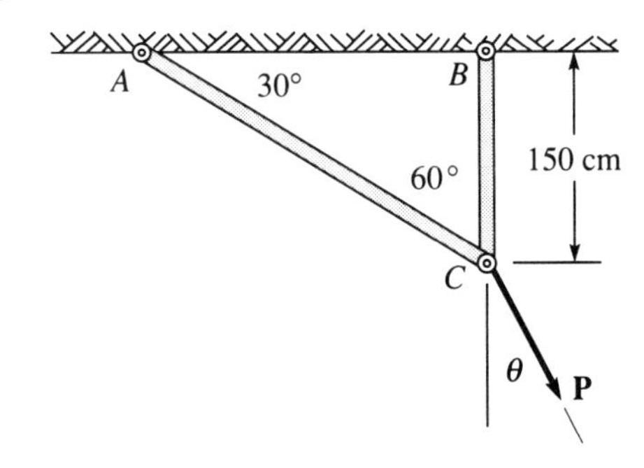

5.65 All members of the truss in Fig. P5.65 are steel and have a cross-sectional area of 5 in^2. Compute the vertical displacement of B when $W = 50{,}000$ lb.

Fig. P5.65

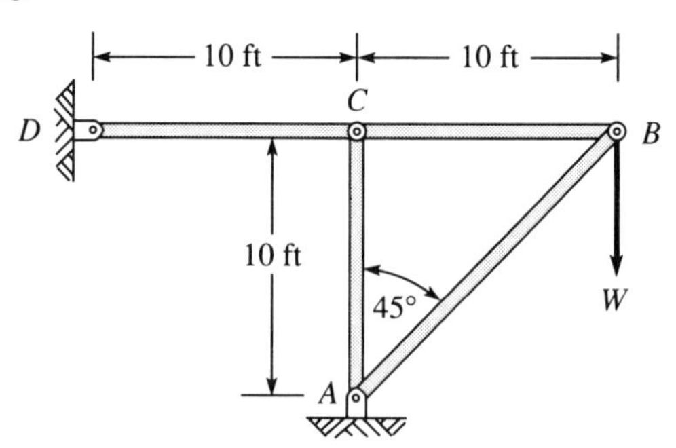

5.66 If the yield stress of the (ideally plastic) steel in Prob. 5.65 is $\sigma_0 = 30{,}000$ lb/in.^{2}, what is the limit load W_0?

5.67 Assume that the members AB, BC, and AC in the truss of Prob. 5.65 are linearly elastic ($E = 10^7$ lb/in.^{2}) but that CD is made of a work-hardening material that has the stress-strain diagram shown in Fig. P5.67. Compute the vertical displacement of joint B when $W = 40{,}000$ lb.

Fig. P5.67

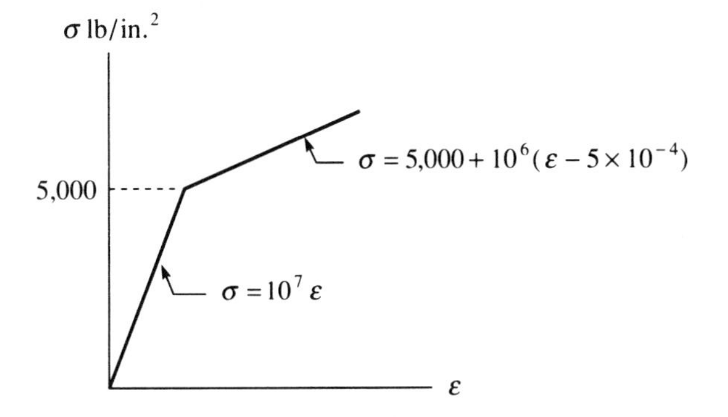

5.68 What is the horizontal displacement of B in the truss shown in Fig. P5.68 if $P = 100/\sqrt{3}$ kN? All the members are aluminum with cross-sectional areas of 10 cm^2.

Fig. P5.68

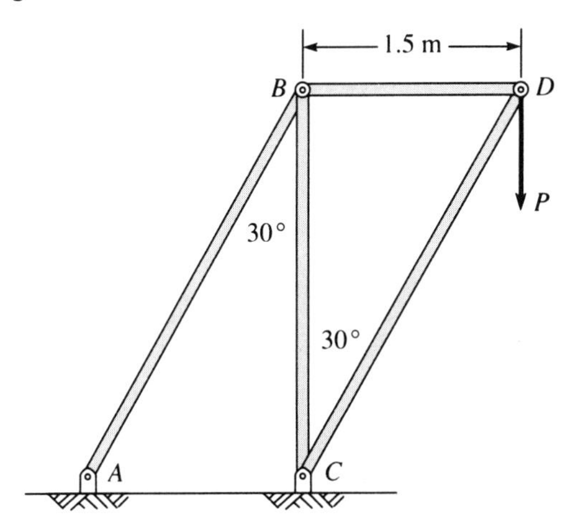

5.69 In the truss of Prob. 5.68, all the members are ideally plastic ($E = 200 \times 10^3$ MPa, $\sigma_0 = 270$ MPa). Compute the limit load P_0.

5.70 If each member of the truss in Prob. 5.68 is ideally plastic, with $\sigma_0 = 340$ MPa, determine the cross-sectional area of each member so that all members yield simultaneously when $P = 500/\sqrt{3}$ kN.

5.71 The members of the truss in Fig. P5.71 are aluminum, each has a cross-sectional area of 1.00 in.^{2}, and each is 20 ft long. Determine the vertical displacement of joint E due to the 5-T load.

Fig. P5.71

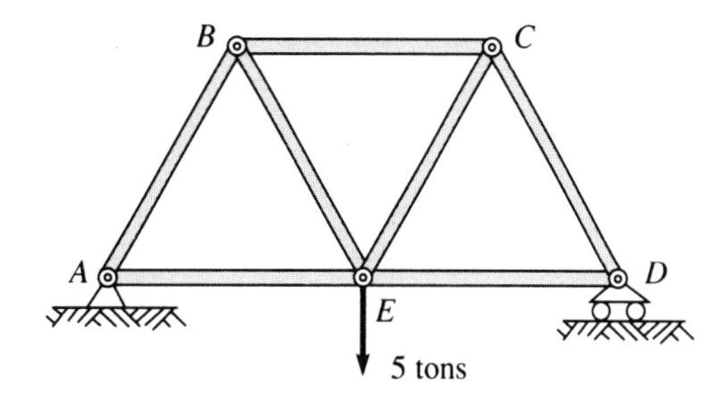

5.72 Each member of the truss shown in Fig. P5.72 is made of steel, is 50 cm long, and has a cross-sectional area of 4 cm^2. Calculate the stresses in each member and the displacement of joint D.

Fig. P5.72

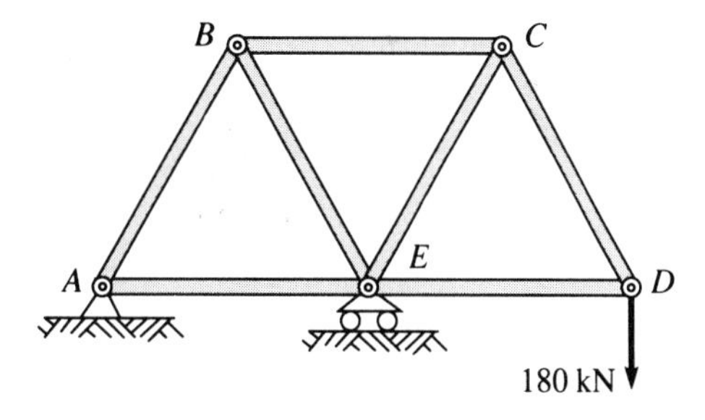

5.73 Each member of the plane symmetric truss in Fig. P5.73 is made of the same steel ($E = 30 \times 10^6$ lb/in.^{2}). Each has the same circular cross-section, with a 0.564-in. diameter. This steel is hookean to the yield stress ($\sigma_0 = 48 \times 10^3$ lb/in.^{2}). Subsequently, it is ideally plastic. The load P acts vertically at joint O along the line of symmetry (DO).

a. Determine the load P_0, the force on each member, and the displacement of joint O when yielding is initiated.

b. Calculate the limit load P_L.

c. Compute the displacement of joint O when the load just reaches P_L.

Fig. P5.73

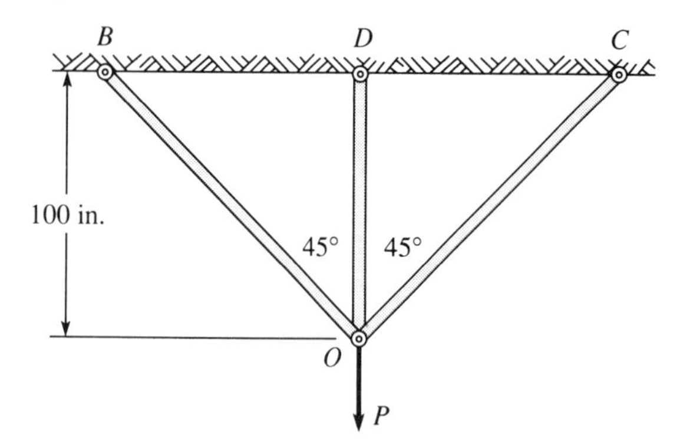

5.74 The truss of Prob. 5.73 is modified by adding two members, as shown in Fig. P5.74. The additional members are made of the same steel ($E = 30 \times 10^6$ lb/in.^{2}, $\sigma_0 = 48 \times 10^3$ lb/in.^{2}) and also have the same circular cross section, with diameter 0.564 in. The truss remains symmetric about the line DO. Load P acts along the line of symmetry.

a. Determine the displacement of joint O and load P_0 that initiates yielding.

b. Calculate the limit load P_L.

c. Sketch the plot of load P versus the deflection of joint O.

Fig. P5.74

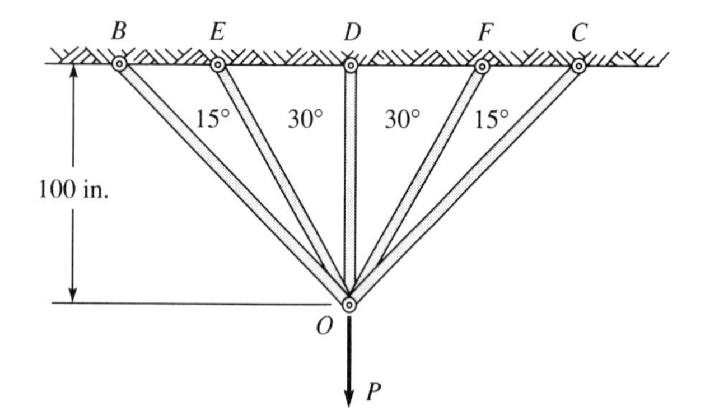

5.75 Each member of the truss shown in Fig. P5.75 is steel, and the cross-sectional areas are each 3.20 cm^2. Calculate the displacement of point A and the forces acting on each member.

Fig. P5.75

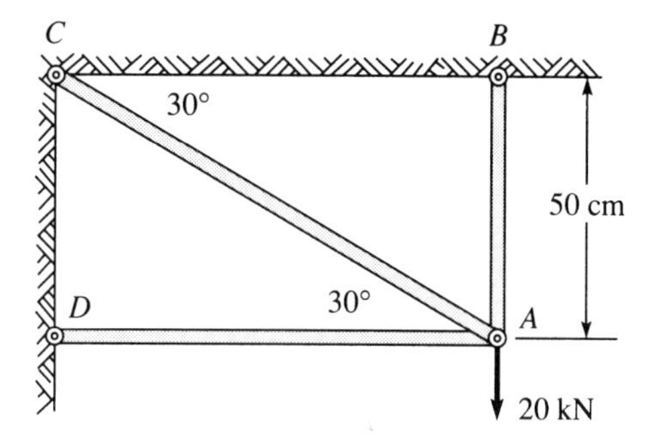

5.76 All members of the truss shown in Fig. P5.76 are of the same material and have the same cross-sectional areas. Express the vertical displacement of joint O in terms of the load W, L, E, and A.

Fig. P5.76

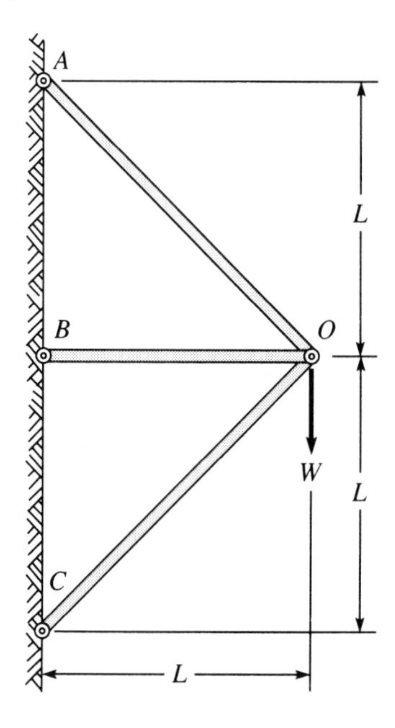

5.77 All members of the truss shown in Fig. P5.77 consist of the same material. Bar OA has a cross-sectional area of 6 cm^2. Choose the areas of the remaining members so that all three sustain equal stresses under the action of load P.

Fig. P5.77

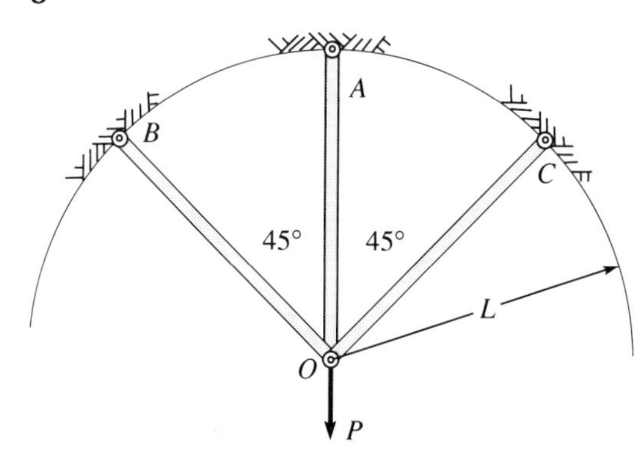

5.78 The three bars of the plane truss in Fig. P5.78 are composed of the same material, hookean ($E = 200 \times 10^3$ MPa) to the yield stress ($\sigma_0 = 280$ MPa), and then ideally plastic. Each rod has the same cross-sectional area ($A = 3.20$ cm^2).

a. Compute the displacement of joint O and the load at the onset of yielding.

b. Calculate the limit load and the displacement of the joint O when the load just attains the limit value.

Fig. P5.78

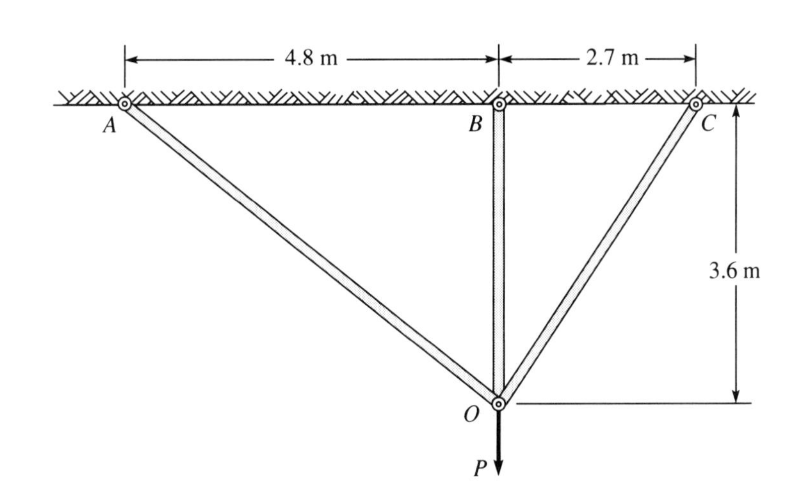

5.79 Each member of the truss shown in Fig. P5.79 is made of the same material and has the same cross-sectional area A. Express the horizontal displacement of B in terms of Q, L, E, and A.

Fig. P5.79

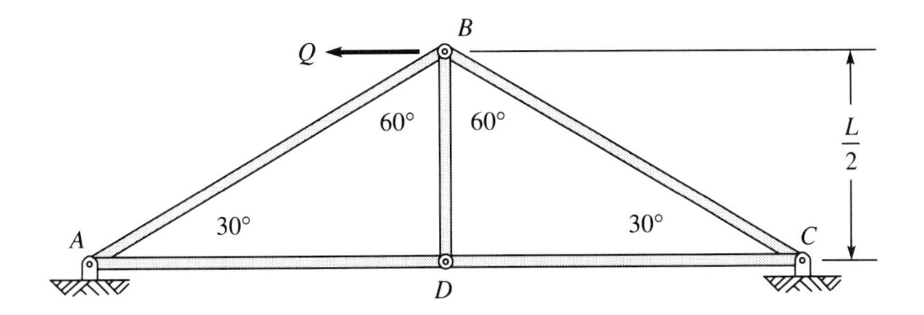

5.80 Each member of the truss shown in Fig. P5.80 is made of structural steel and has a cross-sectional area of 0.80 in^2. If the steel is ideally plastic at the yield stress $\sigma_0 = 50{,}000$ lb/in.^{2}, calculate (a) the limit load P_L and (b) the residual axial stress in each member if the load is increased to a value just under P_L and then removed.

Fig. P5.80

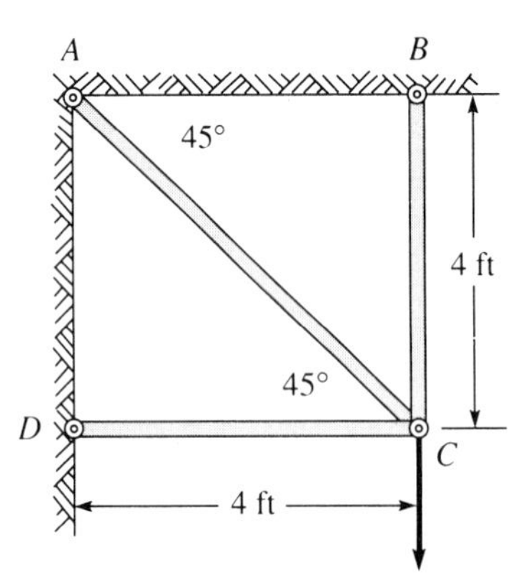

5.81 Each member of the truss shown in Fig. P5.81 is steel. The cross-sectional areas of members AD and DC are each 36 cm^2; the areas of the others are each 12 cm^2.

a. Compute the displacement of joint B.

b. If all of the members of the truss have cross-sectional areas of 36 cm^2, compute the displacement of joint B.

Fig. P5.81

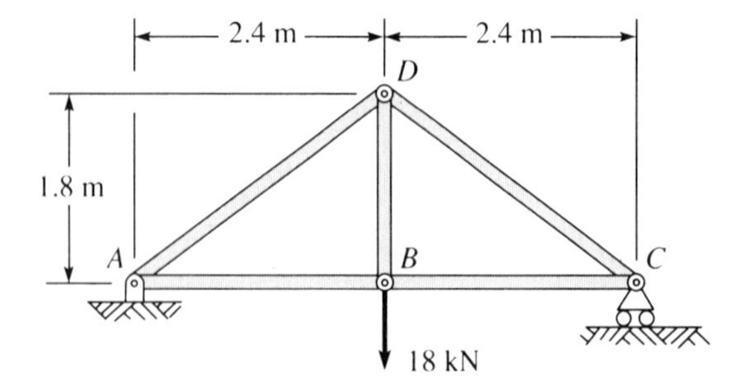

5.82 Each member of the truss illustrated in Fig. P5.82 is made of the same material ($E = 70 \times 10^3$ MPa) and has a cross-sectional area of 0.640 cm^2. Determine the stress in each member.

Fig. P5.82

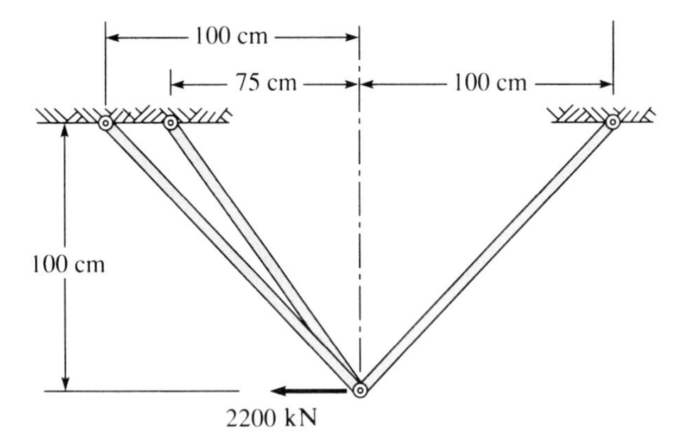

5.83 Each member of the truss in Fig. P5.83 is steel and has a cross-sectional area of 2 in^2. Compute the vertical displacement of joint C.

Fig. P5.83

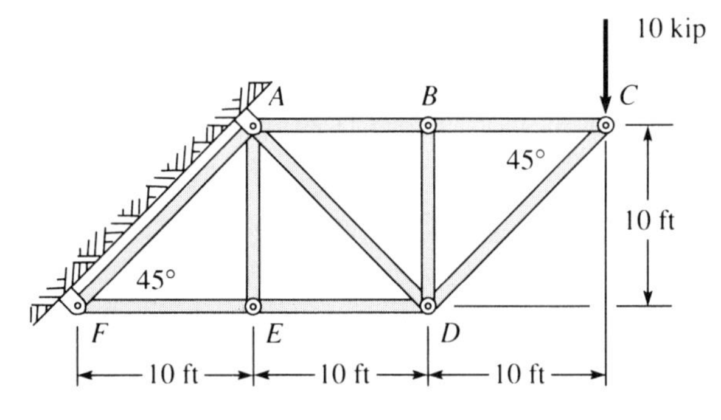

5.84 Calculate the limit load for the truss of Prob. 5.83 if the members are ideally plastic ($\sigma_0 = 40{,}000$ lb/in.^{2}).

5.85 The truss illustrated in Fig. P5.85 is composed of three rigid plastic members, AB, BC, CD ($\sigma_0 = 40{,}000$ lb/in.^{2}) and two ideally elastic-plastic members,

BD and *AC* ($E = 30 \times 10^6$ lb/in.^{2}, $\sigma_0 = 40{,}000$ lb/in.^{2}). All members have a cross-sectional area of $2\sqrt{2}$ in^2. Compute the normal stresses in all members when $P = 80{,}000$ lb.

Fig. P5.85

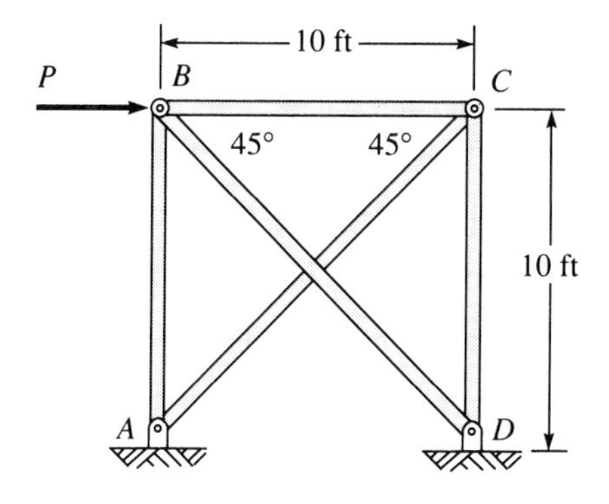

5.86 Compute the limit load for the truss of Prob. 5.85.

5.87 What is the displacement of the joint *B* for the truss shown in Fig. P5.87? All members are steel and all cross-sectional areas are 6 cm^2.

Fig. P5.87

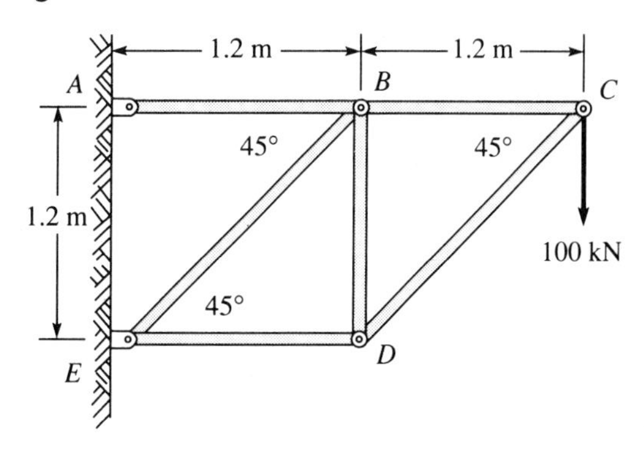

5.88 The yield stress for all members of the truss in Prob. 5.87 is $\sigma_0 = 250$ MPa. Determine the limit load P_L.

5.89 A weight of 36 kN is supported by three steel bars, as shown in Fig. P5.89. Each bar has a circular cross section with diameter 1.3 cm. Determine the stress in each bar and the vertical displacement of joint *O* due to the weight *W*. Treat the connections as ball-and-socket joints; that is, assume that they do not transmit couples.

Fig. P5.89

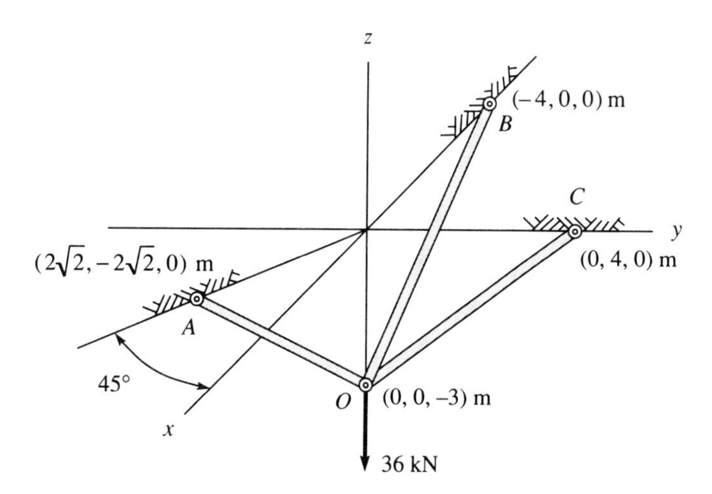

5.90 If $P = 70{,}000$ lb, compute the stress in member AC in Fig. P5.90. All the members of the truss are steel, with cross-sectional areas equal to 4 in^2.

Fig. P5.90

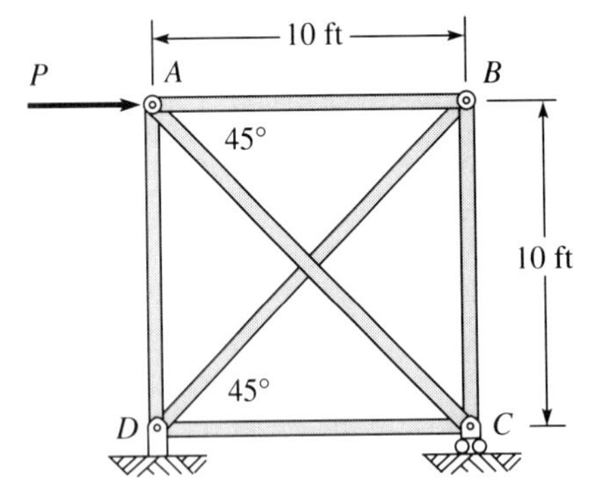

5.91 The members of the truss in Prob. 5.90 are all ideally plastic ($\sigma_0 = 45{,}000$ lb/in.^{2}). What is the limit load P_L for the truss?

5.92 Compute the stress in member AD for the truss shown in Fig. P5.92. (For each member $E = 70 \times 10^3$ MPa, $A = 16$ cm^2).

Fig. P5.92

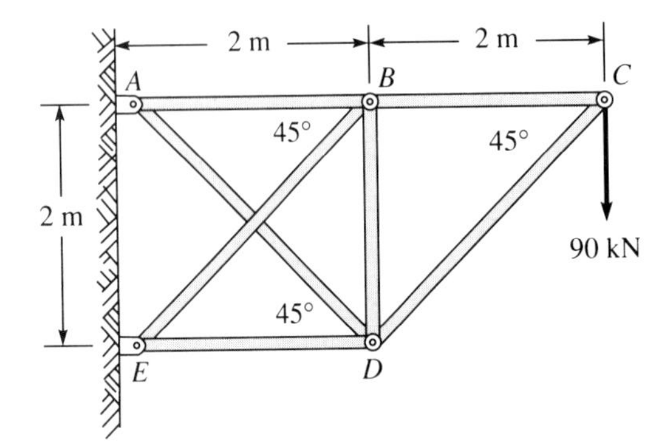

5.93 Three similar HI-HO rods are pinned together at joint O and pinned to the fixed joints at B, C, and D (see Fig. P5.93). The system is assembled at 70°F. It is symmetric about the plane, stress-free but snug when assembled. Each bar is 20.0 in. long. Essential properties are the modulus ($E = 30 \times 10^3$ ksi) and coefficient of thermal expansion ($\alpha = 10 \times 10^{-6}$/°F). Compute the displacement of the joint O and the stress in each rod when the temperature of rod OB only is lowered to -30°F.

Fig. P5.93

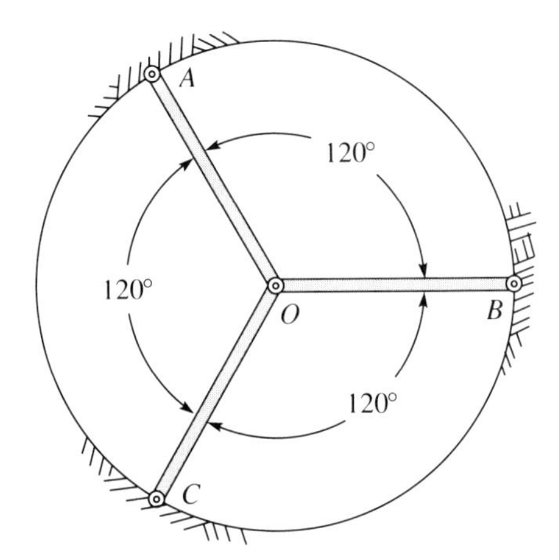

5.8 | Twisting of Circular Shafts

The Problem

Bars of circular cross section are frequently subjected to twisting couples. The drive shaft of an automobile, for example, is subjected to torques that tend to twist one end relative to the other. Another example is the torsion bar found in spring suspension systems. As a specific example, consider the circular cylinder shown in Fig. 5.23. One end is fixed and the free end is subjected to the twisting couple $\bar{T}$.

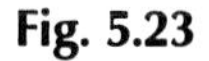
Fig. 5.23

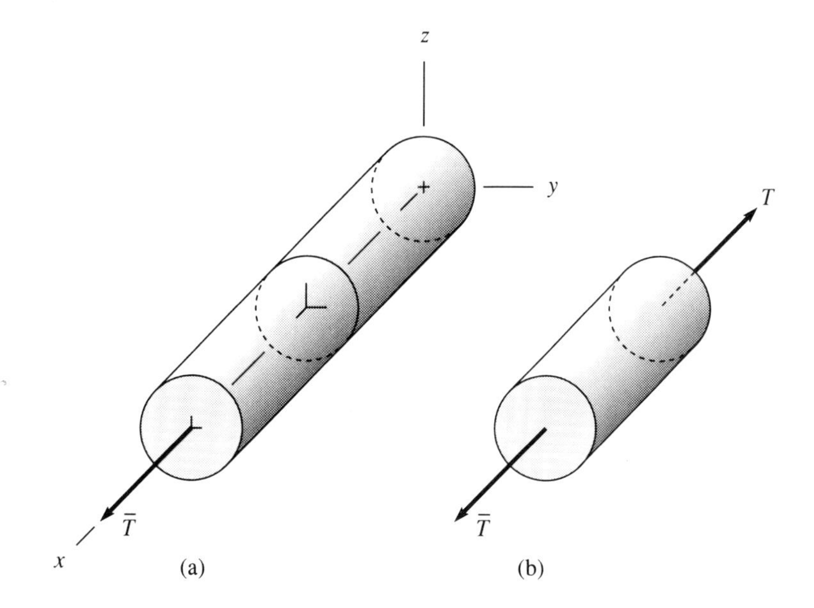

From a free-body diagram of any portion of the cylinder, Fig. 5.23(b), it is apparent that the net internal force system on any cross section is simply the twisting couple $T = \bar{T}$.

Our attention is presently limited to the shaft of circular cross section, wherein all properties are symmetric about the axis. This symmetry of properties and loading implies certain symmetry in the deformation.

Description of the Deformation

Our symmetric shaft appears the same when viewed from either end. This means that the midsection must remain plane; either half is a reflection of the other about that midplane. Cross sections must remain symmetric about the axis, but rotate. After deformation, a straight diameter must appear the same from either end—in other words, straight.

The significant feature of the twisting deformation is the relative rotation of cross sections about the axis of the cylinder. We assume that, when this deformation occurs, (1) the axis of the circular cylinder remains straight and (2) all radii in a cross section remain straight and rotate through the same angle about the axis.

Let $\phi(x)$ represent the rotation of a cross section S at a distance x from the fixed end of the cylinder, as indicated in Fig. 5.24. As a result of the twisting, a nearby cross section will rotate relative to S by the small amount $\Delta\phi$ as shown in Fig. 5.25. If P is a point in the undeformed cylinder located by the polar coordinates (r, θ) in S [Fig. 5.25(a)], then the segment PR in the θ-direction is perpendicular to the segment PQ,

Fig. 5.24

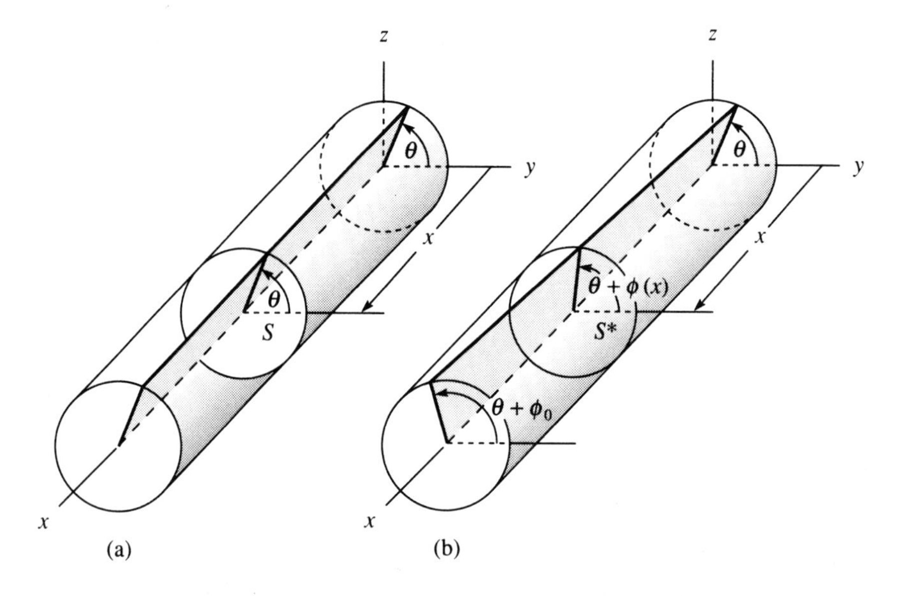

Fig. 5.25

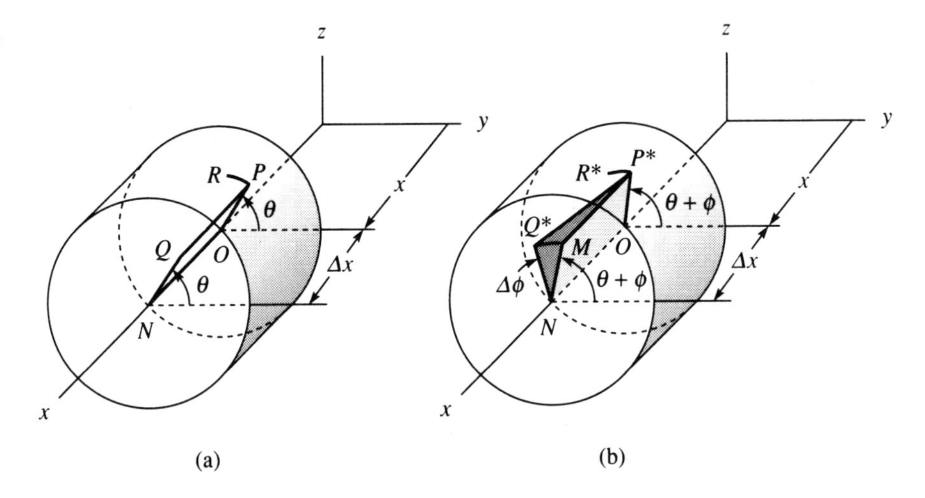

parallel to the axis of the cylinder. After deformation, the radius NQ^* has rotated $\Delta\phi$ more than the radius OP^*, as indicated in Fig. 5.25(b). The angle MP^*Q^* is given by

$$\tan \angle MP^*Q^* = \frac{MQ^*}{P^*M} = r\frac{\Delta\phi}{\Delta x} \tag{5.63}$$

But

$$\gamma_{x\theta} = \lim_{\Delta x \to 0} \angle MP^*Q^* \tag{5.64}$$

Hence, for small strains,

$$\gamma_{x\theta} = r\frac{d\phi}{dx} \tag{5.65}$$

(See Example 7 in Sec. 3.11 (p. 139) for a more detailed analysis.)

Under the prescribed circumstances, the geometry and the loading, the strains of lines in a cross section are neglected; we presume that $\varepsilon_\theta = \varepsilon_r = \gamma_{r\theta} = 0$. Likewise, the components ε_x and γ_{xr} are neglected.

Statics

In view of the loading and the anticipated strain $\gamma_{x\theta}$, we must anticipate a shear stress $\tau_{x\theta}$. Moreover, the symmetry requires that the component can depend only on the radius; that is, $\tau_{x\theta} = \tau_{x\theta}(r)$.

The cross section S is shown in Fig. 5.26. On the element of area ΔA, the shear stress produces a force

$$\Delta V_\theta = \tau_{x\theta}\,\Delta A$$

Fig. 5.26

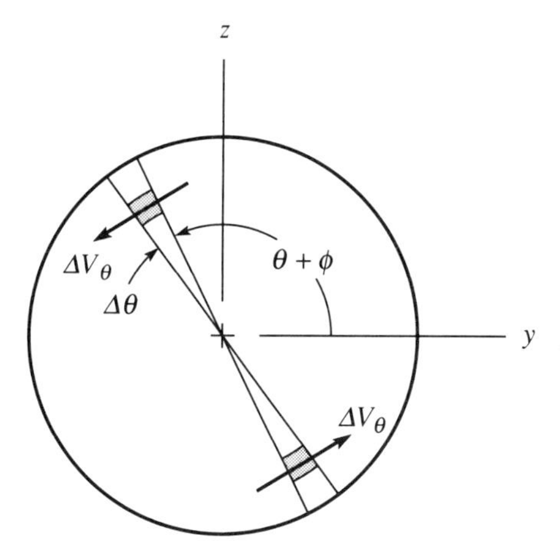

Since $\tau_{x\theta}$ is independent of the angle θ, the distribution is symmetric about the axis x. As shown on the cross section of Fig. 5.26, for each elemental area with the force ΔV_θ, there is a diametrically opposed element with an opposite force ΔV_θ. Such distribution produces only a couple T upon the cross section; in other words, the net shear force vanishes ($V_y = V_z = 0$). The twisting couple is the integral

$$T = \lim_{\Delta A \to 0} \sum r\,\Delta V_\theta$$

$$= \iint r\,\tau_{x\theta}\,dA \tag{5.66}$$

Hookean Isotropic Behavior

If the material is hookean and isotropic, then our assumptions about the deformation—specifically, $\varepsilon_x = \varepsilon_y = \varepsilon_z = \gamma_{xr} = \gamma_{r\theta} = 0$—imply also that $\sigma_x = \sigma_y = \sigma_z = \tau_{xr} = \tau_{r\theta} = 0$. Indeed, the latter conditions are as plausible and equivalent to the previous assumptions. The nonvanishing stress is

$$\tau_{x\theta} = G\gamma_{x\theta} \tag{5.67}$$

The *kinematic* result (5.65) the *stress-strain* equation (5.67), and the *statical* requirement (5.66) can be used to obtain the interrelationships among the stress ($\tau_{x\theta}$), applied torque (T), and rotation (ϕ). Substituting Eq. (5.65) into the right side of Eq. (5.67) and the latter into the right side of Eq. (5.66), we obtain

$$T = \frac{d\phi}{dx} \iint Gr^2 \, dA$$

In polar coordinates, $dA = r \, d\theta \, dr$, so that

$$T = \frac{d\phi}{dx} \iint Gr^3 \, d\theta \, dr \tag{5.68}$$

Our underlying assumptions presume only symmetry about the axis. Indeed, the modulus G might depend on the radius r. Also, the shaft might be tubular with an inner (hollow) portion of radius a and outer radius R. Therefore, Eq. (5.68) takes the general form

$$T = 2\pi \frac{d\phi}{dx} \int_a^R G(r) r^3 \, dr \tag{5.69}$$

If the shaft is composed of HI-HO material (hookean isotropic *and* homogeneous), then G is constant, and Eq. (5.69) assumes the particular form

$$T = G \frac{\pi}{2} (R^4 - a^4) \frac{d\phi}{dx} \tag{5.70}$$

Frequently, engineers employ the abbreviation[3]

$$\frac{\pi}{2}(R^4 - a^4) \equiv J \tag{5.71}$$

Then, the twisting of the HI-HO circular shaft is described by the equation

$$T = GJ \frac{d\phi}{dx} \tag{5.72a}$$

or

$$\frac{d\phi}{dx} = \frac{T}{GJ} \tag{5.72b}$$

---

3 J is sometimes called the polar moment of the cross section S.

The quantity GJ is sometimes called the *torsional rigidity*; the modulus G reflects the stiffness of the material, whereas the quantity J depends solely on the diameter(s) of the shaft. If the twist $(d\phi/dx)$ and strain $(\gamma_{x\theta})$ are eliminated from the three equations (5.65), (5.67), and (5.72), we obtain

$$\tau_{x\theta} = \frac{Tr}{J} = \frac{\overline{T}r}{J} \tag{5.73}$$

The right side of Eq. (5.73) merely acknowledges the equilibrium condition; the couple on the cross section $T = \overline{T}$, the applied torque.

The overall deformation of the cylinder is described by the rotation of one end relative to the other, in this case the rotation of the free end at $x = L$. From Eq. (5.72) the relative rotation of two nearby cross sections is given by

$$\Delta\phi = \frac{d\phi}{dx}\Delta x = \frac{T}{GJ}\Delta x$$

If the rotation of the face at $x = L$ relative to the face at $x = 0$ is denoted by ϕ_0, then

$$\phi_0 = \int_0^L \frac{T}{GJ}\,dx \tag{5.74}$$

In the present case $T = \overline{T}$ and G and J are constant, so that

$$\phi_0 = \frac{\overline{T}}{GJ}\int_0^L dx = \frac{\overline{T}L}{GJ} \tag{5.75}$$

The stress distribution of Eq. (5.73) applies to every cross section of the cylinder. Strictly speaking, the end faces at $x = 0$ and $x = L$ are also included. This would greatly restrict the utility of Eqs. (5.73) and (5.75) if it were not for Saint-Venant's principle. If a circular cylinder is reasonably slender and the loading applied to its ends are twisting couples, Eq. (5.73) is applicable to intermediate cross sections, and Eq. (5.75) gives the overall deformation.

If the end torques are actually applied according to Eq. (5.73), then Eqs. (5.73) and (5.75) are the same as results obtained from more elaborate theories. Furthermore, it can be shown that all the other components of stress and strain vanish; that is,

$$\varepsilon_r = \varepsilon_\theta = \varepsilon_z = \gamma_{r\theta} = \gamma_{rz} = 0$$

$$\sigma_r = \sigma_\theta = \sigma_z = \tau_{r\theta} = \tau_{rz} = 0$$

For cylinders that do not have circular cross sections, the assumptions about the deformation are not valid. In particular, the noncircular cross section will not remain plane.

EXAMPLE 8

Nonhomogeneous Hookean Shaft The foregoing analysis of torsion incorporated certain *kinematics*, *statics*, and *material* properties. The underlying description presumes symmetry about the axis and uniformity along the shaft. Accordingly, cross sections are not distorted, and the kinematic relation (5.65) follows:

$$\gamma_{x\theta} = r\frac{d\phi}{dx} \tag{5.76}$$

In view of the symmetry, the essential statical result is Eq. (5.66); on every cross section the resultant is the couple

$$T = \iint \tau_{x\theta} r\, dA \tag{5.77}$$

These results apply to any circular shaft with properties that are symmetric about the axis. As an example, suppose that the material is hookean and isotropic *but* the modulus is a function:

$$G = G(r)$$

One might conceive of a material (a matrix of metal or plastic) filled with small fibers of stiffer and stronger material. The latter are dispersed more densely in the outer portions to create a more efficient shaft. To illustrate, suppose that

$$G = G_0\left(1 + \frac{r}{R}\right)$$

Then, the material at the outer surface is twice as stiff as the material at the center and

$$\tau_{x\theta} = G\gamma_{x\theta} = G_0\left(1 + \frac{r}{R}\right)\gamma_{x\theta} \tag{5.78}$$

Again, the kinematics, the statics, and the material relations are all needed to effect the solution. Substituting Eq. (5.76) into Eq. (5.78) and Eq. (5.78) into Eq. (5.77), we obtain

$$\begin{aligned} T &= G_0\frac{d\phi}{dx}\int_0^R \int_0^{2\pi} \left(1 + \frac{r}{R}\right) r^3\, d\theta\, dr \\ &= G_0\left(\frac{9}{10}\pi R^4\right)\frac{d\phi}{dx} \end{aligned} \tag{5.79}$$

If $d\phi/dx$ and $\gamma_{x\theta}$ are eliminated from Eqs. (5.76), (5.79), and (5.78), we obtain

$$\tau_{x\theta} = \frac{10r}{9\pi R^4}\left(1 + \frac{r}{R}\right)T \tag{5.80}$$

The reader can compare the results of Eqs. (5.79) and (5.80) with the equations for the Hi-HO shaft, Eqs. (5.70) and (5.73). ◆

EXAMPLE 9

Composite Shaft In keeping with the general description of the symmetric circular shaft, we might form the shaft with a light cylindrical core of hookean isotropic material and an outer cylindrical tube of harder and stiffer material. Such a shaft would be lighter, stiffer, and more wear-resistant. To illustrate, suppose that the core and outer tube are bonded together; in particular, since both are joined at the ends, both undergo the same twist. Kinematic equation (5.65) holds; again,

$$\gamma_{x\theta} = r\frac{d\phi}{dx} \tag{5.81}$$

The torque is again given by (5.66):

$$T = \iint \tau_{x\theta} r\, dA \tag{5.82}$$

Let us denote the radius of the core (also, the inner radius of the surrounding tube) by R_c, the modulus of the core by G_c, and the modulus of the outer shell by G_s. Then

$$\tau_{x\theta} = G_c\gamma_{x\theta}, \qquad r < R_c \tag{5.83a}$$

$$\tau_{x\theta} = G_s\gamma_{x\theta}, \qquad R_c < r \le R_0 \tag{5.83b}$$

Again, utilizing the *kinematics*, the *statics*, and the *description* of the *material* and substituting Eq. (5.81) into Eq. (5.83) and the latter into Eq. (5.82), we have

$$T = \frac{d\phi}{dx}\left[G_c \iint_{A_c} r^2\, dA + G_s \iint_{A_s} r^2\, dA\right] \tag{5.84a}$$

The two distinct moduli for the core and shell require that we split the integral into the two parts. Here, as before, the integrals reflect the size of the sections:

$$\iint_{A_c} r^2\, dA = \frac{\pi R_c^4}{2} \equiv J_c$$

$$\iint_{A_s} r^2\, dA = \frac{\pi}{2}(R_0^4 - R_c^4) \equiv J_s$$

Substituting these abbreviations into Eq. (5.84a), we obtain

$$T = (G_c J_c + G_s J_s)\frac{d\phi}{\delta x} \tag{5.84b}$$

The reader can compare this result with that obtained for the homogeneous (HI-HO) shaft, Eq. (5.72). As might be anticipated, the stiffness of the composite shaft is the sum of the stiffnesses of the core and shell. ◆

PROBLEMS

5.94 A composite shaft has a cylindrical core made from a linearly elastic material, with a 2-in. radius and $G_1 = 10^7$ psi. The outer portion (inner radius 2 in., outer radius 3 in.) is linearly elastic, with $G_2 = 5 \times 10^6$ psi. Make the usual assumptions for simple torsion: The radial lines in a cross section remain straight and rotate through the same angle. Compute the relative rotation of cross sections 10 in. apart when the shaft is subjected to a twisting couple of $T_0 = 20{,}000$ in.-lb.

5.95 A solid aluminum ($G = 30 \times 10^3$ MPa) shaft has an outside diameter of 2.50 cm. What is the length of this shaft if the maximum shear stress is 16 MPa and the relative rotation of the ends is 0.040 (rad)?

5.96 A shaft has a solid circular cross section: diameter = 1.00 in., length = 20.00 in. It is composed of a brittle HI-HO material, which fails if the *normal* stress exceeds $\sigma_u = 96$ ksi; $E = 30 \times 10^3$ ksi and $v = 0.25$. The shaft is subjected to opposing twisting couples applied at the ends. Use our theory of simple torsion; compute the total twist (relative rotations of the ends) when failure occurs.

5.97 A shaft is made of steel ($E = 30 \times 10^3$ ksi, $v = 0.25$). It has an outside diameter of 1.40 in. and a coaxial hole with a diameter of 0.80 in. The shaft is 30.0 in. long and subject to opposing twisting couples at the ends; $T_0 = 30$ kip-in. Calculate (a) the relative rotation of the ends and (b) the maximum *tensile* stress in accordance with our theory of simple torsion.

5.98 A hollow circular shaft is made of steel ($G = 83 \times 10^3$ MPa, $v = 0.25$). The inner and outer diameters are $D_1 = 2.50$ cm and $D_0 = 3.75$ cm, respectively. The shaft is 100 cm long and subject to opposing twisting couples at the ends, $T = 66$ kN-cm. Use the theory of simple torsion to compute (a) the maximum shear stress and (b) the maximum *extensional* strain.

5.99 An aluminum wire (HI-HO behavior) has a diameter of 0.200 in. The modulus and Poisson ratio are $E = 12{,}000$ ksi and $v = 0.30$. Determine the minimum length so that it can be twisted through one complete revolution (relative rotation of the ends) without exceeding the yield stress of 10 ksi in shear.

5.100 A shaft is 50 cm long and composed of an aluminum core (2.50-cm diameter) and a concentric steel tube (2.50-cm inside diameter and 3.00-cm outside diameter). The core and tube are joined so that they twist together when opposing twisting couples are applied at the ends of the composite. The yield stresses of the aluminum and steel are 300 MPa and 450 MPa, respectively, in simple *tension*. The shear moduli are 30×10^3 MPa and 90×10^3 MPa, respectively. Use the maximum shear-stress criterion (Tresca) to determine the relative rotation of the ends when yielding is initiated.

5.101 A solid shaft is built up by laminating very thin concentric cylinders of varying stiffness so that, although the material at any point of the shaft is linearly elastic, the shear modulus varies with the radius in accordance with $G(r) = G_0[1 - (r/2R)]$, where R is the radius of the shaft. If the length of the shaft is L and the ends are subjected to the twisting couple T, compute the shear stress $\tau_{x\theta}$ at any point of the shaft, and compute the relative rotation of the end cross sections.

5.102 Determine the relationship between the applied torque T and the relative rotation ϕ for a shaft whose material behavior is approximated by

$$\tau_{x\theta} = K\sqrt{\gamma_{x\theta}}$$

($\tau_{x\theta}$ and $\gamma_{x\theta} \geq 0$.)

5.103 Repeat Prob. 5.102 for a shaft whose material behavior is approximated by

$$\tau_{x\theta} = K\gamma_{x\theta}^n$$

where $0 \leq n \leq 1$. Show how, by suitable choice of values for K and n, these results reduce to the results for linearly elastic material obtained in this section.

5.104 Repeat Prob. 5.102 for a shaft whose material behavior is approximated by

$$\tau_{x\theta} = \tau_0 \frac{G\gamma_{x\theta}}{\tau_0 + G\gamma_{x\theta}}$$

Is there a limit load for this shaft?

5.105 A shaft is subjected to a distributed torque so that the twisting couple on a cross section at x is given by $T = T_0\sqrt{x/L}$, as shown in Fig. P5.105. If the material is linearly elastic with shear modulus G, compute the rotation ϕ_0 of the cross section at $x = 0$ in terms of T_0, L, G, and J. Assume that the radial lines in a cross section remain straight and rotate through the same angle.

Fig. P5.105

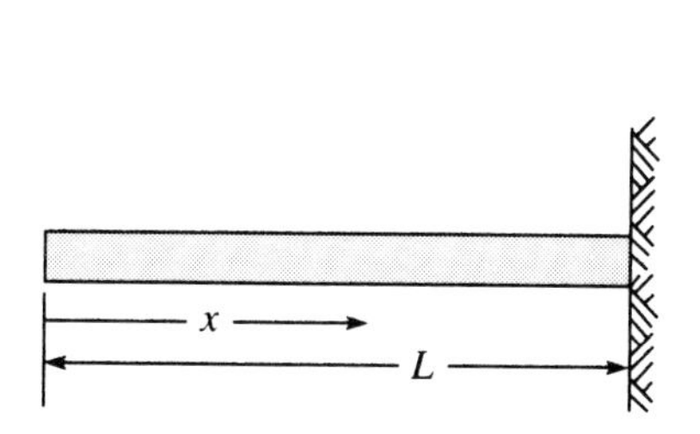

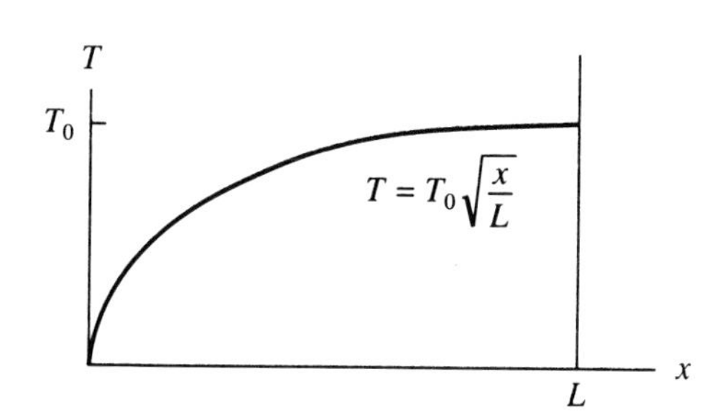

5.106 A linearly elastic shaft 10 in. long and 2 in. in diameter fits snugly in a bored hole in a block, as shown in Fig. P5.106. One end of the shaft is welded to the block at A, and the other end, B, is subjected to a twisting couple of $T = 4000\ \pi$ in.-lb. The block can exert friction forces on the surface of the shaft of 100 psi. If $G = 6 \times 10^6$ psi, what is the rotation of the shaft at B? Make the usual assumption for simple torsion.

Fig. P5.106

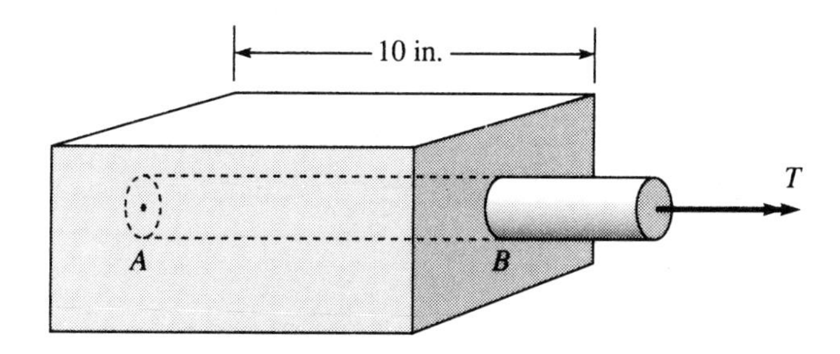

5.107 The shaft shown in Fig. P5.107 is subjected to a distributed torque, in addition to $\overline{T}$ and $2\overline{T}$, so that the twisting couple diagram is linear as shown, and the shaft is in equilibrium. The radius is R_0 at $x = 0$ and varies along the length of the shaft in such a way that the maximum shear stress on any cross section is the same. If the material is linearly elastic, determine how R varies with x. Assume that radial lines in a cross section remain straight and rotate through the same angle.

Fig. P5.107

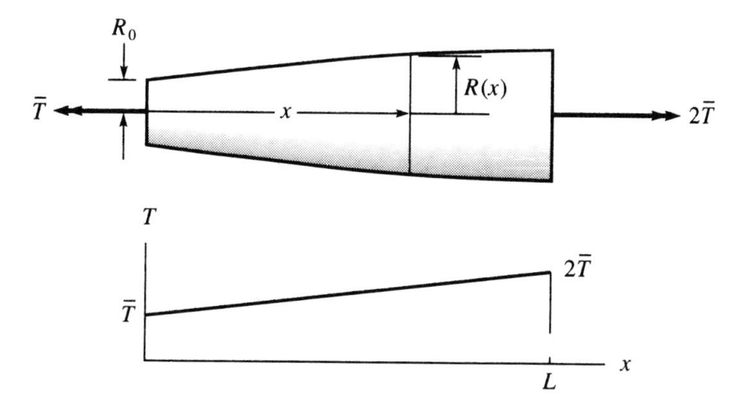

5.108 Compute the rotation of the cross section at A of the slightly tapered linearly elastic shaft shown in Fig. P5.108. Assume that radial lines in a cross section remain straight and rotate through the same angle.

Fig. P5.108

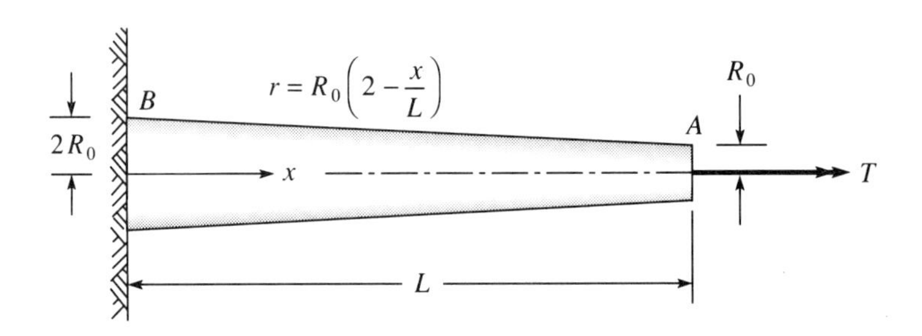

5.109 A cylindrical shaft is fixed at one end and fre of all other externally applied loads. The material of the shaft is set in torsional motion; that is, cross sections of the shaft rotate about the axis but do not displace axially or rotate about a transverse axis. Assume that at all times the axis of the shaft is straight, particles of a cross section always remain the same distance from the fixed end, and the radial lines in a cross section remain straight and rotate through the same angle. In other words, cross sections rotate like rigid discs. The shaft is shown in the unstrained state and also at some subsequent time t in Fig. 5.24.

a. Let $\phi(x, t)$ represent the rotation of the cross section at x and at time t. Show that [see Eq. (5.65)]

$$\gamma_{x\theta} = r\frac{\partial}{\partial x}\phi(x, t) \tag{P5.109a}$$

b. If the material is linearly elastic, $\tau_{x\theta} = G\gamma_{x\theta}$. Show that the twisting couple on the cross section at x is [see Eqs. (5.66) and (5.68)]

$$T(x, t) = GJ\frac{\partial}{\partial x}\phi(x, t) \tag{P5.109b}$$

where J has the usual meaning [see Eq. (5.71)].

c. Denote the mass density of the shaft by ρ so that the mass moment of inertia (about the shaft axis) of the slice between the nearby cross sections at x and $x + \Delta x$ is

$$\Delta I = \tfrac{1}{2}(\rho\pi R^2\,\Delta x)R^2 = \rho J\,\Delta x$$

Since the slice is rotating about the axis with an angular acceleration of $\partial^2/\partial t^2\phi(x, t)$, deduce from the equations of motion that

$$T(x + \Delta x, t) - T(x, t) = \rho J\,\Delta x\frac{\partial^2}{\partial t^2}\phi(x, t)$$

Hence

$$\frac{\partial T}{\partial x} = \rho J\frac{\partial^2}{\partial t^2}\phi(x, t) \tag{P5.109c}$$

d. Show that it follows from Eqs. (P5.109b) and (P5.109c) that the free torsional vibration of a linearly elastic shaft satisfies the wave equation:

$$\frac{\partial^2}{\partial x^2}\phi(x, t) - \frac{\rho}{G}\frac{\partial^2}{\partial t^2}\phi(x, t) = 0$$

5.9 Torsion of an Ideally Elastic-Plastic Shaft

Our description of the twisting depends only upon the axial symmetry of the shaft. In particular, the kinematic result (5.65) is applicable provided only that the properties remain symmetric—that is, dependent only on the radial coordinate r. Restated, the strain is given by the equation:

$$\gamma_{x\theta} = r\frac{d\phi}{dx} \tag{5.85}$$

Now, suppose that the circular shaft of Fig. 5.23 is made of an ideally elastic-plastic material. The stress-strain diagram is shown in Fig. 5.27.

Fig. 5.27

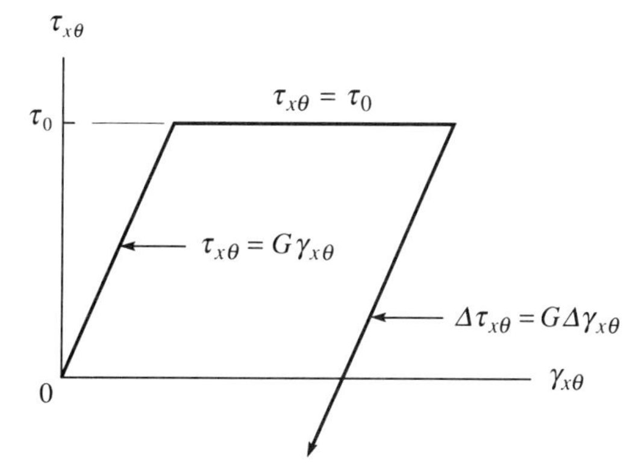

If the shear strain does not exceed $\gamma_0 = \tau_0/G$ anywhere in the shaft, the material behavior is linearly elastic and the results previously obtained are applicable. In particular the shear stress varies linearly along a radius, as shown in Fig. 5.28. As the

Fig. 5.28

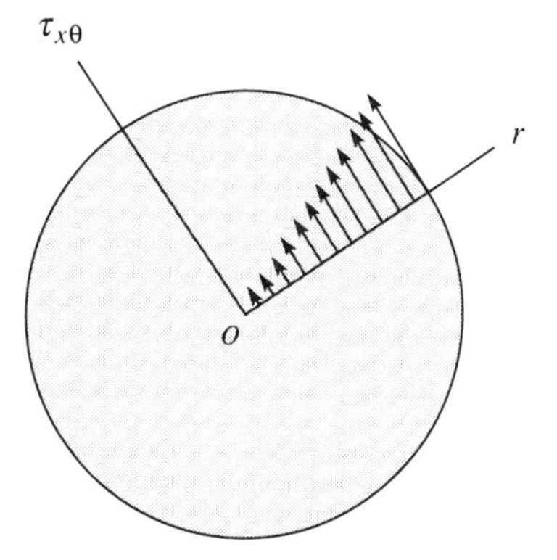

twisting couple is increased, the strain $\gamma_{x\theta}$ exceeds the yield strain γ_0 in the outermost portion of the cross section, as indicated in Fig. 5.29(a). We denote by b the radius at which $\gamma_{x\theta} = \gamma_0$.

The inner core ($r < b$) behaves elastically because $\gamma_{x\theta} < \gamma_0$, whereas the outermost portion ($r > b$) undergoes plastic deformation because $\gamma_{x\theta} > \gamma_0$. In the elastic core the shear stress $\tau_{x\theta}$ is proportional to the shear strain $\gamma_{x\theta}$:

Fig. 5.29

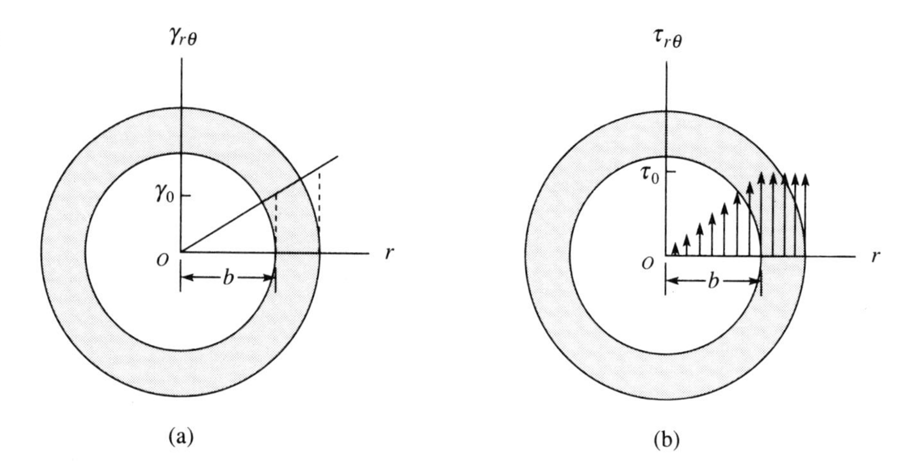

$$\tau_{x\theta} = G\gamma_{x\theta} = G\frac{d\phi}{dx}r, \qquad r \le b \tag{5.86a}$$

In the plastic (outer) portion

$$\tau_{x\theta} = \tau_0, \qquad r \ge b \tag{5.86b}$$

The shear stress $\tau_{x\theta}$ varies along a radius, as shown in Fig. 5.29(b). At the elastic-plastic interface ($r = b$) we have

$$G\frac{d\phi}{dx}b = \tau_0 \tag{5.87}$$

Using this to eliminate $G\,d\phi/dx$ from Eq. (5.86a), we obtain, finally,

$$\tau_{x\theta} = \begin{cases} \dfrac{r}{b}\tau_0, & r \le b \qquad (5.88a) \\ \tau_0, & r \ge b \qquad (5.88b) \end{cases}$$

This is equivalent to Eq. (5.86); we have merely replaced the twist $d\phi/dx$ in Eq. (5.86a) with the unknown radius b in Eq. (5.88a).

Again, we note that all preceding steps are valid as the symmetry about the axis is maintained. Our shaft may be solid or tubular. As before, we admit either by assuming an inner radius a and outer radius R.

The twisting couple T on a cross section is given by

$$T = \iint_S r\tau_{x\theta}\,dA = \int_0^{2\pi}\int_a^R \tau_{x\theta}r^2\,dr\,d\theta$$

and since $\tau_{x\theta}$ does not depend on θ,

$$T = 2\pi\int_a^R \tau_{x\theta}r^2\,dr$$

To evaluate this integral we split it into two integrals, one from a to b and the other from b to R:

$$T = 2\pi \int_a^b \tau_{x\theta} r^2 \, dr + 2\pi \int_b^R \tau_{x\theta} r^2 \, dr$$

In the first integral $\tau_{x\theta}$ is given by Eq. (5.88a); in the second integral, it is given by Eq. (5.88b). After these substitutions, we have

$$T = 2\pi \frac{\tau_0}{b} \int_a^b r^3 \, dr + 2\pi\tau_0 \int_b^R r^2 \, dr$$

$$= \frac{\pi\tau_0}{6}\left[4R^3 - b^3 - 3\frac{a^4}{b}\right] \tag{5.89}$$

The radius of the elastic core can be determined by solving Eq. (5.89) for b (after noting $T = \bar{T}$, the applied couple). For the results to be meaningful, b must lie between a and R. When $b = R$, the material at the outermost portion of the cross section is about to yield. The twisting couple at this point is

$$T_0 = \frac{\pi\tau_0}{2}(R^3 - a^3) \tag{5.90}$$

For T less than T_0, the behavior of the material is linearly elastic, and the previously obtained results can be used.

As b decreases, the elastic core becomes smaller. In the limiting condition $b = a$, we say that the cross section is fully plastic. The twisting couple has the limiting value

$$T_L = \tfrac{2}{3}\pi\tau_0(R^3 - a^3) \tag{5.91}$$

Most shafts are solid; that is, $a = 0$. Then the limit torque is 33% greater than the torque that initiates yield, or

$$\frac{T_L}{T_0} = \frac{4}{3}$$

The inner portion of the ideally plastic shaft provides a measure of safety at the expense of weight and cost. Within the elastic limit, a hollow tube is more efficient than a solid one. Since the solid shaft is far more common, let us examine its behavior in further detail. After yielding is initiated, the radius b is given by Eq. (5.87):

$$b = \frac{\tau_0}{G \, d\phi/dx}$$

Substituting this into (5.89) and setting $a = 0$ for the solid shaft, we obtain

$$T = \frac{\pi\tau_0}{3}\left[2R^3 - \frac{\tau_0^3}{2G^3(d\phi/dx)^3}\right] \tag{5.92}$$

Prior to yielding, the twisting is governed by the *linear* relation (5.70):

$$T = \frac{\pi}{2} R^4 G \frac{d\phi}{dx} \tag{5.93}$$

Linear equation (5.93) applies provided $T < T_0$. Additional loading follows the nonlinear equation (5.92). As the twist increases ($d\phi/dx \to \infty$), the torque approaches asymptotically the limit torque ($T \to T_L$). This progression is depicted graphically in Fig. 5.30.

Fig. 5.30

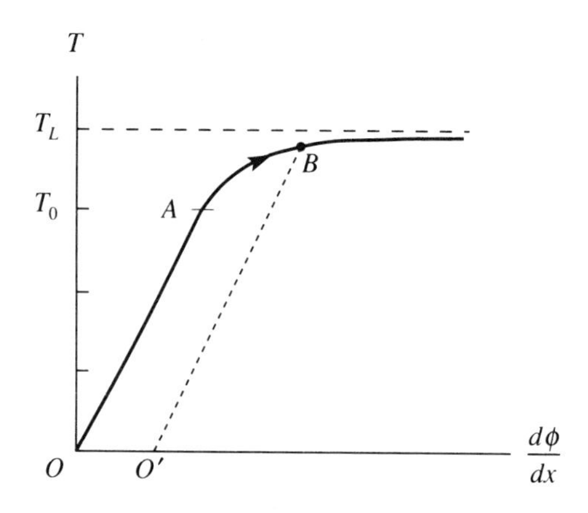

Residual Stress

Suppose that a twisting couple is applied to the solid elastic-plastic shaft, exceeds the yield value T_0, and causes plastic deformation. The history of deformation traces the path OAB of Fig. 5.30. Now suppose that the couple is reduced, changed in amount $\Delta\overline{T}$. This is accompanied by changes in stress, $\Delta\tau_{x\theta}$, in strain, $\Delta\gamma_{x\theta}$, and in twist, $\Delta(d\phi/dx)$. In keeping with the *kinematic* relation (5.65),

$$\Delta\gamma_{x\theta} = r\,\Delta\left(\frac{d\phi}{dx}\right) \tag{5.94}$$

Also, by the *static* condition (5.66)

$$\Delta T = \Delta\overline{T} = \iint r\,\Delta\tau_{x\theta}\,dA \tag{5.95}$$

To examine the *material behavior*, we must consider the regions $r < b$ and $b < r < R$. The former has remained hookean; the state of stress lies on the initial linear segment OA in the diagram of Fig. 5.31. In the outer portion ($b < r < R$), the material has yielded; the state of stress is at a point B of Fig. 5.31. During unloading from this yielded state, as from the elastic state, the changes of stress and strain follow the linear relation

Fig. 5.31

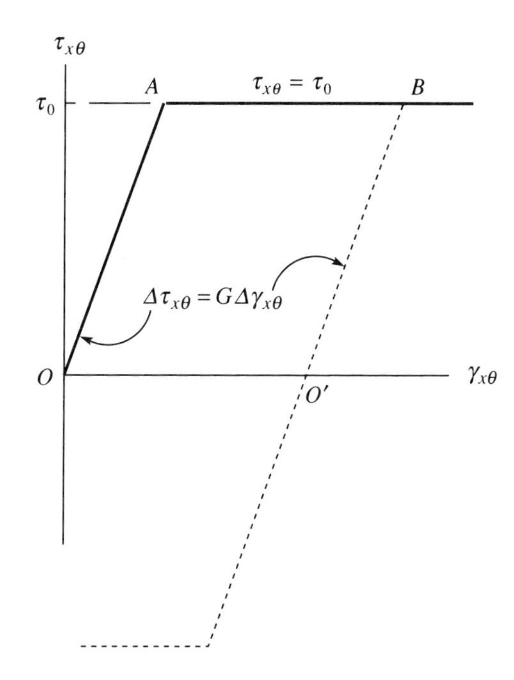

$$\Delta\tau_{x\theta} = G\,\Delta\gamma_{x\theta} \tag{5.96}$$

In other words, unloading from state B traces the dotted curve BO', parallel to OA.

Once again, during this unloading phase, we require the *kinematics*, *statics*, and *material behavior*, which are embodied in Eqs. (5.94), (5.95), and (5.96), respectively. After eliminating the stress and strain increments via Eqs. (5.94) and (5.96), we obtain, from (5.95), the result

$$\begin{aligned}\Delta\overline{T} &= G\,\Delta\left(\frac{d\phi}{dx}\right)\int_0^R\int_0^{2\pi} r^3\,d\theta\,dr \\ &= G\frac{\pi}{2}R^4\,\Delta\left(\frac{d\phi}{dx}\right)\end{aligned} \tag{5.97}$$

Note that this is a linear relation between the changes in torque $\Delta\overline{T}$ and twist $\Delta(d\phi/dx)$. The unloading traces the straight line BO' in Fig. 5.30. If the twist and strain are eliminated from Eqs. (5.94), (5.96), and (5.97), we obtain

$$\Delta\tau_{x\theta} = \Delta\overline{T}\frac{2r}{\pi R^4} = \frac{\Delta\overline{T}r}{J} \tag{5.98}$$

Prior to unloading, the stress distribution was given by Eq. (5.88) and depicted graphically by the plot of Fig. 5.29(b). The distribution $\tau_{x\theta}$ and the change $\Delta\tau_{x\theta}$ are both plotted in Fig. 5.32, together with the residual stress

Fig. 5.32

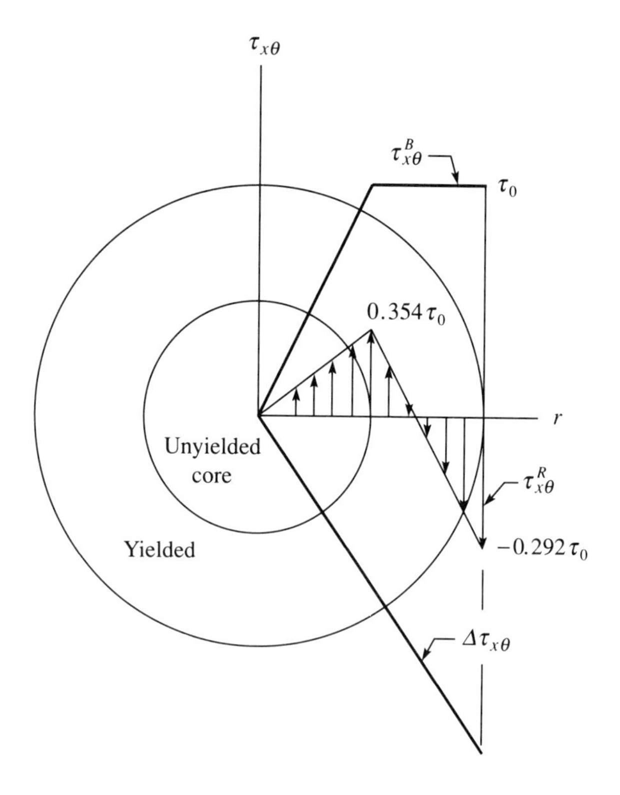

$$\tau_{x\theta}^R = \tau_{x\theta}^B + \Delta\tau_{x\theta}$$

For purposes of illustration, the radius of the elastic core in Fig. 5.32 is $b = R/2$. Then

$$T_B = \frac{31}{48}\pi\tau_0 R^3$$

Upon unloading,

$$\Delta T = -T_B, \qquad \Delta\tau_{x\theta} = -\frac{31}{24}\tau_0\frac{r}{R}$$

The student should verify that the residual distribution gives $T = 0$.

A few words of caution are in order: The ideally elastic-plastic material is merely a mathematical model. It provides a reasonable description of a few materials; the most notable is structural (hot-rolled) steel. It must be understood that all materials are somewhat heterogeneous. Consequently, parts of a member invariably yield at lower stresses and others yield at higher stresses, so that the elastic and plastic regions are not clearly delineated, as suggested by Fig. 5.29. Nonetheless, the model does provide some useful guidelines about and insights into the behavior of elastic-plastic members.

PROBLEMS

5.110 The aluminum and steel in the composite shaft of Prob. 5.100 are both ideally plastic at their respective yield stresses. Determine (a) the torque that initiates yielding, (b) the torque that initiates yielding in the core, and (c) the limit torque.

5.111 A shaft is made of low-carbon steel:

$$E = 30 \times 10^6 \text{ psi}, \qquad \nu = 0.25$$

$$\tau_0 = 24 \times 10^3 \text{ psi} = \text{yield stress in shear}$$

The behavior is HI-HO to the yield stress and then ideally plastic. The shaft is hollow; inner and outer diameters are

$$D_I = 1.00 \text{ in.}, \qquad D_O = 2.00 \text{ in.}$$

The shaft is 60 in. long and is subjected to opposing twisting couples at the ends.

a. Compute the total twist (relative rotation of the opposite ends) when yielding is initiated.

b. Calculate the torque T_O that initiates yielding.

c. Compute the total twist at the limit torque T_L.

d. Calculate the limit torque T_L.

5.112 A hollow steel shaft has inner and outer diameters of 2.00 cm and 3.00 cm. The shaft is 50 cm long and is subjected to opposing twisting couples upon the ends. Modulus, Poisson ratio, and yield stress (in tension-compression) are

$$E = 200 \times 10^3 \text{ MPa}, \quad \nu = 0.25, \quad \sigma_0 = \pm 320 \text{ MPa}$$

Yielding under any state of stress is attributed to the shear stress (Tresca's criterion).

a. Calculate the relative rotation of the ends when yielding is initiated.

b. If the steel is ideally plastic, calculate the torque T_O that initiates yielding and the limit torque T_L.

5.113 A hollow shaft is composed of ideally elastic-plastic material (hookean to the yield stress). Obtain the formula that expresses the ratio T_L/T_O (limit torque/initial-yield torque) in terms of the ratio $r = R_I/R_O$ (inner radius/outer radius). Examine your result for the limiting cases: $r \to 0$, $r \to 1$.

5.114 For the hollow shaft of Prob. 5.113, express the ratio $(d\phi/dx)_L/(d\phi/dx)_O$ (limit twist/initial-yield twist) in terms of $r = R_I/R_O$. Examine this result for the limiting cases: $r \to 0$, $r \to 1$.

5.115 Two hollow cylindrical shafts have the same dimensions:

$$\text{Outside diameter} = 1.00 \text{ in.}$$

$$\text{Inside diameter} = 0.50 \text{ in.}$$

Both are subject to simple torsion.

Shaft I is made of ductile steel. It is HI-HO to the onset of yielding in accordance with Tresca's criterion (maximum shear stress); properties are $\sigma_0 = \pm 48$ ksi (tension, compression).

Shaft II is made of brittle steel, which has the same HI-HO behavior to fracture in tension at an ultimate stress $\sigma_U = 48$ ksi.

Determine the ratio of the maximum torques that can be applied, $T_{\text{I}}/T_{\text{II}}$.

5.116 A solid shaft is composed of ideally elastic-plastic material (HI-HO to the yield stress τ_0). It is subjected to simple torsion until the elastic core has radius $R/2$. Compute each of the following:

a. The ratio $(d\phi/dx)/(d\phi/dx)_O$ (twist/twist at initial yielding)

b. The ratio T/T_L (torque/limit torque)

c. The permanent twist as a ratio $(d\phi/dx)_P/(d\phi/dx)_O$

d. The equations for the distribution of residual stress in terms of radii r and R and yield stress τ_0

5.117 Determine the relationship between the applied twisting couple T and the twist $(d\phi/dx)$ for a solid shaft composed of the linear hardening material:

$$\gamma_{x\theta} = \begin{cases} \dfrac{\tau_{x\theta}}{G}, & \tau \leq \tau_0 \\[2ex] \dfrac{1+k}{kG}\tau_{x\theta} - \dfrac{\tau_0}{kG}, & \tau \geq \tau_0 \end{cases}$$

5.118 The shaft of Prob. 5.117 is loaded by a twisting couple T, which causes yielding in a region $b < r < R$.

Follow the method of Sec. 5.9 and obtain the expression for the distribution of residual stress when the shaft is unloaded. Assume linearly elastic response during unloading (see Fig. 5.23a).

5.10 Torsion of Circular Shafts: Applications

Shafts with circular cross sections are used in machinery to transmit torsional loads. The simplest situation of this kind, a circular shaft subjected to twisting couples applied to its ends, is discussed in Sec. 5.8. We refer to this case as simple torsion and use it as a basis for the solution of other problems involving the torsion of shafts that have circular sections. Let us first recall the essential features and limitations of the simple torsion.

For a shaft composed of a HI-HO material, the twisting couples cause a shear stress on cross sections that is given by Eq. (5.73):

$$\tau = \frac{Tr}{J} \tag{5.99}$$

Also, the relative rotation of the ends is expressed by Eq. (5.75),

$$\phi = \frac{TL}{GJ} \tag{5.100}$$

The simple torsion of a shaft of ideally elastic-plastic material is described in Sec. 5.9. In particular, we note the onset of yielding and the existence of a limit load corresponding to a fully plastic condition.

The results for simple torsion can be applied to portions of a shaft that fit our description in regard to shape, material, and loading. However, the engineer must also appreciate the limitations of the theory: Our physical model is a shaft of circular cross section. Every cross section is presumed to rotate without distortion. Consequently, the only significant strain is the shear component $\gamma_{x\theta} = r(d\phi/dx)$, which varies linearly with the radius r. In the simplest circumstance, a HI-HO material, the one significant stress is, therefore, the shear component $\tau_{x\theta} = Tr/J$, which also varies linearly. In any real application, torque must be transmitted to the shaft via some mechanical connection to gears, pulleys, or levers. Typically, a shaft and gear are joined by splines, which mesh as depicted in Fig. 5.33. Then the torque is the resultant of the numerous forces (f) of contact between the faces of the spline. At the very *end* of the shaft, $\tau_{x\theta} = 0$. Only at some distance from the end does the stress approach the linear distribution. This is another manifestation of Saint-Venant's principle. If the design requires an assessment of the localized stress near the connections (splines or keyways), then more refined analyses are necessary. Our theory of simple torsion is applicable under the special circumstances set forth but *also* in *portions* of actual shafts. The theory gives an adequate description of the overall twist (rotation ϕ) and, by virtue of Saint-Venant's principle, the stress and strain at points sufficiently distant from any connections and any abrupt steps. Our description is then applicable within portions that are relatively slender ($R/L \ll 1$).

Fig. 5.33

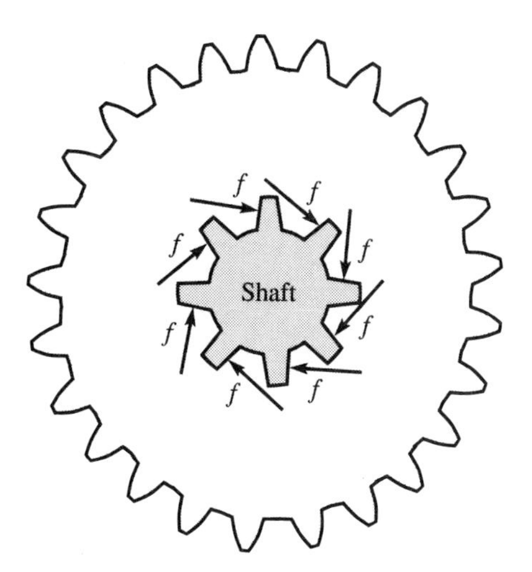

EXAMPLE 10

A Stepped Shaft The shaft of Fig. 5.34(a) consists of two cylindrical portions; part AC has a diameter d_1 and part CD has a diameter d_2. The shaft is clamped at A, subjected to twisting couples applied at locations B and D, and composed of HI-HO material (e.g., steel). Let us calculate the rotation of the end D and the shear stress on cross sections.

We begin by considering the equilibrium of the three parts: AB is part 1, BC is part 2, and CD is part 3. The free-body diagrams of these parts are shown in Fig. 5.34(b). A sectioning plane is placed just to the left of point B so that the couple T_1 is shown acting on part BC. (This was an arbitrary choice; the cut may equally well be to the right of point B, putting couple T_1 on the right end of AB. The net result would be unchanged.) For equilibrium,

$$T_C = T_2, \qquad T_A = T_B = T_1 + T_2 \tag{5.101}$$

Each of the parts in Fig. 5.34(b) is a relatively slender circular shaft subjected to twisting couples acting at its ends; that is, each is a case of simple torsion. Equation (5.100) applies to the HI-HO material; the rotation of end D relative to C (portion 3) is

$$\phi_3 = \frac{L_3 T_C}{GJ_2} = \frac{L_3 T_2}{GJ_2} \tag{5.102a}$$

where

$$J_2 = \frac{\pi}{32}(d_2)^4$$

Similarly, the rotations of C relative to B and B relative to A are

Fig. 5.34

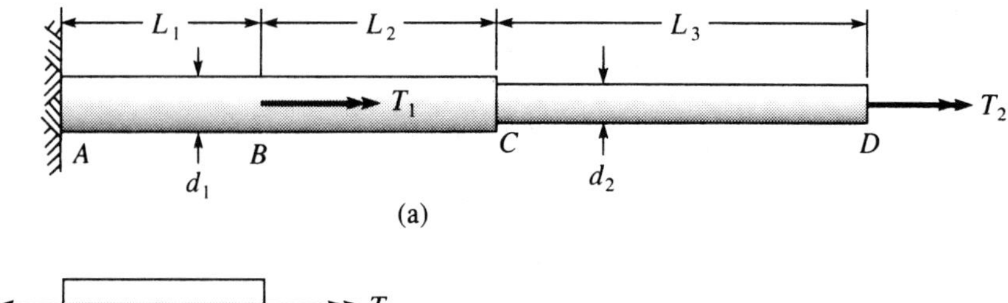

(a)

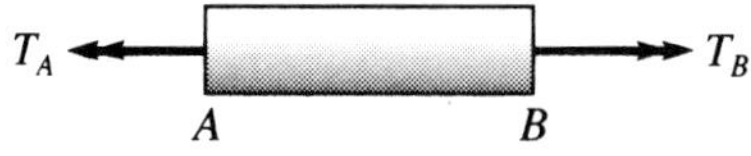

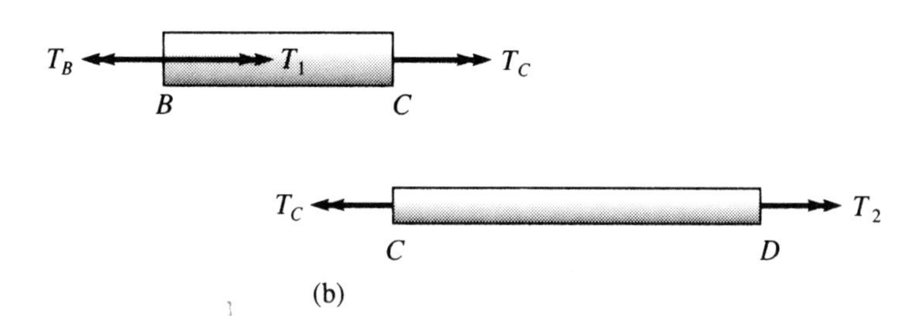

(b)

$$\phi_2 = \frac{L_2 T_C}{GJ_1} = \frac{L_2 T_2}{GJ_1} \tag{5.102b}$$

$$\phi_1 = \frac{L_1 T_B}{GJ_1} = \frac{L_1(T_1 + T_2)}{GJ_1} \tag{5.102c}$$

where

$$J_1 = \frac{\pi}{32}(d_1)^4$$

The rotation ϕ of the right end is simply the sum

$$\begin{aligned} \phi &= \phi_1 + \phi_2 + \phi_3 \\ &= \frac{L_1}{GJ_1} T_1 + \left(\frac{L_1 + L_2}{GJ_1} + \frac{L_3}{GJ_2} \right) T_2 \end{aligned} \tag{5.103}$$

Also, the shear stress on cross sections of each part can be calculated by Eq. (5.99):

$$\begin{aligned} \tau_1 &= \frac{T_1 + T_2}{J_1} r, \qquad 0 \le r \le \frac{d_1}{2} \\ \tau_2 &= \frac{T_2}{J_1} r, \qquad 0 \le r \le \frac{d_1}{2} \\ \tau_3 &= \frac{T_2}{J_2} r, \qquad 0 \le r \le \frac{d_2}{2} \end{aligned} \tag{5.104}$$

In each portion the maximum shear stress occurs at the outer surface, where $r = d_1/2$ or $r = d_2/2$. Of course, Eq. (5.104) does not apply for cross sections near A, B, C, or D. At these locations the localized stresses may be much different than those given by Eq. (5.104).

Again, the complete solution requires the *statics*, the *material properties* embodied in (5.100), and the *kinematics*. ◆

EXAMPLE 11

Statically Indeterminate Consider the shaft described in Example 10, but suppose that it is clamped at both ends, A and D, prior to the application of the torque T_1. Most of the previous relations apply. Certainly the free-body diagrams of Fig. 5.34 and the associated equations of *equilibrium*,

$$T_C = T_2, \qquad T_A = T_1 + T_2 \tag{5.105}$$

apply. Couple T_2 is an unknown, which is imposed by the clamp at D to prevent rotation of that end.

Since this is the same shaft, Eqs. (5.102), which embody the *HI-HO relations*, remain applicable; that is,

$$\phi_3 = \frac{L_3 T_2}{GJ_2}, \qquad \phi_2 = \frac{L_2 T_2}{GJ_1} \tag{5.106a,b}$$

$$\phi_1 = \frac{L_1(T_1 + T_2)}{GJ_1} \tag{5.106c}$$

Indeed, *only* the *kinematic* equation (5.103) is changed: Since the end D is clamped,

$$\phi = \phi_1 + \phi_2 + \phi_3 = 0 \tag{5.107a}$$

$$0 = \frac{L_1}{GJ_1} T_1 + \left(\frac{L_1 + L_2}{GJ_1} + \frac{L_3}{GJ_2} \right) T_2 \tag{5.107b}$$

The couple T_2, imposed by the fixture on the end D, is expressed in terms of the applied torque T_1 by Eq. (5.107b). To obtain some appreciation of the magnitude, let us assume that the shaft is steel with the following lengths and diameters and that the applied torque is

$$T_1 = 2400 \text{ in.-lb } (270 \text{ N}\cdot\text{m})$$

$$G = 12 \times 10^6 \text{ lb/in.}^2 \text{ (83 GPa)}$$

$$L_1 = L_2 = L_3 = 20.00 \text{ in. (51 cm)}$$

$$d_1 = 1.000 \text{ in. (2.54 cm)}, \qquad d_2 = 0.750 \text{ in. (1.91 cm)}$$

The maximum rotation occurs at section B, where

$$\phi_B = 0.0329 \text{ (rad)}, \quad \text{or} \quad 1.9^\circ$$

The maximum stress occurs in the first section AB, where $\tau_{max} = 9900$ lb/in^2. The reader can imagine two plumbers, each with 12-in. pipe wrenches, applying opposing forces of 100 lb. (They are strong, experienced plumbers.) ◆

EXAMPLE 12

Shafts Transmitting Power Since most shafts are used to deliver power, it is useful to relate the shear stress directly to the power transmitted. To obtain such a relation we consider a shaft that is rotating at a constant angular velocity ω and is supplying power to a machine. This power is transmitted to the shaft by means of a torque applied at one end and is delivered by the shaft at the other end. Consequently, there is a torque T tending to twist the shaft between its ends.

During time interval Δt, the shaft rotates through the angle $\Delta\phi = \omega\,\Delta t$. The work done by the torque as the shaft rotates through this angle is $T\,\Delta\phi = T\omega\,\Delta t$. Therefore, the rate at which this work is done is $T\omega$. But this rate of doing work is also the power that the shaft transmits. Suppose the shaft delivers H horsepower. If T is measured in inch-pounds and ω is measured in radians per second, the units of $T\omega$ are inch-pounds per second. Because 1 hp is 6600 lb-in./s, when we equate the power transmitted to the rate at which work is done, we get

$$H = \frac{T\omega}{6600}$$

or

$$T = \frac{6600H}{\omega}$$

Then, if the shaft material is linearly elastic, the shear stress caused by the torque is

$$\tau = \frac{Tr}{J} = \frac{6600Hr}{J\omega} \tag{5.108}$$

where τ is in pounds per square inch when H is in horsepower, ω is in radians per second, r is in inches, and J is in inches to the fourth power.

The corresponding formulas in terms of the SI units of power (W in watts) and torque (T in Newton-meters) follow:

$$T = \frac{W}{\omega}$$

$$\tau = \frac{Wr}{J\omega}$$ ◆

PROBLEMS

Assume HI-HO behavior unless otherwise stated and employ the analysis of simple torsion; that is, discount local effects.

5.119 A steel shaft is 30 in. long and 1.50 in. in diameter. When twisting couples T are applied to its ends, the shear strain $\tau_{x\theta}$ at the surface is 0.0012. Determine the maximum shear stress $\tau_{x\theta}$ and the angle of twist of the shaft.

5.120 A steel shaft is 5 cm in diameter and 300 cm long. It is twisted by end couples of 700 Nm. What is the maximum tensile stress in the shaft?

5.121 Measurements on a bronze shaft 2.50 in. in diameter and 10 ft long show that the total angle of twist of the shaft is 0.040 rad. Compute the maximum shear stress $\tau_{x\theta}$ in the shaft.

5.122 A solid cast-iron drive shaft is 200 cm long and 10 cm in diameter. Determine the angle of twist if the maximum shear stress is 140 MPa.

5.123 A hollow shaft of length L has an inner radius ρ and an outer radius R. A solid shaft of the same material has a radius r. If the cross-sectional areas of the shafts are equal and the maximum shear stress is the same in both, show that

$$\frac{T_S}{T_H} = \frac{\sqrt{1-K}}{1+K}, \qquad K = \frac{\rho^2}{R^2}$$

where T_S and T_H are the twisting couples applied to the solid and hollow shafts, respectively.

5.124 An oil well drill is 3000 ft long. It is a solid steel rod $\frac{3}{4}$ in. in diameter. If the maximum permissible shear stress is 6500 lb/in.^{2}, compute the largest permissible value of the applied couple T. At this stress level what is the angle of twist of the rod?

5.125 For the hollow steel shaft shown in Fig. P5.125, find the maximum possible twisting couple T_1 if it is required that $\tau_{x\theta} \leq 80$ MPa and that the total angle of twist is less than or equal to 0.020 rad.

Fig. P5.125

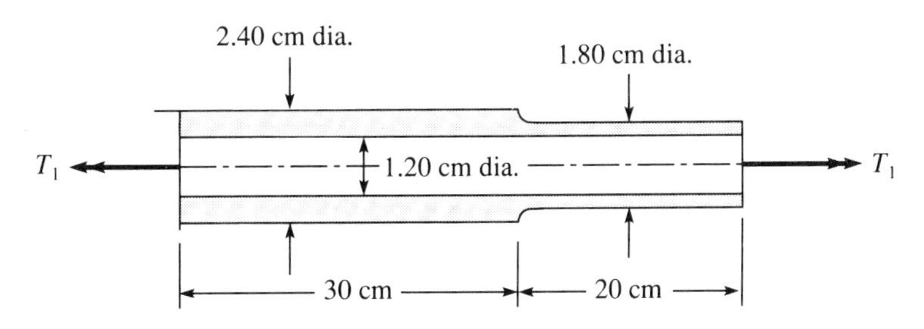

5.126 A solid shaft 2 in. in diameter is made of a brittle material that fractures when the maximum tensile stress reaches a value of 10,000 lb/in^2. What is the maximum twisting couple T_U that can be applied to the shaft?

5.127 A hollow brass shaft is twisted by opposing couples $T = 25$ N·m applied to the ends. Inner and outer diameters are 0.96 cm and 1.60 cm, respectively. Compute the maximum extensional strain.

5.128 A steel shaft is made by welding the hollow tube AB to the solid shaft BC at B. The inside diameter of AB is 1.80 in. The properties of this steel are $E = 30 \times 10^3$ ksi, $\nu = 0.25$. Couples T_1 and T_2 are applied at B and C, respectively, as shown in Fig. P5.128. The resulting rotations at B and C are $\theta_B = 0.020$ rad and $\phi_C = 0$. Determine each of the following.

a. The maximum shear strain

b. The maximum extensional strain

c. The couple T_2

d. The couple T_1

Fig. P5.128

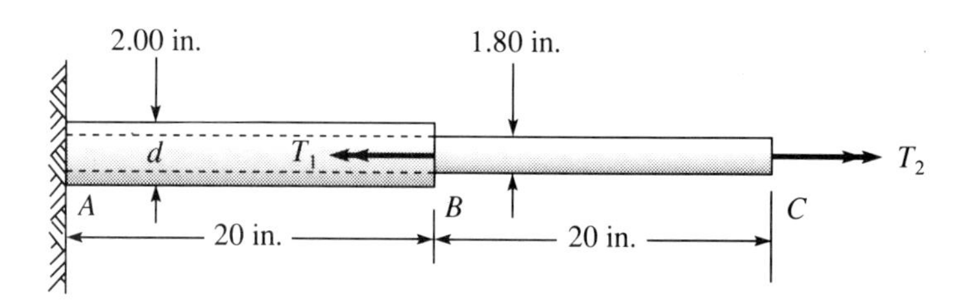

5.129 A hollow aluminum shaft has a 6-cm outside diameter and a 2-cm wall thickness. The material is ideally plastic ($\tau_0 = 200$ MPa). Compute the torque T that causes yielding through one half of the wall thickness.

5.130 When the couples T_1 and T_2 are applied to the shaft shown in Fig. P5.130, the right end undergoes no rotation. What is the relation between T_1 and T_2 in terms of k_1 and k_2?

Fig. P5.130

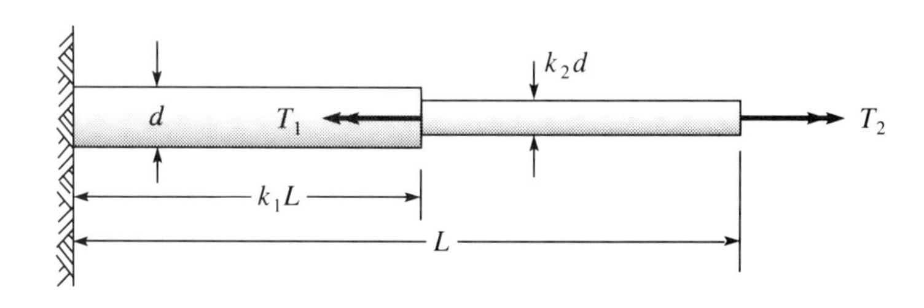

5.131 Determine (a) the shear stress at the outer surface in each portion of the stepped steel shaft shown in Fig. P5.131 and (b) the total angle of twist.

Fig. P5.131

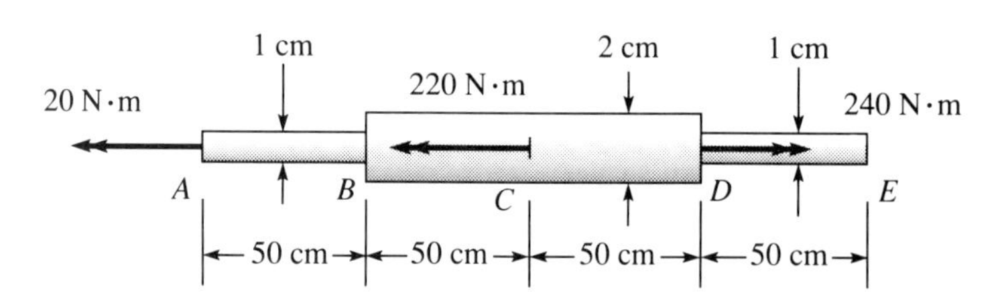

5.132 Two steel shafts are geared together, as shown in Fig. P5.132. Determine the rotation of gear A when the twisting couple $T_A = 200$ in.-lb is applied.

Fig. P5.132

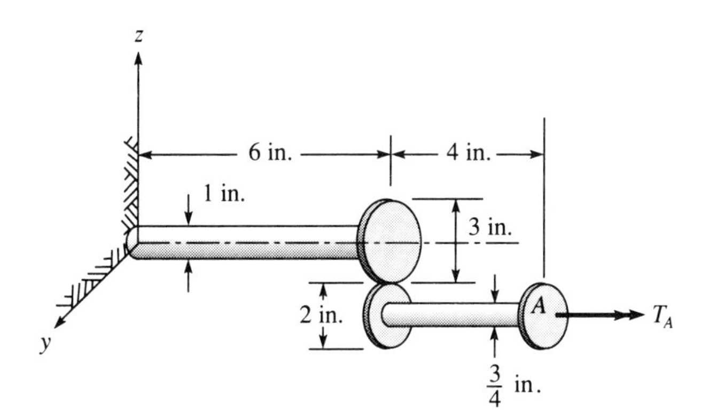

5.133 For the system of Prob. 5.132, determine the maximum normal stress in each shaft.

5.134 Determine the relation between the twisting couple T_1 and the rotation of cross section B if T_1 is applied to the circular shaft as shown at B in Fig. P5.134.

Fig. P5.134

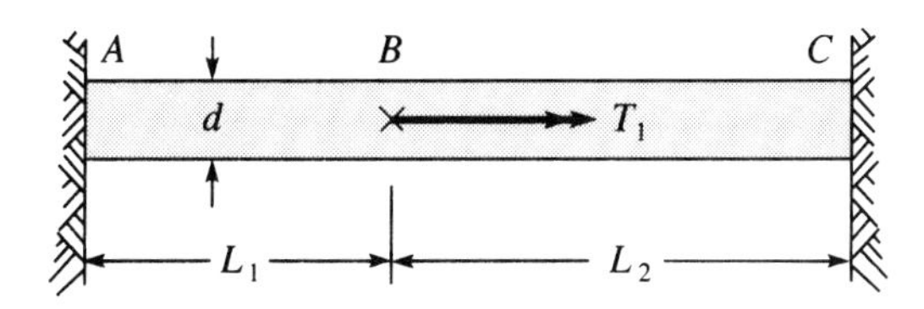

5.135 An aluminum shaft, 6.0 cm in diameter and 60 cm long, is fixed at both ends and has a twisting couple T_1 applied at its center. What is the angle of twist of the central cross section when the maximum shear stress in the shaft is 80 MPa?

5.136 The steel shaft in Fig. P5.136 is made by welding the hollow tube AB to the solid shaft BC at B. Both ends are clamped before the couple T_1 is applied at the juncture B. This couple causes a rotation at B: $\phi_B = 0.0200$ rad. Calculate (a) the maximum shear strain, (b) the maximum extensional strain, and (c) the couple exerted by the clamp at C.

Fig. P5.136

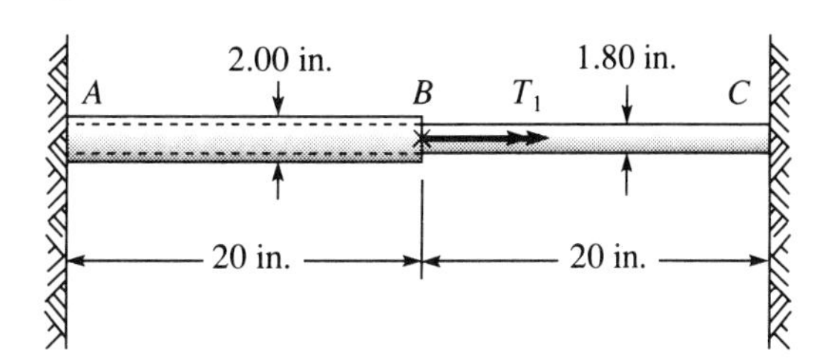

5.137 Shafts AB and BC are made of different materials and are unstressed when welded together at B, as shown in Fig. P5.137. The twisting couples at A and C are equal in magnitude when the couple T_1 is applied at B. Determine the relationship between the elastic moduli of the two materials.

Fig. P5.137

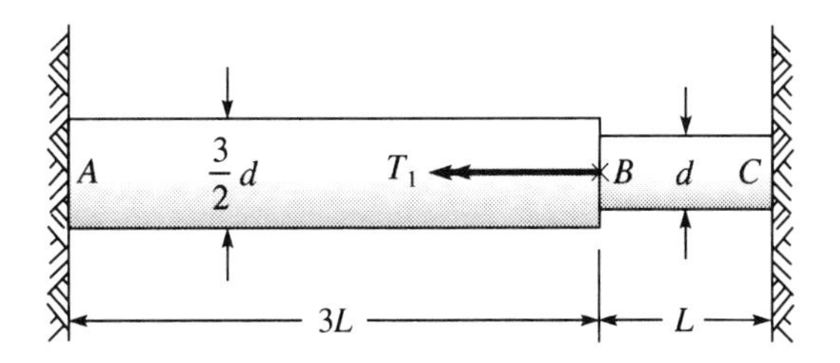

5.138 An aluminum shaft is rigidly restrained at each end and loaded by couples, as shown in Fig. P5.138. If $T_1 = 1200$ N·m and $T_2 = 1000$ N·m, compute the maximum shear stress in each section and the rotations of cross sections at B and C.

Fig. P5.138

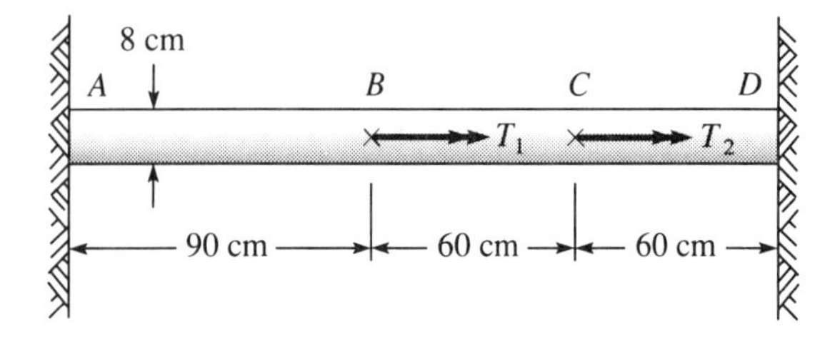

5.139 Solid brass and steel shafts are welded together, as shown in Fig. P5.139. What is the maximum shear stress in the steel part when the couple $T_1 =$ 100 in.-lb?

Fig. P5.139

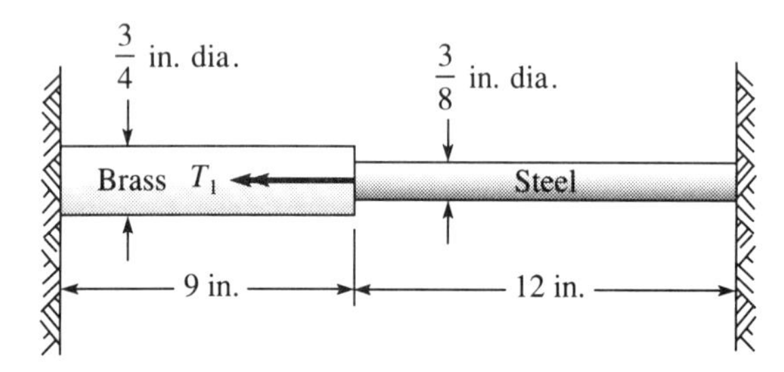

5.140 If the materials of the shaft in Prob. 5.139 are ideally plastic ($\tau_{0s} = 40{,}000$ lb/in.^{2}, $\tau_{0b} = 20{,}000$ lb/in.^{2}), what is the largest possible value for the applied torque?

5.141 A circular cylinder is fabricated with the composite cross section shown in Fig. P5.141. The end cross sections rotate as rigid units. What part of the couple T is carried by each cylindrical layer if the inner core is brass, the middle layer is aluminum, and the outer is steel?

Fig. P5.141

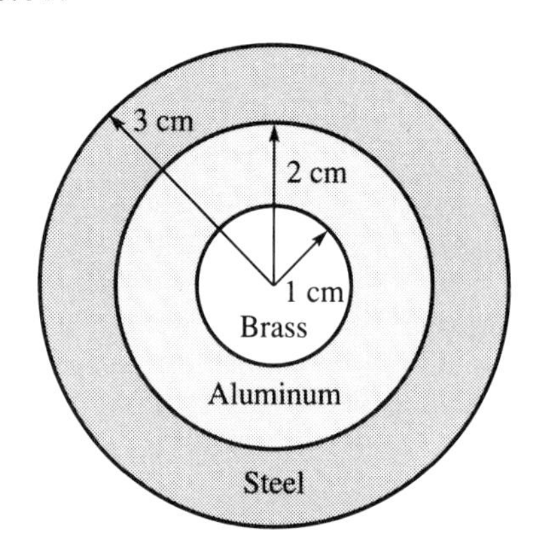

5.142 The steel shaft shown in Fig. P5.142 consists of an ideally plastic material ($\tau_0 = 200$ MPa). The couple T_1 causes yielding of the central portion to a depth equal to one half the radius. Determine T_1.

Fig. P5.142

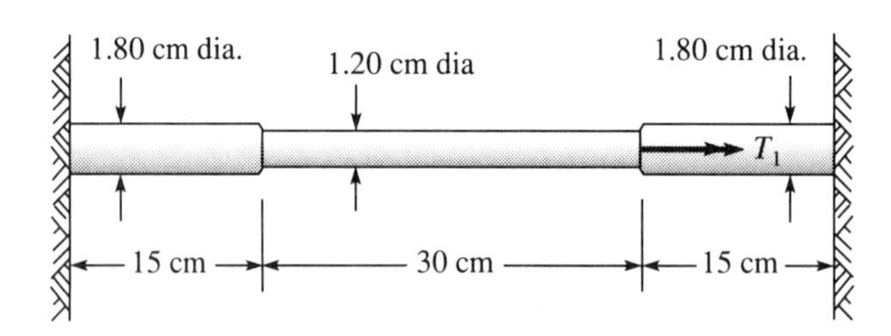

5.143 The shaft AB slips into the end of the hollow shaft BC, as shown in Fig. P5.143. A wrench is used to apply a twisting couple $T = 1000$ in.-lb at the end of the hollow shaft. The two are locked by a set screw and then the couple is removed. Compute the maximum residual shear stress.

Fig. P5.143

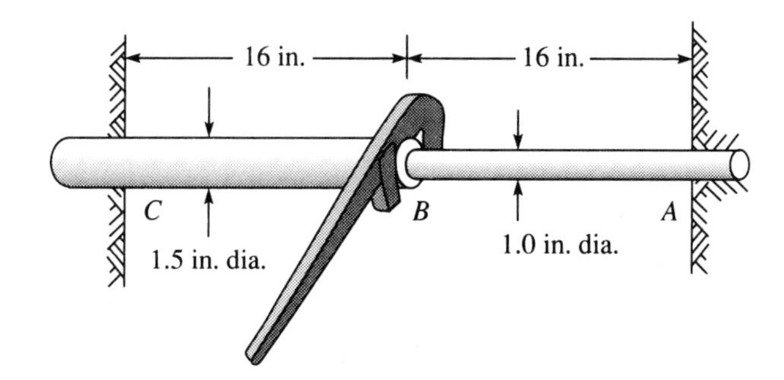

5.144 When gear A is keyed to the shaft AB, as shown in Fig. P5.144, the end of the shaft B must be rotated through a 2° angle in order that the gears will mesh. Compute the residual stresses in each shaft after the gear is installed and released. Both shafts are steel. (Bearings, not shown, prevent bending.)

Fig. P5.144

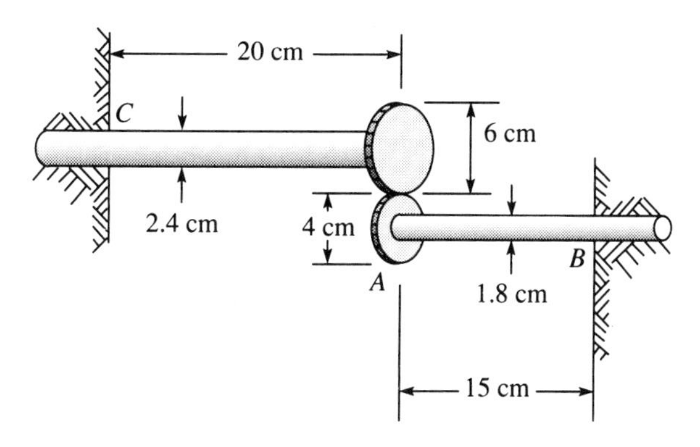

5.145 A tubular steel shaft, shown in Fig. P5.145, is ideally plastic ($\tau_0 = 30{,}000$ lb/in.^{2}) and has a mean diameter of 3 in. and a wall thickness of 0.10 in. The shaft is rigidly restrained at both ends. Determine the relationship between the couple T_1 and the rotation of section B.

Fig. P5.145

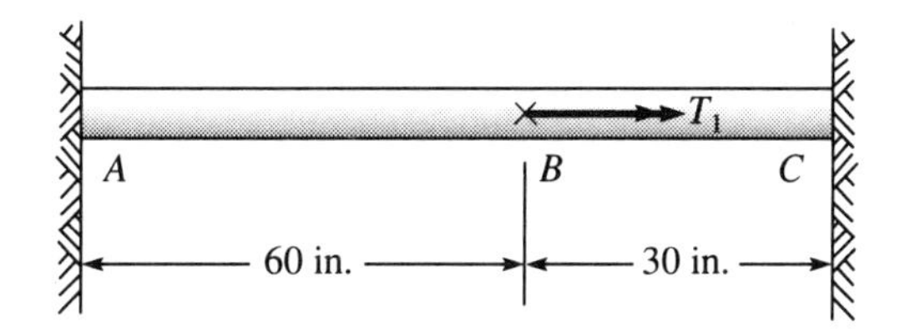

5.146 The steel shaft illustrated in Fig. P5.146 is ideally plastic, $\tau_0 = 200$ MPa. Determine (a) the relation between couple T_1 and the rotation of section B when $\tau < \tau_0$, (b) the value of T_1, when yielding is initiated in section BC, and (c) the limit couple T_L corresponding to the fully plastic condition.

Fig. P5.146

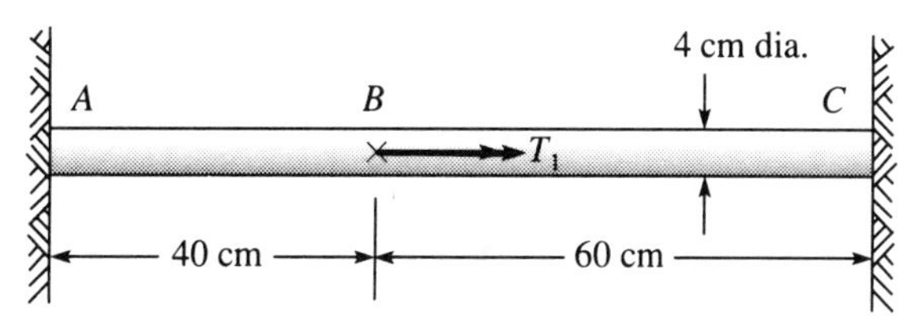

5.147 The shaft shown in Fig. P5.147 is subjected to a twisting couple T_1 at B, which is just sufficient to initiate yielding in part AB. The entire shaft is made of hot-rolled steel. (Assume it is ideally plastic with $\sigma_0 = 40{,}000$ lb/in.^{2}). Use Tresca's yield condition and determine the residual stresses when couple T_1 is removed.

Fig. P5.147

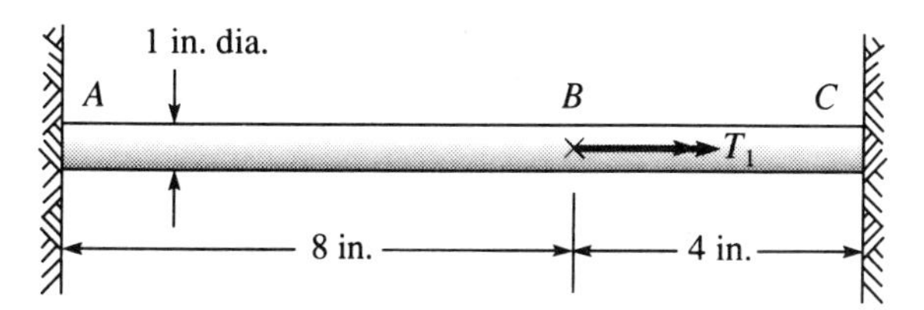

5.148 The solid tapered shaft shown in Fig. P5.148 has a radius given by

$$R = R_0\left(1 - k\frac{x}{L}\right), \qquad k \ll 1$$

Determine the relation between T_1 and the angle of twist at the section where T_1 acts.

Fig. P5.148

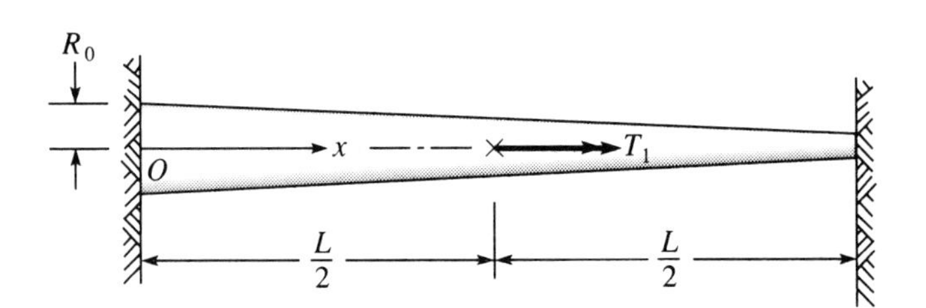

5.149 The shaft illustrated in Fig. P5.149 is made of material with a nonlinear stress-strain relation of the form $\tau = G\gamma^{1/2}$. If the taper is given by

$$R = R_0\left(1 - k\frac{x}{L}\right), \qquad k \ll 1$$

relate T_1 to the rotation of the end $x = L$.

Fig. P5.149

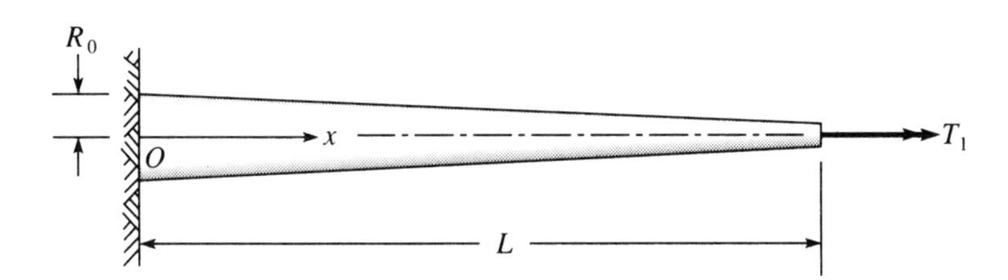

5.150 A uniform circular shaft of radius r and length L is forced into a hole in a rigid block, as shown in Fig. P5.150. The pressure on the lateral surface of the shaft is p and the coefficient of friction between the wall and the cylinder is μ. Compute the angle of twist when the bottom end is on the verge of slipping.

Fig. P5.150

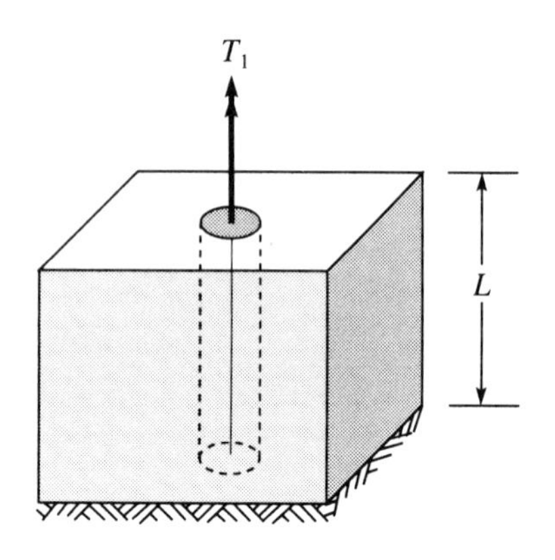

5.151 If the shaft of Prob. 5.134 is ideally plastic, determine the relation between T_1 and the rotation θ of the section at which T_1 acts. Take $L_1 = L_2 = L/2$.

5.152 A composite shaft has a core of radius 4 cm, which is made of material with a shear modulus of 70×10^3 MPa. The outer portion has an inner radius of 4 cm, has an outer radius of 6 cm, and is composed of material with a shear modulus of 35×10^3 MPa. These two portions are bonded together. If the shaft is 20 cm long, determine the total angle of twist when the shaft is subjected to a twisting couple of 2000 N · m.

5.153 A slightly tapered shaft is composed of an ideally plastic material. If the radius is given by

$$R = R_0\left(1 - k\frac{x}{L}\right), \qquad k \ll 1$$

show that yielding is initiated in the cylinder when

$$T_0 = \frac{\pi\tau_0 R_0^3}{2}(1 - k)^3$$

Compute the total angle of twist when the couple attains the value T_0.

5.154 A thin-walled, slightly tapered tube has a mean diameter that varies according to

$$D = D_0\left(1 - \frac{x}{2L}\right)$$

where x is the distance from one end and L is the total length. The wall thickness is a constant t, $t \ll D$. Determine the angle of twist when the tube is subjected to a twisting couple T applied to its ends. If the yield stress (in shear) for the material is τ_0, compute the value of the twisting couple that initiates yielding in the tube.

5.155 A motor delivers k horsepower to a hollow steel shaft rotating at 1750 rev/min. The outside diameter of the shaft is 2 in. and its wall thickness is $\frac{1}{4}$ in. At what horsepower is yielding initiated if $\tau_0 =$ 20,000 lb/in.^{2}?

5.156 A solid steel shaft rotates at 8000 rev/min and delivers 750×10^3 W. If the maximum shear stress is not to exceed 55 MPa, compute the minimum possible diameter for the shaft.

5.157 What is the angle of twist of a 2.00-cm-diameter solid steel shaft 1 m long, which is delivering 11×10^3 watts (W) at 1750 rpm?

5.158 A cast-iron drive shaft is 60 in. long and is slightly tapered. The radius is given by $R = 4(1 - x/240)$ in., where x is the distance from one end in inches. What is the angle of twist of the shaft when it delivers 237 hp at 1000 rev/min?

5.11 Bending of Beams

The Problem: The Symmetric Beam

A slender member subjected to transverse loads is called a *beam*. However, depending on the shape of the cross section and the nature of the loads and supports, twisting can accompany bending. At present we consider only situations in which the beam initially is a straight prismatic member that has a longitudinal plane of symmetry, both geometrically and materially. We also restrict our attention to beams that are loaded and supported symmetrically with respect to this longitudinal plane of symmetry (abbreviated LPS). These restrictions eliminate the possibility of the beam twisting as it bends.

If the loads (in the LPS) are perpendicular to the axis of the beam, the normal-force component on a cross section is zero. Then the net internal force system consists of a shear force and a bending couple, as indicated in Fig. 5.35. The portion of the beam to one side of a cross section S is shown with a coordinate system established at one end. The cross section is located by its distance x from the end of the beam. If the xy-plane is the longitudinal plane of symmetry, the only components of the net internal force system on S that are not identically zero are V_y and M_z. We therefore drop the subscripts and write V for V_y and M for M_z.

Fig. 5.35

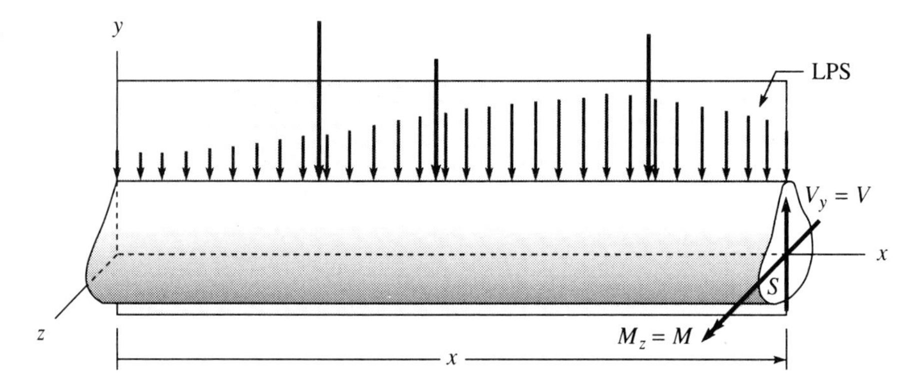

Simple Bending

Let us consider a particular situation. The beam of Fig. 5.36 is held fixed at the left end and is subjected to a bending couple M applied to the right end. The couple M acts in the LPS. The coordinate system is located so that the xy-plane is coincident with the LPS and the x-axis is parallel to the axis of the member. A cross section S is located by its distance x from the left end.

Fig. 5.36

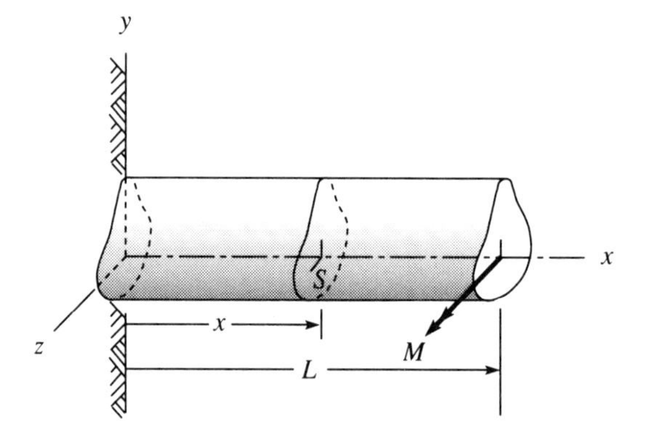

A free-body diagram of the portion to the right of cross section S is shown in Fig. 5.37. From equilibrium considerations, the only nonzero component of the net internal force system on S is the bending couple M.

Fig. 5.37

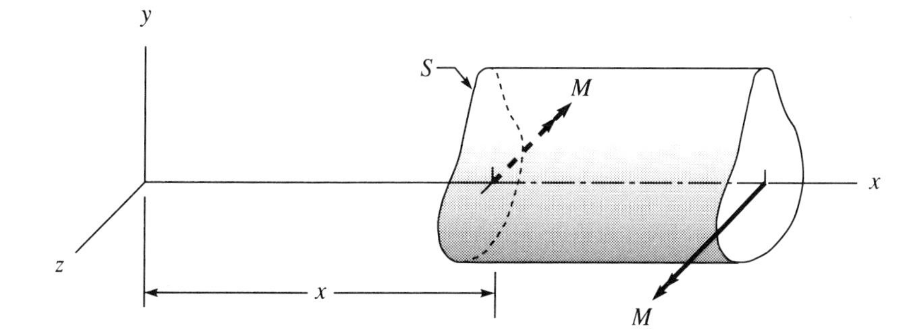

Description of the Deformation

Because the beam and the loading are symmetric with respect to the LPS, we expect the deformed beam to be symmetric with respect to the LPS. The effect of the bending couple is to compress the upper portions and stretch the lower portions of the beam (assuming the sense of M shown in Fig. 5.37). This description suggests that some axial line in the LPS experiences no extension but deforms into a plane curve in the LPS.

To characterize the main features of the deformation, we assume the following:

1. There is a line of particles in the LPS parallel to the axis of the beam that experiences no extensional strain and deforms into a plane curve in the LPS. This line is called the *neutral axis* of the beam.

2. Every cross section remains plane and perpendicular to the neutral axis after deformation.[3]

In Fig. 5.38 the x-axis is placed along the neutral axis of the beam. The position of the x-axis in the cross section is yet to be determined.

Consider the extension of an axial line element PQ. Point P lies in a cross section S located at a distance x from the left end, and point Q lies in the nearby cross section at $x + \Delta x$. The slice between these cross sections is shown before and after deformation in the profile view of Fig. 5.39.

The y-coordinate is the initial distance to particle P. After the deformation, this distance is

3 We can offer logical arguments to justify assumption 2, the so-called Bernoulli hypothesis. We observe that every cross section is subject to the same action, the couple M, so that every slice is expected to deform similarly. Indeed, under this uniform bending, the x-axis must be deformed to a circular arc. The beam of Fig. 5.36 could also be extended to $x = -L$. Then such an extended beam would be entirely equivalent and symmetric about the yz-plane; opposing couples at the ends $x = \pm L$ would preserve that symmetry. Then left and right halves would be reflections about the midsection ($x = 0$). It follows that the midsection remains plane and perpendicular to the bent axis. But, because of the uniformity of properties and loading, it is plausible to assume that all cross sections remain plane and perpendicular to the bent axis, a circular arc.

Fig. 5.38

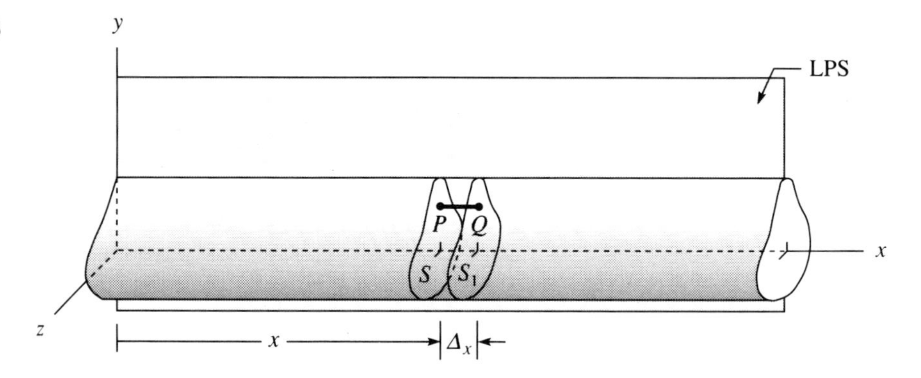

Fig. 5.39

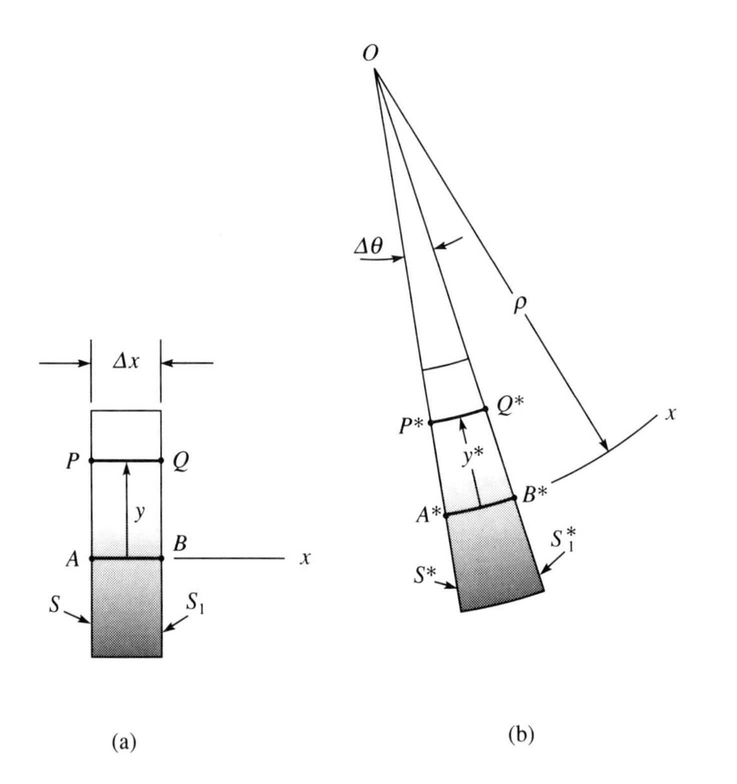

$$y^* = (1 + \bar{\varepsilon}_y)y$$

where $\bar{\varepsilon}_y$ denotes the mean value of the extensional strain. Point O denotes the center of curvature of the neutral axis at A^*, and ρ denotes the radius of curvature at A^*. Because AB is a part of the neutral axis, its length is unchanged; consequently

$$PQ = AB = A^*B^* = \Delta x = \rho\,\Delta\theta$$

However, the length of the deformed line element P^*Q^* is

$$P^*Q^* = (\rho - y^*)\,\Delta\theta$$

Then, from the definition of extensional strain, we have

$$\begin{aligned}\varepsilon_x(P) &= \lim_{Q\to P} \frac{P^*Q^* - PQ}{PQ} \\ &= \lim_{\Delta\theta\to 0} \frac{(\rho - y^*)\,\Delta\theta - \rho\,\Delta\theta}{\rho\,\Delta\theta} \\ &= -\frac{y^*}{\rho} = -(1 + \bar{\varepsilon}_y)\frac{y}{\rho}\end{aligned} \tag{5.109}$$

In most practical problems $\bar{\varepsilon}_y \ll 1$, so that we can neglect the extension strain compared to unity ($\bar{\varepsilon}_y \ll 1$) and employ the approximation

$$\varepsilon_x(P) \doteq -\frac{y}{\rho} \tag{5.110}$$

Equation (5.110) is the basic result of the *kinematic* analysis. The result presumes symmetry with respect to the xy- (LPS) plane; otherwise it is independent of the material.

Statics

We must anticipate a normal stress σ_x. In general, that stress produces a normal force ΔF on an elemental area ΔA, as depicted in Fig. 5.40:

$$\Delta F = \sigma_x\,\Delta A$$

Fig. 5.40

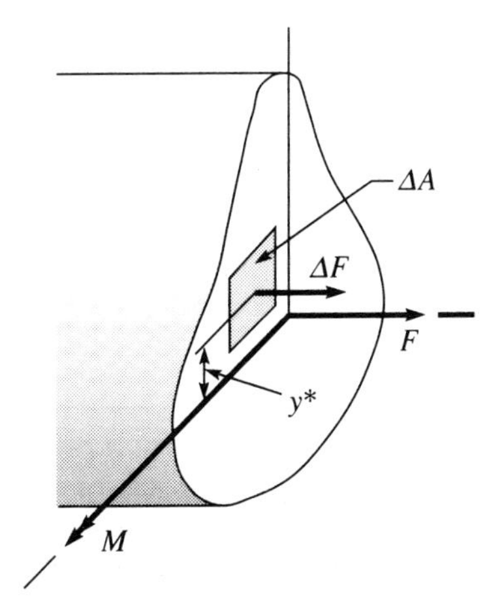

This contributes to the couple

$$\Delta M = -y^* \Delta F = -y^* \sigma_x \Delta A$$

Summing the contributions and passing to the limit, we obtain the normal force and couple:

$$F = \iint_S \sigma_x \, dA \tag{5.111}$$

$$M = -\iint_S \sigma_x y^* \, dA \doteq -\iint \sigma_x y \, dA \tag{5.112}$$

The final approximation is consistent with Eq. (5.110), wherein $y^* = (1 + \bar{\varepsilon}_y)y \doteq y$.

Equations (5.111) and (5.112) are also basic results that are independent of the material.

5.12 Hookean Isotropic Beam

Suppose that the beam is composed of hookean isotropic material. Then

$$\begin{aligned} \varepsilon_x &= \frac{1}{E}[\sigma_x - \nu(\sigma_y + \sigma_z)] \\ \varepsilon_y &= \frac{1}{E}[\sigma_y - \nu(\sigma_z + \sigma_x)] \\ \varepsilon_z &= \frac{1}{E}[\sigma_z - \nu(\sigma_x + \sigma_y)] \end{aligned} \tag{5.113}$$

From the foregoing description of the kinematics, only ε_x is described explicitly, and only σ_x is anticipated. Some additional assumptions are needed.

If we bend a deformable strip, transverse extensions and contractions are observed. The reader can do this with a piece or rubber, though not hookean. The upper portion, compressed by the couple, extends in the z-direction ($\varepsilon_z > 0$), whereas the lower contracts ($\varepsilon_z < 0$). On the other hand, the absence of any stresses on the lateral surfaces suggest that the components σ_y and σ_z are negligible, if not zero. Then

$$\varepsilon_x = \frac{\sigma_x}{E} \tag{5.114}$$

$$\varepsilon_y = \varepsilon_z = -\nu \frac{\sigma_x}{E} \tag{5.115}$$

The kinematics, the statics, and the description of the material now provide the basis for a solution. If we substitute Eq. (5.110) into the left side of (5.114) and use that result in the integrals (5.111) and (5.112), we obtain

$$F = -\frac{1}{\rho}\iint_S Ey\,dA \tag{5.116}$$

$$M = \frac{1}{\rho}\iint_S Ey^2\,dA \tag{5.117}$$

Case I: Doubly Symmetric Cross Section

Suppose that our beam is symmetric with respect to the xz-plane as well as the xy-plane. The I-section of Fig. 5.41(a) is a common example. If the property E is an even function of y (and z), then the integral (5.116) vanishes *if* the x-axis is at the center of the section; stated otherwise, the neutral (unstretched) axis is the central line. Note that the modulus E might be variable, but $E(y)$ must be an even function. Indeed, all the foregoing assumptions and equations would apply to the laminated beam of Fig. 5.41(b), provided that the lamina are symmetrically positioned.

Fig. 5.41

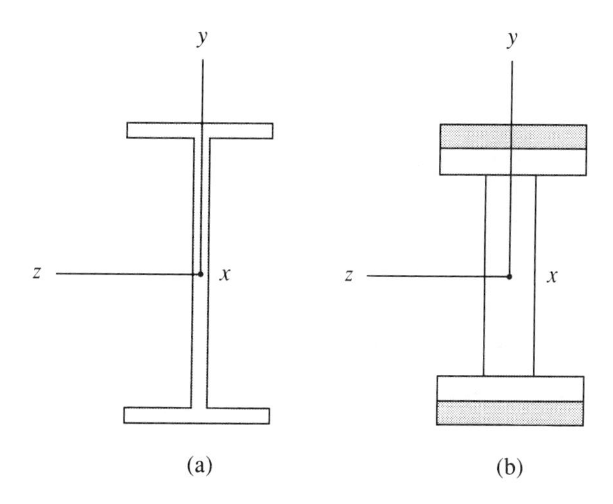

Case II: Hookean, Isotropic, and Homogeneous Beam

Suppose now that the beam is composed of an HI-HO material, that is, E is constant. This is essentially the case for most metals within the yield condition.

Then, according to Eq. (5.116), $F = 0$ if

$$\iint_S y\,dA = 0 \tag{5.118}$$

In other words, the coordinate y must be measured from the centroid of the cross section. The neutral axis (x) pierces the centroid of the section.

The couple M is given by Eq. (5.117):

$$M = \frac{E}{\rho} \iint_S y^2 \, dA$$

Following usual conventions, we denote the integral by I:

$$I \equiv \iint_S y^2 \, dA$$

Then, the curvature and bending couple are linearly related:

$$\kappa \equiv \frac{1}{\rho} = \frac{M}{EI} \tag{5.119}$$

Equation (5.119) is a fundamental relation in elementary beam theory, sometimes called the *Bernoulli-Euler curvature formula.* It relates the bending couple on a cross section to the curvature of the neutral axis at that location if the material is hookean, isotropic, and homogeneous (HI-HO).[4]

The extensional strain ε_x is expressed in terms of the couple by means of Eqs. (5.110) and (5.119):

$$\varepsilon_x = -\frac{My}{EI} \tag{5.120}$$

The stress σ_x is then given by Eq. (5.114):

$$\sigma_x = -\frac{My}{I} \tag{5.121}$$

This linear distribution of normal stress is depicted in Fig. 5.42.

Fig. 5.42

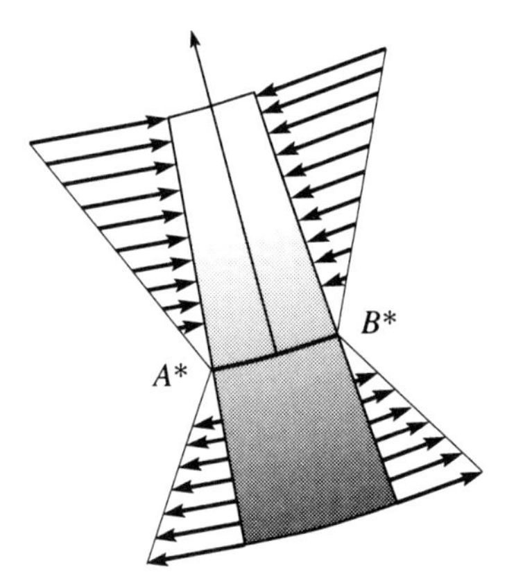

4 If the material is hookean (linearly elastic) but not isotropic and homogeneous, curvature and moment are proportional, *but* the constant of proportionality is not that (EI) of Eq. (5.119).

The strain and stress distributions of Eqs. (5.120) and (5.121) apply to all cross sections of the beam. In particular, this includes the end faces where the load is applied and where the beam is supported. Fortunately, Saint-Venant's principle enables us to use these results in many other cases if the beam is slender. However, if the end couples are applied in accordance with Eq. (5.121), the results obtained here are in agreement with the solutions of more elaborate theories. In fact, it can then be shown that, for a linearly elastic material, $\sigma_y = \sigma_z = \tau_{xy} = \tau_{xz} = \tau_{yz} = 0$. Then $\gamma_{xy} = \gamma_{yz} = \gamma_{xz} = 0$, and, by Hooke's law, $\varepsilon_y = \varepsilon_z = -\nu\varepsilon_x = \nu My/EI$.

Other Materials

The significant features of the bending deformation of straight beams loaded in a longitudinal plane of symmetry (LPS) do not depend on the material. We still assume the existence of a neutral axis that experiences no extension strain (although it need not be the centroidal axis). Furthermore, cross sections are assumed to remain plane and perpendicular to the neutral axis. Under these conditions, Eq. (5.110) holds for all materials.

EXAMPLE 13

Simple Bending of a HI-HO Beam Suppose that the beam has the T-shaped cross section shown in Fig. 5.43, the material is HI-HO, and the beam is subjected to a simple bending couple. Then the net force vanishes and, in according with Eqs. (5.116) and (5.118),

Fig. 5.43

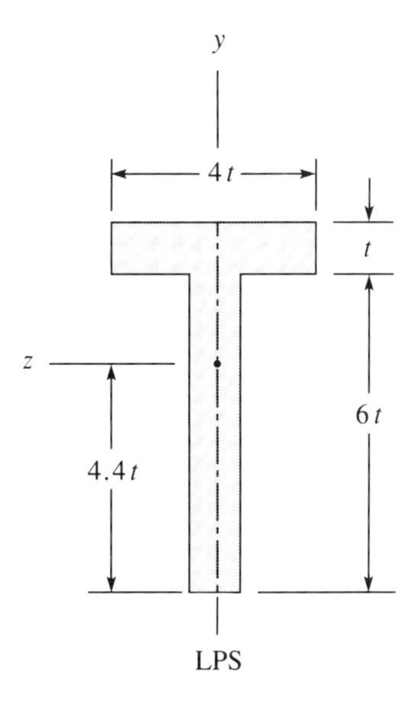

$$\iint_S y\,dA = 0$$

The neutral line is at the centroid, a distance $4.4t$ from the bottom. The bending couple M is related to the curvature in accordance with Eq. (5.117):

$$M = \frac{E}{\rho}\iint_S y^2\,dA = \frac{EI}{\rho}$$

where

$$\begin{aligned} I &= \iint_S y^2\,dA \\ &= t\int_{-4.4t}^{1.6t} y^2\,dy + 4t\int_{1.6t}^{2.6t} y^2\,dy \\ &= (47.733\cdots)t^4 \end{aligned}$$

Here the stress varies linearly, in accordance with Eq. (5.121):

$$\sigma_x = -\frac{M}{I}y$$

The maximum and minimum stresses occur at the bottom and top, respectively:

$$\sigma_{\max} = -\frac{M(-4.4t)}{47.7t^4} = \frac{M}{10.8t^3}$$

$$\sigma_{\min} = -\frac{M(2.6t)}{47.7t^4} = -\frac{M}{18.4t^3}$$

To assign some reasonable values, let us suppose that the member is a rolled-steel beam with $t = 0.25$ in. (0.64 cm) and σ_0 (yield stress) = 48 ksi (330 MPa). Then

$$\sigma_0 = 48{,}000\left(\frac{\text{lb}}{\text{in.}^2}\right) = \frac{M}{10.8(0.25)^3\ \text{in.}^3}$$

$$M = 8140\ \text{in.-lb} \quad (920\ \text{N}\cdot\text{m})$$

Our steel beam is 1.75 in. deep. Two strong students could exert the couple through a 48-in. (1.2 m) crossbar (wrench) by applying forces of 170 lb (760 N). ◆

EXAMPLE 14

A Laminated Beam A beam is built up by laminating *very* thin layers of HI-HO materials, each slightly denser and stiffer, so that the modulus effectively varies according to the equation

$$E = E_0\left(1 + \frac{y}{h}\right)$$

Here the origin ($y = 0$) is arbitrarily placed at the bottom of the rectangular cross section, as depicted in Fig. 5.44; the modulus is E_0 at the bottom ($y = 0$) and $2E_0$ at the top ($y = h$). The beam is subjected to a simple bending couple. In accordance with the foregoing arguments the strain varies linearly, but the neutral axis ($\varepsilon_x = 0$) is not known *a priori*. Since the material is not homogeneous, the position of the neutral line is not necessarily at the centroid. The linear function $\varepsilon_x(y)$ can be written with the origin ($y = 0$) arbitrarily placed at the bottom where $\varepsilon_0(0) = \varepsilon_0$:

$$\varepsilon_x = \varepsilon_0 - \frac{y}{\rho}$$

Fig. 5.44

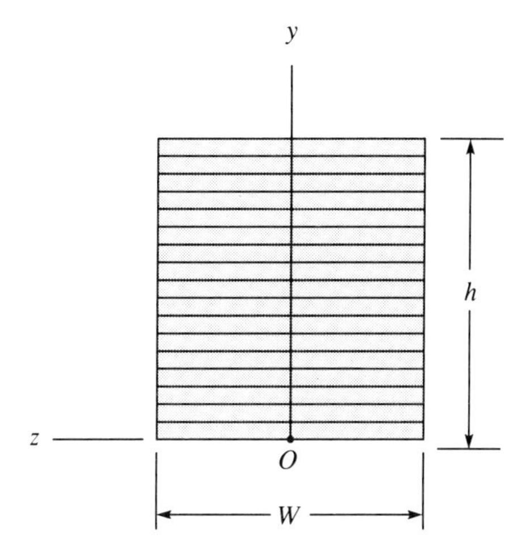

Laminated beam

The force and couple are given by Eqs. (5.111) and (5.112):

$$F = \iint_S \sigma_x \, dA$$

$$M = -\iint_S \sigma_x y \, dA$$

The former is zero in the case of simple bending. With the given material and the linear strain, we have

$$\sigma_x = E_0\left(1 + \frac{y}{h}\right)\left(\varepsilon_0 - \frac{1}{\rho}y\right)$$

$$F = W\int_0^h E_0\left(1 + \frac{y}{h}\right)\left(\varepsilon_0 - \frac{1}{\rho}y\right)dy = 0$$

$$M = -W\int_0^h E_0\left(1 + \frac{y}{h}\right)\left(\varepsilon_0 - \frac{1}{\rho}y\right)y\,dy$$

Upon integrating these two equations, we obtain

$$\left(\frac{3}{2}h\right)\varepsilon_0 - \left(\frac{5}{6}h\right)\frac{h}{\rho} = O$$

$$-\left(\frac{5}{6}h^2\right)\varepsilon_0 + \left(\frac{7}{12}h^2\right)\frac{h}{\rho} = \frac{M}{E_0 W}$$

The solution follows:

$$\varepsilon_0 = \frac{5}{9}\frac{h}{\rho} = \frac{60}{13}\frac{M}{E_0 W h^2}, \qquad \frac{h}{\rho} = \frac{108}{13}\frac{M}{E_0 W h^2}$$

The neutral axis is at the location (y_0) of zero strain:

$$\varepsilon_x(y_0) = 0 = \varepsilon_0 - \frac{y_0}{\rho}$$

$$y_0 = \frac{5}{9}h$$

Since the material is stiffer at the top, the neutral line lies above the midline. Strain and stress distributions are depicted in Fig. 5.45. Note that the stress distribution is *not* linear. ◆

Fig. 5.45

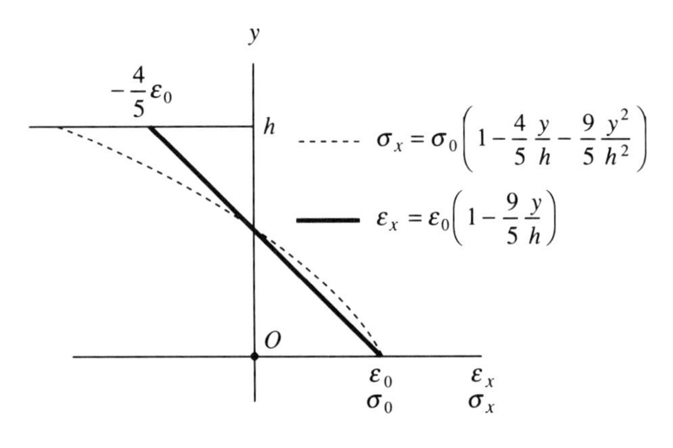

PROBLEMS

5.159 A beam of depth h is bent by end couples so that the extensional strain at the top of the beam is ε_t; at the bottom it is ε_c.

a. What is the radius of curvature of the middle surface (the surface midway between the top and bottom of the beam)?

b. What is the extensional strain in the middle surface?

c. What is the relationship between ε_t and ε_c if the middle surface contains the neutral axis?

5.160 A beam has the square cross section shown in Fig. P5.160. The material is HI-HO and brittle; it fractures if the normal stress reaches 400 MPa. Compare the maximum bending couple M_U when acting in the direction of M with the maximum couple M'_U when acting in the direction of M'; that is, compute M_U/M'_U.

Fig. P5.160

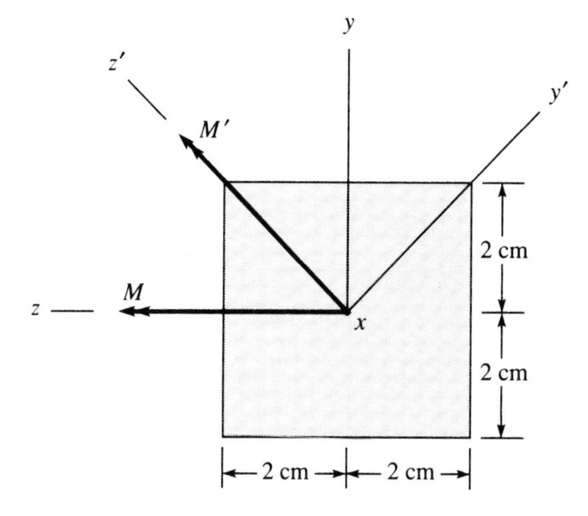

5.161 A beam has the I-section shown in Fig. P5.161. It is composed of hot-rolled steel, which yields if the shear stress attains the value $\tau_0 = 24$ ksi. If the beam is to remain elastic (HI-HO), what is the maximum value of the bending couple when applied in the direction M (only) and M' (only)? Compute the ratio M'_0/M_0.

Fig. P5.161

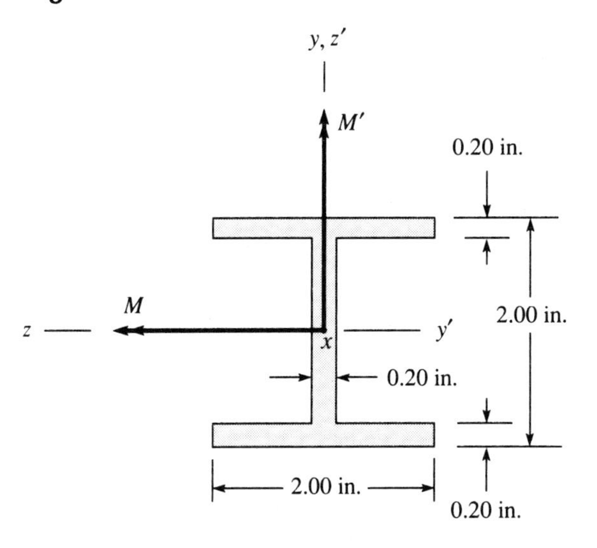

5.162 The product EI is often called the flexural rigidity, since it is the constant of proportionality between moment M and curvature $\kappa = 1/\rho$. Compare the rigidities for bending in the planes (LPS) xy and xy' if the beam has the cross section shown in (a) Fig. P5.160 and (b) Fig. P5.161; that is, calculate EI/EI' in each case.

5.163 A composite I-beam is built up of two materials, as shown in Fig. P5.163. Both are linearly elastic, the web having a modulus of elasticity E^W and the flanges, E^F. Obtain the bending couple-curvature relation, assuming that the cross sections remain plane and perpendicular to the neutral axis.

Fig. P5.163

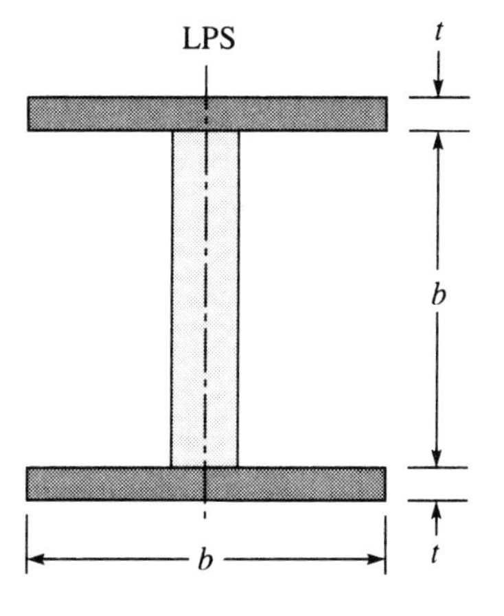

5.164 A composite beam is made by bonding together rectangular beams of two different linearly elastic materials, as shown in Fig. P5.164. Locate the neutral axis if the cross sections remain plane and perpendicular to it during bending. Choose a *convenient*, but arbitrary, origin to impose the condition $F = 0$.

Fig. P5.164

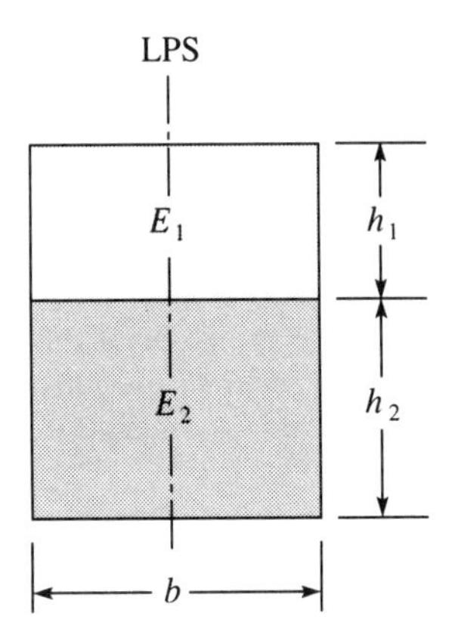

5.165 A beam has the cross section shown in Fig. P5.165. It is HI-HO to the yield stress $\sigma_0 = \pm 300$ MPa. Compute the bending couple that would initiate yielding.

Fig. P5.165

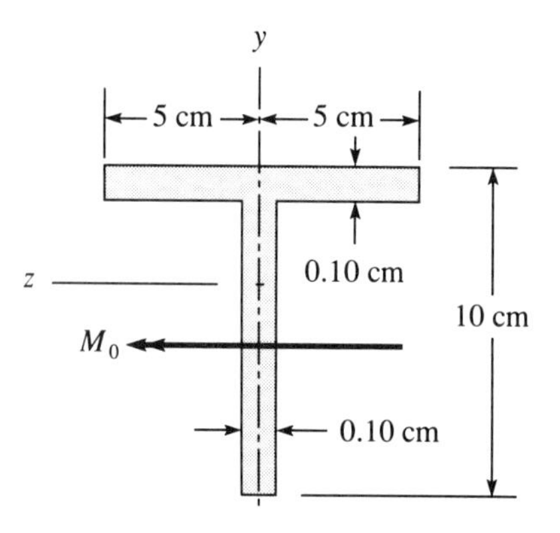

5.166 A beam has the trapezoidal cross section depicted in Fig. P5.166. The material is HI-HO. Express the maximum and minimum normal stress in terms of the applied couple M and the dimension h.

Fig. P5.166

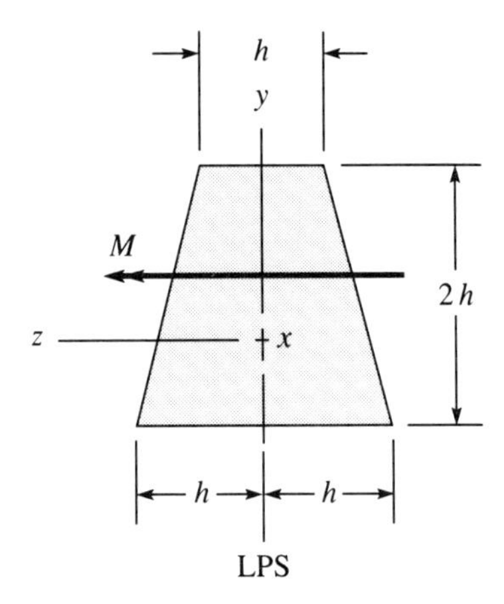

5.167 For a linearly elastic beam subjected to end couples M, the stress distribution is given by

$$\sigma_x = -\frac{My}{I}, \qquad \sigma_y = \sigma_z = \tau_{xy} = \tau_{xz} = \tau_{yz} = 0$$

The beam has a rectangular cross section of width b and depth h, as shown in Fig. 5.47. The neutral surface (originally the plane perpendicular to the LPS and containing the neutral axis) will have a curvature of $1/\rho = M/EI$ along the x-axis. Because $\varepsilon_z = -\nu\varepsilon_x$ is not zero, the neutral surface also has a curvature along the z-axis. Assume that the planes initially parallel to the xy-plane remain plane and perpendicular to the neutral surface. Show that this transverse curvature is given by

$$\frac{1}{\rho^*} = \frac{\nu}{\rho} = \frac{\nu M}{EI}$$

5.168 A linearly elastic beam has a constant depth h but is slightly tapered so that the width decreases linearly from b_0 to b_1 in the length L, as shown in Fig. P5.168. Compute the normal stress distribution on a cross section at x. Assume that the cross sections remain plane and perpendicular to the neutral axis.

Fig. P5.168

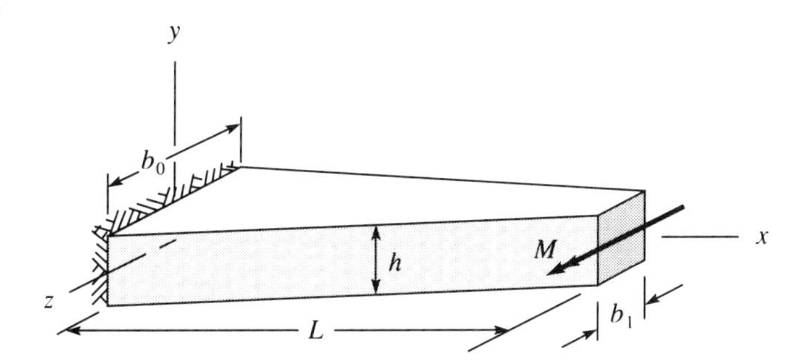

5.169 A linearly elastic beam has a constant width b but is slightly tapered so that the depth decreases linearly from h_0 to h_1 in the length L, as shown in Fig. P5.169. Compute the normal stress distribution on a cross section at x. Assume that the cross sections remain plane and perpendicular to the neutral axis.

Fig. P5.169

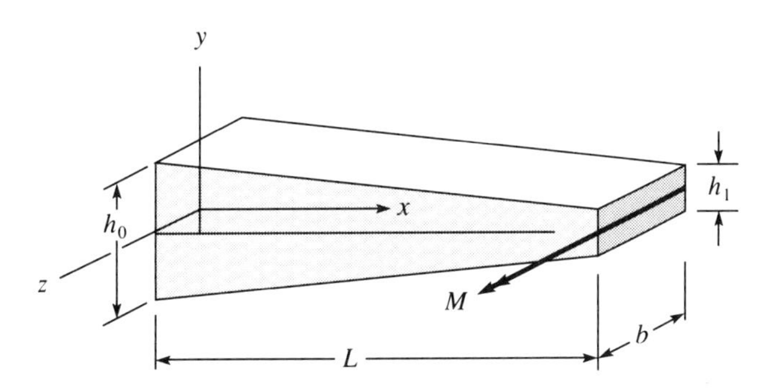

5.13 Ideally Elastic-Plastic Material

Suppose the beam of Fig. 5.36 is made of an ideally elastic-plastic material. The stress-strain diagram is shown in Fig. 5.46. (We again assume $\sigma_y = \sigma_z = 0$).

The stress-strain relationship is expressed by

$$\sigma_x = \begin{cases} -\sigma_0, & \varepsilon_x \leq -\varepsilon_0 \quad (5.122a) \\ E\varepsilon_x, & -\varepsilon_0 \leq \varepsilon_x \leq \varepsilon_0 \quad (5.122b) \\ \sigma_0, & \varepsilon_x \geq \varepsilon_0 \quad (5.122c) \end{cases}$$

The preceding description, specifically the *kinematic* result (5.110), the *dynamic* relations (5.111) and (5.112), and the stress-strain equations (5.122), applies to a beam of arbitrary cross section but symmetric about the xy-plane. Some further simplification results when the shape of the cross section is also symmetric about the xz-plane. To illustrate the analyses, we consider a few examples.

Fig. 5.46

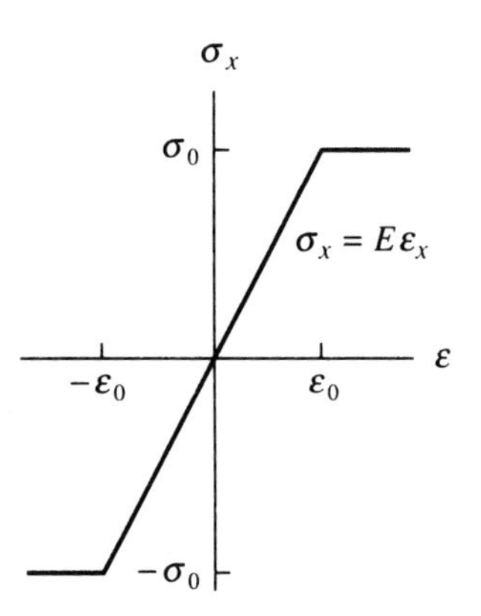

EXAMPLE 15

Rectangular Cross Section Suppose that the cross section is a rectangle of width b and depth h, as shown in Fig. 5.47. We will tentatively assume that the neutral axis is the central axis. This seems reasonable because the axis is the horizontal line of symmetry and the plot of Fig. 5.43 shows an odd function $\sigma_x(\varepsilon_x)$.

Fig. 5.47

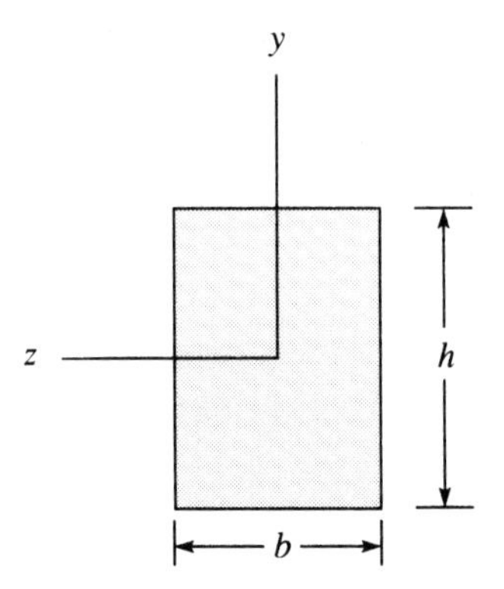

According to Eq. (5.110) the strain ε_x is proportional to the distance y. Hence, the maximum and minimum extensional strains occur at the bottom and top of the beam. These are

$$\varepsilon_{x_{\max}} = -\varepsilon_{x_{\min}} = \frac{h/2}{\rho} \tag{5.123}$$

As long as the extension strain $|\varepsilon_x|$ is everywhere less than the yield strain $\varepsilon_0 = \sigma_0/E$, the behavior is linearly elastic throughout the beam, and the results of the previous analysis are applicable. In particular, the normal stress σ_x is given by Eq. (5.121) and is linearly distributed on a cross section, as indicated in Fig. 5.48 (for positive M). As the bending couple is increased, the extensional strain $|\varepsilon_x|$ exceeds the yield strain ε_0

Fig. 5.48

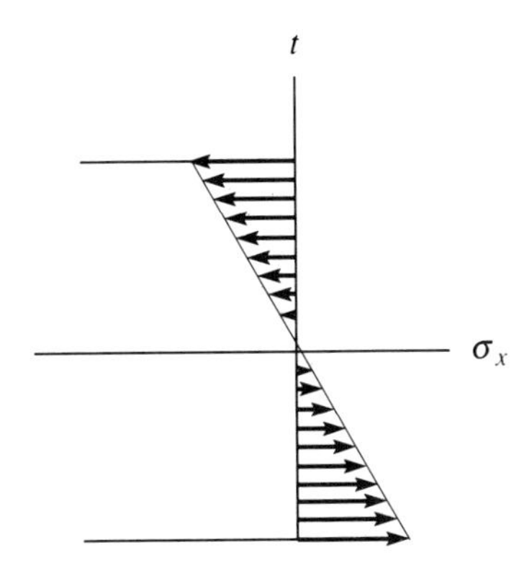

in the top and bottom portions of the beam, as indicated in Fig. 5.49(a). Let c denote half the depth of the elastic core. At $y = \pm c$ we have $\varepsilon_x = \mp \varepsilon_0$; hence, from Eq. (5.110),

$$\varepsilon_0 = \frac{c}{\rho}, \quad \text{or} \quad \frac{1}{\rho} = \frac{\varepsilon_0}{c} = \frac{\sigma_0}{Ec} \tag{5.124}$$

Fig. 5.49

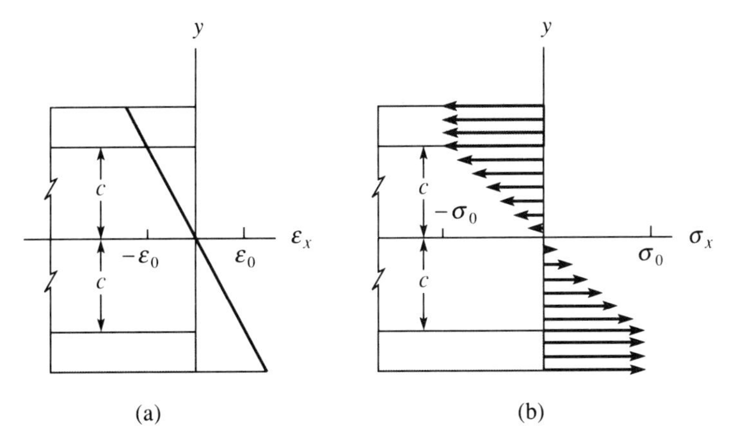

The normal stress in the elastic core $|\varepsilon| \leq \varepsilon_0$ is proportional to the extensional strain:

$$\sigma_x = E\varepsilon_x = -\frac{Ey}{\rho}$$

or, with Eq. (5.124),

$$\sigma_x = -\frac{y}{c}\sigma_0, \qquad |y| \le c \tag{5.125a}$$

In the plastic portions of the beam $|\varepsilon| \ge \varepsilon_0$, the material is subjected to the yield stress

$$\sigma_x = \begin{cases} -\sigma_0, & c \le y \le \dfrac{h}{2} & (5.125\text{b}) \\[2ex] \sigma_0, & -\dfrac{h}{2} \le y \le -c & (5.125\text{c}) \end{cases}$$

The distribution of normal stress on a cross section is as indicated in Fig. 5.49(b).

The normal force upon the cross section is given by Eq. (5.111), where the stress is given by Eq. (5.125a, b, c) and depicted in Fig. 5.49(b). The integral of this odd function $\sigma_x(y)$ over the rectangular region vanishes; that is, $F = 0$. Then the neutral axis is along the central axis *in this case.*

The bending couple is given by Eq. (5.112), wherein the stress is now given by Eq. (5.125a, b, c). Accordingly, the integral must be evaluated in three parts, as follows:

$$\begin{aligned} M &= -b\int_{-h/2}^{-c} \sigma_0 y\,dy + b\int_{-c}^{c} \frac{\sigma_0}{c} y^2\,dy + b\int_{c}^{h/2} \sigma_0 y\,dy \\ &= \sigma_0 b\left[\frac{h^2}{4} - \frac{c^2}{3}\right] \end{aligned} \tag{5.126}$$

The half-depth of the elastic core can be determined by solving Eq. (5.126) for c. However, the results are meaningless unless c is between 0 and $h/2$. When $c = h/2$, the outermost portion of the cross section is about to yield. The bending couple, from Eq. (5.126), is

$$M_0 = \sigma_0 \frac{bh^2}{6} \tag{5.127}$$

For M less than M_0 the behavior is linearly elastic, and the previous results apply. As c decreases, more and more of the cross section deforms plastically. In the limiting condition, $c = 0$, the cross section is said to be fully plastic. For this condition the bending couple has the value

$$M_L = \lim_{c \to 0} \sigma_0 b\left[\frac{h^2}{4} - \frac{c^2}{3}\right] = \sigma_0 \frac{bh^2}{4} \tag{5.128}$$

For future reference let us note the relative magnitudes of the limit couple M_L and initial-yield couple M_0:

$$\frac{M_L}{M_0} = \frac{3}{2} \tag{5.129}$$

Prior to yielding, $M < M_0$, the moment M and curvature κ follow the linear relation (5.119). After yielding, $M_0 < M < M_L$, the stress is distributed according to Eq. (5.125). At the elastic-plastic interface, $y = \pm c$, the yielding is initiated, so that

$$\sigma_x = \pm\sigma_0 = E\varepsilon_x(\mp c) \tag{5.130}$$

According to Eq. (5.110) and the definition $\kappa = 1/\rho$,

$$\varepsilon_x(\mp c) = \pm\frac{c}{\rho} \equiv \pm c\kappa \tag{5.131}$$

It follows that

$$c = \frac{\sigma_0}{E\kappa} \tag{5.132}$$

Substituting (5.132) into (5.126), we obtain

$$M = \sigma_0 b\left[\frac{h^2}{4} - \frac{\sigma_0^2}{3E^2\kappa^2}\right], \qquad M_0 < M < M_L \tag{5.133}$$

This nonlinear equation governs after yielding. As the curvature increases, the second term in the bracket diminishes, and, in the limit $\kappa \to \infty$, the couple M approaches the *limit* value M_L.

The complete load-deformation curve is drawn in Fig. 5.50. From O to A the curvature κ and load M are linearly related by Eq. (5.119). As the load M is increased beyond M_0, plastic deformation occurs in the beam. As M is increased, the curve approaches a horizontal asymptote at M_L; this is the limit load.

Fig. 5.50

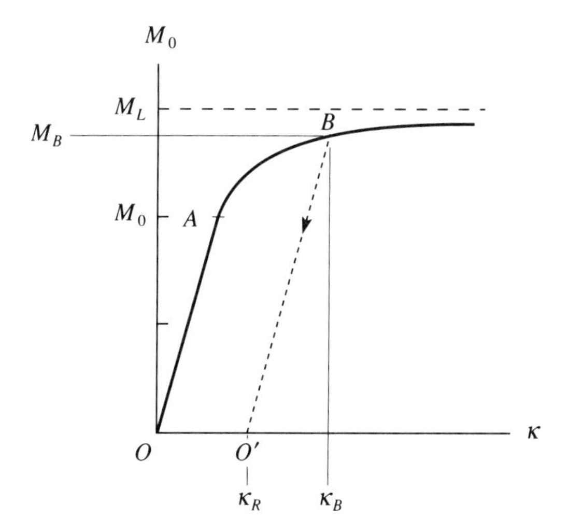

Let us now consider the behavior if the couple is removed *after* yielding has occurred—that is, from the point B of Fig. 5.47. Let ΔM, $\Delta\sigma_x$, $\Delta\varepsilon_x$, and $\Delta\kappa$ denote the subsequent *changes* in moment, stress, strain, and curvature, respectively. In accordance with our previous description of this material (see Sec. 4.11), the changes in stress and strain follow a linear relation (see Fig. 4.22) whether the material has yielded (regions $|y| > c$ in the beam) or not (region $-c < y < c$). Then, *throughout* the beam, during unloading,

$$\Delta\sigma_x = E\Delta\varepsilon_x \tag{5.134}$$

The kinematic relation (5.110) also holds, so that

$$\Delta\varepsilon_x = -y\,\Delta\kappa \tag{5.135}$$

The change in the couple ΔM is related to the change in stress $\Delta\sigma_x$ in the same way as M and σ_x in Eq. (5.112):

$$\Delta M = -\iint \Delta\sigma_x y\,dA \tag{5.136}$$

Again, Eqs. (5.134) and (5.135) hold throughout; their substitution into Eq. (5.136) gives the result

$$\Delta M = EI\,\Delta\kappa \tag{5.137}$$

where, in the present example, $I = bh^3/12$. In short, the unloading follows the linear relation (5.137), wherein the constant of proportionality is the same (EI) that appears in Eq. (5.119), which governs the initial elastic response. The unloading traces the straight line BO', parallel to AO in Fig. 5.50. If the moment M_B is removed, the curvature changes in amount

$$\Delta\kappa = \kappa_R - \kappa_B = -\frac{M_B}{EI} \tag{5.138}$$

Here κ_R is the residual curvature (see Fig. 5.50). ◆

Residual Stress

During removal of the moment, the *change* in the stress is given by Eq. (5.134) and, according to Eq. (5.135), that *change* is linear in the coordinate y:

$$\Delta\sigma_x = -E\,\Delta\kappa y \tag{5.139}$$

Because the strains are presumed small, this change can be superposed upon the stress prior to unloading. To illustrate, let us suppose that yielding has progressed to one-half the depth. Prior to unloading the stress distribution is depicted in Fig. 5.51(a):

$$\sigma_x^B = \begin{cases} -\sigma_0, & \dfrac{h}{4} \le y \\[2ex] -\sigma_0\dfrac{4y}{h}, & -\dfrac{h}{4} \le y \le \dfrac{h}{4} \\[2ex] +\sigma_0, & y \le -\dfrac{h}{4} \end{cases} \tag{5.140}$$

Fig. 5.51

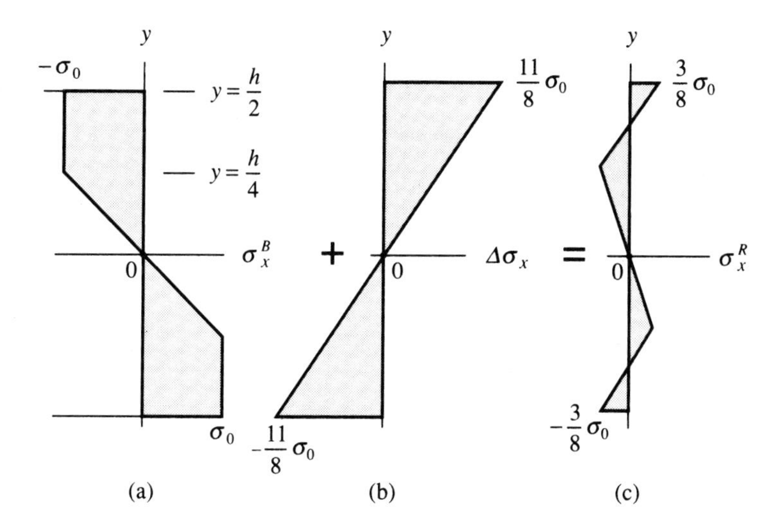

At $y = \pm h/4$, $\sigma_x = \mp E\kappa_B h/4 = \mp\sigma_0$, so that

$$\kappa_B = \frac{4}{h}\frac{\sigma_0}{E} \tag{5.141}$$

During unloading the change of stress is given by Eqs. (5.139), (5.138), and (5.141):

$$\begin{aligned}\Delta\sigma_x &= -E\,\Delta\kappa y = -E(\kappa_R - \kappa_B)y \\ &= -E\left(\kappa_R - \frac{4}{h}\frac{\sigma_0}{E}\right)y\end{aligned} \tag{5.142}$$

The unloaded state implies a vanishing moment:

$$M = -b\int_{-h/2}^{h/2} (\sigma_x^B + \Delta\sigma_x)y\,dy = 0 \tag{5.143}$$

The latter equation, (5.143), is expressed in terms of one unknown, κ_R(σ_0, E, and h are known properties) in accordance with Eqs. (5.140) and (5.142). After integrating we have

$$\kappa_R = \frac{5}{4}\frac{\sigma_0}{Eh}, \qquad \Delta\sigma_x = \frac{11}{4}\sigma_0\frac{y}{h}$$

The residual stress distribution, $\sigma_x^R = \sigma_x^B + \Delta\sigma_x$, is depicted in Fig. 5.51(c); it is the sum of the distributions in (a) and (b) $[\sigma_x^B + \Delta\sigma_x]$. Again, we observe a residual stress when loading causes yielding in some regions of the body while other regions remain elastic.

EXAMPLE 16

T-Shaped Cross Section Suppose that a beam has a T-shaped cross section, as illustrated in Fig. 5.52. For simplicity the width of the flange ($4t$) and depth of the web ($6t$) are given in terms of their common thickness (t). The material is assumed ideally elastic-plastic, as in Example 15.

Fig. 5.52

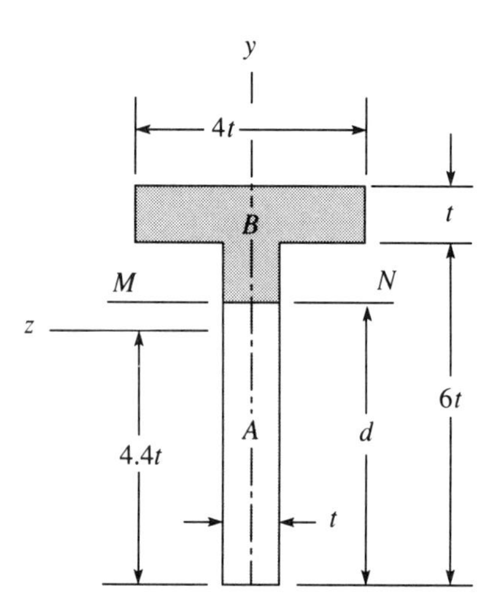

Prior to yielding the behavior is elastic (hookean), and the centroidal line is also the neutral line. It is located at distance $4.4t$ from the bottom. Since the yield stress in tension and compression have the same magnitude, yielding occurs initially in tension at the bottom, where $|y| = 4.4t$. In accordance with Eq. (5.121), the moment is

$$M_0 = \frac{I\sigma_0}{4.4t} = 10.8t^3\sigma_0$$

If the moment is then increased, yielding occurs at the bottom, and a region of inelastic (plastic) deformation progresses inward. Eventually yielding is also initiated (in compression) at the top. As the moment is increased, the regions of inelastic deformation at the bottom and top grow and eventually coalesce. Within a region A, below some line MN, the stress is $\sigma_x = +\sigma_0$; within a region B, above line MN, the stress is $\sigma_x = -\sigma_0$. Throughout the deformation the net force is zero; according to Eq. (5.111),

$$F = \iint_A \sigma_0 \, dA - \iint_B \sigma_0 \, dA = 0 \tag{5.144a}$$

or

$$\text{Area } A - \text{area } B = 0 \tag{5.144b}$$

Equation (5.144b) serves to locate the line MN at $d = 5t$. The corresponding limit couple is $M_L = 19\sigma_0 t^3$, and the relative magnitudes of the limit couple and initial-yield couple follow:

$$\frac{M_L}{M_0} = \frac{19}{10.8} = 1.75 \tag{5.145}$$

Several features are noteworthy: Firstly, the neutral axis moves during the progressive yielding of the beam. Secondly, the relative magnitudes of the limit couple and initial yield couple depend upon the shape of the cross section. Finally, note that (5.144b) is but a special form of the more general requirement $F = 0$; (5.144b) results only when the magnitudes of the tensile and compressive yield stresses are the same. ◆

*5.14 Double-Modulus Material

As another illustration, consider a material that is stiffer in tension than in compression. The stress-strain diagram for such a material is shown in Fig. 5.53. The stress-strain law is

$$\sigma_x = \begin{cases} E^T \varepsilon_x, & \varepsilon_x \geq 0 \\ E^C \varepsilon_x, & \varepsilon_x \leq 0 \end{cases} \qquad E^T \geq E^C \tag{5.146}$$

Fig. 5.53

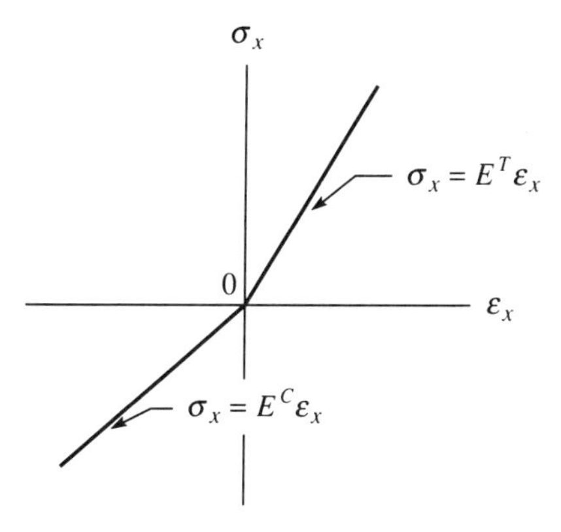

Again let us work with a rectangular cross section of width b and depth h. From Eq. (5.110) we have

$$\varepsilon_x = -\frac{y}{\rho} \tag{5.147}$$

If ρ is positive at a cross section,

$$\sigma_x = \begin{cases} -\dfrac{E^T y}{\rho}, & y \leq 0 \\ -\dfrac{E^C y}{\rho}, & y \geq 0 \end{cases} \tag{5.148}$$

The neutral axis in this case will not be located at the centroid of the cross section. Suppose it is a distance c from the bottom of the beam, as shown in Fig. 5.54. Then the normal force on the cross section is given by

$$\begin{aligned} F &= \int_{-c}^{h-c} \sigma_x b\,dy \\ &= b\int_{-c}^{0} -\frac{E^T}{\rho} y\,dy + b\int_{0}^{h-c} -\frac{E^C}{\rho} y\,dy \\ &= \frac{b}{2\rho}[E^T c^2 - E^C(h-c)^2] \end{aligned} \tag{5.149}$$

Fig. 5.54

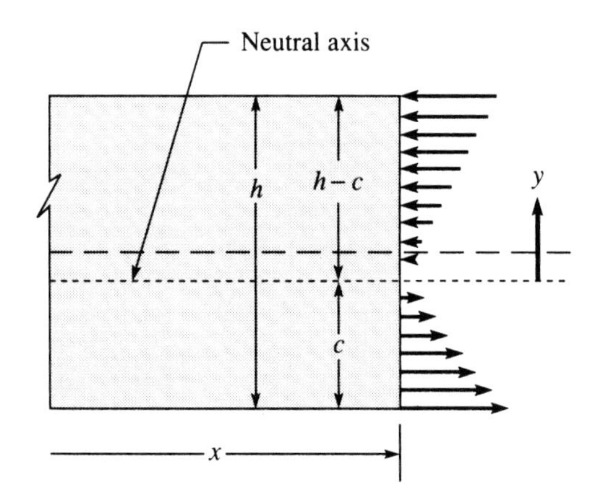

Now the normal force F will be zero if

$$\frac{c^2}{(h-c)^2} = \frac{E^C}{E^T}$$

or

$$\frac{c}{h-c} = \sqrt{\frac{E^C}{E^T}} \tag{5.150}$$

(We take the positive root because c must be positive but less than h.) Then

$$c = \frac{h\sqrt{\dfrac{E^C}{E^T}}}{1 + \sqrt{\dfrac{E^C}{E^T}}} = \frac{h\sqrt{E^C}}{\sqrt{E^T} + \sqrt{E^C}} \tag{5.151}$$

The bending couple on this cross section is

$$
\begin{aligned}
M &= -\int_{-c}^{h-c} y\sigma_x b\,dy \\
&= \int_{-c}^{0} \frac{E^T b}{\rho} y^2\,dy + \int_{0}^{h-c} \frac{E^C b}{\rho} y^2\,dy \\
&= \frac{b}{3\rho}[E^T c^3 + E^C (h-c)^3] \\
&= \frac{bc^3}{3\rho} E^T \left[1 + \frac{E^C}{E^T}\left(\frac{h-c}{c}\right)^3\right]
\end{aligned}
$$

From Eq. (5.150) it follows that

$$
M = \frac{bc^3}{3\rho} E^T \left(\frac{\sqrt{E^C} + \sqrt{E^T}}{\sqrt{E^C}}\right) \tag{5.152}
$$

Substituting for c from Eq. (5.151), we obtain

$$
M = \frac{bh^3}{3\rho} \frac{E^T E^C}{(\sqrt{E^T} + \sqrt{E^C})^2} \tag{5.153}
$$

Let us *define* a reduced modulus E^R as follows:

$$
E^R = \frac{4E^T E^C}{[\sqrt{E^T} + \sqrt{E^C}]^2} \tag{5.154}
$$

Then, upon noting that

$$
\frac{bh^3}{12} = I
$$

is the moment of inertia of this rectangular cross section about the horizontal line through the centroid, we finally obtain, from Eqs. (5.153) and (5.154),

$$
M = \frac{E^R I}{\rho} \tag{5.155}
$$

This is exactly the same as the Bernoulli-Euler curvature formula with E replaced by E^R.

Placing Eq. (5.155) into Eqs. (5.147) and (5.148) we get

$$
\varepsilon_x = -\frac{My}{E^R I}
$$

and

$$
\sigma_x = \begin{cases} -\dfrac{E^T}{E^R}\dfrac{My}{I}, & y \le 0 \\[2ex] -\dfrac{E^C}{E^R}\dfrac{My}{I}, & y \ge 0 \end{cases} \tag{5.156}
$$

Note that these results were derived for a rectangular cross section only. For a differently shaped cross section, E^R will *generally not be defined by Eq.* (5.154). Also, y is measured from the neutral axis defined by Eq. (5.151), but I is computed with respect to the centroidal axis.

PROBLEMS

5.170 An ideally plastic beam ($E = 30 \times 10^6$ lb/in.^{2}, $\sigma_0 = 30{,}000$ lb/in.^{2}) has a square cross section 4 in. on a side. The beam is bent so that the extensional strain at the top of the beam is $\varepsilon_t = 0.004$, under the action of a bending couple.

a. Compute the depth of yielding.

b. Compute the bending couple.

5.171 A beam of ideally plastic material (modulus of elasticity E, yield stress σ_0) has a circular cross section. Compute the ratio of the fully plastic bending couple to the bending couple at which yielding is initiated, M_L/M_E.

5.172 Repeat Problem 5.171 for the I-shaped cross section shown in Fig. P5.172.

Fig. P5.172

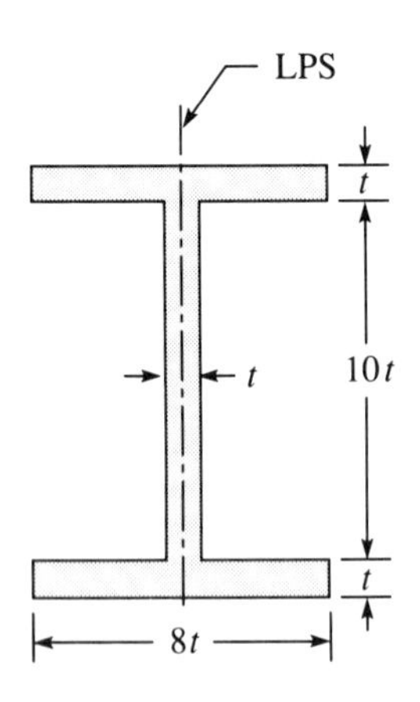

5.173 Repeat Problem 5.171 for the diamond-shaped cross section shown in Fig. P5.173.

Fig. P5.173

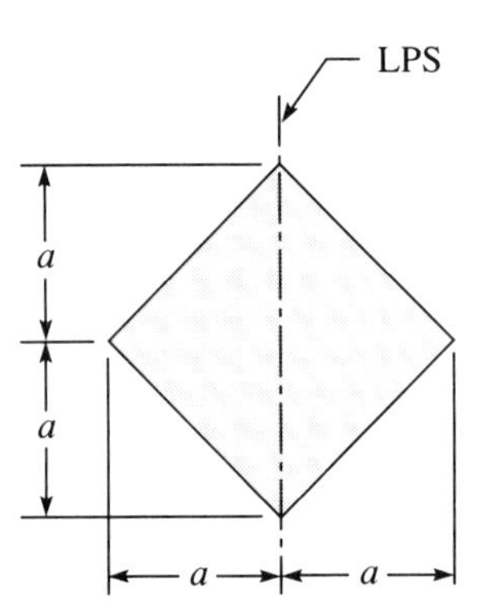

5.174 Repeat Problem 5.171 for the dumbell-shaped cross section shown in Fig. P5.174.

Fig. P5.174

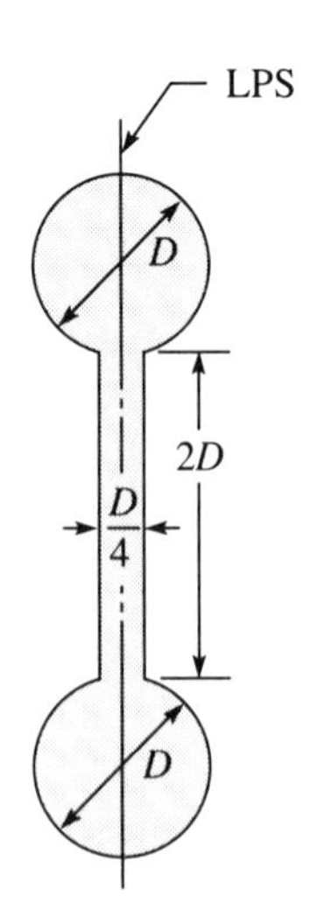

5.175 A beam with a *T*-shaped cross section, as shown in Fig. P5.175 consists of an ideally plastic material (modulus of elasticity E, yield stress σ_0). Locate the neutral axis before any yielding occurs. Where is the neutral axis when the cross section is subjected to the fully plastic bending couple?

Fig. P5.175

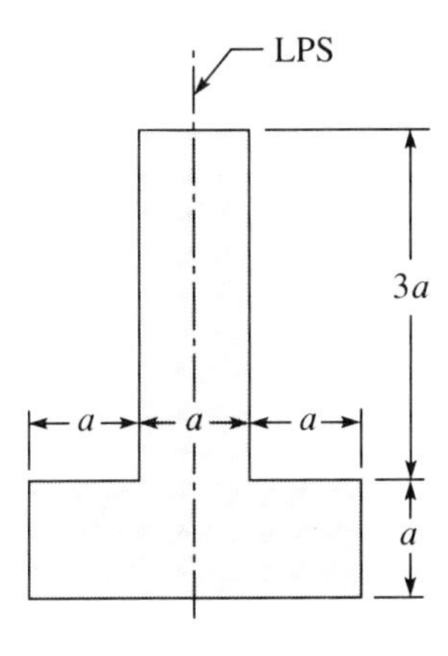

5.176 Repeat Problem 5.170 if the beam is made of a linear-hardening material with $E = 30 \times 10^6$ psi, $\sigma_0 =$ 30,000 psi, and $k = 0.2$. [See Eq. (5.14).]

5.177 Two beams, one with a circular cross section and the other with a square cross section, are made of the same ideally plastic material. If both beams have the same cross-sectional area, which can carry the greater bending couple before yielding is initiated? Consider both planes of symmetry (LPS) for the square.

5.178 A beam has the cross section shown in Fig. P5.178. The material exhibits linear hardening according to the equations

$$\varepsilon = \begin{cases} 2.2 \times 10^{-4}\sigma - 0.05, & \sigma \geq 250 \text{ MPa} \\ 2.0 \times 10^{-5}\sigma, & |\sigma| \geq 250 \text{ MPa} \\ 2.2 \times 10^{-4}\sigma + 0.05, & \sigma \leq -250 \text{ MPa} \end{cases}$$

If the couple is increased so that the maximum stress increases from $\sigma_0 = 250$ MPa to 300 MPa (20% increase), what is the corresponding (percentage) increase in the couple?

Fig. P5.178

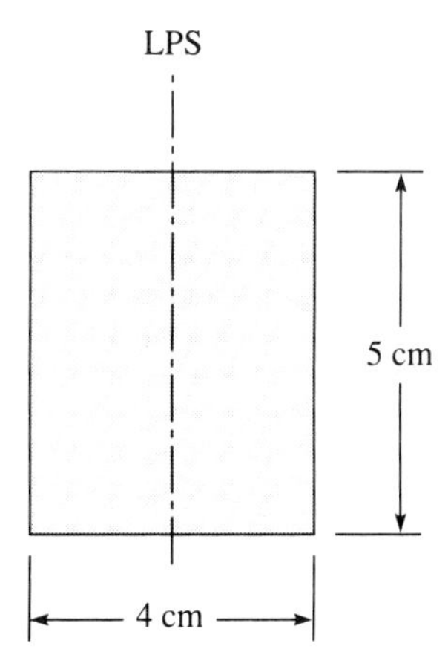

5.179 A beam with a rectangular cross section is made of material whose behavior can be described by

$$\sigma = \begin{cases} K\sqrt{\varepsilon}, & \varepsilon \geq 0 \\ -K\sqrt{-\varepsilon}, & \varepsilon \leq 0 \end{cases}$$

The bending couple on a cross section of width b and depth h is M. Obtain the normal stress-bending couple relation and the curvature-bending couple relation, assuming that the cross sections remain plane and perpendicular to the neutral axis.

5.180 Repeat Problem 5.179 for a material whose behavior is described by

$$\sigma = \begin{cases} K|\varepsilon|^n, & \varepsilon \geq 0 \\ -K|\varepsilon|^n, & \varepsilon \leq 0 \end{cases}$$

where $0 \leq n \leq 1$. Show how, by proper choice of K and n, these results reduce to the results for a linearly elastic material obtained in the text.

5.181 Repeat Problem 5.179 for a material whose behavior is described by

$$\varepsilon = \frac{\sigma}{E}\left[1 + \left|\frac{\sigma}{\sigma_0}\right|\right]$$

5.182 Repeat Problem 5.179 for a material whose behavior is described by

$$\sigma = \begin{cases} \sigma_0 \dfrac{E\varepsilon}{\sigma_0 + E\varepsilon}, & \varepsilon \geq 0 \\ -\sigma_0 \dfrac{E\varepsilon}{-\sigma_0 + E\varepsilon}, & \varepsilon \leq 0 \end{cases}$$

5.183 An ideally plastic beam (modulus of elasticity E, yield stress σ_0) has a rectangular cross section of width b and depth h. The beam is loaded by end couples M, which cause yielding in all but an elastic core of depth $2c$. Compute the residual stresses in the beam when it is subsequently unloaded. Assume that, during unloading, the cross sections remain plane and perpendicular to the neutral axis and that the material behavior is linearly elastic.

5.184 Repeat Problem 5.183 if the material is linear-hardening:

$$\varepsilon = \begin{cases} \dfrac{\sigma}{E}, & |\sigma| \leq \sigma_0 \\ \dfrac{(1+k)\sigma}{kE} - \dfrac{\sigma_0}{kE}, & \sigma \geq \sigma_0 \\ \dfrac{(1+k)}{kE}\sigma + \dfrac{\sigma_0}{kE}, & \sigma \leq -\sigma_0 \end{cases}$$

5.185 A beam of rectangular cross section consists of a linearly elastic core (modulus of elasticity E) and top and bottom plates of a rigid plastic material (yield stress σ_0) bonded to the core, as shown in Fig. P5.185. Obtain the bending couple-curvature relation. (Assume that cross sections remain plane and perpendicular to the neutral axis.) What is the least value of bending couple for which the beam deforms?

Fig. P5.185

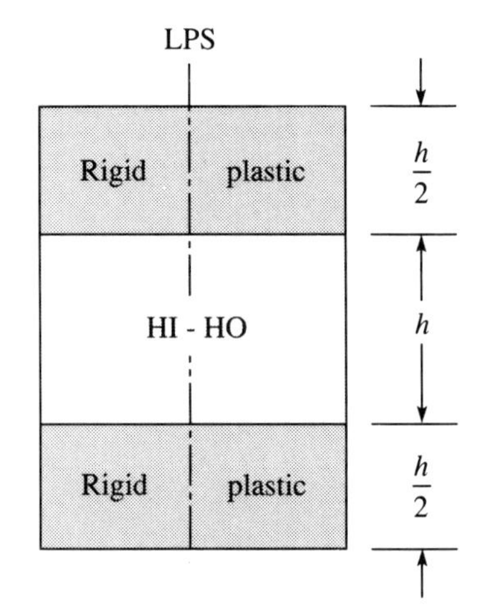

5.15 Extension and Bending of Symmetric Beams

The kinematic equation (5.110) requires that the origin ($y = 0$) be at the neutral axis; that is, $\varepsilon_x(0) = 0$. Suppose that we were to shift the origin, that is, install another system $(x, \bar{y}, z)$ with origin at $y = a$, so that

$$y = a + \bar{y}$$

In this new system,

$$\varepsilon_x(\bar{y}) = -\frac{\bar{y}}{\rho} - \frac{a}{\rho}$$

If $\varepsilon_x(0) = \varepsilon_0$ and $1/\rho = \kappa$, then

$$\varepsilon_x(\bar{y}) = \varepsilon_0 - \kappa\bar{y} \tag{5.157}$$

Evidently, the essential feature of the expression for strain is linearity in the coordinate $\bar{y}$ (or y). From another viewpoint, Eq. (5.157) describes the superposition of small uniform extensional strain (see Sec. 5.4) and uniform bending strain [Eq. (5.110)]. Provided only that the extension (ε_0) and curvature (κ) are constant along the beam, the strain ε_x is linear in the coordinate $\bar{y}$ (or y). In many practical problems, such as the example of the preceding section, we do not know a priori the location of the neutral (unstretched) line. In such cases, we can put the origin at any convenient point on the plane of symmetry, provided that we acknowledge the extensional strain of the axis—that is, we use the more general form

$$\varepsilon_x(y) = \varepsilon_0 - \kappa y \tag{5.158}$$

EXAMPLE 17

Bimetallic Strip Consider a beam composed of two rectangular strips of different materials that are bonded together as depicted in Fig. 5.55. To be more specific, suppose that the top strip S is steel; the bottom strip B is a bronze. Both behave as HI-HO materials within their respective proportional limits. Let us distinguish their moduli (E) and coefficients (α) of thermal expansion by subscripts S and B, respectively. The mechanical and geometric properties are

$$E_S = 30 \times 10^6 \text{ psi} \quad (200 \text{ GPa}), \qquad E_B = 15 \times 10^6 \text{ psi} \quad (100 \text{ GPa})$$

$$\alpha_S = 6.5 \times 10^{-6}/^\circ\text{F} \quad (12 \times 10^{-6}/^\circ\text{C}), \qquad \alpha_B = 13 \times 10^{-6}/^\circ\text{F} \quad (23 \times 10^{-6}/^\circ\text{C})$$

$$h = 0.015 \text{ in.} \quad (0.038 \text{ cm}), \qquad H = 0.060 \text{ in.} \quad (0.152 \text{ cm})$$

$$\text{Length} = l = 2.00 \text{ in.} \quad (5.08 \text{ cm})$$

Fig. 5.55

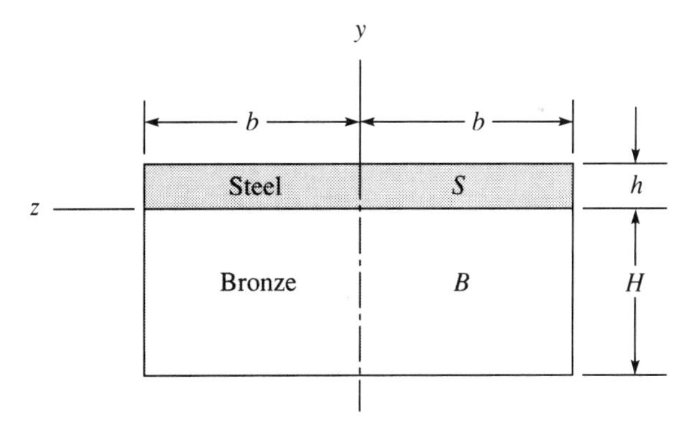

Suppose now that the strip is clamped at one end ($x = 0$) but otherwise free. It is initially straight at room temperature and is then heated so that the temperature is uniformly increased an amount

$$\Delta T = 400^\circ\text{F} \quad (220^\circ\text{C})$$

Since the coefficient α_B is greater than the coefficient α_S, the bronze tends to expand more, so that the composite strip bends when heated. Such a bimetallic strip might be used as a thermostatic device to detect temperature change or activate an electrical switch. Our task is to determine the deflection of the free end ($x = l$) and the stresses that accompany the rise in temperature.

We observe that the strip is a slender beam, uniform along its length, and symmetric about the middle plane (xy). Then, according to all previous arguments, each slice (Δx) behaves as every other, with the possible exception of regions near the ends ($x = 0, l$) and edges ($z = \pm b$). Again, according to Saint-Venant's principle, such end effects are expected to be localized. Then, throughout the beam, except near the ends, the kinematical equation (5.158) is invoked

$$\varepsilon_x = \varepsilon_0 - \kappa y \tag{5.159}$$

Again, we assume a simple state of stress; we neglect all but the normal component σ_x. Then, the HI-HO materials (S and B) are described by the stress-strain-temperature equations:

$$\varepsilon_x = \frac{\sigma_x}{E_B} + \alpha_B \Delta T \qquad \text{(in region } B\text{)} \tag{5.160a}$$

$$\varepsilon_x = \frac{\sigma_x}{E_S} + \alpha_S \Delta T \qquad \text{(in region } S\text{)} \tag{5.160b}$$

Because no external loads act on the strip, the resultant on every cross section vanishes. By Eqs. (5.111) and (5.112),

$$F = \iint_B \sigma_x \, dA + \iint_S \sigma_x \, dA = 0 \tag{5.161}$$

$$M = -\iint_B \sigma_x y \, dA - \iint_S \sigma_x y \, dA = 0 \tag{5.162}$$

We recall that the origin of the coordinates can be located at any convenient point on the plane of symmetry. What is the most convenient? The expressions for σ_x are different in regions B and S. According to Eqs. (5.159) and (5.160),

$$\sigma_x = E_B(\varepsilon_0 - \kappa y) - E_B \alpha_B \Delta T \qquad \text{(in region } B\text{)} \tag{5.163a}$$

$$\sigma_x = E_S(\varepsilon_0 - \kappa y) - E_S \alpha_S \Delta T \qquad \text{(in region } S\text{)} \tag{5.163b}$$

Since the integrations of Eqs. (5.161) and (5.162) must be taken in the two parts (B and S), as indicated, a convenient position for the origin is at the interface between the regions B and S, as shown in Fig. 5.55.

When the expressions (5.163a, b) for the stress are substituted into the integrals (5.161) and (5.162), the latter take the forms:

$$F = 2b \int_{-H}^{0} [E_B(\varepsilon_0 - \kappa y) - E_B \alpha_B \Delta T] \, dy$$
$$+ 2b \int_{0}^{h} [E_S(\varepsilon_0 - \kappa y) - E_S \alpha_S \Delta T] \, dy \tag{5.164}$$

$$M = -2b \int_{-H}^{0} [E_B(\varepsilon_0 y - \kappa y^2) - E_B \alpha_B \Delta T y]\, dy$$
$$- 2b \int_{0}^{h} [E_S(\varepsilon_0 y - \kappa y^2) - E_S \alpha_S \Delta T y]\, dy = 0 \tag{5.165}$$

Now we note that the properties $(E_B, E_S, \alpha_B, \alpha_S)$ and temperature ΔT are constants; also, by our assumptions, ε_0 and κ are constants. The integrands are simply quadratic in y. These provide two linear equations in the two unknowns ε_0 and κ. They are put in nondimensional form by dividing F by bHE_B and M by $-bH^2E_B$. The results follow:

$$\left(1 + \frac{E_S h}{E_B H}\right)\varepsilon_0 + \frac{1}{2}\left(1 - \frac{E_S h^2}{E_B H^2}\right)\kappa H = \left(1 + \frac{E_S}{E_B}\frac{h}{H}\frac{\alpha_S}{\alpha_B}\right)\alpha_B \Delta T \tag{5.166}$$

$$\left(1 - \frac{E_S}{E_B}\frac{h^2}{H^2}\right)\varepsilon_0 + \frac{2}{3}\left(1 + \frac{E_S}{E_B}\frac{h^3}{H^3}\right)\kappa H = \left(1 - \frac{E_S}{E_B}\frac{h^2}{H^2}\frac{\alpha_S}{\alpha_B}\right)\alpha_B \Delta T \tag{5.167}$$

With the numerical values given, we obtain

$$\varepsilon_0 = 36.0 \times 10^{-4}, \qquad \kappa = 41.8 \times 10^{-3}$$

The deflected form of the axis is a circular arc (κ = constant), as depicted in Fig. 5.56, where the deflection is exaggerated.

Fig. 5.56

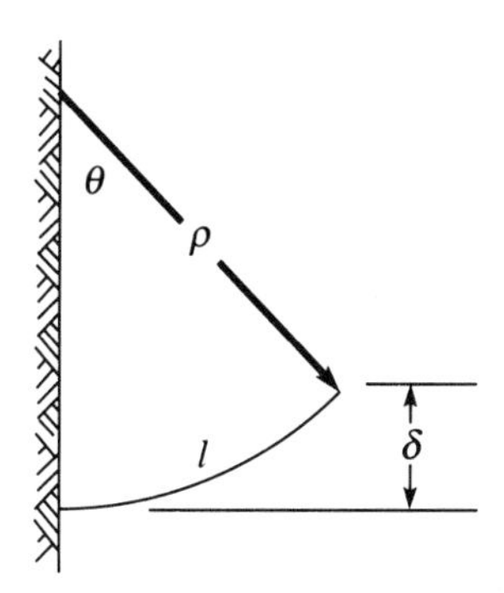

The deflection at the end is

$$\delta = \rho - \rho \cos \theta$$

However, $\theta = l/\rho = \kappa l$:

$$\delta = \rho(1 - \cos \kappa l)$$

As noted, the radius of curvature ρ is relatively large and the angle θ is small. In our case

$$\frac{1}{\rho} = \kappa l = 0.0837$$

For small θ, $\cos\theta \doteq 1 - \frac{1}{2}\theta^2$; therefore,

$$\delta \doteq \frac{1}{\kappa}\left(1 - 1 + \frac{1}{2}\kappa^2 l^2\right) = \frac{1}{2}\kappa l^2 \doteq 0.0837 \text{ in.} \quad (0.213 \text{ cm})$$

The stress distributions are linear in each region according to Eq. (5.163a, b). These are plotted in Fig. 5.57.

Fig. 5.57

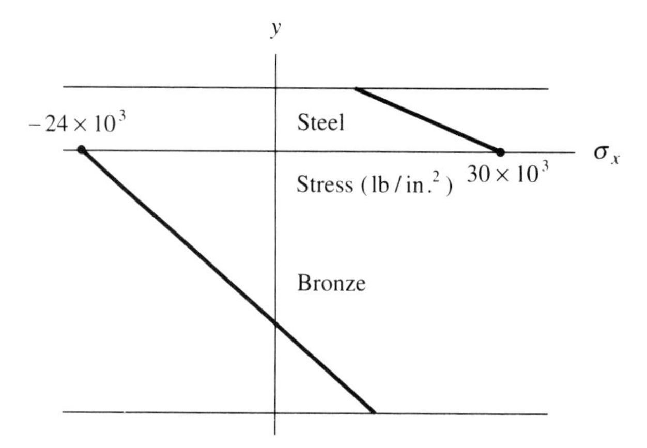

The foregoing analysis is founded on several sweeping assumptions. Firstly, we assumed $\sigma_y = \sigma_z = 0$; neither is true throughout. A better result is obtained if we do not assume $\sigma_z = 0$, but only that the resultant upon the edges ($z = \pm b$) vanishes. Secondly, we assumed that the strains and stresses are independent of x and z. Indeed, this is reasonably valid at a distance from the ends and the edges. Actually very large, but localized, stresses occur in narrow zones near the ends and edges. More complicated analyses are required to determine such concentrated stresses. Our prediction of the deflection is quite good. ◆

PROBLEMS

In each of the following problems the properties vary through the depth; $E = E(y)$. Because of the nonhomogeneity and asymmetry, the position of the neutral axis is not evident. We can simply employ linear equation (5.158) and place the origin at a *convenient* point; see Examples 14 and 17.

5.186 A beam has the composite cross section shown in Fig. P5.186. Both parts are HI-HO, but the modulus of one is twice that of the other. (a) Locate the neutral line, (b) obtain the moment-curvature relation in terms of the modulus E and length h, and (c) determine the stress distribution.

Fig. P5.186

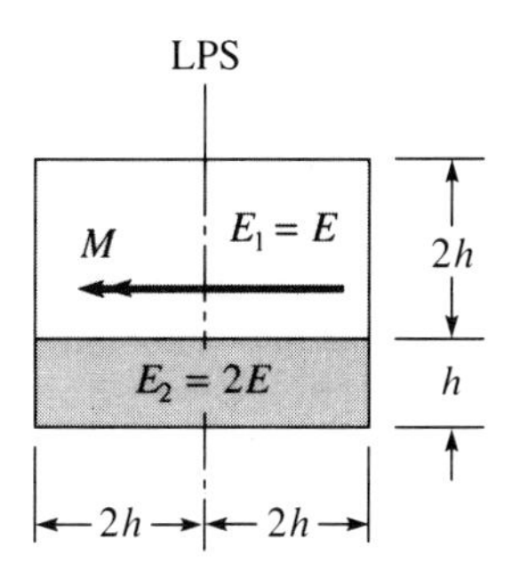

5.187 A beam has the composite cross section depicted in Fig. P5.187. The HI-HO materials are bonded together. The flange and web have different moduli:

$$E_1 = 20 \times 10^3 \text{ MPa}, \qquad E_2 = 40 \times 10^3 \text{ MPa}$$

These materials yield at their respective yield stresses:

$$\sigma_1 = \pm 20 \text{ MPa}, \qquad \sigma_2 = \pm 40 \text{ MPa}$$

Determine the bending couple M_0 that initiates yielding.

Fig. P5.187

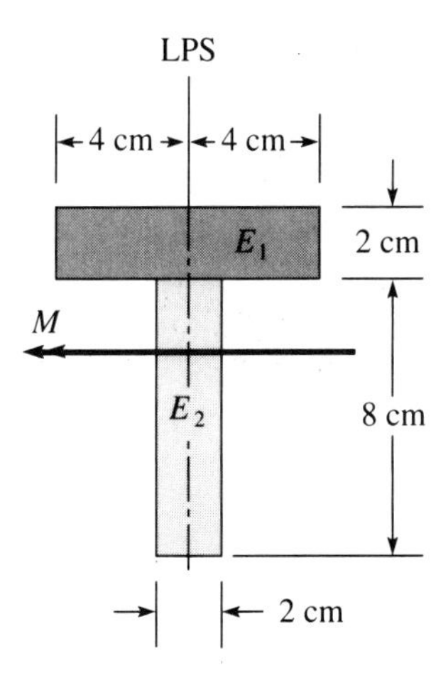

5.188 A reinforced concrete beam has the rectangular cross section shown in Fig. P5.188. The modulus for this concrete (assumed HI-HO) is $E_c = 4 \times 10^6$ psi; the modulus for the steel is 30×10^6 psi. The four steel rods are 1 in. in diameter. Note that the cross-sectional area of the rods is small compared to the area of the concrete. If the bending couple is 86×10^3 ft-lb, compute the maximum and minimum normal stresses.

Fig. P5.188

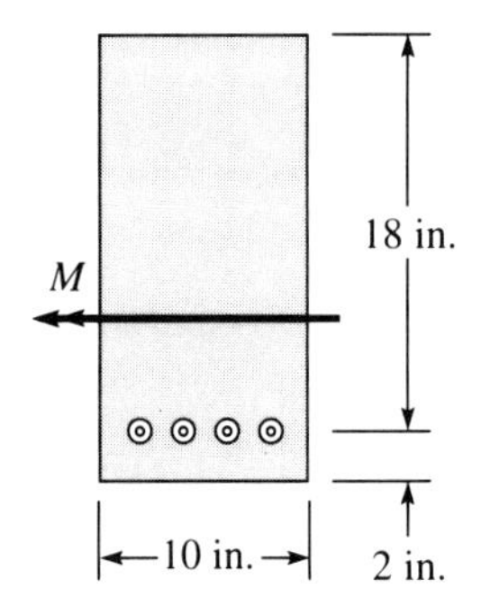

5.189 If the materials of Prob. 5.186 are ideally plastic at their respective yield stresses, determine the limit couple: $\sigma_1 = \pm 200$ MPa, $\sigma_2 = \pm 300$ MPa, and $h = 4$ cm.

5.190 A beam has a square cross section 2 in. × 2 in. It is composed of a fiber-reinforced material that is hookean, but, because of the distribution of the fibers, the modulus varies:

$$E = 4 \times 10^6 \left(1 + \frac{y^2}{2}\right) \text{psi}$$

Here y is the distance (inches) from one face, as depicted in Fig. P5.190.

a. Locate the neutral line when the beam is subjected to the bending couple M.

Fig. P5.190

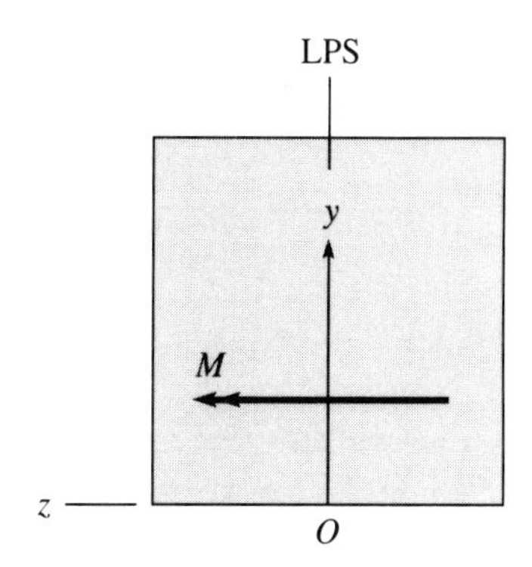

b. Determine the modulus of rigidity, that is, the constant of proportionality $K = M/\rho$.

5.191 A bow has a composite cross section of three different HI-HO materials. The backing (bottom in Fig. P5.191) is a *thin* layer of fiberglass reinforced plastic ($E_B t = 2 \times 10^5$ lb/in.^{2}). The core is an impregnated wood ($E_C = 2 \times 10^6$ lb/in.^{2}), and the facing (top layer) is a dense plastic ($E_F = 4 \times 10^6$ lb/in.^{2}).

a. Determine the flexural rigidity $K = M\rho$.

b. Locate the neutral line.

c. Compute the maximum and minimum normal stresses when $M = 14 \times 10^2$ in.-lb.

Fig. P5.191

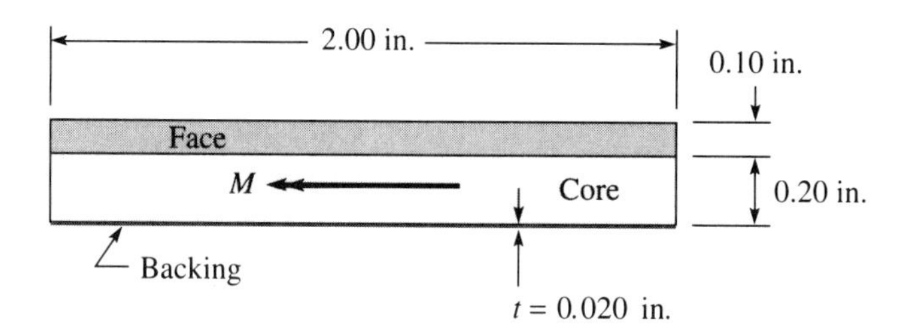

5.16 Thin-Walled Vessels

The Spherical Shell

Many vessels are used in practice to contain gases or fluids under pressure. Frequently, the wall thickness of these vessels is small compared to their lineal dimensions; in such a case we call the vessel a *thin-walled shell.* The simplest shapes of such vessels are cylinders and spheres. We begin with the spherical shell.

Equilibrium Figure 5.58 shows a hemispherical portion of a vessel that contains a fluid under pressure. The force against the inner wall of the container, therefore, is due to a hydrostatic pressure, p. We suppose that the wall thickness is constant throughout the spherical shell. If R denotes the mean radius of the sphere, then t is usually so small compared to R that t/R can be regarded as a small number. In this case R is approximately the inner radius as well as the outer radius of the shell.

Fig. 5.58

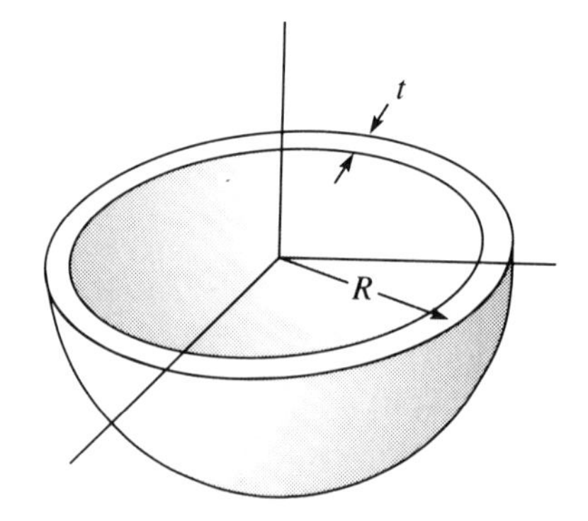

Figure 5.59 depicts a free-body diagram of half a spherical shell and the fluid contained in that half. The fluid pressure p causes a normal force F^f to act on the exposed surface area of the fluid. For equilibrium it must be balanced by a normal force F acting on the exposed area of the shell:

$$F + F^f = 0$$

Fig. 5.59

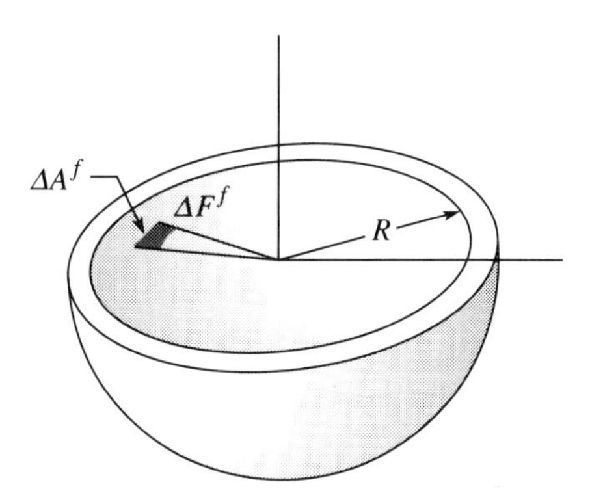

On the element of fluid surface area ΔA^f, we have a normal force ΔF^f, due to the pressure p:

$$\Delta F^f = -p\,\Delta A^f$$

Then, over the entire fluid surface,

$$F^f = \iint_{A^f} -p\,dA^f = -\pi R^2 p \tag{5.168}$$

Hence

$$F = -F^f = \pi R^2 p \tag{5.169}$$

is the required normal force on the exposed portion of the shell.

Description of the Deformation The loading is symmetric about the center of the spherical shell, and the wall of the vessel is of constant thickness. If the material is homogeneous and isotropic, the deformation, as well, should be symmetric about the center of the shell. Consequently, each particle at radius R undergoes the same radial displacement u. A small portion of the unstrained shell is shown in Fig. 5.60. The middle surface, $PABC$, lies in a sphere of radius R. In Fig. 5.61 the portion of the middle surface is shown in its deformed state, $P^*A^*B^*C^*$. This lies in a sphere of radius $R + u$, where u is the radial displacement of all points of $PABC$. If the segment PA subtends the angle $\Delta\theta$, then P^*A^* subtends the same angle. Thus,

$$PA = R\,\Delta\theta, \qquad P^*A^* = (R + u)\,\Delta\theta$$

Fig. 5.60

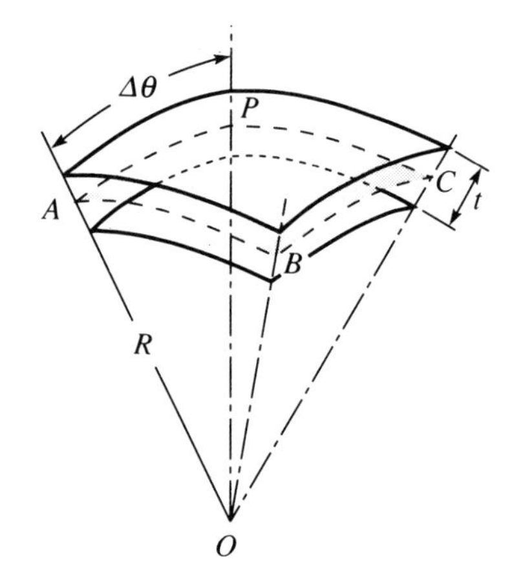

Fig. 5.61

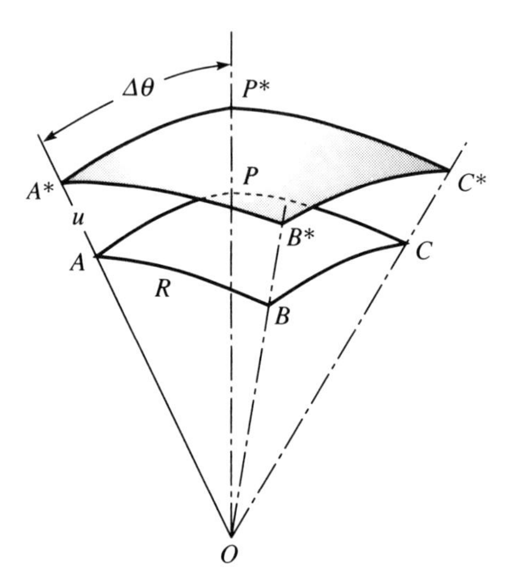

and the extensional strain in the θ-direction is simply

$$\varepsilon_\theta(P) = \lim_{A \to P} \frac{P^*A^* - PA}{PA} = \lim_{\Delta\theta \to 0} \frac{(R+u)\Delta\theta - R\,\Delta\theta}{R\,\Delta\theta} = \frac{u}{R} \tag{5.170}$$

Because of radial symmetry, the arc PA could be in any tangential direction. It therefore follows that the extensional strain in any direction tangent to the middle surface is the same:

$$\varepsilon_t(P) = \varepsilon_\theta(P) = \frac{u}{R} \tag{5.171}$$

If expansion of the thin spherical shell is the dominant feature of this deformation, any changes in thickness of the shell should be small compared with the displacement of the middle surface. This means that we can approximate the increase in radius of any tangential arc in the shell as u, the same as for the middle surface. Furthermore, for thin shells ($t \ll R$), the radius of any arc in the shell is approximately R. Thus, the tangential strain at any point in the spherical shell is approximately the same:

$$\varepsilon_t = \frac{u}{R} \tag{5.172}$$

Linearly Elastic Material If the shell is composed of HI-HO material, we use Hooke's law to obtain

$$\begin{aligned} \varepsilon_t &= \frac{1}{E}[\sigma_t - \nu(\sigma_t + \sigma_r)] = \frac{1-\nu}{E}\sigma_t - \frac{\nu}{E}\sigma_r \\ \varepsilon_r &= \frac{1}{E}[\sigma_r - \nu(\sigma_t + \sigma_t)] = \frac{\sigma_r}{E} - \frac{2\nu}{E}\sigma_t \end{aligned} \tag{5.173}$$

where ε_r and σ_r are, respectively, the extensional strain and the normal stress in the radial direction. We do not have sufficient information at this point to use Eq. (5.173); some additional assumptions are needed. The radial stress σ_r must vary from the value $-p$ at the inner surface of the shell to the value zero at the outer surface. It is, therefore, reasonable to suppose that, in the small thickness of the shell, σ_r will always lie within these bounds. We now assume that σ_r is everywhere small compared with σ_t, and we will therefore neglect it in Eq. (5.173). The basis for such an assumption is not immediately clear. We can, however, accept it tentatively and compare the resulting value of σ_t with p to see how well the assumption holds. Proceeding on this basis, we neglect σ_r in Eq. (5.173), thereby obtaining the relation

$$\varepsilon_t = \frac{1-\nu}{E}\sigma_t, \quad \text{or} \quad \sigma_t = \frac{E}{1-\nu}\varepsilon_t \tag{5.174}$$

With Eq. (5.172), this becomes

$$\sigma_t = \frac{E}{1-\nu}\frac{u}{R} \tag{5.175}$$

which is approximately constant throughout the shell.

(These results are applicable even if the sphere is also subjected to a moderate external pressure p_e in addition to an internal pressure p_i. In this case we take the equivalent pressure p as the difference $p_i - p_e$. Although σ_r now varies between $-p_i$ and $-p_e$, if p is comparable to p_i or p_e, we still neglect σ_r in comparison to σ_t.)

In Fig. 5.62, the normal force ΔF on the element of area ΔA of the shell is

$$\Delta F = \sigma_t \Delta A = \left(\frac{E}{1-\nu}\frac{u}{R}\right)(tR\,\Delta\theta) = \frac{Et}{1-\nu}u\,\Delta\theta$$

Then the normal force on the exposed shell area is

Fig. 5.62

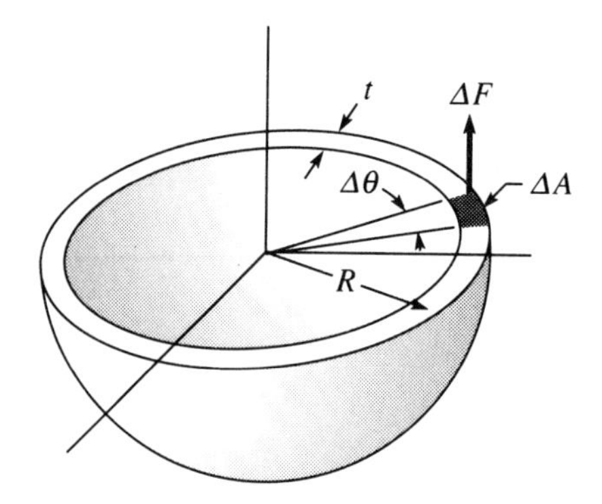

$$F = \int_0^{2\pi} \frac{Et}{1-\nu} u \, d\theta = \frac{2\pi Etu}{1-\nu} \tag{5.176}$$

But, from Eq. (5.169)

$$F = \pi R^2 p$$

Combining this with Eqs. (5.175) and (5.176), we have, finally,

$$\sigma_t = \frac{E}{1-\nu} \frac{u}{R} = \frac{pR}{2t} \tag{5.177}$$

For thin shells $t \ll R$, so that $\sigma_t \gg p$, which supports our presumption that $\sigma_r \ll \sigma_t$.

The extensional strain ε_t is computed by substituting Eq. (5.177) into Eq. (5.174):

$$\varepsilon_t = \frac{1-\nu}{E} \sigma_t = \frac{1-\nu}{E} \frac{pR}{2t} \tag{5.178}$$

The increase in radius of the middle surface is u. Equation (5.177) gives

$$u = \frac{1-\nu}{E} \frac{pR^2}{2t} \tag{5.179}$$

Once again we emphasize that Eqs. (5.177), (5.178), and (5.179) are based on the presumption that the shell is thin. Additionally, we have neglected the radial normal stress σ_r and its variation through the thickness of the shell. The results are also based on a complete spherical container without irregularities, such as pipe inlets or welds. Again, Saint-Venant's principle enables us to use Eq. (5.177) at points not near such irregularities. Also, from equilibrium only, the *mean* value of the stress σ_t is given by Eq. (5.177). Also, note that radial symmetry implies that the strain γ_{tr} and stress τ_{tr} must vanish.

Ideally Elastic-Plastic Material Suppose that the spherical shell is made from an ideally elastic-plastic material. In the present case the material of the shell is not subjected to simple tension or simple shear; an element cut from the shell, as in Fig. 5.63, is subjected to a biaxial state of stress (neglecting σ_r in comparison to σ_t). To

Fig. 5.63

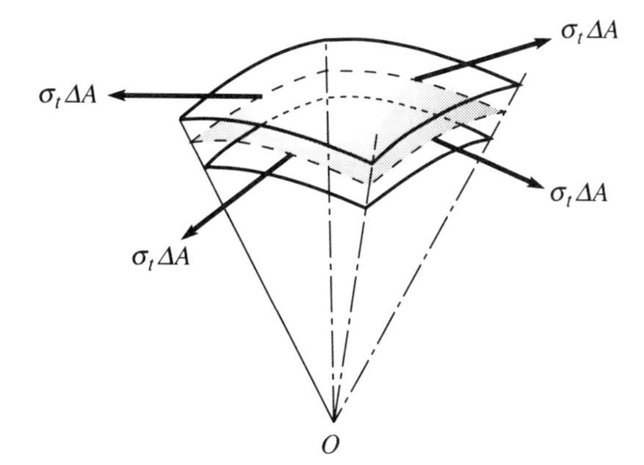

determine when yielding occurs, we resort to a theory of failure; in this case let us use the maximum shear stress theory of Tresca (Sec. 4.15).

Suppose the material, in a simple tension test, yields when the normal stress reaches the value σ_0. At this point, the maximum shear stress in the specimen is

$$\tau_0' = \tfrac{1}{2}\sigma_0 \tag{5.180}$$

For the element of shell in Fig. 5.63, the maximum shear stress occurs on planes that intersect the radius at 45°, bisecting the right angle between a radial plane carrying the maximum stress σ_t and a tangential plane carrying the minimum stress σ_r $(= 0)$. The maximum shear stress in the shell is, therefore,

$$\tau_{\max} = \frac{\sigma_t - \sigma_r}{2} = \frac{1}{2}\sigma_t \tag{5.181}$$

The maximum shear stress theory of Tresca then states that yielding in the shell occurs when

$$\tau_{\max} = \tau_0'$$

or, from Eqs. (5.180) and (5.181), when

$$\sigma_t = \sigma_0 \tag{5.182}$$

As the pressure p is increased from zero, the behavior is linear and elastic until the stress σ_t satisfies the yield criterion, Eq. (5.182). Then by Eq. (5.177), the pressure p has the value p_0 defined by

$$\sigma_0 = \frac{p_0 R}{2t}$$

or

$$p_0 = \frac{2t\sigma_0}{R} \tag{5.183}$$

When p attains the value p_0, unrestricted deformation takes place; p_0 is the limit load.

The Cylindrical Shell

The analysis of a linearly elastic thin-walled cylindrical pressure vessel will only be outlined here; the details are left for the reader to develop. A portion of a cylindrical vessel containing fluid under pressure is shown in Fig. 5.64. On a transverse cut, as in Fig. 5.65, the fluid pressure p produces a normal force F_x^f on the exposed fluid:

$$F_x^f = -\pi R^2 p$$

Fig. 5.64

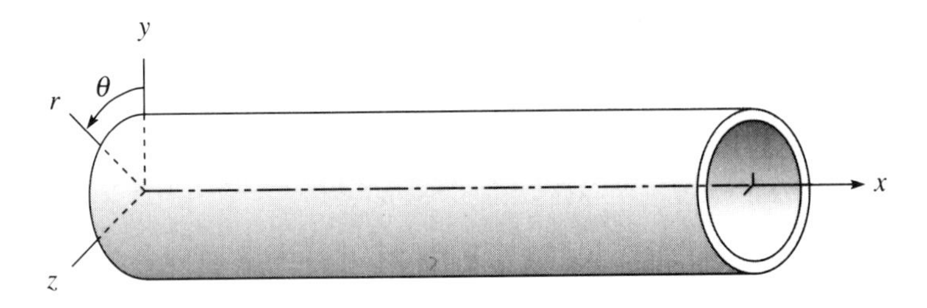

Fig. 5.65

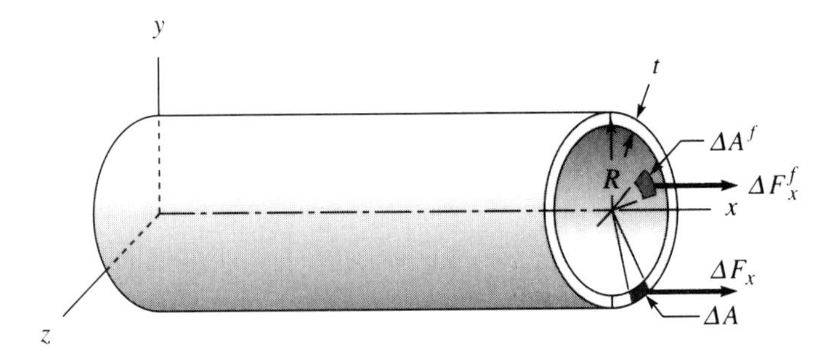

For equilibrium, this must be balanced by a normal force F_x on the exposed shell surface:

$$F_x = -F_x^f = \pi R^2 p \tag{5.184}$$

On the longitudinal cut of a slice of the cylinder, as in Fig. 5.66, the fluid pressure produces normal force F_θ^f on the exposed fluid:

$$F_\theta^f = -2Rlp$$

The normal force on the longitudinal portion of the shell must then be

$$F_\theta = -F_\theta^f = 2Rlp \tag{5.185}$$

In Fig. 5.67(a), $PABC$ is the middle surface of an element of unstrained shell. We assume that the deformation is symmetric about the axis of the cylinder; also a cross section (perpendicular to the axis) remains a cross section, and the axis remains straight (as in axial extension). Then u is the increase in radius to every point of

Fig. 5.66

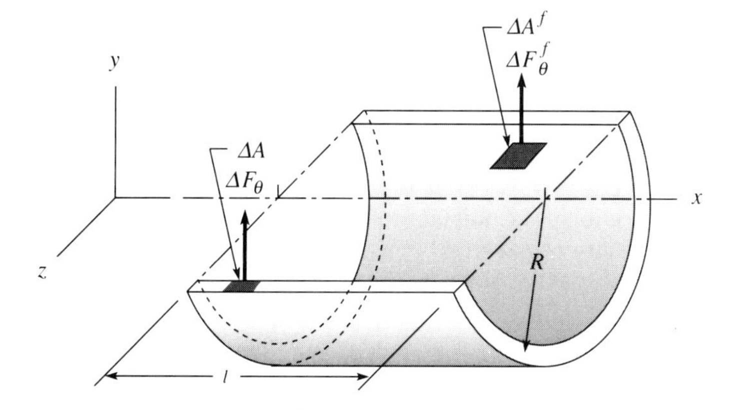

Fig. 5.67

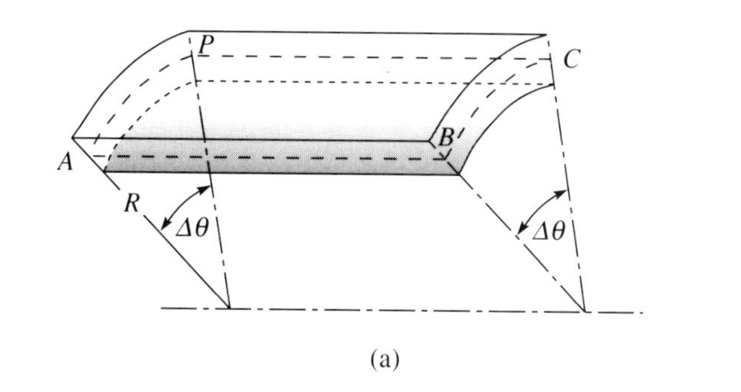

(a)

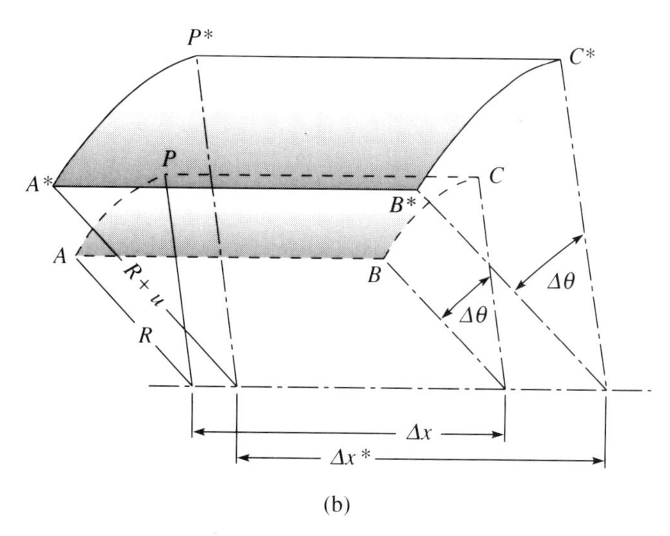

(b)

$P^*A^*B^*C^*$, the deformed portion of middle surface. The angle subtended by P^*A^* is $\Delta\theta$, the same as the angle subtended by PA. Thus,

$$\varepsilon_\theta(P) = \frac{u}{R}$$

We now assume that any change in thickness of the shell is insignificant compared with the increase in radius. It then follows that for any point in the shell,

$$\varepsilon_\theta = \frac{u}{R} \tag{5.186}$$

As in the axial extension problem, the extensional strain ε_x is constant on every cross section perpendicular to the axis of the shell.

If normal stress σ_r is neglected, as before, then extensional strains of the HI-HO material are given by the equations

$$\begin{aligned} \varepsilon_\theta &= \frac{1}{E}[\sigma_\theta - \nu\sigma_x] \\ \varepsilon_x &= \frac{1}{E}[\sigma_x - \nu\sigma_\theta] \end{aligned} \tag{5.187}$$

Inversely,

$$\begin{aligned} \sigma_\theta &= \frac{E}{1-\nu^2}[\varepsilon_\theta + \nu\varepsilon_x] \\ \sigma_x &= \frac{E}{1-\nu^2}[\varepsilon_x + \nu\varepsilon_\theta] \end{aligned} \tag{5.188}$$

The forces F_x and F_θ are related to these stresses as follows:

$$F_x = \int_0^{2\pi} \sigma_x Rt\,d\theta = \frac{2\pi RtE}{1-\nu^2}\left[\varepsilon_x + \nu\frac{u}{R}\right]$$

$$F_\theta = 2\int_x^{x+l} \sigma_\theta t\,dx = \frac{2ltE}{1-\nu^2}\left[\frac{u}{R} + \frac{\nu}{l}\int_x^{x+l} \varepsilon_x\,dx\right]$$

From Eqs. (5.184) and (5.185) we have

$$\varepsilon_x + \nu\frac{u}{R} = \frac{1-\nu^2}{E}\frac{R}{2t}p$$

$$\frac{\nu}{l}\int_x^{x+l} \varepsilon_x\,dx + \frac{u}{R} = \frac{1-\nu^2}{E}\frac{R}{t}p$$

When $l \to 0$, the last equation becomes

$$\nu\varepsilon_x + \frac{u}{R} = \frac{1-\nu^2}{E}\frac{R}{t}p$$

so that

$$\begin{aligned} \frac{u}{R} &= \frac{2-\nu}{E}\frac{R}{2t}p \\ \varepsilon_x &= \frac{1-2\nu}{E}\frac{R}{2t}p \end{aligned} \tag{5.189}$$

and

$$\sigma_\theta = \frac{R}{t}p$$
$$\sigma_x = \frac{R}{2t}p \qquad (5.190)$$

If the material is ideally plastic and yields according to the maximum shear stress yield criterion, the limit load is given by

$$p_0 = \frac{t}{R}\sigma_0$$

where σ_r has again been neglected.

5.192 Long cylindrical pipes carrying fluids under pressure cannot be treated as close-ended pressure vessels, since the ends of the pipe are usually restrained from axial motion. It is, therefore, reasonable to assume that the axial forces applied to the ends of the pipe are such as to insure $\varepsilon_x = 0$. Under these conditions, show that

$$\sigma_\theta = \frac{R}{t}p, \qquad \sigma_x = \frac{\nu R}{t}p$$

$$\frac{u}{R} = \varepsilon_\theta = \frac{1-\nu^2}{E}\sigma_\theta = \frac{1-\nu^2}{Et}Rp$$

5.193 A steel water pipe has a 40-in. diameter and a 1-in. wall thickness. The pipe carries water under 100 psi pressure and lies at the bottom of a lake where the external pressure is 40 psi. Since the pipe is long, assume that the axial strain is zero (see Prob. 5.192). Compute the circumferential and axial stress in the pipe and also the percentage increase in pipe diameter. Assume linearly elastic behavior: $E = 30 \times 10^6$ psi, $\nu = 0.3$.

5.194 An aluminum alloy tube (closed ends), 2.50 cm in diameter and 0.125 cm in wall thickness, bursts at an internal pressure of 8.4 MPa. Estimate the ultimate strength of the aluminum alloy using a maximum shear stress criterion and neglecting the radial component of stress.

5.195 A long aluminum pipe (no axial strain, as in Prob. 5.192), 4 in. in diameter and 0.062 in. in wall thickness, bursts at an internal pressure of 1040 psi. Estimate the ultimate strength of this aluminum alloy using a maximum shear stress criterion and neglecting the radial component of stress. Assume hookean behavior to the point of rupture, with $\nu = 0.25$. Also estimate what the bursting pressure would be if the pipe had closed ends and was free to expand axially.

5.196 A cylindrical thin-walled pressure vessel (closed ends) is made of HI-HO material. Compare the percent increase in length of the cylinder to the percent increase in diameter when an internal pressure is introduced.

5.197 An aluminum pipe (70×10^3 MPa, $\nu = 0.25$) is compressed exactly 1 in. and placed between two smooth, but rigid, walls, as shown in Fig. P5.197. The internal pressure is then increased until the limit pressure p_0 is reached. (At p_0 the pipe has shortened enough so that leakage occurs at the smooth walls if an attempt is made to increase the pressure.) Compute p_0.

Fig. P5.197

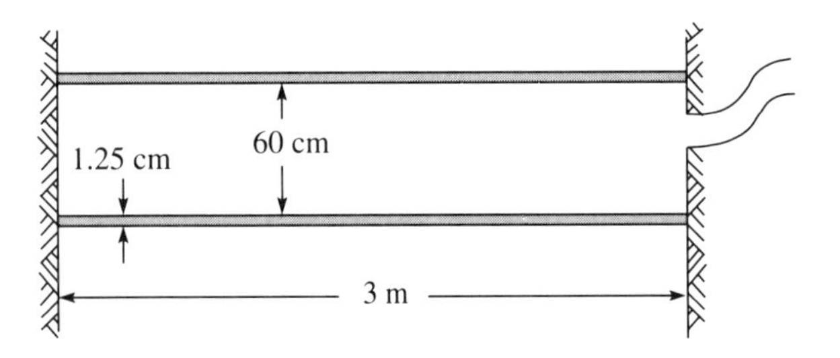

5.198 A steel stand pipe has a 1-ft diameter and a $\frac{1}{3}$-in. wall thickness. It is filled to a height of 100 ft with water (density 62.4 lb/ft^3). Assume that in any short length of the pipe, the radial displacement of every point is the same, although the variation in radial displacement from the top to the bottom is significant. Also neglect the weight of the pipe itself and the radial stress in it. Compute the circumferential stress in the pipe at any height x feet above the bottom of it.

5.199 A thin-walled cylinder of inside radius R and thickness t is forced over another cylinder of the same wall thickness but of outside radius $R + \delta$, where δ is small. Both cylinders are of the same HI-HO material. Compute the circumferential stress in each cylinder. (They are free to expand axially.)

5.200 A composite cylinder is made by sliding a steel sleeve (E_s, ν_s) over an aluminum cylinder (E_a, ν_a), as shown in Fig. P5.200. The ends of this composite cylinder are then closed and an internal pressure p is introduced. The closure is affixed to both layers of the composite so that both experience the same axial extension. Make the same basic assumptions as are made if the cylinder is of one material, and relate the circumferential stress and axial stress in each material to the pressure.

Fig. P5.200

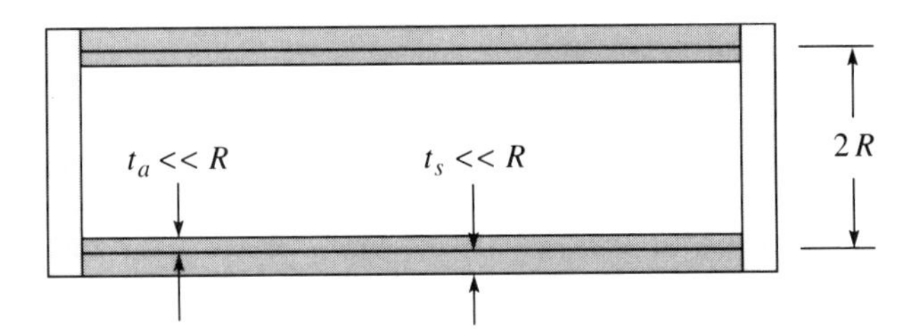

5.17 Miscellaneous Problems

The following problems are intended as further exercise in relating the fundamental concepts of stress, strain, and material behavior.

5.201 A thin plate of linearly elastic material is bent by couples uniformly distributed along two opposite edges, as shown in Fig. P5.201. Assume as for a beam, that there is a neutral axis (in fact, the x-axis) and that the cross sections remain plane and perpendicular to the neutral axis. However, because of constraints along the edges, it is better to assume $\varepsilon_z = 0$, rather than to assume $\sigma_z = 0$. (However, the assumption $\sigma_y = 0$ is still acceptable.) Obtain the bending couple-curvature relation for this plate.

Fig. P5.201

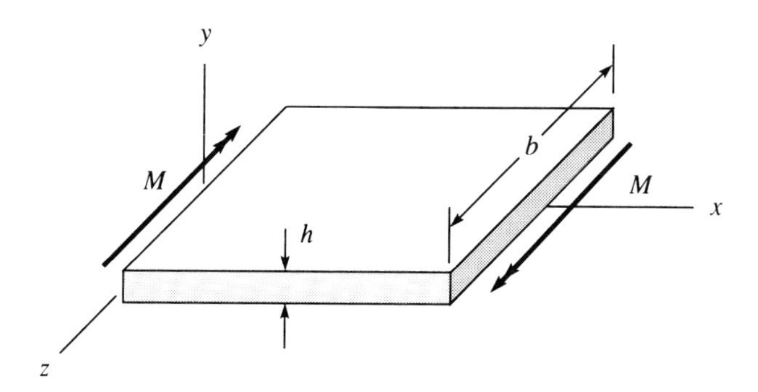

5.202 A flexible mount is shown in Fig. P5.202. Assume that the post A and the frame C are rigid and the rubber pads B deform so that planes remain plane. Relate the (small) displacement of the post, δ, to the load P. Assume that the rubber is HI-HO with a shear modulus G.

Fig. P5.202

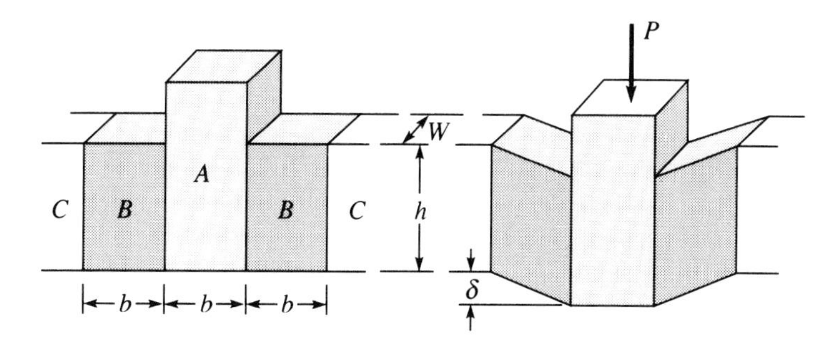

5.203 Repeat Prob. 5.202 for the mount shown in Fig. P5.203. Use $\frac{1}{2}$ for the value of Poisson's ratio for the rubber.

Fig. P5.203

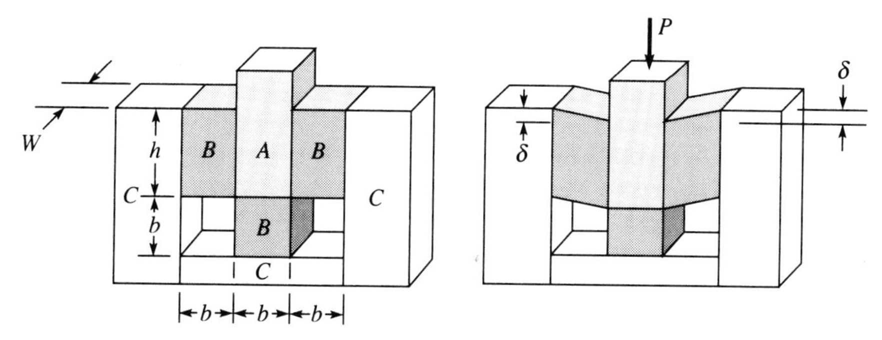

5.204 The flexible mounting shown in Fig. P5.204 consists of a rigid cylindrical plug in a thick-walled rubber cylinder whose outer surface is held fixed. When the load P is applied to the plug, it displaces a small amount δ. Assume that the deformation in the rubber cylinder is symmetric about the axis and that every cross section (perpendicular to the axis) deforms to a similar surface. Also assume that the rubber is HI-HO with a shear modulus G. Show that

$$\tau_{rz} = \frac{P}{2\pi r L}$$

and hence that

$$\delta = \int_b^R \gamma_{rz}\, dr = \frac{P}{2\pi L G} \log \frac{R}{b}$$

Fig. P5.204

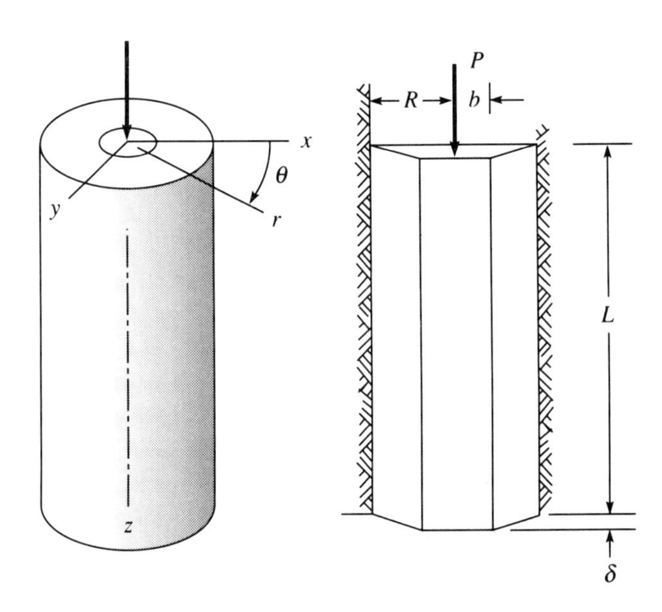

5.205 The rigid plug of the flexible mount described in the preceding problem is twisted through a small angle ϕ by a twisting couple T, as indicated in Fig. P5.205. Assume that the deformation in the rubber cylinder is symmetric about the axis and that every cross section (perpendicular to the axis) remains plane and deforms like every other. Show that

$$\tau_{r\theta} = \frac{T}{2\pi r^2 L}$$

and hence that

$$\phi = \int_b^R \frac{\gamma_{r\theta}}{r}\, dr = \frac{T}{4\pi L G}\left(\frac{1}{b^2} - \frac{1}{R^2}\right)$$

Fig. P5.205

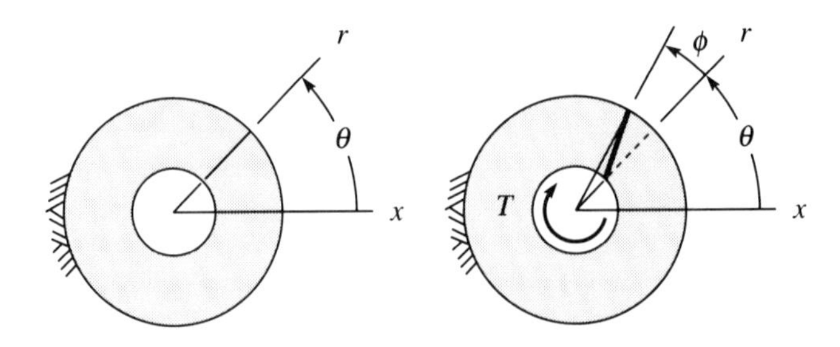

5.206 A segment of a circular ring (with an LPS) is subjected to end couples M and bends so that (a) there is an arc of particles (called the neutral axis) that deforms into another arc without stretching and (b) every cross section of the ring remains plane and perpendicular to the neutral axis. Referring to Fig. P5.206, R is the initial (undeformed) radius of curvature of the neutral axis, and ρ is the final (deformed) radius of curvature. For a linearly elastic material, assume $\sigma_y = \sigma_z = 0$ and show that

$$\sigma_\theta = -\frac{My}{I^*(1 - y/R)}$$

$$\frac{1}{\rho} = \frac{1}{R} + \frac{M}{EI^*}$$

where the neutral axis is located from

$$\iint_S \frac{y\,dA}{1 - y/R} = 0$$

and

$$I^* = \iint_S \frac{y^2\,dA}{1 - y/R}$$

Fig. P5.206

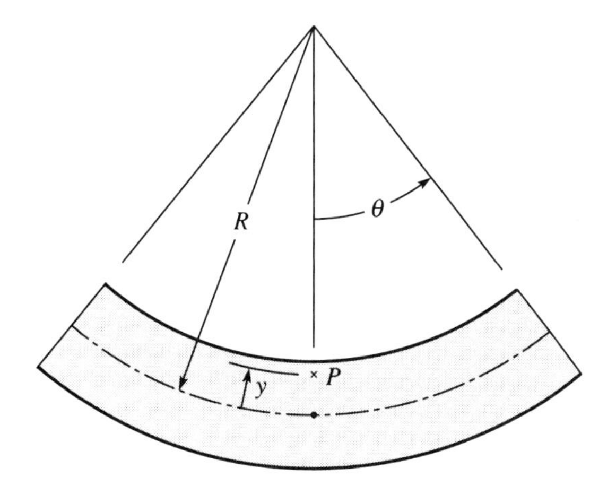

(a) Undeformed

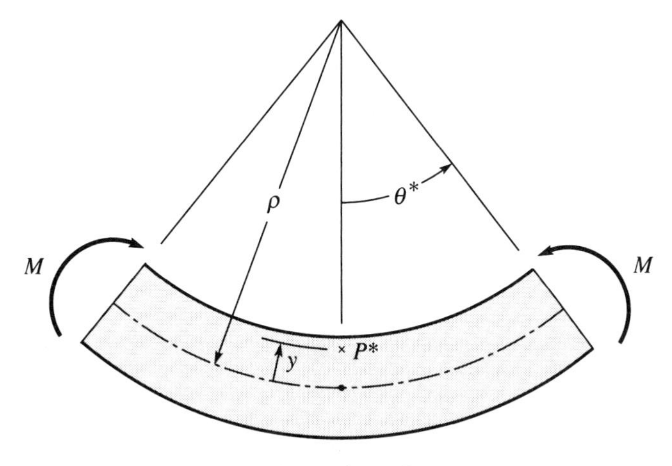

(b) Deformed

5.207 A thin rectangular plate (thickness t), held fixed on two adjacent edges, is bonded to a slender rod along a third edge (the cross-sectional area of the rod is A). When an axial load P is applied to the rod, it extends according to the usual assumptions for axial extension. The plate deforms in such a way that initially horizontal planes remain plane, as shown in Fig. P5.207. The plate and rod are made of HI-HO materials, E is the modulus of elasticity of the rod, and G is the modulus of rigidty of the plate.

a. Show that the shear strain in the plate is given by

$$\gamma_{xy}(x) = \frac{u(x)}{b}$$

and the normal strain in the rod is

$$\varepsilon_x(x) = \frac{d}{dx} u(x)$$

where $u(x)$ is the displacement of the cross section in the rod at x.

b. Show that consideration of the equilibrium of a slice from the rod leads to

$$A \frac{d}{dx} \sigma_x(x) - t\tau_{xy}(x) = 0$$

Hence

$$\frac{d^2}{dx^2} u(x) - \frac{Gt}{AEb} u(x) = 0$$

c. The solution of the preceding equation is

$$u(x) = C_1 \sinh \lambda \frac{x}{L} + C_2 \cosh \lambda \frac{x}{L}$$

where $\lambda \sqrt{\frac{GtL^2}{AEB}}$ and C_1 and C_2 are constants.

Since we must have

$$\tau_{xy}(0) = 0, \qquad \sigma_x(L) = \frac{P}{A}$$

show that

$$C_1 = \frac{PL}{AE} \frac{1}{\lambda \cosh \lambda}, \qquad C_2 = 0$$

Consequently,

Fig. P5.207

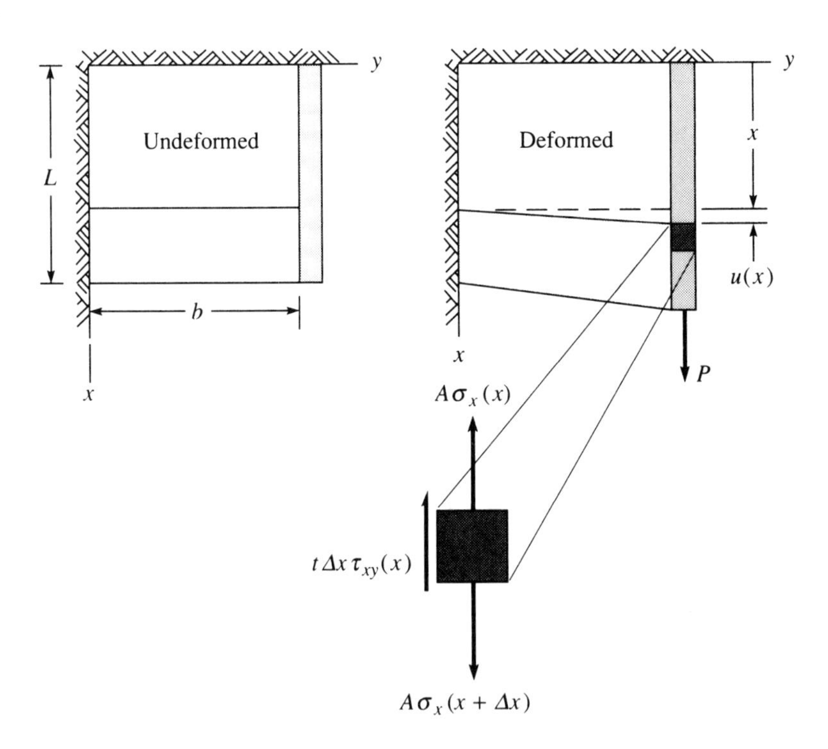

$$u(x) = \frac{PL}{AE} \frac{\sinh \lambda \frac{x}{L}}{\lambda \cosh \lambda}$$

5.208 A thick-walled cylinder is subjected to an internal pressure p_i and an external pressure p_0, as shown in Fig. P5.208(a). For convenience we assume that there is no axial stress in the cylinder; that is, $\sigma_z = 0$. Assume also that the deformation is symmetric about the axis of the cylinder; each point displaces radially an amount $u(r)$.

a. Show that

$$\varepsilon_r = \frac{du}{dr} \tag{a}$$

$$\varepsilon_\theta = \frac{u}{r} \tag{a'}$$

b. Remove a slice from the cylinder (say of thickness t) and remove an element from this slice, as shown in Fig. P5.208(b). Show that a consideration of the equilibrium of this element leads to

$$\sigma_\theta = \sigma_r + r\frac{d\sigma_r}{dr} \tag{b}$$

c. If the material is HI-HO, show that

$$\frac{du}{dr} = \frac{1-\nu}{E}\sigma_r - \frac{\nu}{E} r \frac{d\sigma_r}{dr} \tag{c}$$

Fig. P5.208

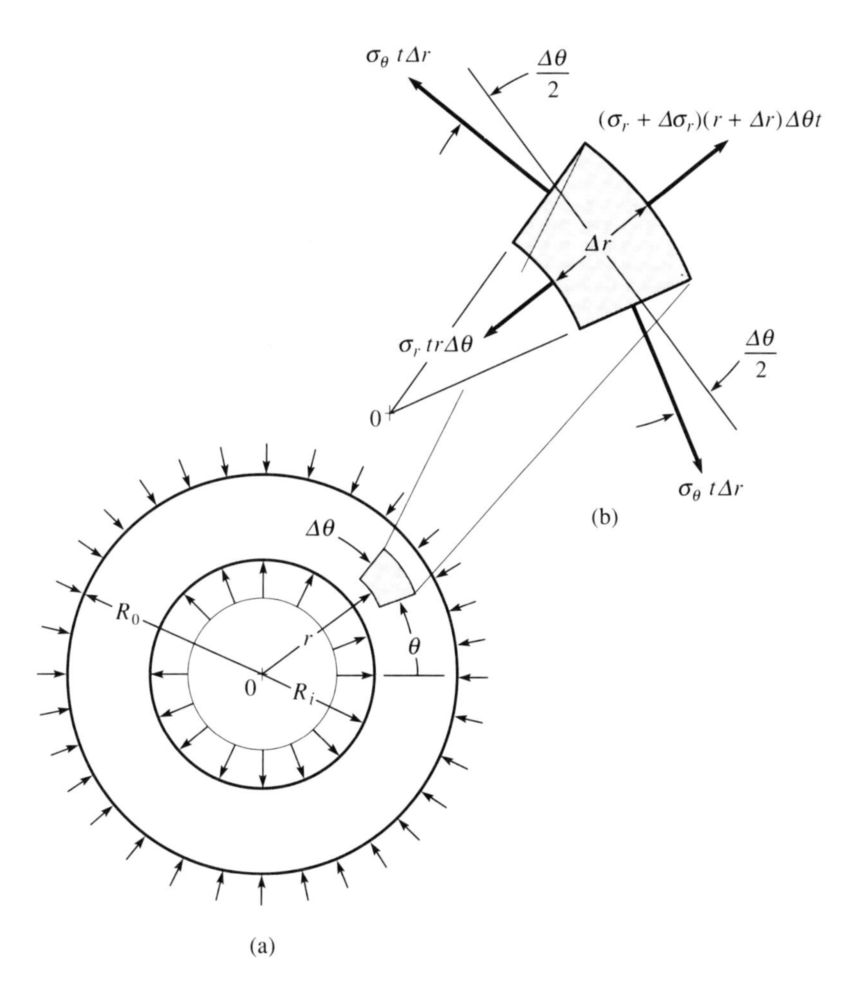

$$\frac{u}{r} = \frac{1-\nu}{E}\sigma_r + \frac{1}{E}r\frac{d\sigma_r}{dr} \tag{c'}$$

d. Eliminate u between Eqs. (c) and (c′) to obtain

$$\frac{d^2\sigma_r}{dr^2} + \frac{3}{r}\frac{d\sigma_r}{dr} = 0$$

e. The solution of Eq. (d) is

$$\sigma_r = C_1 + \frac{C_2}{r^2}$$

where C_1 and C_2 are constants. Show that

$$C_1 = \frac{R_i^2 p_i - R_0^2 p_0}{R_0^2 - R_i^2}, \qquad C_2 = -\frac{R_0^2 R_i^2 (p_i - p_0)}{R_0^2 - R_i^2}$$

Hence,

$$\sigma_r = -\frac{R_0^2 R_i^2}{R_0^2 - R_i^2}\left[\frac{p_0}{R_0^2} - \frac{p_i}{R_i^2} + \frac{p_i - p_0}{r^2}\right] \tag{e}$$

$$\sigma_\theta = \frac{R_0^2 R_i^2}{R_0^2 - R_i^2}\left[-\frac{p_0}{R_0^2} + \frac{p_i}{R_i^2} + \frac{p_i - p_0}{r^2}\right] \tag{e'}$$

and

$$u = \frac{1}{E}\left(\frac{R_0^2 R_i^2}{R_0^2 - R_i^2}\right)\left[(1-\nu)\left(\frac{p_i}{R_i^2} - \frac{p_0}{R_0^2}\right)r + (1+\nu)\frac{p_i - p_0}{r}\right] \tag{e''}$$

5.209 For the thick-walled cylinder of Prob. 5.208, assume $\varepsilon_z \equiv 0$ rather than $\sigma_z \equiv 0$, and obtain results for this case that are identical to Eqs. (c), (c′), and (e″), except that E is replaced everywhere by $E^* = E/(1-\nu^2)$ and ν is replaced everywhere (except in E^*) by $\nu^* = \nu/(1-\nu)$.

6 Bending of Symmetric Beams

In Secs. 5.11 through 5.14, we described the bending of symmetric beams under the action of couples only. Then every slice behaves the same. Certain results of simple bending are also applicable to nonuniform distributions of symmetric loads, wherein each segment experiences different actions and deformation. The extension of the earlier analyses and results entails a degree of approximation and also additional mathematical tools. As we extend the early analysis, we must be mindful of the limitations inherent in the approximation. Still, the resulting theory of beams remains an important and useful tool in the hands of the practicing engineer.

6.1 The Symmetric Beam

The beam, symmetric about a plane (LPS), was described in Sec. 5.11 as a prelude to the special circumstance of simple bending. In practice, the engineer seldom encounters *simple* bending but often employs beams to support various transverse loads, which cause bending, strains, and stresses. We now consider such problems but limit our immediate attention to circumstances wherein the beam, supports, and loadings are symmetric about the plane.

6.2 Loads, Internal Forces, and Equilibrium

Symmetric Loads and Supports

Loads and reactions are addressed in Sec. 2.1. Recall now that the reaction of a support upon any body can be represented by a resultant consisting of a force acting at a point and a couple; each can be represented by three components. Also recall that such representation provides no specific information about the distribution of the forces and therefore does not admit an analysis of the local behavior—that is, strains or stresses in the immediate neighborhood. Once again, by virtue of St. Venant's principle (Sec. 5.2), the overall response, as well as strains and stresses at remote locations, is little affected by the actual distribution. With an awareness of such limitations, let us consider the nature of the reactions that arise in the present circumstance. To be specific, let us denote the reactions of a support by a force **R** and couple **M**:

$$\mathbf{R} = F_x\hat{\mathbf{i}} + R_y\hat{\mathbf{j}} + R_z\hat{\mathbf{k}} \tag{6.1}$$

$$\mathbf{M} = T_x\hat{\mathbf{i}} + M_y\hat{\mathbf{j}} + M_z\hat{\mathbf{k}} \tag{6.2}$$

Since the loading is entirely symmetric about the plane (xy LPS),

$$R_z = T_x = M_y = 0$$

In other words, a transverse (shear) force R_z, the twisting couple T_x, and a bending couple M_y vanish, as shown in Fig. 6.1(a). The nonzero components are depicted in Fig. 6.1(b). The subscripts can now be deleted from the reactive force **R** and couple **M**:

$$\mathbf{R} = F\hat{\mathbf{i}} + R\hat{\mathbf{j}} \tag{6.3}$$

$$\mathbf{M} = M\hat{\mathbf{k}} \tag{6.4}$$

Fig. 6.1

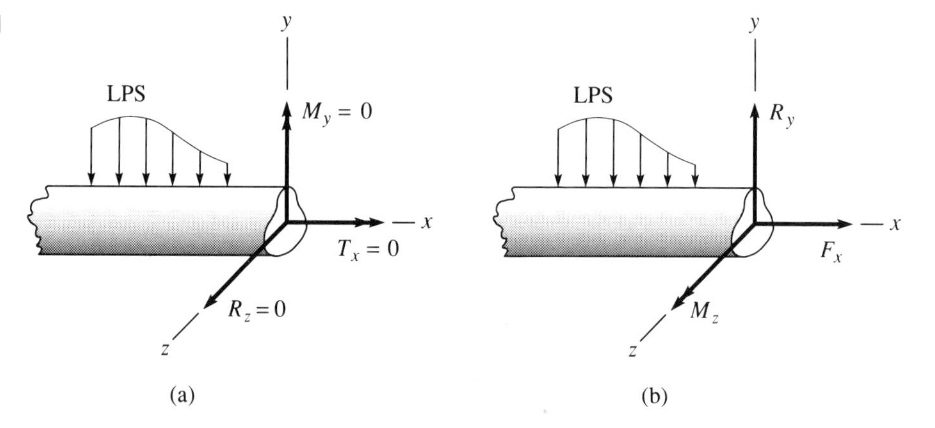

The purpose of the roller support at B in Fig. 6.2(a) is to prevent vertical movement. It does this by applying the necessary reaction R_B. (For clarity, the load on the beam is not shown.) Such a support cannot prevent rotation; the bending couple at B is zero. Hence the reactive force system is only a vertical force, as shown in Fig. 6.2(b). The pin at support A also prevents horizontal movement, since the reactive force system includes a horizontal component, as in Fig. 6.2(b). A beam supported as shown in the figure is said to be *simply supported*.

The beam of Fig. 6.3(a) has a fixed, built-in, or clamped connection at A. Such a support prevents both translation and rotation. This is accomplished by the force system shown in Fig. 6.3(b). The force components and the bending couple assume the necessary magnitudes to inhibit completely displacement at A. This type of end connection is frequently called a *fixed* end. At B the beam is free of support; this is called a *free* end. The name obviously implies that all the components of the force

Fig. 6.2

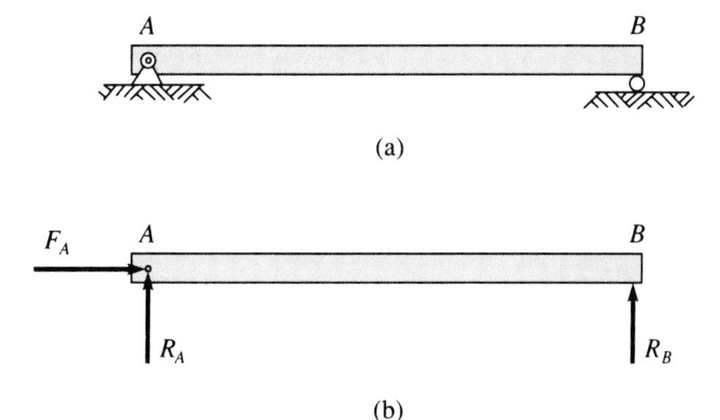

Fig. 6.3

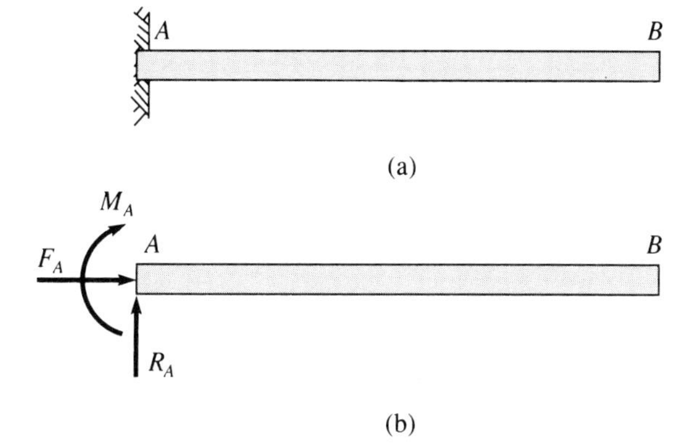

system at B are zero. A beam that is fixed at one end and free at the other is said to be *cantilevered.*

The simple supports of Fig. 6.2 and the fixed and free ends of Fig. 6.3 are the most commonplace, but other, more complicated circumstances arise when beams are employed as structural members. For example, the floor joists of a building might be supported by vertical columns that are also deformable—that is, the column resists the movement, displacement, and rotation of the joist (beam) but cannot entirely prevent movement. The situation is illustrated in Fig. 6.4(a), wherein the horizontal joist (beam) is affixed to the vertical column. Under transverse (vertical) loading, the flexible beam bends; the end tends to rotate. Such rotation is restrained by the column, which exerts the couple M upon the beam; the opposite action upon the column tends to bend the column. The restraint cannot entirely prevent the rotation θ. If the column is hookean, then the rotation is proportional to the couple and the condition at the end of the beam can be expressed by the equation

$$M = k\theta \tag{6.5}$$

Fig. 6.4

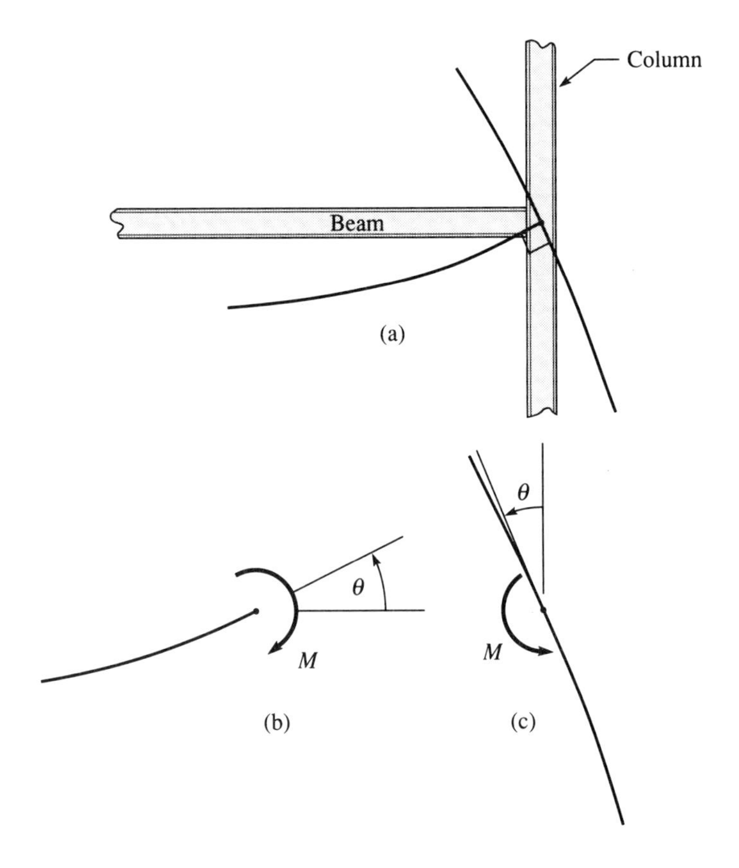

where the coefficient k depends upon the stiffness, material, and size of the column. The beam is neither fixed nor simply supported; the constraint is intermediate.

Net Internal Force System

With the loading (in the LPS) perpendicular to the axis of the beam, the normal force component on any cross section is zero. Consequently, the net internal force system on any cross section consists of, at most, a shear force and bending couple. Figure 6.5 shows a portion of a beam to one side of a cross section S. This cross section is located by its distance x from the end of the beam. Since V_y and M_z are the only components of the net internal force system on S that are not identically zero, we drop the subscripts and write V for V_y and M for M_z.

At this point it is advantageous to adopt a sign convention for the bending couple M and the shear force V. A bending couple that tends to compress the upper portions of the beam is regarded as positive. If the couple tends to stretch the upper portions, it is negative. This is illustrated in Fig. 6.6(a). A shear force is regarded as positive if it tends to rotate the free body on which it acts in a counterclockwise sense. If the

Fig. 6.5

Fig. 6.6

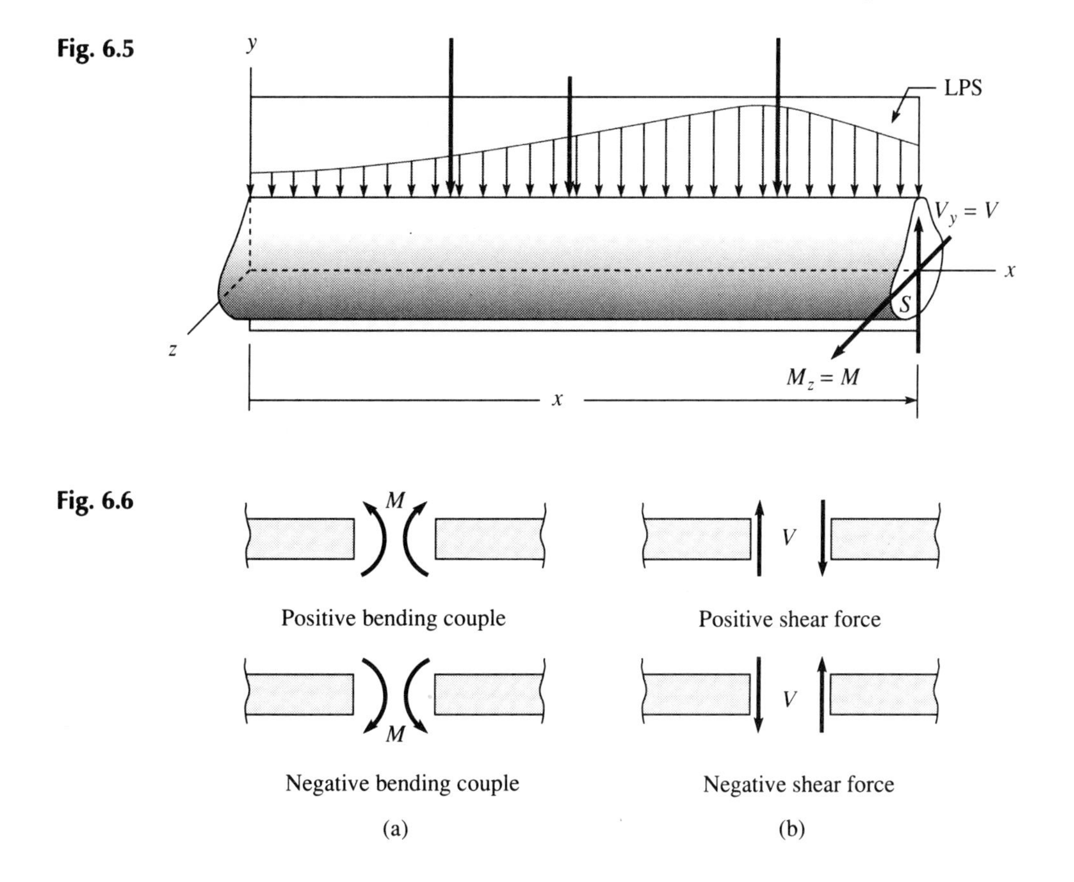

shear force tends to rotate the free body in a clockwise sense, then it is negative. This situation is illustrated in Fig. 6.6(b).

The shear force and bending couple are related to the loading. For example, consider a beam subjected to a distributed load $p(x)$ (force/length), as shown in Fig. 6.7(a). Additionally, one or more concentrated forces may act. For clarity we have shown only a typical concentrated force F_i. Also, the end supports have been replaced by their reactions (R_A, M_A, R_B, M_B).

The cross section S is located by the coordinate distance x. To relate the shear force and bending couple on S to the loading, we consider the free-body diagram of the portion of beam to the left of S shown in Fig. 6.7(b). Our choice of this part rather than the other is arbitrary; the portion to the right of S would serve as well. The conditions for equilibrium of the free body determine V and M:

$$V(x) = -R_A + F_i + \int_0^x p(\xi)\,d\xi \tag{6.6}$$

$$M(x) = M_A + R_A x - F_i(x - L_i) - \int_0^x (x - \xi)p(\xi)\,d\xi \tag{6.7}$$

Fig. 6.7

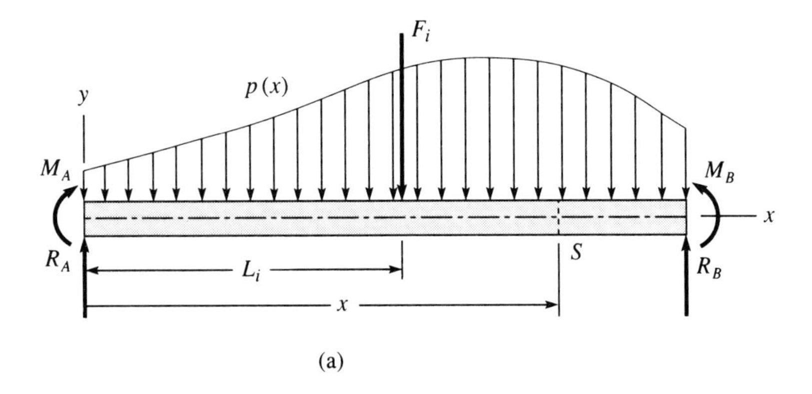

(a)

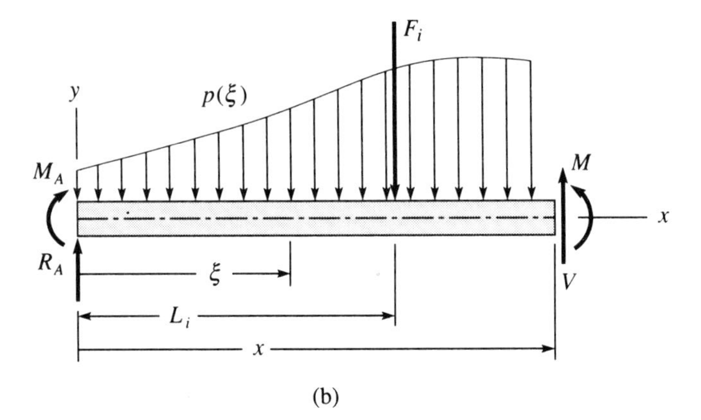

(b)

If there is more than one concentrated load, the F_i term in each of these equations is replaced by a sum of similar terms ($i = 1, 2, \ldots$).

Differential Equations of Equilibrium

We could eliminate the integrals in Eqs. (6.6) and (6.7) by differentiating with respect to x and so relate the derivatives of M and V to $p(x)$. Instead, let us obtain the equations by considering the equilibrium of a slice from the beam.

In Fig. 6.8 we have isolated a slice of the beam between the cross section S at x and one nearby at $x + \Delta x$. The distributed load $p(x)$ exerts a force $\bar{p}\,\Delta x$, where $\bar{p}$ denotes the mean value over the segment Δx. For the slice to be in equilibrium, $\Sigma F_y = 0$, so that

$$V(x + \Delta x) - V(x) - \bar{p}\,\Delta x = 0$$

or

$$\frac{V(x + \Delta x) - V(x)}{\Delta x} = \bar{p}$$

Fig. 6.8

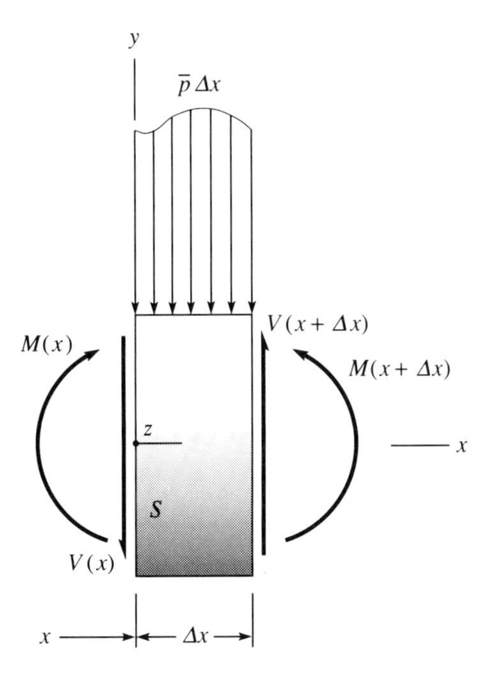

As Δx approaches zero, $\bar{p}$ approaches the actual load $p(x)$ at x, and in the limit we obtain

$$\frac{dV}{dx} = p(x) \tag{6.8}$$

Equilibrium of the slice in Fig. 6.8 also requires that $\Sigma M_z = 0$; hence,

$$M(x + \Delta x) - M(x) + V(x + \Delta x)\,\Delta x + \bar{p}\,\Delta x\left(\frac{\Delta x}{2}\right) = 0$$

or

$$\frac{M(x + \Delta x) - M(x)}{\Delta x} = -V(x + \Delta x) - \bar{p}\frac{\Delta x}{2}$$

Again, as Δx approaches zero, we obtain

$$\frac{dM}{dx} = -V(x) \tag{6.9}$$

The equilibrium equations (6.8) and (6.9) are immediately useful in two ways. Firstly, the equations can be integrated to determine $V(x)$ and $M(x)$. If these equations are satisfied everywhere (every slice is in equilibrium), then the internal force system (V and M) satisfies equilibrium (every piece of the beam is in equilibrium). Secondly, the equations are helpful in plotting the diagrams of shear V and bending couple M. According to Eq. (6.8), the *slope* of the shear diagram *is* the load $p(x)$. According to Eq. (6.9) the *slope* of the moment diagram *is* the negative shear $-V(x)$. It follows, too, that the maximum and minimum values of the bending couple $M(x)$ occur where the shear V is zero. The derivations of Eqs. (6.8) and (6.9) and their applications presuppose that the limits exist and that the load, shear, and moment are continuous and integrable. Other circumstances, such as concentrated loads, must be considered, but first let us apply the equations to a straightforward example.

EXAMPLE 1

The beam of Fig. 6.9(a) is simply supported at ends A $(x = 0)$ and B $(x = \ell)$. The load is distributed so that $p(x) = \bar{p}x/\ell$, where $\bar{p}$ is a constant, the magnitude at the right end $(x = \ell)$. Substituting the load $p(x)$ into the right side of Eq. (6.8) and integrating, we obtain

$$dV = p(x)\,dx = \bar{p}\frac{x}{\ell}\,dx \tag{6.10a}$$

$$\int_{V_0}^{V(x)} dV = \int_0^x \bar{p}\frac{x}{\ell}\,dx \tag{6.10b}$$

$$V(x) - V_0 = \bar{p}\frac{x^2}{2\ell} \tag{6.10c}$$

Note that the limits on the left and right sides of (6.10b) are in accord: $V_0 \equiv V(0)$ is the value of the shear force at $x = 0$; the upper limit is indefinite, because we need the function $V(x)$ in order to proceed with Eq. (6.9), namely,

$$dM = -V(x)\,dx = -\left(V_0 + \bar{p}\frac{x^2}{2\ell}\right)dx \tag{6.11a}$$

$$\int_0^{M(x)} dM = -\int_0^x \left(V_0 + \bar{p}\frac{x^2}{2\ell}\right)dx \tag{6.11b}$$

$$M(x) = -\left(V_0 x + \bar{p}\frac{x^3}{6\ell}\right) \tag{6.11c}$$

Note that the condition of simple support, $M(0) = 0$, has been introduced into the left side of (6.11b). Similarly,

$$M(\ell) = 0 = -\left(V_0\ell + \bar{p}\frac{\ell^2}{6}\right) \tag{6.12a}$$

or

$$V_0 = -\bar{p}\frac{\ell}{6} \tag{6.12b}$$

Fig. 6.9

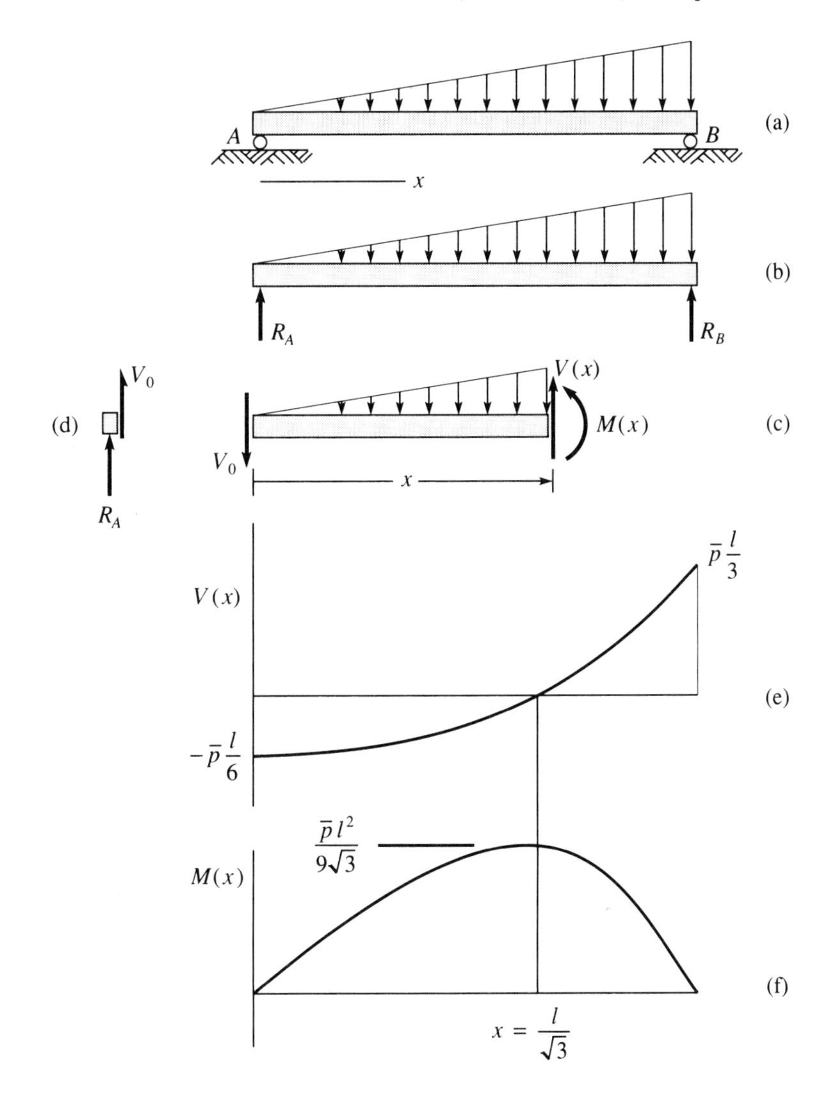

A free body of a slice above the left support is pictured in Fig. 6.9(d), which shows that

$$R_A = -V_0 = \bar{p}\frac{\ell}{6} \tag{6.13a}$$

Likewise, by Eq. (6.10c)

$$R_B = V(\ell) = \bar{p}\frac{\ell}{3} \tag{6.13b}$$

Note that Eq. (6.10c) is an equilibrium condition, $\Sigma F = 0$, for the free-body diagram of Fig. 6.9(c); Eq. (6.11c) is another equilibrium equation, $\Sigma M = 0$, for the same free-body diagram. These were obtained by integrating the equations of equilibrium for a slice (summing) from the end $x = 0$ to $x = \ell$, which gives the corresponding equilibrium equations for the finite body lying in the segment $(0, x)$.

Finally, the shear force and bending moments are plotted in Fig. 6.9(c) and (d). Note that the slope of the shear vanishes at the end $x = 0$, where $p(0) = 0$; the moment is a maximum at $x = \ell/\sqrt{3}$, where $V(\ell/\sqrt{3}) = 0$. ◆

Concentrated Loads—Forces and Couples

In reality, a concentrated load is one that is distributed, but over a very small region compared with other lengths, for instance, much less than the length and usually less than the depth. Again, we invoke the St. Venant principle, which enables us to utilize the concept of the load concentrated at a point, but we acknowledge that our theory *and* our descriptions of loads and supports do not admit an analysis of the local behavior at points of concentrated loading. With this in mind, let us examine the implications of a concentrated load.

Figure 6.10 shows a free body of a slice immediately beneath a concentrated force P. The shear forces upon the left and right sections are denoted by V_- and V_+, respectively. Equilibrium requires that $\Sigma F = 0$. Accordingly, we obtain

$$V_+ - V_- = P \tag{6.14}$$

In other words, the shear force exhibits a jump of amount P in the small distance of the loading. If the load were actually concentrated (it isn't) at a point, then the shear force would be discontinuous at the point of application.

Fig. 6.10

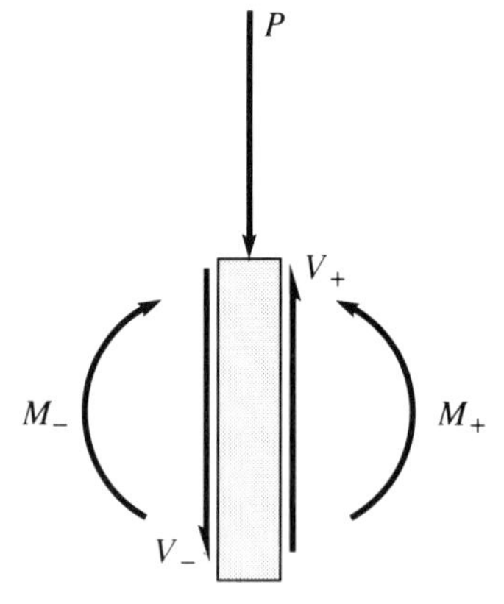

A concentrated couple is also conceivable. For example, two equal but opposing forces might be applied to a bracket, as shown in Fig. 6.11(a). The net effect upon the beam is a concentrated couple, which acts at a point, as depicted in Fig. 6.11(b). The

Fig. 6.11

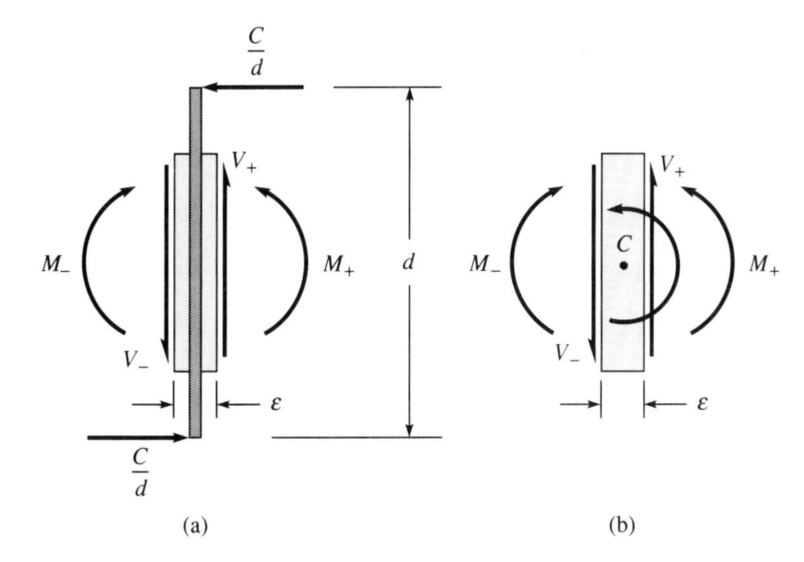

free-body diagrams of Fig. 6.11(a) and (b) show the force system that must satisfy equilibrium, in particular, $\Sigma M = 0$. The width of the slice is presumed small (compared with the other dimensions of the beam), so that the moment of the shear forces is neglected (in the limit $\varepsilon \to 0$, it vanishes). Then,

$$M_{+} - M_{-} = -C \tag{6.15}$$

The bending moment $M(x)$ exhibits a jump in the small distance ε. If the couple were concentrated, then the function $M(x)$ would be discontinuous at the point.

The reader can compare the conditions in (6.14) and (6.15) with their counterparts, (6.8) and (6.9), which hold where the functions $p(x)$ and $V(x)$ exist as smooth (continuous and bounded) functions.

EXAMPLE 2

The beam of Fig. 6.12(a) serves to illustrate the relationships between load, shear force, and bending couple. The free-body diagram of the entire beam is shown in Fig. 6.12(b). For equilibrium, the reactions must be 333 lb and 1167 lb, as shown. To determine V and M on a cross section between the supports, we consider the free-body diagram of Fig. 6.12(c). For equilibrium,

$$V(x) = -333 + 100x, \qquad M(x) = 333x - 50x^2, \qquad \text{for } 0 < x < 10 \text{ ft}$$

To obtain expressions for V and M on a cross section in the overhang, we consider the free-body diagram shown in Fig. 6.12(d). For equilibrium,

$$V(x) = -5(20 - x)^2, \qquad M(x) = -\tfrac{5}{3}(20 - x)^3, \qquad \text{for } 10 \text{ ft} < x < 20 \text{ ft}$$

Fig. 6.12

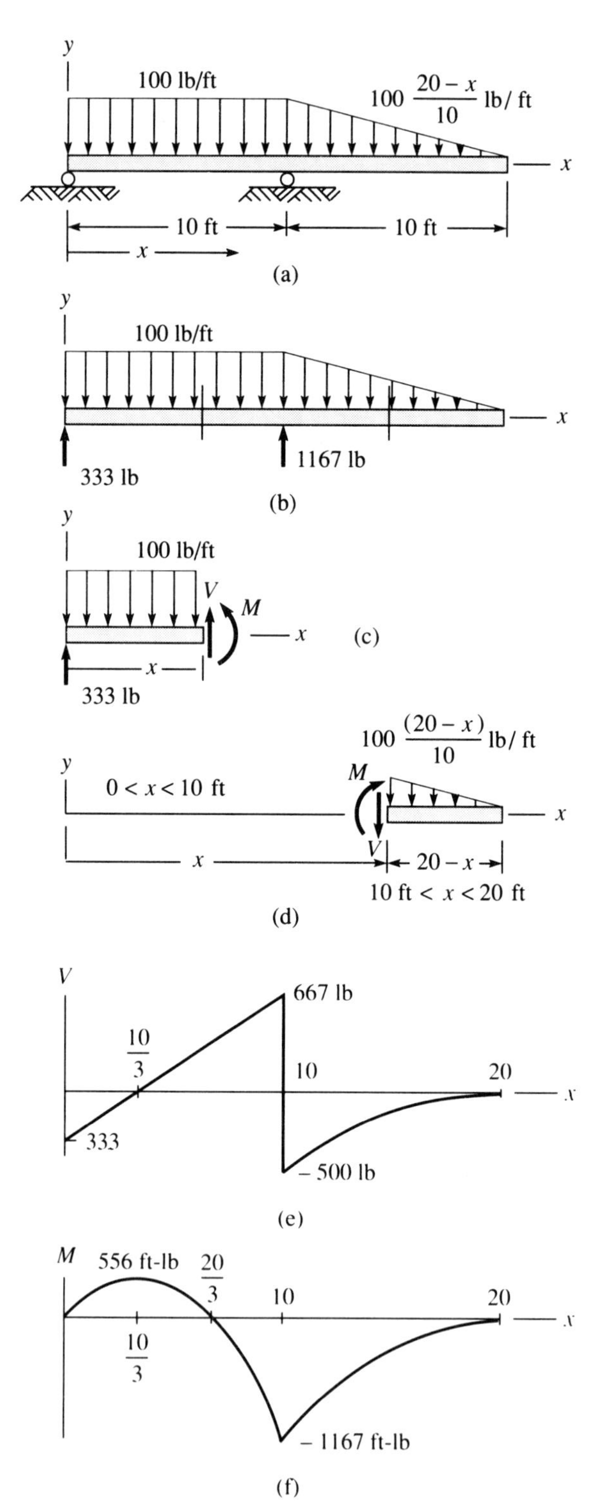
y
100 lb/ft
100 (20 − x)/10 lb/ft
x
10 ft
10 ft
x
(a)
y
100 lb/ft
x
333 lb
1167 lb
(b)
y
100 lb/ft
V
M
x
x
(c)
333 lb
100 (20 − x)/10 lb/ft
y
0 < x < 10 ft
M
x
V
x
20 − x
10 ft < x < 20 ft
(d)
V
667 lb
10/3
10
20
x
333
− 500 lb
(e)
M
556 ft-lb
20/3
10
20
x
10/3
− 1167 ft-lb
(f)

Now we can plot V and M over the entire length of the beam. However, it is helpful to use Eqs. (6.8) and (6.9). According to Eq. (6.8), the slope of the shear diagram is constant at 100 lb/ft from $x = 0$ to $x = 10$ ft; then it diminishes to zero at $x = 20$ ft. Also, there is a discontinuity in the shear diagram at $x = 10$ ft. The shear force jumps by the amount $\Delta V = -1167$ lb, due to the concentrated force acting there. These observations aid in sketching the shear diagram shown in Fig. 6.12(e).

From Eq. (6.9), the slope of the bending couple diagram can be read from the shear diagram. The slope is $+333$ lb at $x = 0$, vanishes at $x = \frac{10}{3}$ ft, and diminishes to -667 lb at $x = 10$ ft; here the slope changes abruptly to $+500$ lb, decreasing to zero at $x = 20$ ft. This information is used in plotting the bending couple diagram in Fig. 6.12(f).

The bending couple is a relative maximum (556 ft-lb) at $x = \frac{10}{3}$ ft, where the shear force vanishes. However, the largest bending couple (in absolute value) occurs at $x = 10$ ft, where the shear force has a discontinuity. The bending couple here is -1667 ft-lb. ◆

PROBLEMS

6.1 Draw the shear and bending couple diagrams for each beam in Fig. P6.1. Label all important values on each diagram.

Fig. P6.1

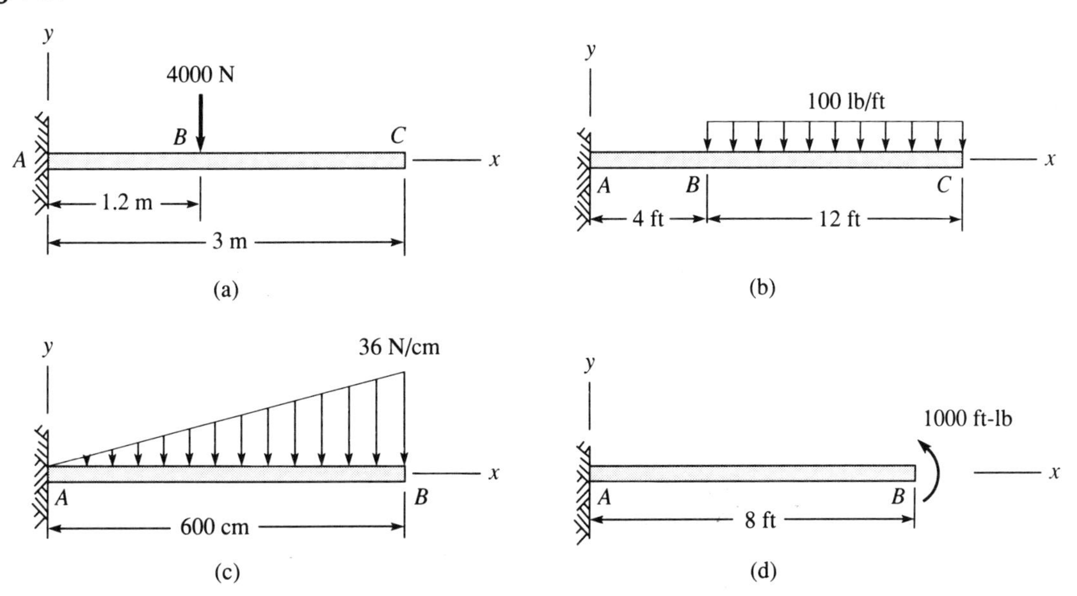

Fig. P6.1
(*cont.*)

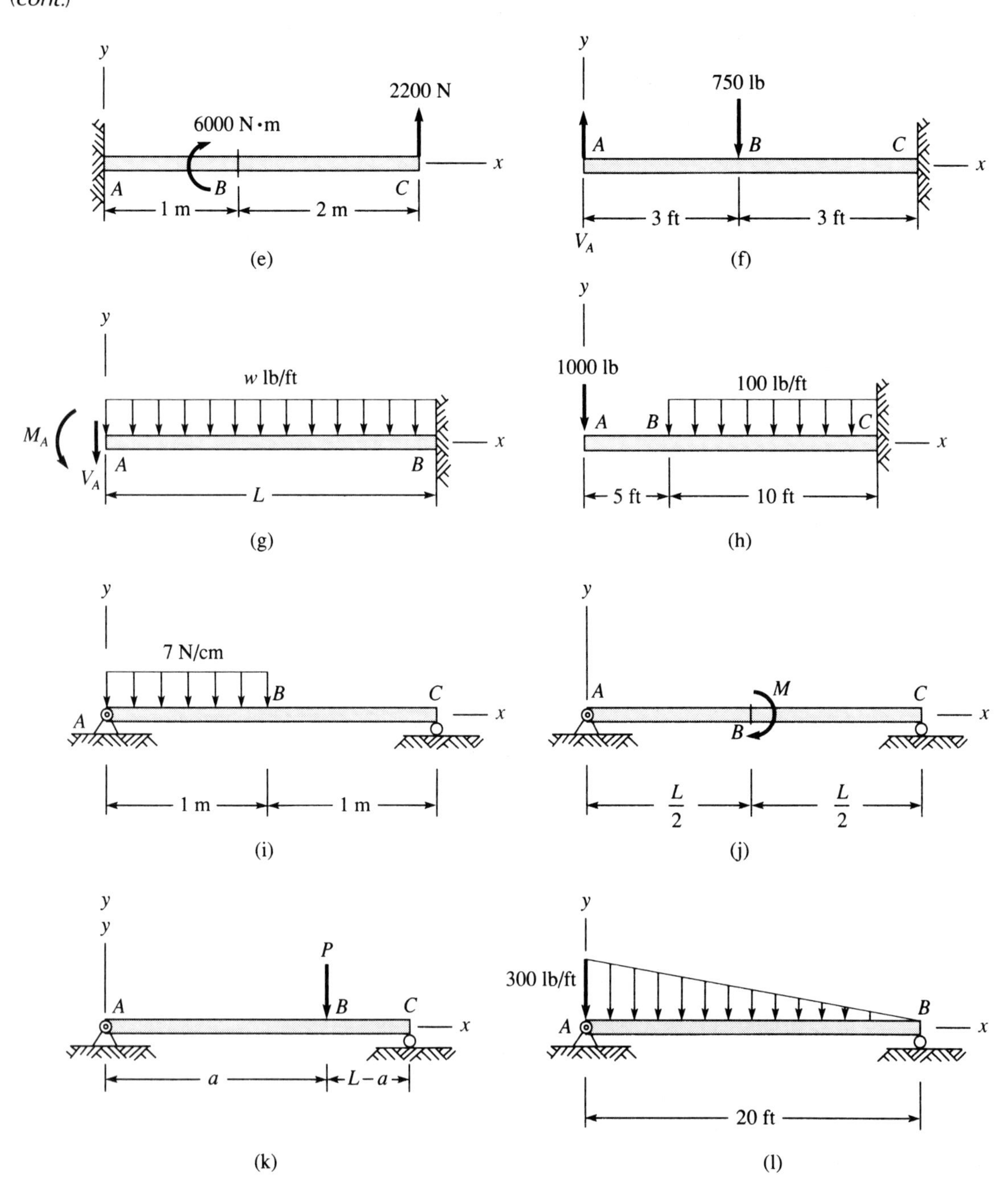
y
2200 N
6000 N·m
A
B
C
x
1 m
2 m
(e)
y
750 lb
A
B
C
x
3 ft
3 ft
V_A
(f)
y
w lb/ft
M_A
V_A
A
B
x
L
(g)
y
1000 lb
100 lb/ft
A
B
C
x
5 ft
10 ft
(h)
y
7 N/cm
A
B
C
x
1 m
1 m
(i)
y
A
M
B
C
x
L/2
L/2
(j)
y
y
P
A
B
C
x
a
L−a
(k)
y
300 lb/ft
A
B
x
20 ft
(l)

Fig. P6.1
(cont.)

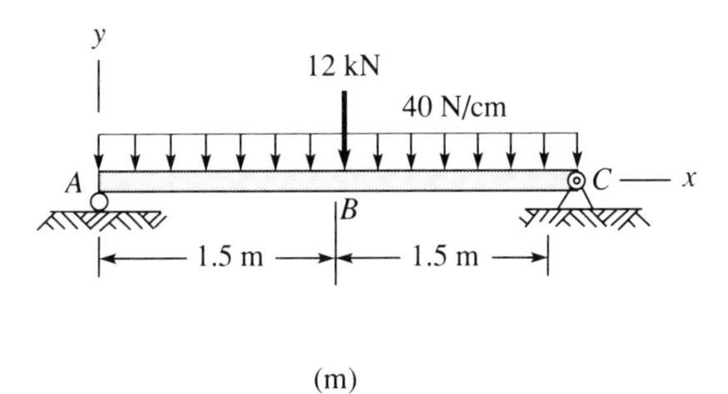

(m)

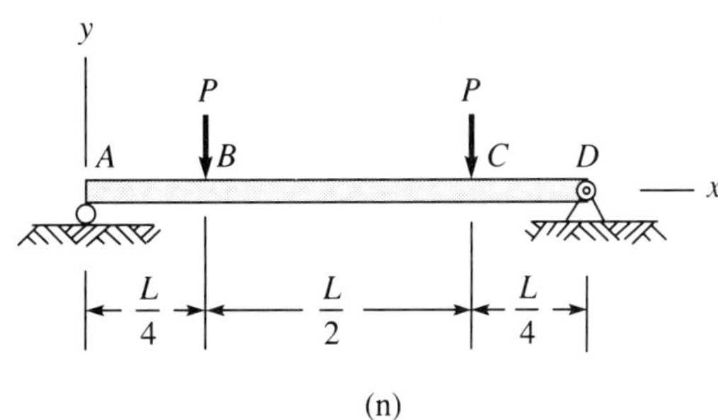

(n)

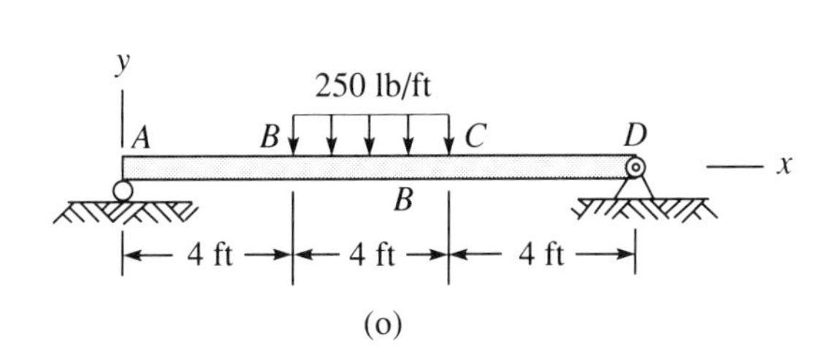

(o)

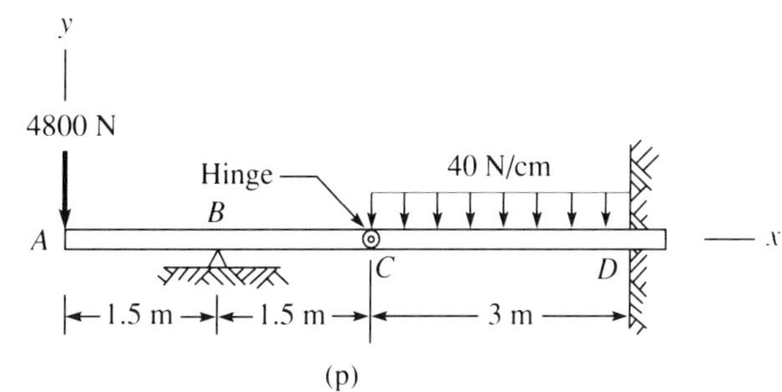

(p)

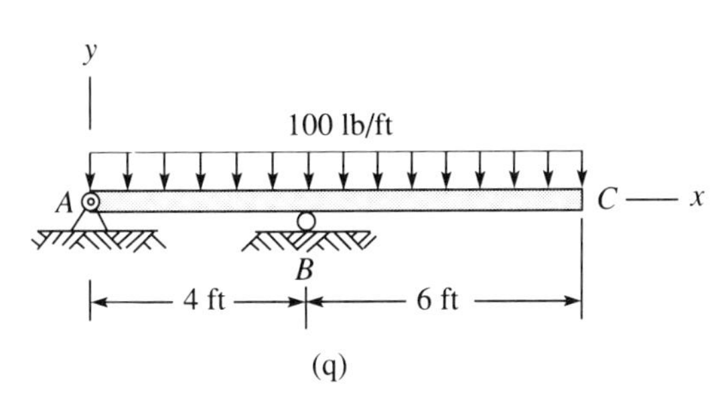

(q)

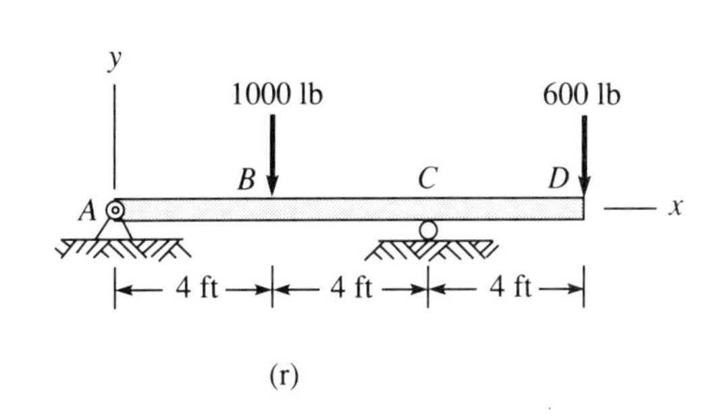

(r)

6.2 Determine the distribution of the load $p(x)$ on a beam of length $3L$ $(0 < x < 3L)$, which is free at both ends $(x = 0, 3L)$ and has the following shear force:

$$V = 0 \qquad 0 < x < L$$

$$V = \frac{\overline{p}}{2}(3L - 2x) \qquad L < x < 2L$$

$$V = 0 \qquad 2L < x < 3L$$

Fig. P6.3

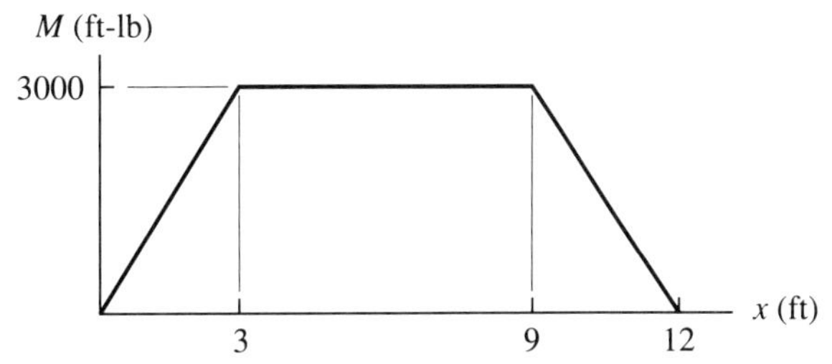

6.3 A beam 12 ft long is simply supported at each end. Determine the loading on the beam from the bending couple diagram shown in Fig. P6.3.

6.4 Determine the loading on a beam of length L that is simply supported at each end if the bending couple in the beam is that shown in Fig. P6.4.

Fig. P6.4

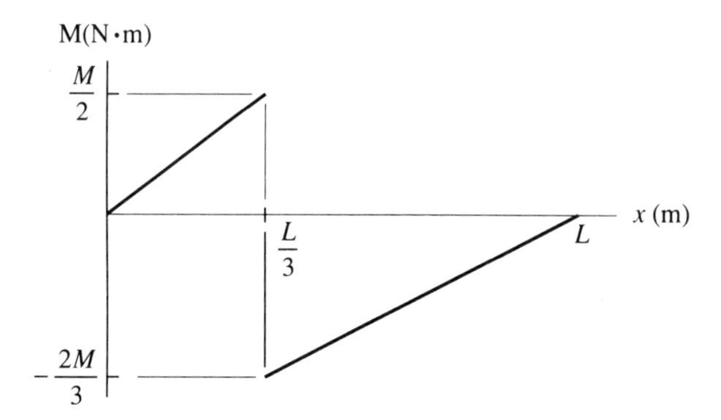

6.5 The shear force diagram for a cantilevered beam is shown in Fig. P6.5. Determine the manner in which the beam is loaded. It is fixed at $x = 0$ and free at $x = 15$.

Fig. P6.5

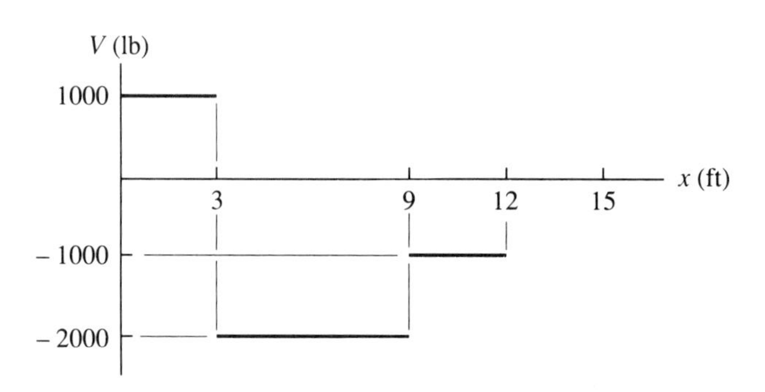

6.6 If the shear force diagrams for two cantilevered beams differ by a constant over the entire lengths of the beams, what is the difference in the loading on the beams?

6.7 Two beams are simply supported at their ends. The difference in their shear diagrams is a constant over the lengths of the beams. How do the loads on the two beams differ?

6.8 A beam of length L has no supports. Show that if such a beam carries a distribution load $p(x)$ pounds/foot, then the following conditions on $p(x)$ must be satisfied for the loaded beam to be in equilibrium:

$$\int_0^L p(x)\,dx = \int_0^L xp(x)\,dx = 0$$

6.3 Normal Stresses in HI-HO Beams

In Sec. 5.12 we obtained a relation [Eq. (5.121)] between the normal stress on a cross section and the bending couple for the case of simple bending of linearly elastic beams:

$$\sigma_x = -\frac{My}{I} \tag{6.16}$$

This relation is based on the assumption that plane cross sections of the beam remain plane and perpendicular to the neutral axis after bending. For simple bending the only component of the net internal force system on a cross section is the bending couple M. However, with transverse loading, a shear force V must also act to maintain equilibrium. The presence of this shear force raises a question regarding the validity of the hypothesis about the behavior of cross sections. If the cross sections remain plane and perpendicular to the neutral axis, the shear strain must be zero there. Then the shear stress must also be zero, but this is unlikely because of the presence of the shear force V. The dilemma is resolved by taking the view that the influence of the shear strain is small enough to be ignored in obtaining Eq. (6.16). This is generally true for a slender beam, one whose depth is small compared to its length. In most applications Eq. (6.16) gives excellent results in general, although $V \neq 0$.

A word of caution is appropriate: Although transverse shear is neglected in the derivation of Eq. (6.16) and, indeed, the stress component τ_{xy} is usually also negligible, shear stress on other surfaces can be very significant. If the component σ_x is the *only* significant stress on the *coordinate* surfaces (x, y, z), the state is one of simple stress, wherein the maximum shear stress is $\tau_{\max} = \sigma_x/2$ (see Sec. 2.9) and that maximum shear stress might initiate yielding (see Sec. 4.15).

EXAMPLE 3

Suppose that the beam of Fig. 6.12 has the cross section shown in Fig. 6.13(a) and is composed of HI-HO material. Then the normal stress σ_x is given by Eq. (6.16). The maximum and minimum values can occur at $x = \frac{10}{3}$ ft or at $x = 10$ ft, depending upon

Fig. 6.13

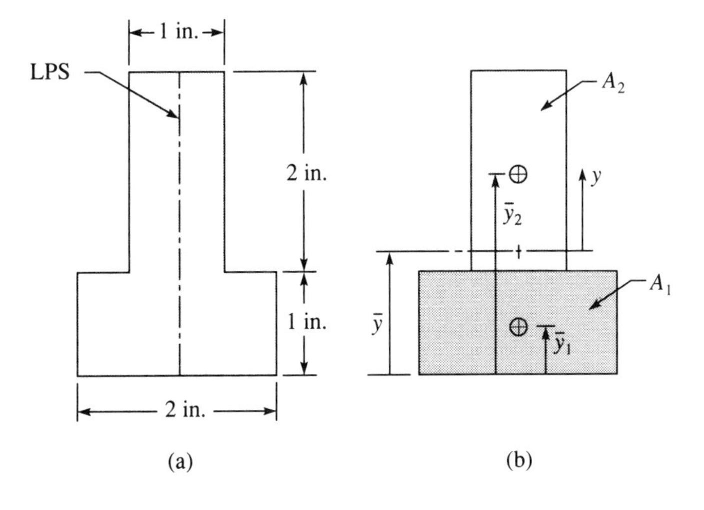

the maximum and minimum values of the variable y, that is, the distance from the neutral line. We must first locate the centroidal line: To that end, we divide the cross section into two 1-in. × 2-in. rectangles, as shown in Fig. 6.13(b). Then if $\bar{y}$ locates the centroid, we have

$$\bar{y} = \frac{A_1\bar{y}_1 + A_2\bar{y}_2}{A_1 + A_2} = \frac{(2)(\frac{1}{2}) + (2)(2)}{2 + 2} = \tfrac{5}{4} \text{ in.}$$

With the origin at the centroid, $-\frac{5}{4}$ in. $\leq y \leq \frac{7}{4}$ in.

The moment of inertia of the cross section (about the line through the centroid) is also needed. Using the parallel-axis theorem[1] for each of the rectangular areas, we get

$$I_1 = \tfrac{1}{12}(2)(1)^3 + (2)(\tfrac{3}{4})^2 = \tfrac{31}{24} \text{ in.}^4$$

$$I_2 = \tfrac{1}{12}(1)(2)^3 + (2)(\tfrac{3}{4})^2 = \tfrac{43}{24} \text{ in.}^4$$

Hence,

$$I = I_1 + I_2 = \tfrac{37}{12} \text{ in.}^4$$

At $x = \frac{10}{3}$ ft, $M = 556$ ft-lb, so that the maximum and minimum normal stresses *at that location* are

$$\sigma_x = -\frac{(556 \times 12)(-\frac{5}{4})}{\frac{37}{12}} = +2700 \text{ lb/in.}^2$$

$$\sigma_x = -\frac{(556 \times 12)(\frac{7}{4})}{\frac{37}{12}} = -3790 \text{ lb/in.}^2$$

At $x = \frac{10}{3}$ ft the maximum (tensile) stress occurs at the bottom ($y = -\frac{5}{4}$ in.); the minimum (compressive) stress occurs at the top ($y = +\frac{7}{4}$ in.).

At $x = 10$ ft, $M = -1167$ ft-lb, so that the maximum and minimum normal stresses *at that location* are

$$\sigma_x = -\frac{(-1167 \times 12)(\frac{7}{4})}{\frac{37}{12}} = +7950 \text{ lb/in.}^2$$

$$\sigma_x = -\frac{(-1167 \times 12)(-\frac{5}{4})}{\frac{37}{12}} = -5680 \text{ lb/in.}^2$$

From the computations, we must conclude that the beam is likely to fail near the simple support ($x = 10$ ft). The maximum tensile stress is 7950 lb/in.^{2} (55 MPa) and the maximum shear stress is (7950 lb/in.^{2})/2 = 3980 lb/in.^{2} (27 MPa). Of course, these values are based upon the effects of bending. They do not account for the local conditions—for example, "stress concentrations"—at the supports. ◆

1 The parallel-axis theorem is reviewed in Appendix B.

PROBLEMS

Assume HI-HO behavior unless otherwise stated.

6.9 The cross sections of the beams of Figs. P6.1(a) through P6.1(f) are shown in Figs. P6.9(a) through P6.9(f). Compute the maximum normal stress in each beam.

Fig. P6.9

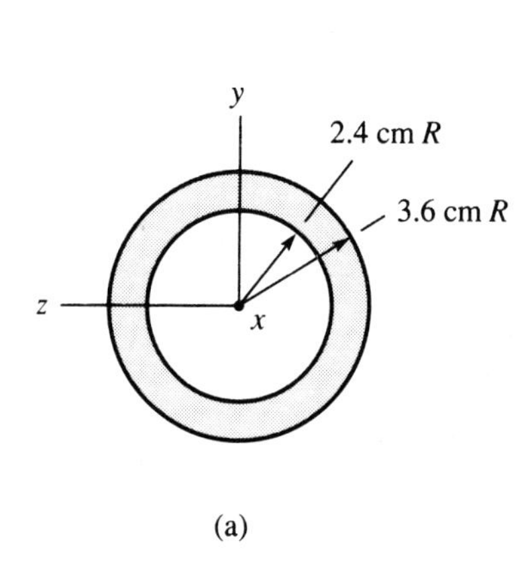

(a)

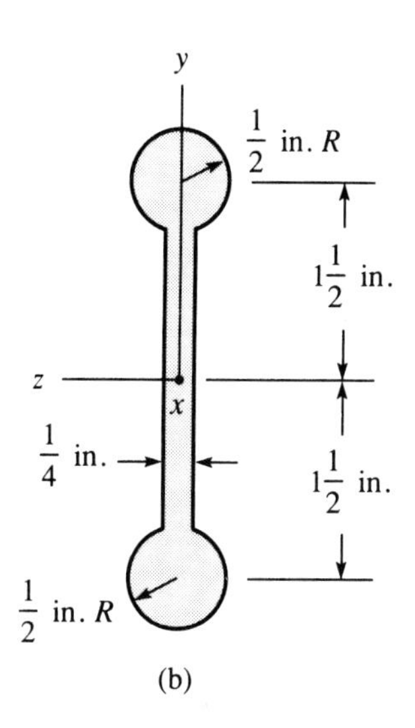

(b)

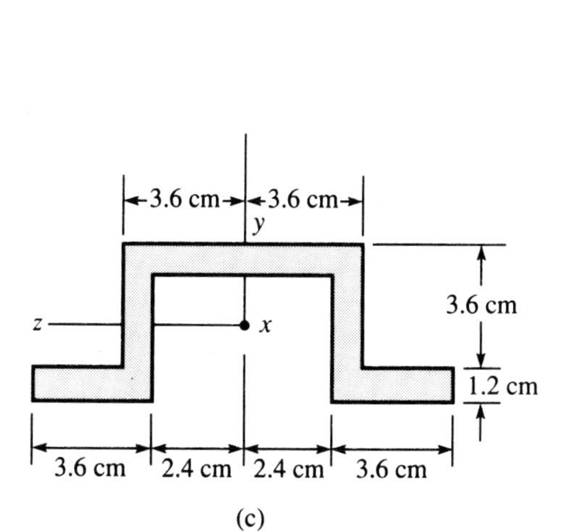

(c)

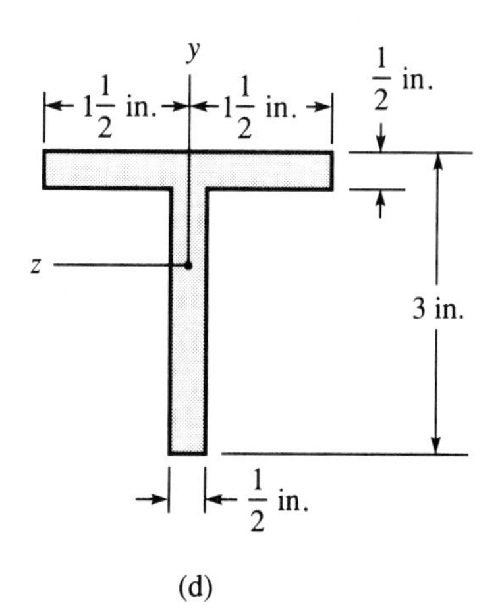

(d)

Fig. P6.9
(*cont.*)

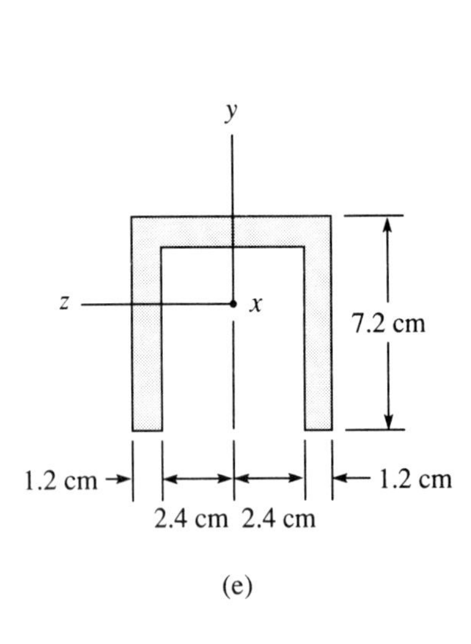

(e)

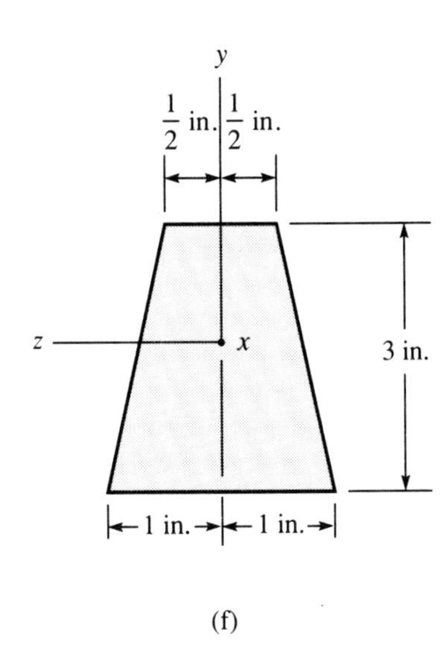

(f)

6.10 A solid cylindrical beam 3 m long with a radius of 5 cm is simply supported at its ends. If the beam carries a uniform load of 2.40 N/cm, compute the maximum normal stress in the beam.

6.11 A cantilevered strip of steel 0.020 in. thick, 0.25 in. wide, and 1.50 in. long carries an end load $P = \frac{1}{4}$ lb. What is the maximum normal stress in the strip?

6.12 The cross section of the cantilevered beam shown in Fig. P6.12 is semicircular ($r = 4$ cm). If the vertical shear in the beam is

$$V = \begin{cases} (4800 - 4000x)\text{ N}, & 0 < x < 1.2\text{ m} \\ -4800\text{ N}, & 1.2\text{ m} < x < 1.8\text{ m} \\ 0, & 1.8\text{ m} < x < 2.4\text{ m} \end{cases}$$

find the maximum and minimum values of the normal stress.

Fig. P6.12

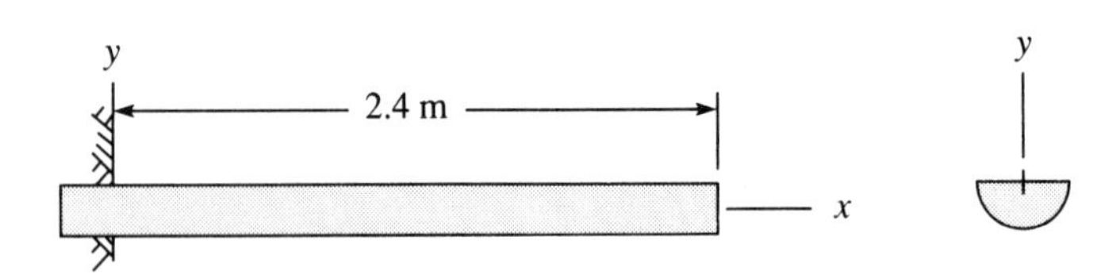

6.13 The depth of a tapered cantilevered beam varies linearly as illustrated in Fig. P6.13. Compute the maximum normal stress in the beam.

Fig. P6.13

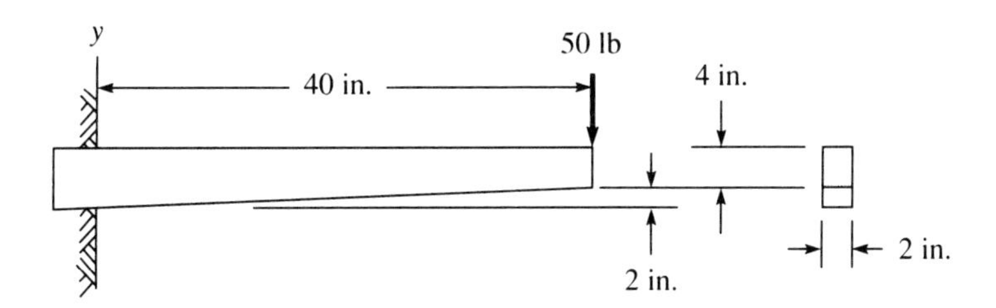

6.14 If the beam of Fig. P6.13 is steel, compute the maximum normal stress in the beam when the weight is included in the loading on the beam.

6.15 Find the maximum normal stress in the beam shown in Fig. P6.15.

Fig. P6.15

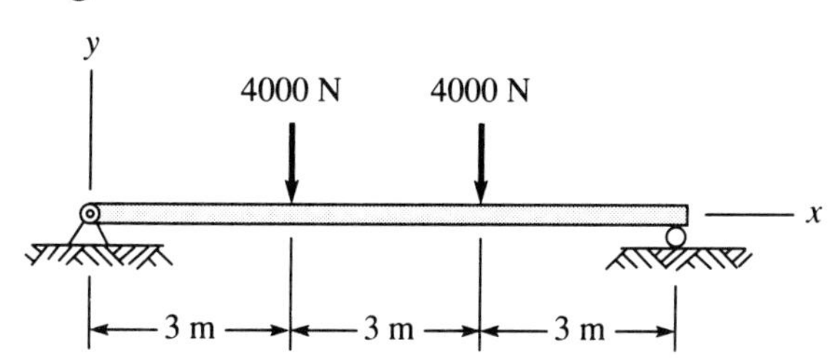

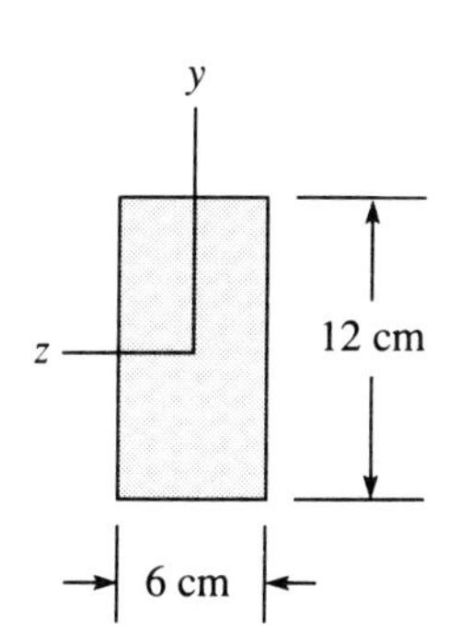

6.16 A beam with the cross section shown in Fig. P6.16 is 12 ft long and simply supported at each end. The bending couple in the beam is

$$M_z = \begin{cases} 250x \text{ ft-lb}, & 0 < x < 4 \text{ ft} \\ \frac{1000}{32}x(12 - x) \text{ ft-lb}, & 4 \text{ ft} < x < 12 \text{ ft} \end{cases}$$

Determine the maximum normal stress in the beam.

Fig. P6.16

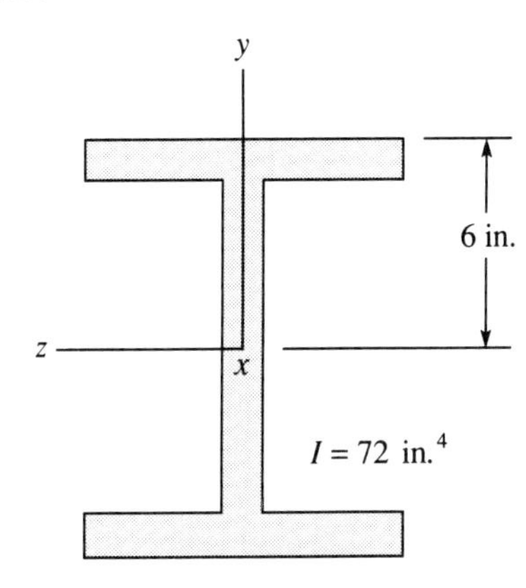

6.17 A beam 6.0 m long with the cross section shown in Fig. P6.17 is supported by rollers 1.2 m from each end. The beam carries concentrated loads of 8000 N at

Fig. P6.17

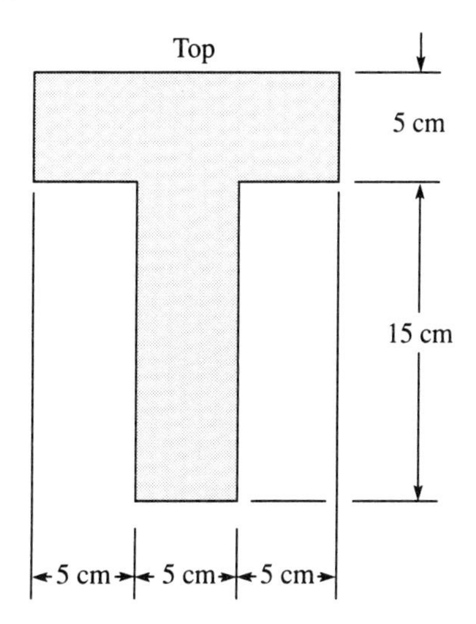

each end and 20,000 N at the center (all downward). Compute the maximum and minimum values of the normal stress in the beam.

6.18 Repeat Prob. 6.17 if the beam is of cast iron and the effects of its weight are included.

6.19 The cantilevered beam shown in Fig. P6.19 has a constant thickness. How should the depth vary in order that the maximum value of the normal stress on each cross section be constant over the length of the beam?

Fig. P6.19

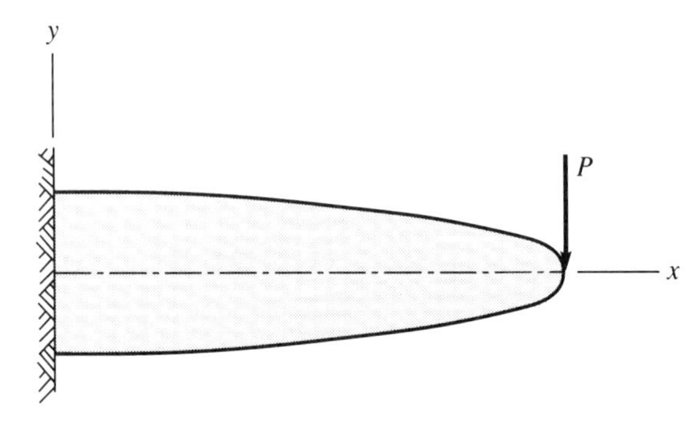

6.20 A composite beam has the cross section shown in Fig. P6.20. Compute the maximum normal stress in the beam if $M_z = 320$ in.-lb.

Fig. P6.20

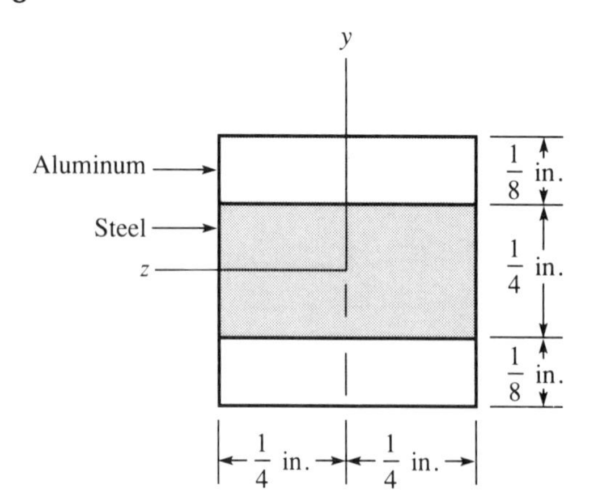

6.21 The beam shown in Fig. P6.21 consists of a steel jacket bonded to an aluminum core, as shown. What is the maximum normal stress in the beam if $M_z = 36$ N·m?

Fig. P6.21

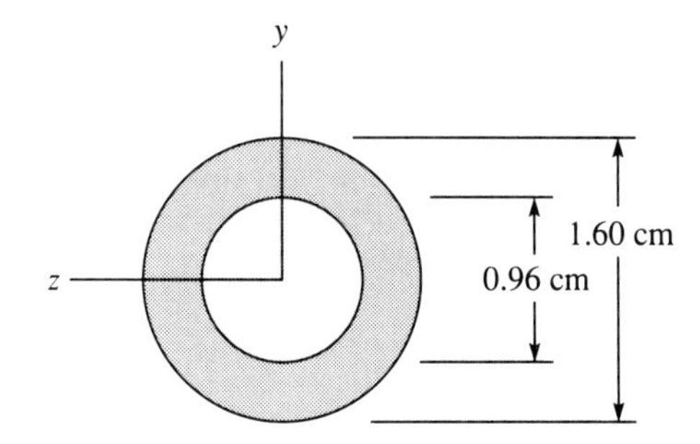

6.22 A steel reinforced concrete beam 12 ft long is simply supported at each end and carries a uniform load of 4800 lb/ft. If $E_{\text{concrete}} = 4 \times 10^6$ lb/in.^{2}, compute the maximum normal stress in the beam if it has the cross section shown in Fig. P6.22.

Fig. P6.22

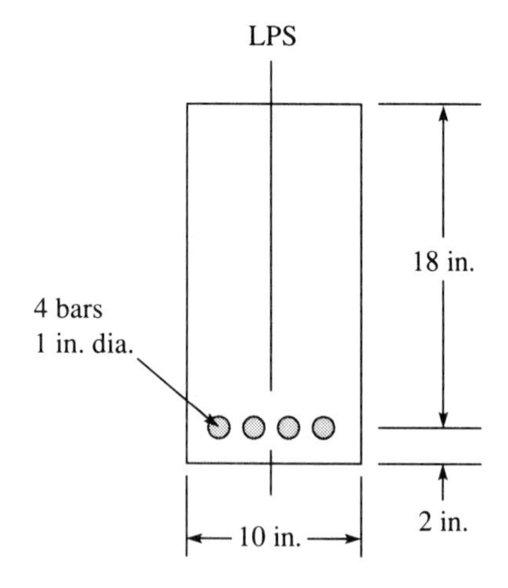

6.23 Repeat Prob. 6.22, assuming that the normal stress in the concrete is zero wherever the extensional strain is positive.

6.24 In the prestressed concrete beam shown in Fig. P6.24, the steel bar (centered 2 in. from the top of the beam) is held in tension until the concrete has hardened, and then the external forces on the bar are removed. What initial force must be applied to the bar if the concrete is to experience no tensile stress when the beam is subjected to a uniform downward load of 400 lb/ft? ($E_{concrete} = 1.5 \times 10^6$ lb/in.^{2})

Fig. P6.24

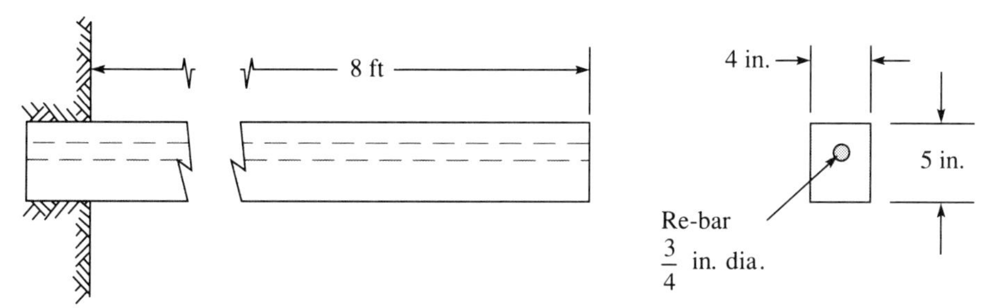

6.25 A beam has the doubly symmetric cross section shown in Fig. P6.25. (a) Determine the bending couple M_z that initiates yielding, (b) find the bending couple M_y that initiates yielding, and (c) compute the ratio M_y/M_z.

Fig. P6.25

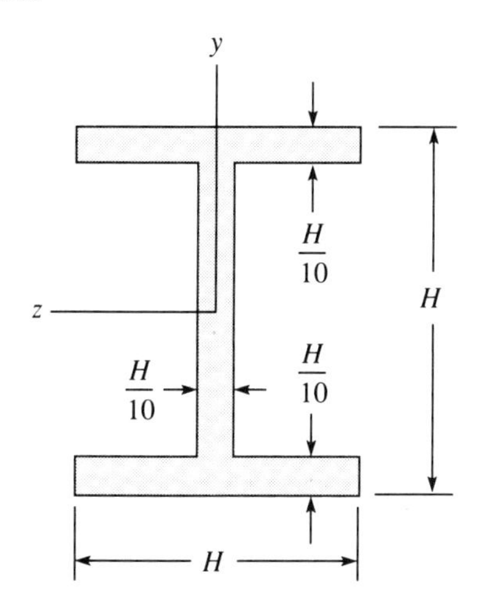

6.26 The beam AB in Fig. P6.26 is simply supported at A and on the end B of the cantilever BC. Both beams are American Standard steel I-beams (12-I-35); depth is 12 in. and, from the tabulated values,

$$I = 227.0 \text{ in.}^4$$

Determine the maximum and minimum normal stresses in each beam.

Fig. P6.26

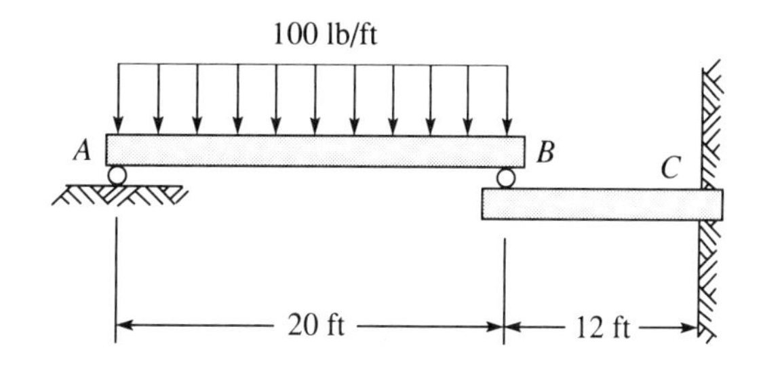

6.4 Shear Stresses in HI-HO Beams

As previously noted, the shear strain γ_{xy} is neglected in the derivation of the equation for the normal stress σ_x. Still, a shear stress τ_{xy} must exist whenever the force V is present, since

$$V = \iint_x \tau_{xy}\, dA$$

In some circumstances the shear component can be important. Specific situations are noteworthy: Firstly, the beam might be composed of anisotropic material that has little resistance to that stress. The common wooden beam is an example; the grain is commonly parallel to the axis and shearing along the grain is a possible mode of failure. Secondly, the beam might have a cross section like the I-beam shown in Fig. 6.14. Such cross sections are desirable because they provide much stiffness and strength [the property I in Eq. (6.16) is large] with less material and weight, and they are also easily formed. I-beams are so commonplace that we use the I-section to develop a procedure for estimating the shear stress.

Fig. 6.14

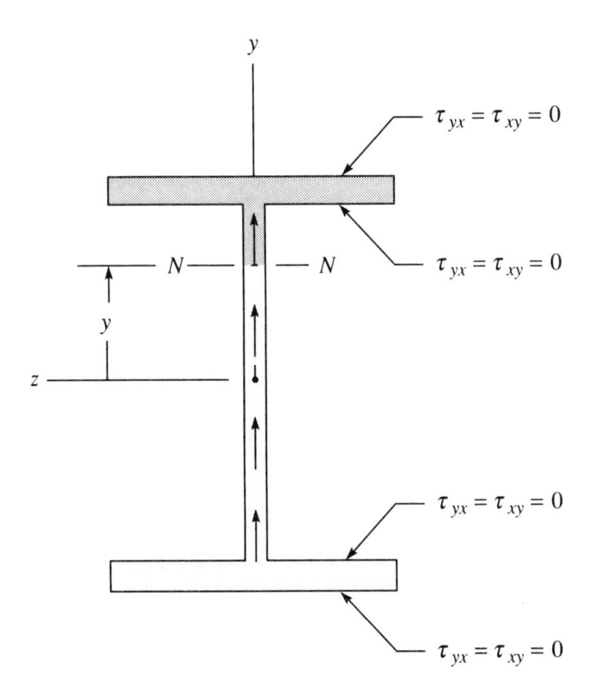

Note first that the shear stress must vanish at the upper and lower surfaces of the flanges, because these are free surfaces; that is, $\tau_{yx} = \tau_{xy} = 0$. Then it is unlikely that the shear stress τ_{xy} is appreciable in the flanges. It follows that most of the shear force, and the significant shear stress, acts upon the web. This is typical of "thin-walled" members; shear stresses upon the thin-walled sections act effectively tangent to the walls, as τ_{xy} acts in the y-direction upon the web. This means that most of the shear force is carried by the web.

To estimate the shear stress τ_{xy} at the distance y, we accept the normal stress σ_x given by Eq. (6.16) and appeal to the conditions for equilibrium of a piece of the beam. In particular, we isolate a piece of the beam as follows. Firstly, we consider a slice lying between sections at x and at $x + \Delta x$; secondly, we cut the slice by the plane N-N

at distance y above the neutral line in Fig. 6.14. The free body isolated by these cuts is shown in Fig. 6.15. On the free body, we show only the forces acting in the direction of the axis x: On the shaded portion A of the cross section, the *net* force is

$$\iint \Delta\sigma_x \, dA$$

On the surface cut by plane N-N the force is

$$\bar{\tau}_{yx} \Delta x \, t$$

where $\bar{\tau}_{yx} = \bar{\tau}_{xy}$ denotes the mean value of the shear stress and t denotes the thickness of the web. The sum of the components must vanish for equilibrium:

$$\Sigma F_x = \iint_A \Delta\sigma_x \, dA - \bar{\tau}_{yx} \Delta x \, t = 0 \tag{6.17a}$$

The normal stress σ_x is given by Eq. (6.16), and the change $\Delta\sigma_x$ is due to the change in the moment ΔM, so that Eq. (6.17a) takes the form

$$\bar{\tau}_{yx} = -\frac{\Delta M}{\Delta x} \frac{\iint_A y' \, dA}{It} \tag{6.17b}$$

Fig. 6.15

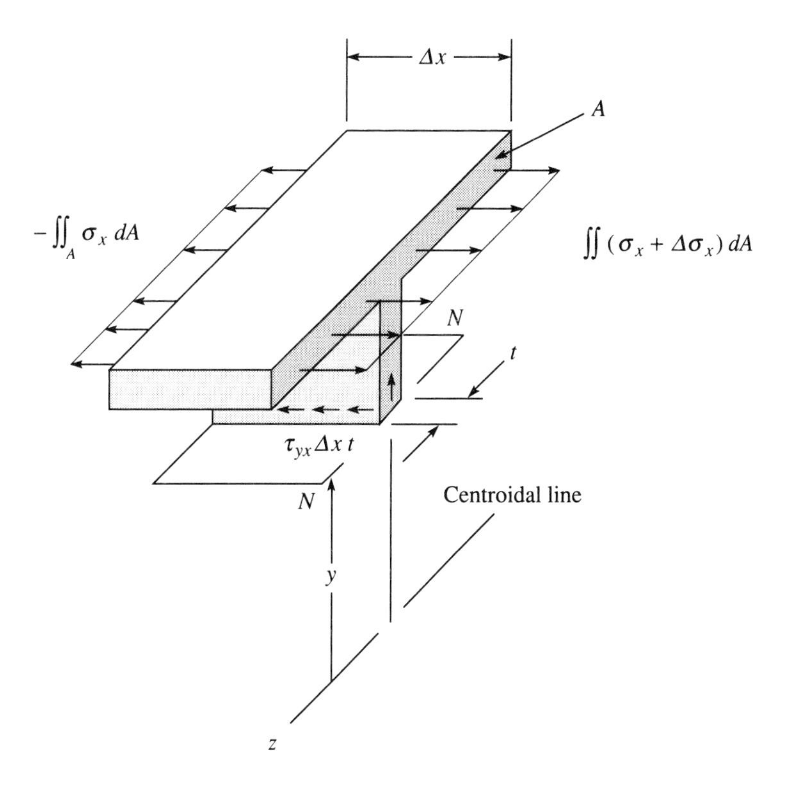

For brevity, we denote the integral in Eq. (6.17b) by

$$Q(y) \equiv \iint_A y' \, dA \tag{6.18}$$

Note that y' denotes the variable in the integral, since y denotes the distance to the cut N-N. The integral $Q(y)$ is sometimes called the *moment* of the area A (above $y' = y$) with respect to the centroidal line ($y = y' = 0$).

In the limit ($\Delta x \to 0$), according to equilibrium equation (6.9),

$$\lim_{\Delta x \to 0} \frac{\Delta M}{\Delta x} = \frac{dM}{dx} = -V(x) \tag{6.19}$$

Therefore, Eq. (6.17) takes the form

$$\bar{\tau}_{yx} = \bar{\tau}_{xy} = \frac{V(x)Q(y)}{It} \tag{6.20}$$

The value $\bar{\tau}_{yx} = \bar{\tau}_{xy}$ is the mean value (across the width t). This vanishes at the top and bottom as $Q(y)$ vanishes. Note that $Q(y)$ changes little over the web of the cross section. The area of the flanges provides the greater contribution; also, in most conventional I-beams, the web is thinner than the flanges. An approximation is obtained by dividing the force V by the area of the flange.

EXAMPLE 4

Shear Stress in a Standard I-Beam An American Standard wide-flange I-beam has the cross section shown in Fig. 6.16; this is an 18-WF-50 (18-in. depth—wide-flange beam—50 lb/ft).[2] The beam is 30.0 ft long, is simply supported at both ends, and carries a uniform load of 240 lb/ft. The maximum (absolute) value of the component τ_{xy} occurs near end supports, where the shear force is

$$|V_0| = \frac{240 \times 30.0}{2} = 3600 \text{ lb}$$

The property I in Eq. (6.20) can be calculated from the given dimensions (I = 793 in.^{4}); however a table gives the value:

$$I = 800 \text{ in.}^4$$

Note that the actual cross section of the hot-rolled beam does not have the precise shape illustrated; for example, corners have fillets and flanges exhibit slight taper. We use the tabulated value, since it is based on actual cross-sectional shape.

To compute the maximum value of τ_{xy}, we require the integral $Q(0)$ for the darker shaded region of Fig. 6.16. From the given dimensions,

2 The weight of the beam is neglected in our examples but can be included if circumstances warrant.

Fig. 6.16

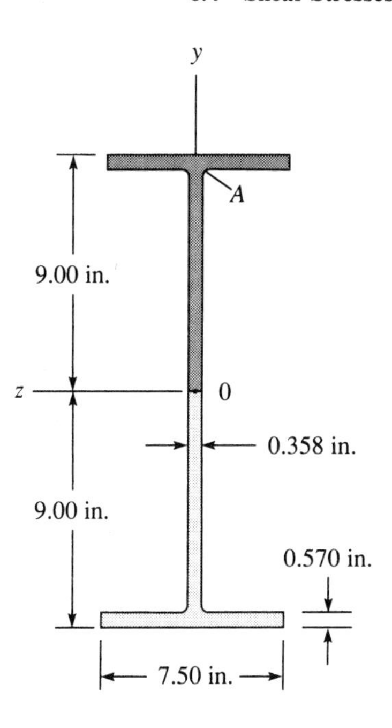

$$Q(0) = 0.358 \int_0^{8.43} y\,dy + 7.50 \int_{8.43}^{9.00} y\,dy$$
$$= 50.0 \text{ in.}^3$$

The shear $\bar{\tau}_{xy}$ can now be computed by means of Eq. (6.20):

$$\bar{\tau}_{xy}(0) = \frac{3600 \times 50.0}{800 \times 0.358} = 628 \text{ lb/in.}^2 \quad (4.33 \text{ MPa})$$

Let us also calculate the value at the point A—that is, at the top of the web. There

$$Q(8.43) = 37.3 \text{ in.}^3$$

so that

$$\bar{\tau}_{xy}(A) = \frac{3600 \times 37.3}{800 \times 0.358} = 469 \text{ lb/in.}^2 \quad (3.23 \text{ MPa})$$

We note again that the component $\bar{\tau}_{xy}$ does not vary greatly over the web; it is 25% less at A. Also, we note that the web supports most of the shear force. An approximation is obtained by dividing the shear force V by the area of the web:

$$\tau_{xy} \text{ (approx.)} = \frac{3600}{(16.86 \times 0.358)} = 596 \text{ lb/in.}^2 \quad (4.11 \text{ MPa})$$

This is only 5% less than the maximum value at the centroid.

Note that this value of $\bar{\tau}_{xy}(0)$ is *not* the maximum shear stress in this beam. At the center the bending moment is

$$M_{\max} = 27{,}000 \text{ ft-lb.} \quad (36{,}600 \text{ N} \cdot \text{m})$$

Then the maximum value of the component σ_x occurs at the bottom ($y = -9.00$) and at the middle where, according to Eq. (6.16),

$$\sigma_x = -\frac{27{,}000 \times 12 \times (-9.00)}{800}$$

$$= 3645 \text{ lb/in.}^2 \quad (25.1 \text{ MPa})$$

At this location the maximum shear stress occurs on planes inclined 45° to the axis (see Sec. 2.9). That maximum value is $3645/2 = 1820$ lb/in.^{2} (12.5 MPa), approximately three times greater than the maximum value of the component $\bar{\tau}_{xy}$. Yielding would probably occur at the location of the maximum shear stress (see Sec. 4.15). ◆

EXAMPLE 5

Built-Up Beam Another example serves to illustrate the basic approach to the computation of longitudinal shear stress. Consider a small member made by riveting together five strips 0.500 cm × 2.50 cm × 150 cm. Suppose that the five strips were laid together and joined only by the left clamp at one end, as shown in Fig. 6.17(a). Under end load P the assembly bends as depicted in Fig. 6.17(b); the individual strips are free to slip, one upon the other. In effect, the assembly acts merely as five beams of thickness 0.500 cm, each supporting one-fifth of the load P. When joined by equally spaced rivets, the individual strips cannot slip relative to one another. The rivets provide the shear resistance (as the longitudinal stress $\bar{\tau}_{xy}$ in a homogeneous beam). Now the assembly is unified; it acts much as a single homogeneous beam of thickness 2.50 cm. The action is depicted in Fig. 6.17(c), and the cross section of the composite is illustrated in Fig. 6.17(d). The reader can well imagine that the composite Fig. 6.17(c) is much stiffer (deflects much less) than the assembly of disjoint strips in Fig. 6.17(b). The former acts as the leaf-spring in an older automobile. The differences in stiffness are considered in Sec. 6.5.

Let us now pose the problem: The rivets are spaced at 2.50-cm intervals along the axis. The rivets are 0.625 cm in diameter and are made of steel. Specifications require that the mean transverse shear on the rivets cannot exceed $\bar{\tau} = 70$ MPa.

We approach this problem as we approached the homogeneous beam, *but* this beam is not quite homogeneous. We isolate not an *arbitrary* slice but a typical slice of width $\Delta x = 2.50$ cm, as shown in Fig. 6.18. From our previous analysis, we can anticipate that the maximum shear occurs at the interfaces nearest the centroidal axis. Accordingly, the cut (counterpart of N-N in Fig. 6.15) is made at that interface in Fig. 6.18. On this free-body diagram, we show the forces acting in the axial (x) direction: the unequilibrated normal force ΔF_n and the shear force $\bar{\tau} A_R$ upon the rivet.

Fig. 6.17

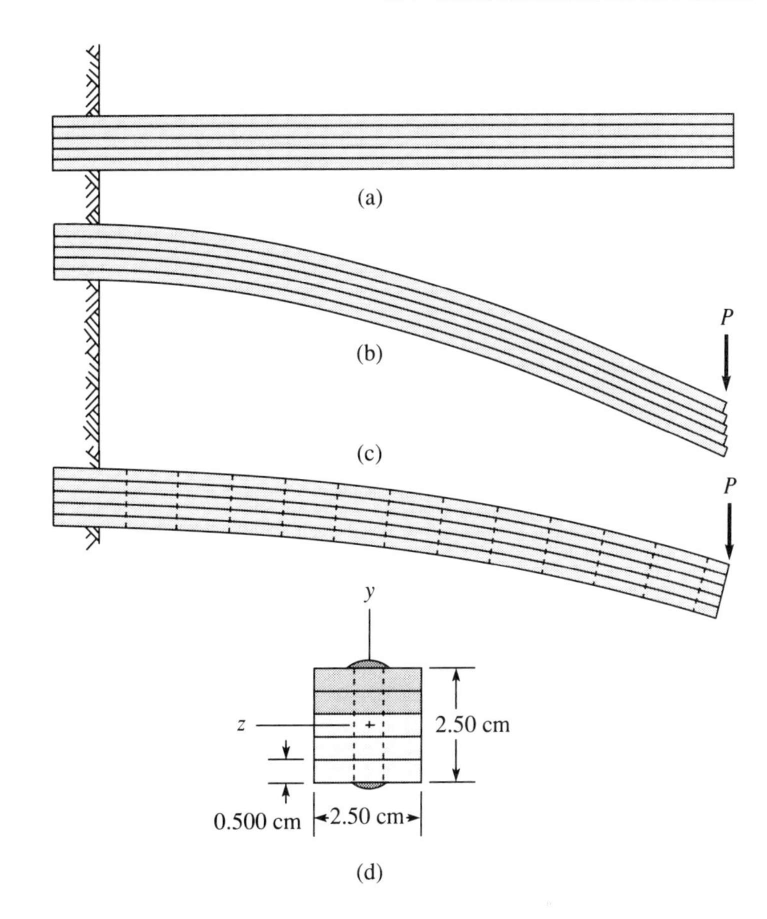

Fig. 6.18

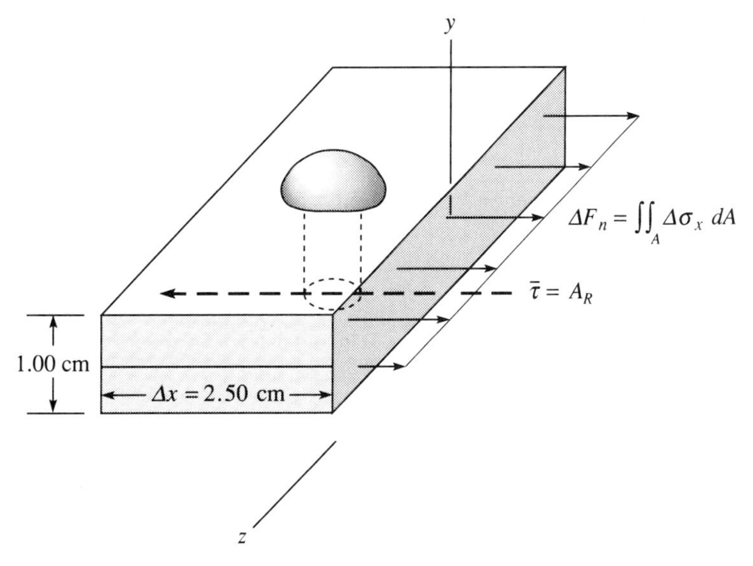

Equilibrium requires that

$$\bar{\tau} A_R = + \iint \Delta\sigma_x \, dA \qquad (6.21)$$

As before, the increment $\Delta\sigma_x$ is caused by the increment ΔM, in accordance with Eq. (6.16):

$$\Delta\sigma_x = -\frac{\Delta M y}{I} \qquad (6.22)$$

Now the increment ΔM is related to the shear V by equilibrium. The condition differs from Eq. 6.8 in that we cannot pass to the limit. Instead, we have the condition for our typical slice:

$$\Delta M = -V \Delta x \qquad (6.23)$$

By substituting Eq. (6.23) into (6.22) and that result into Eq. (6.21), we obtain

$$\bar{\tau} = +\frac{V \Delta x}{I A_R} \iint_A y \, dA \qquad (6.24)$$

As before, the integral in Eq. (6.24) extends over the shaded area of Fig. 6.17 and Fig. 6.18. We need only calculate that integral and I and substitute the numerical values into Eq. (6.24):

$$Q = \iint_A y \, dA = 2.5 \int_{0.25}^{1.25} y \, dy = 1.875 \text{ cm}^3$$

$$I = \frac{2.5 \times (2.5)^3}{12} = 3.255 \text{ cm}^4$$

$$A_R = \frac{\pi}{4}(.625)^2 = 0.3068 \text{ cm}^2$$

$$\bar{\tau} = 70 \times 10^2 \text{ N/cm}^2$$

$$V = \frac{I A_R \bar{\tau}}{\Delta x Q} = \frac{3.255 \times 0.3068 \times 70 \times 10^2}{2.5 \times 1.875} \text{N} = 1491 \text{ N}$$

Note that the foregoing analysis entails approximations and provides only a mean value, $\bar{\tau}$. Also, the use of Eq. (6.22) presumes HI-HO behavior of the steel. ◆

6.27 A beam with a rectangular cross section 20 cm wide and 25 cm deep is 6.0 m long and simply supported at its ends. Determine the vertical shear stress at a point 5 cm from the top and 1.5 m from the left support if the beam carries a 36-kN load at midspan.

6.28 Compute the maximum vertical shear stress in the beam shown in Fig. P6.28.

Fig. P6.28

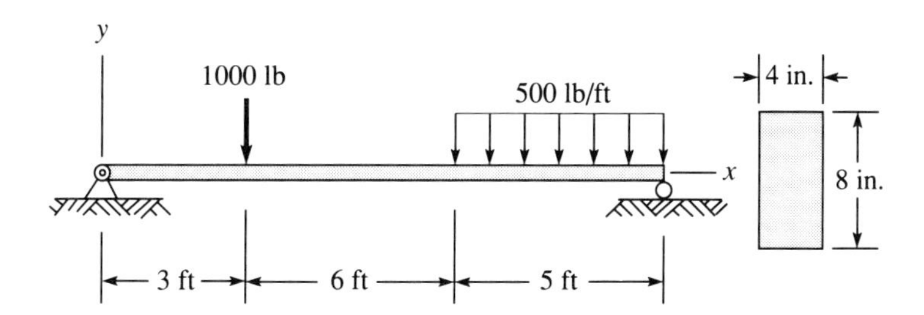

6.29 The beam in Fig. P6.29 has a rectangular cross section 10 cm wide. Determine the minimum depth d if the vertical shear stress is not to exceed 1400 kPa.

Fig. P6.29

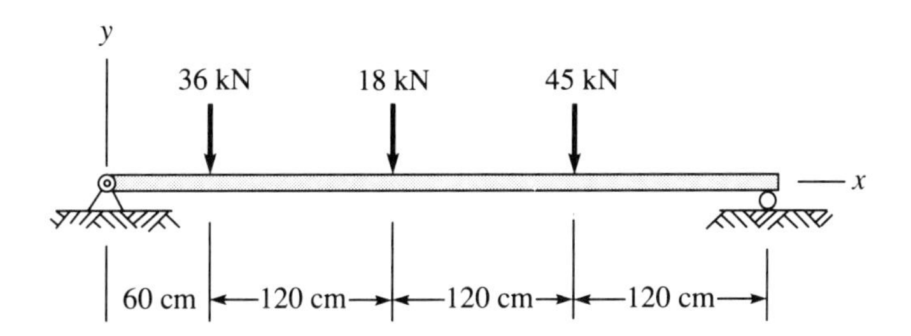

6.30 Find the maximum value of shear stress τ_{xy} for the beam shown in Fig. P6.30.

Fig. P6.30

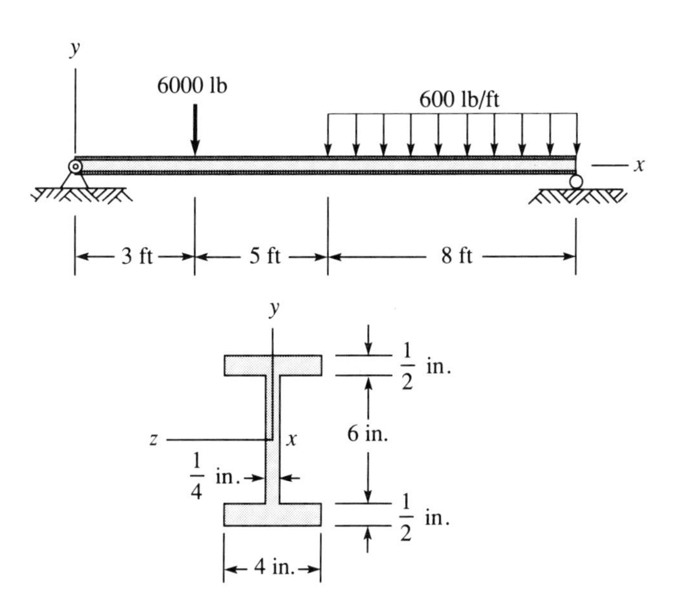

6.31 Determine the maximum value of shear stress τ_{xy} for the beam in Fig. P6.31.

Fig. P6.31

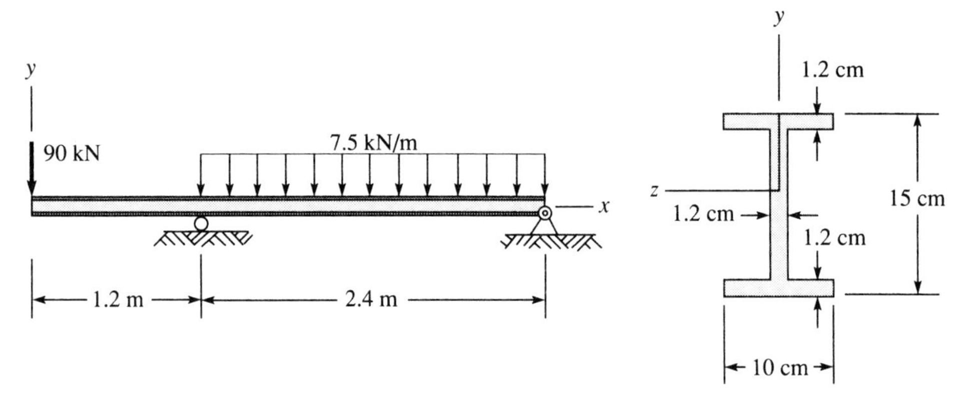

6.32 Obtain an expression for the value of shear stress τ_{xy} in the web of the beam of Fig. P6.32.

Fig. P6.32

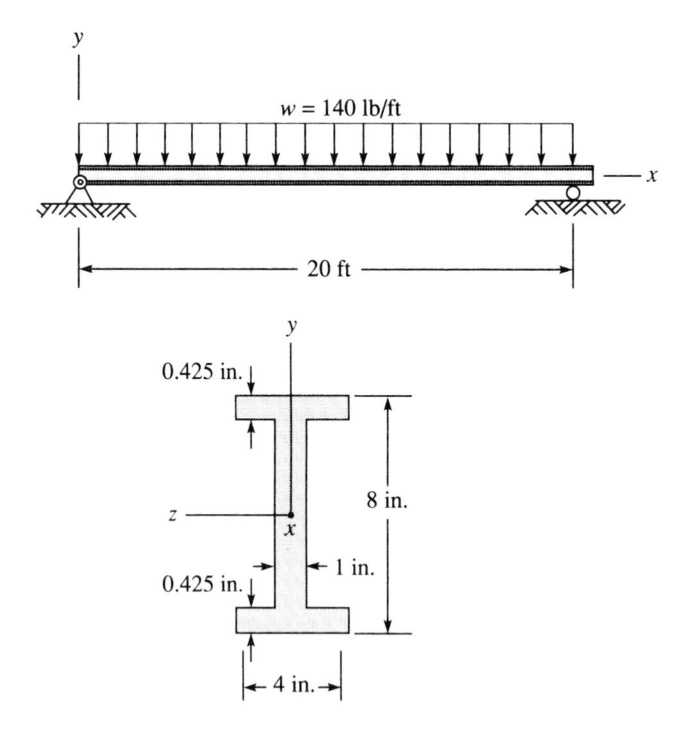

6.33 Obtain an expression for the shear stress in the top flange of the beam of Fig. P6.32.

6.34 A beam is made by welding two steel plates to form the T-section shown in Fig. P6.34. The beam is used as a cantilever, 12 ft long, which is subjected to the concentrated load $P = 2000$ lb at the unsupported end. The weld is continuous along the two reentrant corners. Determine (a) the shear force per unit length of each weld and (b) the maximum shear stress in the plate.

Fig. P6.34

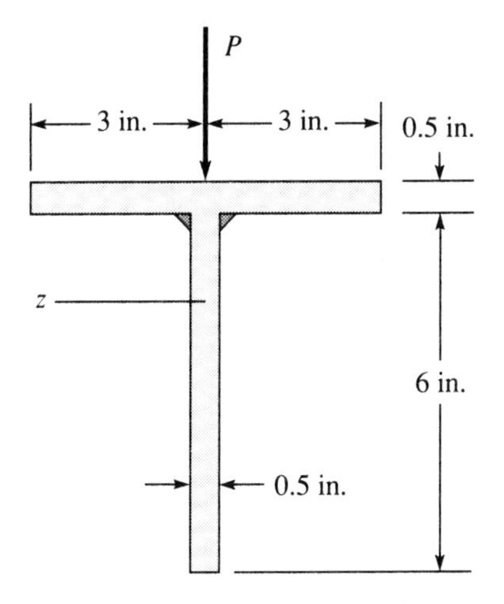

6.35 A beam is composed of two steel strips joined by rivets that are 0.30 cm in diameter (Fig. P6.35). The beam must sustain a transverse shear force $V = 200$ N. If the rivets are placed at equal intervals along the axis and each can sustain a mean shear stress of 50 MPa, determine the maximum spacing (the distance between centers).

Fig. P6.35

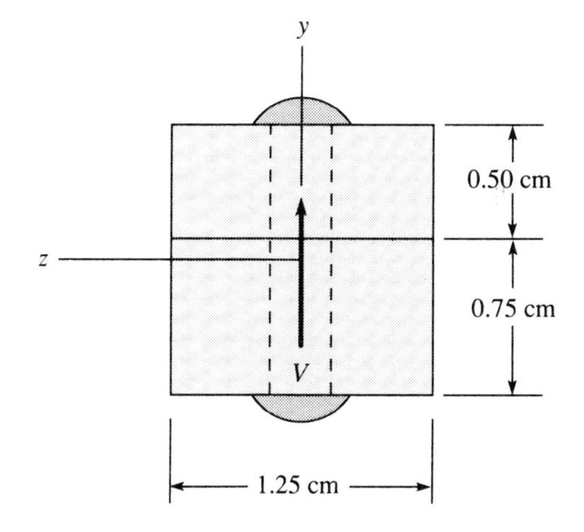

6.36 A simple beam is made of three strips of steel $1/4 \times 3/4$ in., which are joined by equally spaced rivets, as depicted in Fig. P6.36. The rivets are 1/4 in. in diameter and can sustain a maximum shear stress of 10 ksi. The beam must sustain a transverse shear $V = 100$ lb. Determine the maximum allowable spacing of the rivets (the distance between centers).

Fig. P6.36

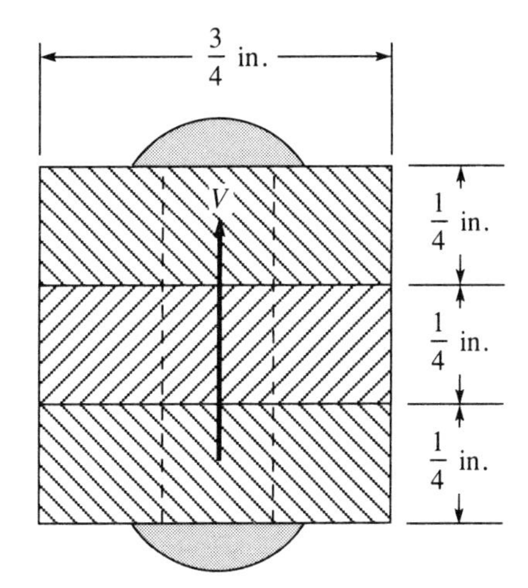

6.37 A wide-flange I-beam is 24 ft long, is simply supported at both ends, and supports a concentrated load of 6 kips at midspan. The beam is reinforced by bolting 8.00×0.250-in. plates on the top and bottom flanges, as shown in Fig. P6.37. The shanks of these

Fig. P6.37

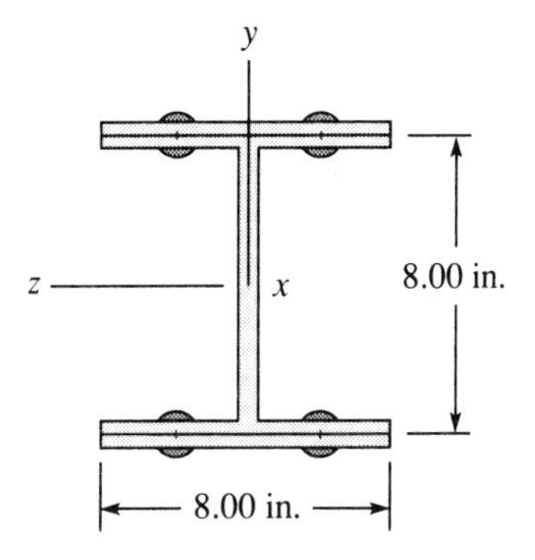

bolts are 0.500 in. in diameter; two bolts (one on each side) are spaced at equal intervals, 12.00 in. between centers. For the I-beam *only* (does not include the plates)

$$I = 178 \text{ in.}^4$$

Calculate by the theory for HI-HO beams (a) the maximum normal stress σ_x in the beam and (b) the mean shear stress upon each bolt.

6.38 A wide-flange I-beam (21 × 13 in., 112 lb/ft) is 30 ft long and is simply supported at both ends. It supports a concentrated load of 25 kips, which acts 10 ft from one end. Dimensions of the cross section are given in Fig. P6.38; from tabulated values $I = 2621$ in.^{4} Use the theory of HI-HO beams, neglect the weight, and calculate (a) the maximum normal stress, (b) the maximum transverse shear stress (τ_{xy}), and (c) the maximum shear stress. Repeat these calculations with the weight included.

6.39 The beam shown in Fig. P6.39 is a 12 × 5-in., 31.8-lb/ft American Standard I-beam. From tabulated properties, the depth is 12 in., the width (flanges) is 5 in., the thicknesses of flanges and web are 0.544 in. and 0.350 in., respectively, and

$$I = 215.8 \text{ in.}^4$$

Neglect the weight of the beam and (a) sketch the shear force and moment diagrams, (b) determine the maximum and minimum values of the shear force and moment, (c) compute the maximum transverse shear stress (τ_{xy}), and (d) compute the maximum shear stress.

Fig. P6.38

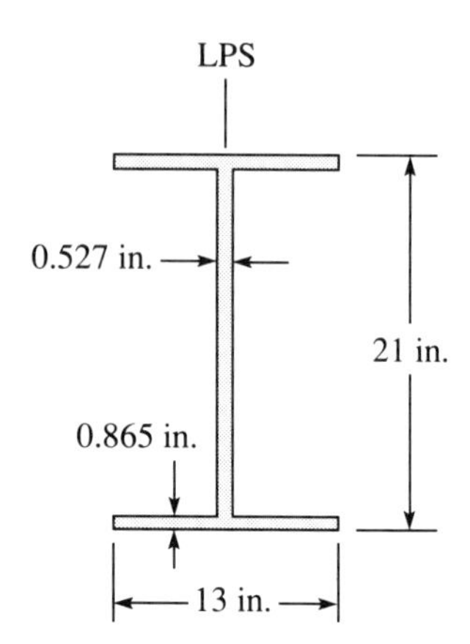

Fig. P6.39

300 lb/ft
5000 lb
10 ft
4 ft
6 ft

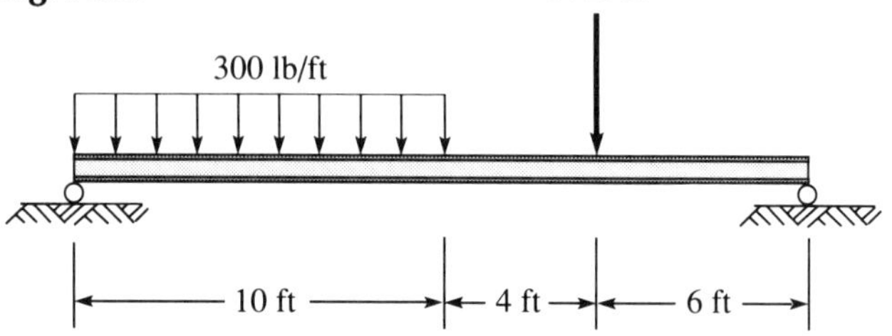

***6.40** A beam has the cross section shown in Fig. P6.40 ($t \ll h$). Show that, if the beam bends according to the description for symmetric beams, then the resultant shear on the cross section must pass through *P*.

Fig. P6.40

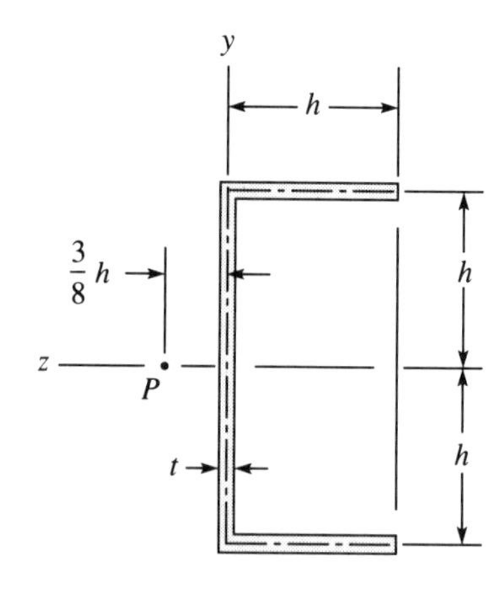

***6.41** A thin-walled cylindrical tube is used as a beam. Obtain an expression for the shear stress perpendicular to the radial direction anywhere in a cross section.

6.5 Deflection of Symmetric Hookean Beams

The Deflection Equation

The Bernoulli-Euler curvature formula was obtained [Sec. 5.12, Eq. (5.119)] for the case of simple bending of a linearly elastic beam:

$$\kappa \equiv \frac{1}{\rho} = \frac{M}{EI} \tag{6.25}$$

As in the discussion of bending stresses just concluded, we extend the use of Eq. (6.25) to cases where the shear stress is not zero on every cross section. Again, this leads to very useful results for slender beams, since the effects due to shear are generally insignificant. We simply treat κ and M as functions of the coordinate x that locates a cross section. Note, too, that the curvature κ and moment M are proportional for any hookean beam; only the coefficient (the flexural rigidity EI) is different for anisotropic or nonhomogeneous beam.

The deformation of the beam is determined by the deformed shape of the neutral axis. Initially, the neutral axis is along the x-axis, as in Fig. 6.19. After the beam deforms, the neutral axis becomes a curve in the xy-plane, as shown in Fig. 6.19. The equation of this curve determines the transverse displacement of particles on the neutral axis. This equation is therefore called the deflection function, or, more simply, the deflection. In Fig. 6.19, Δs is an element of arc length along the deformed neutral axis and ρ is the radius of curvature of the neutral axis at x. From the figure, we see that $\Delta s \doteq \rho\,\Delta\theta$, and in the limit, as Δs approaches zero,

$$\kappa \equiv \frac{1}{\rho} = \frac{d\theta}{ds} \tag{6.26}$$

The angle θ is the rotation of the tangent to the neutral axis. If we substitute Eq. (6.25) into Eq. (6.26), the result is

$$\kappa = \frac{d\theta}{ds} = \frac{M}{EI} \tag{6.27}$$

The relation (6.27) is applicable to symmetric HI-HO beams wherein the strains are small. Indeed, because the strains are small,

$$\Delta s = (1 + \varepsilon_0)\,\Delta x \doteq \Delta x \tag{6.28}$$

The use of this approximation is entirely consistent with previous approximations and enables us to replace the length s by the initial length x in Eqs. (6.26) and (6.27), to obtain

$$\kappa = \frac{d\theta}{dx} = \frac{M}{EI} \tag{6.29}$$

Fig. 6.19

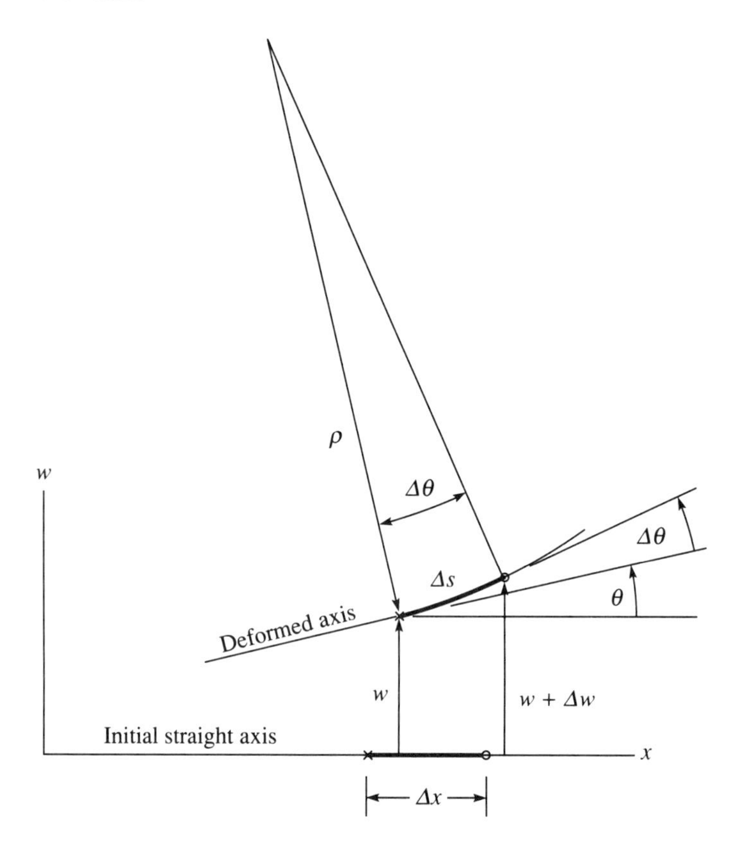

It remains to relate the rotation θ to the displacement w, shown in Fig. 6.19. We note that, as $\Delta x \to 0$, $\Delta\theta \to 0$ as well, and

$$\frac{dw}{ds} \doteq \frac{dw}{dx} = \sin\theta \tag{6.30a}$$

Let us now collect the differential equations that govern the four variables: dynamic variables V and M and kinematic variables θ and w. The dynamic variables are governed by the equilibrium equations (6.8) and (6.9). The kinematic variables are governed by (6.29) and (6.30). The first three equations are linear differential equations; the fourth is not. We note, however, that the dynamic equations (6.8) and (6.9) presumed very small rotations, so small that rotation was neglected in the free-body diagrams and hence in the equations. Thus, it is now consistent to treat the rotation as very small and use the first-order *approximation*:

$$\sin\theta \doteq \theta \tag{6.30b}$$

Then the four linear equations follow:

$$\frac{dV}{dx} = p(x) \tag{6.31}$$

$$\frac{dM}{dx} = -V(x) \tag{6.32}$$

$$\frac{d\theta}{dx} = \frac{M(x)}{EI} \tag{6.33}$$

$$\frac{dw}{dx} = \theta(x) \tag{6.34}$$

To obtain the functions $V(x)$, $M(x)$, $\theta(x)$, and $w(x)$, we must successively integrate the four first-order linear equations. This introduces some arbitrariness in the form of constants of integration. The evaluation of these constants is accomplished by requiring the force V, couple M, rotation θ, and/or deflection w to satisfy the conditions imposed by the supports. Let us therefore examine the support conditions at the more common types of supports.

Support Conditions

At a clamped or fixed support, the axis cannot rotate, nor can it displace. For example, the cantilever of Fig. 6.20(a) is clamped at $x = L$; the rotation θ and deflection w must satisfy the conditions

$$\theta(L) = 0, \qquad w(L) = 0$$

The cantilever of Fig. 6.20(a) is free at the end $x = 0$; the force V and couple M must satisfy the conditions

$$V(0) = 0, \qquad M(0) = 0$$

Fig. 6.20

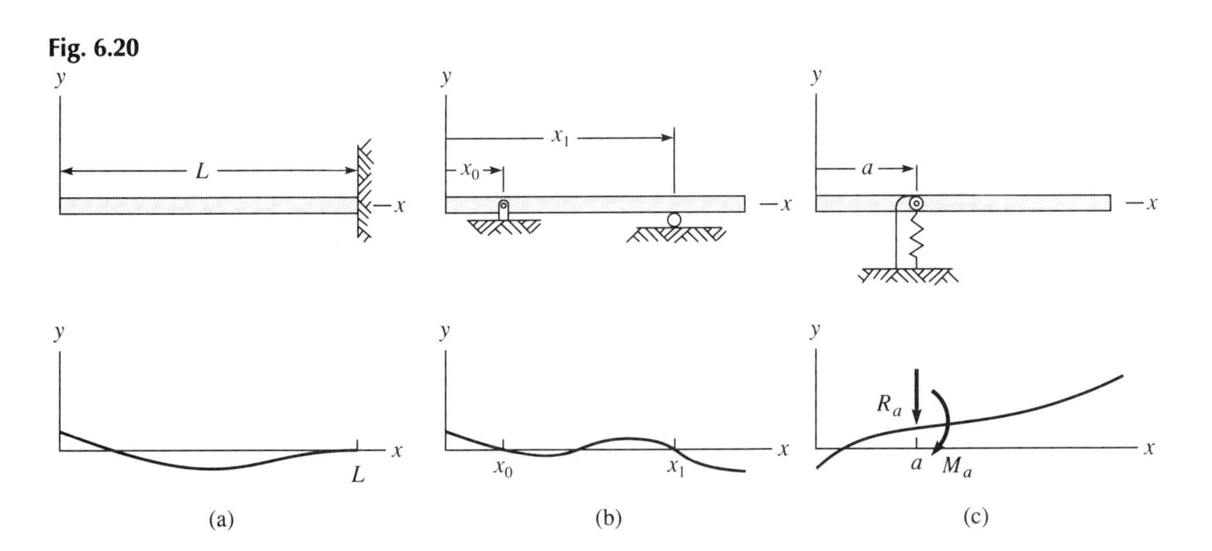

A pin or roller prevents transverse displacement w but does not inhibit rotation θ; that is, it can exert no couple. For example, the beam of Fig. 6.20(b) is supported by a pin at $x = x_0$ and a roller at $x = x_1$; the deflection must satisfy the conditions

$$w(x_0) = 0, \qquad w(x_1) = 0$$

The rotation θ and/or the deflection w can be partially constrained when joined to an elastic supporting structure. Such a situation is depicted schematically in Fig. 6.20(c). Here the couple M_A and force R_A are exerted by a hookean support; hence, these actions are proportional to the rotation $\theta(a)$ and displacement $w(a)$, respectively. These conditions follow:

$$\theta(a) = \frac{M_A}{\beta}, \qquad w(a) = \frac{R_A}{K}$$

Here the constants, β and K, depend upon the properties of the adjoining structure. The following examples illustrate the application of the differential equations, (6.31), (6.32), (6.33), and (6.34), and the support conditions in order to determine the variables, V, M, θ, and w.

EXAMPLE 6

The beam in Fig. 6.21(a) is fixed at the left end ($x = 0$) and simply supported at the right end ($x = L$). The conditions at the left end are

$$\theta(0) = 0, \qquad w(0) = 0 \tag{6.35a, b}$$

At the right end, the simple support can exert no couple but prevents any displacement:

$$M(L) = 0, \qquad w(L) = 0 \tag{6.36a, b}$$

The beam is subjected to the linearly varying load

$$p(x) = \bar{p}\left(1 - \frac{x}{L}\right) \tag{6.37}$$

where $\bar{p}$ is a constant (240 lb/ft), the magnitude of the load at the left end ($x = 0$).

We begin the solution by substituting the load $p(x)$, Eq. (6.37), into the first differential, Eq. (6.31), and integrating between the corresponding limits $(0, x)$:

$$dV = p\,dx = \bar{p}\left(1 - \frac{x}{L}\right) dV \tag{6.38a}$$

$$\int_{V_0}^{V(x)} dV = \bar{p} \int_0^x \left(1 - \frac{x}{L}\right) dx \tag{6.38b}$$

$$V = V_0 + \bar{p}\left(x - \frac{x^2}{2L}\right) \tag{6.38c}$$

Fig. 6.21

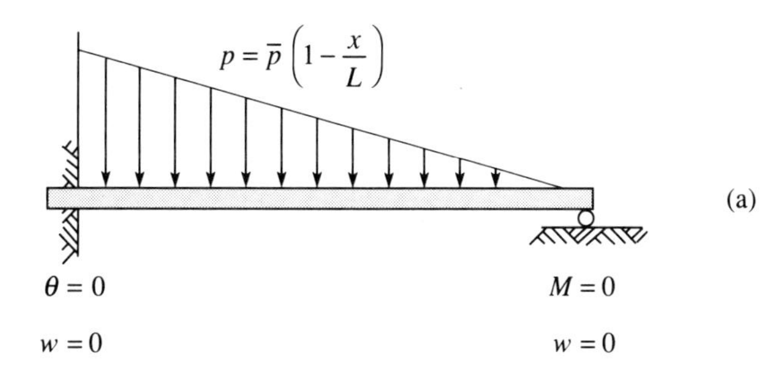

$p = \overline{p}\left(1 - \frac{x}{L}\right)$
(a)
$\theta = 0$
$w = 0$
$M = 0$
$w = 0$

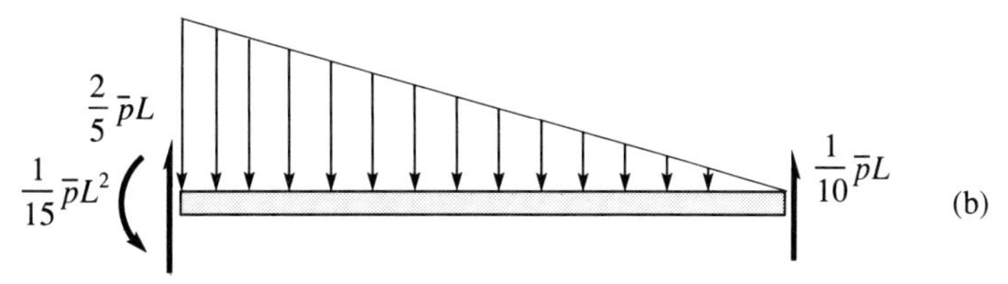

$\frac{2}{5}\overline{p}L$
$\frac{1}{15}\overline{p}L^2$
$\frac{1}{10}\overline{p}L$
(b)

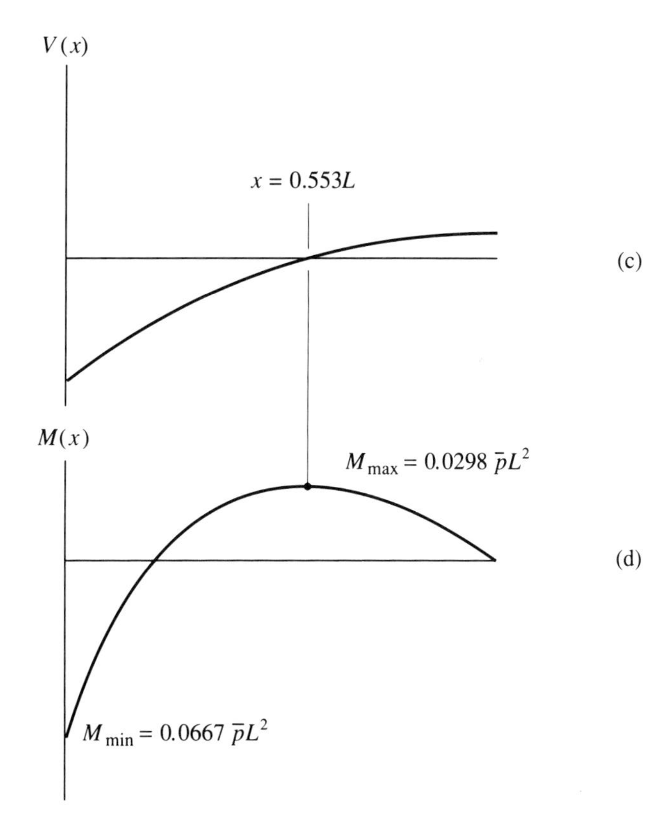

$V(x)$
$x = 0.553L$
(c)
$M(x)$
$M_{max} = 0.0298\ \overline{p}L^2$
(d)
$M_{min} = 0.0667\ \overline{p}L^2$

Note that the limits correspond: $V(0) = V_0$, the unknown force exerted on the left end, $x = 0$. The resulting Eq. (6.38c) is now substituted into the second differential equation, Eq. (6.32), which is integrated between the corresponding limits $(0, x)$:

$$dM = -V\,dx = -\left[V_0 + \bar{p}\left(x - \frac{x^2}{2L}\right)\right]dx \tag{6.39a}$$

$$\int_{M_0}^{M} dM = -\int_0^x \left[V_0 + \bar{p}\left(x - \frac{x^2}{2L}\right)\right]dx \tag{6.39b}$$

$$M = M_0 - V_0 x - \bar{p}\left(\frac{x^2}{2} - \frac{x^3}{6L}\right) \tag{6.39c}$$

Here $M(0) = M_0$, the unknown couple exerted upon the left end, $x = 0$. This last result, Eq. (6.39c), is now substituted into the third differential equation, Eq. (6.33), which is integrated between the corresponding limits $(0, x)$:

$$EI\,d\theta = M\,dx = \left[M_0 - V_0 x - \bar{p}\left(\frac{x^2}{2} - \frac{x^3}{6L}\right)\right]dx \tag{6.40a}$$

$$EI\int_0^{\theta(x)} d\theta = \int_0^x \left[M_0 - V_0 x - \bar{p}\left(\frac{x^2}{2} - \frac{x^3}{6L}\right)\right]dx \tag{6.40b}$$

$$EI\theta(x) = M_0 x - V_0\frac{x^2}{2} - \bar{p}\left(\frac{x^3}{6} - \frac{x^4}{24L}\right) \tag{6.40c}$$

Note that the end condition, $\theta(0) = 0$, has been inserted in the lower limit of Eq. (6.40b, c). Finally, the result, Eq. (6.40c), is substituted into the fourth differential equation, Eq. (6.34), which is integrated between the corresponding limits $(0, x)$:

$$dw = \theta(x)\,dx \tag{6.41a}$$

$$EI\int_0^{w(x)} dw = \int_0^x \left[M_0 x - V_0\frac{x^2}{2} - \bar{p}\left(\frac{x^3}{6} - \frac{x^4}{24L}\right)\right]dx \tag{6.41b}$$

$$EIw(x) = M_0\frac{x^2}{2} - V_0\frac{x^3}{6} - \bar{p}\left(\frac{x^4}{24} - \frac{x^5}{120L}\right) \tag{6.41c}$$

Here, too, an end condition, $w(0) = 0$, has been enforced.

The functions $V(x)$, $M(x)$, $\theta(x)$, and $w(x)$ are now determined, except for the two unknown constants, M_0 and V_0. We presume that the dimensions of the beam (L and I), the material (E), and the load ($\bar{p}$) are given; also, we have assumed that the properties (E and I) are constant. It remains only to determine the two constants by means of the remaining conditions at the right end, Eq. (6.36a, b); these are the only requirements not yet satisfied. They provide two linear algebraic equations in the two unknowns:

$$M(L) = M_0 - V_0 L - \bar{p}\frac{L^2}{3} = 0$$

$$EIw(L) = M_0 \frac{L^2}{2} - V_0 \frac{L^3}{6} - \bar{p}\frac{L^4}{30} = 0$$

The second is simplified when multiplied by $2/L^2$; then the two equations take the forms:

$$M_0 - V_0 L = \bar{p}\frac{L^2}{3}$$

$$M_0 - V_0 \frac{L}{3} = \bar{p}\frac{L^2}{15}$$

The solution of these provides the two actions, V_0 and M_0, at the left end:

$$V_0 = -\frac{2}{5}\bar{p}L, \qquad M_0 = -\bar{p}\frac{L^2}{15}$$

Also, by means of Eq. (6.38c),

$$V(L) = \tfrac{1}{10}\bar{p}L$$

The free body of the entire beam is shown in Fig. 6.21(b), and the plots of force $V(x)$ and couple $M(x)$ are shown in Fig. 6.21(c) and (d).

So that we can appreciate the orders of magnitude, let us compute the maximum normal stress σ_x at $x = 0$ and the deflection $w(L/2)$ at midspan. For this purpose, we use an American Standard wide-flange (12-WF-40) beam:

$$I = 310 \text{ in.}^4 \qquad \text{depth} = 12.0 \text{ in.}$$

$$L = 20.0 \text{ ft}$$

Then, by Eq. (6.16) the maximum normal stress occurs at the left end and at the top ($y = 6.00$ in.):

$$\sigma_{\max} = -\frac{M_{\min}(6.00)}{310}$$

$$= -\frac{(-76{,}800)(6.00)}{310} = 1490 \text{ lb/in.}^2 \quad (10.3 \text{ MPa})$$

The deflection at midspan is given by Eq. (6.41c) with $x = 10$ ft. The modulus for steel is approximately $E = 30 \times 10^6$ lb/in^2. Then,

$$w\left(\frac{L}{2}\right) = \frac{1}{EI}\left[\frac{M_0}{2} - \frac{V_0 L}{12} - \bar{p}\left(\frac{L^2}{96} - \frac{L^2}{960}\right)\right]\frac{L^2}{4}$$

$$= \frac{10^{-6}}{30 \times 310}\left[-\frac{1}{30} + \frac{1}{30} - \frac{9}{960}\right]\frac{(240)^4 \times 20}{4}$$

$$= -0.0167 \text{ in.} \quad (-0.0424 \text{ cm}) \ \blacklozenge$$

EXAMPLE 7

The cantilever of Fig. 6.22 is the same beam employed in the preceding example, but the simple support (roller) has been removed from the right end. As before, we can integrate the equilibrium equations to obtain Eqs. (6.38c) and (6.39c):

$$V = V_0 + \bar{p}\left(x - \frac{x^2}{2L}\right) \tag{6.42}$$

$$M = M_0 - V_0 x - \bar{p}\left(\frac{x^2}{2} - \frac{x^3}{6L}\right) \tag{6.43}$$

Fig. 6.22

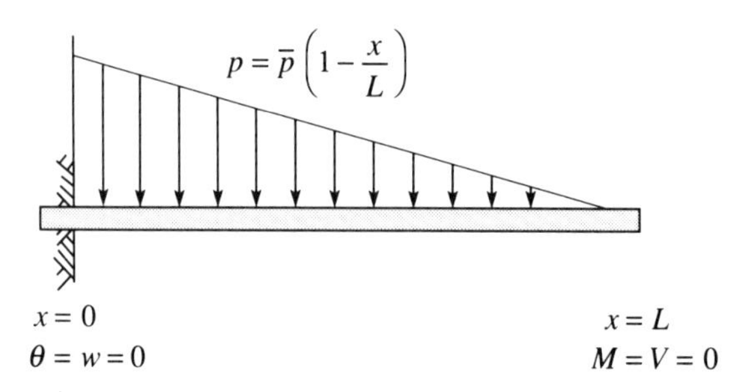

The conditions at the free end ($x = L$) are dynamic requirements, namely,

$$V(L) = 0, \qquad M(L) = 0$$

Substituting Eqs. (6.42) and (6.43) into these conditions, we obtain

$$V_0 = -\frac{\bar{p}L}{2}, \qquad M_0 = -\frac{\bar{p}L^2}{6}$$

Note that the force $V(x)$ and couple $M(x)$ are fully determined by these equilibrium conditions. The problem is said to be statically determinate; the internal force system can be determined without recourse to kinematic conditions. However, the problem is not completely solved. We still need the kinematic equations to determine the rotation $\theta(x)$ and the deflection $w(x)$. The two additional differential equations and their solutions are Eqs. (6.40c) and (6.41c). With the present values of the constants V_0 and M_0, Eq. (6.41c) takes the form

$$w(x) = -\frac{\bar{p}L^4}{12EI}\left(\frac{x^2}{L^2} - \frac{x^3}{L^3} + \frac{x^4}{2L^4} - \frac{x^5}{10L^5}\right) \tag{6.44}$$

With the previous dimensions and properties, we compute the maximum normal stress σ_x and deflection $w(L)$. The former occurs at the fixed end, where

$$M_0 = -\frac{240 \times 20 \times 240}{6} = -192{,}000 \text{ in.-lb} \quad (21.7 \times 10^3 \text{ N}\cdot\text{m})$$

$$\sigma_x = \frac{192{,}000 \times 6.00}{310} = 3720 \text{ lb/in.}^2 \quad (25.6 \text{ MPa})$$

The deflection at the free end is

$$w(L) = -\frac{\bar{p}L^4}{30EI} = 0.238 \text{ in.} \quad (0.605 \text{ cm})$$

This deflection is 14 times as great as the midspan deflection of the beam with the support at the right end. ◆

EXAMPLE 8

Recall the example of the beam in Sec. 6.4 (Fig. 6.17). The beam is a cantilever, fixed at the left end ($x = 0$) and subjected to the concentrated load at the right ($x = L$). The conditions at the left end are

$$\theta(0) = 0, \qquad w(0) = 0$$

The conditions at the right end can be stated as follows:

$$V(L) = -P, \qquad M(L) = 0$$

As before, four successive integrations lead to the expressions $[p(x) = 0]$:

$$\begin{aligned} V(x) &= V_0 = -P \\ M(x) &= M_0 - V_0 x = -PL\left(1 - \frac{x}{L}\right) \\ \theta(x) &= -\frac{PL^2}{EI}\left(\frac{x}{L} - \frac{x^2}{2L}\right) \\ w(x) &= -\frac{PL^3}{2EI}\left(\frac{x^2}{L^2} - \frac{x^3}{3L^3}\right) \\ w(L) &= -\frac{PL^3}{3EI} \end{aligned} \tag{6.45}$$

Let us examine the difference in flexibility between the assembly of Fig. 6.17(b) and the unified composite of Fig. 6.17(c). In the former, each strip acts independently. If t denotes the thickness and b is the width of the individual strip, then the property I of the rectangular cross section is

$$I = \frac{bt^3}{12}$$

Since each strip of the assembly supports one-fifth of the load P, the deflection is obtained by substituting the property I and $P/5$ into Eq. (6.45):

$$w(L) = -\frac{PL^3 12}{15Ebt^3} = -\frac{4}{5}\frac{PL^3}{Ebt^3} \tag{6.46}$$

The property I of the composite beam is

$$I = \frac{b(5t)^3}{12}$$

The deflection of the composite beam is obtained by substituting this expression and the total load P into Eq. (6.45):

$$w(L) = -\frac{PL^3}{3Eb}\frac{12}{(125)t^3} = -\frac{4}{125}\frac{PL^3}{Ebt^3} \tag{6.47}$$

Comparison of the deflections, Eqs. (6.46) and (6.47), indicates that the assembly is 25 times as flexible as the composite. ◆

EXAMPLE 9

The determination of the deflection is not as straightforward if concentrated loads act at *intermediate* locations on a beam. For example, consider the simply supported beam subjected to the concentrated force as shown in Fig. 6.23. The load $p(x)$ is zero everywhere but at the location $x = a$. The differential equations, Eqs. (6.31), (6.32), (6.33), and (6.34), apply in the interval $0 < x < a$ and in the interval $a < x < L$, but a discontinuity in shear occurs at point $x = a$. The expressions for each of the variables (V, M, θ, w) are different in the two segments. Accordingly, we identify the expressions in the first $(0 < x < a)$ by the subscript 1 and in the second $(a < x < L)$ by the subscript 2. Integrating successively the four equations and using the conditions $w(0) = M(0)$ in the first, we obtain in $0 < x < a$:

$$V_1(x) = V_0 \tag{6.48a}$$

$$M_1(x) = -V_0 x \tag{6.48b}$$

$$EI\theta_1(x) = EI\theta_0 - V_0\frac{x^2}{2} \tag{6.48c}$$

$$EIw_1(x) = EI\theta_0 x - V_0\frac{x^3}{6} \tag{6.48d}$$

In the interval $(a < x < L)$ our lower limit is $x = a$, where we denote the values by $V_2(a)$, $M_2(a)$, $\theta_2(a)$, and $w_2(a)$. Then the successive integrations lead to the expressions in $a < x < L$:

Fig. 6.23

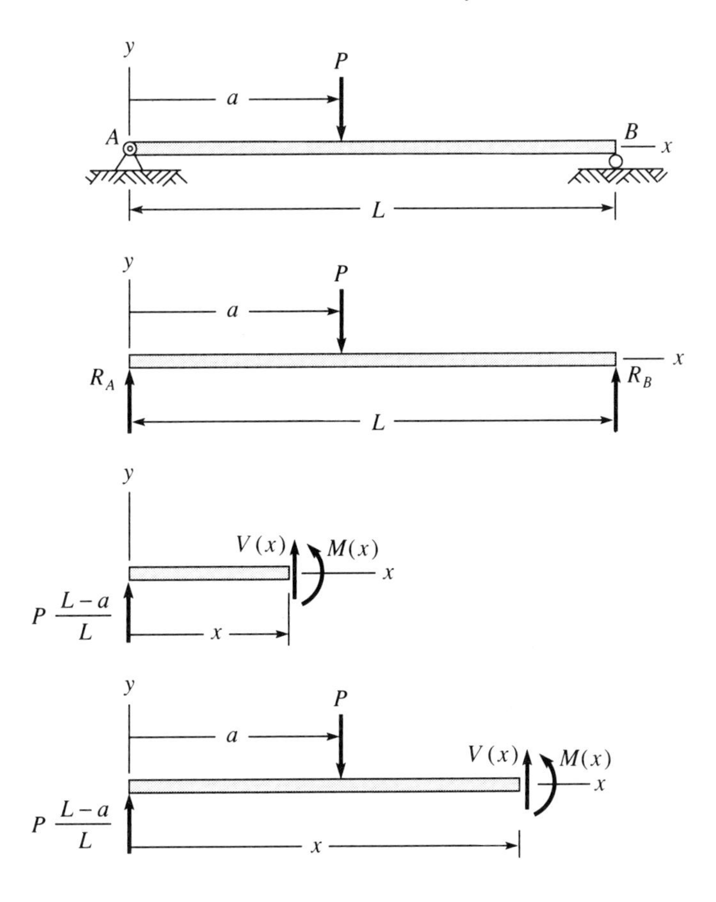

$$V_2(x) = V_2(a) \tag{6.49a}$$

$$M_2(x) = M_2(a) - V_2(a)(x - a) \tag{6.49b}$$

$$EI\theta_2(x) = EI\theta_2(a) + M_2(a)(x - a) - V_2(a)\frac{(x - a)^2}{2} \tag{6.49c}$$

$$EIw_2(x) = EIw_2(a) + EI\theta_2(a)(x - a) + M_2(a)\frac{(x - a)^2}{2} - V_2(a)\frac{(x - a)^3}{6} \tag{6.49d}$$

The complete deflection is given by the two equations (6.48d) and (6.49d), which contain six unknown constants, V_0, θ_0, $V_2(a)$, $M_2(a)$, $\theta_2(a)$, and $w_2(a)$. We presume that the load (P) and properties (E, I) are known.

As before, the two conditions at the right end provide two linear algebraic equations in the unknown constants:

$$M_2(L) = 0, \qquad w_2(L) = 0 \tag{6.50a,b}$$

Four additional equations are needed. These are obtained by acknowledging the conditions of continuity or discontinuity, which must hold at the location of the concentrated load ($x = a$).

Firstly, let us consider the dynamic requirements at concentrated loads, as set forth in Sec. 6.2. By Eq. (6.14) the shear force must exhibit a jump at the concentrated load; therefore, we have

$$V_2(a) - V_1(A) = P \tag{6.50c}$$

Since there is no concentrated couple, the moment is everywhere continuous; therefore,

$$M_2(a) = M_1(a) \tag{6.50d}$$

Secondly, let us consider the kinematic requirements. Underlying all our work has been the assumption that the materials we are considering are continuous and cohesive media. The distinguishing characteristic of a cohesive media is that lines and surfaces of particles in the body cannot be torn apart or made to pass through one another. Then the deflection of the neutral axis to either side of the concentrated load must be the same; otherwise, we will have a break in the neutral axis. Furthermore, the rotation of the neutral axis to either side of the concentrated load must be the same. This follows from our assumption that cross sections remain perpendicular to the neutral axis. If there were a corner in the deformed shape of the neutral axis, then two cross sections near this corner would have to pass through each other, as illustrated in Fig. 6.24.

Fig. 6.24

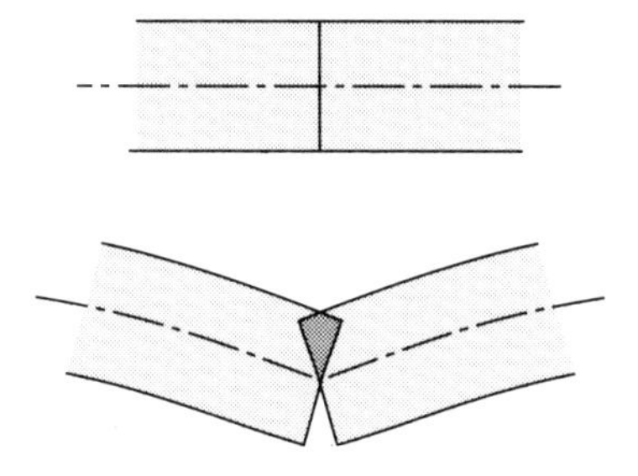

The required continuity of rotation θ and deflection w provides the remaining equations:

$$\theta_2(a) = \theta_1(a) \tag{6.50e}$$

$$w_2(a) = w_1(a) \tag{6.50f}$$

With the expressions in Eqs. (6.48) and (6.49) for the variables in the respective segments, the six equations of Eq. (6.50) assume the forms [Eqs. (6.48) are used to eliminate $V_1(a), M_1(a), \theta_1(a)$, and $w_1(a)$]:

$$M_2(L) = M_2(a) - V_2(a)(L-a) = 0$$

$$EIw_2(L) = EIw_2(a) + EI\theta_2(a)(L-a) + M_2(a)\frac{(L-a)^2}{2} - V_2(a)\frac{(L-a)^3}{6} = 0$$

$$V_2(a) = P + V_1(a) = P + V_0$$

$$M_2(a) = M_1(a) = -V_0 a$$

$$\theta_2(a) = \theta_1(a) = \theta_0 - \frac{V_0 a^2}{2EI}$$

$$w_2(a) = w_1(a) = \theta_0 a - \frac{V_0 a^3}{6EI}$$

The solution follows:

$$V_1 = V_0 = -\frac{P(L-a)}{L}, \qquad V_2 = \frac{Pa}{L}$$

$$M_1(a) = M_2(a) = +\frac{P(L-a)a}{L}$$

$$\theta_0 = -\frac{Pa(L-a)(2L-a)}{6EIL}, \qquad \theta_1(a) = \theta_2(a) = -\frac{Pa(L-a)(L-2a)}{3LEI}$$

$$w_2(a) = w_1(a) = -\frac{Pa^2(L-a)^2}{3LEI}$$

The greatest deflection occurs where the rotation is zero. This is generally *not* under the load. In fact, $\theta(a)$ vanishes only if the load acts at the middle ($x = L/2$).

This example illustrates an apparent complication when concentrated loads act at intermediate locations. By the procedure described, each load requires an additional expression for each of the four variables and four additional constants, which are determined by the four conditions of continuity and/or discontinuity at the site of each concentrated load. Operational procedures are available to simplify the treatment of such problems. Since these are essentially mathematical and procedural tools, we relegate our presentation to Appendix C. ◆

PROBLEMS

Apply the theory of symmetric HI-HO beams to the following problems. Those marked by a dagger (†) involve discontinuous or concentrated loads; as such they are most readily solved by the methods of Appendix C.

6.42 Determine all reactions and the deflection $w(x)$; express them in terms of the given parameters ($\bar{p}, P, M, L$) as shown in Fig. P6.42(a) through (p).

Fig. P6.42

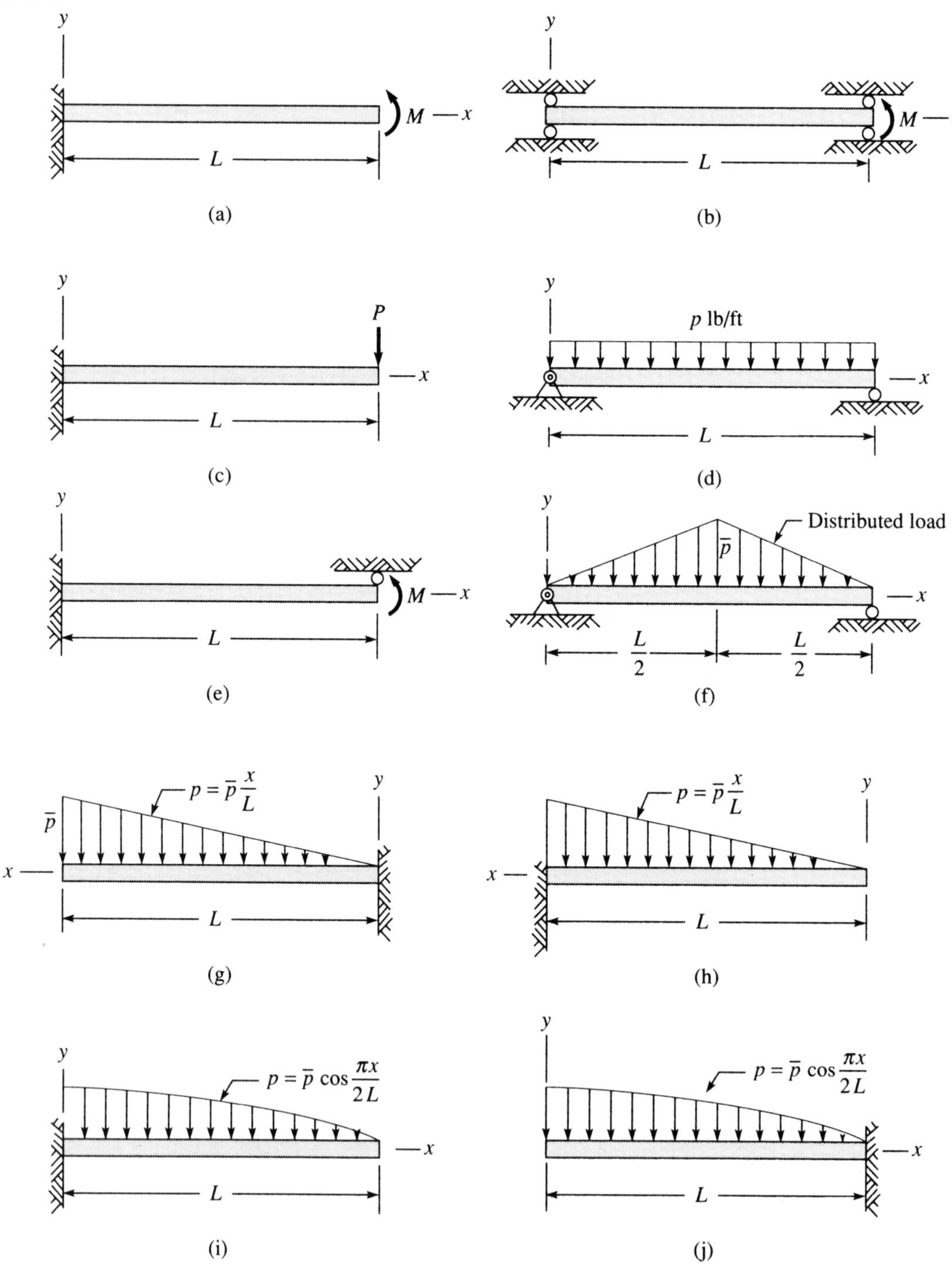
y
M
x
L
(a)
y
M
x
L
(b)
y
P
x
L
(c)
y
p lb/ft
x
L
(d)
y
M
x
L
(e)
y
Distributed load
$\bar{p}$
x
$\frac{L}{2}$
$\frac{L}{2}$
(f)
$p = \bar{p}\frac{x}{L}$
y
$\bar{p}$
x
L
(g)
$p = \bar{p}\frac{x}{L}$
y
x
L
(h)
y
$p = \bar{p}\cos\frac{\pi x}{2L}$
x
L
(i)
y
$p = \bar{p}\cos\frac{\pi x}{2L}$
x
L
(j)

Fig. P6.42
(*cont.*)

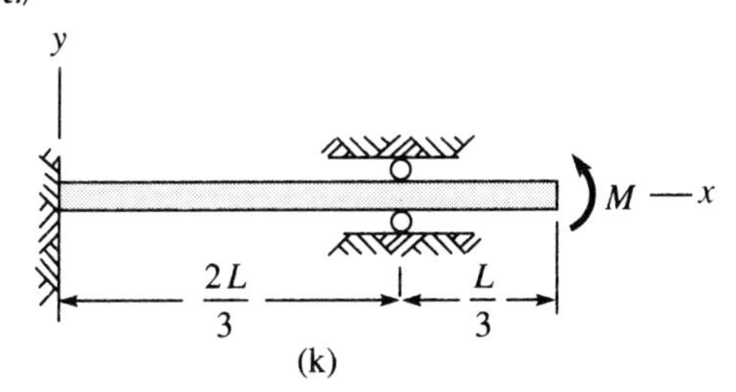

(k)

(l)

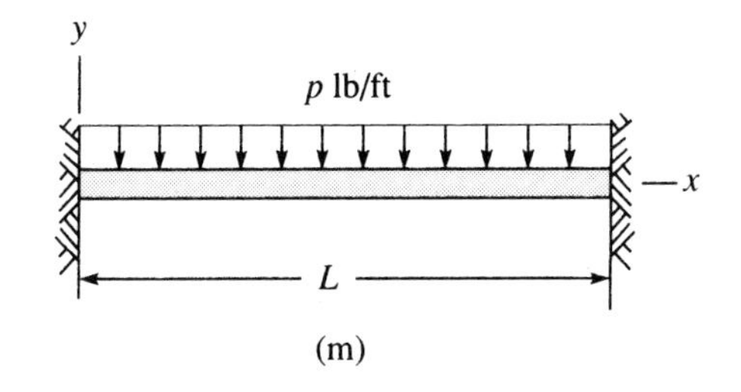

(m)

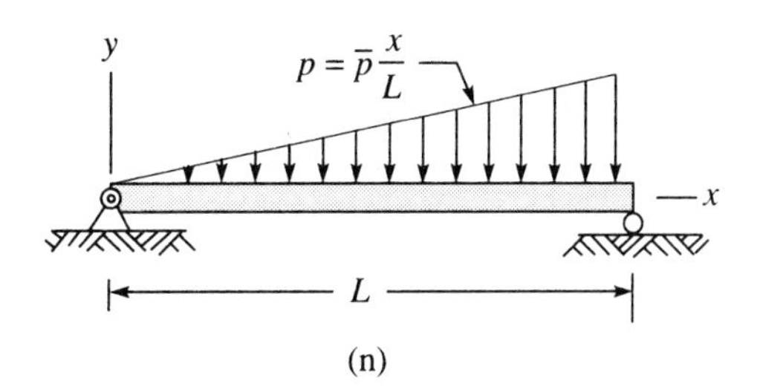

(n)

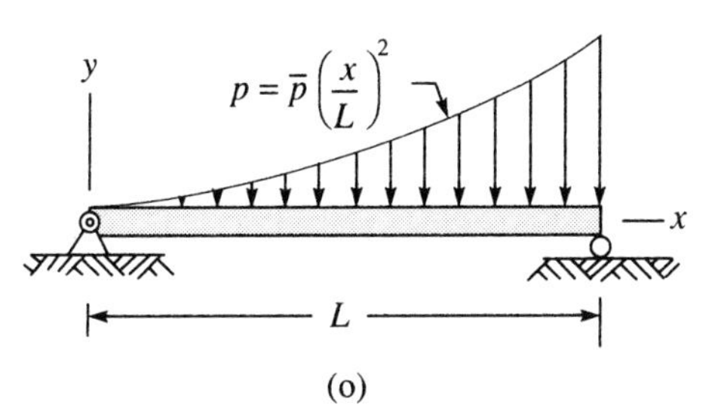

(o)

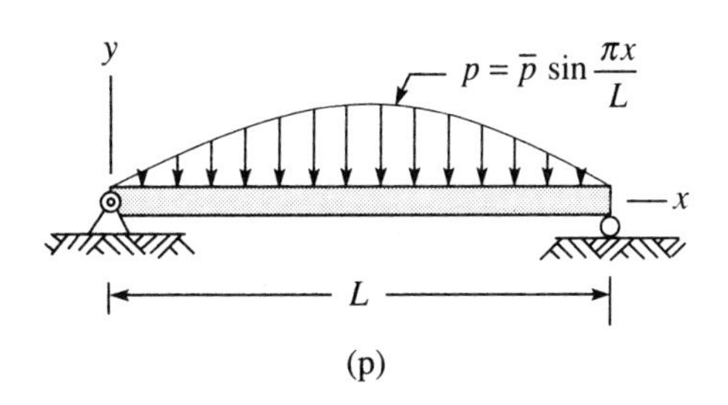

(p)

6.43 The beams of Fig. P6.43(a) and (b) are restrained by linear springs at the right ends. Constants K and β are the moduli of the linear springs, which resist lateral displacement (*a*) and rotation (*b*) by exerting force ($F = Kw$) and couple ($M = \beta\theta$), respectively. Obtain the equations of the deflection $w(x)$ and the reactions at the left end. What are the implications of passing to the limits $K \to 0$, $K \to \infty$ and $\beta \to 0$, $\beta \to \infty$?

Fig. P6.43

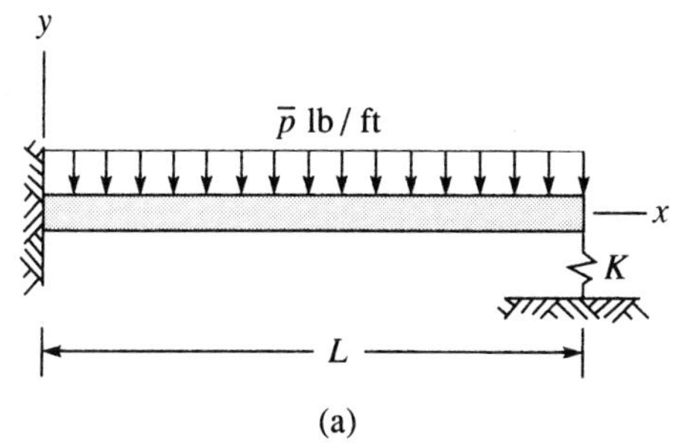

(a)

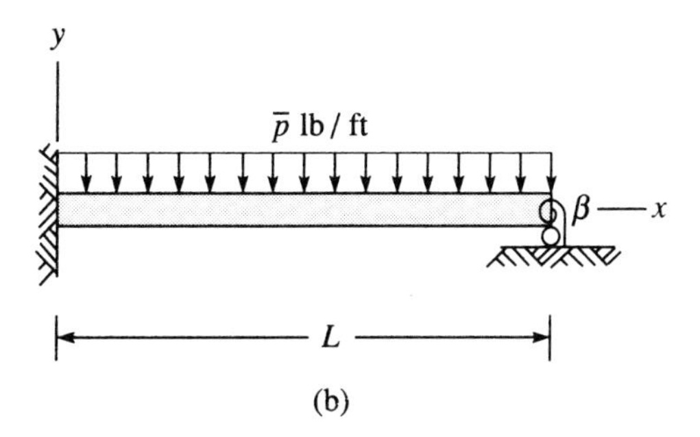

(b)

[†]**6.44** The beams of Fig. P6.44(a) through (m) are subject to concentrated loads and/or reactions. Consequently, the shear force and/or bending couple exhibit discontinuity, and the solution must also exhibit the appropriate discontinuity. Determine all reactions and the deflection; express them in terms of the given parameters ($\bar{p}$, P, M, L, a, b, K, β) as shown.

Fig. P6.44

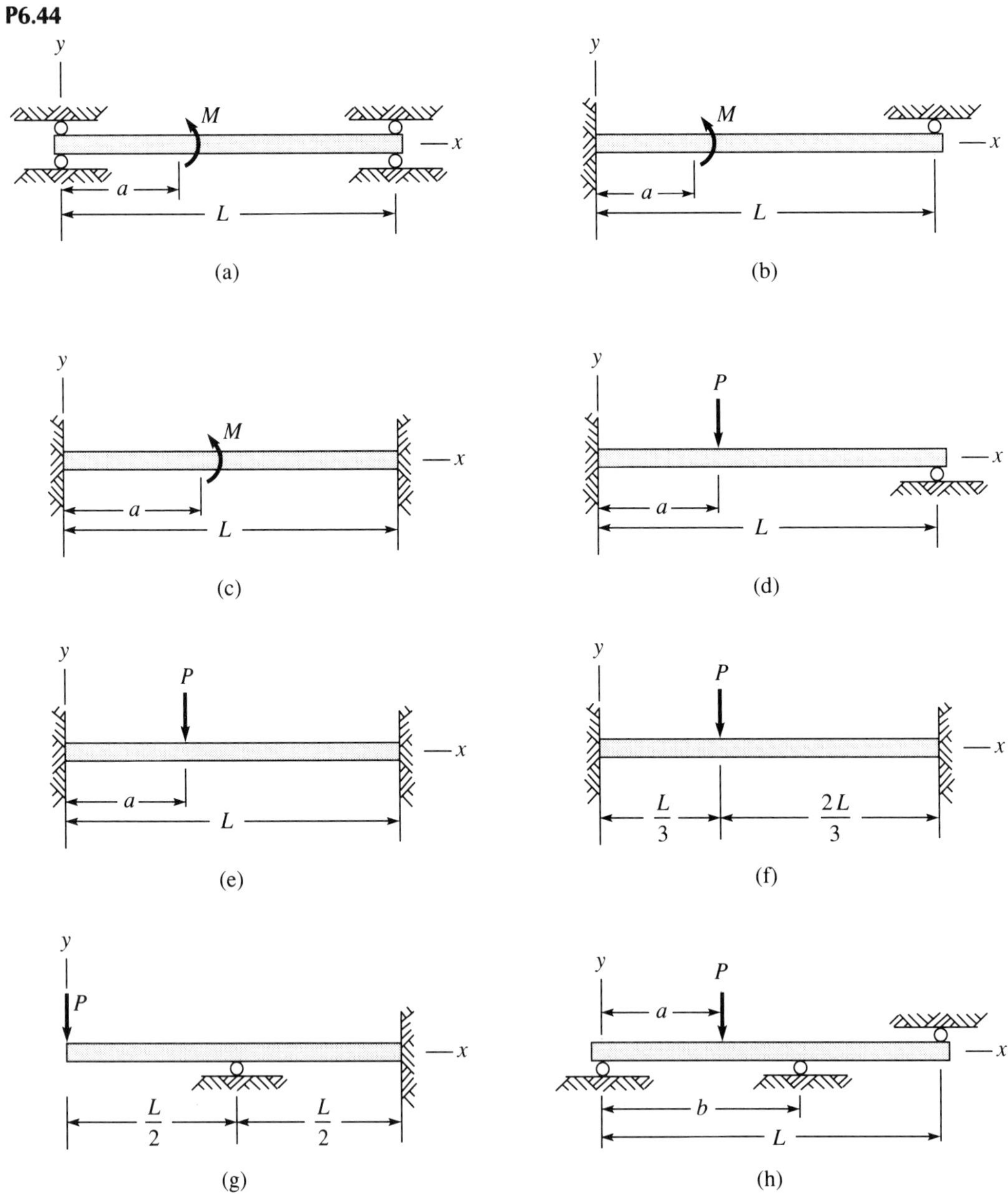

Fig. P6.44
(cont.)

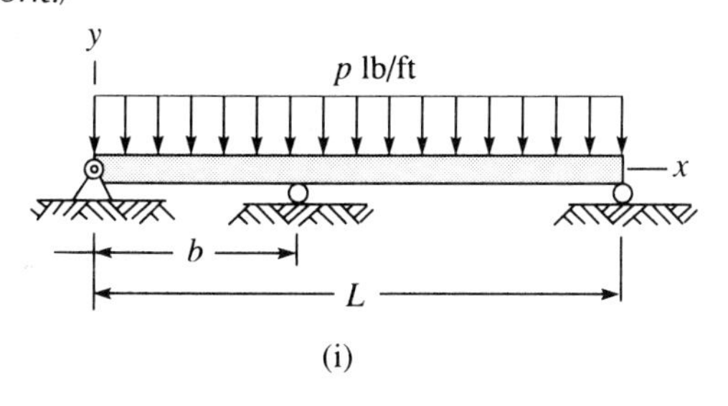

(i)

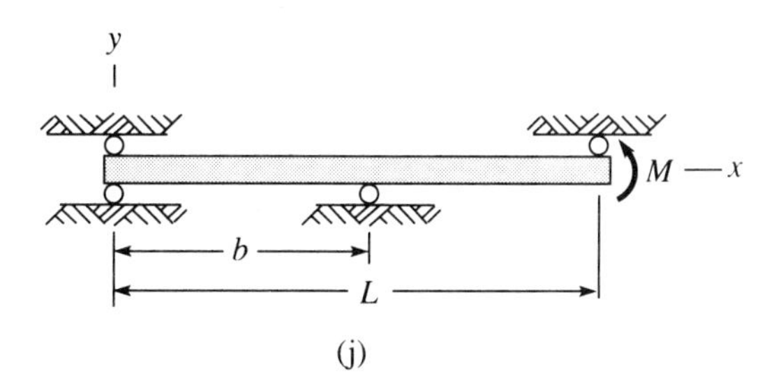

(j)

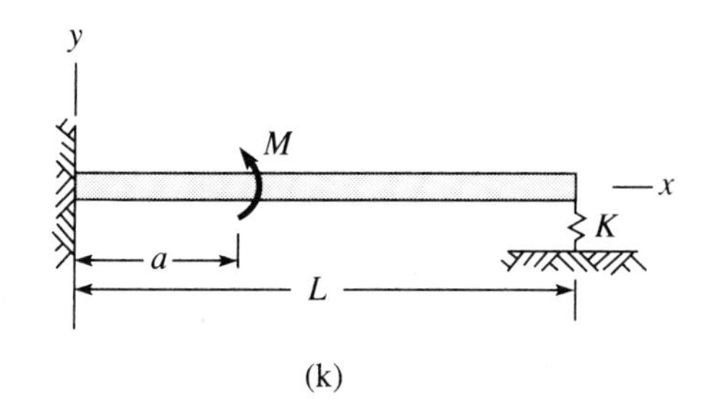

(k)

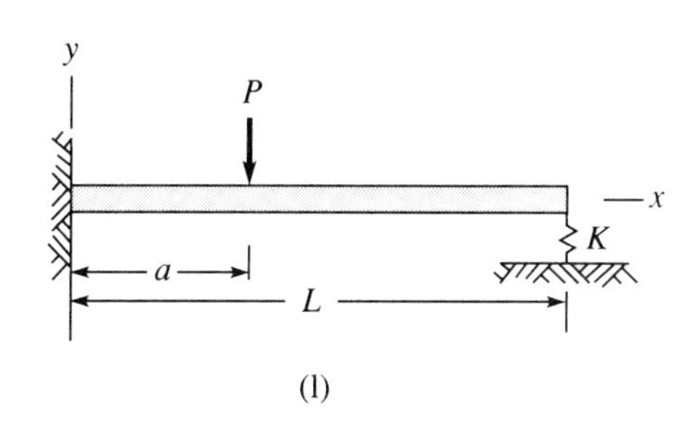

(l)

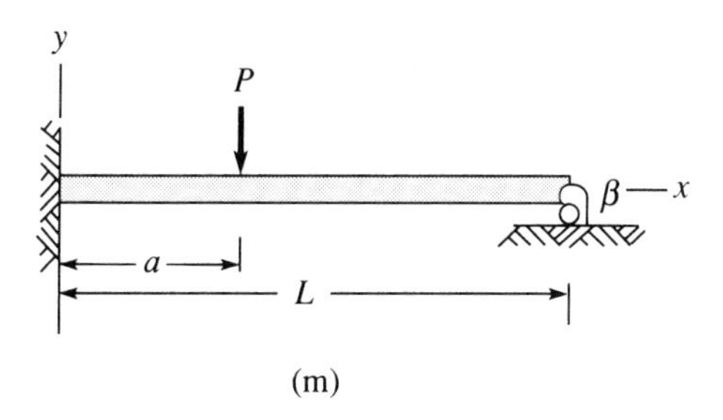

(m)

6.45 A steel strip is used as a cantilever and subjected to the concentrated load at the midpoint, as depicted in Fig. P6.45. The width of the strip is 2.50 cm and the depth h is 1.25 cm; the modulus is 21×10^4 MPa. Compute the deflection at the free end C.

Fig. P6.45

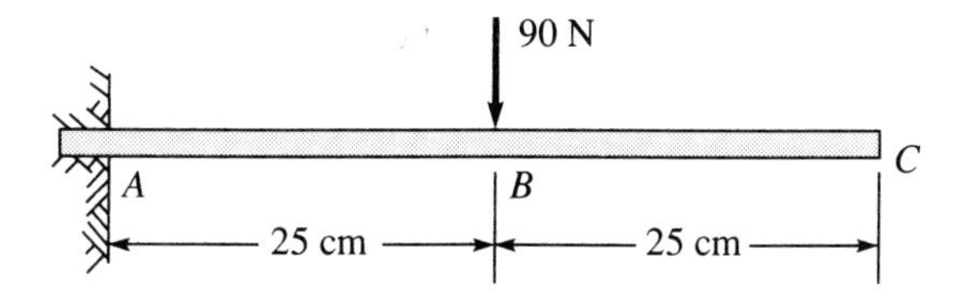

6.46 The steel strip of Prob. 6.45 is subjected to a couple at the midpoint, as depicted in Fig. P6.46. The

Fig. P6.46

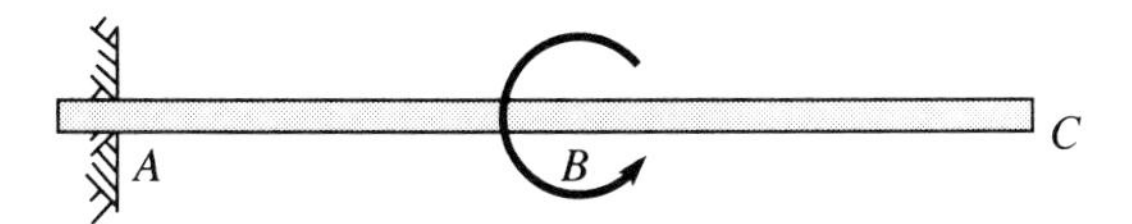

couple is just sufficient to cause a maximum normal stress of 340 MPa. Compute the deflection of the free end C.

6.47 Find the displacement of the tip of the pointer when the uniform load $\bar{p}$ is applied to the cantilever beam in Fig. P6.47.

Fig. P6.47

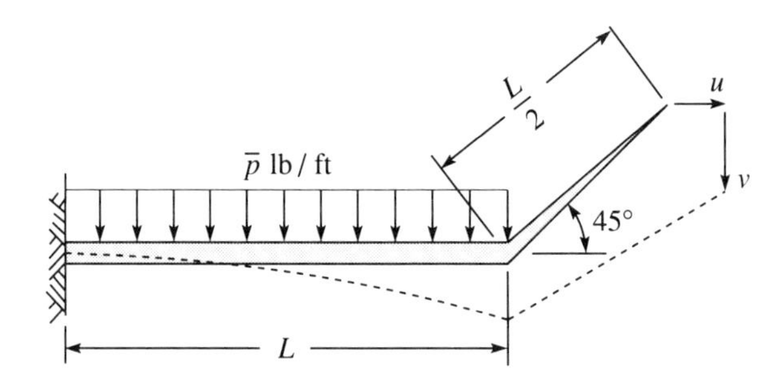

6.48 Pointers are attached at the ends and perpendicular to the axis of the simply supported beam of Fig. P6.48. Compute the distance L^* between the tips of the pointers when the load P is applied.

Fig. P6.48

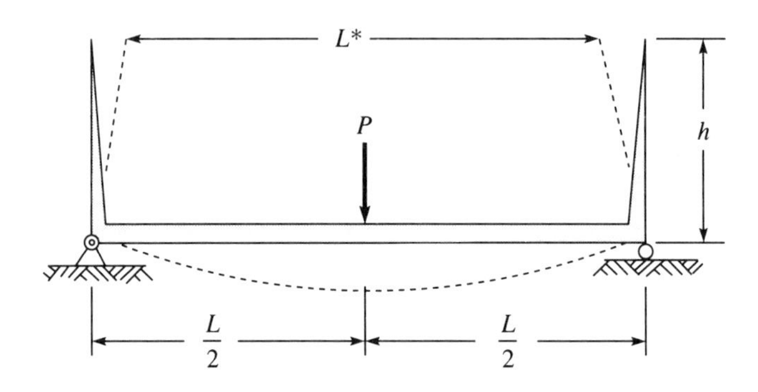

6.49 A weight rests on the end of a cantilevered beam. Determine the minimum value of the coefficient of friction between the weight and the beam in order that the weight does not slide off the beam.

6.50 The simply supported, uniformly loaded beam of Fig. P6.50 has a cross section that is a 4-in. by 6-in. rectangle. Determine its maximum deflection if (a) the 6-in. dimension is the depth of the beam and (b) the 4-in. dimension is the depth of the beam. Use $E = 4 \times 10^6$ lb/in^2.

Fig. P6.50

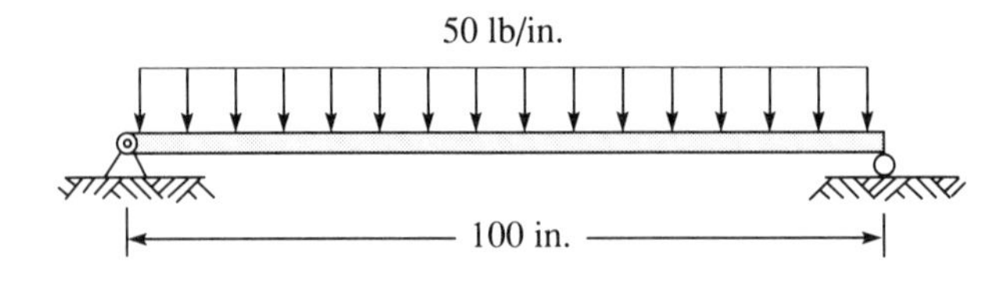

6.51 A round steel shaft is clamped at one end and subjected to a bending couple, as shown in Fig. P6.51. Determine the deflection of point A.

Fig. P6.51

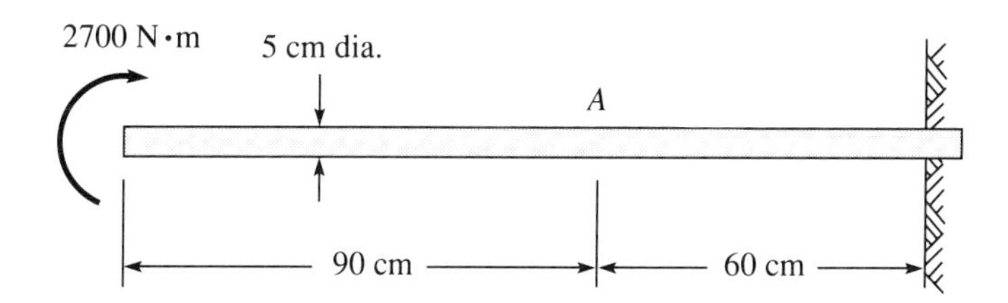

6.52 How much does the steel beam of Prob. 6.51 deflect if the effect of its weight is taken into account?

6.53 The cantilever shown in Fig. P6.53 is a standard steel (10-WF-49) I-beam. From tables, $I = 272.9$ in.^{4} The uniform load on the indicated portion is 4000 lb/ft. Determine the slope and deflection at the free end A.

Fig. P6.53

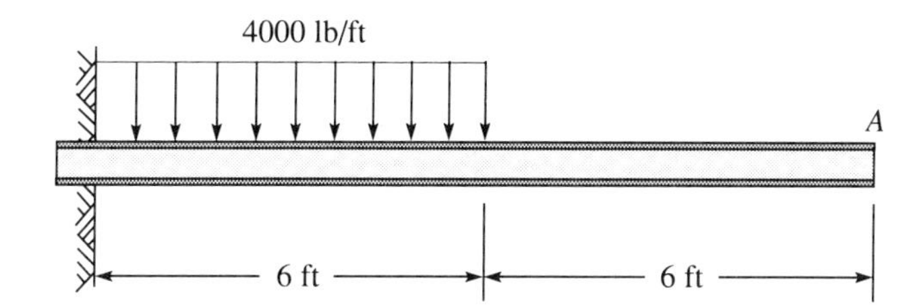

6.54 The beam BC in Fig. P6.54 is simply supported at end C and on the cantilever AB at B. Both are American Standard steel I-beams (12 × 5-in., 35 lb/ft); $I = 227.0$ in.^{4} Determine the deflection of the intermediate support B and then the deflection at the midpoint of the beam BC caused by the uniform load on BC.

Fig. P6.54

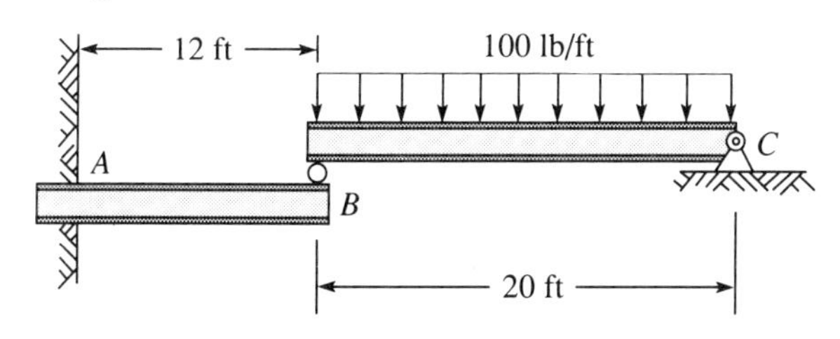

6.55 A steel shaft 30 cm long is cantilevered at one end. It must support a 900-N transverse load concentrated at the free end. Specifications require that the deflection is not to exceed 0.125 cm. Determine the minimum diameter that will serve the purpose.

6.56 Repeat Prob. 6.55, including the effect of the weight of the shaft.

6.57 What minimum cross-sectional moment of inertia is required for a 10-in. (depth) structural steel I-beam, simply supported at each end and carrying a uniformly distributed load of 500 lb/ft, if specifications require that the maximum deflection is not to exceed 1/360 of the span length and the maximum normal stress is not to exceed 16,000 psi?

6.58 The torque wrench shown in Fig. P6.58 is to be designed so that the pointer moves 2 cm on the scale to indicate a 4500-cm · N couple acting on bolthead A. If the wrench is made of steel, determine the proper width h.

Fig. P6.58

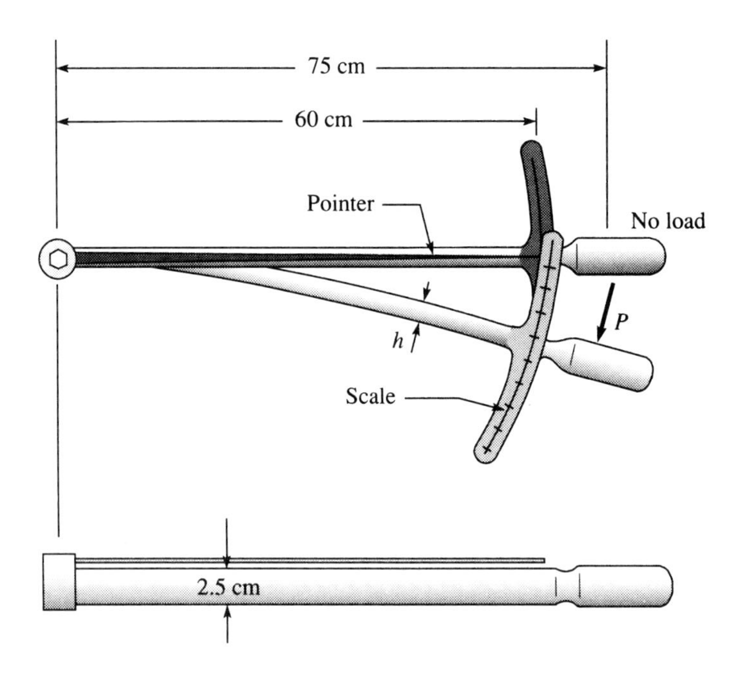

6.59 A uniformly loaded cantilevered beam is supported by a wire, as shown in Fig. P6.59. The wire has a cross-sectional area A and is linearly elastic. Determine the deflection of the right end as a function of load intensity $\bar{p}$.

Fig. P6.59

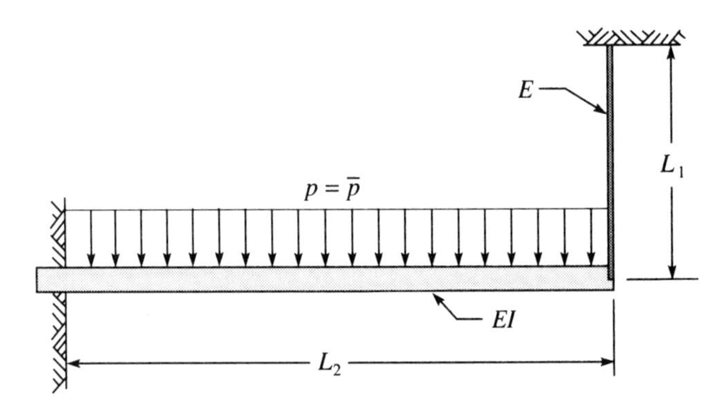

6.60 The free end of the cantilever beam in Fig. P6.60 is a small distance Δ above a roller support. Determine the reaction of the roller as a function of $\bar{p}$.

Fig. P6.60

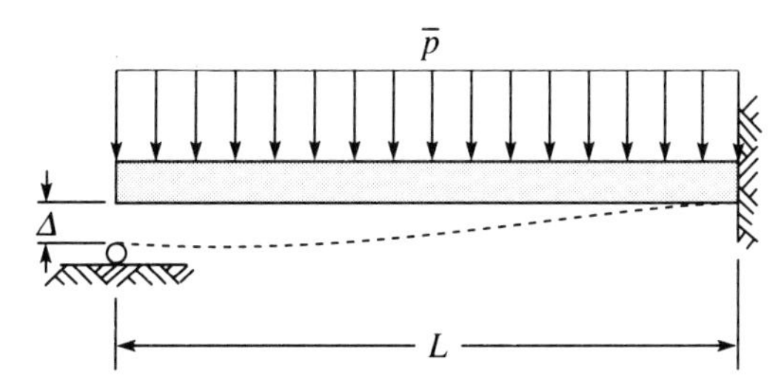

6.61 A straight beam of Douglas fir ($E = 1.5 \times 10^6$ lb/in.^{2}) is simply supported at the ends. When it is subjected to a uniform loading $\bar{p} = 20$ lb/ft, it deflects enough to contact the roller support B (shown in Fig. P6.61), which was originally 0.40 in. below the beam. Determine the reactions of supports A, B, and C.

Fig. P6.61

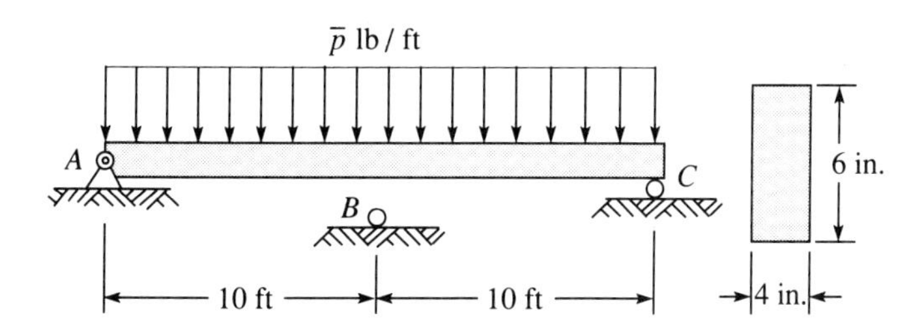

6.62 A beam has a symmetric I cross section and length L. It is simply supported at the ends and carries a concentrated load P at midspan. The maximum deflection is equal to the depth h of the beam. If the maximum normal stress on a cross section is σ, obtain an expression for the depth h in terms of the length L, stress σ, and modulus of elasticity E.

6.63 Two strips of metal have the same width and thickness. They are used together as a cantilever beam. In the case of Fig. P6.63(a) and (b), the strips are welded together at point A, and in the case of Fig. P6.63(c) and (d), they are not. Determine the ratio of the end deflections in the two cases.

Fig. P6.63

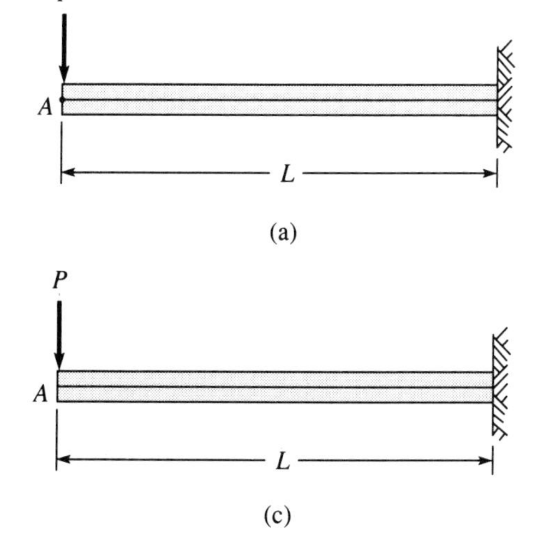

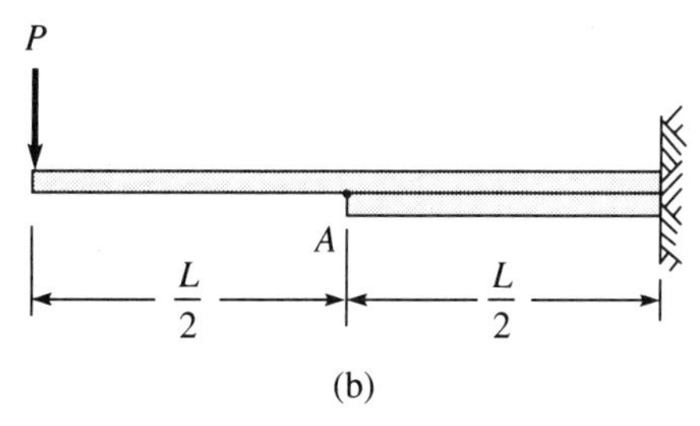

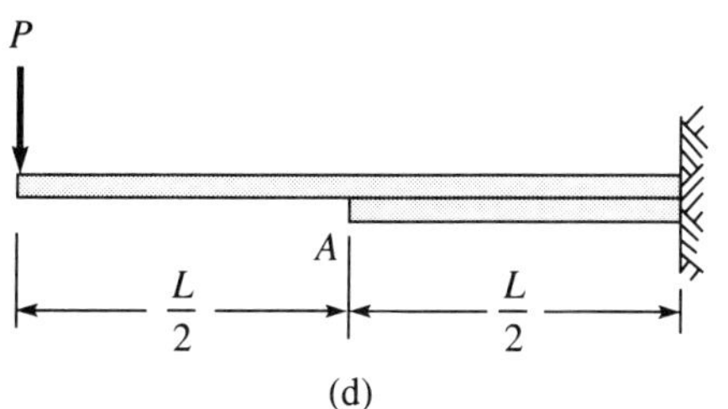

6.64 The continuous beam BCD in Fig. P6.64 forms a right angle at C; it is clamped at B and is simply supported at C. The force $P = 1000$ N acts at D parallel to BC. Determine the reactions at B and C.

Fig. P6.64

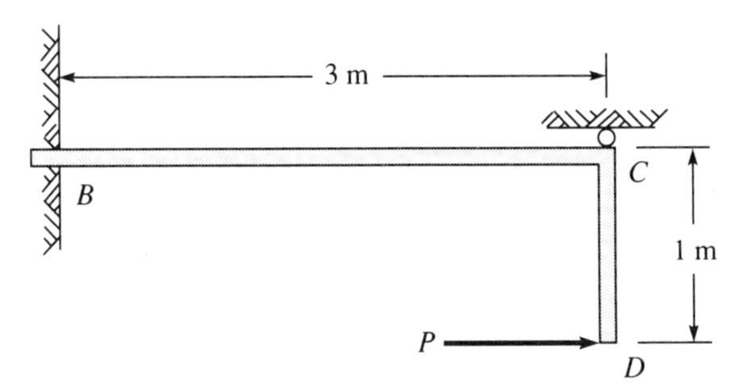

6.65 The beam of Fig. P6.65 is simply supported at A and clamped at B; it is subjected to the uniformly distributed load, $P = 800$ N/m. Determine the reactions at both ends, sketch the shear and moment diagrams, and obtain the maximum and minimum values of each.

Fig. P6.65

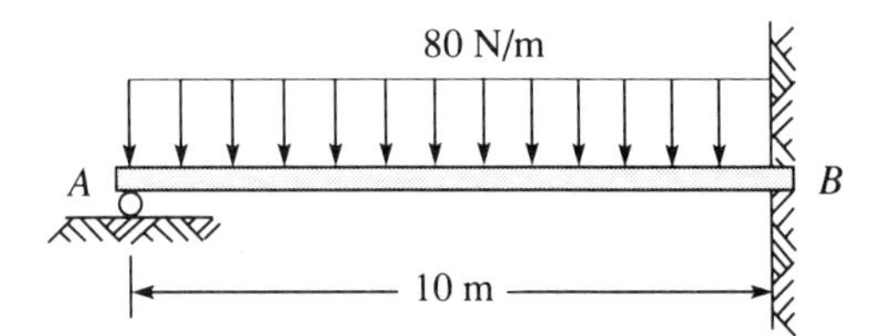

6.66 A symmetric HI-HO beam is clamped at one end ($x = 0$) and simply supported at the other ($x = \ell$) It supports a distributed load:

$$P = \left(1 - \frac{x}{\ell}\right)\bar{p}, \qquad \bar{p} = 2000 \text{ N/m}, \qquad \ell = 10 \text{ m}$$

Determine the reactions upon both ends, and sketch the shear and moment diagrams.

6.67 A symmetric HI-HO beam is fixed at one end A ($x = 0$) and simply supported at an intermediate site B ($x = 20$ ft). It carries the uniform load over the portion AB and the concentrated load at end C, as shown in Fig. P6.67. (a) Determine the bending moment at B, (b) find the reactions at A and B, and (c) sketch the shear and moment diagrams.

Fig. P6.67

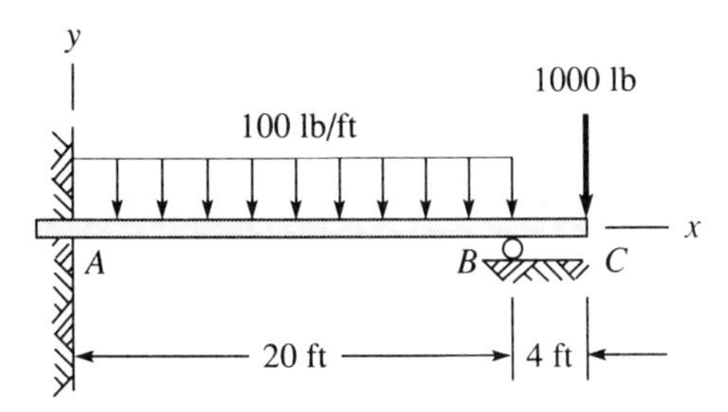

6.68 The steel beam of Fig. P6.68 is clamped at A and simply supported at B; $I = 42.7$ in.^{4} and $E = 30 \times 10^3$ ksi. Determine the force exerted by the support at B and the deflection of end C. $P = -20x$ (lb/ft).

Fig. P6.68

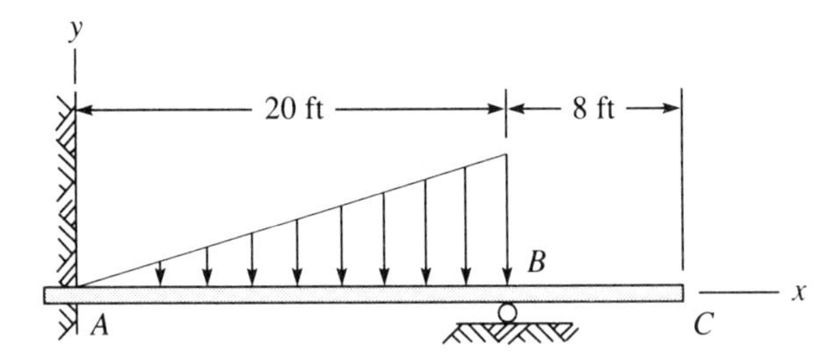

6.69 A symmetric HI-HO beam is clamped at the left end ($x = 0$) and simply supported at the right ($x = 2a$). In terms of the constants $(\bar{p}, a)$, determine the reactions at both ends and sketch the shear and moment diagrams if

$$p = \bar{p}(2a - x)\frac{x}{a}$$

6.70 A symmetric HI-HO beam is fixed at the left end ($x = 0$) and simply supported at the right ($x = L$). It is subjected to the load

$$p(x) = \bar{p} \sin \pi \frac{x}{L}$$

In terms of the constants, $\bar{p}$ and L, determine the reactions at both ends and sketch the shear and moment diagrams.

6.71 A cantilever beam is tapered as shown in Fig. P6.71. Compute the deflection of the free end.

Fig. P6.71

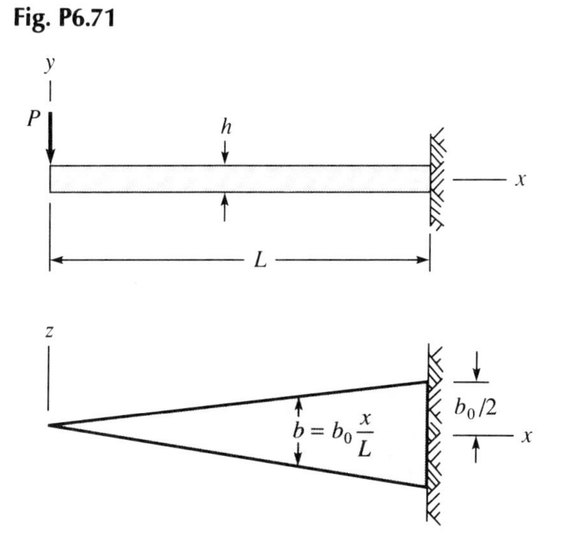

6.72 For the cantilevered beam tapered as illustrated in Fig. P6.72, obtain the equation for the deflection curve.

Fig. P6.72

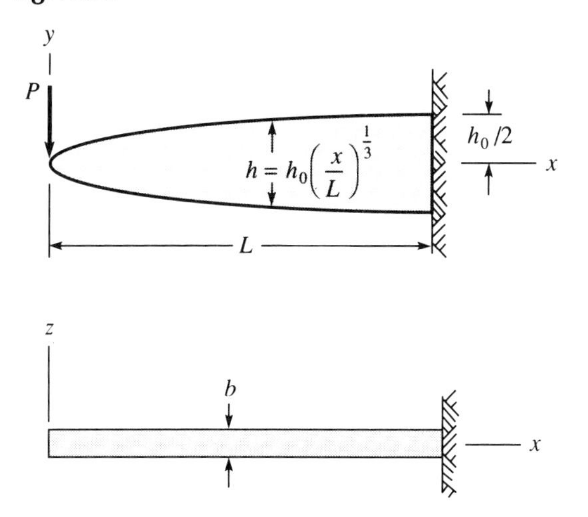

6.73 The depth of the beam shown in Fig. P6.73 varies such that the maximum normal stress σ at every cross section is the same. Compute the deflection of the free end.

Fig. P6.73

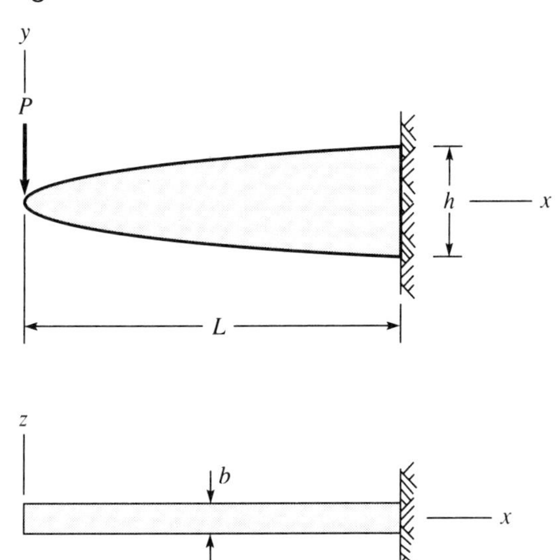

6.74 A cantilevered beam is tapered as shown in Fig. P6.74. Determine the deflection of the free end caused by the concentrated load $P = 40$ lb. $L = 20$ in.

Fig. P6.74

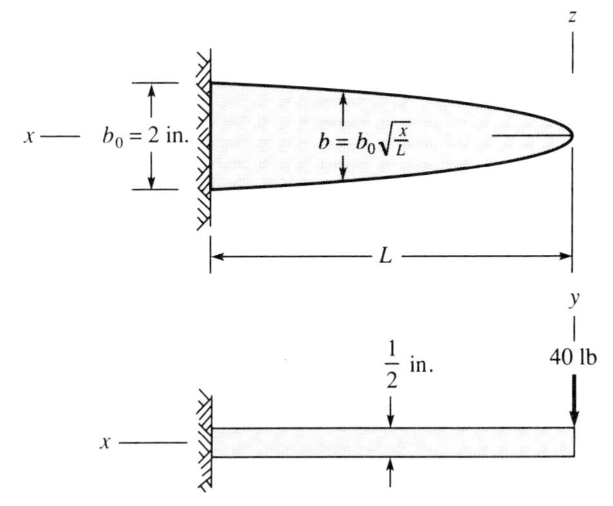

6.75 The beam in Fig. P6.75 has a rotational spring at $x = 0$ which resists rotation. Determine the slope of the beam at its free end in terms of the parameters $(P, \beta, E, I, a, \text{and } L)$.

Fig. P6.75

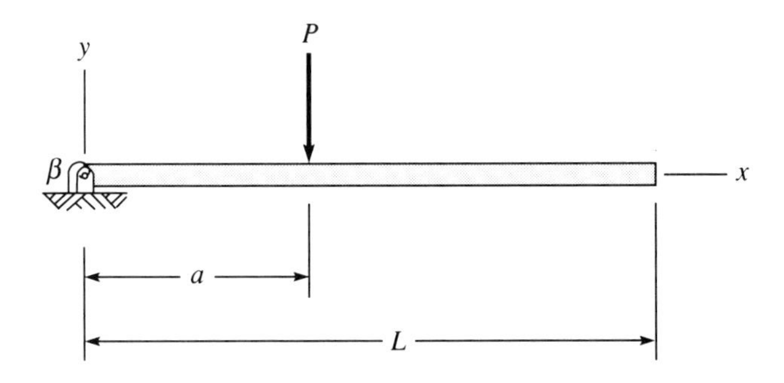

[†]**6.76** Obtain the deflection curve for the stepped beams of Fig. P6.76(a) through (d). Express the result in terms of the given parameters.

Fig. P6.76

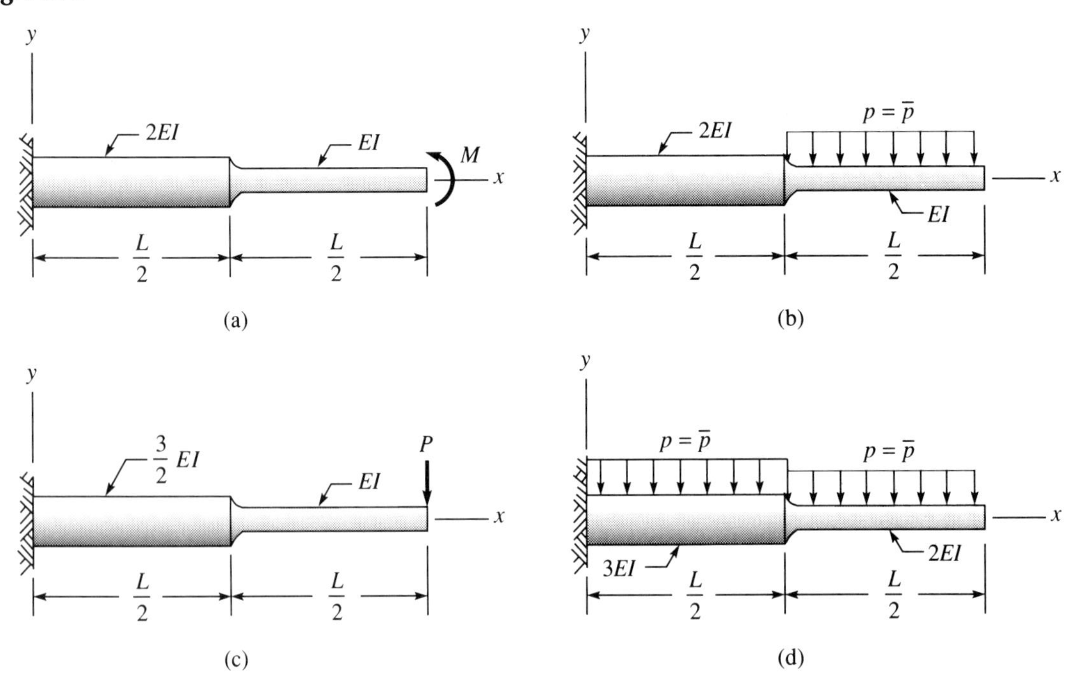

[†]**6.77** Calculate the deflection at point A for the stepped steel beams of Fig. P6.77(a) through (d).

Fig. P6.77

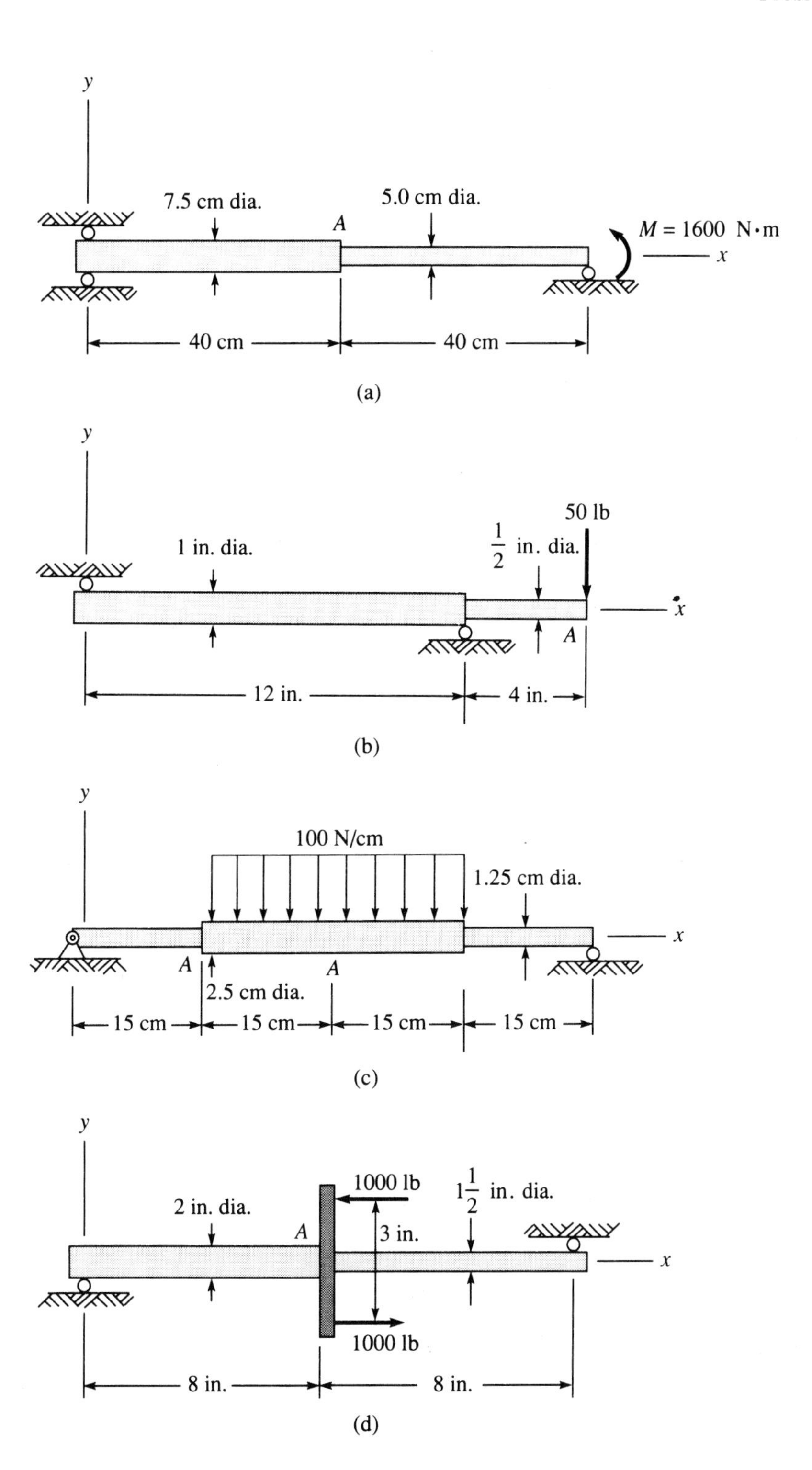
y
7.5 cm dia.
A
5.0 cm dia.
M = 1600 N·m
x
40 cm
40 cm
(a)
y
50 lb
1/2 in. dia.
1 in. dia.
x
A
12 in.
4 in.
(b)
y
100 N/cm
1.25 cm dia.
x
A
A
2.5 cm dia.
15 cm
15 cm
15 cm
15 cm
(c)
y
1000 lb
1 1/2 in. dia.
2 in. dia.
A
3 in.
x
1000 lb
8 in.
8 in.
(d)

†**6.78** For the beam of Fig. P6.78, determine the length a in terms of L so that the slope of the beam is zero over the supports.

Fig. P6.78

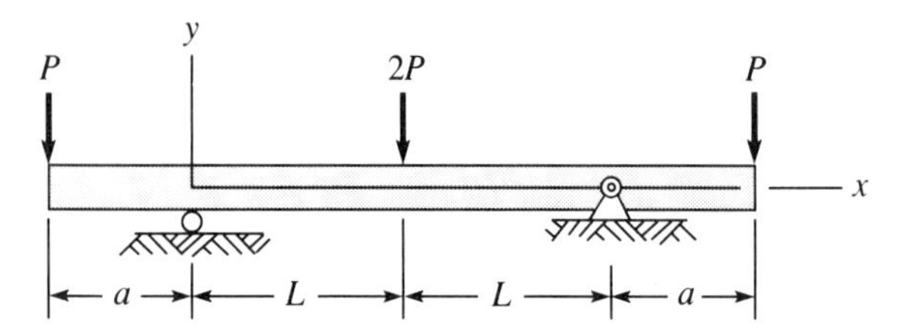

†**6.79** Determine the deflection at C for the beams of Fig. P6.79(a) through (c).

Fig. P6.79

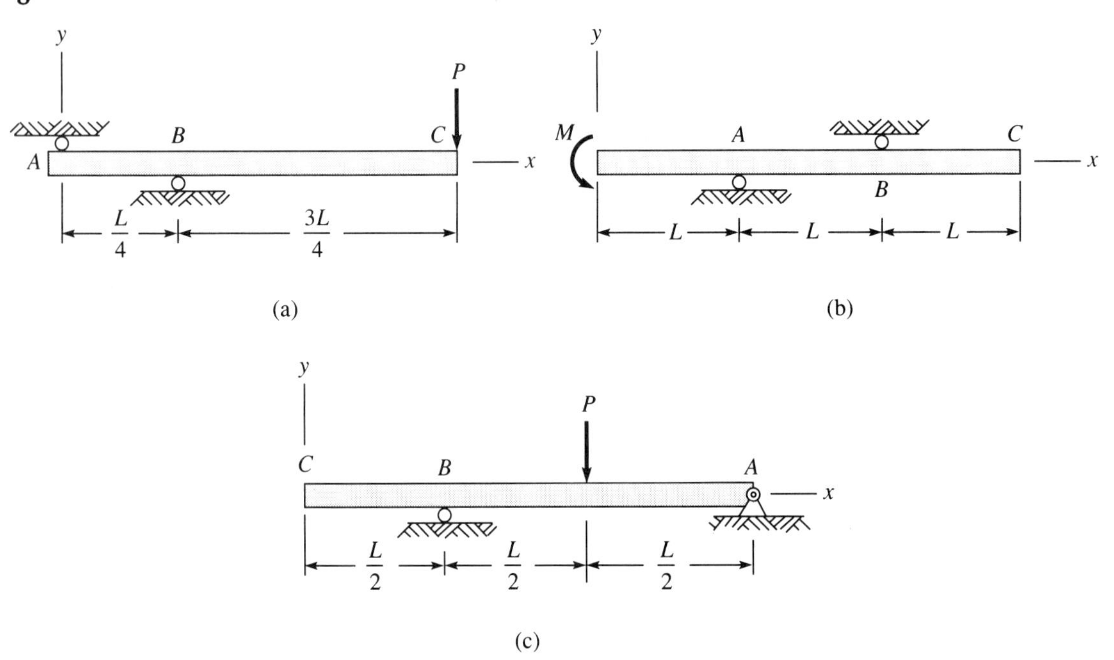

†**6.80** The beam of Fig. P6.80 is clamped at C, rests on a roller at B, and is subjected to the couple M_A. Find the reaction of the roller at B on the beam.

Fig. P6.80

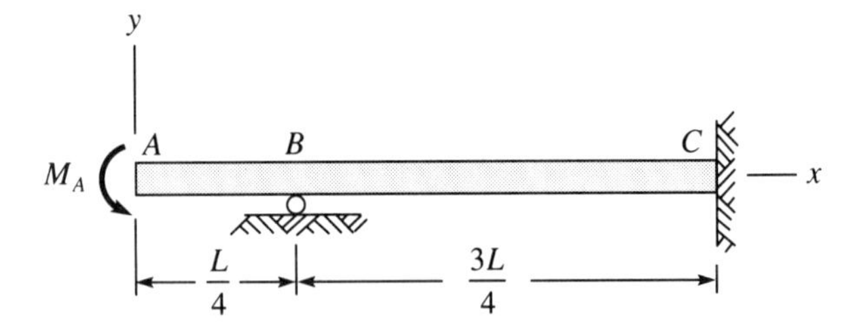

†**6.81** The two beams of Fig. P6.81 are composed of the same material and have the same cross sections. Determine the vertical displacement of the roller A caused by the load P.

Fig. P6.81

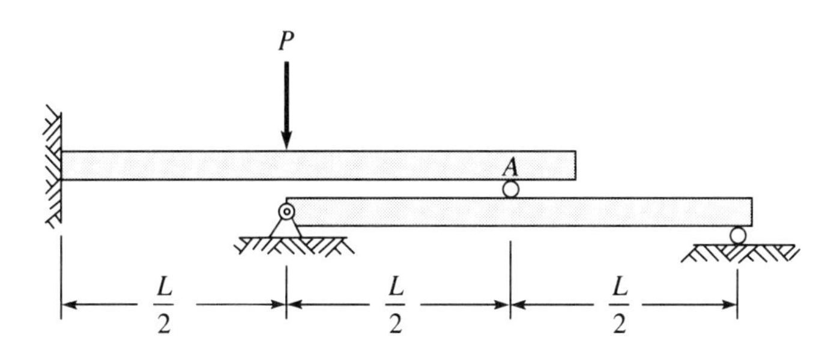

6.82 A steel I-beam supports two equal loads, as shown in Fig. P6.82. If I = 60 in.^{4}, calculate the deflection of points A and B.

Fig. P6.82

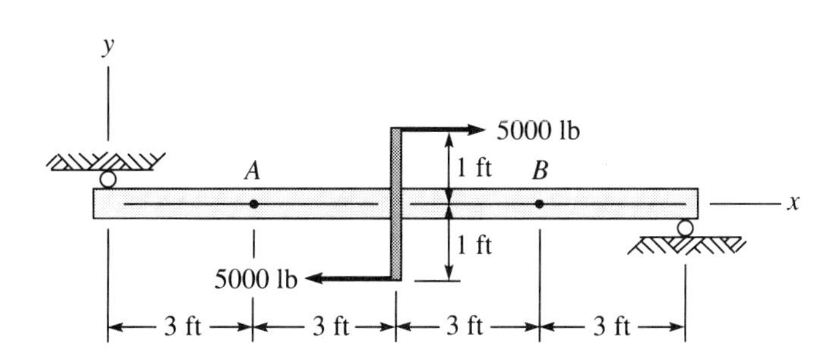

6.83 A steel I-beam is loaded as shown in Fig. P6.83. The maximum normal stress is 100 MPa. Find the maximum deflection.

Fig. P6.83

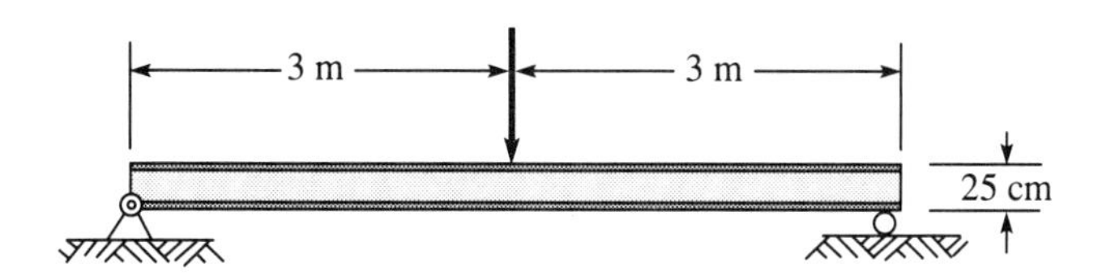

†**6.84** Show that the deflection at b in the beam (a) is equal to the deflection at a in the beam (b) of Fig. P6.84.

Fig. P6.84

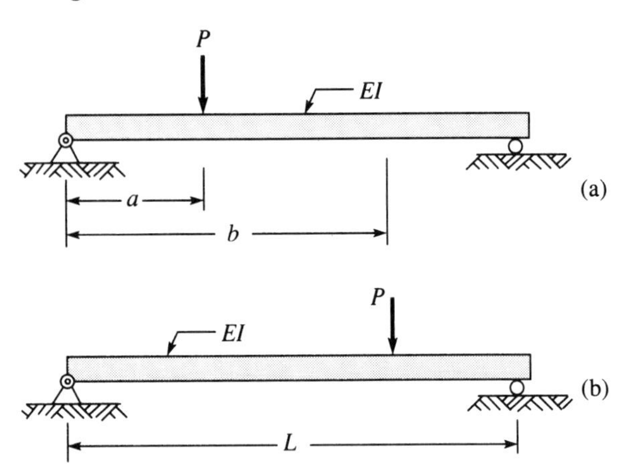

†**6.85** Show that the slope at B in the beam (a) is equal to the slope at A in the beam (b) of Fig. P6.85.

Fig. P6.85

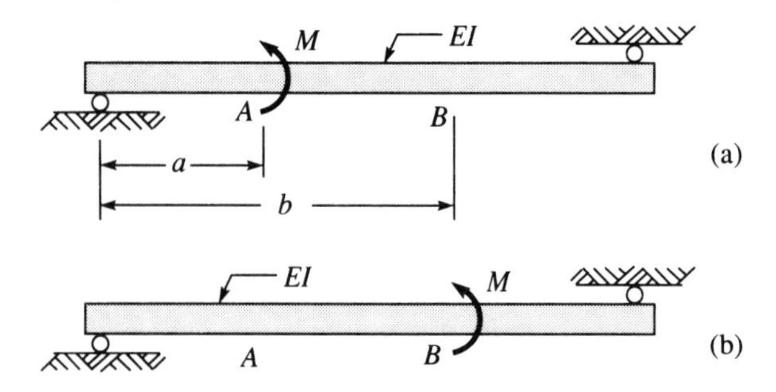

6.86 If a uniform beam has a distributed load $p(x)$ (force per unit length positive downward), show that the deflection satisfies the following equation:

$$EI\frac{d^4w}{dx^4} = -p(x)$$

6.87 A beam has a longitudinal plane of symmetry but is slightly tapered so that $I = I(x)$. Assume that the Bernoulli-Euler theory holds as for prismatic beams, and show that

$$E\frac{d}{dx}\left(I\frac{d^2w}{dx^2}\right) = -V(x)$$

Hence,

$$E\frac{d^2}{dx^2}\left(I\frac{d^2w}{dx^2}\right) = -p(x)$$

where V and p denote the transverse shear force and load intensity.

6.88 If M/EI is plotted as a function of x, as shown in Fig. P6.88, show that $\theta_2 - \theta_1$ = shaded area ($\bar{A}$), where θ_1 and θ_2 are the slopes at x_1 and x_2 of the neutral axis. Then show also that

$$w_2 - w_1 = \theta_1(x_2 - x_1) + \bar{A}\bar{x}$$

where w_1 and w_2 are the deflections at x_1 and x_2 of the neutral axis and $\bar{x}$ is the centroidal distance of the shaded area shown in the M/EI diagram.

Fig. P6.88

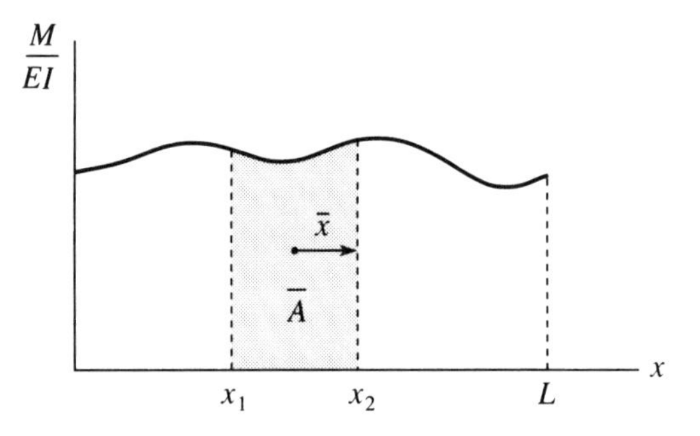

6.89 A uniform beam rests on a continuous bed of springs with spring constant k lb/ft/ft of length. Show that the appropriate equation is

$$EI\frac{d^2w}{dx^2} = M_p(x) - k\int_0^x (x - \xi)w\,d\xi$$

where M_p is the bending couple due to $p(x)$.

6.90 A continuous beam rests on a number of supports. Only three are shown in Fig. P6.90. Show that

$$M_{n-1}L_{n-1} + 2M_n(L_{n-1} + L_n) + M_{n+1}L_{n+1}$$
$$= -\tfrac{1}{4}p_{n-1}L_{n-1}^3 - \tfrac{1}{4}p_nL_n^3$$

where M_{n-1}, M_n, and M_{n+1} are the bending couples at $n - 1$, n, and $n + 1$, respectively.

Fig. P6.90

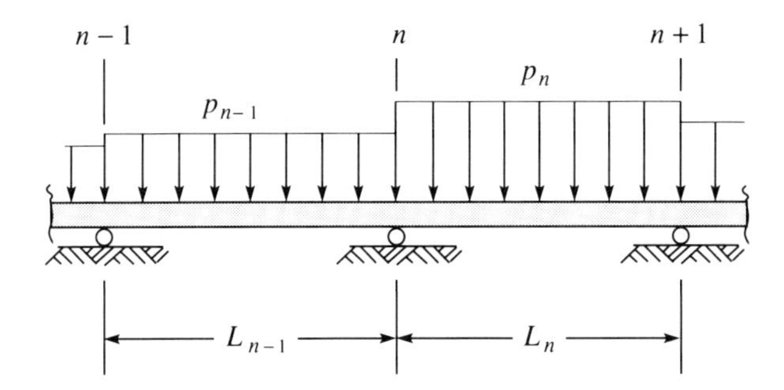

6.91 A leaf spring consists of n individual layers, which are fixed at the middle, as depicted in Fig. P6.91(a). Each contacting layer has the same thickness t and width w, and each slips freely along the adjoining layer but maintains contact. Each layer is made of the same HI-HO material. Show that the spring acts essentially as the tapered beam of Fig. P6.91(b). Use the theory of HI-HO beams to show that the curvature is essentially constant and the load deflection relation is

$$\delta = \frac{6Pl^3}{nEwt^3}$$

Fig. P6.91

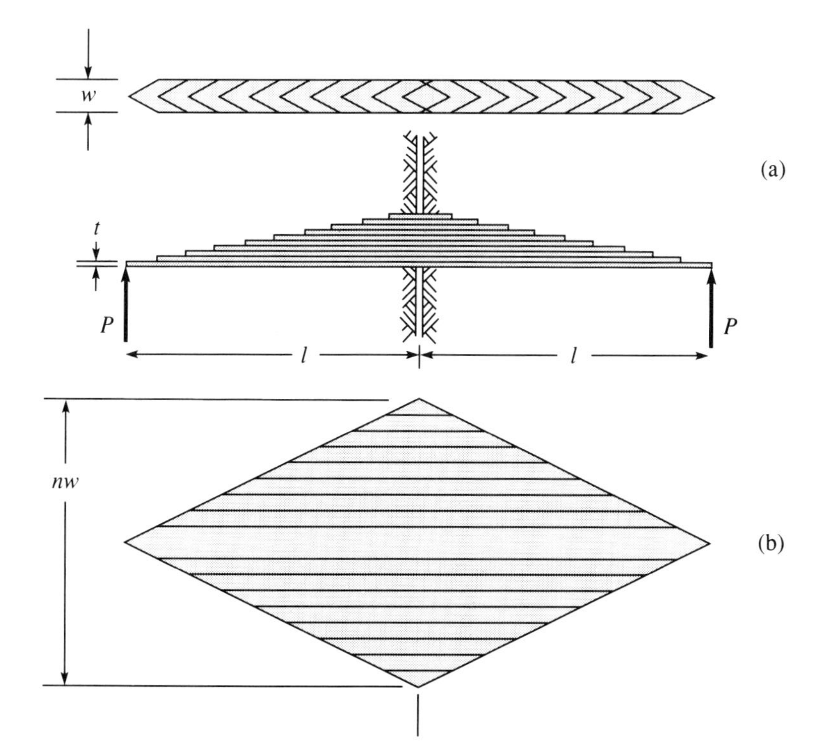

6.92 A beam has a *slight* initial curvature in its plane of symmetry. If the depth h is small compared with the initial radius of curvature $\rho_0 (h \ll \rho_0)$ and the strains and relative rotations are small, show that

$$\frac{M}{EI} = \frac{1}{\rho} - \frac{1}{\rho_0} = \frac{d\theta}{ds} - \frac{1}{\rho_0}$$

6.93 A round wire forms a semicircular arc, as shown in Fig. P6.93. Neglect the extension caused by normal forces on a cross section. Since $r \ll R$, each small element of length $R\,\Delta\theta$ deforms much as a straight beam; that is,

$$\frac{M}{EI} = \frac{d}{ds}(\theta^* - \theta)$$

By considering the rotations of each small element, show that the vertical deflection of end B is given by

$$v = \frac{3\pi PR^3}{2EI}$$

Fig. P6.93

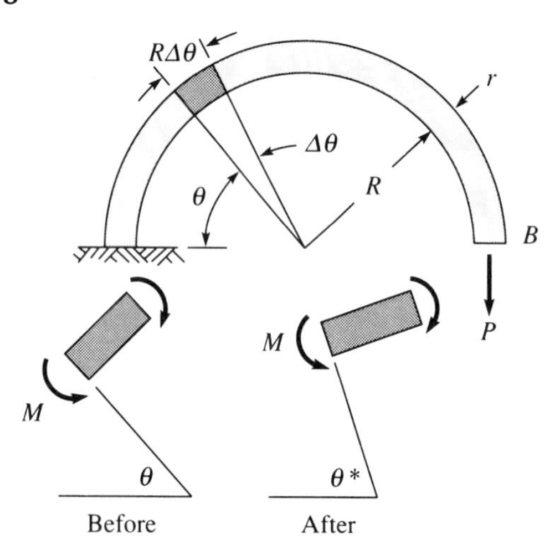

6.94 A *thin* steel strip is clamped at one end and subjected only to a bending couple at the unsupported end. The strip is HI-HO, 0.040 in. thick, 1.000 in. wide and 20.00 in. long. The maximum normal stress is 42 ksi. Show that the moment-curvature relation Eq. (6.29) is valid, but the approximation Eq. (6.30b) is *not* valid. Then calculate the deflection of the unsupported end.

6.95 A uniform beam rests on a continuous bed of springs, with spring constant density k. Show that the deflection of the neutral axis satisfies the equation

$$EI\frac{d^4w}{dx^4} + kw = -p(x)$$

where $p(x)$ is a distributed load (force per unit length positive downward).

6.6 Deflection Due to Shear

The Bernoulli-Euler theory of hookean beams ($\kappa = M/EI$) neglects the transverse shear strain (γ_{xy}). Actual beams under transverse loads are subjected to shear force V and stress τ_{xy}; the stress in a HI-HO beam is estimated, according to Eq. (6.20):

$$\bar{\tau} = \frac{VQ(y)}{It}$$

The corresponding strain in the HI-HO beam is

$$\bar{\gamma}_{xy} = \frac{VQ(y)}{GIt}$$

The shear stress and the strain vanish at the top and bottom and are a maximum at the centroidal axis:

$$\bar{\gamma}_{xy}(0) \equiv \bar{\gamma}_0 = \frac{VQ_0}{GIt} \tag{6.51}$$

where $Q_0 = Q(0)$.

From Fig. 6.25 it is evident that the shear strain (which is assumed to be small) causes additional deflection of the neutral axis

$$\Delta w_s = \bar{\gamma}_0 \, \Delta x$$

or in the limit

$$\frac{dw_s}{dx} = \bar{\gamma}_0$$

Fig. 6.25

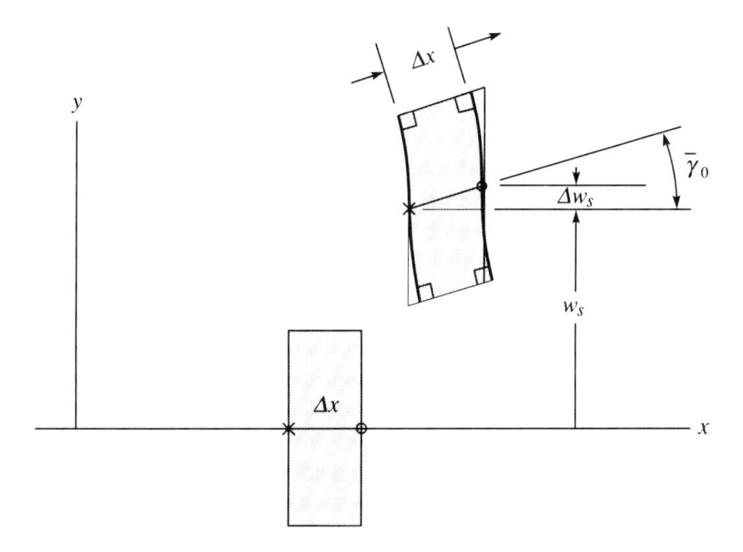

In view of Eq. (6.51) the *additional* rotation due to shear is

$$\frac{dw_s}{dx} = \frac{V(x)Q_0}{GIt} \tag{6.52}$$

The additional deflection can now be computed by integrating Eq. (6.52). This additional deflection is generally quite small compared to that caused by bending and is usually neglected. The shear deflection is significant when the beam is relatively deep compared to its length. Even then the accuracy of such an estimate is dubious, since the basis neglected the shear strain. Nevertheless, in the absence of a more precise analysis, such a calculation provides a correction to the deflection equation when the shear is not insignificant.

EXAMPLE 10

As an illustration, consider the cantilever shown in Fig. 6.26. The same beam and loading used in Example 7 are used here. Recall that

$$V = -\frac{\bar{p}L}{2}\left(1 - \frac{x}{L}\right)^2$$

Fig. 6.26

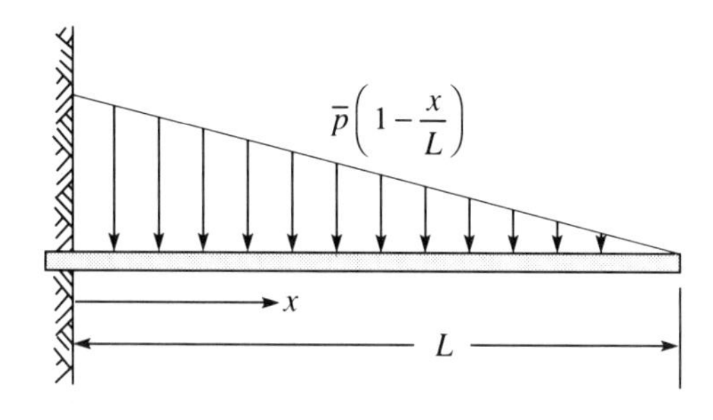

Then, by Eq. (6.52)

$$\frac{dw_s}{dx} = -\frac{Q_0}{GIt}\frac{\bar{p}L}{2}\left(1 - \frac{x}{L}\right)^2$$

Integrating, we obtain

$$w_s = +\frac{Q_0}{GIt}\frac{\bar{p}L^2}{6}\left[\left(1 - \frac{x}{L}\right)^3 - 1\right]$$

At the end $x = L$,

$$w_s(L) = -\frac{Q_0\bar{p}L^2}{6GIt}$$

From Example 6.7, the deflection caused by bending, according to the Bernoulli-Euler theory, is

$$w_B(L) = -\frac{\bar{p}L^4}{30EI}$$

The ratio at the end $x = L$ is

$$\frac{w_s(L)}{w_B(L)} = \frac{5EQ_0}{GtL^2}$$

The cross section of the American Standard WF (12-40) beam is shown in Fig. 6.27. The integral Q_0 follows:

Fig. 6.27

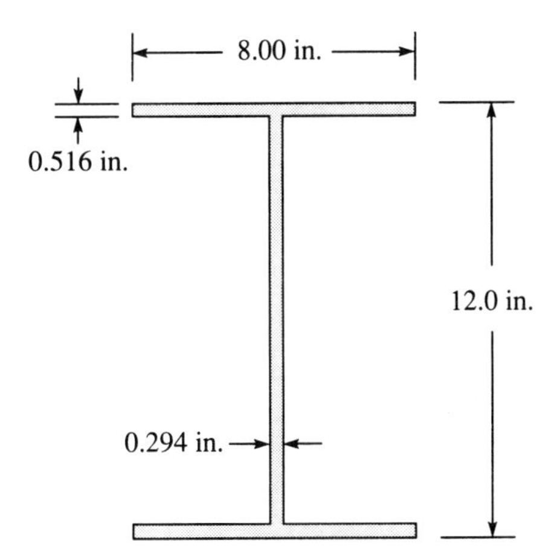

$$Q_0 = (8 \times 0.516 \times 5.742) + \frac{(5.484)^2}{2} \times (.294) = 28.1 \text{ in.}^3$$

Also,

$$I = 310 \text{ in.}^4, \qquad t = 0.294 \text{ in.}$$

The beam is steel, so that

$$\frac{E}{G} \doteq 2.50$$

As before, let us take the length

$$L = 20.0 \text{ ft}$$

The ratio follows:

$$\frac{w_s(L)}{w_B(L)} = 0.0207$$

We see that the deflection due to shear is approximately 2%. ◆

PROBLEMS

6.96 A cantilever beam is fixed at one end and subjected only to a transverse load at the opposite end. Obtain the ratio $w_s(L)/w_B(L)$ for the cross sections shown in Fig. P6.96. Use $L/h = 10$ and $L/h = 20$ and $E/G = 2.5$. $w_B(L)$ denotes the deflection at the load according to Bernoulli-Euler theory of Sec. 6.5, and $w_s(L)$ is the deflection attributed to shear, as given in Sec. 6.6.

Fig. P6.96

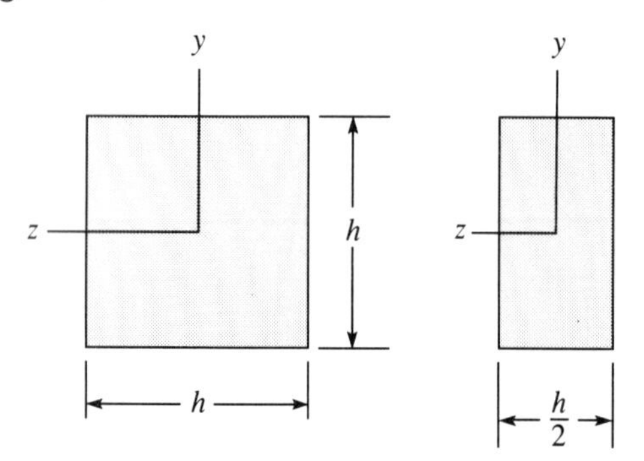

Fig. P6.97

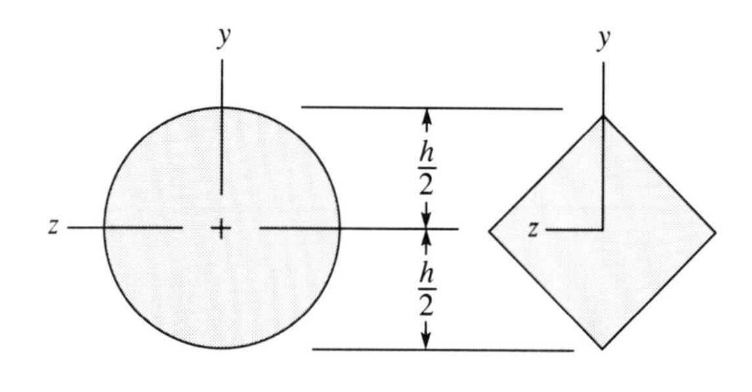

6.97 Repeat Prob. 6.96 for each cross section shown in Fig. P6.97.

6.98 Repeat Prob. 6.96 for each cross section shown in Fig. P6.98.

Fig. P6.98

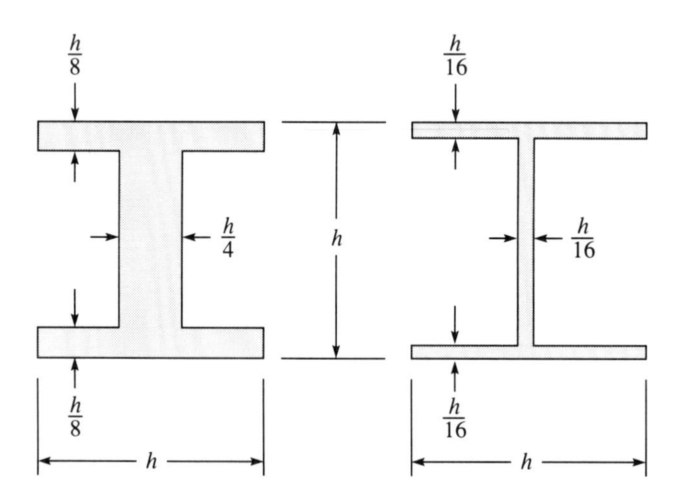

***6.99** A composite beam is made of two HI-HO materials. The inner core of the square cross section in Fig. P6.99 has modulus $E_c = E_F/8$, where E_F is the modulus of the upper and lower facings. For both materials, $E/G = 2.5$. For a simple cantilever, subject to a concentrated load upon the free end, and proportions $L/h = 20$, determine the ratio $w_s(L)/w_B(L)$. *Note*: The normal stress σ_x is relatively small in the core.

Fig. P6.99

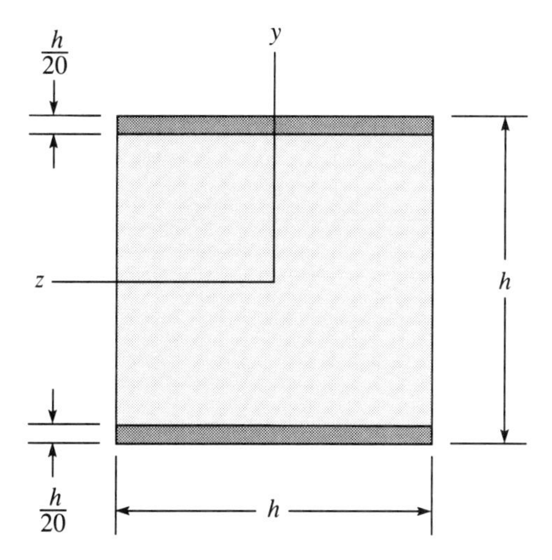

6.7 Limit Loads for Ideally Plastic Beams

In Sec. 5.13 we considered the behavior of a beam composed of an ideally plastic material when subjected to simple bending. On the assumption that plane cross sections remain plane and perpendicular to the neutral axis, we obtained a relationship between the bending couple and the curvature. A distinctive feature of this relation is the existence of a limiting condition, which occurs when the bending couple attains a value M_L. The beam cannot support a greater moment, since the action of this couple causes unrestricted rotations of nearby cross sections.

We now apply this concept to ideally plastic beams subjected to transverse loads, ignoring the effects of shear. When the bending couple at any cross section of the beam reaches the limiting value, a plastic "hinge" forms, and unrestricted rotation can occur. Here, we make no attempt to calculate the deflection after yielding but concern ourselves with the load-carrying capacity of the beam as it is limited by the occurrence of plastic hinges and subsequent plastic collapse. The method of calculating the limit load for an ideally plastic beam is illustrated by the following examples.

EXAMPLE 11

The cantilever beam of Fig. 6.28 is made of structural steel, which can be regarded as ideally plastic with $\sigma_0 = 40{,}000$ lb/in.^{2} (the same magnitude in tension and in compression). Its cross section is shown in Fig. 6.28(b). A single concentrated load P is applied to the end of the beam so that the bending couple on any cross section is given by

$$M(x) = P(L - x)$$

Fig. 6.28

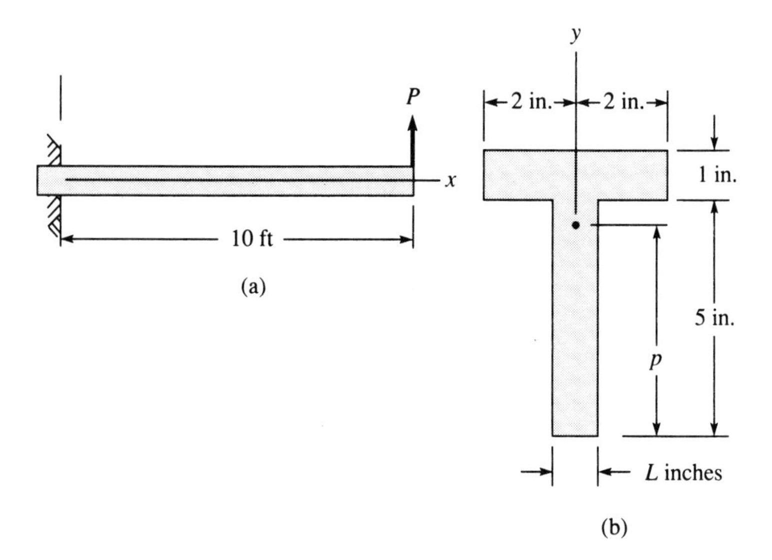

For a given load, the greatest bending couple occurs at $x = 0$. We therefore expect a plastic hinge to form there when $M(0) = M_L$.

To calculate M_L we suppose that the cross section is in the fully plastic state illustrated in Fig. 6.29. The entire cross section has yielded; the portion above the neutral axis is subjected to the normal stress $\sigma = -\sigma_0$, whereas the remainder is acted upon by the stress $\sigma = \sigma_0$.

Fig. 6.29

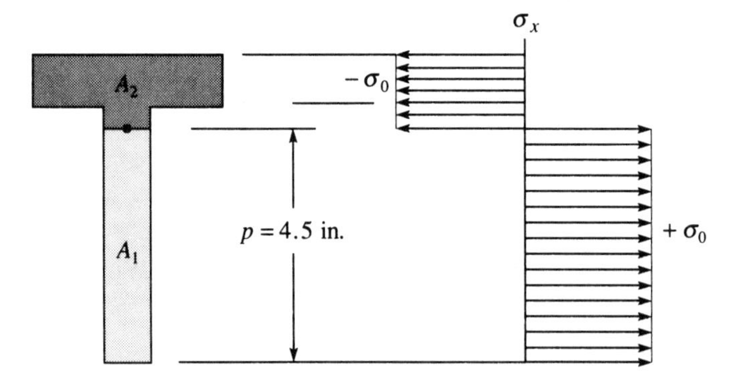

The position of the neutral axis is determined by the condition that the normal force on the cross section is zero;

$$F = \iint_A \sigma_x \, dA = 0$$

If p locates the neutral axis from the bottom edge of the cross section, as in Fig. 6.29, then

$$F = \sigma_0 A_1 - \sigma_0 A_2 = \sigma_0 p - \sigma_0[4 + (5 - p)] = 0$$

so that $p = 4.5$ in. The fully plastic bending couple is then

$$M_L = -\iint_A \sigma_x y\, dA = -\sigma_0 \iint_{A_1} y\, dA + \sigma_0 \iint_{A_2} y\, dA$$

where A_2 and A_1 are the areas above and below the neutral axis, respectively. When these integrals are evaluated, the result is $M_L = 570{,}000$ in.-lb. The load causes that this bending couple at $x = 0$ is the limit load,

$$P_L = \frac{M_L}{L} = 4750 \text{ lb} \quad (21.1 \times 10^3 \text{ N})$$

As a matter of interest let us compute the yield couple M_0. This couple is the bending couple that initiates yielding. For HI-HO behavior, the neutral axis is at the centroid of the cross section; in Fig. 6.28(b) it is located at distance $\bar{p} = 3.833$ in. Accordingly, the largest (absolute value) stress occurs at the bottom, where, for incipient yielding, $\sigma_0 = M_0 \bar{p}/I$. In this case $I = 30.75$ in.^{4}, so yielding is initiated when

$$M_0 = \frac{\sigma_0 I}{\bar{p}} = 321{,}000 \text{ in.-lb}$$

The behavior of the beam is linearly elastic as long as the bending couple is everywhere less than M_0. Where the couple exceeds M_0, plastic deformation occurs.

Since the first cross section to yield is at $x = 0$, the load that initiates yielding in the beam is

$$P_0 = \frac{M_0}{L} = 2675 \text{ lb} \quad (11.9 \times 10^3 \text{ N})$$

As the load increases beyond P_0, the bending couple at $x = 0$ increases. When $P = P_L$, the region adjacent to the wall behaves as a plastic hinge. The beam is then, effectively, a mechanism that can rotate about the plastic hinge, as indicated in Fig. 6.30. This constitutes plastic collapse. The load P_L is, therefore, the limit load.

Fig. 6.30

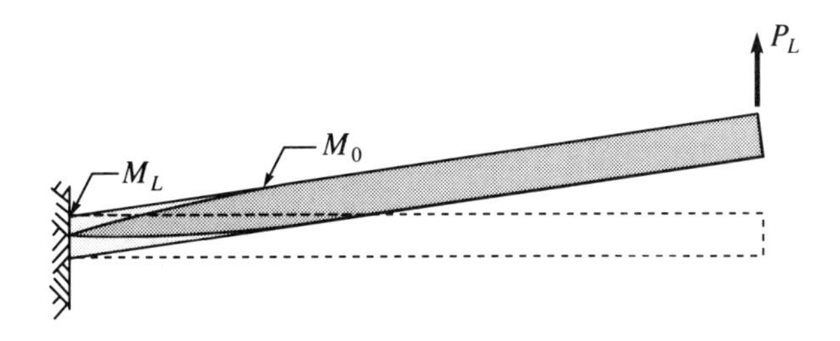

Note that yielding is not restricted to the immediate vicinity of the hinge. In Fig. 6.30 some yielding occurs throughout the region $0 < x < 52.4$ in., since $M > M_0$ in that part of the beam when $P = P_L$. ◆

EXAMPLE 12

Suppose that the cantilevered beam of the preceding example is provided with an additional support at the right end, as shown in Fig. 6.31(a), and is then subjected to a uniform load W. For equilibrium of the free body in Fig. 6.31(b), we obtain

$$M(x) = -\frac{W}{2}(L - x)^2 + R_B(L - x) \tag{6.53}$$

Fig. 6.31

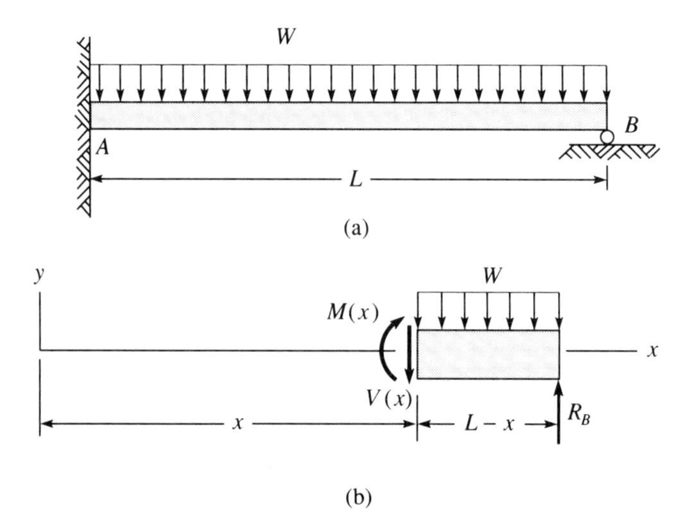

where R_B is the reaction of the support at B. The bending couple diagram of Fig. 6.32 indicates, in a general way, the variation of the bending couple.

At the outset we cannot assign values, because R_B remains unknown. However, it is clear from Fig. 6.32 that the bending couple may have its largest values at the built-in end $x = 0$ or at some intermediate location $x = a$. Furthermore, a single plastic hinge at any location will not convert the beam to a mechanism. Because of the additional support, two hinges are needed to transform the beam into a collapse mechanism. One of these hinges must, therefore, develop at the fixed end $x = 0$ and the other, at $x = a$. Thus, at the limit load W_L, we must have

$$M(0) = -M_L, \qquad M(a) = M_L \tag{6.54}$$

Putting the first of these conditions in Eq. (6.53) yields

$$-M_L = -\frac{W_L}{2}L^2 + R_B L$$

Solving this equation for R_B and eliminating it from Eq. (6.53), we obtain (at the limit load)

Fig. 6.32

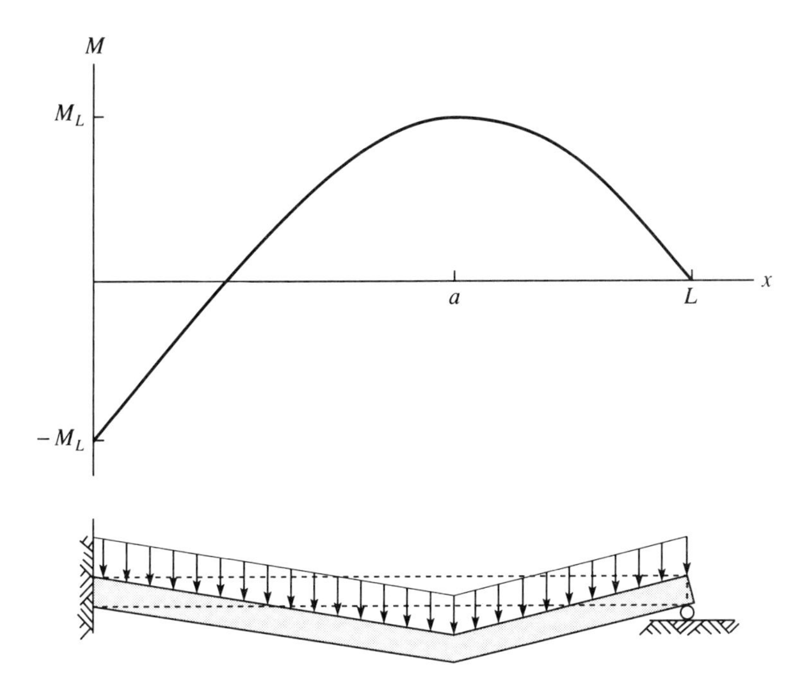

$$M(x) = -\frac{W_L}{2}(L-x)^2 + \frac{W_L}{2}(L-x) - \frac{M_L}{L}(L-x) \tag{6.55}$$

The second plastic hinge is at the intermediate location $x = a$ where M is a maximum—that is, where $dM/dx = 0$. Differentiating Eq. (6.55) we have

$$\left.\frac{dM}{dx}\right|_{x=a} = W_L(L-a) - \frac{W_L L}{2} + \frac{M_L}{L} = 0$$

or

$$L - a = \frac{L}{2} - \frac{M_L}{W_L L} \tag{6.56}$$

Returning to the second of Eqs. (6.54), when $x = a$, M takes on the value M_L. Substituting this in Eq. (6.55), with the aid of Eq. (6.56), we obtain

$$M_L = -\frac{W_L}{2}\left(\frac{L}{2} - \frac{M_L}{W_L L}\right)^2 + \frac{W_L L}{2}\left(\frac{L}{2} - \frac{M_L}{W_L L}\right) - \frac{M_L}{L}\left(\frac{L}{2} - \frac{M_L}{W_L L}\right)$$

which reduces to

$$W_L^2 - \frac{12M_L}{L^2}W_L + \frac{4M_L^2}{L^4} = 0$$

Solving this equation for W_L gives

$$W_L = \frac{M_L}{L^2}(6 \pm 4\sqrt{2}) \tag{6.57}$$

By Eq. (6.56) we have

$$L - a = \frac{L}{2} - \frac{L}{6 \pm 4\sqrt{2}}$$

or

$$a = \frac{2 \pm \sqrt{2}}{3 \pm 2\sqrt{2}} L$$

The negative sign yields a value $a > L$, which has no physical meaning in the present problem. Therefore, the hinge is at

$$a = \frac{2 + \sqrt{2}}{3 + 2\sqrt{2}} L$$

Then, taking the positive sign in Eq. (6.57), the limit load is

$$W_L = (6 + 4\sqrt{2})\frac{M_L}{L^2}$$

Using the value of M_L calculated in the preceding example, we arrive finally at

$$W_L = 461 \text{ lb/in.} \quad (807 \text{ N/cm})$$

This example suggests that limit load analyses of indeterminate beams and frames can be very laborious. This is generally true, because we cannot usually foretell the correct collapse mechanism—that is, the location of the plastic hinges. ◆

PROBLEMS

Assume ideally plastic behavior unless otherwise stated.

6.100 A beam has the cross section shown in Fig. P6.100. Compare M_L for loading in the xy-plane to M_L for loading in the xz-plane.

Fig. P6.100

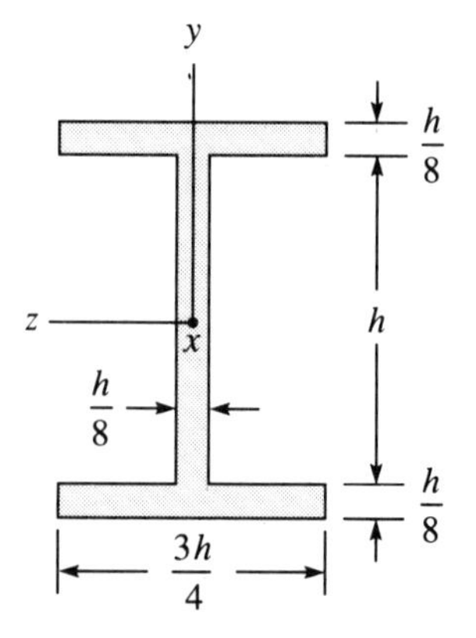

6.101 Determine the ratio M_L/M_0 for the cross section in Fig. P6.101 and evaluate

$$\lim_{t/h \to 0} \left(\frac{M_L}{M_0}\right)$$

Fig. P6.101

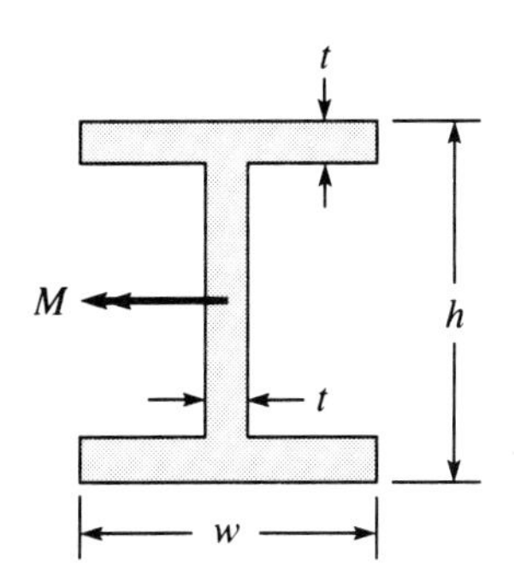

6.102 A beam has the cross section shown in Fig. P6.102 and carries transverse loads in the xy-plane. Compute the width of the $\frac{1}{2}$-in. plates (dotted) of the same material that need to be added to the flanges in order to double M_L.

Fig. P6.102

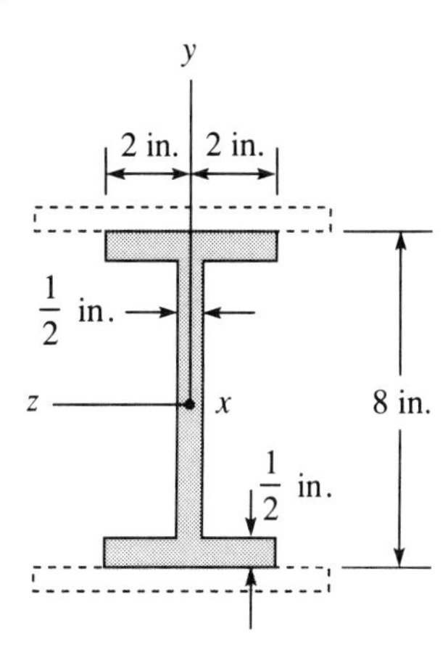

6.103 Compute the ratio M_L/M_0 for the cross sections shown in Fig. P6.9(a) through (f).

6.104 The beams illustrated in Fig. P6.104(a) through (f) are all structural steel with $\sigma_0 = 40{,}000$ psi. Compute the ratio of the limit load to the load at initial yielding.

Fig. P6.104

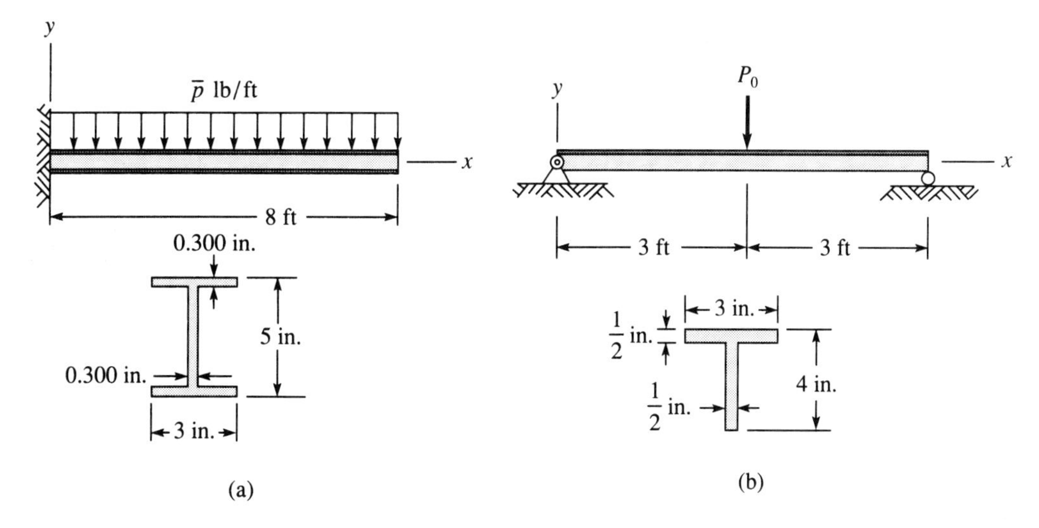

Fig. P6.104
(*cont.*)

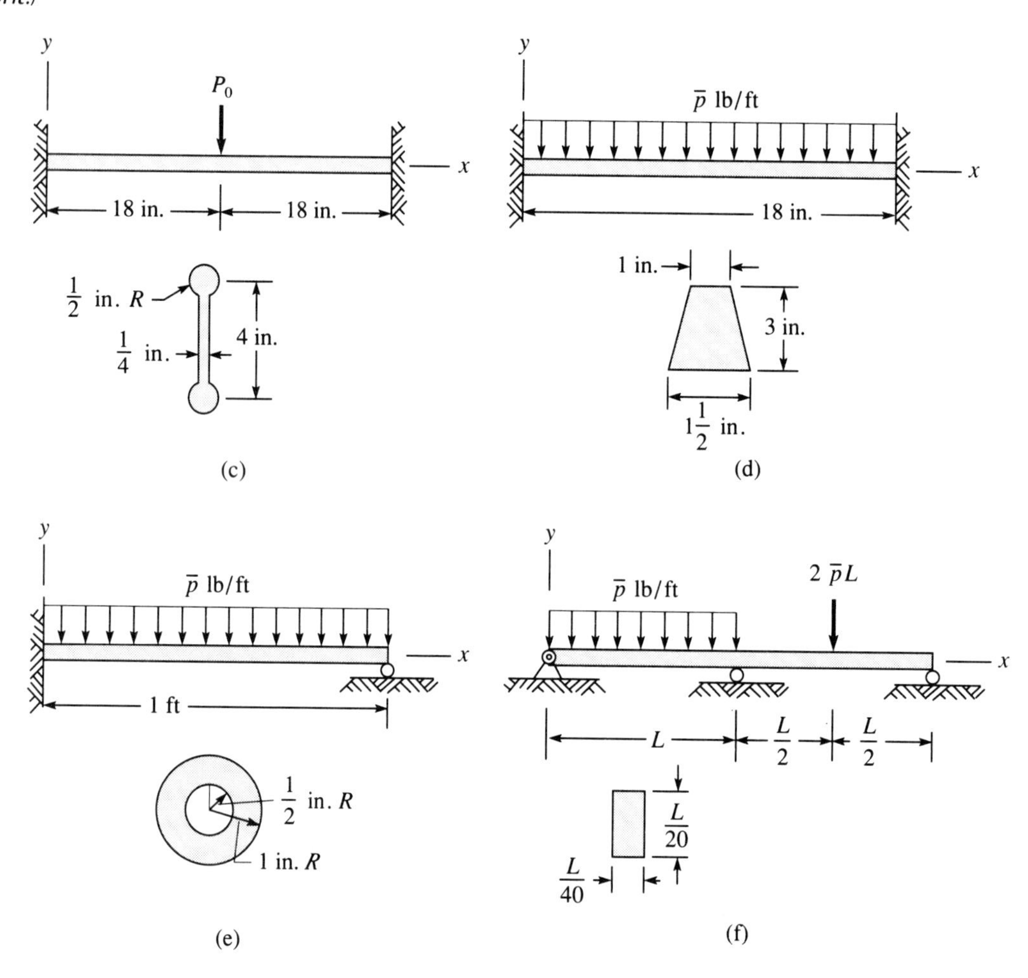

6.105 For the beam shown in Fig. P6.105, $\sigma_0 =$ 60,000 psi in compression and 40,000 psi in tension. Should the T be up (as shown) or inverted in order to carry the maximum limit load? Calculate that maximum load.

Fig. P6.105

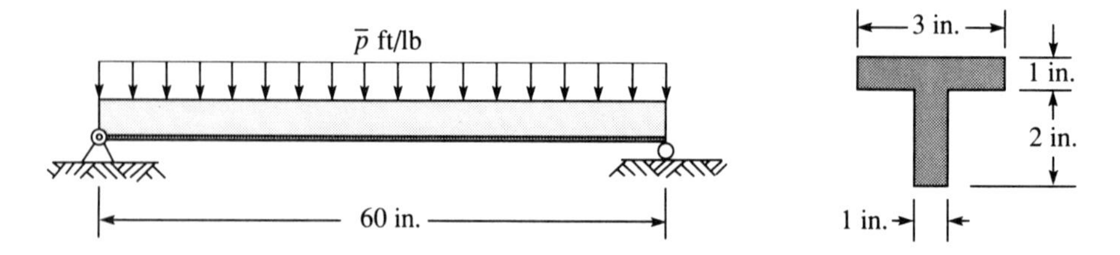

6.106 When estimating the collapse load of a reinforced concrete beam, we may assume that the concrete does not resist tensile stresses and that it behaves as ideally plastic in compression with $\sigma_0 = 0.8\sigma_u$. At collapse the steel is at the yield stress σ_0'. Use the notation indicated in Fig. P6.106 (A is the area of the steel) and obtain the limit couple

$$M_u = dA\sigma_0'\left[1 - \frac{A\sigma_0'}{1.6bd\sigma_u}\right]$$

Fig. P6.106

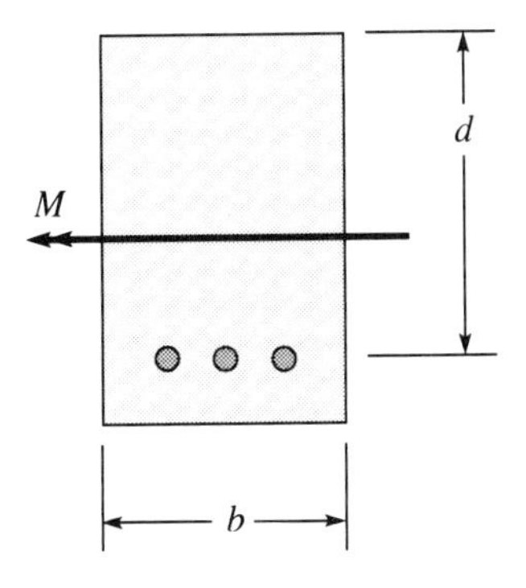

6.107 Assume that the materials in a reinforced concrete beam are ideally plastic at collapse. The beam is shown in Fig. P6.107. In compression the concrete has a yield stress $\sigma_0 = 2000$ lb/in.^{2} and does not resist tensile stresses. The steel reinforcing bars are fully plastic at collapse with $\sigma_0 = 40{,}000$ lb/in^2. What is the diameter of the steel bars if the upper 8 in. of the beam is in compression at collapse? What is the corresponding bending couple M_L?

Fig. P6.107

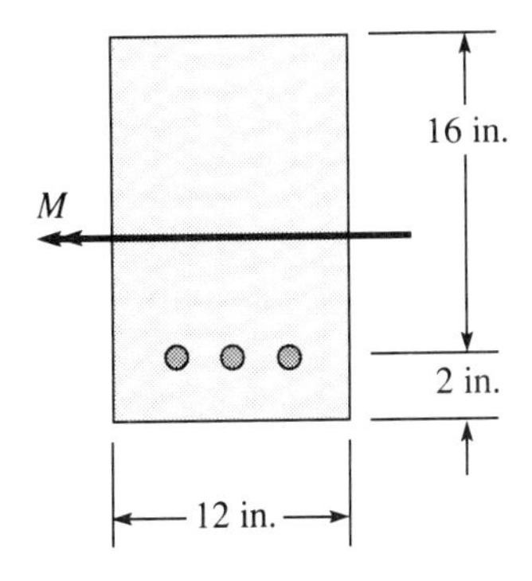

7 Combined Loads; Superposition

With two exceptions (Sections 5.15 and 6.6), the preceding problems involved a single type of loading, such as axial force, twisting couple, *or* bending couple; each entailed a simple deformation, such as extension, twist, *or* flexure. Here, we recognize that most actual members are subjected to a combination of loadings. For example, a driveshaft is intended to transmit torque but is also subjected to bending couples imposed by attachments and supporting bearings. In reality, most of the geometric changes are *relatively* small; specifically, strains and rotations are small. Consequently, the geometric relations are *linear*. If the description of the material is also linear, that is, the material is hookean, then all relevant equations are linear. The important consequence is the superposition of solutions. Stated otherwise, we can add the effects of two (or more) loads to obtain the effects of the combined loading. The purpose of the present chapter is to extend greatly the utility of the foregoing results and to provide some exercise in the process of superposing.

7.1 Combined Loads

In many instances structural members and machine parts are subjected to a variety of loads acting simultaneously. These loads may be of the same type; for example, a bar may be acted upon by several axial loads at the same time. On the other hand the same member may support axial, twisting, and bending loads simultaneously. For example, an automobile axle carries loads that cause it to twist and bend. Loading of this kind is referred to as *combined loading*.

We have already discussed methods for determining the response of a member to axial loads, twisting couples, and transverse loads. Now let us turn our attention to a procedure that enables us to treat problems involving combined loadings by superposing the effects of each individual load. The following examples serve to illustrate the concept, the procedure, and the limitations of *superposition*.

The linearly elastic cylinder of Fig. 7.1(a) is stretched by an axial force of magnitude $P_1 + P_2$. If the strains are small, we have

Fig. 7.1

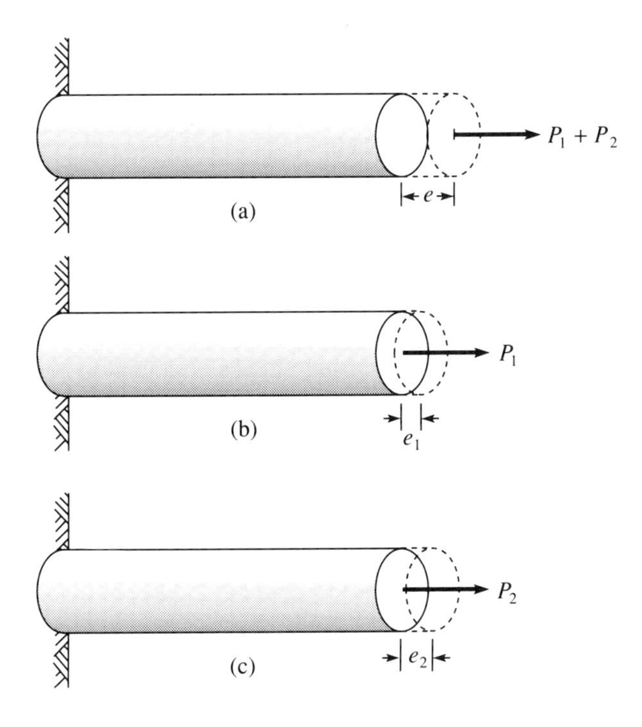

$$\sigma_1 = \frac{P_1 + P_2}{A}, \qquad e = \frac{(P_1 + P_2)L}{EA}$$

where L is the length of the cylinder, A is its cross-sectional area, and e is the total extension. Now consider the same bar subjected to the load P_1 alone, as in Fig. 7.1(b), and P_2 alone, as in Fig. 7.1(c). For P_1 alone

$$\sigma_1 = \frac{P_1}{A}, \qquad e_1 = \frac{P_1 L}{EA}$$

whereas for P_2 alone,

$$\sigma_2 = \frac{P_2}{A}, \qquad e_2 = \frac{P_2 L}{EA}$$

From the form of these relations we see that

$$\sigma = \sigma_1 + \sigma_2$$

and

$$e = e_1 + e_2$$

Hence the response of the bar loaded by the axial force $P_1 + P_2$ can be obtained by adding, or *superposing*, the effects of the loads P_1 and P_2 acting individually. *If* the force P_1 were applied first and *if* the initial area A were significantly changed (contracted), then the true stress would not be simply $\sigma_1 = P_1/A$. If the actual deformed area were A^*, then the true stress would be $\sigma_1 = P_1/A^*$. Moreover, the second force P_2 would also cause further change. If the final area were A^{**}, then the final stress would be

$$\sigma = \frac{P_1 + P_2}{A^{**}} \neq \sigma_1 + \sigma_2$$

The resulting stress is not precisely the sum of the stresses associated with each force acting alone. The difference is caused by the geometric change; the change of area invalidates the simple *linear* relation. Fortunately, such geometric changes are usually so small that they are negligible. The linear equations are adequate and the stresses can be added.

As another example consider the problem illustrated in Fig. 7.2. The cylindrical shaft is composed of a linearly elastic material. It is simultaneously subjected to an axial force P and a twisting couple T_0. We already know the effects caused by each force system when it is separately applied. The force P acting alone causes an axial extension and a normal stress of amounts

$$e = \frac{PL}{AE}, \qquad \sigma = \frac{P}{A}$$

Fig. 7.2

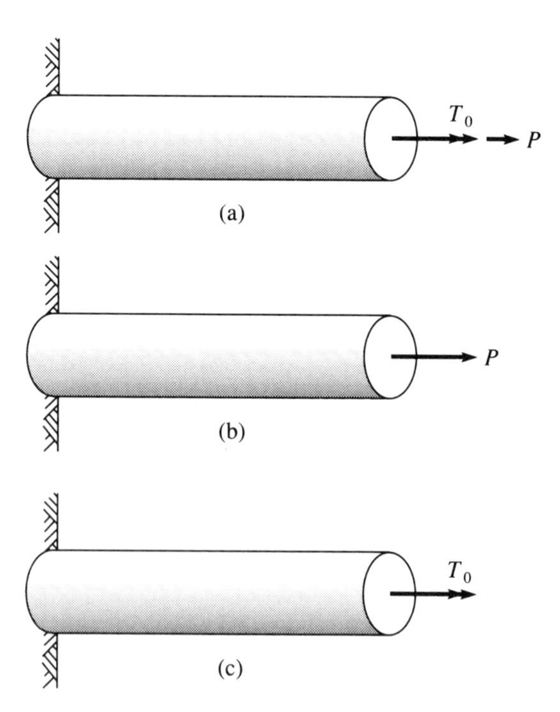

The effect of the twisting couple T is a relative rotation of end sections and a shear stress. These are

$$\phi = \frac{T_0 L}{GJ}, \qquad \tau = \frac{T_0 r}{J}$$

The combined-loading problem could be solved in the same manner as the individual problems in Chapter 5. We could begin by assuming that cross sections of the cylinder displace axially and also rotate about the axis. Since both deformations are *small*, the presence of one does not significantly affect the other. Consequently, the axial strain and shear strain are essentially the same as those in the individual cases of axial extension and torsion. However, both are present when P and T_0 act simultaneously, and the effect of the combined loading is therefore the sum of the effects that result when each load acts separately.

These examples illustrate the superposition of effects. An inspection of these two examples reveals that the addition of effects has its origin in the fact that one system of loads does not cause any change that will modify the effects of a second system of loads, and vice versa. In other words, there is no coupling between one cause and the effects of another cause. The following counterexamples show that the procedure is not always applicable.

Let the cylindrical bar of Fig. 7.1 be composed of a material that has a nonlinear stress-strain relation of the form

$$\sigma = k\varepsilon^{1/2}$$

The elongations of a bar loaded as in Fig. 7.1(a), (b), and (c), respectively, are

$$e = L\left(\frac{P_1 + P_2}{kA}\right)^2 = L\left(\frac{P_1}{kA}\right)^2 + L\left(\frac{P_2}{kA}\right)^2 + 2L\left(\frac{P_1}{kA}\right)\left(\frac{P_2}{kA}\right)$$

$$e_1 = L\left(\frac{P_1}{kA}\right)^2, \qquad e_2 = L\left(\frac{P_2}{kA}\right)^2$$

It is evident that in general $e \neq e_1 + e_2$, and so the principle cannot be employed to obtain extensions in this case. This can be seen graphically in Fig. 7.3. Comparing Fig.

Fig. 7.3

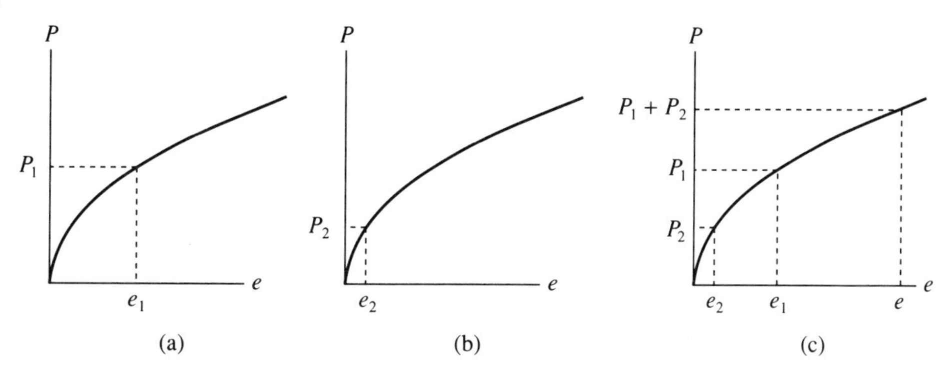

7.3(b) and (c) we see that the effect of load P_1 is to change the response of the member to the second load P_2. If P_2 is applied after P_1 has been applied, it causes the elongation e shown in Fig. 7.3(c). If P_2 is applied alone, the elongation is e_2, shown in Fig. 7.3(b). The nonlinear behavior of the material is responsible for the difference in effects.

If a body undergoes extensive changes in size or shape, superposition is not generally applicable. To illustrate this we consider the structure of Fig. 7.4(a), which consists of two identical linearly elastic wires OA and OB. A vertical load P applied at O causes the structure to deform as shown in Fig. 7.4(b). If δ denotes the vertical displacement of the point O, the elongation of AO or BO is

$$e = \sqrt{L^2 + (L + \delta)^2} - \sqrt{2}L$$

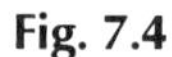

Fig. 7.4

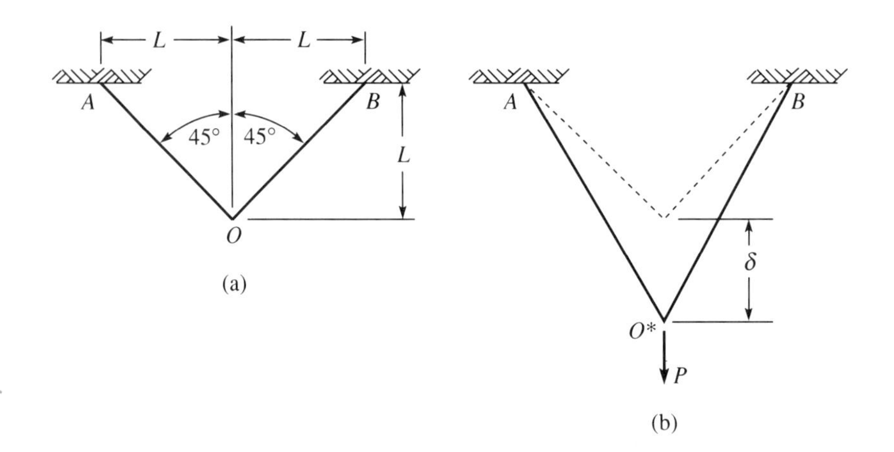

Denoting the axial force in each wire by F, we find from equilibrium that

$$F = P\frac{\sqrt{L^2 + (L + \delta)^2}}{2(L + \delta)}$$

The elongation e is related to the force F by

$$e = \frac{F\sqrt{2}L}{AE}$$

where A is the cross-sectional area of the wire. Eliminating e and F between these equations, we find that

$$P = \frac{2(L + \delta)}{\sqrt{L^2 + (L + \delta)^2}} \frac{AE}{\sqrt{2}L}[\sqrt{L^2 + (L + \delta)^2} - \sqrt{2}L] \tag{7.1}$$

Although the material is linearly elastic, the load-displacement relation embodied in Eq. (7.1) is nonlinear, as illustrated in Fig. 7.5. The effect of a combined load $P = P_1 + P_2$ is the displacement δ in Fig. 7.5(c). This is not the sum of the displacements

Fig. 7.5

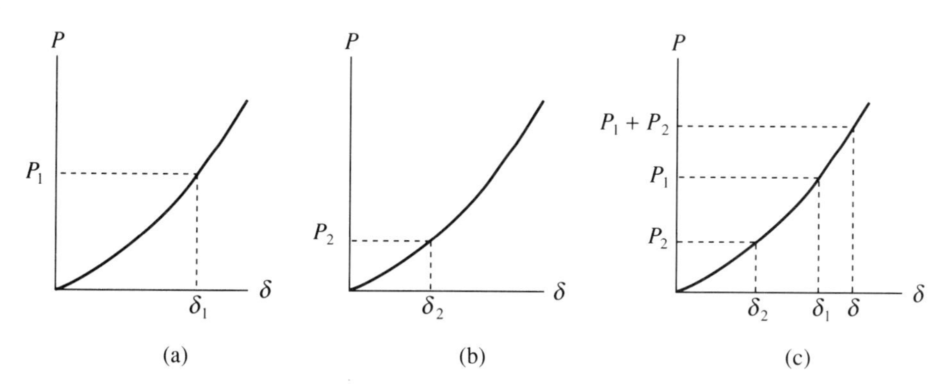

δ_1 and δ_2 caused by P_1 and P_2 acting separately. When the loads are applied consecutively, the first load changes the geometry of the structure significantly (if δ_1 is large) and so alters the response to the second load.

Another counterexample is illustrated in Fig. 7.6(a). A cantilevered beam is stretched by an axial load P and is bent by a couple M_0. According to the theory of Sec. 6.5,

$$\kappa = \frac{d^2w}{dx^2} = \frac{M}{EI} = \frac{M_0}{EI} - \frac{P}{EI}(w_0 - w) \tag{7.2}$$

Fig. 7.6

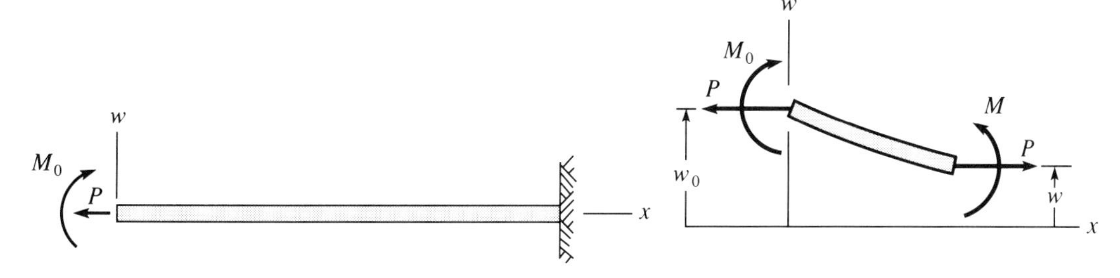

In this case the presence of the axial load P causes an addition to the bending couple in the amount $-P(w_0 - w)$. However, this additional bending couple exists only because the beam bends and gives a transverse deflection so that $w_0 - w \neq 0$. On the other hand, the axial load P acting alone causes no transverse deflection and therefore cannot contribute a bending couple. The difficulty stems from coupling between the deformation due to the bending load M_0 and the effects of the axial load P. In other words, because the bending couple M_0 causes deflection w, the axial load P causes a bending couple $-P(w_0 - w)$, which, through Eq. (7.2), also affects the transverse deflection.

Each of the preceding examples illustrates a type of situation in which the superposition procedure cannot be applied. Each example involves a nonlinear relationship: The example of Fig. 7.3 involves a nonlinear material; the stress-strain relation is *nonlinear*. The example of Fig. 7.4 involves a relatively large displacement; δ/L is *not* small compared to unity. This is a geometric *nonlinearity*. The example of Fig. 7.6 involves a *nonlinearity* in the equilibrium equation, namely, a *product* of the applied force and the displacement $[P(w_0 - w)]$. This is essentially a geometric *nonlinearity*. Superpositions (additivity of effects) are invalidated by nonlinearities that might be attributed to the behavior of the *material* or to relatively large changes of *geometry*. Material nonlinearities arise in the stress-strain relations. Geometric nonlinearities manifest themselves in kinematic relations (e.g., strain-displacement equations) and also in dynamic relations (e.g., equilibrium equations). These cases are excluded in the following applications of superposition.

7.2 Method of Superposition

If a body is composed of a linearly elastic material, the stresses can be expressed as linear functions of the strains. If the body is subjected independently to each of two different equilibrated loads, then the sum of the two effects is essentially the same as the effect caused by the simultaneous application of the two loads, provided that the following conditions hold:

1. Neither force system creates strains or relative rotations that significantly alter the geometric configuration of the body.
2. Neither force system contributes an effect dependent on the presence of the effects of the other force system.

Mathematically speaking, superposition requires linear equations in the dependent variables; internal forces, stresses, strains, and displacements must be all *linearly* related in the equations governing the system (the material, equilibrium, and kinematics).

7.3 Superposition of Stresses

EXAMPLE 1

Combined Axial Force and Bending Couple The bracket depicted in Fig. 7.7(a) is composed of steel, HI-HO to the yield stresses 280 MPa (tensile) and −350 MPa (compressive). It must support a force F, as shown, but must sustain no permanent deformation. It has the T-shaped cross section shown at (b). We require the force F_0 that would initiate yielding.

Fig. 7.7

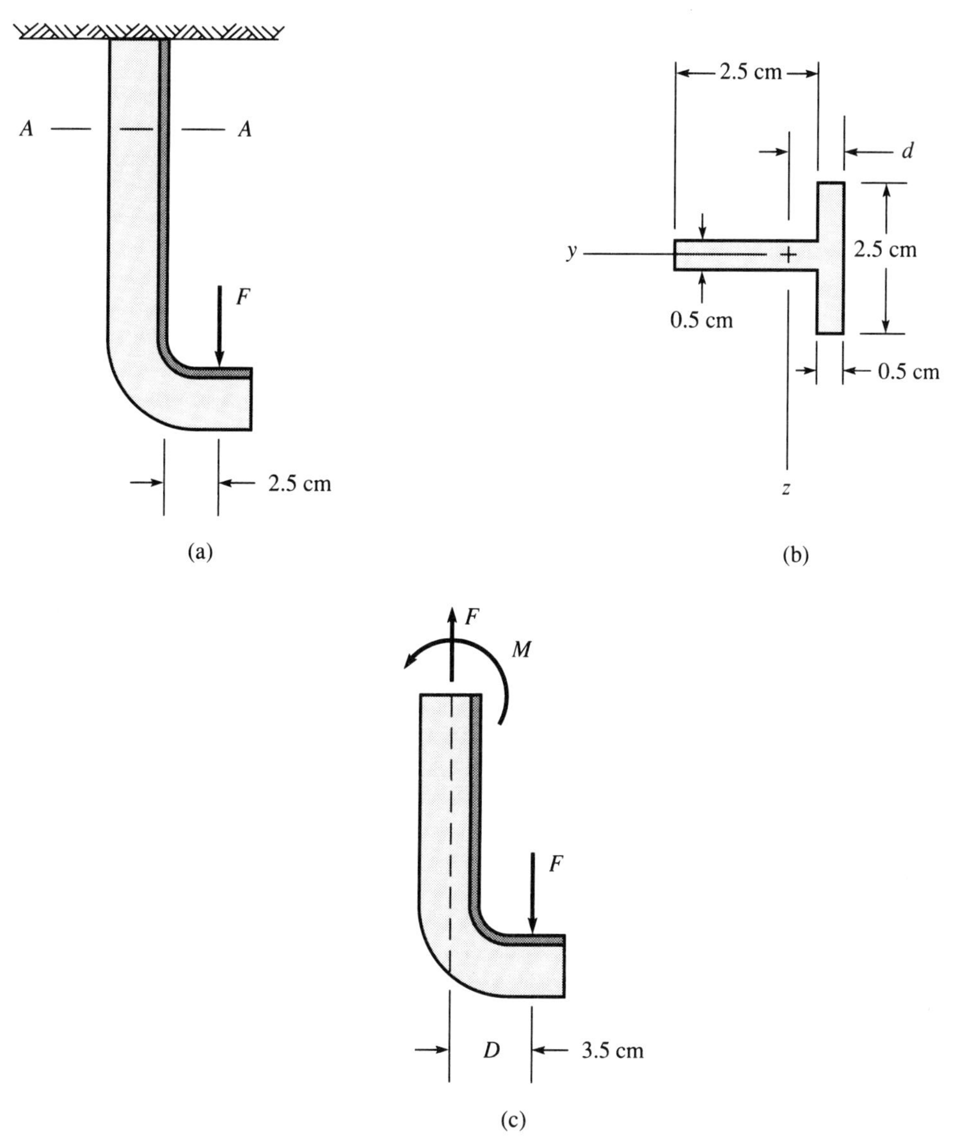

From the free body at (c) we find that the action upon a cross section (A-A) consists of a tensile force F at the centroid and a couple $M = FD$, where D denotes the distance from the centroid to the line of action.

From the given dimensions of Fig. 7.7(a), we determine the area A, distance d to the centroid, and then the integral I with respect to the centroidal axis:

$$A = 2.5 \text{ cm}^2, \qquad d = \frac{1.25 \times 1.75 + 1.25 \times 0.25}{2.5} = 1.00 \text{ cm}$$

$$I = \iint y^2 \, dA = 2.5 \int_{-1}^{-0.5} y^2 \, dy + 0.5 \int_{-0.5}^{2.0} y^2 \, dy = 2.08333\ldots \text{ cm}^4$$

The stress at any point of the cross section can be obtained by superposing that attributed to the axial force F ($\sigma_x = F/A$) *and* the couple M ($\sigma_x = -My/I$):

$$\sigma_x = \frac{F}{A} - \frac{3.5Fy}{I}$$

The maximum and minimum values occur at $y = -1$ cm and $y = +2$ cm, respectively.

$$\sigma_x(\text{max}) = \frac{F}{2.5} + \frac{3.5F}{2.083} = 2.08F$$

$$\sigma_x(\text{min}) = \frac{F}{2.5} - \frac{3.5 \times 2F}{2.083} = -2.96F$$

The limiting values (yielding in tension and compression) follow:

$$F = \frac{280 \times 10^2}{2.08} \text{N} = 13.5 \text{ kN}$$

$$F = \frac{350 \times 10^2}{2.96} \text{N} = 11.8 \text{ kN}$$

To avoid permanent deformation the load must be less than 11.8 kN. ◆

EXAMPLE 2

Combined Bending in Two Planes of Symmetry The cantilever beam of Fig. 7.8(a) is linearly elastic and supports an end load P, which acts perpendicular to the axis of the beam and along a diagonal of the rectangular cross section. The line of action is not in a longitudinal plane of symmetry but passes through the central axis. To determine the effects of the load we superpose the effects of the components, as shown in Fig. 7.8(b) and (c); each acts in a plane of symmetry:

$$P_y = \left(\frac{2}{\sqrt{5}}\right) 100 \text{ lb}, \qquad P_z = \left(\frac{1}{\sqrt{5}}\right) 100 \text{ lb}$$

The maximum stress σ'_x, caused by P_y alone, is at end A and at the top, where $y =$ 2 in. and $M_z = -P_y L$:

$$\sigma'_x = -\frac{M_z y}{I_{yy}} = \frac{P_y L(2)}{I_{yy}}$$

Fig. 7.8

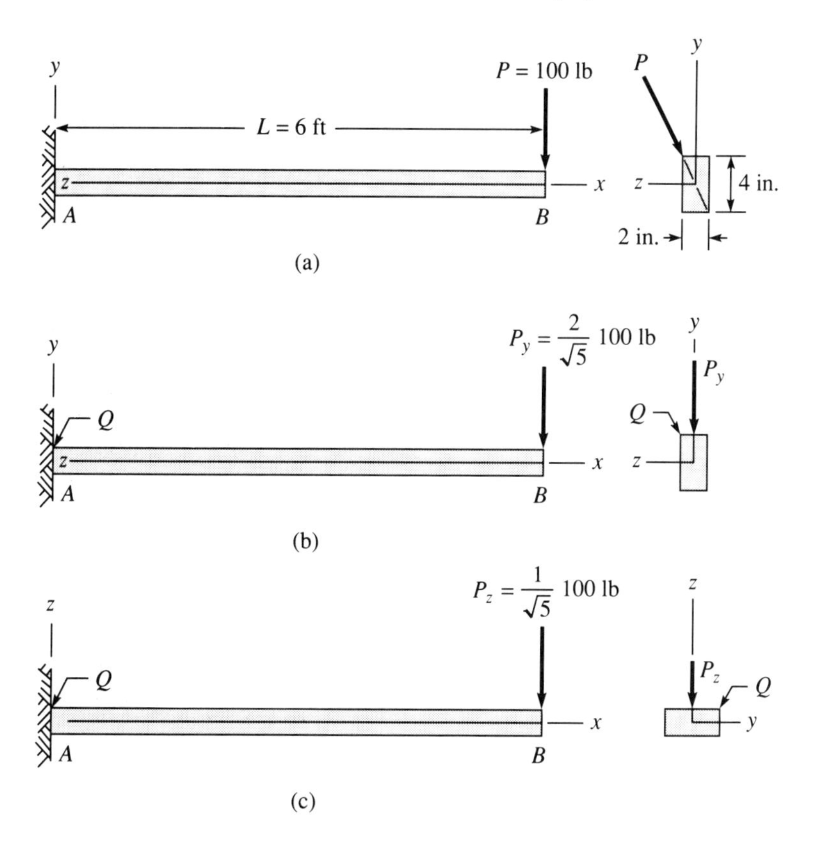

The maximum normal stress σ_x'', produced by P_z alone, is at end A and at the front, where $z = 1$ in. and $M_y = P_z L$:

$$\sigma_x'' = \frac{M_y z}{I_{zz}} = \frac{P_z L(1)}{I_{zz}}$$

For the rectangular cross section,

$$I_{yy} \equiv \iint y^2 \, dA = \frac{32}{3} \text{ in.}^4, \qquad I_{zz} \equiv \iint z^2 \, dA = \frac{8}{3} \text{ in.}^4$$

Both contributions, σ_x' and σ_x'', are a maximum at point Q of Fig. 7.8. By superposition, the maximum normal stress is at Q, where

$$\begin{aligned}\sigma_x &= \sigma_x' + \sigma_x'' \\ &= \frac{P_y L(2)}{I_{yy}} + \frac{P_z L(1)}{I_{zz}} \\ &= 1207 + 1207 \\ &= 2415 \text{ psi} \; \blacklozenge\end{aligned}$$

EXAMPLE 3

Stresses in Bending and Torsion A rigid arm is attached at a right angle to a circular cylinder ("torsion bar"), as shown in Fig. 7.9. The cylinder is made of steel that is HI-HO to the onset of yielding in accordance with the Mises criterion [Eq. (4.73)]:

$$(\sigma_1 - \sigma_2)^2 + (\sigma_2 - \sigma_3)^2 + (\sigma_1 - \sigma_2)^2 = C^2$$

Here, $\sigma_1, \sigma_2, \sigma_3$ are the principal stresses at the point of yielding and C is a property of the material that must be obtained experimentally. If σ_0 denotes the yield stress in a simple tensile test, then $\sigma_1 = \sigma_0$, $\sigma_2 = \sigma_3 = 0$ and

$$2\sigma_0^2 = C^2 \quad \text{or} \quad C = \sqrt{2}\sigma_0$$

We require the load P_0 that would initiate yielding.

Fig. 7.9

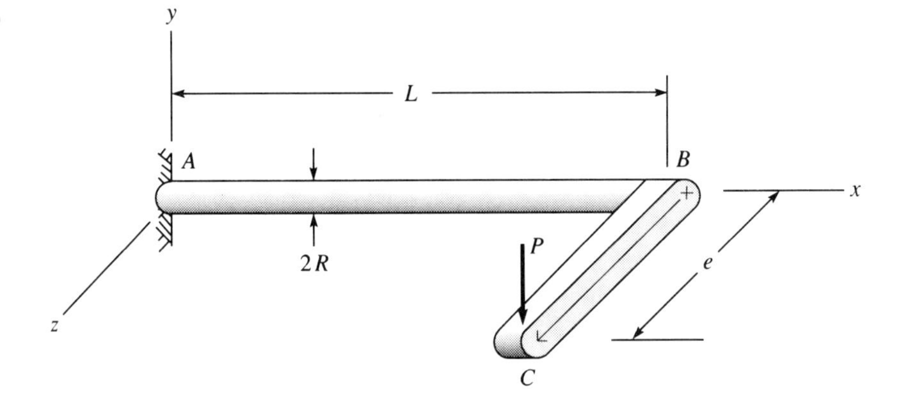

From the free bodies we observe that the cylinder is subject to a constant twisting couple $T = Pe$ *and* a bending couple M_z, which has the maximum absolute value at the fixed end ($x = 0$) where $M_z = -PL$. These are shown on the free body of Fig. 7.10(a).

To determine the state of stress, we can superpose the stress caused by the couple $M_z = -PL$ of Fig. 7.10(b) and that caused by the couple $T = Pe$ of Fig. 7.10(c). *Note:* We can do this because the equations are linear to the yield condition. The bending couple causes only a normal stress at point A:

$$\sigma_x = -\frac{M_z R}{I} = \frac{PLR}{I}$$

Fig. 7.10

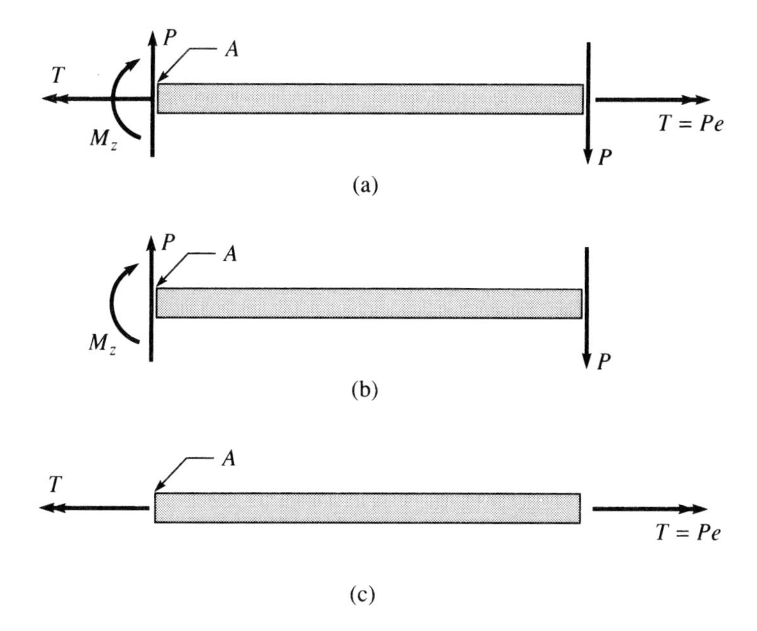

The twisting couple causes only a shear stress at point A:

$$\tau_{xz} = \frac{TR}{J} = \frac{PeR}{J}$$

An element at A is depicted in Fig. 7.11 with the stresses, σ_x and τ_{zx}, superposed. At this location a principal stress is $\sigma_3 = \sigma_y = 0$. By the theory of Chapter 3 the other principal directions are in the xz-plane. The directions (angles β) are determined by Eq. (2.72b):

Fig. 7.11

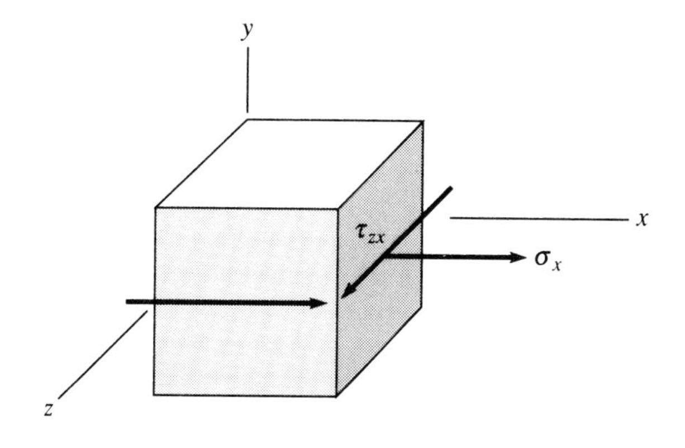

$$\tan 2\beta = \frac{2\tau_{xz}}{\sigma_x}$$

$$= -\frac{2Ie}{JL} = -\frac{e}{L},$$

since $2I = J$ for the circular cross section. The principal stresses (σ_1 and σ_2) are then given by

$$\sigma = \frac{\sigma_x}{2}(1 + \cos 2\beta) - \tau_{xz} \sin 2\beta$$

Recall that the principal directions are orthogonal. Since $e/L > 0$, the angles $2\beta_1$ and $2\beta_2$ lie in the second and fourth quadrants:

$$\cos 2\beta = \mp \frac{L}{\sqrt{e^2 + L^2}}, \qquad \sin 2\beta = \pm \frac{e}{\sqrt{e^2 + L^2}}$$

$$\left.\begin{matrix}\sigma_1 \\ \sigma_2\end{matrix}\right\} = \frac{\sigma_x}{2}\left(1 \mp \frac{L}{h}\right) \mp \tau_{xz}\frac{e}{h}$$

where $h \equiv \sqrt{e^2 + L^2}$ for brevity.

$$\sigma_1 - \sigma_2 = -\left(\sigma_x \frac{L}{h} + 2\tau_{xz}\frac{e}{h}\right)$$

$$(\sigma_1 - \sigma_2)^2 = \frac{\sigma_x^2 L^2 + 4\sigma_x \tau_{xz} eL + 4\tau_{xz}^2 e^2}{h^2}$$

$$\sigma_1^2 + \sigma_2^2 = \frac{\sigma_x^2}{2} + \frac{\sigma_x^2 L^2 + 4\sigma_x \tau_{xz} eL + 4\tau_{xz}^2 e^2}{2h^2}$$

Substituting the expressions for σ_x and τ_{xz}, using $I = J/2$ and $h^2 = e^2 + L^2$ we obtain

$$(\sigma_1 - \sigma_2)^2 + \sigma_1^2 + \sigma_2^2 = 2\frac{P_0^2 R^2}{J^2}\left(\frac{4L^4 + 7e^2L^2 + 3e^4}{e^2 + L^2}\right) = 2\sigma_0^2$$

or

$$P_0 = \frac{J\sigma_0}{R\sqrt{4L^2 + 3e^2}} = \frac{\pi R^3}{2}\frac{\sigma_0}{\sqrt{4L^2 + 3e^2}}$$

The normal stress σ_x and shear stress τ_{xz} are a maximum at the point A; the yield condition is attained there, and yielding is indicated at the load P_0.

The reader can verify that the result holds for bending only ($e = 0$) and torsion only ($L = 0$, $\tau_0 = \sigma_0/\sqrt{3}$). ◆

PROBLEMS

Assume linearly elastic behavior unless otherwise stated.

7.1 Find the maximum and minimum normal stresses at cross section A of the round steel bar in Fig. P7.1.

Fig. P7.1

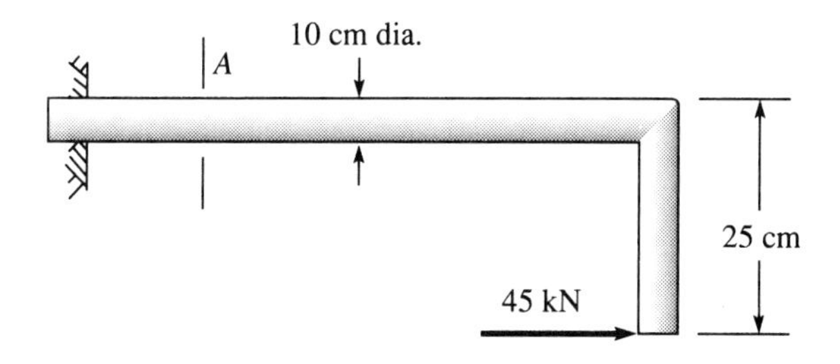

7.2 A shaft 1 in. in diameter is subjected to a bending couple M and an axial tensile load P at its ends. The maximum tensile stress in the shaft is 3500 psi, and the minimum tensile stress is 1500 psi. Compute P and M.

7.3 An aluminum shaft 3.75 cm in diameter is subjected to an axial tensile load of 135 kN and a twisting couple T at its ends. What is the largest possible value for T if the maximum shear stress in the shaft is not to exceed 200 MPa?

7.4 A circular steel shaft has a radius of 1 in. and a length of 4 ft. It is stretched by an axial load of 10,000 lb and twisted by a couple of 20,000 in.-lb. Determine the principal stresses at a point in the surface of the shaft.

7.5 A cylindrical brick chimney has a 6-ft diameter and a 1-ft wall thickness. The weight of the material is 160 lb/ft^3. If the maximum horizontal wind pressure is 40 lb/ft of height, determine the maximum height to which the chimney can be built and have only compressive stress on all cross sections.

7.6 A block 20 cm square and 60 cm long is loaded as shown in Fig. P7.6. By what ratio will the maximum compressive stress be changed if 3.8 cm of material is removed from the top of the block?

Fig. P7.6

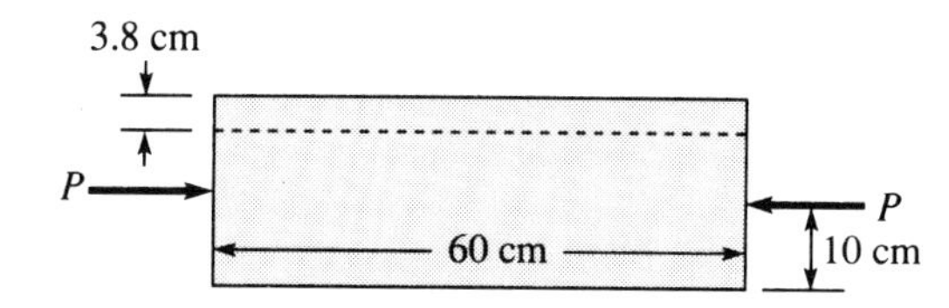

7.7 A plane frame of uniform cross section is loaded as shown in Fig. P7.7. Compute the principal stresses at point A.

Fig. P7.7

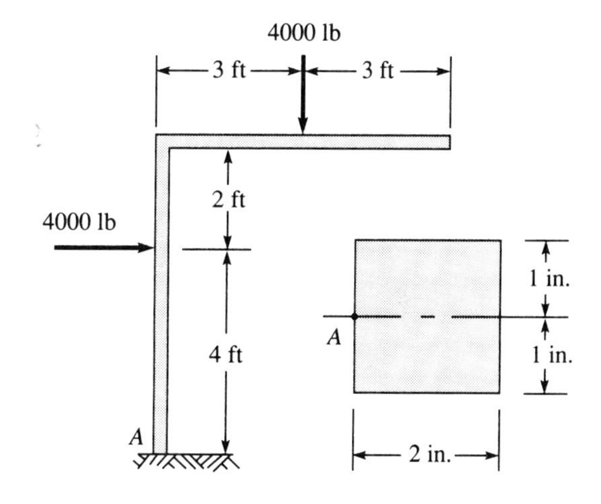

7.8 A thin tube has an outside diameter of 10 cm with a 0.30-cm wall thickness. The maximum allowable shear stress is 100 MPa. What combinations of axial load P and twisting couple T will produce this shear stress?

7.9 A bending couple M of magnitude 20,000 in.-lb acts on the beam cross section shown in Fig. P7.9. Calculate the normal stress on the section at points A and B.

Fig. P7.9

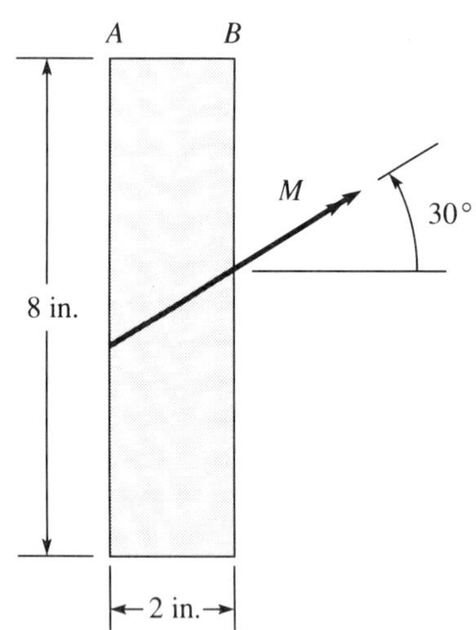

Fig. P7.10

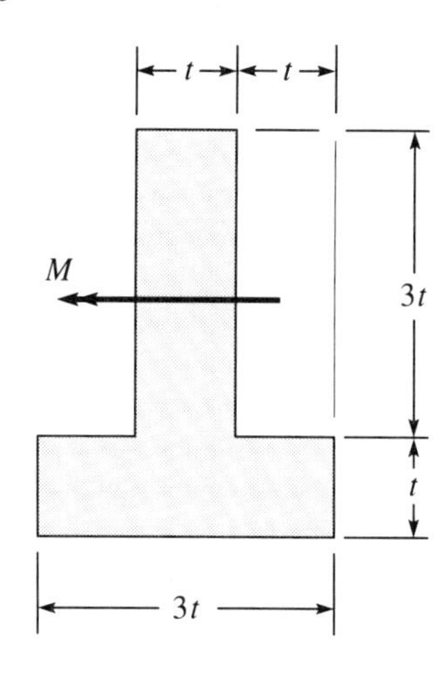

7.10 A beam with the cross section illustrated in Fig. P7.10 is subjected to a bending couple M and an axial centroidal load P. Determine the ratio M/P that causes equal normal stress magnitudes at the top and bottom of the beam.

7.11 What is the maximum principal stress at point A of the shaft shown in Fig. P7.11?

Fig. P7.11

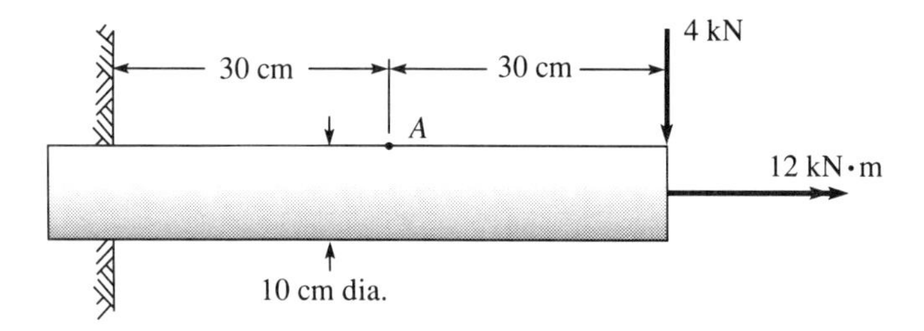

7.12 A hollow aluminum tube has a 20-in. outside diameter and a $\frac{1}{4}$-in. wall thickness. It is stretched by a force P pounds and expanded by an internal pressure of $(P/100)$ pounds per square inch (lb/in.^{2} or psi). Determine P so that the maximum shear stress in the cylindrical wall is 25,000 lb/in.^{2}

7.13 For the beam in Fig. P7.13, compute P so that the normal stress on the cross section at $x = y = 0$ is -28 MPa.

Fig. P7.13

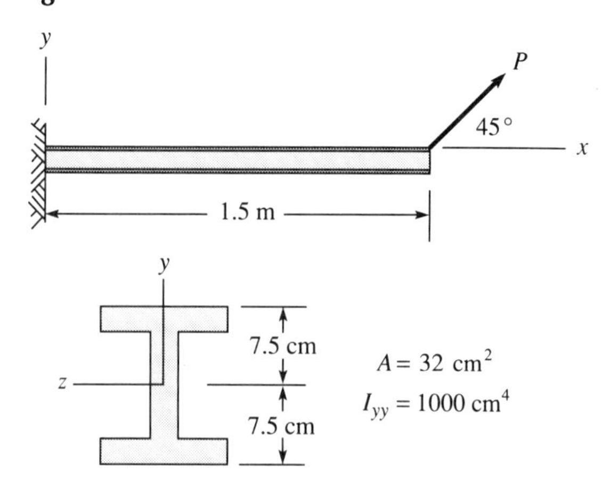

7.14 Calculate the maximum and minimum normal stresses on cross section A of the prismatic bar loaded in the LPS, as shown in Fig. P7.14.

Fig. P7.14

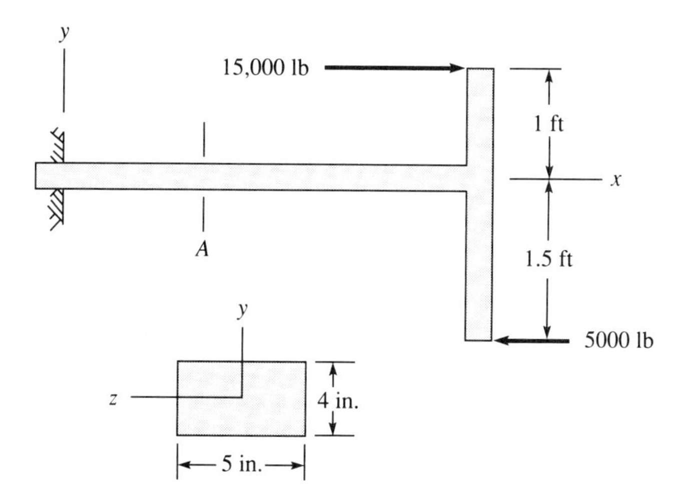

7.15 A circular steel bar is stretched by a force of 40 kN and is bent by end couples of magnitude 100 cm·kN. What must the radius of the bar be in order that the maximum normal stress in the bar does not exceed 210 MPa?

7.4 Combined Loads on Symmetric Beams

EXAMPLE 4

Deflection of a Beam A simply supported beam with $EI = 10^8$ lb-in.^{2} is loaded as shown in Fig. 7.12(a). To find the deflection at point A by the method of superposition, we consider the beams of Fig. 7.12(b) and 7.12(c). Note that we have two beams in these figures with different loads but the *same end conditions* as the original beam. From Example 6.9, the deflection at x of a simply supported beam carrying a concentrated load P at a distance a from the left end of the beam is

$$w = \begin{cases} -\dfrac{P}{6EIL}(L-a)[(2L-a)ax - x^3], & x \le a \\ -\dfrac{P}{6EIL}(L-x)[(2L-x)ax - a^3], & x \ge a \end{cases}$$

Therefore, the deflection of the beam in Fig. 7.12(b) at $x = 6$ ft is

$$w_{1|A} = 0.18 \text{ in.}$$

Fig. 7.12

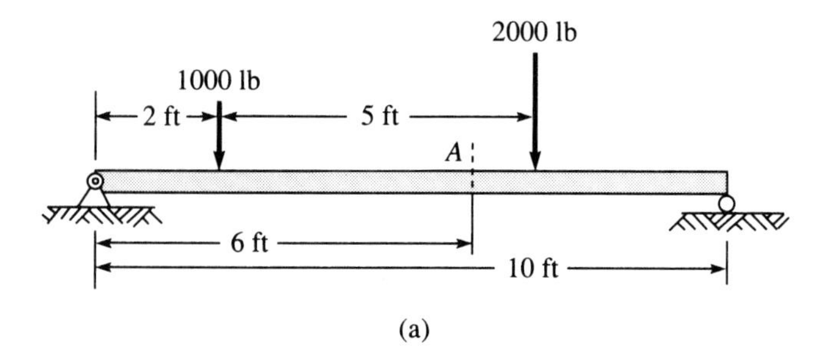

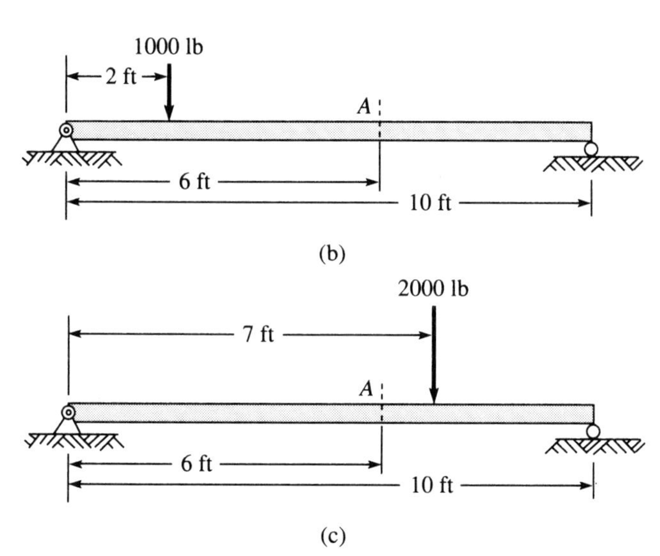

and for the beam in Fig. 7.7(c),

$$w_{2|A} = 0.57 \text{ in.}$$

On adding these two, we obtain the deflection at $x = 6$ ft for the beam of Fig. 7.12(a):

$$w_{|A} = w_{1|A} + w_{2|A} = 0.75 \text{ in.} \quad \blacklozenge$$

EXAMPLE 5

A Statically Indeterminate Problem The beam of Fig. 7.13(a) is statically indeterminate. We can employ the superposition principle to obtain the reaction at B. To do this consider the two statically determinate beams of Figs. 7.13(b) and 7.13(c). The first is a cantilevered beam loaded by the uniform load q. The second cantilevered beam is subjected only to an unknown force R acting at B. (This force is created by the roller in the original problem.) For the first beam,

Fig. 7.13

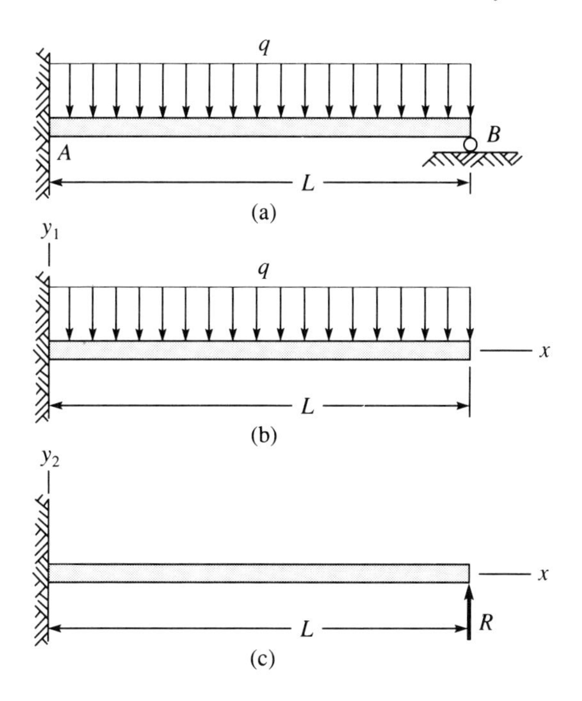

$$w_1(L) = -\frac{qL^4}{8EI}$$

For the second beam,

$$w_2(L) = \frac{RL^3}{3EI}$$

Since $w_1(L) + w_2(L) = 0$ (due to the roller at B), we have

$$-\frac{qL^4}{8EI} + \frac{RL^3}{3EI} = 0$$

or

$$R = \tfrac{3}{8}qL \quad \blacklozenge$$

EXAMPLE 6

Combined Bending in Two Planes of Symmetry Let us revisit the cantilevered beam of Fig. 7.8(a). The end load P acts perpendicular to the axis of the beam and along a diagonal of the rectangular cross section. This load is not in a longitudinal plane of symmetry; however, it does pass through the central axis. Suppose that the material of the beam (wood) is linearly elastic with $E = 1.5 \times 10^6$ lb/in^2.

To determine the effect of the load P, we superimpose the effects of its components $P_y = (2/\sqrt{5})100$ lb and $P_z = (1/\sqrt{5})100$ lb shown in Fig. 7.8(b) and 7.8(c). Notice that each component does act in a longitudinal plane of symmetry. The load P_y acting alone would cause bending in the xy-plane. The deflection of end B that would be caused by the load P_y is

$$w_y(B) = -\frac{P_y L^3}{3EI_{yy}}$$

Similarly, the deflection that would be caused by the load P_z acting alone is

$$w_z(B) = -\frac{P_z L^3}{3EI_{zz}}$$

in the xz-plane. Here, I_{zz} and I_{yy} denote the area moments of inertia with respect to coordinates z and y:

$$I_{zz} = \tfrac{8}{3} \text{ in.}^4, \qquad I_{yy} = \tfrac{32}{3} \text{ in.}^4$$

Substituting numerical values, we have

$$w_y(B) = -0.696 \text{ in.}$$

$$W_z(B) = -1.39 \text{ in.} \quad \blacklozenge$$

Solve the following problems of symmetric HI-HO beams by superposing the solutions for simple loadings. Consult the table in Appendix D.

7.16 The cantilever beam in Fig. P7.16 is loaded by a force P and a bending couple M. Find the value of M (in terms of P, EI, and L) that will cause the rotation of the end A to be zero.

Fig. P7.16

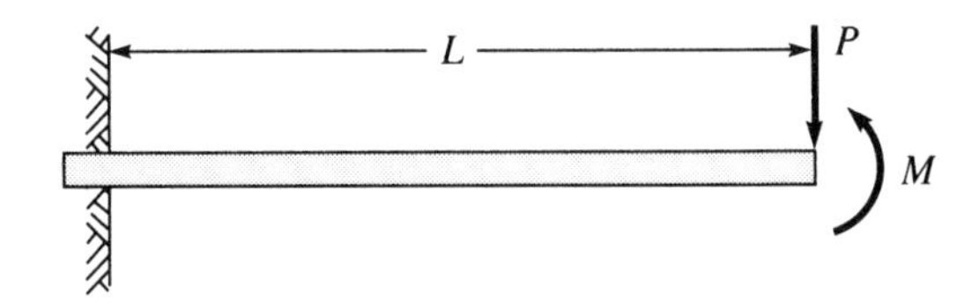

7.17 What is the reaction of the roller B on the beam shown in Fig. P7.17?

Fig. P7.17

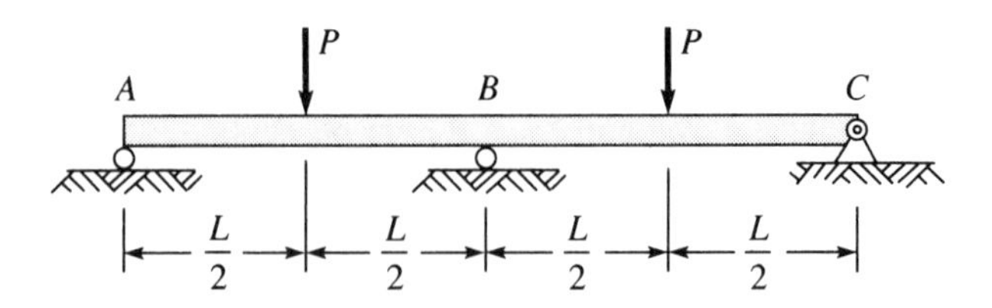

7.18 Express the deflection at end A of the beam in Fig. P7.18 in terms of the parameters (K, E, I, L, and $\bar{p}$), where K denotes the modulus of the spring.

Fig. P7.18

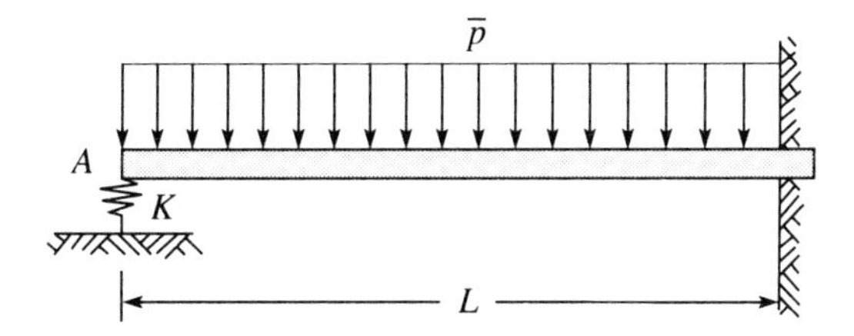

7.19 Determine the reactions of all supports on the beams illustrated in Fig. P7.19(a) and (b). (Assume all members have the same cross section and modulus).

Fig. P7.19

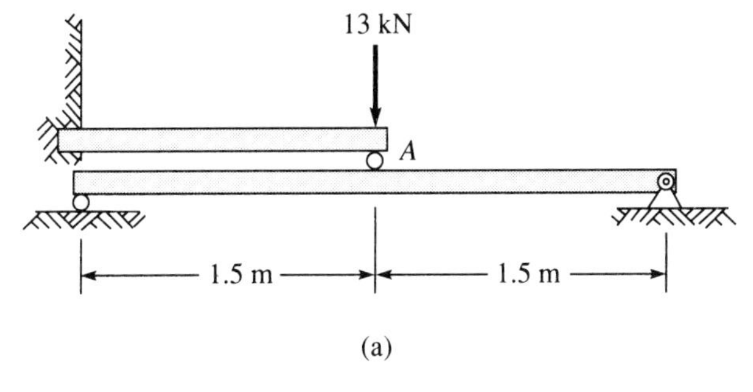

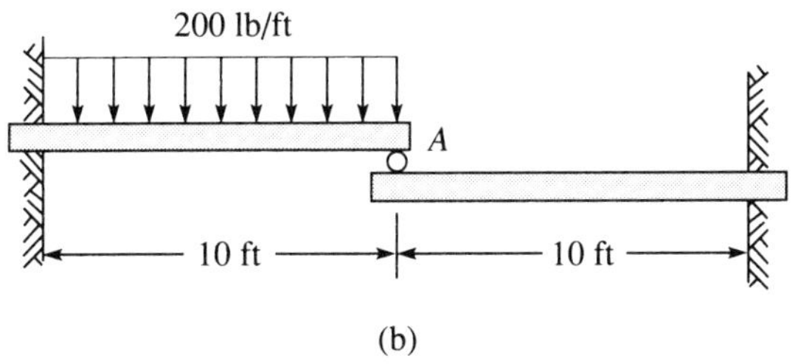

7.20 The two steel cantilever beams in Fig. P7.19(b) have the same cross sections ($I = 20$ in.^{4}). Each just touches the roller A before the uniform load w is applied. Calculate the displacement of the roller.

7.21 The two beams of Fig. P7.19(a) are steel and have the same cross sections ($I = 830$ cm^4). Both are in contact with the roller A when load P is applied. Calculate (a) the load transmitted by the roller and (b) the displacement of the roller.

7.22 An elastic wire supports a cantilever beam, as shown in Fig. P7.22. The wire is straight and unstressed when the load P is applied. Determine the force in the wire after the load is applied; express the answer in terms of the parameters (E, I, P, L, A), where A denotes the cross-sectional area of the wire and E is the modulus of the HI-HO wire *and* beam.

Fig. P7.22

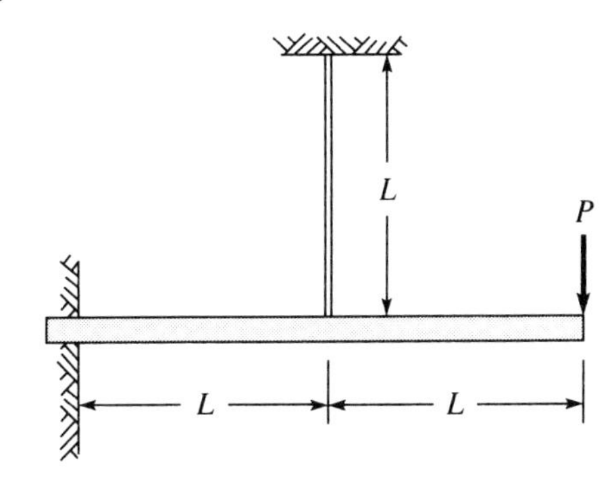

7.23 The left end of the steel I-beam in Fig. P7.23 is fixed; the right end is supported by a steel rod. The cross-sectional area of the rod is $A = 0.040$ in.^{2}, and for the beam $I = 192$ in^4. The stop-nut at the end of the rod is just tight against the bottom flange of the beam when the uniform load is applied. Calculate the tensile force in the rod after the load is applied; also

Fig. P7.23

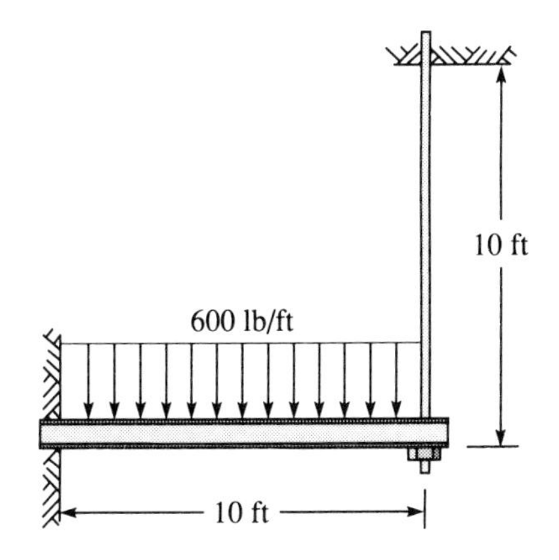

determine the maximum and minimum bending moment in the beam.

7.24 Repeat Prob. 7.23 if there is a gap of 0.100 in. between the nut and the flange when the load is applied.

7.25 Obtain the deflection equations for the beams shown in Fig. P7.25 by superposing the effects of simple loadings. Give the answers in terms of the parameters E, I, L, and $\bar{p}$.

Fig. P7.25

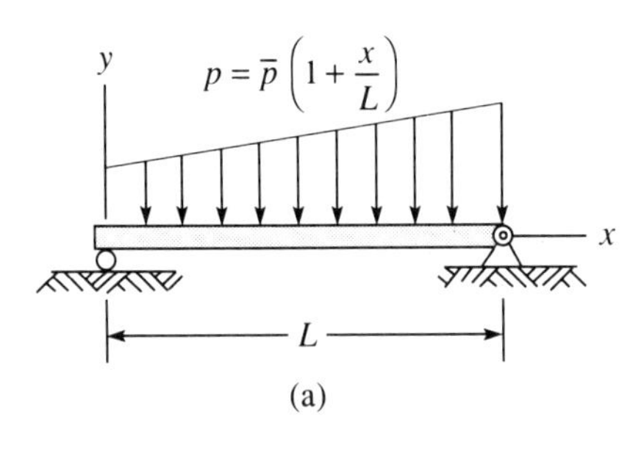

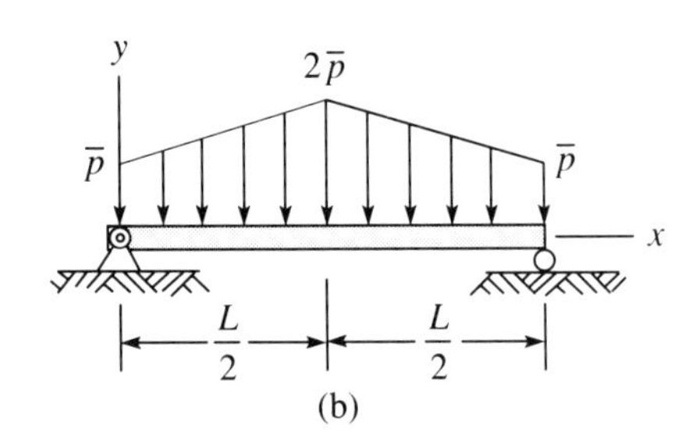

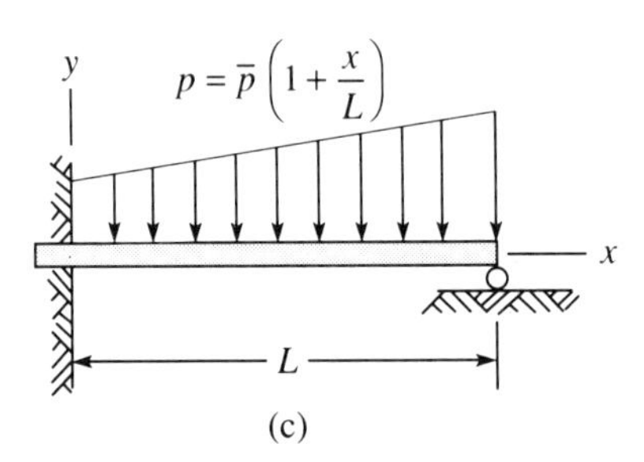

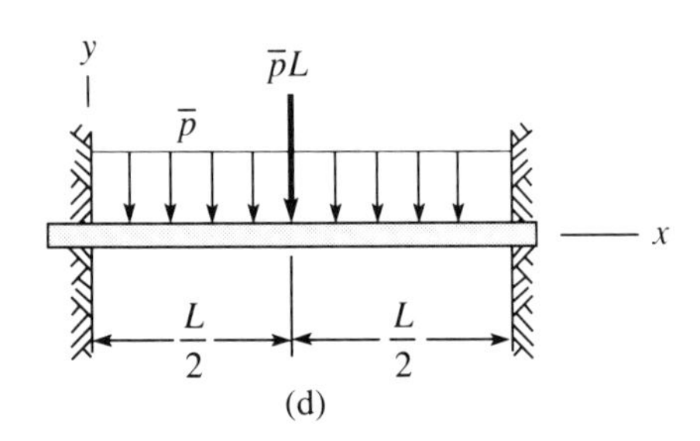

7.26 A steel beam has the rectangular cross section shown in Fig. P7.26. It is fixed at one end ($x = 0$) and subjected to a couple $M = 9000$ N·cm at the other end ($x = 50$ cm). Determine the deflection (vector) at the end ($x = 50$ cm).

Fig. P7.26

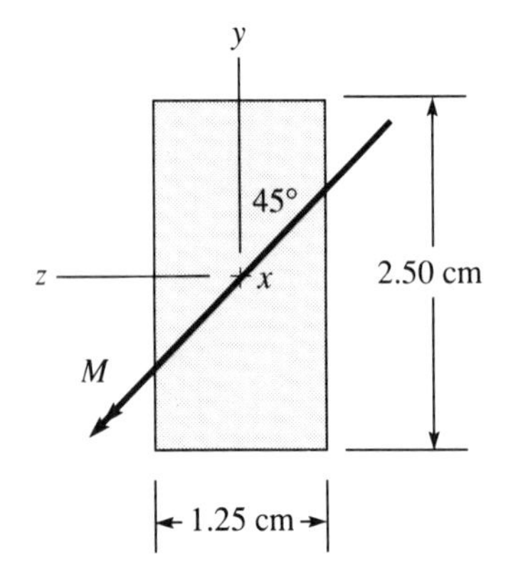

7.27 A cantilever beam has the rectangular cross section shown in Fig. P7.27. It is 1 m long, composed of steel, and subjected to the load $P = 1300$ N at the un-

Fig. P7.27

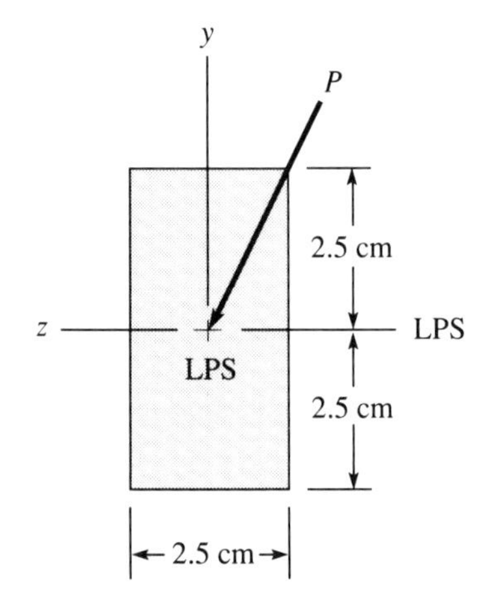

supported end. Determine the deflection of the loaded end.

7.28 A standard wide-flange steel I-beam is used as a cantilever. It is 20 ft long and is subjected to the load $P = 1000$ lb upon the unsupported end, as shown in Fig. P7.28. Properties of the beam are

$$I_{yy} = 310 \text{ in.}^4; \qquad I_{zz} = 44.0 \text{ in.}^4$$

Calculate the deflection at the end.

Fig. P7.28

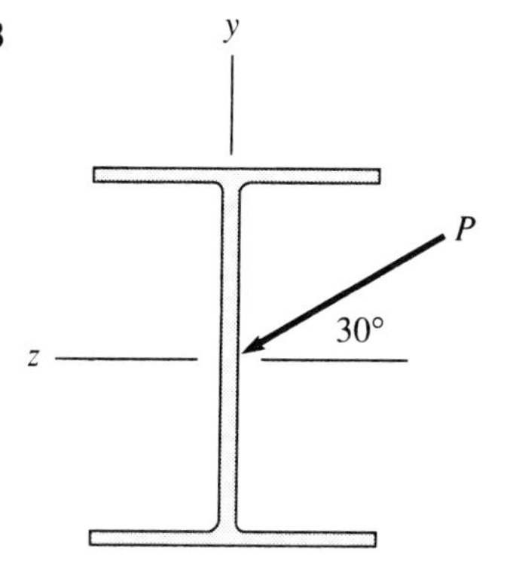

7.29 Compute the deflection at end B of the steel frame of Prob. 7.7. Note that axial extension of the members is negligible compared to the transverse deflections.

7.30 The steel truss illustrated in Fig. P7.30 has a turnbuckle on the steel rod OA. Before the truss is loaded, it is prestressed by turning the turnbuckle. If the turnbuckle has 20 threads per inch, how much should it be turned in order to equalize the stresses in all three rods for $P = 8000$ lb?

Fig. P7.30

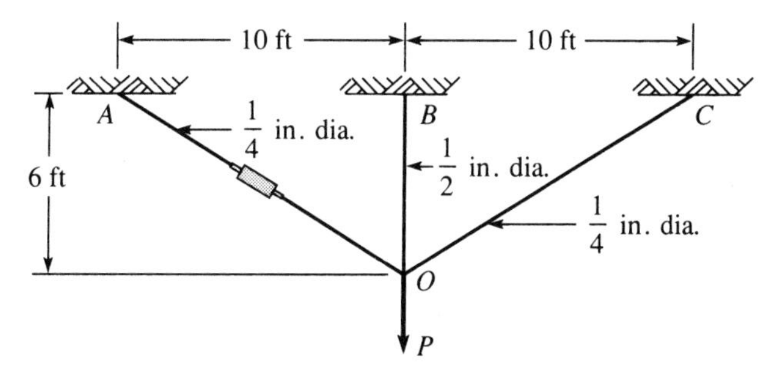

7.31 What is the maximum normal stress on the cross section at B in the beam shown in Fig. P7.31? What is the maximum normal stress in the beam?

Fig. P7.31

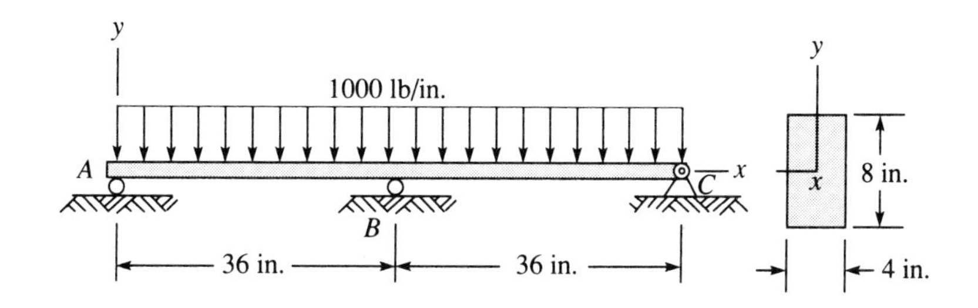

7.32 Compute the maximum normal stress on the cross section at A for the beam shown in Fig. P7.32.

Fig. P7.32

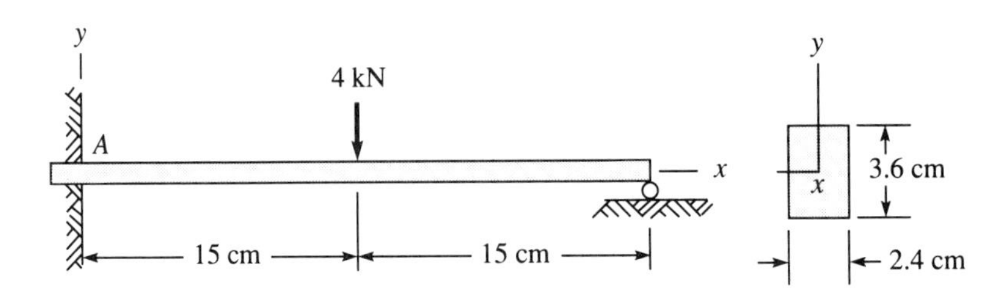

7.33 The load P, shown acting on the cantilever beam of Fig. P7.33(a), causes the free-end deflection

$$-\frac{P\xi^2}{6EI}(3L - \xi)$$

Use this result and the principle of superposition to obtain the end deflection for the distributed loads shown in the remaining figures. (*Hint:* Consider the deflection caused by the load on a slice of width $\Delta\xi$; then superpose the effects of all such loads.)

Fig. P7.33

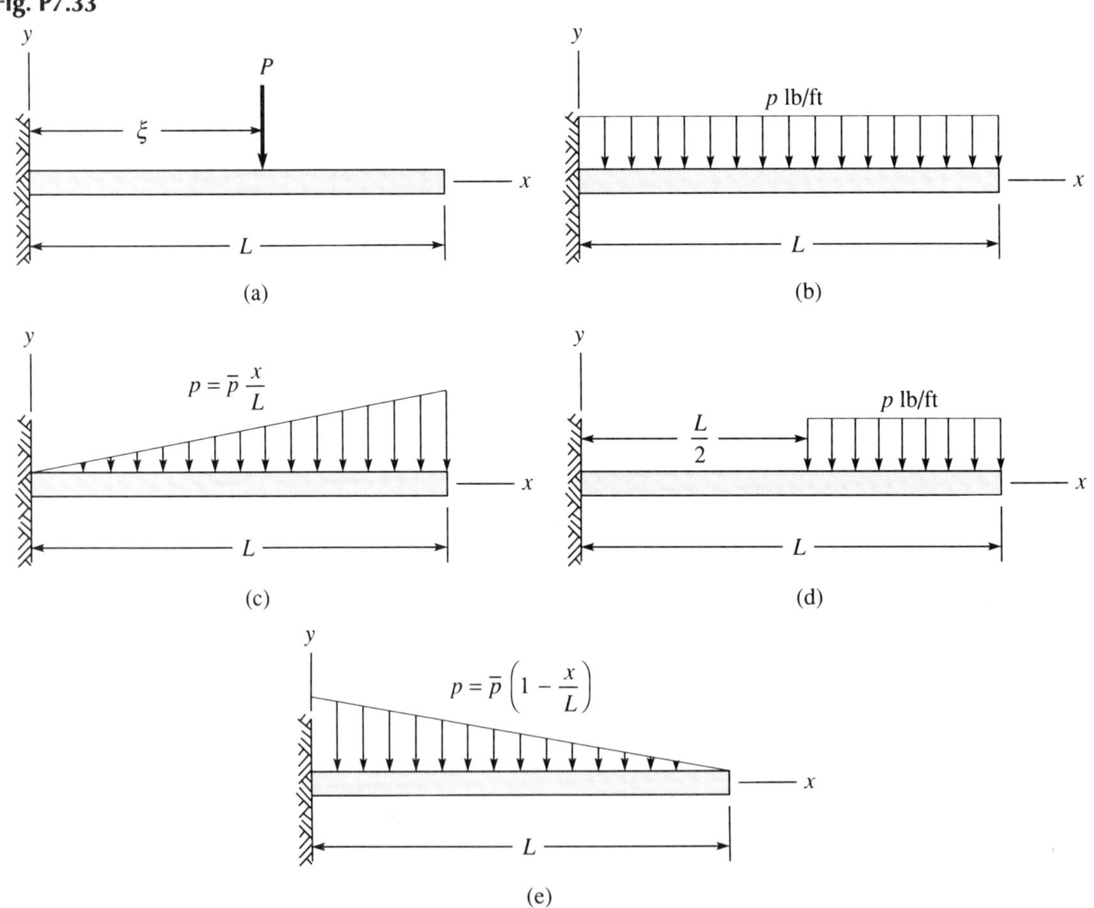

7.34 The load P on the cantilever beam of Fig. P7.34(a) causes a deflection at any point given by

$$w(x) = \begin{cases} -\dfrac{P}{6EI}(3\xi x^2 - x^3), & x \leq \xi \\ -\dfrac{P}{6EI}(3\xi^2 x - \xi^3) & x \geq \xi \end{cases}$$

Use this result and the principle of superposition to obtain the deflection equation for the distributed load in Fig. P7.34(b).

Fig. P7.34

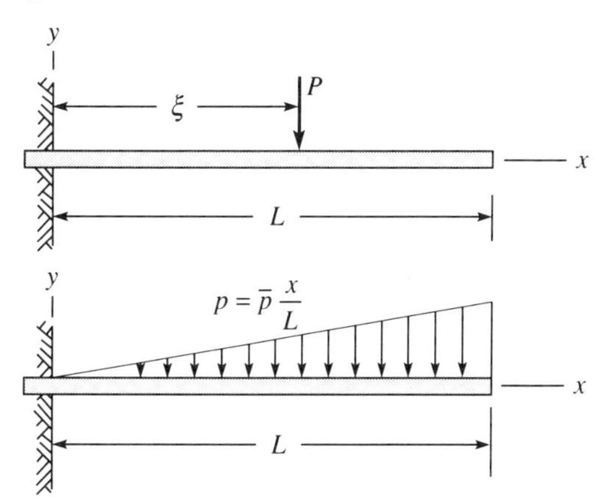

7.5 | Combined Bending and Torsion

EXAMPLE 7

Stresses in Bending and Torsion The member OBC in Fig. 7.14(a) is formed of standard $\frac{3}{4}$-in. steel pipe. Segment OB is joined to OC by a right elbow at B (assume a rigid joint) and clamped to a fixed support at O. A concentrated force Q is applied perpendicular to the plane (xz) at end C. We can assume that the steel is HI-HO to the yield condition. Tension and torsion tests provide the yield stresses and moduli:

$$\sigma_0 = 48 \times 10^3 \text{ lb/in.}^2, \qquad \tau_0 = 24 \times 10^3 \text{ lb/in.}^2, \qquad E = 30 \times 10^6 \text{ lb/in.}^2$$

$$G = 12 \times 10^6 \text{ lb/in.}^2$$

Inner and outer diameters of the pipe are

$$\text{I.D.} = 0.824 \text{ in.}, \qquad \text{O.D.} = 1.050 \text{ in.}$$

The cross-sectional integrals and areal moments of inertia follow:

$$I = \iint_A y^2\, dA = 0.0370 \text{ in.}^4$$

$$J = \iint (y^2 + z^2)\, dA = 2I = 0.0741 \text{ in.}^4$$

Since the steel appears to yield in shear—that is, according to the Tresca criterion [see Sec. 4.15, Eqs. (4.71) and (4.72)]—we want to compute the maximum shear stress.

We can determine the internal force system on any cross section by considering the equilibrium of a free body such as that depicted in Fig. 7.14(b). Only the nonzero

Fig. 7.14

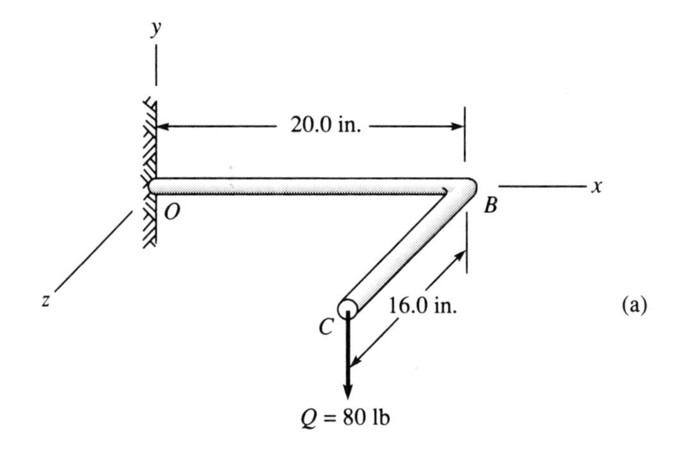

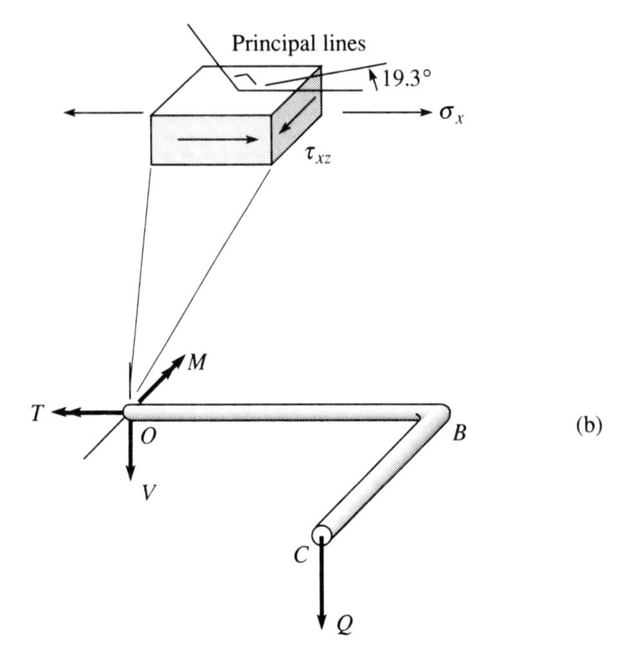

components are shown on the cross section through O, the clamped end. At O these assume their greatest values. From the equilibrium equations, we have

$$\sum M_x = 0: \quad T = 80 \times 16 = 1280 \text{ in.-lb}$$

$$\sum M_z = 0: \quad M = -80 \times 20 = -1600 \text{ in.-lb}$$

$$\sum F_y = 0: \quad V = -80 \text{ lb}$$

On this cross section the maximum value of the normal stress σ_x occurs at the top $(0, 0.525, 0)$. By Eq. (6.16)

$$\sigma_x = -\frac{My}{I} = -\frac{(-1600)(0.525)}{0.0370} = 22{,}700 \text{ lb/in.}^2$$

At this cross section the maximum value of the shear stress (τ_{xz}) also occurs at the same location. That stress is caused by the twisting couple T and is given by Eq. (5.73):

$$\tau_{xz} = \frac{Ty}{J} = \frac{(1280)(0.525)}{0.0741} = 9{,}070 \text{ lb/in.}^2$$

We note that the transverse shear force V causes a small shear stress τ_{xy} at the central line ($y = 0$). However, at that location the dominant stress component σ_x is zero. Therefore, the maximum shear stress occurs at the site $(0, 0.525, 0)$, where the state of stress has the two nonzero values just computed, namely,

$$\begin{bmatrix} \sigma_x & \tau_{xy} & \tau_{xz} \\ & \sigma_y & \tau_{yz} \\ & & \sigma_z \end{bmatrix} = \begin{bmatrix} 22{,}700 & 0 & 9{,}070 \\ & 0 & 0 \\ & & 0 \end{bmatrix} \text{ lb/in.}^2$$

The free upper surface is a principal surface (y-direction). The other principal directions are determined according to Eq. (2.72b):

$$\tan 2\theta = \frac{2\tau_{xz}}{\sigma_x} = 0.799$$

$$\theta = 19.3^\circ, 109.3^\circ$$

The corresponding normal (principal) stresses are given by Eq. (2.69):

$$\sigma\begin{pmatrix}\max\\ \min\end{pmatrix} = \frac{\sigma_x}{2} + \frac{\sigma_x}{2}\cos 2\theta + \tau_{xy}\sin 2\theta = \begin{pmatrix}+25{,}900\\ -3{,}200\end{pmatrix} \text{ lb/in.}^2$$

Recall that

$$\tau_{\max} = \frac{\sigma_{\max} - \sigma_{\min}}{2} = 14{,}600 \text{ lb/in.}^2$$

The data for this material indicate that it yields at $\tau_{\max} = 24{,}000$ lb/in^2. Since the behavior is linear to the onset of yielding, the stress increases linearly with the load. Therefore, we can anticipate yielding at the load

$$Q_0 = \frac{24{,}000}{14{,}600} \times 80 = 130 \text{ lb} \; \blacklozenge$$

EXAMPLE 8

Deflections from Bending and Twisting Let us now compute the deflection of the end C of the member in Fig. 7.14. We do this by superposing (adding) the deflections caused by the bending and the twisting. We first divide the member at B and consider separately the segments OB and BC, as in Fig. 7.15(a) and (b), respectively. The

Fig. 7.15

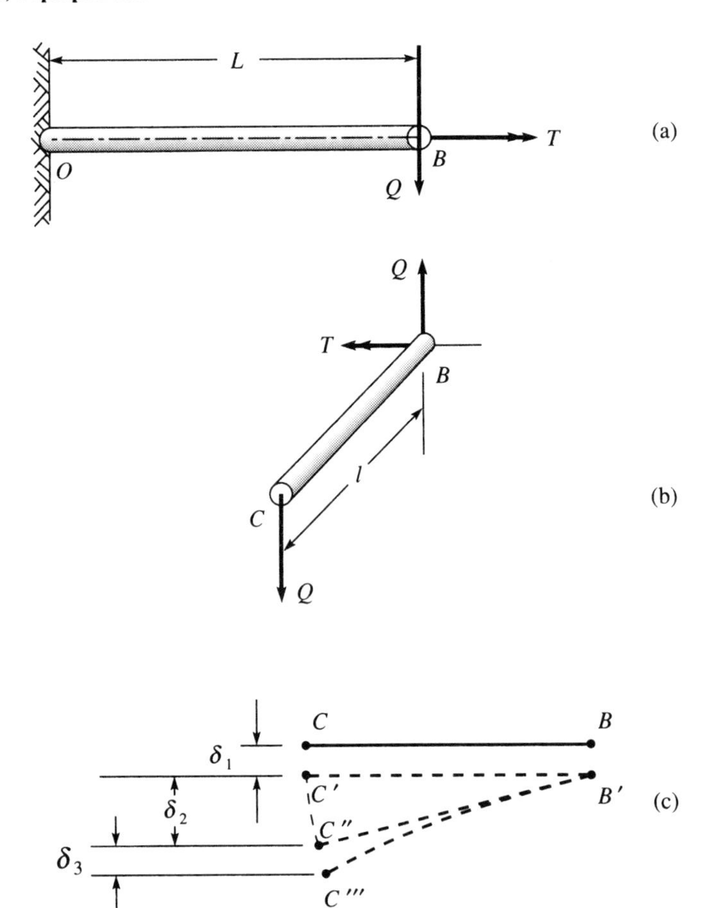

segment OB is a cantilever subjected to an end load $Q = 80$ lb *and* a twisting couple $T = 1600$ in.-lb. The former causes a deflection of the end B, according to the flexure theory of Chapter 6:

$$\delta_1 = \frac{QL^3}{3EI} = \frac{80 \times (20)^3}{3 \times 30 \times 10^6 \times 0.0370} = 0.192 \text{ in.}$$

The twisting couple T causes a rotation of the cross section B, according to the torsion theory of Chapter 5:

$$\phi = \frac{TL}{GJ} = \frac{1280 \times 20}{12 \times 10^6 \times 0.0741} = 0.0288$$

The segment BC is also a cantilever but subject *only* to the bending caused by the end load Q. Disregarding (temporarily) any movement of the end B, the deflection of end C relative to end B is again given by the flexure theory:

$$\delta_3 = \frac{QL^3}{3EI} = \frac{80 \times 16^3}{3 \times 30 \times 10^6 \times 0.0370} = 0.0984 \text{ in.}$$

Next, we add the displacements caused by the bending of segment OB, the twisting of segment OB, and the bending of segment BC. The bending of part OB alone causes B and C to displace the amount 0.192 in. *If* no other deformation occurred, the portion BC would translate in the $-y$-direction, carrying BC to $B'C'$, as shown in Fig. 7.15(c). However, the end B rotates the additional amount $\phi = 0.0288$, which would displace C' to position C''; that displacement is given by the first-order approximation ($\phi \ll 1$):

$$\delta_2 = \phi L = 0.0288 \times (16) = 0.461 \text{ in.}$$

Finally, the bending of the segment BC causes the additional deflection $\delta_3 = 0.0984$, so that the total deflection of end C is

$$\begin{aligned}\delta &= \delta_1 + \delta_2 + \delta_3 \\ &= 0.192 + 0.461 + 0.098 \\ &= 0.751 \text{ in.}\end{aligned}$$

The successive movements, corresponding to the three terms, are depicted in Fig. 7.15(c). ◆

PROBLEMS

Assume linearly elastic behavior unless otherwise stated.

7.35 An L-shaped bar is loaded perpendicular to its plane of symmetry (xy-plane), as shown in Fig. P7.35.

Fig. P7.35

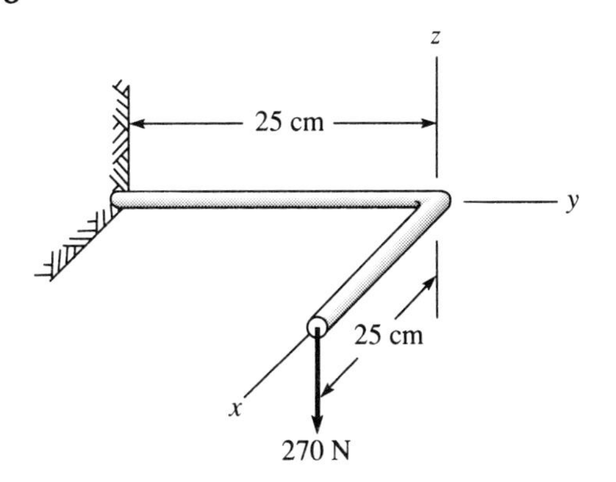

The bar is steel and has a circular cross section with a 2.5-cm diameter. What is the deflection of end A?

7.36 A plumber applies a wrench at location B of a pipe, which is clamped at A. The pipe is a standard 1-in. (nominal diameter) steel pipe; inside diameter = 1.049 in., outside diameter = 1.315 in., and $I = 0.087$ in^4. The plumber exerts the normal force of 120 lb on the handle and the normal force of 20 lb against the pipe, as shown in Fig. P7.36. Determine, in turn, (a) all components of stress at cross section A on the top and on the bottom ($y = \pm 0.658$, $x = z = 0$), (b) the deflection at B, and (c) the deflection at the plumber's hand, assuming that the wrench is rigid.

Fig. P7.36

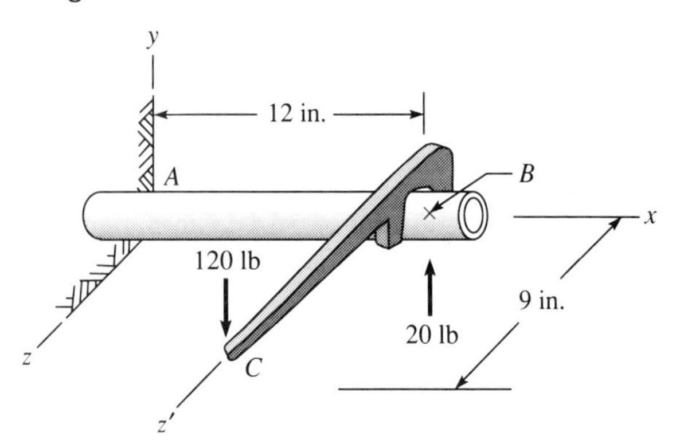

7.37 Two pieces of standard $\frac{3}{4}$-in. steel pipe are joined at a right angle, as shown in Fig. P7.37. Properties of this pipe follow:

$$\text{Outside diameter} = 1.050 \text{ in.}$$

$$\text{Inside diameter} = 0.824 \text{ in.}$$

$$I = 0.037 \text{ in.}^4; \qquad E = 30 \times 10^6 \text{ psi}; \qquad \nu = 0.25$$

The pipe is clamped at section O and the normal force $P = 20$ lb acts at B. Calculate (a) all components of stress at the top and bottom of section O (at $y = \pm 0.525$ in., $x = z = 0$), (b) the deflection of end C, and (c) the deflection of end B.

Fig. P7.37

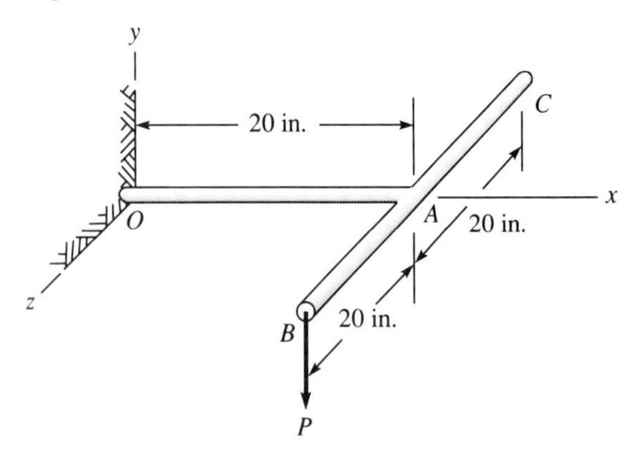

7.38 The member shown in Fig. P7.38 is made of standard $\frac{3}{4}$-in. steel pipe joined by a right elbow at B. Forces P and Q act in the directions of the $\mp y$ axis. Properties of the pipe follow:

$$\text{Outside diameter} = 1.050 \text{ in.}$$

$$\text{Inside diameter} = 0.824 \text{ in.}$$

$$J = 0.0741 \text{ in.}^4$$

If $P = Q = 60$ lb, calculate (a) the maximum shear stress in portion OB, (b) the maximum shear stress in portion BC, and (c) the deflection of end C.

Fig. P7.38

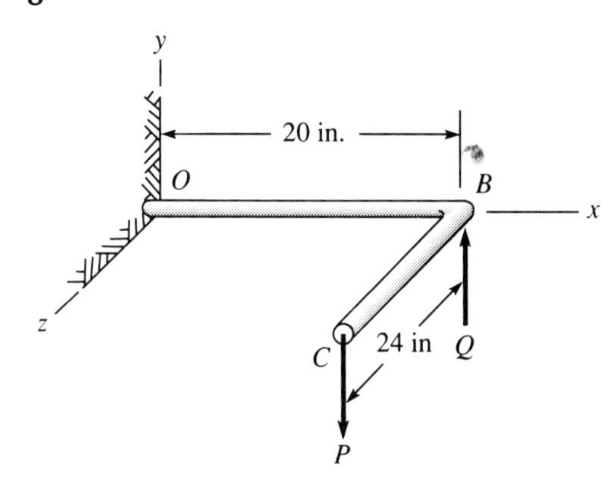

7.39 Repeat Prob. 7.38 if $P = 60$ lb and $Q = 0$.

7.40 All segments of the T-shaped member in Fig. P7.40 are made of the same solid steel rod:

$$\text{Outside diameter} = 2.50 \text{ cm}$$

$$E = 210 \times 10^3 \text{ MPa}, \qquad \nu = 0.25$$

Segment ED is welded to segment BDC at a right angle. The latter is clamped at B and C, stress free, when

Fig. P7.40

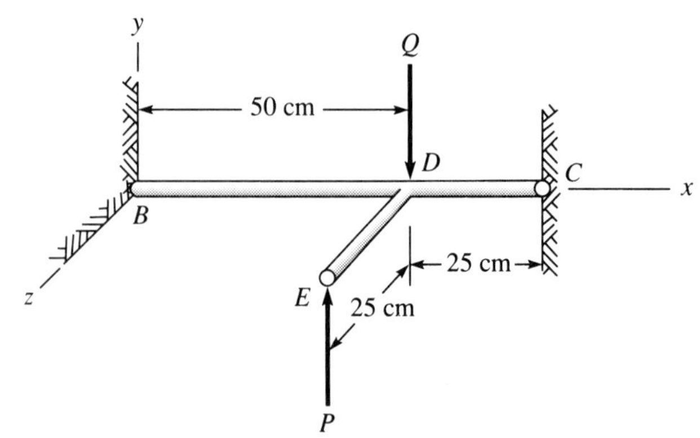

the normal forces P and Q are applied in the $\pm y$ directions. If $P = Q = 400$ N, determine, in order, (a) the reactions at B and C, (b) the displacement of end E, and (c) the maximum shear stress in segment DE.

7.41 Repeat Prob. 7.40 if $P = 400$ N and $Q = 0$.

7.42 The T-shaped member of Prob. 7.40 is depicted again in Fig. P7.42, but loads P and Q are removed. Instead the member is subjected to a couple $M = 10$ kN cm, which acts on end E in the direction $-x$. Determine (a) the actions on the ends B and C, (b) the deflection of end E, and (c) the maximum shear stress in segment DE.

Fig. P7.42

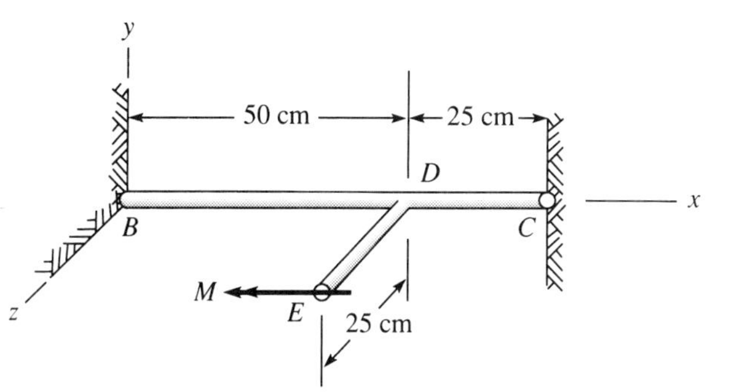

7.43 The L-shaped member of Prob. 7.38 is shown again in Fig. P7.43. Here the member is subjected to the couples ($M = 600$ in.-lb), which act at C and B in the directions $\pm x$, as shown. Calculate (a) the maximum shear stress in segment OB, (b) the maximum shear stress in segment BC, and (c) the deflection of end C.

Fig. P7.43

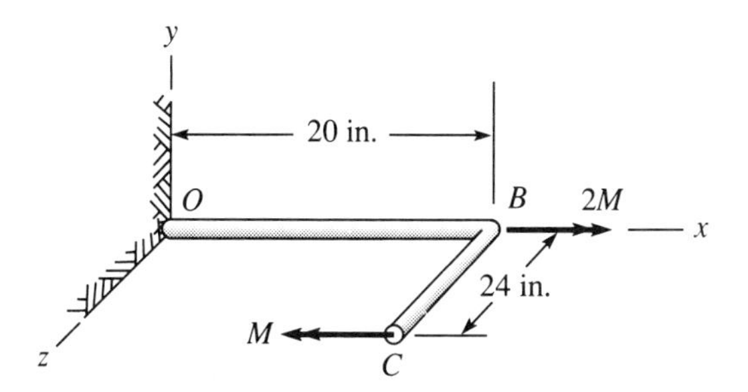

7.44 Repeat Prob. 7.43 if the couple is removed at B, that is, the only load is the couple M at C.

7.45 Replace the couple 2M at B of Prob. 7.43 by a couple kM and choose K such that end C is not displaced when the couples are applied.

7.46 Repeat Prob. 7.38 if $P = 60$ lb $= -Q$; both act in the direction of P.

7.47 The T-shaped member of Fig. P7.47 is made by welding two identical steel rods of circular cross section. The rods form a right angle at the weld, the midpoint B of rod CBD. The forces P and Q act at the ends C and D, respectively, in the $\mp y$ direction. Determine the ratio P/Q such that end D does not displace when the forces are applied ($E = 2.5\ G$).

Fig. P7.47

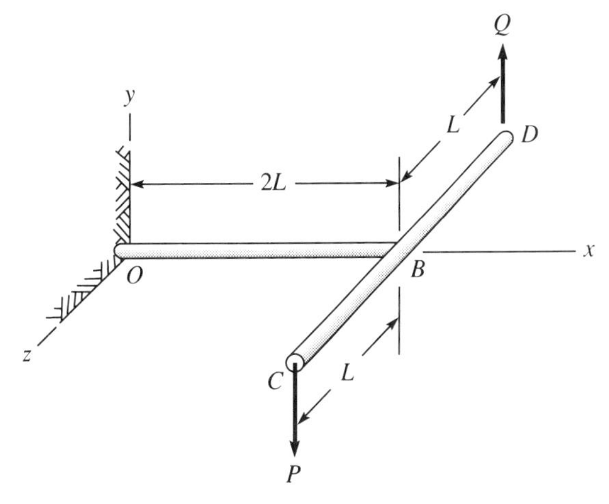

7.48 The L-shaped bracket of Fig. P7.48 is formed from a steel rod of circular cross section; ABC is a right angle.

$$I = 40\ \text{cm}^4, \qquad J = 80\ \text{cm}^4$$

$$E = 200 \times 10^3\ \text{MPa}, \qquad \nu = 0.25$$

The bracket is clamped at A and simply supported at C. It is unstressed when the transverse force $P = 4$ kN is applied, normal to the xy plane. Compute all reactions at the ends A and C.

Fig. P7.48

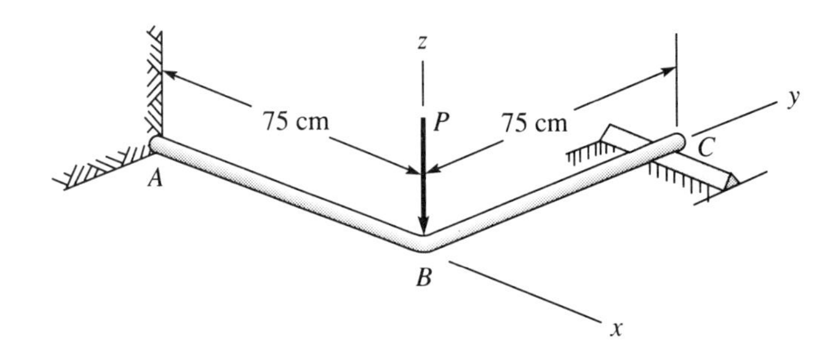

7.49 In Fig. P7.49 the member OBC is made from a solid steel rod with a circular cross section; diameter = 2.5 cm. The rod forms a right angle at B in the xy plane. A simple support is located 0.450 cm below the member at B. Determine the reaction of the simple support *and* the deflection at C when the load P = 900 N ($E = 2.5\,G = 200 \times 10^3$ MPa).

Fig. P7.49

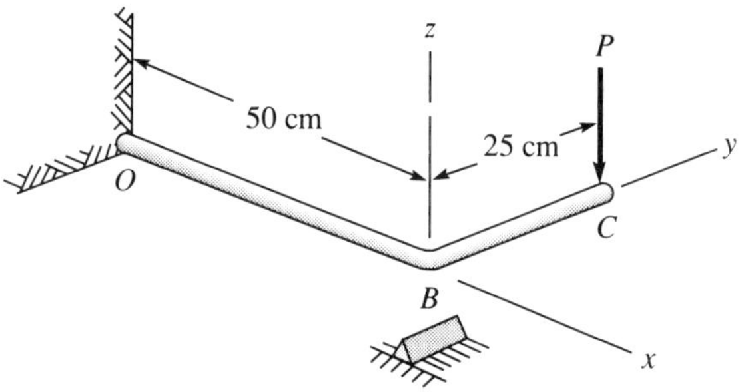

z
P
50 cm
25 cm
y
O
C
B
x

7.6 Thermal Stresses

Previously (Sec. 4.7) *small* strains from changes of temperature were superposed upon other *small* strains of hookean solids. Those linear strain-stress-temperature relations were subsequently employed (Secs. 5.5 and 5.15) to illustrate procedures of analysis. We now consider some additional problems in which the stresses and strains originate with changes of temperature.

EXAMPLE 9

A steel pipe carries hot water. The pipe passes through two fixed walls 20 ft apart, as shown in Fig. 7.16(a). The pipe is rigidly fastened to the walls at A and B. When the pipe was installed (stress free) the temperature was 60°F. Because of poor insulation the temperature of the water entering at A drops from 210° at A to 160° at B in a linear manner. Let us determine the normal stress on a cross section of the pipe. The temperature at x (x is measured in inches) is

$$Q(x) = 160 + 50\left(1 - \frac{x}{240}\right)$$

Fig. 7.16

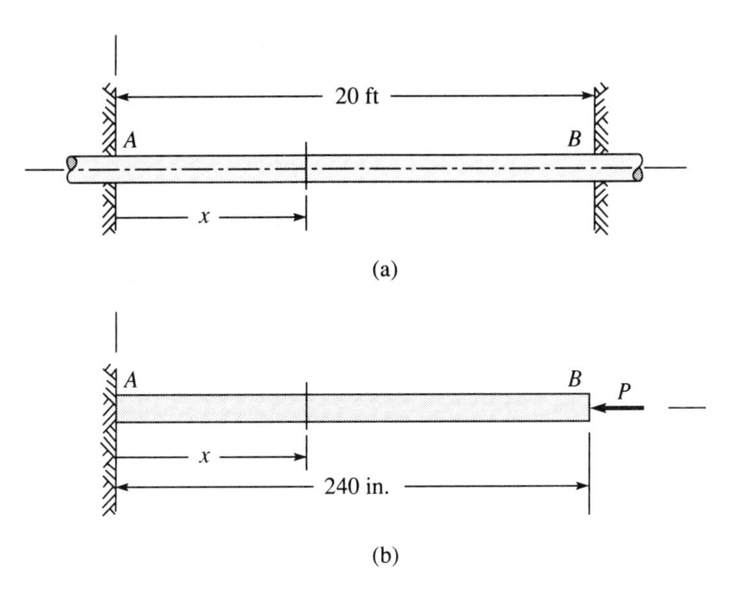

Then, the temperature rise is

$$\Delta Q(x) = 50\left(3 - \frac{x}{240}\right)$$

Since the pipe tends to expand, the walls will exert an axial force P, as shown in Fig. 7.16(b). This must be accompanied by a stress σ_x. By the assumptions previously used in axial extension, the stress, σ_x, is constant on a cross section. Again, we assume that $\sigma_y = \sigma_z = 0$. Then, the stress-strain-temperature relation (see Sec. 4.7) yields

$$\varepsilon_x = \frac{\sigma_x}{E} + \alpha\,\Delta Q$$

For this steel the modulus E and coefficient α are

$$E = 30 \times 10^6 \text{ lb/in.}^2, \qquad \alpha = 6.5 \times 10^{-6}/^\circ\text{F}$$

Then

$$\varepsilon_x = \frac{1}{3} \times 10^{-7}\sigma_x + 3.25 \times 10^{-4}\left(3 - \frac{x}{240}\right)$$

The length of pipe between A and B cannot change; therefore,

$$e = \int_0^{240} \varepsilon_x\,dx = 0$$

or

$$\int_0^{240} 3.25 \times 10^{-4}\left(3 - \frac{x}{240}\right)dx + 8 \times 10^{-6}\sigma_x = 0.195 + 8 \times 10^{-6}\sigma_x = 0$$

Hence, the normal stress in the pipe is

$$\sigma_x = -24{,}400 \text{ lb/in.}^2$$

Note that σ_x does not depend on the cross-sectional area of the pipe. ◆

EXAMPLE 10

A wire is attached to a cantilevered beam, as shown in Fig. 7.17(a). At a temperature Q_1 the wire is of length ℓ, and the cantilevered beam is straight. If the temperature drops to a value Q_0, the wire AB tends to shorten. Let us determine the vertical displacement of the right end of the beam when the temperature changes. Consider the wire and beam separately, as in Fig. 7.17(b) and (c). The force P acting on the beam is applied by the wire when it tends to contract. The beam tends to stretch the wire with an axial force P, as indicated in Fig. 7.17(c). Although P is unknown, the continuity of the original structure requires that the vertical displacement of the beam at B be equal to the amount of contraction of the wire. Since the beam of Fig. 7.17(b) is a cantilever carrying a load at is end, the deflection at this point is

$$w(L) = \frac{PL^3}{3EI}$$

Fig. 7.17

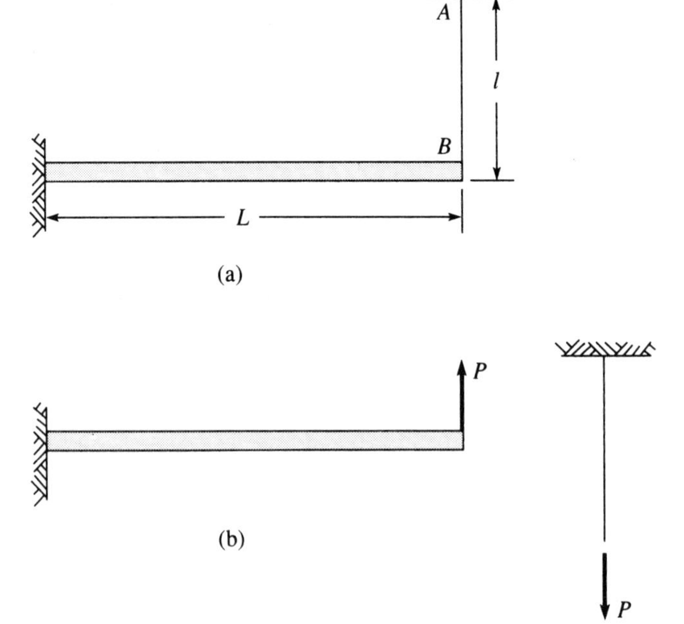

The contraction, $-e$, of the wire consists of two parts: that due to the temperature drop and that due to the axial load P. Thus

$$-e = -\frac{P\ell}{A_1 E_1} - \alpha(Q_0 - Q_1)\ell$$

(α is the coefficient of thermal expansion for the wire, E_1 is its modulus of elasticity, and A_1 is the cross-sectional area of the wire.) Since $w(L) = -e$, we find that

$$\frac{PL^3}{3EI} = -\frac{P\ell}{A_1 E_1} - \alpha(Q_0 - Q_1)\ell$$

or

$$P = \frac{\alpha(Q_1 - Q_0)\ell}{\dfrac{L^3}{3EI} + \dfrac{\ell}{A_1 E_1}}$$

Therefore, the vertical displacement of the tip of the beam is

$$w(L) = \frac{\alpha(Q_1 - Q_0)\ell}{1 + \dfrac{3EI\ell}{A_1 E_1 L^3}} \quad ◆$$

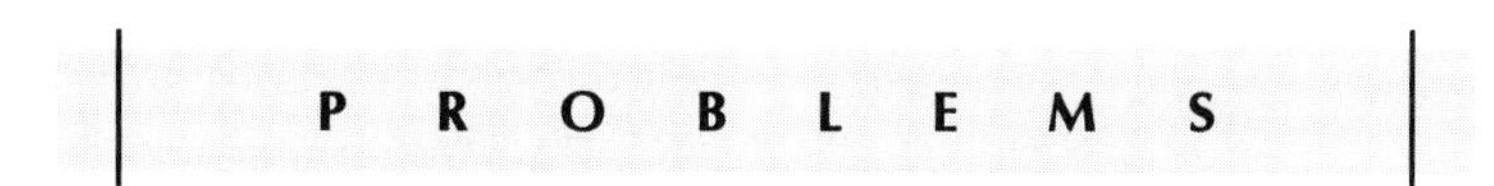

PROBLEMS

Assume linearly elastic behavior unless otherwise noted.

7.50 The rod in Fig. P7.50 is composed of aluminum and steel and is unstressed when fastened to the walls at 21°C. Compute the normal stress in each member when the temperature is 132°C.

Fig. P7.50

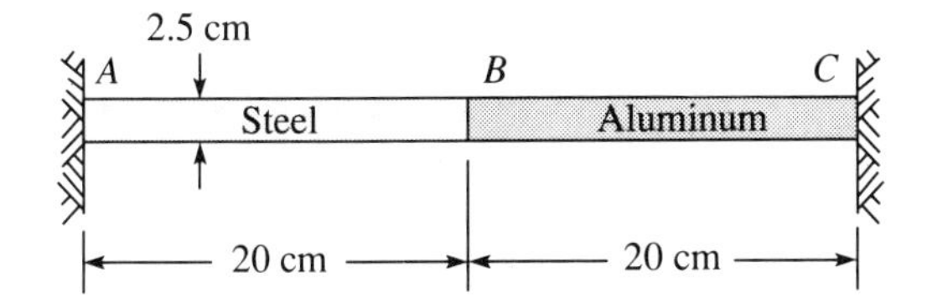

7.51 If each material in Fig. P7.50 is ideally plastic, with $\sigma_0 = 200$ MPa for the steel and $\sigma_0 = 370$ MPa for the aluminum, determine the displacement of the cross section at B as a function of the temperature change.

7.52 The composite rod of Fig. P7.50 is unstressed at the uniform temperature 60°F. It is then heated nonuniformly so that the temperature varies linearly from 60°F at A to 260°F at B. What is the stress in the rod?

7.53 In Fig. P7.50 the connection between the two materials fails in tension at a stress of 10,000 psi. If the connection is unstressed at 70°F, at what temperature will the connection fail? Assume uniform temperature throughout.

7.54 The steel bolt B in Fig. P7.54 is passed through a cast-iron tube T. At 80°F the nut is just tight enough so there is contact but no stress. Find the normal stresses in the steel and cast iron when the temperature is 255°F.

Fig. P7.54

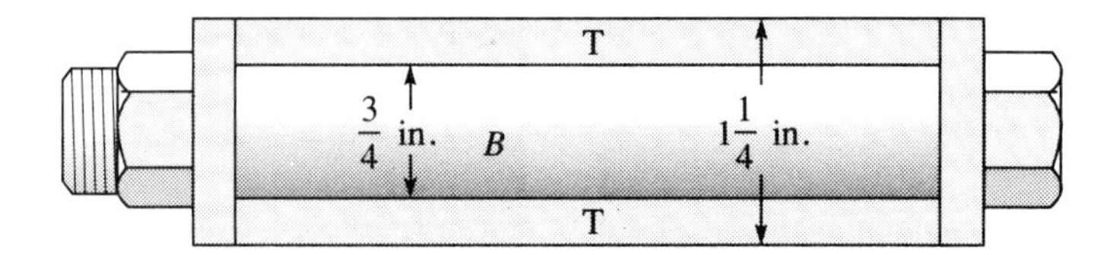

7.55 If the bolt in Fig. P7.54 has 16 threads per inch, calculate the stresses in the tube and bolt when the nut is tightened 1/10 turn and then the temperature is raised 100°F.

7.56 A thin-walled tube of mild steel just fits inside a rigid cylinder at temperature Q_0 degrees Fahrenheit, as shown in Fig. P7.56. The cylinder is filled with a fluid at temperature Q_1 and pressure p_0. If the von Mises yield criterion is used ($\sigma_0 = 200$ MPa), what relation between p_0 and $(Q_1 - Q_0)$ exists when failure occurs in the cylindrical wall? (Assume Q_1 is also the final temperature of the cylinder. Also, neglect friction between the tube and the enclosure.)

Fig. P7.56

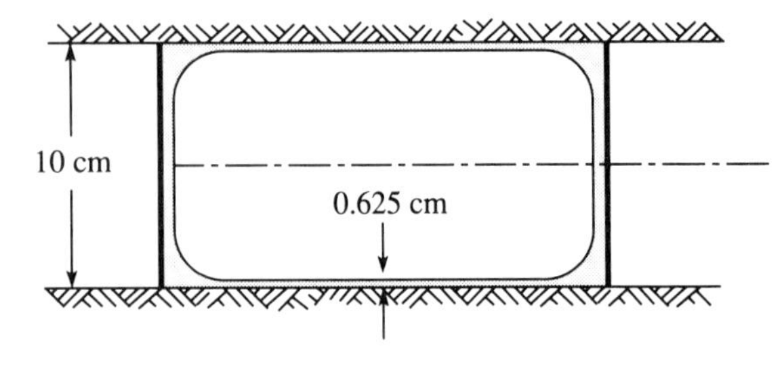

7.57 A cast-iron pipe 4 in. in diameter, $\frac{1}{2}$ in. in wall thickness, and 10 ft. long is rigidly restrained at each end. If cast iron fails in tension at a stress of 20,000 lb/in.^{2}, at what temperature does the pipe fail? Assume the pipe to be unstressed at 100°F.

7.58 A circular tube is composed of an inner layer of aluminum and an outer layer of steel, as illustrated in Fig. P7.58. Initially, the tube is stress-free. Compute the circumferential stress in each material when the tube is heated 200°F.

Fig. P7.58

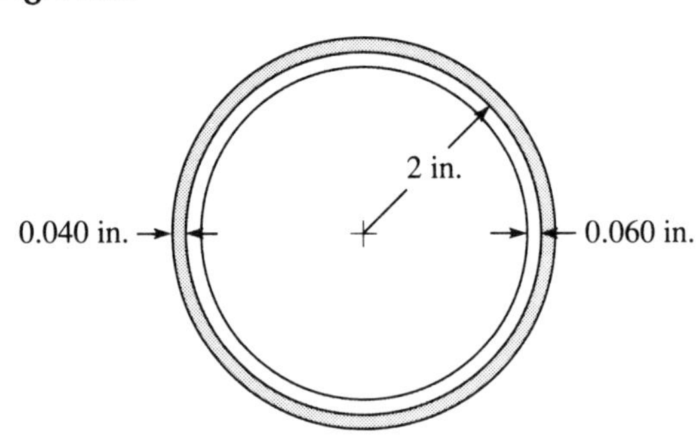

7.59 Steel, brass, and aluminum bars support the rigid platform AC, as shown in Fig. P7.59. Each bar has a cross-sectional area of 6 cm^2. The platform is horizontal, and the bars are unstressed at 21°C. If the temperature is increased to 65°C, where must the 44,000-N load act to keep the platform horizontal?

Fig. P7.59

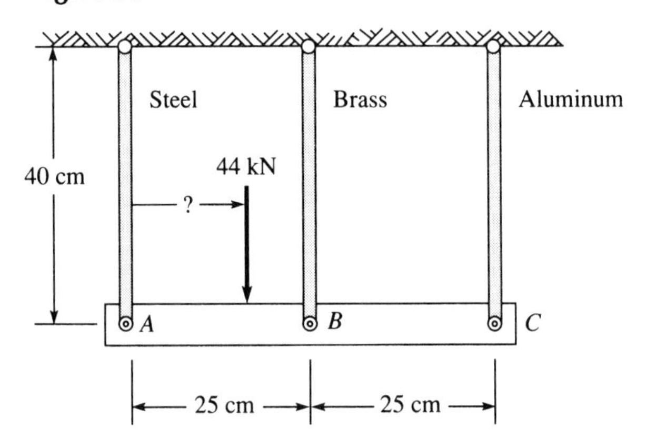

7.60 In Fig. P7.60, AB and CD have equal cross-sectional areas and have the same length at a certain temperature. Determine the ratio x/a such that the rigid bar BC remains horizontal at any temperature.

Fig. P7.60

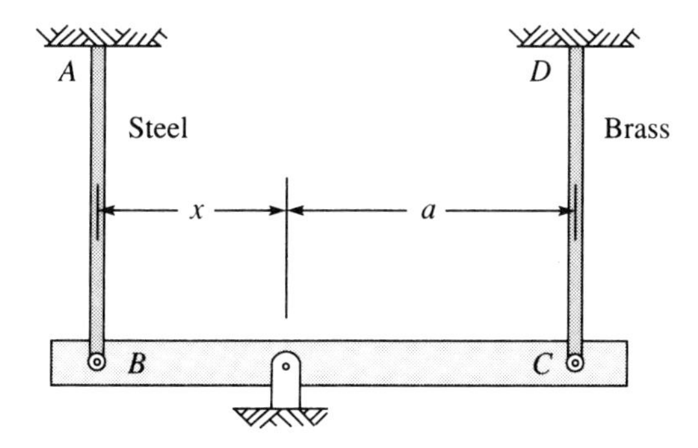

7.61 A cylindrical rod just fits between two rigid walls at temperature Q_0. If E (the modulus of elasticity) is a function of temperature such that

$$E = E_0 - k(Q - Q_0), \qquad Q_0 < Q < \frac{E_0 + kQ_0}{k}$$

and if the bar is uniformly heated, determine the temperature at which the stress in the bar is a maximum.

7.62 A steel cylinder (inside diameter 4 in., wall thickness $\frac{1}{16}$ in.) just fits over an aluminum cylinder of outside diameter 4 in. and wall thickness $\frac{1}{16}$ in. at 32°F. What is the stress in the steel cylinder when the temperature is raised to 150°F?

7.63 A strip of steel, 0.32 cm thick, is plated on one side with aluminum, 0.16 cm thick. If the strip is uniformly heated to 28°C, find the change in curvature of the strip.

7.64 A titanium tube, $\frac{1}{8}$ in. thick, slips over an aluminum tube, $\frac{1}{8}$ in. thick and 3 in. in outside diameter, with 0.0010 in. clearance. Calculate the stress in each tube when (a) the inner tube is heated to 200°F and (b) both tubes are heated to 200°F.

7.65 The top flange of the steel I-beam in Fig. P7.65 is heated so that its temperature is 200°F greater than that of the lower flange. Determine the maximum stress, assuming that the web offers no resistance to bending.

Fig. P7.65

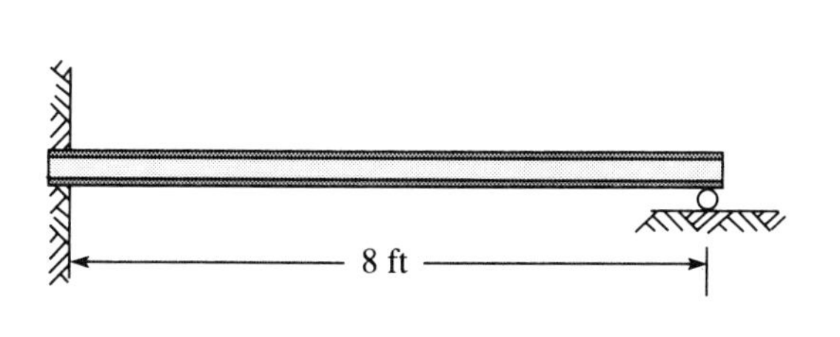

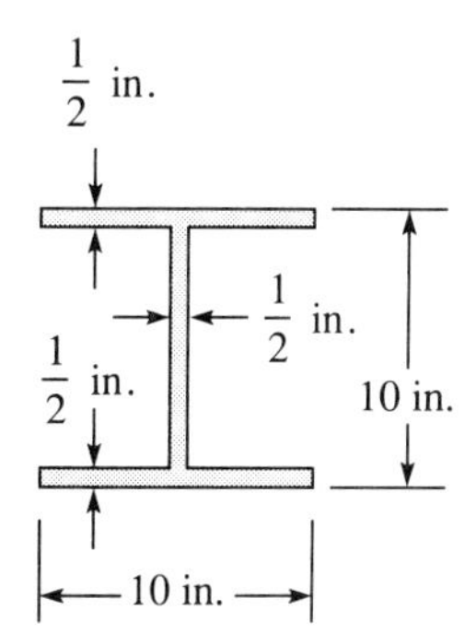

7.66 The members of the truss in Fig. P7.66 are steel and unstressed at 20°C. Calculate the normal stress in each member if the temperature of bar OB *only* is increased to 130°C. Each member has the same cross section.

Fig. P7.66

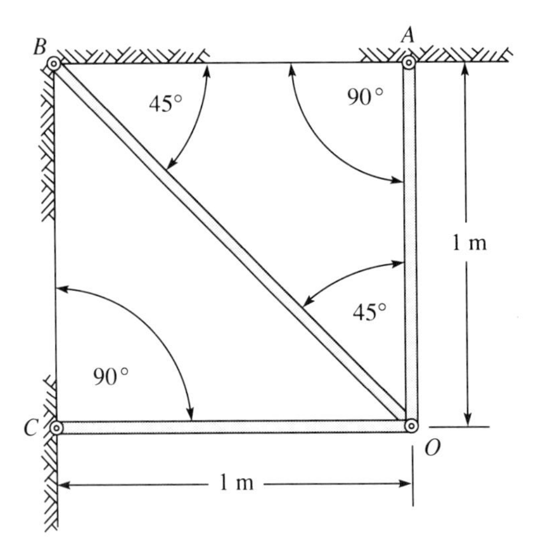

7.7 Bending of Nonsymmetric Beams

In some circumstances the engineer employs members that do not have symmetric cross sections. Often these are required to support loads that cause bending. Then the axis is not generally deformed to a plane curve but to a three-dimensional spatial curve. Still, the curvature and strains are small enough to admit superposition.

Suppose that an axial line Ox lies initially along the axis of Fig. 7.18. Under loads that line is deformed to the curve $O'x'$; it is stretched and bent slightly. Projections

Fig. 7.18

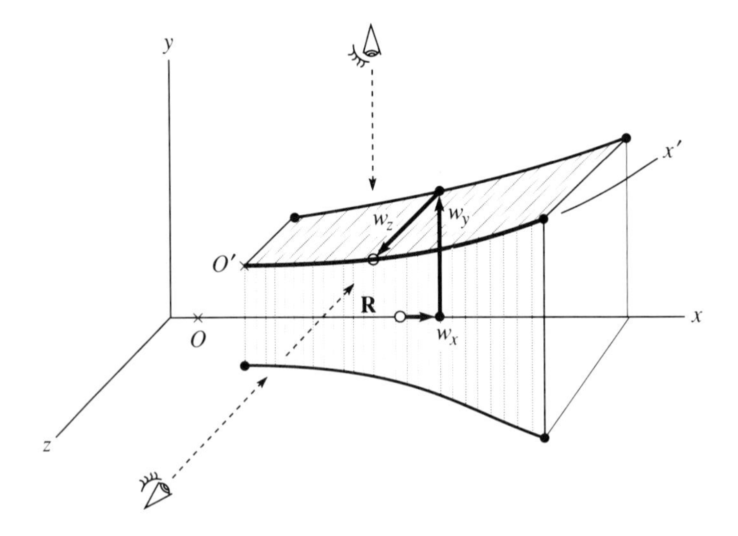

upon the xy and xz planes are plane curves with curvatures κ_y and κ_z, respectively. The reader can imagine viewing these curves, as suggested in Fig. 7.18. The small strain of the beam is then obtained by superposition. Assuming (see Sec. 5.15) that plane cross sections remain plane, the extensional strain ε_x is obtained by superposing the strains associated with the extension (ε_0) and the flexures in the xy-plane (κ_y) and in the xz-plane (κ_z). In accordance with Eq. (5.158),

$$\varepsilon_x = \varepsilon_0 - \kappa_y y - \kappa_z z \tag{7.3}$$

The relations between the curvatures and the small rotations and between the rotations and the small deflections are similar to the earlier results [see Sec. 6.5, Eqs. (6.26) and (6.34)]. We then have the curvatures κ_y and κ_z and rotations θ_y and θ_z, as shown in Fig. 7.19.

$$\frac{d\theta_y}{dx} \doteq \kappa_y, \qquad \frac{d\theta_z}{dx} \doteq \kappa_z \tag{7.4a, b}$$

$$\frac{dw_y}{dx} \doteq \theta_y, \qquad \frac{dw_z}{dx} \doteq \theta_z \tag{7.5a, b}$$

These linear equations relate the kinematic variables.

Fig. 7.19

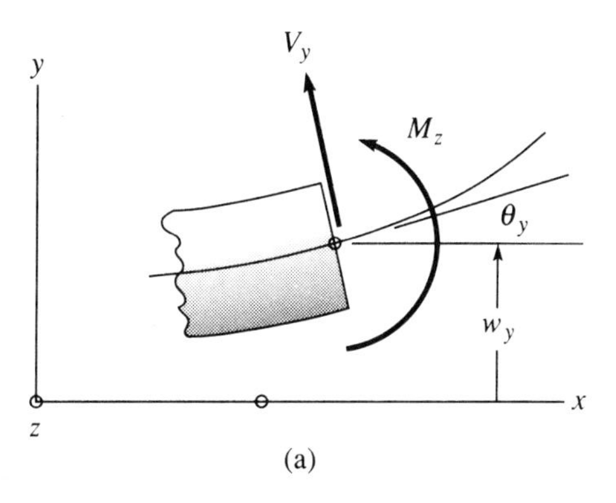

(a)

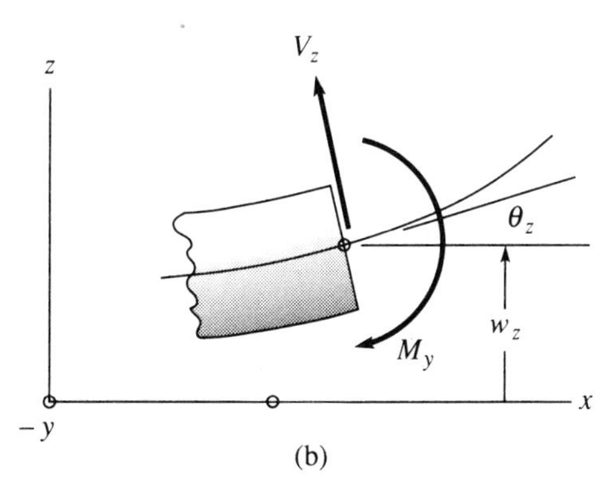

(b)

Let us now turn to the equations of equilibrium. The transverse shear forces, V_y and V_z, and the bending couples, M_y and M_z, are shown in Fig. 7.19. Now, as before, let us presume small rotations, so that a slice of the beam appears as shown in Fig. 7.20. Components of the internal forces and couples are shown only on the positive section (viewed from the positive x-axis). The notations and sign conventions are those indicated and are consistent with the previous development (see Fig. 6.8). The complete free-body diagram of a slice and the equilibrium conditions provide four differential equations, two like Eq. (6.8) and two like Eq. (6.9):

$$\frac{dV_y}{dx} = p_y(x), \qquad \frac{dV_z}{dx} = p_z(x) \tag{7.6a, b}$$

$$\frac{dM_z}{dx} = -V_y(x), \qquad \frac{dM_y}{dx} = +V_z(x) \tag{7.7a, b}$$

Fig. 7.20

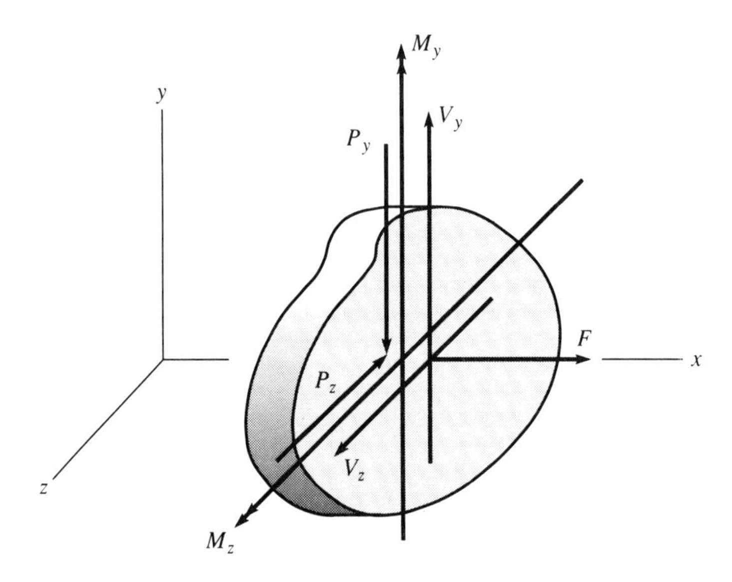

Note that the difference in signs on the right sides of Eqs. (7.7a) and (7.7b) is a consequence of the convention; M_y has the direction of the positive y-axis but stretches fibers on the positive side ($z > 0$). These equations of equilibrium are linear and uncoupled; the variables V_y and M_z are related only to the load p_y, whereas V_z and M_y are related only to p_z. We must now turn to the relations between the kinematic variables ($\varepsilon_0, \kappa_z, \kappa_y$) and the internal actions (F, M_z, M_y), which depend upon the behavior of the material.

Let us assume the simplest behavior, a HI-HO material. As before, we assume that the transverse normal stresses, σ_y and σ_z, are negligible. Then

$$\sigma_x = E\varepsilon_x \tag{7.8a}$$

Substituting Eq. (7.3) into the right side of Eq. (7.8a), we obtain the distribution

$$\sigma_x = E(\varepsilon_0 - \kappa_y y - \kappa_z z) \tag{7.8b}$$

The net force F and the couples M_y and M_z are the integrals

$$F = \iint_A \sigma_x \, dA \tag{7.9}$$

$$M_z = -\iint_A \sigma_x y \, dA \tag{7.10a}$$

$$M_y = +\iint_A \sigma_x z \, dA \tag{7.10b}$$

The first two equations are the same as Eqs. (5.111) and (5.112); only the subscript has been added to the couple M_z. The difference in the signs on the right sides of the Eq. (7.10a) and (7.10b) is again a consequence of the convention. When the distribution σ_x of Eq. (7.8b) is substituted into Eqs. (7.9) and (7.10a, b), these results follow:

$$F = (EA)\varepsilon_0 - E\kappa_y \iint_A y \, dA - E\kappa_z \iint_A z \, dA \tag{7.11}$$

$$M_z = -E\varepsilon_0 \iint_A y \, dA + E\kappa_y \iint_A y^2 \, dA + E\kappa_z \iint_A yz \, dA \tag{7.12a}$$

$$M_y = E\varepsilon_0 \iint_A z \, dA - E\kappa_y \iint_A yz \, dA - E\kappa_z \iint_A z^2 \, dA \tag{7.12b}$$

These equations contain five unnamed integrals, which depend on the size and shape of the cross section *and* on the choice of the coordinates (y, z). Let us give them names:

$$\iint_A y \, dA \equiv B_y, \qquad \iint_A z \, da \equiv B_z \tag{7.13a, b}$$

$$\iint_A y^2 \, dA \equiv I_{yy}, \qquad \iint_A yz \, dA \equiv I_{yz}, \qquad \iint_A z^2 \, dA \equiv I_{zz} \tag{7.13c, d, e}$$

With these notations Eqs. (7.11) and (7.12a, b) assume the forms

$$F = (EA)\varepsilon_0 - (EB_y)\kappa_y - (EB_z)\kappa_z \tag{7.14}$$

$$M_z = -(EB_y)\varepsilon_0 + (EI_{yy})\kappa_y + (EI_{yz})\kappa_z \tag{7.15a}$$

$$M_y = (EB_z)\varepsilon_0 - (EI_{yz})\kappa_z - (EI_{zz})\kappa_z \tag{7.15b}$$

These are three linear equations relating the three dynamic variables (F, M_z, and M_y) to the three kinematic variables (ε_0, κ_y, and κ_z). In the absence of axial force ($F = 0$ if only transverse loads are present), the strain ε_0 can be eliminated by means of Eq. (7.14). Then a system of ten linear equations [(7.4a), (7.4b), (7.5a), (7.5b), (7.6a), (7.6b), (7.7a), (7.7b), (7.15a), and (7.15b)] relate ten dependent variables ($V_y, V_z, M_z, M_y, \kappa_y, \kappa_z, \theta_y, \theta_z, w_y$, and w_z). This is little more than an elaboration of the five equations for the symmetric hookean beam, which govern five dependent variables [$V = V_y$, $M = M_z$, $\kappa = \kappa_y$, $\theta = \theta_y$, and $w = w_y$], as presented in Chapter 6. Indeed, except for the coupling between the two equations (7.15a) and (7.15b), the present case would be just two separate problems; each would entail only five variables, uncoupled from one another.

We recall that the integrals B_y and B_z vanish if the origin is located at the centroid. Then the extension (ε_0) is decoupled from the flexure; the variable ε_0 does not appear in Eqs. (7.15a) and (7.15b). These two equations remain coupled; κ_y and κ_z appear in both. However, it is always possible to orient the axes y and z such that $I_{yz} = 0$. The axes so oriented are called the principal axes of the cross section. Proof of the existence and means for locating principal axes are given in Appendix B. In many practical problems a cross section does have one line of symmetry; then, that line and the perpendicular line are the principal lines. The proof is as follows: If the y line is a line of symmetry, as in Fig. 7.21, then each elemental area, such as A at $z = \ell$, has a counterpart, such as A' at $z = -\ell$; their contributions to the integral I_{yz} cancel.

Fig. 7.21

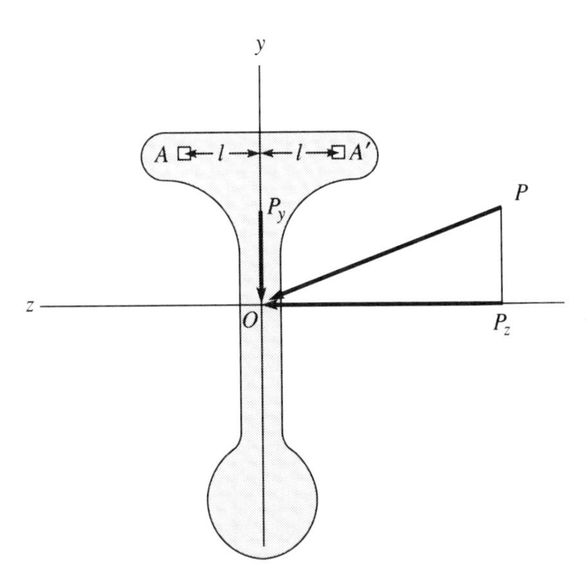

If the y- and z-axes are principal axes, then the equations are fully uncoupled. For convenience, let us reiterate them. Firstly, we have the uncoupled equations of equilibrium (7.6a, b) and (7.7a, b):

$$\frac{dV_y}{dx} = p_y(x), \qquad \frac{dV_z}{dx} = p_z(x) \tag{7.16a, b}$$

$$\frac{dM_z}{dx} = -V_y(x), \qquad \frac{dM_y}{dx} = +V_z(x) \tag{7.17a, b}$$

Secondly, we have the uncoupled version of the equations (7.15a, b); κ_y and κ_z can be eliminated by means of the kinematic equations (7.4a, b):

$$M_z = EI_{yy}\frac{d\theta_y}{dx}, \qquad M_y = -EI_{zz}\frac{d\theta_z}{dx} \tag{7.18a, b}$$

Finally, we have the kinematic equations (7.5a, b):

$$\frac{dw_y}{dx} = \theta_y, \qquad \frac{dw_z}{dx} = \theta_z \tag{7.19a, b}$$

We see that four first-order differential equations (designated by *a*) govern four variables $(V_y, M_z, \theta_y, w_y)$ and four (designated by *b*) govern the other four $(V_z, M_y, \theta_z, w_z)$. Each system is just like the system of Chapter 6 ([Eqs. (6.31), (6.32), (6.33), and (6.34)]. Each can be solved independently of the other.

In addition to the equations governing flexure, we also obtain, by Eqs. (7.9) and (7.8b),

$$F = (EA)\varepsilon_0 \tag{7.20}$$

where ε_0 is the extensional strain of the centroidal axis. When Eqs. (7.18a, b) and (7.20) are substituted into (7.8b), we obtain

$$\sigma_x = \frac{F}{A} - \frac{M_z y}{I_{yy}} + \frac{M_y z}{I_{zz}} \tag{7.21}$$

Here the origin is at the centroid and the coordinates (y, z) are in principal directions. Equation (7.21) is again an example of superposition; the three terms on the right represent the contributions from extension and bending in each of the principal planes.

Limitations of the foregoing theory must be noted. Equations (7.8a, b) and (7.18a, b) apply only to HI-HO materials. The linear equations of equilibrium, (7.16a, b) and (7.17a, b), neglect the rotation and curvature of the element. Also, the linear kinematic equations (7.19a, b) are valid only for small rotations.

One further note of caution is necessary: Any transverse load, whether distributed $p(x)$ or concentration P, can be decomposed into components in the principal directions so that bending in these directions is uncoupled; then superposition of the bending effects is possible. However, if the principal lines are not also lines of symmetry, the loads could cause twisting. For example, the transverse load P in Fig. 7.21 can be decomposed into components P_y and P_z. The component P_y causes bending in the xy-plane (LPS), *but* the xz-plane is not a longitudinal plane of symmetry. Without additional theory and analysis, we do not know whether the loading P_z causes twisting about the x-axis. There exists a point O such that loads acting through the point cause no twist. That point is termed the *shear center*; in general, the point is *not* the centroid. A limited consideration of the question is contained in the following section.

PROBLEMS

7.67 A standard steel angle is employed as a cantilever beam with a force P applied at the unsupported end. Assume that the wall is thin compared to the length of the equal legs; $t \ll h$ in Fig. P7.67. Determine the ratio P_y/P_z if the end deflects in the direction AO of the one leg.

Fig. P7.67

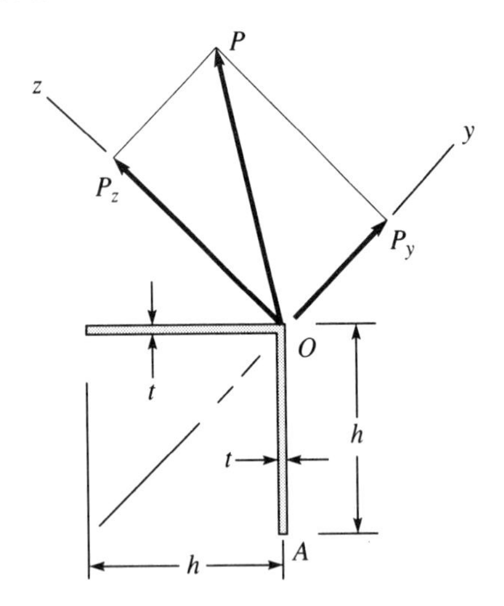

7.68 A wide-flange steel I-beam is used as a cantilever. It is 10 ft long. A concentrated force $P = 1000$ lb acts at the unsupported end, as shown in Fig. P7.68. With respect to centroidal axes

$$I_{yy} = 100 \text{ in.}^4, \qquad I_{zz} = 36 \text{ in.}^4$$

Calculate the maximum tensile stress, the maximum shear stress, and the deflection of the unsupported end.

Fig. P7.68

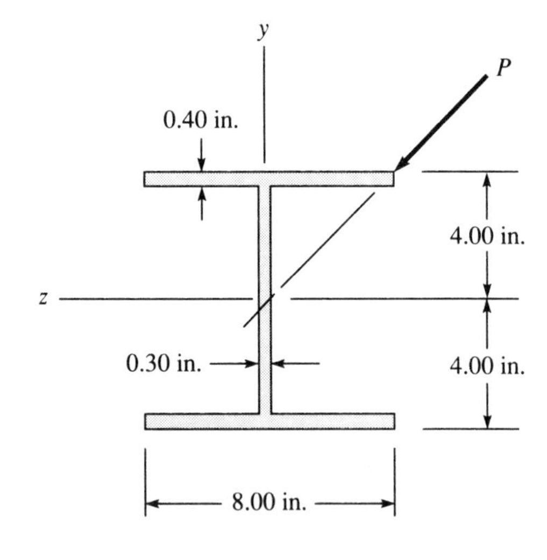

7.69 A beam has the thin-wall semicircular cross section depicted in Fig. P7.69. It is subjected to the bending couple of magnitude 180 Nm. Compute the maximum normal stress.

Fig. P7.69

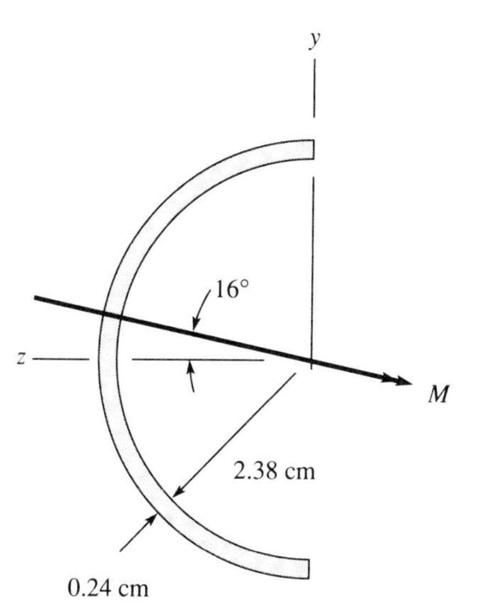

Fig. P7.70

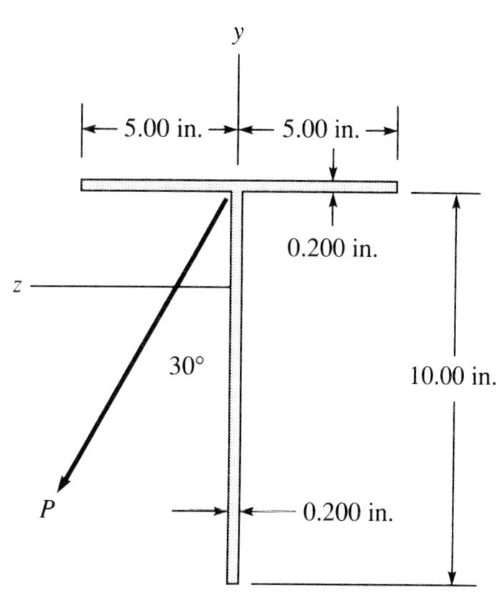

7.70 A cantilever beam is 12 ft long. It has the T-section shown and is made of structural steel that exhibits yield stresses of 40 ksi and 48 ksi in tension and compression, respectively. The transverse load P acts at the unsupported end, as shown in Fig. P7.70. Determine the load P_0 that would initiate yielding.

7.71 The beam shown in Fig. P7.71 is made of a steel angle with equal legs oriented in the y- and z-directions. It is fixed at one end ($x = 0$) and supported at the other ($x = L$). The support provides simple support in the y-direction but none in the z-direction (the end slips freely in the z-direction). The beam is subjected to the uniform load $\bar{p}_y$, as depicted. Obtain the equations for the components of the deflection and compare them at the midpoint; that is, obtain $w_y(L/2)/w_z(L/2)$.

Fig. P7.71

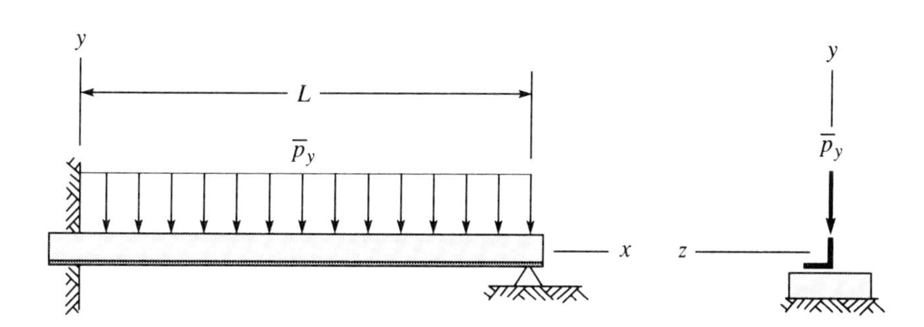

7.72 A simple cantilever beam is made of aluminum alloy with the following properties:

$$E = 80 \times 10^3 \text{ MPa}, \qquad \nu = 0.25$$

The beam is 250 cm long and is subjected to a concentrated load applied at the free end, as illustrated on the cross section of Fig. P7.72. Compute the maximum normal stress and the deflection of the unsupported end. Dimensions are given to the midlines of the thin web and flanges. Thicknesses of web and flange are 0.500 cm.

Fig. P7.72

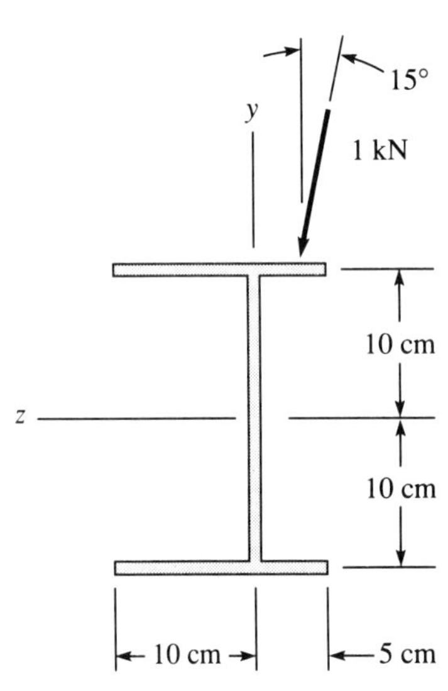

***7.73** A steel angle is used as a beam to support loading in the xy-plane. Determine the angle θ in Fig. P7.73 such that the beam does not deflect in the z-direction. Note that $t \ll a$.

Fig. P7.73

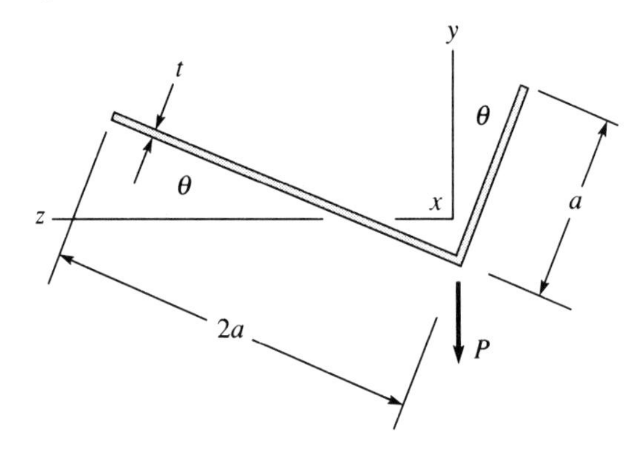

***7.74** A beam is made of two *thin* plates welded to form the cross section depicted in Fig. P7.74. This beam acts as a cantilever with the couple $M = M_z$ ($M_y = 0$) acting upon the unsupported end. Determine the direction of the consequent deflection w.

Fig. P7.74

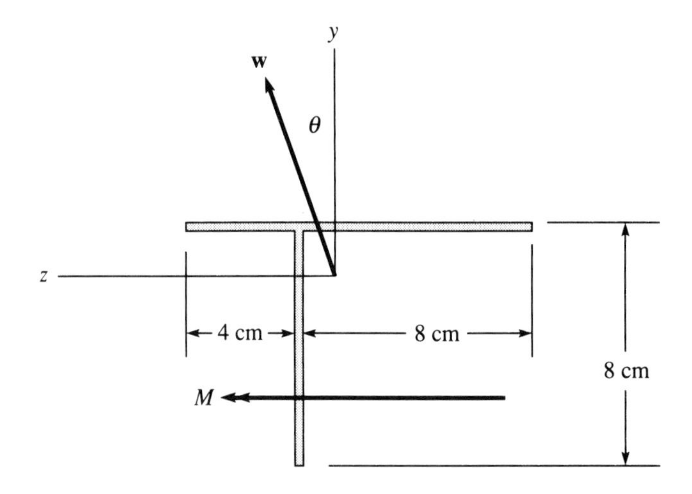

7.8 Shear Stresses in Thin-Walled Beams

As in the earlier discussion (Sec. 6.4) of shear stress in symmetric beams, we accept the normal stress given by the Bernoulli theory for HI-HO beams and appeal to equilibrium. With respect to principal lines of the cross section, the normal stress from bending only ($F = 0$) is given by the final two terms of Eq. (7.21):

$$\sigma_x = -\frac{M_z y}{I_{yy}} + \frac{M_y z}{I_{zz}} \tag{7.22}$$

Also, we recall the equilibrium equations (7.17a, b):

$$\frac{dM_z}{dx} = -V_y, \qquad \frac{dM_y}{dx} = +V_z$$

Let us now consider a nonsymmetrical thin-walled cross section, the channel depicted in Fig. 7.22. The transverse loads and, therefore, the shear forces can be decomposed. The component V_z must act on the axes of symmetry or the beam will twist. The line of action for the component V_y, which causes no twist, is not apparent. To locate the line we presume bending in the xy-plane under the action of transverse

Fig. 7.22

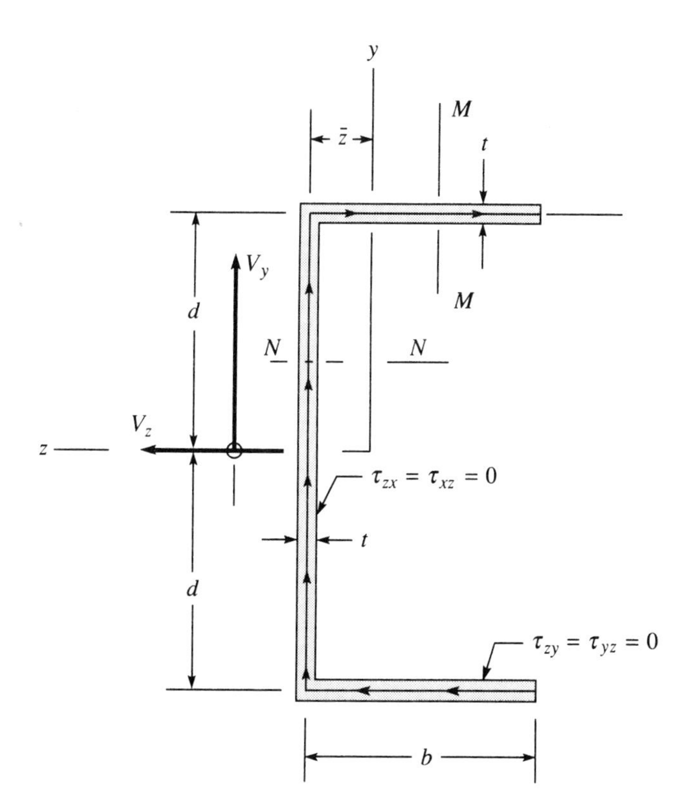

loads p_y, shear force V_y, and bending couple M_z; according to the Bernoulli theory:

$$\sigma_x = -\frac{M_z y}{I_{yy}} \tag{7.23}$$

For simplicity, let us assume that the web and flange have the same thickness $t \ll b$. On the lateral surfaces of the web, $\tau_{zx} = \tau_{xz} = 0$; therefore, the stress τ_{xz} is assumed negligible throughout the *thin* web. Likewise, the component $\tau_{yx} = \tau_{xy}$ is neglected in the *thin* flange. In general, the shear stress on the thin-walled section is everywhere tangent to the wall; here, the significant components are τ_{xy} in the web and τ_{xz} in the flange. To determine their distributions we appeal to the equilibrium of a free body, as in Sec. 6.4. We slice the beam by sections at x and $x + \Delta x$. To determine the stress $\bar{\tau}_{xy}$ on the web, we cut the cross section by a plane N-N at coordinate y. To determine the stress $\bar{\tau}_{xz}$ on the flange, we cut the cross section by a plane at M-M at coordinate z. The cuts (N-N and M-M) are shown in Fig. 7.22; the respective free bodies and the net force components in the x-direction are shown in Fig. 7.23(a) and (b).

Fig. 7.23

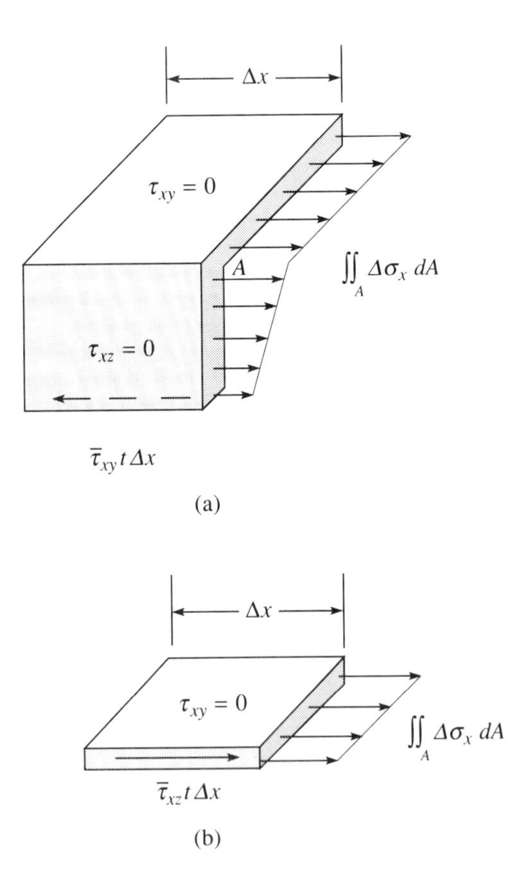

For equilibrium of the body in Fig. 7.23(b), we require

$$\sum F_x = \bar{\tau}_{xz} \Delta x t + \iint_A \Delta \sigma_x \, dA = 0 \tag{7.24a}$$

The integral extends over the shaded area A on the portion of the section shown on the body. The change $\Delta \sigma_x$ is caused by the change of moment ΔM_z in accordance with Eq. (7.23), so that

$$\bar{\tau}_{xz} = \frac{\Delta M_z}{\Delta x} \frac{\iint_A y' \, dA}{I_{yy} t} \tag{7.24b}$$

Passing to the limit ($\Delta x \to 0$) and using Eq. (7.17a), we obtain

$$\bar{\tau}_{xz} = -\frac{V_y}{I_{yy} t} \iint_A y' \, dA \tag{7.24c}$$

Let us evaluate the stress for the channel of Fig. 7.22. Here, we recognize that $t \ll y'$ and take $y' = d$ throughout the upper flange. Then

$$\iint_A y' \, dA \equiv Q(z) = td \int_{\bar{z}-b}^{z} dz = td(z - \bar{z} + b)$$

The distribution of the shear stress varies linearly from $\bar{\tau}_{xz} = -V_y bd/I_{yy}$ at the corner $z = \bar{z}$ to $\bar{\tau}_{xz} = 0$ at the end $z = \bar{z} - b$. The distribution is similar in the lower flange. The entire distribution is plotted adjacent to the walls in Fig. 7.24.

Let us now locate the line of action. The resultant V_y must have the same moment as the distribution. Since any axis will suffice, we choose the most convenient axis at point A; the shear stress acting along the web acts through point A. The moments of the shear stresses on the upper and lower flanges are the same; the total is equal to the moment of the force V_y (clockwise):

$$-2 \int_{\bar{z}-b}^{\bar{z}} \bar{\tau}_{xz} y \, dA = V_y \bar{c}$$

$$+2V_y \frac{d}{I_{yy}} \int_{\bar{z}-b}^{\bar{z}} (z - \bar{z} + b) \, dt \, dz = V_y \bar{c}$$

$$\bar{c} = \frac{tb^2 d^2}{I_{yy}} = \frac{b^2}{2(b + d/3)}$$

Note: In the limit ($b \to 0$), $\bar{c} \to 0$; also, if $b \geq d$, $0.500b > \bar{c} \geq 0.375b$. For a typical channel, American Standard 12 × 30, $b = 0.528d$, $\bar{c} = 0.162b$. The point O at $z = \bar{z} + \bar{c}$ is the shear center. We anticipate little or no twisting if the shear force acts through this point.

The foregoing procedure applies only to open, thin-walled cross sections. If the cross section is unsymmetric and closed, as the section of Fig. 7.25(a), then statics

Fig. 7.24

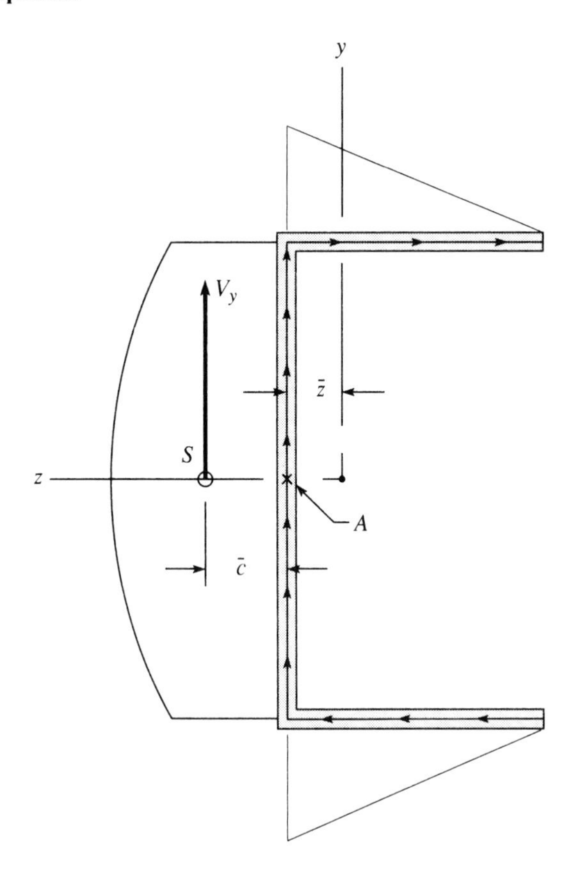

alone is not sufficient to locate the shear center. Additional kinematics is needed to describe the torsion.

From statics we can glean some additional information about the torsion of thin-walled symmetric sections, such as those of Fig. 7.25(b) and (c). The shear centers for sections such as those of Fig. 7.25(b) and (c) can be identified from symmetry. A transverse shear force upon the section of Fig. 7.25(b) can be decomposed into components in directions y and z, which are axes of symmetry. A shear force upon the section of Fig. 7.25(c) can be decomposed into components in two of the directions y, y', or y'', which are axes of symmetry. In each case the actions upon the cross section can be represented by a force at the shear center and a couple. Then, we can superpose the shear stresses caused by the transverse shear force and the twisting couple.

Let us suppose that the location of the shear center is known, or perhaps the cross section is subject only to a twisting couple T. In either case, we want to express the shear stress upon the thin wall in terms of the couple T. As noted previously, the significant shear stress upon the cross section is tangent to the wall. The component

Fig. 7.25

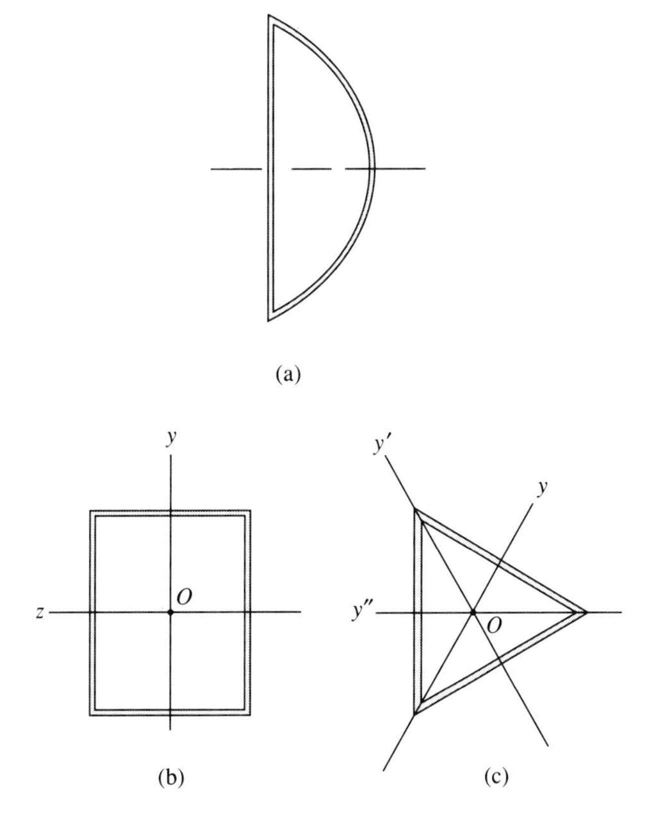

$\tau_{nx} = \tau_{xn}$ is negligible on a thin wall, as shown in Fig. 7.26. The couple caused by the component $\bar{\tau}_{xt}$ (tangent to the midline) upon a segment of arc length Δs is

$$\Delta T = h\bar{\tau}_{xt} t \, \Delta s$$

We note that the product $h \, \Delta s$ is twice the area ΔA of the small shaded triangle. Therefore,

$$\Delta T = 2t\bar{\tau}_{xt} \, \Delta A$$

Integrating around the entire midline, we obtain the net twisting couple:

$$T = 2At\bar{\tau}_{xt} \tag{7.25a}$$

or

$$\bar{\tau}_{xt} = \frac{T}{2At} \tag{7.25b}$$

Equation (7.25b) expresses the shear stress upon the simply connected thin-walled section in terms of the applied torque T and the geometric properties, the area A

Fig. 7.26

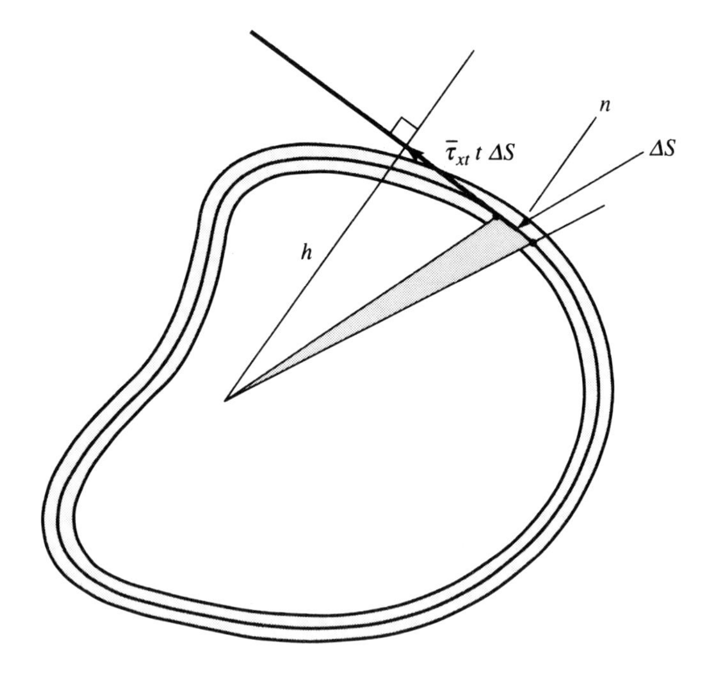

enclosed by the midline and the wall thickness t. Note that the result applies only to a simply connected cross section. The shear stress distribution upon a multiply connected section, as that of Fig. 7.27, requires additional considerations of the deformation.

Fig. 7.27

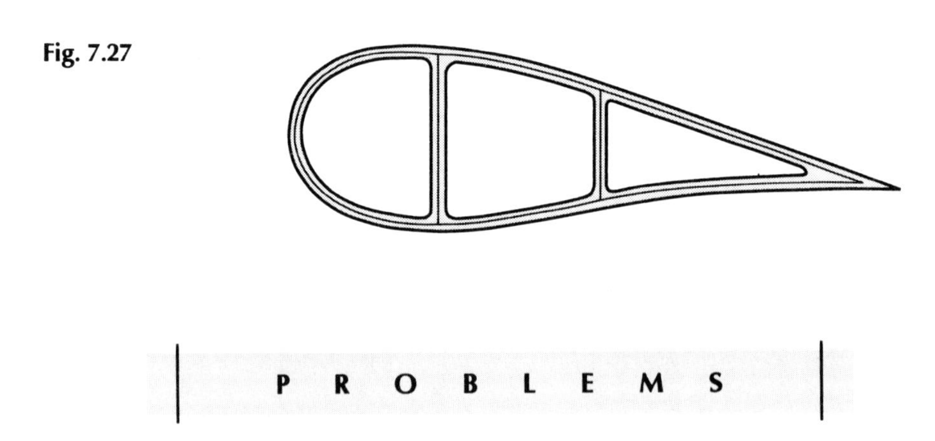

PROBLEMS

7.75 Locate the shear center for each of the thin-walled cross sections in Fig. P7.75; $t/a \ll 1$.

Fig. P7.75

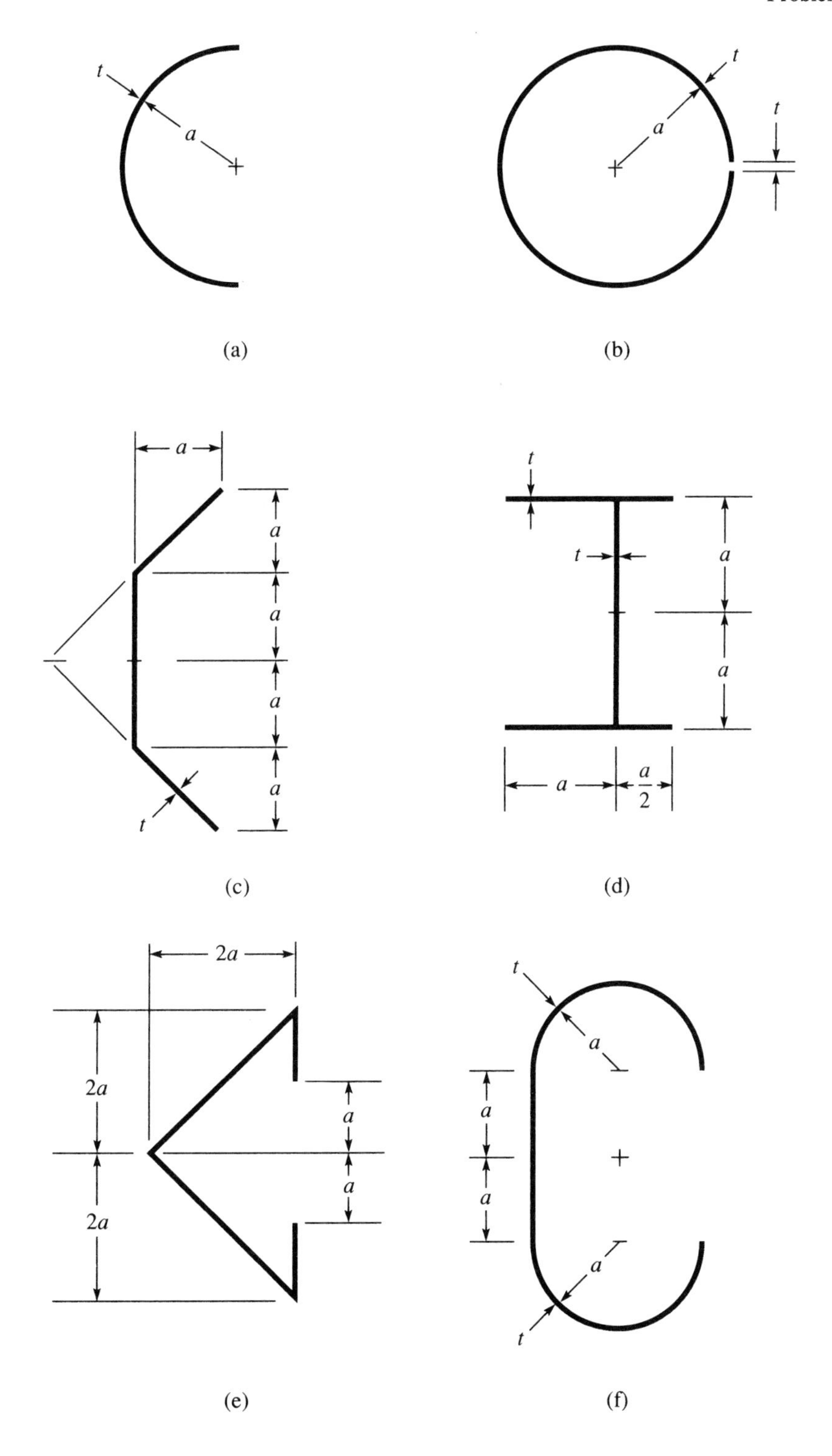

(a) (b) (c) (d) (e) (f)

7.76 A beam has a thin-walled rectangular cross section, as shown in Fig. P7.76. It is acted upon by a twisting couple $T = 20{,}000$ in.-lb and a bending couple $M = 10{,}000$ in.-lb. (Assume τ_{xy} and τ_{xz} are constant through a wall.) (a) Compute the stresses $\sigma_x, \tau_{xy}, \tau_{xz}$ at $y = 2$, $z = 0$. (b) Determine the maximum shear stress in the beam.

Fig. P7.76

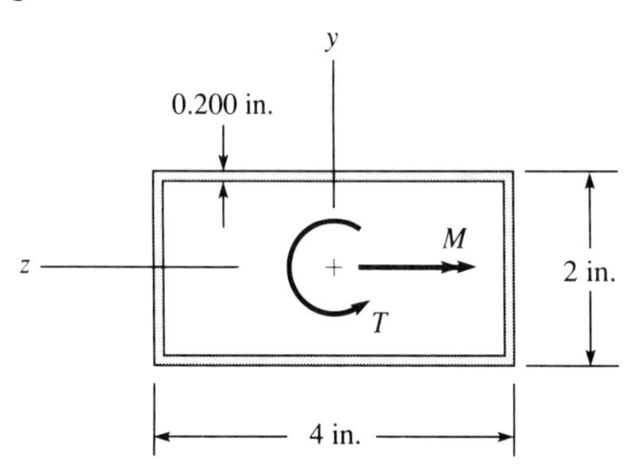

7.77 The cross section of the beam in Fig. P7.77 is an equilateral triangle. The sides are very thin webs 0.150 cm thick; the stringers at the vertices have equal areas of 3.00 cm^2. Assume that the webs carry no normal force and compute the shear stress in each web.

Fig. P7.77

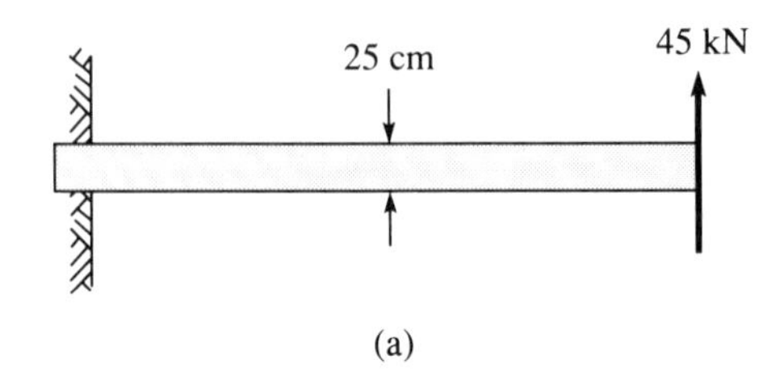

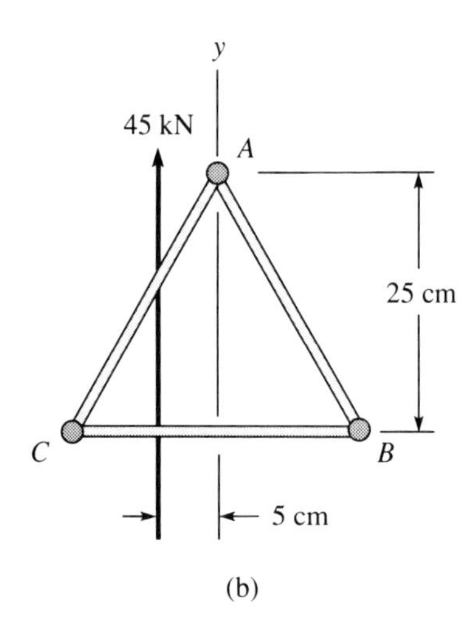

7.78 The net shear force on a cross section of a box beam acts as indicated in Fig. P7.78. The cross-sectional area of stringers A and B is 0.50 in.^{2} and the area of D and C is 1.00 in.^{2} The webs are 0.06 in. thick. Assume that the webs carry no normal force (parallel to the axis of the beam). Compute the shear stress in each web.

Fig. P7.78

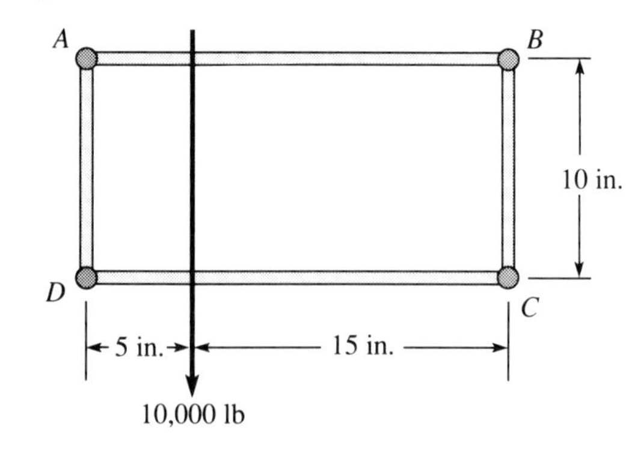

7.79 A beam has the thin-walled rectangular cross section illustrated in Fig. P7.79. The thickness is constant $t \ll h$. Determine the line of action (distance d)

for the shear force V, such that no shear stress acts at point A.

Fig. P7.79

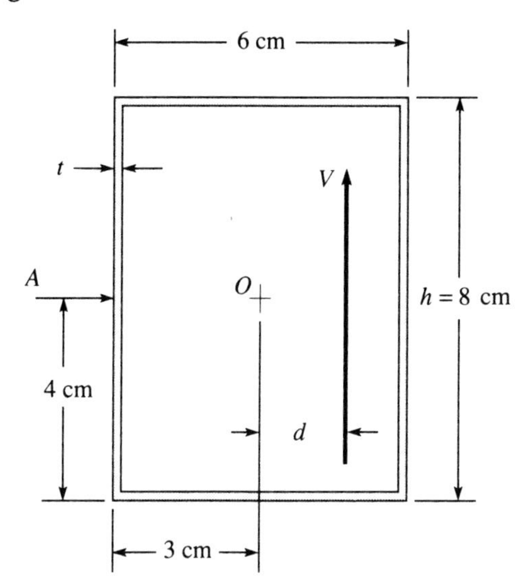

7.80 A beam has the thin-walled rectangular cross section depicted in Fig. P7.80. The top and bottom walls are 0.200 in.; the sides are 0.100 in. When subjected to the transverse shear force $V = 1$ kip, determine the maximum value of the stress $\bar{\tau}_{xy}$.

Fig. P7.80

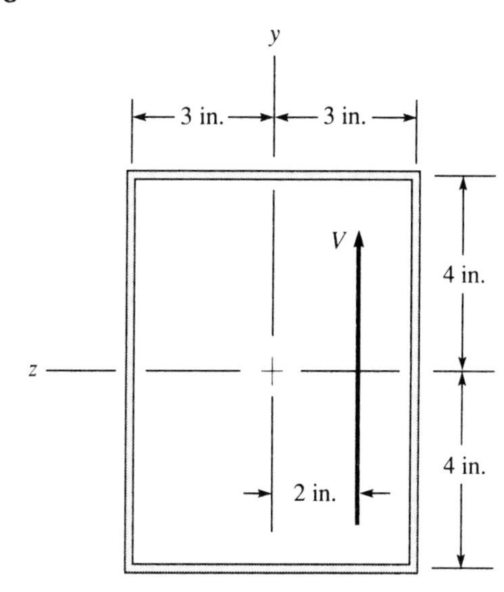

8 Methods of Work and Energy

Throughout the preceding text our approach has been the mechanics of Newton,[1] wherein we isolate a body (free-body diagram) and impose Newton's "laws" to enforce equilibrium. In the present chapter we adopt the very different approach of Lagrange,[2] wherein equilibrium is imposed via criteria placed upon the work expended or energy changed during an incremental movement of a mechanical system. The Lagrangian approach entails significant practical differences: The free-body diagram is no longer essential, since only active (working) forces play a role. Instead, the *systems*, the work expended and energy transferred, play the key roles. In some instances (conservative systems), internal forces and stresses play no roles. This has distinct advantages in complicated systems of many components. For example, the displacement at the joint of a truss can be determined without attention to any of the *internal* forces. They do no work explicitly, though their work is accommodated by changes of internal energy.

Considerations of work and energy provide not only a different avenue to equilibrium, but a means to assess the stability of equilibrium. A body may be in a state of equilibrium but unstable in the sense that a minor disturbance would cause severe disfiguration, as in the case of a ball resting at the topmost point of a hill (see Fig. 8.16a).

The intent of the present chapter is to set forth the basic concepts, which provide different and powerful methods not employed in the preceding chapters. Only enough examples and exercises are provided to reinforce the concepts and to establish a firm foundation for further study.

---

1 I. Newton, *Principia Philosophiae*, 1686.

2 J. L. Lagrange, *Mécanique analytique* (Paris, 1788). For additional historical remarks, see: C. Lanczos, *The Variational Principles of Mechanics* (Toronto: University of Toronto Press, 1949); H. L. Langhaar, *Energy Methods in Applied Mechanics* (New York: John Wiley, 1962); G. A. E. Oravas and L. McLean, "Historical Development of Energetical Principles in Elastomechanics," *Applied Mechanics Review* 19, nos. 8 and 11 (1966).

8.1 Principle of Virtual Work

Our previous approach to equilibrium involved the free-body diagram and the balance of force and moment. The notion of *virtual work* provides a very different viewpoint and a method to assure a state of equilibrium. The concept of virtual work relies on a consideration of small changes in the configuration of the mechanical system and the work expended to effect such changes.

A *configuration* is defined by the collective positions of all particles that comprise the system. Any change of configuration involves kinematics, that is, a change in the geometry. Of course, kinematics always plays a role in the study of deformable bodies. In the principle and method of virtual work, kinematics plays the major role, whereas the free body is not essential.

Virtual Displacement

A virtual displacement is any infinitesimally small change (first order in the geometric variables) in the configuration of a mechanical system, *but* a change that is consistent with physical constraints. For example, the teeter-totter (or seesaw) illustrated in Fig. 8.1 is a simple mechanical system consisting of a board (assumed rigid) that can rotate only about a pivot (assumed frictionless). The configuration of this system is fully determined by specifying the angle θ. Here, the constraint is only the pivot. A virtual displacement that moves that point is *not* admissible.

Fig. 8.1

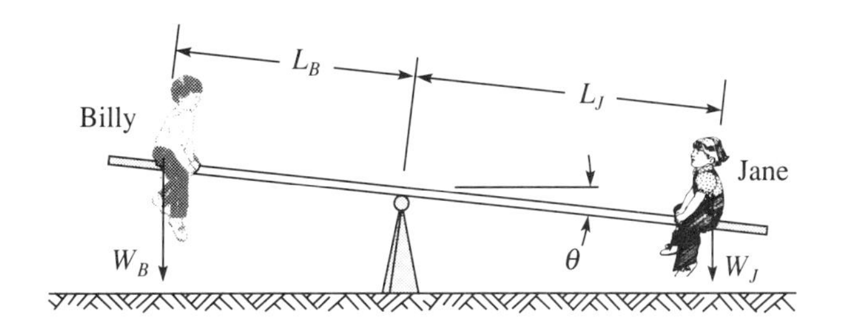

Another familiar mechanical system is the block and tackle of Fig. 8.2(a) or (b). A block and tackle is a system consisting of pulleys (assumed frictionless) and ropes (assumed inextensible). In both cases (for simplicity) we assume that the pulleys are small and the intermediate distances are large; then the slight inclinations of the ropes can be neglected. Each is a system with one degree of freedom; if the displacement S_A of the knot A is prescribed, then the displacement S_O of axle O follows. The relationship is an exercise in kinematics: In case (a) $S_A = 2S_O$; in case (b) $S_A = 3S_O$.

The system depicted in Fig. 8.3(a) consists of two rigid links joined by pins at A and at O. Torsional springs resist but do not prevent relative rotations at these

Fig. 8.2

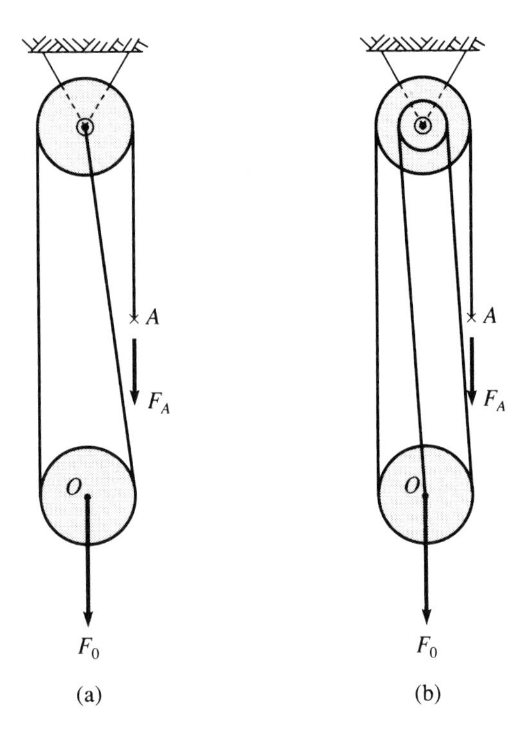

joints. This system has two degrees of freedom, since two angles, θ_1 and θ_2, define an admissible configuration. Similarly, the system in Fig. 8.3(b) consists of four rigid links, pinned at O, A, B, and C; again, torsional springs resist relative rotations. This system has four degrees of freedom, since four angles, θ_1, θ_2, θ_3, and θ_4, are needed to define a configuration. We can elaborate on the linkages of Fig. 8.3(a) and (b) and create systems with any number of degrees of freedom. The limit of the process is a continuous flexible beam, as shown in Fig. 8.3(c). At each point the rotation can be different, but continuous; we could say that the beam has infinitely many degrees of freedom. A virtual displacement of the two-link system [Fig. 8.3(a)] consists of any *two* independent, but infinitesimally small, changes in the angles θ_1 and θ_2. Here, *independent* is a key word; the virtual displacement can be *any* change consistent with the constraints. In other words, any movement of these systems (Fig. 8.3) is an admissible virtual displacement if it does not break a physical constraint, a pivot in Fig. 8.3(a) or (b) or a kink in the beam of Fig. 8.3(c).

Fig. 8.3

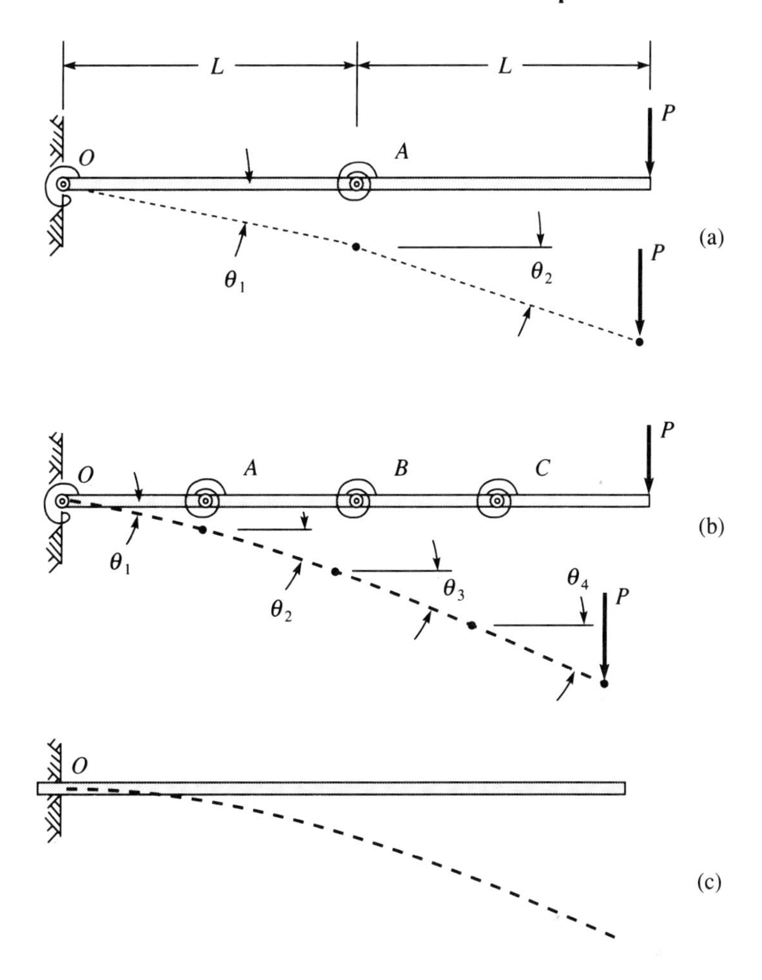

Virtual Work

We define the virtual work as the work that we would expend in the virtual displacement.[3] This is hypothetical, since we only imagine the motion and the work expended. For example, a small (positive) change in the angle θ in the teeter-totter in Fig. 8.1 produces a small upward displacement δ_B of Billy's center of gravity and a small downward displacement δ_J of Jane's.

3 Others might give a definition as the work done by the imposed forces. Our definition facilitates subsequent treatment of conservative systems and changes in potential. More specifically, our definition is such that the virtual work on a conservative mechanical system *is* the variation of its potential.

$$\delta_B = L_B[\sin(\theta + \Delta\theta) - \sin\theta]$$
$$\doteq L_B \cos\theta\,\Delta\theta$$
$$\delta_J \doteq L_J \cos\theta\,\Delta\theta$$

Note: These equations are first-order approximations consistent with the notion of a virtual displacement. If W_B and W_J are the weights of Billy and Jane, respectively, then the total virtual work (we would do) is

$$\Delta w = (W_B L_B - W_J L_J)\cos\theta\,\Delta\theta \tag{8.1}$$

The relationships between the motions of knot A and axle O in Fig. 8.3(a) and (b) were noted previously. If a small downward displacement ΔS_O is given to the axle O and downward forces F_O and F_A act at axle O and knot A, then the virtual work in case (a) is

$$\Delta w = (2F_A - F_O)\,\Delta S_O \tag{8.2}$$

In case (b),

$$\Delta w = (3F_A - F_O)\,\Delta S_O \tag{8.3}$$

Now, consider the virtual work for the two-degree-of-freedom system of Fig. 8.3(a). Under the action of the end load P, the system is deformed to a configuration, as indicated by the dotted lines. An admissible virtual displacement corresponds to any infinitesimally small changes in the *two* angles θ_1 and θ_2. Such (positive) changes, $\Delta\theta_1$ and $\Delta\theta_2$, produce a downward displacement of the loaded end:

$$\delta = L[\sin(\theta_1 + \Delta\theta_1) + \sin(\theta_2 + \Delta\theta_2)$$
$$- \sin\theta_1 - \sin\theta_2]$$
$$\doteq (L\cos\theta_1)\,\Delta\theta_1 + (L\cos\theta_2)\,\Delta\theta_2$$

The virtual work (we do) upon this *external* force P is

$$\delta w_E = -LP[(\cos\theta_1)\,\Delta\theta_1 + (\cos\theta_2)\,\Delta\theta_2] \tag{8.4}$$

The number of *independent* variables is always equal to the number of degrees of freedom; for example, the system of four links [Fig. 8.3(b)] has four independent changes ($\Delta\theta_1$, $\Delta\theta_2$, $\Delta\theta_3$, and $\Delta\theta_4$).

When the system of Fig. 8.3(a) is displaced, one or both links must rotate. Because our principle requires the work for *any* admissible virtual displacement, $\Delta\theta_1$ and $\Delta\theta_2$ must be arbitrary. Then relative rotations are admissible, indeed required, at both joints. When this occurs, we expend work against the springs. Virtual work upon *internal* forces is inherent in a displacement of any deformable body. That work depends upon the physical attributes of the springs (the properties of the material). If these are linear (hookean) springs, then the resisting couples are proportional to the relative rotations. For simplicity, let us assume the same torsional stiffness β for both. The virtual work upon these internal couples is

$$\delta w_I = \beta[\theta_1\Delta\theta_1 + (\theta_2 - \theta_1)(\Delta\theta_2 - \Delta\theta_1)] \tag{8.5}$$

Principle of Virtual Work

With the foregoing definitions of virtual displacement and virtual work, we can now state, quite simply, the principle of virtual work. *A mechanical system is in equilibrium if and only if the virtual work vanishes for arbitrary admissible virtual displacements.*

Let us apply the principle to each of the mechanical systems in Figs. 8.1, 8.2, and 8.3. The teeter-totter of Fig. 8.1 is in equilibrium if and only if the virtual work of Eq. (8.1) vanishes for the arbitrary displacement (nonzero) $\Delta\theta$: The criterion is met if and only if the parenthetical term is zero. Therefore, we obtain the equilibrium equation

$$W_B L_B = W_J L_J$$

We recognize that this is equivalent to the condition of vanishing moment of the active forces about the pivot.

The block and tackle of Fig. 8.2(a) is in equilibrium if and only if the virtual work of Eq. (8.2) vanishes for nonzero ΔS_O. Again, equilibrium requires that the parenthetical term vanish:

$$2F_A = F_O$$

Likewise, the equilibrium of the system in Fig. 8.2(b) requires the vanishing of the virtual work of Eq. (8.3), which yields

$$3F_A = F_O$$

The two-degree-of-freedom system in Fig. 8.3(a) requires also that the virtual work vanish for *arbitrary* admissible virtual displacements. That virtual work consists of the work against external *and* internal forces, that is, the sum of the work given by Eqs. (8.4) and (8.5):

$$\begin{aligned}\delta w_E + \delta w_I = {} & [-LP\cos\theta_1 + \beta(2\theta_1 - \theta_2)]\,\Delta\theta_1 \\ & + [-LP\cos\theta_2 + \beta(\theta_2 - \theta_1)]\,\Delta\theta_2 = 0\end{aligned}$$

Because the virtual work must vanish for arbitrary virtual displacements ($\Delta\theta_1$ and $\Delta\theta_2$ are independent), both bracketed terms must vanish for equilibrium. Accordingly, we obtain two equations in the variables θ_1 and θ_2:

$$\begin{aligned}2\beta\theta_1 - \beta\theta_2 &= LP\cos\theta_1 \\ -\beta\theta_1 + \beta\theta_2 &= LP\cos\theta_2\end{aligned}$$

These are nonlinear equations; however, viewing this as a very simplified model of a cantilever beam, we might assume that the angles θ_1 and θ_2 are small. Then a first-order approximation ($\cos\theta_1 \doteq \cos\theta_2 \doteq 1$) provides the linear equations and their solution:

$$\begin{aligned}2\beta\theta_1 - \beta\theta_2 &\doteq LP \\ -\beta\theta_1 + \beta\theta_2 &\doteq LP\end{aligned}$$

$$\theta_1 = \frac{2LP}{\beta}, \qquad \theta_2 = \frac{3LP}{\beta}$$

Note that the method requires no free body. Moreover, no interactions (forces and/or couples) at interconnections (pivots, axles, or joints) are involved in the solutions of the foregoing examples. In many problems such interactions are not required. The avoidance of interactions via the principle of virtual work is then a simplification and is, therefore, an advantage.

A word of caution is in order: The preceding description of virtual displacement presumes that the paths of particles are smooth. Likewise, the description of virtual work presumes that the forces vary continuously. Special circumstances do arise. For example, the roller of Fig. 8.4 is at rest, in equilibrium, in the V-shaped groove. The only admissible displacement of the center is parallel to one or the other plane. In either case the virtual work cannot vanish, because the normal force exerted by the other surface vanishes abruptly. Any admissible displacement, however small, abruptly violates equilibrium.

Fig. 8.4

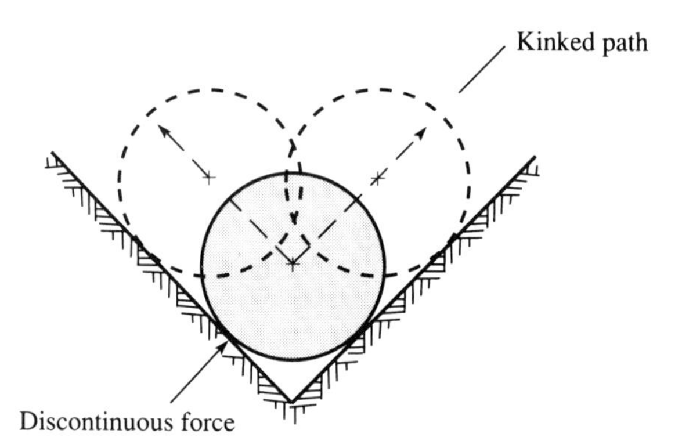

Let us now apply the principle of virtual work to a familiar problem of deformable bodies.

EXAMPLE 1

Simple Planar Truss Let us reconsider the truss of Fig. 5.20 (Example 7, Sec. 5.7) which is redrawn in Fig. 8.5. According to the earlier analysis, the behavior of the individual members is described in terms of the extension (e) and force (F) upon the member. The kinematic analysis expresses each extension in terms of the two displacements (u and v) of the joint A:

$$e_D = v, \qquad e_C = u, \qquad e_B = \frac{u}{\sqrt{2}} + \frac{v}{\sqrt{2}} \tag{8.6a,b,c}$$

Therefore, this structure is reduced to a mechanical system with two degrees of freedom.

Fig. 8.5

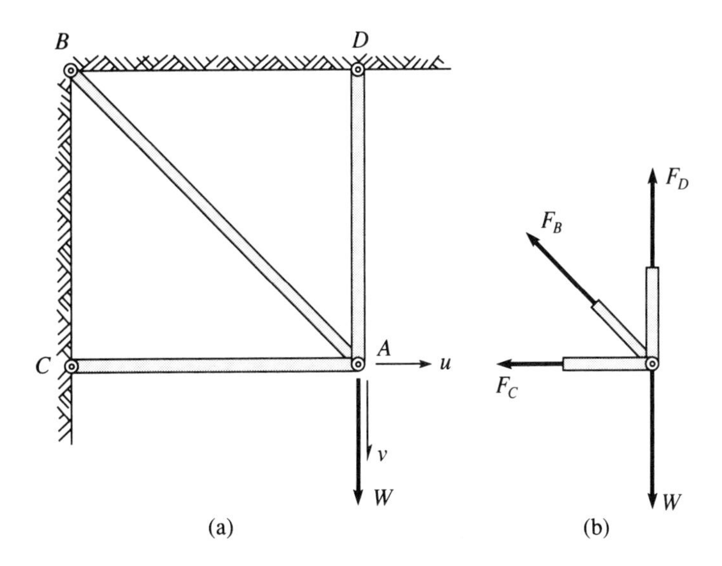

Under the action of the one external load, the virtual work (we do) upon a virtual displacement is

$$\delta w_E = -W\delta v$$

The total virtual work expended against the internal forces (F_D, F_C, F_B) is

$$\begin{aligned}\delta w_I &= F_D\delta e_D + F_C\delta e_C + F_B\delta e_B \\ &= F_D\delta v + F_C\delta u + \frac{F_B(\delta u + \delta v)}{\sqrt{2}}\end{aligned}$$

We require that the total virtual work vanish for arbitrary virtual displacements:

$$\delta w_E + \delta w_I = \left(-W + F_D + \frac{F_B}{\sqrt{2}}\right)\delta v + \left(F_C + \frac{F_B}{\sqrt{2}}\right)\delta u = 0$$

Because the virtual displacements are arbitrary, δv and δu are independent, and both parenthetical terms must vanish:

$$F_D + \frac{F_B}{\sqrt{2}} = W \tag{8.7a}$$

$$F_C + \frac{F_B}{\sqrt{2}} = 0 \tag{8.7b}$$

These equilibrium equations are precisely Eq. (5.50) but derived via the principal of virtual work. The complete solution of the problem requires the introduction of the behavior of the material and follows the steps of Example 7 in Chapter 5. ◆

PROBLEMS

Use the principle of virtual work, not the free-body diagram, to solve the following problems.

8.1 Determine the mechanical advantage of the block and tackle system of Fig. P8.1, that is, the ratio W/P. Neglect the small inclination of the ropes; that is, assume that the segments between the pulleys are parallel.

Fig. P8.1

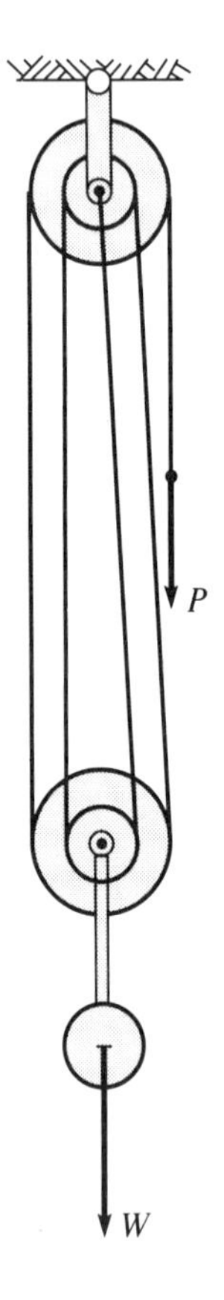

8.2 Obtain the equations of equilibrium for each of the plane trusses in Fig. P8.2, and determine the limit load in terms of the limit force $F_0 = \sigma_0 A$ for the individual members. The yield stress σ_0 and cross-sectional area A are the same for all members.

Fig. P8.2

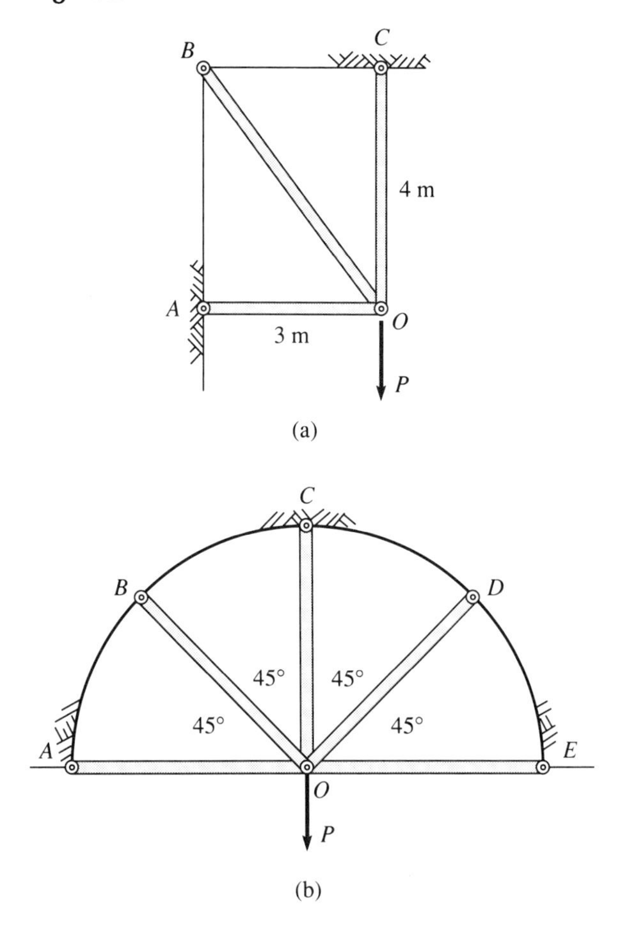

8.3 An assembly of four rigid rods is illustrated in Fig. P8.3. The rods are pinned together at their ends and support equal weights at the interconnections A, B, and C. Determine the equilibrium configuration, that is, the angles θ_1 and θ_2. The pins O and P are immovable. The length of each rod is the same l.

Fig. P8.3

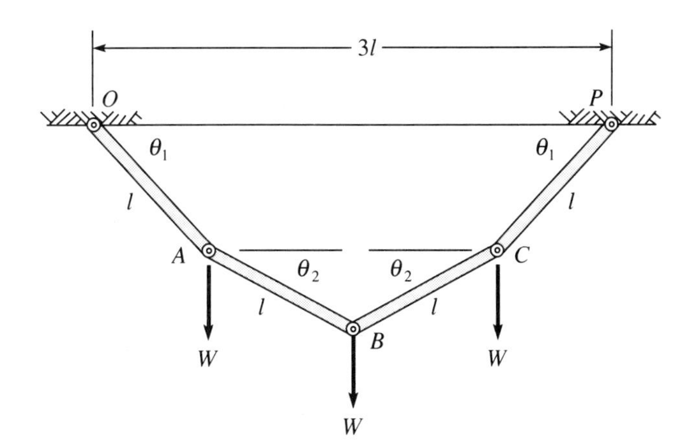

8.4 The "rigid" boom CO of Fig. P8.4 is supported by the two cables OA and OB. Express the forces upon each cable in terms of the load W. Both cables are made from the same stock, which fails at 5 kN. Calculate the load W_M that would cause failure.

Fig. P8.4

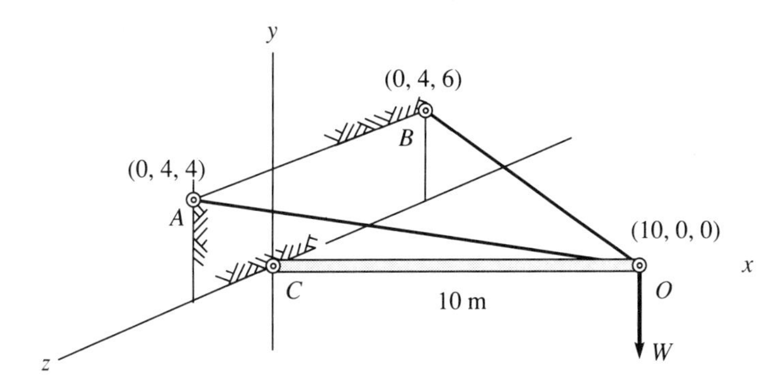

8.5 The deflection of a simply supported beam can be represented by the Fourier series:

$$w(x) = \sum a_n \sin n\frac{\pi x}{L}$$

where L is the length of the beam. Now a_n represents a generalized coordinate, that is, the values determine the deformed configuration. If the beam carries a continuous load $p(x)$, show that the components of the generalized force P_n are proportional to the coefficients in the Fourier (sine) series for $p(x)$ and determine the constant of proportionality. The virtual work has the form $\sum P_n \delta a_n$.

8.6 Do Prob. 8.5 if the load on the beam has the following explicit distributions:

a. A concentrated load at $x = L/2$

b. A uniform load $p(x) = \bar{p}$, constant

c. A linearly varing load $p(x) = \bar{p}x/L$

*8.2 Equilibrium of a Symmetric Beam

In Chapter 6 (Sec. 6.2) we examined the equilibrium of symmetric beams. Now let us revisit that problem via the principle of virtual work.

EXAMPLE 2

Equilibrium of an Element Let us first apply the criterion to a small deformed segment, as illustrated in Fig. 8.6(a). The initial length is Δx, and the deformed length is $(1 + \varepsilon_0)\Delta x$. Fig. 8.6(b) shows the corresponding element of the beam with the forces and couples that act on it. If we discount a virtual *deformation*, the element has three degrees of freedom; it can displace in each of two directions and rotate in the plane of symmetry. Let us assume the virtual displacements δv and δu are normal and tangent to the middle line, as depicted in Fig. 8.6(b). Additionally, we take a virtual rotation $\delta\theta$ in the sense of the angle θ. The virtual work (we do) in the arbitrary displacement $(\delta v, \delta u)$ and rotation $(\delta\theta)$ is

$$\begin{aligned}\delta w = &\left[-\Delta F\cos\frac{\Delta\theta}{2} + (2V + \Delta V)\sin\frac{\Delta\theta}{2}\right]\delta u \\ &- \left[(2F + \Delta F)\sin\frac{\Delta\theta}{2} + \Delta V\cos\frac{\Delta\theta}{2} - \bar{p}\,\Delta x\right]\delta v \\ &- \left[\Delta M + 2\rho\sin\frac{\Delta\theta}{2}\left(V + \frac{\Delta V}{2}\right)\cos\frac{\Delta\theta}{2}\right]\delta\theta \qquad (8.8)\end{aligned}$$

The final term incorporates the distance OA between the ends: $OA = 2\rho\sin\Delta\theta/2$.

Again, the virtual work must vanish for arbitrary virtual displacements. Therefore, each bracketed term must vanish independently. We suppose a practical application wherein the length Δx is relatively small, $\varepsilon_0 \ll 1$, and forces F and V and couple M are continuous. Then $\Delta\theta \doteq \kappa\,\Delta x \ll 1$, $\Delta F \ll F$, and $\Delta V \ll V$,

$$\cos\frac{\Delta\theta}{2} \doteq 1, \qquad \text{and} \qquad \sin\frac{\Delta\theta}{2} \doteq \frac{\Delta\theta}{2}$$

With these approximations the three equations of equilibrium (bracketed terms vanish) have the approximate forms

$$\frac{\Delta F}{\Delta x} - \kappa V \doteq 0 \qquad (8.9a)$$

$$\frac{\Delta V}{\Delta x} + \kappa F \doteq \bar{p} \qquad (8.9b)$$

$$\frac{\Delta M}{\Delta x} + V \doteq 0 \qquad (8.9c)$$

Fig. 8.6

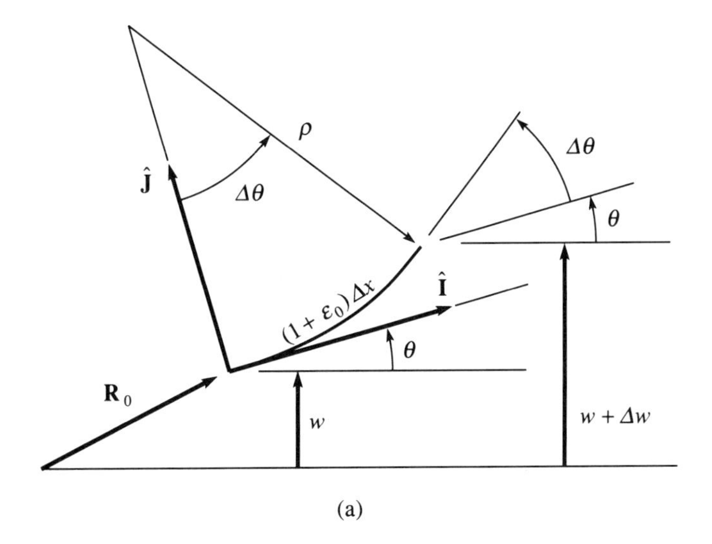

(a)

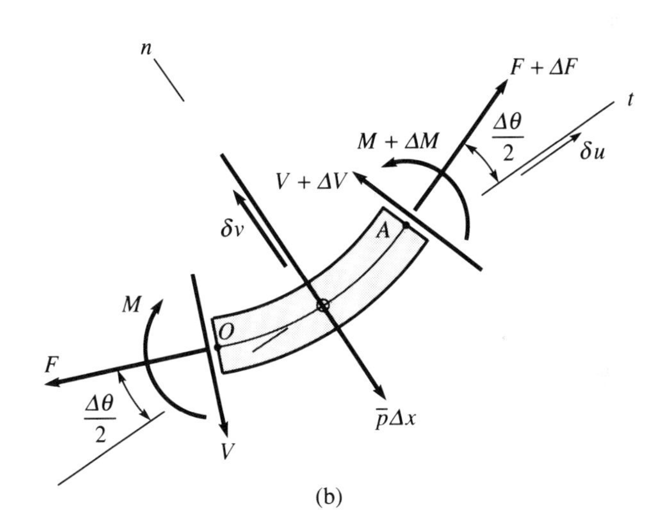

(b)

These three equations are valid for a small, but *finite*, element; that is, Δx is relatively small compared to the length of the beam. In other words, these might be used in an approximation of the beam by a finite number of small discrete elements. However, to describe the continuous beam we pass to the limit, $\Delta x \to 0$, and obtain the differential equations [see Eqs. (6.8) and (6.9)]

$$\frac{dF}{dx} = \kappa V, \qquad \frac{dV}{dx} + \kappa F = \bar{p}, \qquad \frac{dM}{dx} = -V \tag{8.10a,b,c}$$

Note that Eq. (8.10b) is a refinement of the previous approximation [Eq. (6.8)], which did not account for the curvature. These equations apply to the symmetric beam, which might be composed of any material. Finally, we note that our virtual displacements $(\delta v, \delta v, \delta\theta)$ constitute only a rigid motion of the element. Additionally, the element can experience deformation. The virtual work expended against those forces depends upon the material. ◆

EXAMPLE 3

Limiting Condition of an Ideally Plastic Beam Let us consider the ideally plastic beam shown in Fig. 8.7. The cantilever is fixed at the left end, is simply supported at the right, and carries the concentrated load P. We presume that the beam behaves according to the theory of Sec. 5.13. The load approaches the limit value P_L, the inelastic strains dominate; the elastic deformations are negligible. Very simply, the "collapse" corresponds to a plastic hinge at the fixed end and at the load, where the bending couple attains the limiting values

$$M(0) = -M_L, \qquad M(2L) = +M_L$$

Fig. 8.7

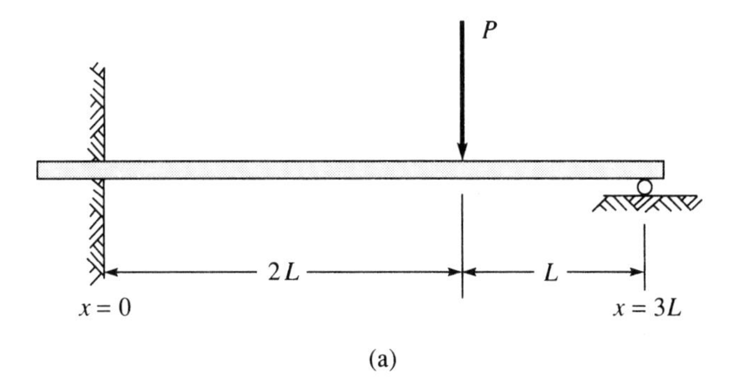

(a)

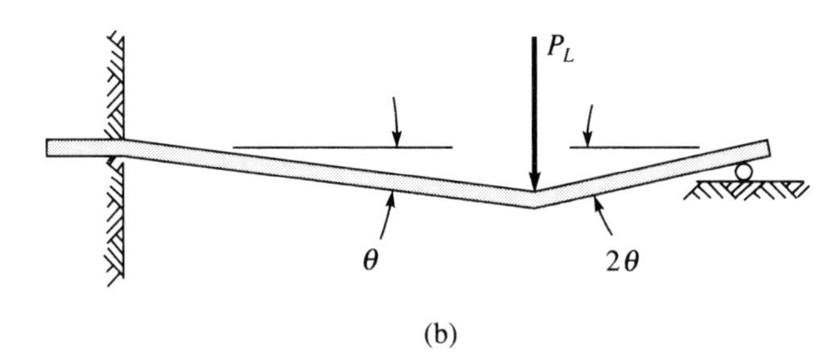

(b)

At this limiting state *no* appreciable deformations occur except at the two hinges. Therefore, if $\Delta\theta$ denotes the small virtual rotation (change in θ), the virtual work of the load is

$$\Delta w_E = -2LP_L\Delta\theta$$

We also expend work Δw_I upon the internal couples ($\mp M_L$ at $x = 0, 2L$), which undergo relative rotations ($\Delta\theta, -3\,\Delta\theta$):

$$\Delta w_I = 4M_L\,\Delta\theta$$

By the principle of virtual work, the *total* work must vanish.

$$(-2LP_L + 4M_L)\,\Delta\theta = 0$$

or

$$P_L = 2\frac{M_L}{L}$$

The limit couple M_L is determined as before (see Secs. 5.13 and 6.7) by the material and shape of the cross section. ◆

EXAMPLE 4

Equilibrium of a Symmetric Beam In Example 2 we examined a discrete element of a beam and, in fact, assumed only a rigid virtual displacement. The principle applies to any mechanical system and so to a continuous beam. To be specific, let us apply the method of virtual work to the entire beam illustrated in Fig. 8.8. All quantities and notations follow our previous conventions, but transverse displacement is denoted by $\bar{v}(x)$.

Fig. 8.8

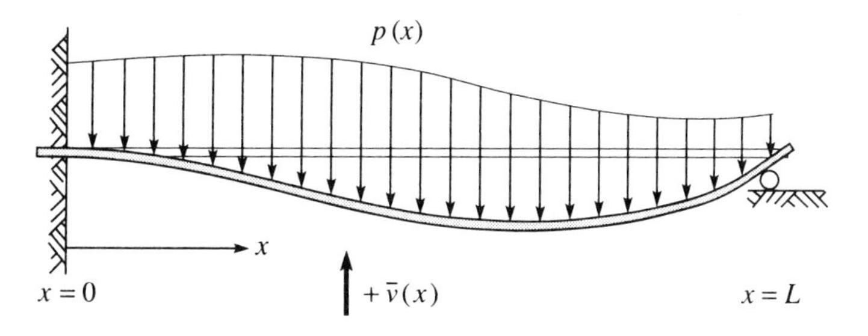

The only active external force is the load $p(x)$, so that the virtual work (we do) upon the external forces is

$$\delta w_E = +\int_0^L p(x)\,\delta\bar{v}\,dx \tag{8.11}$$

Here $\delta\bar{v}$ (a function of position x) is an arbitrary virtual displacement, normal to the *initial* position of the axis.

The virtual work expended against the internal forces, according to the Bernoulli hypothesis, consists of the work of normal force F in the virtual extension $\delta\varepsilon_0$ and the bending couple M in the virtual curvature $\delta\kappa$.

$$\delta w_I = \int_0^L (F\delta\varepsilon_0 + M\delta\kappa)\,dx \tag{8.12}$$

The major task now is to express the strain and curvature increments, $\delta\varepsilon_0$ and $\delta\kappa$, in terms of the virtual displacements. We refer to Fig. 8.6 and employ the components tangent and normal to the bent axis (δu and δv). Then the virtual displacement (vector) of a particle on the axis is [see Fig. 8.6(a)]

$$\delta\mathbf{R}_0 = \delta u\mathbf{I} + \delta v\mathbf{J} \tag{8.13}$$

Also,

$$\frac{d\mathbf{R}_0}{dx} = (1 + \varepsilon_0)\mathbf{I} \tag{8.14}$$

$$\frac{d}{dx}\delta\mathbf{R}_0 = \delta\varepsilon_0\mathbf{I} + (1 + \varepsilon_0)\delta\theta\mathbf{J} \tag{8.15a}$$

$$\doteq \delta\varepsilon_0\mathbf{I} + \delta\theta\mathbf{J} \tag{8.15b}$$

The virtual change of the unit vector $\mathbf{I}$ is $\delta\theta\mathbf{J}$, caused by rotation $\delta\theta$. Approximation (8.15b) is based on the assumption of small strain ($\varepsilon_0 \ll 1$). Now, we observe that

$$\frac{d\mathbf{I}}{dx}\cdot\mathbf{J} \equiv \kappa = -\frac{d\mathbf{J}}{dx}\cdot\mathbf{I} \tag{8.16a, b}$$

When we differentiate Eq. (8.13), use Eqs. (8.16a, b), and then equate the result to Eq. (8.15b), we obtain

$$\frac{d}{dx}(\delta\mathbf{R}_0) = \frac{d\delta u}{dx}\mathbf{I} + \kappa\delta u\mathbf{J} + \frac{d\delta v}{dx}\mathbf{J} - \kappa\delta v\mathbf{I}$$

$$= (\delta\varepsilon_0)\mathbf{I} + \delta\theta\mathbf{J}$$

It follows that

$$\delta\varepsilon_0 = \frac{d\delta u}{dx} - \kappa\delta v \tag{8.17}$$

$$\delta\theta = \frac{d\delta v}{dx} + \kappa\delta u$$

$$\delta\kappa = \frac{d\delta\theta}{dx} = \frac{d^2\delta v}{dx^2} + \frac{d}{dx}(\kappa\delta u) \tag{8.18}$$

Substituting Eqs. (8.17) and (8.18) into Eq. (8.12) and adding that result to Eq. (8.11), we obtain

$$\delta w_I + \delta w_E = \int_0^L \left[F\frac{d}{dx}\delta u - F\kappa\delta v + M\frac{d^2}{dx^2}\delta v + M\frac{d}{dx}(\kappa\delta u) + p\delta\bar{v}\right]dx \tag{8.19a}$$

The first, third, and fourth terms of the integrand can be integrated by parts, the third twice, to obtain

$$\delta w_I + \delta w_e = F\delta u\Big|_0^L + M\left(\frac{d}{dx}\delta v + \kappa\delta u\right)\Big|_0^L - \frac{dM}{dx}\delta v\Big|_0^L + \int_0^L \left[-\left(\frac{dF}{dx} + \kappa\frac{dM}{dx}\right)\delta u + \left(\frac{d^2M}{dx^2} - \kappa F\right)\delta v + p\delta\bar{v}\right]dx \quad (8.19b)$$

We are now in a position to enforce the following principle: The virtual work, Eq. (8.19), must vanish for *arbitrary* virtual displacements, the independent variations of the components $\delta u(x)$ and $\delta v(x)\,[\delta\bar{v} = \delta v\cos\theta + \delta u\sin\theta]$. This means that either may be zero over all but some small region. Also, the values at the ends can assume any values consistent with the constraints. In the specific example depicted in Fig. 8.8, δu and δv must be zero at the fixed end $(x = 0)$. δu can have any value at the right end $(x = L)$; slip along the roller is admissible. Also, the rotation at the right end $(x = L)$ can assume any value, specifically:

$$\delta\theta(L) = \left(\frac{d}{dx}\delta v + \kappa\delta u\right)(x = L)$$

If we are to have no virtual work for the arbitrary virtual displacement, then the coefficients of the components $\delta u(x)$ and $\delta v(x)$ in the integrand must vanish everywhere. Throughout $(0 < x < L)$,

$$\frac{dF}{dx} + \kappa\frac{dM}{dx} - p\sin\theta = 0 \quad (8.20)$$

$$\frac{d^2M}{dx^2} - \kappa F + p\cos\theta = 0 \quad (8.21)$$

These are the differential equations of equilibrium. Also, because the tangential displacement and rotation are independently variable at the right end, their coefficients must vanish:

$$F(L) = 0 \quad (8.22)$$

$$M(L) = 0 \quad (8.23)$$

The additional end conditions are the prescribed kinematic constraints imposed by the fixture at the left $(x = 0)$ and the simple support at the right $(x = L)$, namely,

$$\bar{v}(0) = \frac{d\bar{v}}{dx}(0) = 0$$

$$v(L) = 0$$

Note that the shear force V does not appear explicitly in the foregoing equations. This happens because the Bernoulli hypothesis is a constraint against transverse shear strain; hence, the force V is inactive. We recall Eq. (8.10c):

$$V = -\frac{dM}{dx}$$

With this equation the shear can be eliminated from Eqs. (8.10a) and (8.10b) to obtain the equivalents of Eqs. (8.20) and (8.21). The differences are only in the loading: The previous load $\bar{p}$ is *normal* to the deformed axis, so that $\bar{p} = p\cos\theta$; the tangential component, $p\sin\theta$, is absent from Eq. (8.10a). ◆

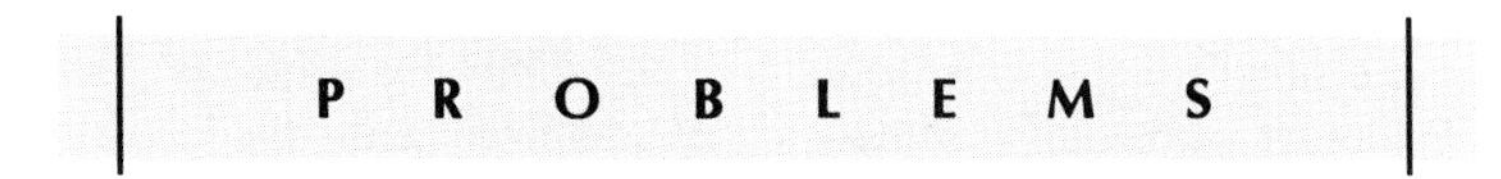

8.7 Suppose that a continuous HI-HO beam is approximated by n straight rigid finite elements, which are pinned together as depicted in Fig. P8.7. Relative rotations between elements is resisted by linear torsional springs with equal moduli β:

$$\beta\,\Delta\theta_n = \beta(\theta_{n+1} - \theta_n) = M_n$$

Treat the rotations (θ_n) as small and neglect any axial force (F) on the elements. From equilibrium (criteria of zero virtual work) obtain the *linear* versions of Eqs.

Fig. P8.7

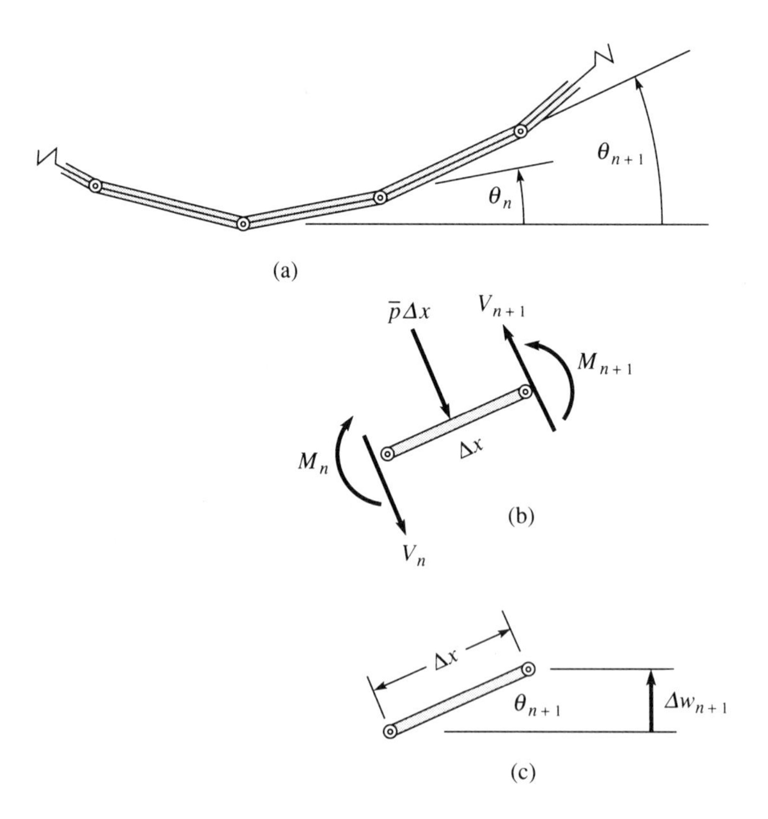

(8.9b, c). From the geometry of Fig. P8.7(c) and the preceding equation, obtain the remaining (linear) equations that collectively govern the discrete model of the hookean beam. By successive elimination of the variables θ_n, M_n, and V_n, obtain the one difference equation:

$$\beta \frac{(w_{n+2} - 4w_{n+1} + 6w_n - 4w_{n-1} + w_{n-2})}{\Delta x^4} = \bar{p}$$

Choose the moduli β such that the foregoing model approaches the continuous beam in the limit $\Delta x \to 0$.

8.8 A symmetric beam is composed of a HI-HO material. Then, under the Bernoulli hypothesis (cross sections remain plane and normal to the axis):

$$M = EI\kappa$$

If axial extension and force are neglected, then the virtual work of internal and external actions has the form

$$\delta w = \delta w_I + \delta w_E = \int_0^L (EI\kappa\delta\kappa + p\delta\bar{v})\, dx$$

Use the additional geometric equations, which govern small rotations, and do the necessary integrations by parts to obtain

$$\delta w = EI\frac{d^2\bar{v}}{dx^2}\delta\theta\Big|_0^L - EI\frac{d^3\bar{v}}{dx^3}\delta\bar{v}\Big|_0^L + \int_0^L \left(EI\frac{d^4\bar{v}}{dx^4} + p\, d\bar{v}\right)\delta\bar{v}\, dx$$

What is then implied by the principle of virtual work? That is, equilibrium requires that $\delta w = 0$ for *arbitrary* virtual displacement, which is consistent with geometric constraints such as end supports.

8.9 Obtain the limit load (P_L or $\bar{p}_L$) for each ideally plastic beam depicted in Fig. P8.9. Express your answers in terms of the given parameters (M_L, L).

Fig. P8.9

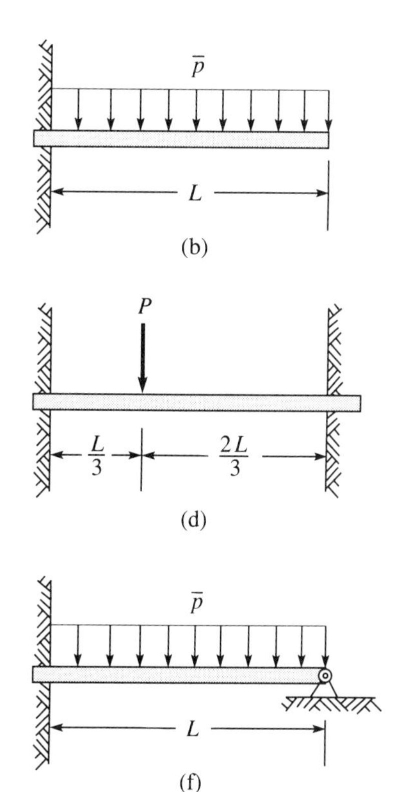

8.3 Principle of Stationary Potential

Conservative Force

A force is conservative if the work done is independent of the path traced by its point of application. To elaborate, let us examine the work of the conservative force acting at point O as it moves along the path OAP of Fig. 8.9. By definition, the same work is done if the force acts upon the point traversing path $OA'P$. Likewise, the work done by the force must be the same if the point travels along any of the closed loops $OABO$, $OA'B'O$, or OCO. The loop can be arbitrarily small. It follows that no work is expended in *any* closed loop. The last observation implies that work is not dissipated, but is *conserved*, as the action completes a closed path.

Fig. 8.9

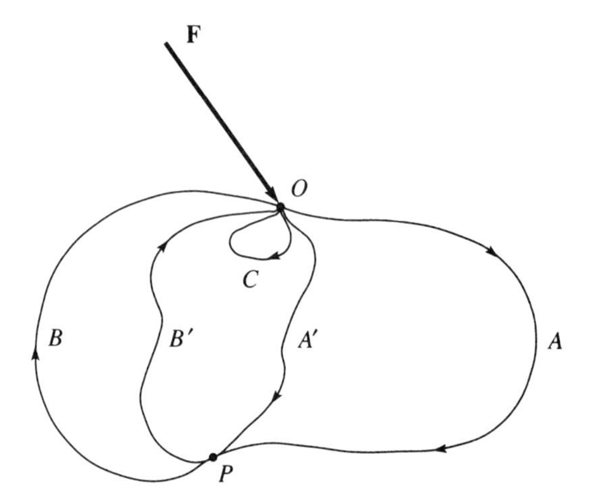

To illustrate conservative forces, consider the examples of Fig. 8.10. The weight W of any object, such as the block of Fig. 8.10(a), is a conservative force exerted by the earth's gravitational field. The work performed by the weight in any small displacement (ds) along the path is

$$dw = -W \cos\theta \, ds$$
$$= -W \, dh$$

Along any path the net work is $-Wh$, where h is the vertical distance shown. If we physically transport the weight along any path from O to O', *we* do positive work. In a virtual displacement of a weight or any constant force (so-called dead load), the virtual work (we do) constitutes a change in potential. The change of potential (available energy) is

$$\Delta w = Wh$$

Fig. 8.10

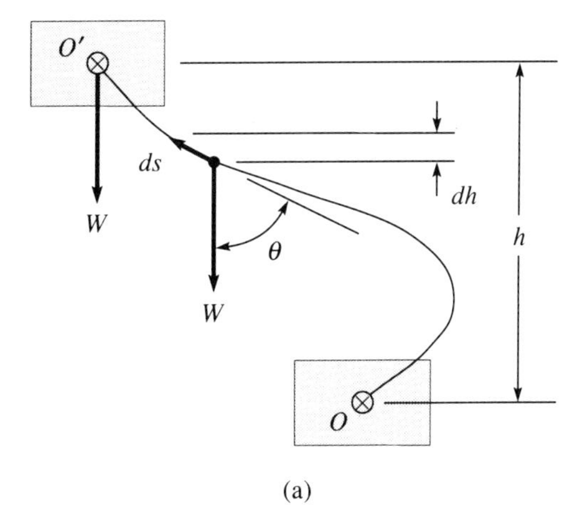

(a)

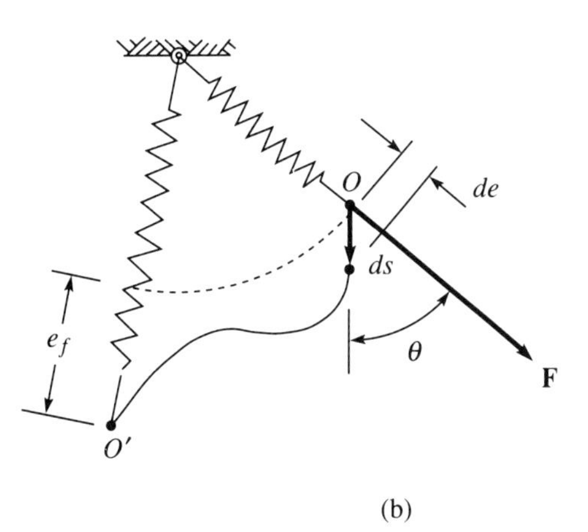

(b)

The elastic spring of Fig. 8.10(b) is also a conservative mechanical system. The work we do in moving end O a small amount ds is

$$dw = F \cos\theta \, ds = F \, de$$

where de is the small extension of the spring. Since the spring is elastic, the force F depends only on the extension; that is, $F = F(e)$. Therefore, the work expended upon the spring in traversing *any* path from O to O' is

$$w = \int_0^{e_f} F(e) \, de$$

Again, the work we do constitutes a change of potential; the potential is energy stored in the spring. No work is dissipated.

Principle of Stationary Potential

A mechanical system is conservative if all the forces, internal and external, are conservative. Constant loads, such as weights, are the most common form of conservative external forces. The internal forces (stresses) in elastic bodies are conservative internal forces. If a system is conservative, then the virtual work (as defined in Sec. 8.1) is the change in potential. It is a small change associated with an arbitrary infinitesimal (first-order in the kinematic variables) change of the configuration. If a function (or functional) is unchanged by such small variations of its variables, it is said to be stationary. Accordingly, the principle of virtual work, when applied to a conservative system, is called the principle of stationary potential: *A conservative mechanical system is in equilibrium if and only if the total potential is stationary.*

The teeter-totter of Fig. 8.1 is a conservative mechanical system. The potential of the system is the gravitational potential of the weights, W_B and W_J. Because we are concerned only with changes of potential, we can refer the potential to any state. Let us regard the potential as zero when the board is horizontal. Then the total potential is

$$V(\theta) = W_B L_B \sin\theta - W_J L_J \sin\theta$$

The criterion of stationary potential requires that the smallest (first-order) change dV vanish for arbitrary change $d\theta$; that is,

$$dV = \frac{dV}{d\theta}\,d\theta = 0 = (W_B L_B - W_J L_J)\cos\theta\,d\theta = 0$$

Again,

$$W_B L_B = W_J L_J$$

The mechanical system of Fig. 8.3(a) is also conservative if the force P is a constant and the springs are elastic. As before, we assume that the springs are undeformed when the links are horizontal; also, we assume that these springs are linear (hookean). If the horizontal position of the links is taken to be the state of zero potential of the external force P, then the total potential is

$$\begin{aligned} V(\theta_1, \theta_2) &= V_E + V_I \\ &= -[PL(\sin\theta_1 + \sin\theta_2)] + \frac{\beta}{2}[\theta_1^2 + (\theta_2 - \theta_1)^2] \end{aligned}$$

The stationary condition requires that

$$dV = \frac{\partial V}{\partial\theta_1}\,d\theta_1 + \frac{\partial V}{\partial\theta_2}\,d\theta_2 = 0$$

This must vanish for *arbitrary* variations; $d\theta_1$ and $d\theta_2$ are independent. Therefore, we obtain the two equations of equilibrium:

$$\frac{\partial V}{\partial\theta_1} = -PL\cos\theta_1 + \beta\theta_1 - \beta(\theta_2 - \theta_1) = 0$$

$$\frac{\partial V}{\partial \theta_2} = -PL\cos\theta_2 + \beta(\theta_2 - \theta_1) = 0$$

These are precisely the equations obtained previously (Sec. 8.1).

EXAMPLE 5

Hookean Truss Let us revisit the planar truss of Fig. 8.5 and assume now that the members are uniform hookean bars. For simplicity, suppose that all are composed of the same steel and have the same cross sections. Lengths of members AC and AD are denoted by L. Then the potential of the internal forces is (see Sec. 4.9)

$$V_I = \frac{EA}{2}\left[L\left(\frac{e_C}{L}\right)^2 + \sqrt{2}L\left(\frac{e_B}{\sqrt{2}L}\right)^2 + L\left(\frac{e_D}{L}\right)^2\right]$$

Recall that the elongations are expressed in terms of the displacement at joint A by Eqs. (8.6a, b, c). Substituting the latter, we obtain

$$V_I = \frac{EA}{2L}\left[v^2 + \frac{1}{2\sqrt{2}}(u+v)^2 + u^2\right]$$

Referred to the initial position of joint A, the potential of the load is

$$V_E = -Pv$$

The stationary criterion for the system is

$$dV = dV_I + dV_E = 0$$

$$= \frac{\partial}{\partial v}(V_I + V_E)\,dv + \frac{\partial}{\partial u}(V_I + V_E)\,du = 0 \tag{8.24}$$

Again, the variation (virtual displacement) is arbitrary; dv and du are independent. It follows that

$$\frac{\partial}{\partial v}(V_I + V_E) = 0, \qquad \frac{\partial}{\partial u}(V_I + V_E) = 0$$

Stated otherwise,

$$\frac{\partial V_I}{\partial v} = -\frac{\partial V_E}{\partial v}, \qquad \frac{\partial V_I}{\partial u} = -\frac{\partial V_E}{\partial u} \tag{8.25a, b}$$

There are two equations in the unknowns v and u obtained previously (Example 7, Sec. 5.7), namely,

$$\left(1 + \frac{1}{2\sqrt{2}}\right)v + \frac{1}{2\sqrt{2}}u = \frac{PL}{EA}$$

$$\frac{1}{2\sqrt{2}}v + \left(1 + \frac{1}{2\sqrt{2}}\right)u = 0$$

Several features of this formulation are noteworthy, since every problem of a discrete hookean body with small displacements shares these attributes: The governing equations are linear in the kinematic variables (u, v); the internal potential is quadratic in those variables. The matrix of coefficients is symmetric. The external potential of the constant load(s) is linear. In general, if the system (e.g., a truss) has n degrees of freedom (e.g., displacements v_i of joints, $i = 1, \ldots, n$), then

$$V_I = \sum_i^n \sum_j^n \tfrac{1}{2} a_{ij} v_i v_j \tag{8.26}$$

$$V_E = -\sum_i^n P_i v_i \tag{8.27}$$

Alternatively, we might employ customary matrix notations: **a** denotes the square matrix with symmetrical elements a_{ij}, **v** denotes the column (matrix) with elements v_i, and **P** denotes the column (matrix) with elements P_i. Of course, some elements might be zero (e.g., some joints may not be loaded). Then

$$V_I = \tfrac{1}{2} \mathbf{v}^T \mathbf{a} \mathbf{v} \tag{8.28}$$

$$V_E = -\mathbf{P}^T \mathbf{v} \tag{8.29}$$

The linear equations governing such systems have the alternative forms

$$\frac{\partial V_I}{\partial v_i} = \sum_j a_{ij} v_j = -\frac{\partial V_E}{\partial v_i} = P_i \tag{8.30a}$$

or

$$\mathbf{a}\mathbf{v} = \mathbf{P} \tag{8.30b}$$ ◆

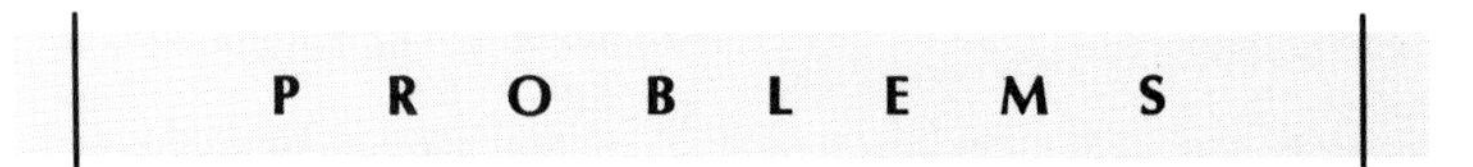

PROBLEMS

8.10 Each member of the plane truss in Fig. P8.10 is made from the same steel ($E = 30 \times 10^3$ ksi), which yields in tension and compression at $\sigma = \pm 50$ ksi. The cross-sectional areas are all $A = 0.800$ in^2. Use the principle of stationary potential to compute the displacement of joint O when $P = 4$ kip. Which member yields first? What is the limit load if the steel is ideally plastic?

Fig. P8.10

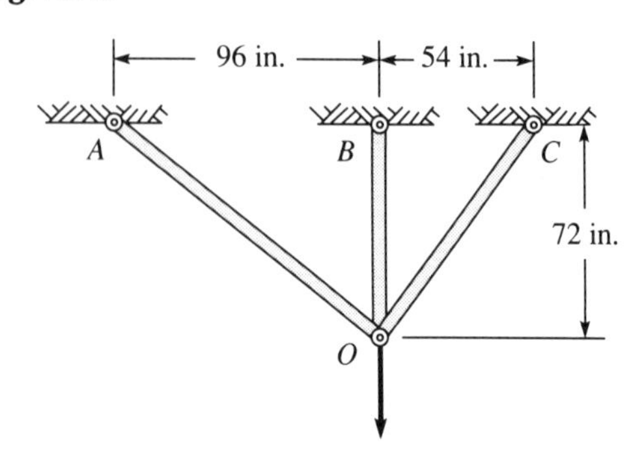

8.11 Members of the rectangular frame in Fig. P8.11 are pinned at A, B, C, and D. HI-HO cables BD, BE, and BF have the same properties ($E = 200 \times 10^3$ MPa) and cross-sectional areas $A = 0.100$ cm^2. The members of the frame are relatively rigid. Determine the displacement of joint D under the action of the load $P = 5$ kN. Compute the maximum tensile stress in the wires.

Fig. P8.11

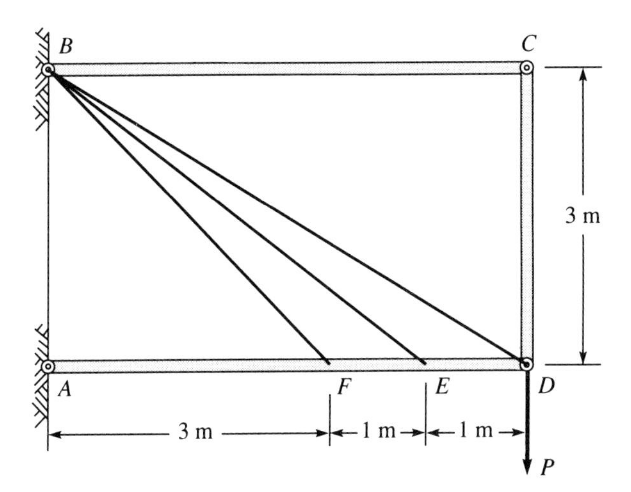

8.12 The three equal legs of the tripod in Fig. P8.12 are pinned to the fixed base at A, B, and C; ABC forms an equilateral triangle with sides 10 ft long. The pin joint O is 20 ft above the base and subjected to the force $P = 1000$ lb acting parallel to the x-axis. The cross-sectional area of the HI-HO legs is 0.500 in.^{2}, the modulus is $E = 10^7$ psi. Determine the displacement of the joint O.

Fig. P8.12

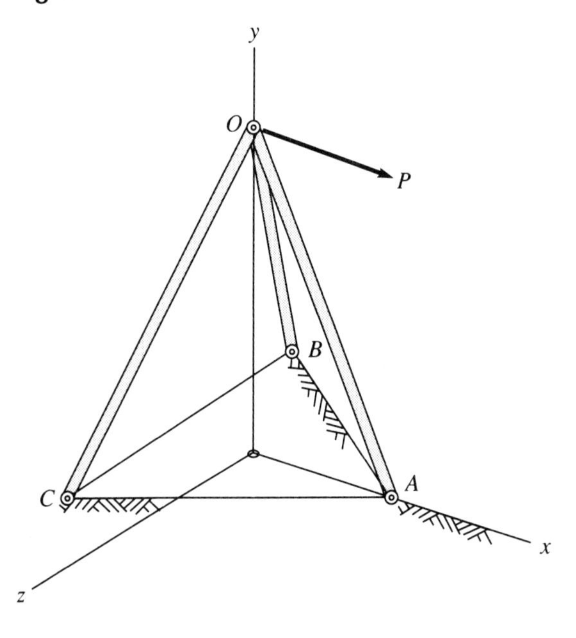

8.13 All members in the plane truss of Fig. P8.13 are made of the same stock, steel bars:

$$E = 30 \times 10^3 \text{ ksi}, \qquad \text{dia.} = 1.00 \text{ in.}$$

$$\sigma_0 = \pm 48 \text{ ksi}$$

Calculate the horizontal and vertical components of the displacement at joint O when the load $P = 10$ kip is applied. Calculate the load that would initiate yielding.

Fig. P8.13

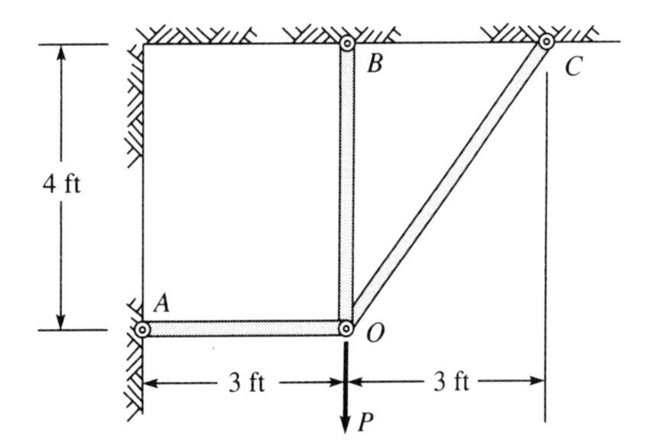

8.14 The members of the plane truss in Fig. P8.14 are made of the same steel stock (modulus E and cross-sectional area are the same). Members I, II, and III form an equilateral triangle. The truss is subject to the vertical load P. Denote the horizontal displacements of joints 1 and 2 by u_1 and u_2 and the vertical displacements by v_1 and v_2. Obtain the 4×4 stiffness matrix (the coefficients of the components u_1, u_2, v_1, and v_2) and verify that the matrix is symmetric.

Fig. P8.14

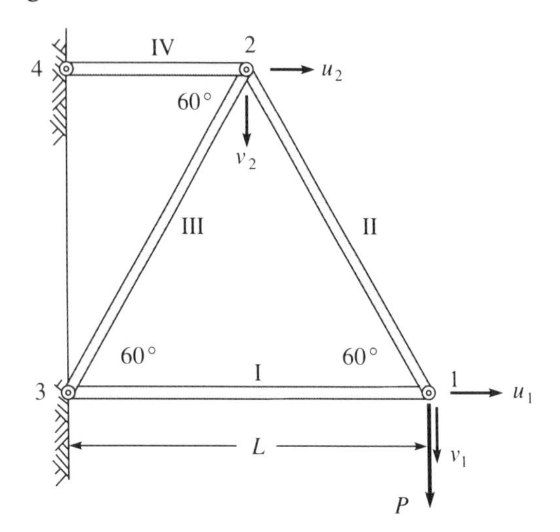

8.15 The three-dimensional truss shown in Fig. P8.15 forms a regular octahedron (faces, such as ABC, are equilateral triangles). All members, including the cross braces in the central square, have the same modulus and cross-sectional areas (EA is the same). Determine the forces in all members by the principle of stationary potential. Note that symmetries reduce the number of degrees of freedom.

Fig. P8.15

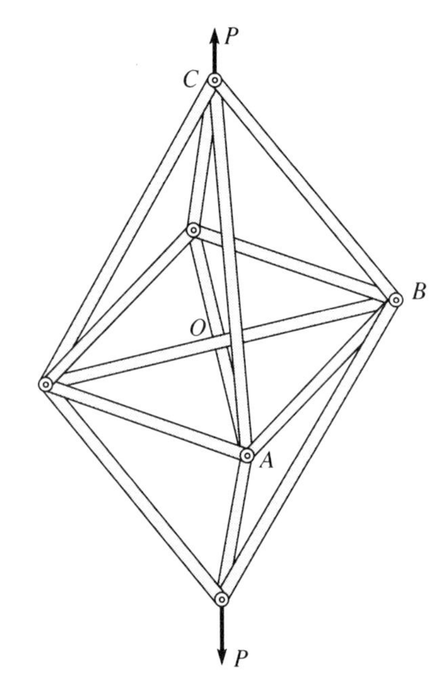

*8.4 Bernoulli-Euler Theory of Symmetric Beams

If a symmetric beam—for example, the beam of Fig. 8.8—is a uniform hookean beam, then the internal potential U_I (per unit length) is obtained in accordance with the theory of Bernoulli-Euler [see Eqs. (5.158) and (4.59)]:

$$U_I = \iint_A \frac{E}{2}(\varepsilon_0 - \kappa y)^2 \, dA$$

With the axis at the centroid,

$$U_I = \frac{EA}{2}\varepsilon_0^2 + \frac{EI}{2}\kappa^2$$

$$= \frac{EA}{2}\varepsilon_0^2 + \frac{EI}{2}\left(\frac{d\theta}{dx}\right)^2$$

The total internal potential is obtained by integrating throughout the beam:

$$V_t = \frac{E}{2}\int_0^L \left[A\varepsilon_0^2 + I\left(\frac{d\theta}{dx}\right)^2\right] dx \tag{8.31}$$

If u and v denote the displacements tangent and normal to the deformed axis (see Fig. 8.6), then by Eqs. (8.17) and (8.18),

$$\delta\varepsilon_0 = \frac{d}{dx}(\delta u) - \kappa\delta v \tag{8.32}$$

$$\delta\kappa = \frac{d^2}{dx^2}(\delta v) + \frac{d}{dx}(\kappa\delta u) \tag{8.33}$$

The potential of the constant load $p(x)$ is

$$V_E = +\int_0^L p(x)\bar{v}\, dx \tag{8.34}$$

The variation ("virtual" change of potential) is

$$\delta V = \delta V_I + \delta V_E = \int_0^L [EA\varepsilon_0\delta\varepsilon_0 + EI\kappa\delta\kappa + p\delta\bar{v}]\, dx \tag{8.35a}$$

$$\delta V = \int_0^L \left\{EA\varepsilon_0\left[\frac{d}{dx}(\delta v) - \kappa\delta u\right] + EI\kappa\left[\frac{d^2}{dx^2}(\delta v) + \frac{d}{dx}(\kappa\delta u)k\right] + p[\delta v\cos\theta + \delta u\sin\theta]\right\} dx \tag{8.35b}$$

This variation δV differs from Eq. (8.19a) only because the force F and couple M are conservative actions. They are now determined by the deformation; specifically,

$$F = EA\varepsilon_0, \qquad M = EI\kappa$$

Let us now simplify our analysis by restricting it to small rotations and deleting nonlinear terms. Then, the component v ($\doteq \bar{v}$) is normal to the initial axis ($\cos\theta \doteq 1$). Instead of Eqs. (8.32) and (8.33) we have the approximations:

$$\varepsilon_0 \doteq \frac{du}{dx}, \qquad \kappa = \frac{d\theta}{dx} \doteq \frac{d^2v}{dx^2}$$

The potential follows from Eqs. (8.31) and (8.34):

$$V = \int_0^L \left[\frac{EA}{2}\left(\frac{du}{dx}\right)^2 + \frac{EI}{2}\left(\frac{d^2v}{dx^2}\right)^2 + pv\right] dx \tag{8.36}$$

Now, this integral is a *functional*; this potential depends on two *functions.* Note that, as in Eq. (8.26), the internal energy is a quadratic in the functions $u(x)$ and $v(x)$, and, as in Eq. (8.27), the external potential is linear in the function $v(x)$. The enforcement of the principle requires that this functional V be stationary ($\delta V = 0$) for arbitrary variations ($\delta u, \delta v$) of the functions $u(x)$ and $v(x)$. Specifically,

$$\delta V = \int_0^L \left[EA\left(\frac{du}{dx}\right)\frac{d}{dx}(\delta u) + EI\left(\frac{d^2v}{dx^2}\right)\frac{d^2}{dx^2}(\delta v) + p\delta v \right] dx = 0$$

As before, the first term and second term must be integrated by parts to obtain

$$\delta V = EA\frac{du}{dx}\delta u\bigg|_0^L + EI\left(\frac{d^2v}{dx^2}\right)\frac{d}{dx}(\delta v)\bigg|_0^L$$

$$- EI\frac{d}{dx}\left(\frac{d^2v}{dx^2}\right)\delta v\bigg|_0^L$$

$$+ \int_0^L \left[-\left(EA\frac{d^2u}{dx^2}\right)\delta u + \left(EI\frac{d^4v}{dx^4} + p\right)\delta v \right] dx = 0$$

For the beam illustrated in Fig. 8.8, we have geometric constraints at the ends:

$$\delta u\bigg|_0 = \delta v\bigg|_0^L = \frac{d}{dx}(\delta v)\bigg|_0 = 0$$

No constraint is imposed upon the displacement $u(L)$ or upon the rotation $dv/dx(L)$. Since these are independent, it follows that

$$EA\frac{du}{dx}\bigg|_L = EI\left(\frac{d^2v}{dx^2}\right)\bigg|_L = 0$$

These are the conditions that the axial force F and couple M vanish at $x = L$. Throughout the beam the displacements $\delta u(x)$ and $\delta v(x)$ are arbitrary continuous functions. Therefore, the stationary criterion is satisfied if and only if each parenthetical term of the integrand vanishes:

$$EA\frac{d^2u}{dx^2} = \frac{dF}{dx} = 0 \tag{8.37}$$

$$EI\left(\frac{d^4v}{dx^4}\right) = -\frac{dV}{dx} = -p(x) \tag{8.38}$$

These two linear differential equations govern the stretching (u) and small transverse deflection (v) of the uniform hookean beam. Two end conditions on u and/or du/dx accompany Eq. (8.37). Four end conditions (two at each end) on v, dv/dx, d^2v/dx^2, or d^3v/dx^2 (deflection, rotation, bending couple, or shear force) accompany Eq. (8.38).

Whether the mechanical system is *discrete* (the configuration is defined by a number of discrete variables as the truss of the preceding section) or *continuous* (the configuration is defined by continuous functions as the beam), the potential is expressed as a function (or functional) of the kinematic variables (e.g., displacements), which define the configuration. In any case the internal forces do not enter explicitly. Even the behavior of the material is expressed by an internal energy (e.g., a hookean member of a truss has energy $U_I = EAe^2/2L$).

The principle of stationary potential provides a means to formulate the mathematical description of a continuous body, because the potential of the beam (8.36) leads to Eqs. (8.37) and (8.38). In practice the principle provides the basis for approximating a *continuous* body by a *discrete* model. Such is the method of finite elements, wherein the continuous body is subdivided into small but finite elements that are joined at interelement nodes. The displacements within each element are approximated, usually by simple polynomials, which, in turn, are prescribed by the nodal displacements. In this way the continuous body is replaced by a discrete mechanical system; the *functional*, which gives the potential [such as Eq. (8.36)], is replaced by a *function* of the discrete variables. Long before the advent of the finite element (circa 1956) Lord Rayleigh (1877) introduced the notion of approximating the displacements of a continuous body by functions that reduced the potential to a function of discrete variables. The variables in any such scheme are obtained by the stationary condition for the approximated potential. Such approximations are tantamount to physical constraints; they serve to inhibit the continuous body from assuming its most natural configuration of least potential. Accordingly, the discrete models (approximations) tend to be excessively stiff. We illustrate Rayleigh's method with a Bernoulli-Euler beam.

EXAMPLE 6

A Cantilever Beam The cantilever beam in Fig. 8.11 is clamped at the left end ($x = 0$) and simply supported at the right ($x = L$). It is subjected to the uniform load $p(x) = \bar{p}$ and the concentrated load P at the midpoint ($x = L/2$).

Fig. 8.11

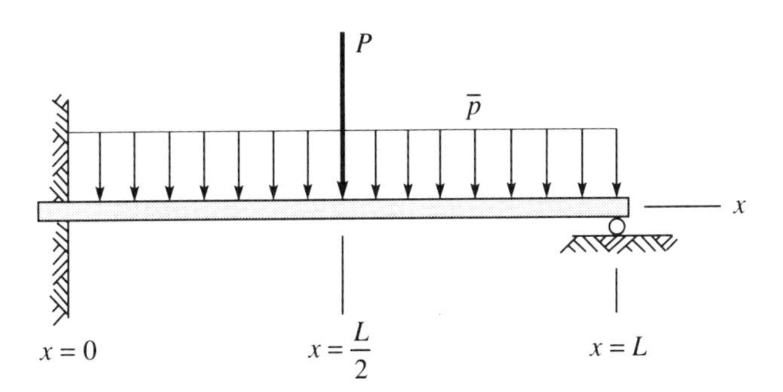

Cantilever Beam

At best, we choose an approximation of $v(x)$ that satisfies the geometric constraints, $v(0) = v'(0) = 0$ and $v(L) = 0$ ($v' = dv/dx$), and the natural conditions at the end, $M(L) = EIv''(L) = 0$ ($v'' = d^2v/dx^2$). At least, we must adopt a function $v(x)$ that satisfies the geometric constraints. To illustrate, let

$$v(x) \doteq Ax^2 + Bx^3 + Cx^4$$

To satisfy the geometric requirements,

$$v(L) = 0, \qquad C = -\frac{A}{L^2} - \frac{B}{L}$$

$$v(x) = A\left(1 - \frac{x^2}{L^2}\right)x^2 + B\left(1 - \frac{x}{L}\right)x^3$$

Omitting the extensional energy of Eq. (8.36), we obtain our approximation to the potential of this cantilever:

$$V = \int_0^L \left[\frac{EI}{2}(v'')^2 + \bar{p}v\right]dx + Pv(L/2)$$

$$\doteq \frac{EI}{5}(42LA^2 + 44L^2AB + 12L^3B^2) + \frac{\bar{p}}{60}(8L^3A + 3L^4B) + \frac{P}{16}(3L^2A + L^3B)$$

The stationary criteria provide the two equations that determine constants A and B:

$$\frac{\partial V}{\partial A} = \frac{EI}{5}(84LA + 44L^2B) + \frac{4\bar{p}L^3}{30} + \frac{3PL^2}{16} = 0$$

$$\frac{\partial V}{\partial B} = \frac{EI}{5}(44L^2A + 24L^3B) + \frac{\bar{p}L^4}{20} + \frac{PL^3}{16} = 0$$

or

$$84A + 44LB = -\frac{2\bar{p}L^2}{3EI} - \frac{15PL}{16EI}$$

$$44A + 24LB = -\frac{\bar{p}L^2}{4EI} - \frac{5PL}{16EI}$$

The solution follows:

$$A = -\frac{\bar{p}L^3}{16EI} - \frac{7}{64}\frac{PL}{EI}$$

$$B = \frac{5\bar{p}L^2}{48EI} + \frac{3P}{16EI}$$

To compare with the exact solution of the continuum, let us examine the values at the midpoint ($x = L/2$):

$$(L/2) = -\frac{\bar{p}L^4}{192EI} - \frac{9}{1024}\frac{PL^3}{EI}$$

The exact value is

$$(L/2) = -\frac{\bar{p}L^3}{192} - \frac{7}{764}\frac{PL^3}{EI}$$

Note that the contribution from the distributed load $\bar{p}$ is the same, because the quartic chosen embraces the exact solution. The contribution from the concentrated load P is 4% less; the approximate system is too stiff. The exact solution of the latter is a function with discontinuity in the third derivative ($V = -EIv'''$) at $x = L/2$ (see Appendix C). ◆

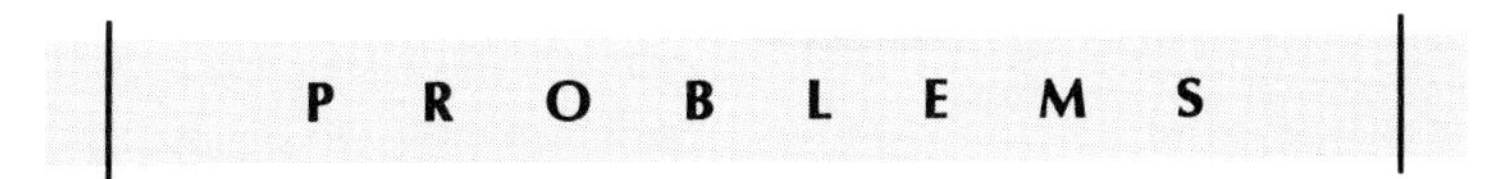

PROBLEMS

8.16 Treat by Rayleigh's method the problem of a simple cantilever fixed at one end ($x = 0$) and subject only to a concentrated load at the unsupported end ($x = L$). Use the approximation (a) $w(x) \doteq ax^2$, then (b) $ax^2 + bx^3$, and, finally, (c) $ax^2 + bx^3 + cx^4$. Compare your answers with the exact solution of the differential equations and the prescribed end conditions.

8.17 The HI-HO beam of Fig. P8.17(a) is subject only to a bending couple at the unsupported end. Since every section is subject to the same moment ($M = EI\kappa$), the rotation of the end is simply $\theta = L\kappa$. The mechanical system of Fig. P8.17(b) consists of N rigid links of length $\ell = L/N$, joined by smooth hinges and interconnected by torsional springs that resist relative rotations; that is, $\beta(\theta_n - \theta_{n-1}) = M$. Choose the spring modulus β such that the internal energies are the same for a typical segment $\Delta x = \ell$ of each system. Compare the governing equations in the limit $N = L/\ell \to \infty$. *Note:*

$$\int_a^{a+l} \kappa\, ds = \kappa l = \theta_n - \theta_{n-1}$$

Fig. P8.17

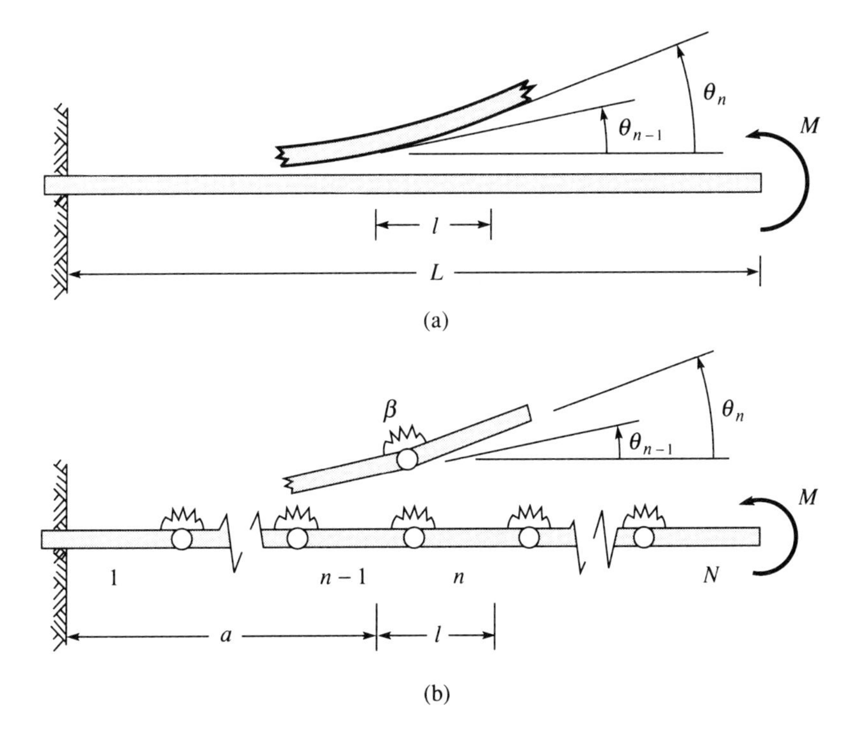

8.18 A HI-HO beam is subjected to a uniform load $p(x) = \bar{p}$. It is clamped at both ends; that is, $v(0) = v'(0) = v(L) = v'(L) = 0$. A simple function that satisfies these conditions is $v(x) = A(1 - \cos 2\pi x/L)$. Use Rayleigh's method to determine A and compare this approximation with the exact solution.

8.19 The deflection of a simply supported beam can be represented by a Fourier series of the form

$$v(x) = \sum a_n \sin n\frac{\pi}{L}x, \qquad n = 1, 2, \dots$$

Likewise, the distribution of load $p(x)$ can be represented by a Fourier series:

$$p(x) = \sum p_n \sin n\frac{\pi}{L}x$$

Note the orthogonality of the functions in the interval (O, L), the length of the beam:

$$\int_0^L \sin n\frac{\pi x}{L} \sin m\frac{\pi x}{L}\,dx = \begin{cases} 0, & n \neq m \\ \dfrac{L}{2}, & n = m \end{cases}$$

Write the expression for the potential of the HI-HO symmetric beam with simply supported ends, use the Fourier series, impose the conditions of stationary potential $(dV/da_n = 0)$, and show that

$$a_n = \frac{p_n L^4}{n^4 \pi^4 EI}$$

8.20 Use the leading terms of the Fourier expansion (see Prob. 8.19) and obtain approximations $v(x)$ with one term (a_1) and then with two terms $(a_1$ and $a_3)$, and compare the approximate values for $v(L/2)$ with the exact solutions of the governing equations for a simply supported beam subject to the following loads:

a. $p(x) = \bar{p}$ (constant)

b. $p(x) = \bar{p} \sin \pi x/L$

c. A concentrated load P at $x = a$

d. $p(x) = \bar{p}x/L$

*8.5 Complementary Potential and Castigliano's Theorem

In Sec. 4.9 we discussed the strain energy of an elastic body. This internal energy (per unit volume) is a function of the six components of strain:

$$u(\varepsilon_x, \varepsilon_y, \varepsilon_z, \gamma_{xy}, \gamma_{xz}, \gamma_{yz}) \tag{8.39}$$

The stress components derive from this function according to Eqs. (4.62) and (4.63):

$$\begin{aligned} \sigma_x &= \frac{\partial u}{\partial \varepsilon_x}, & \sigma_y &= \frac{\partial u}{\partial \varepsilon_y}, & \sigma_z &= \frac{\partial u}{\partial \varepsilon_z} \\ \tau_{xy} &= \frac{\partial u}{\partial \gamma_{xy}}, & \tau_{xz} &= \frac{\partial u}{\partial \gamma_{xz}}, & \tau_{yz} &= \frac{\partial u}{\partial \gamma_{yz}} \end{aligned} \tag{8.40}$$

These equations are an implicit form of the stress-strain relations. For example, in the case of simple stress σ_x and hookean behavior

$$u = E\frac{\varepsilon_x^2}{2}, \qquad \sigma_x = \frac{\partial u}{\partial \varepsilon_x} = E\varepsilon_x$$

We presume that Eqs. (8.40) can be inverted to express the strains in terms of the stresses; for example, in the simple case, $\varepsilon_x = \sigma_x/E$.

We now define an internal complementary energy (per unit volume):

$$u_c \equiv \sigma_x \varepsilon_x + \sigma_y \varepsilon_y + \sigma_z \varepsilon_z + \tau_{xy}\gamma_{xy} + \tau_{xz}\gamma_{xz} + \tau_{yz}\gamma_{yz} - u \tag{8.41}$$

Since the strains can (presumably) be expressed in terms of the stresses, the complementary function can be regarded as a function of the components of stress:

$$u_c = u_c(\sigma_x, \sigma_y, \sigma_z, \tau_{xy}, \tau_{xz}, \tau_{yz}) \tag{8.42}$$

In accordance with Eqs. (8.39) and (8.41) and the chain rule, the differential of the function (8.42) has the forms:

$$\begin{aligned} du_c = {} & d\sigma_x \varepsilon_x + d\sigma_y \varepsilon_y + d\sigma_z \varepsilon_z \\ & + d\tau_{xy}\gamma_{xy} + d\tau_{xz}\gamma_{xz} + d\tau_{yz}\varepsilon_{yz} \\ & + \Bigg[\sigma_x d\varepsilon_x + \sigma_y d\varepsilon_y + \sigma_z d\varepsilon_z \\ & + \tau_{xy} d\gamma_{xy} + \tau_{xz} d\gamma_{xz} + \tau_{yz} d\varepsilon_{yz} \\ & - \frac{\partial u}{\partial \varepsilon_x} d\varepsilon_x - \frac{\partial u}{\partial \varepsilon_y} d\varepsilon_y - \frac{\partial u}{\partial \varepsilon_z} d\varepsilon_z \\ & - \frac{\partial u}{\partial \gamma_{xy}} d\gamma_{xy} - \frac{\partial u}{\partial \gamma_{xz}} d\gamma_{xz} - \frac{\partial u}{\partial \gamma_{yz}} d\gamma_{yz} \Bigg] \end{aligned} \tag{8.43a}$$

and

$$\begin{aligned} du_c = {} & \frac{\partial u}{\partial \sigma_x} d\sigma_x + \frac{\partial u_c}{\partial \sigma_y} d\sigma_y + \frac{\partial u_c}{\partial \sigma_z} d\sigma_z \\ & + \frac{\partial u_c}{\partial \tau_{xy}} d\tau_{xy} + \frac{\partial u_c}{\partial \tau_{xz}} d\tau_{xz} + \frac{\partial u_c}{\partial \tau_{yz}} d\tau_{yz} \end{aligned} \tag{8.43b}$$

By virtue of Eqs. (8.40), the entire bracketed term of Eq. (8.43a) vanishes. We presume that the six components of strain uniquely define the strained state and, therefore, the internal energy, which, in turn, uniquely determines the six components of stress. These components of stress are likewise independent; any one can be subject to a variation (e.g., $d\sigma_x \neq 0$) while the remaining components remain unchanged (e.g., $d\sigma_y = d\sigma_z = d\tau_{xy} = d\tau_{xz} = d\tau_{yz} = 0$). Then the right sides of Eqs. (8.43a) and (8.43b) are equivalent if—but only if—

$$\begin{aligned} \varepsilon_x &= \frac{\partial u_c}{\partial \sigma_x}, & \varepsilon_y &= \frac{\partial u_c}{\partial \sigma_y}, & \varepsilon_z &= \frac{\partial u_c}{\partial \sigma_z} \\ \gamma_{xy} &= \frac{\partial u_c}{\partial \tau_{xy}}, & \gamma_{xz} &= \frac{\partial u_c}{\partial \tau_{xz}}, & \gamma_{yz} &= \frac{\partial u_c}{\partial \tau_{yz}} \end{aligned} \tag{8.44}$$

The complementary character of the energies u and u_c is evident on comparison of Eqs. (8.40) and (8.44).

Let us now apply the foregoing notions to a discrete system (e.g., a truss). Let the discrete kinematic variables v_i $(i = 1, \ldots, n)$ (e.g., displacements of a joint) define the configuration. Then the internal energy of the system is a function of those discrete variables:

$$V_I = V_I(v_1, \ldots, v_n) \tag{8.45}$$

By the principle of stationary potential [see Eq. (8.30a)], the associated "forces" are

$$P_i = \frac{\partial V_I}{\partial v_i} \qquad (i = 1, \ldots, n) \tag{8.46}$$

Note that the variables v_i need not be displacements, but any kinematic variable that serves to define the configuration—for example, the angle θ of Fig. 8.1. Then the quantity P_i is not necessarily a force, but the conjugate dynamic variable. The latter is defined by the requirement that $P_i dv_i$ is the increment of work done by this "generalized force." For example, if v_i is a rotation, then P_i is a couple.

Here, too, we presume that the n equations (8.46) between "forces" (P_i) and "displacements" (v_i) can be inverted to express the displacements in terms of the forces.

As before [see Eqs. (8.41) and (8.42)], we define a complementary "energy":

$$V_c(P_1, \ldots, P_n) = \sum_{i=1}^{n} P_i v_i - V_I(v_1, \ldots, v_n) \tag{8.47}$$

Again, the differential of the function V_c has the forms

$$dV_c = \sum_{i=1}^{n} dP_i v_i + \left[\sum_{i=1}^{n} P_i \, dv_i - \sum_{i=1}^{n} \frac{\partial V_I}{\partial v_i} dv_i \right] \tag{8.48}$$

$$= \sum_{i=1}^{n} \frac{\partial V_c}{\partial P_i} dP_i \tag{8.49}$$

By virtue of Eq. (8.46) the bracketed term vanishes. Again, we presume that the n forces (P_i) are independent, just as are the n displacements (v_i). Then the two equations, Eqs. (8.48) and (8.49), are equivalent if—but only if—

$$v_i = \frac{\partial V_c}{\partial P_i} \tag{8.50}$$

Much as Eq. (8.46) expresses the force P_i in terms of the displacements v_i, so Eq. (8.50) expresses the displacement v_i in terms of the forces P_i.

In general, it is difficult to obtain an explicit expression of the complementary energy $V_c(P_1, \ldots, P_n)$. But, in case of small deflections of hookean bodies, this is possible. Then the internal energy V_I is a quadratic form, such as Eq. (8.26) or (8.28):

$$V_I = \tfrac{1}{2} \sum_{i=1}^{n} \sum_{j=1}^{n} a_{ij} v_i v_j \tag{8.51a}$$

or

$$V_I = \tfrac{1}{2} \mathbf{v}^T \mathbf{a} \mathbf{v} \tag{8.51b}$$

Then the forces are given by the linear equations

$$P_i = \sum_{j=1}^{n} a_{ij} v_j \tag{8.52a}$$

or

$$\mathbf{P} = \mathbf{a}\mathbf{v} \tag{8.52b}$$

The inverse is

$$v_i = \sum_{j=1} a_{ij}^{-1} P_j \tag{8.53a}$$

or

$$\mathbf{v} = \mathbf{a}^{-1}\mathbf{P} \tag{8.53b}$$

In accordance with definition (8.47), the complementary energy is

$$V_c = \sum_{i=1}^{n} P_i v_i - V_I \tag{8.54a}$$

or

$$V_c = \mathbf{P}^T\mathbf{v} - V_I \tag{8.54b}$$

We can substitute Eq. (8.53) into Eq. (8.51) and also into Eq. (8.54) to obtain, respectively,

$$V_I = \tfrac{1}{2} \sum_{i=1}^{n} \sum_{j=1}^{n} a_{ij}^{-1} P_i P_j \tag{8.55a}$$

$$= \tfrac{1}{2}\mathbf{P}^T\mathbf{a}^{-1}\mathbf{P} \tag{8.55b}$$

and

$$V_c = V_I \tag{8.56}$$

In other words, the internal energy V_I and its complement V_c are equal in the case of linear behavior. This is true only if the material is hookean and the displacements and especially the rotations are small. Then the relations between forces (P_i) and displacements (v_i) are linear, and, according to Eq. (8.56), the complementary energy *is* the internal energy *but* expressed in terms of the forces. Then, in keeping with Eqs. (8.50), (8.55), and (8.56),

$$v_i = \frac{\partial V_I}{\partial P_i} = \sum_{j=1}^{n} a_{ij}^{-1} P_j \tag{8.57a}$$

or

$$\mathbf{v} = \mathbf{a}^{-1}\mathbf{P} \tag{8.57b}$$

In this form, the result [Eq. (8.57a)] is known as *Castigliano's theorem.* Let us state the theorem: If $V_I(P_1, \ldots, P_n)$ is the internal energy (potential) expressed in terms of the discrete external "forces" (P_i), then the *conjugate* "displacement" (v_i) is given by Eq. (8.57a). Again, a force and a displacement are conjugate in the sense that the increment of work done by said force is expressed by $P_i dv_i$.

In general, it is difficult to express the internal energy in terms of the applied forces. The internal energy is variable throughout a body, depending on the state of strain. Although we can usually invert the strain-stress relations and express the internal energy in terms of stresses, the stresses are readily expressed in terms of external forces only in certain simple statically determinate problems. To illustrate the theorem, we consider some simple problems of this type.

EXAMPLE 7

A Statically Determinate Truss The truss of Fig. 8.12 is the system examined previously (Example 6 in Chapter 5, p. 287). The hookean members, AB and AC, are made of the same steel and have the same cross-sectional areas.

Fig. 8.12

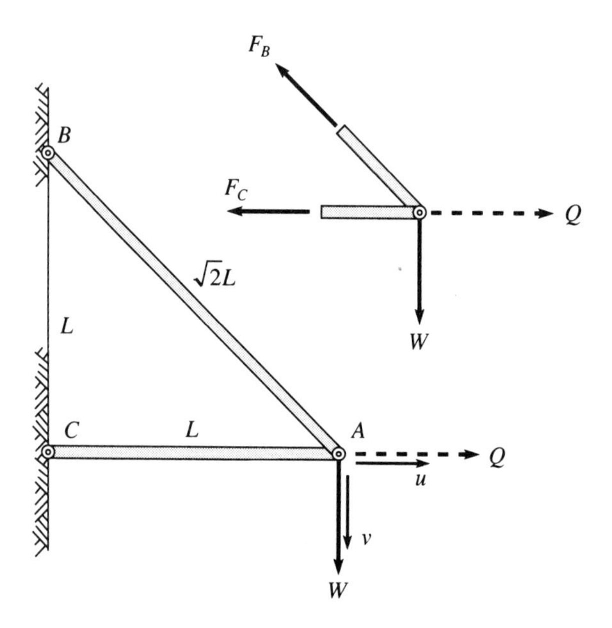

From the free-body diagram and equilibrium,

$$F_B = \sqrt{2}W, \qquad F_C = -W \tag{8.58a, b}$$

Since these are hookean members, the complementary energy is the strain energy, but expressed in terms of the forces: The energies in the individual members B and C are

$$V_B = \frac{EA}{2\sqrt{2}L} e_B^2, \qquad V_C = \frac{EA}{2L} e_C^2$$

However,

$$e_B = \frac{F_B\sqrt{2}L}{EA}, \qquad e_C = \frac{F_C L}{EA}$$

Therefore,

$$V_B = \frac{F_B^2 L}{\sqrt{2}EA} = \frac{\sqrt{2}W^2 L}{EA} \tag{8.59a}$$

$$V_C = \frac{F_C^2 L}{2EA} = \frac{W^2 L}{2EA} \tag{8.59b}$$

The total internal energy is the sum

$$V = V_B + V_C = \frac{W^2 L}{2EA}(2\sqrt{2} + 1) \tag{8.60}$$

In accordance with Eq. (8.57a), the displacement v in the direction of the load W is

$$v = \frac{\partial V}{\partial W} = \frac{WL}{EA}(2\sqrt{2} + 1) \tag{8.61}$$

To obtain the component u of the displacement at joint A, we introduce a "dummy" load Q in that direction. Then the free-body diagram and the equations of equilibrium also include that force, shown dotted in Fig. 8.12. In place of Eqs. (8.58a,b), (8.59a,b), and (8.60), we obtain

$$F_B = \sqrt{2}W, \qquad F_C = -W + Q$$

$$V_B = \frac{\sqrt{2}W^2 L}{EA}, \qquad V_C = \frac{(W^2 - 2WQ + Q^2)L}{2EA}$$

$$V = \frac{L}{2EA}[(2\sqrt{2} + 1)W^2 - 2WQ + Q^2]$$

The dispalcement u is also obtained in accordance with Eq. (8.57a):

$$u = \frac{\partial V}{\partial Q} = \frac{L}{EA}(-W + 2Q) \tag{8.62}$$

In our problem the only load is the force W; the dummy load Q is introduced merely as an artiface to obtain displacement u. Accordingly, in Eq. (8.62) we set $Q = 0$ to obtain the actual displacement under load W.

$$u = -\frac{WL}{EA}$$

The reader can compare the solution and results with those presented previously (Example 6 in Chapter 5, p. 288). ◆

EXAMPLE 8

Statically Indeterminate Truss The truss of Fig. 8.13 (also Fig. 5.20, Sec. 5.7) has the additional member AD. This truss is statically indeterminate; the forces in the members F_C, F_B, and F_D cannot be determined by statics. We temporarily remove member AD and impose the force F_D on the modified system. Equilibrium of the modified system provides the two equations

$$F_B = \sqrt{2}(W - F_D), \qquad F_C = -(W - F_D) \tag{8.63a, b}$$

Fig. 8.13

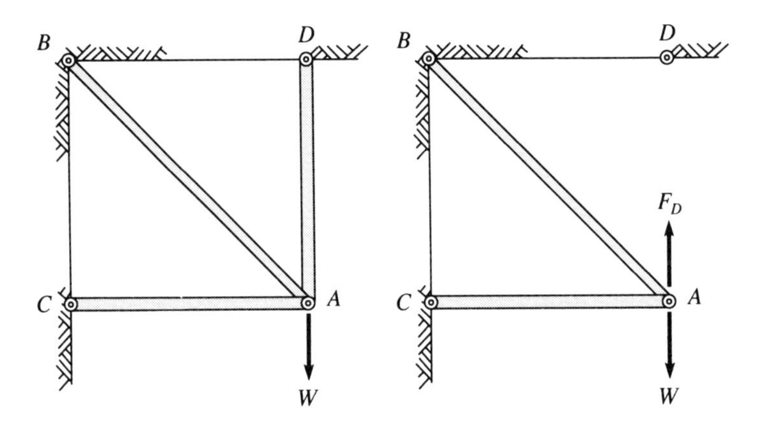

Following the steps of the preceding example, we have in place of Eq. (8.60)

$$\begin{aligned} V &= V_B + V_C \\ &= \frac{(W - F_D)^2 L}{2EA}(2\sqrt{2} + 1) \end{aligned} \tag{8.64}$$

Now the displacement associated with the unknown force F_D is $-v$. By Eq. (8.57a) this displacement is

$$-v = \frac{\partial V}{\partial F_D} = -\frac{(W - F_D)L}{EA}(2\sqrt{2} + 1) \tag{8.65a}$$

Also, in reality,

$$v = e_D = \frac{F_D L}{EA} \tag{8.65b}$$

Equating Eq. (8.65a) and (8.65b) and solving for the unknown F_D, we obtain

$$F_D = \frac{(4 + \sqrt{2})W}{2(2 + \sqrt{2})} \tag{8.66}$$

With the result, Eq. (8.66), the forces of Eqs. (8.63a, b) are expressed in terms of the load W as well as the energy V of Eq. (8.64) and the displacement v of Eq. (8.65).

$$V = V_B + V_C + V_D = \frac{(4+\sqrt{2})}{2(4+2\sqrt{2})}\frac{W^2L}{EA}$$

$$v = \frac{\partial V}{\partial W} = \frac{(4+\sqrt{2})}{(4+2\sqrt{2})}\frac{WL}{EA} \quad \blacklozenge$$

EXAMPLE 9

Symmetric Hookean Beam For expediency, Castigliano's theorem [Eq. (8.57a)] is derived for a discrete system. However, the theorem applies equally to a continuous body, such as the symmetric hookean beam of Fig. 8.14 (see Example 6 and Fig. 6.21 on pp. 416–417). If the internal energy is expressed in terms of a discrete external force, then the conjugate displacement is given by Eq. (8.57a).

The reactions upon the beam of Fig. 8.14(b), like the internal forces of the truss in Fig. 8.13, are statically indeterminate. Again, we can temporarily express the internal bending couple and then the energy in terms of an unknown reaction R_B. The equilibrium condition for the free-body diagram of Fig. 8.14(c) gives

$$M = R_B x - \frac{\bar{p}x^3}{6L} \tag{8.67}$$

According to the Bernoulli theory, the internal energy is attributed solely to the bending. The energy per unit length is

$$u_I = \iint_A \frac{E\varepsilon^2}{2}\,dA = \iint_A \frac{E}{2}(-\kappa y)^2\,da = \frac{EI}{2}\kappa^2$$

Since $\kappa = M/EI$,

$$u_I = \frac{M^2}{2EI}$$

Then, by means of Eq. (8.67), the internal energy for the entire beam is

$$V_I = \int_0^L u_I\,dx = \frac{1}{2EI}\int_0^L \left(R_B x - \bar{p}\frac{x^3}{6L}\right)^2 dx$$

$$= \frac{L^3}{2EI}\left[\frac{R_B^2}{3} - \frac{R_B\bar{p}L}{15} + \frac{\bar{p}^2L^2}{252}\right]$$

The simple support at the right end requires that the displacement vanish. That displacement v_B is conjugate to the force R_B; therefore, by Eq. (8.57a),

$$v_B = \frac{\partial V_I}{\partial R_B} = \frac{L^3}{2EI}\left[\frac{2}{3}R_B - \frac{\bar{p}L}{15}\right] = 0$$

Fig. 8.14

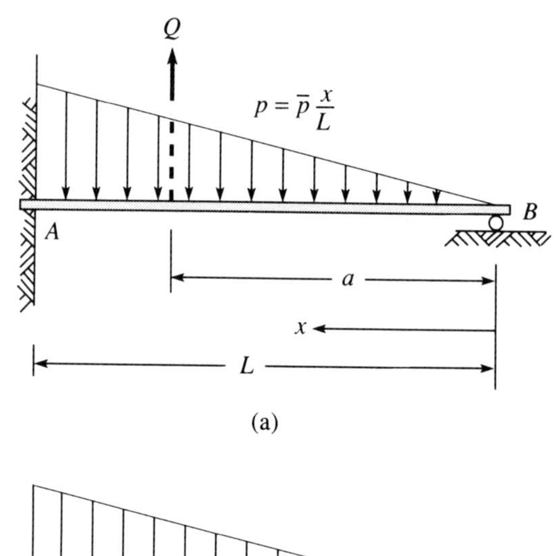

(a)

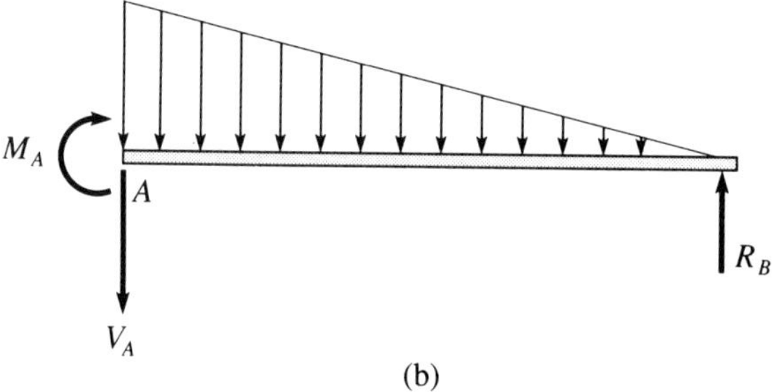

(b)

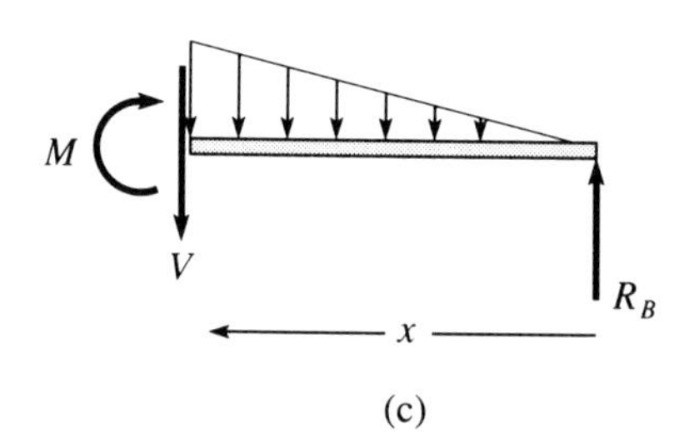

(c)

or

$$R_B = +\frac{\bar{p}L}{10}$$

With this result the bending couple and then the internal energy are expressed throughout the beam in terms of the load $\bar{p}$. The displacement $w(a)$ at any location $(x = a)$ can be obtained by introducing a "dummy" force Q, as shown by the dashed line in Fig. 8.14(a). The conjugate displacement is given by

$$w(a) = \left.\frac{\partial V_I}{\partial Q}\right|_{Q=0} \quad \blacklozenge$$

EXAMPLE 10

Bending and Twisting of a Curved Rod If a straight rod is composed of HI-HO material and has a circular cross section, then the combined action of a bending couple M and twisting couple T causes axial extensional strain ε_x and shear strain $\gamma_{x\theta}$ in accordance with the theories of Secs. 5.11 and 5.8. Specifically,

$$\varepsilon_x = -\kappa_y, \qquad \gamma_{x\theta} = r\frac{d\phi}{dx} \tag{8.68a, b}$$

where

$$\kappa = \frac{M}{EI}, \qquad \frac{d\phi}{dx} = \frac{T}{GJ} \tag{8.69a, b}$$

According to Eqs. (4.59) and (4.60), the strain energy (per unit volume) is

$$u = \frac{E}{2}\varepsilon_x^2 + \frac{G}{2}\gamma_{x\theta}^2 \tag{8.70}$$

Then, the strain energy (per unit length) of the circular rod is

$$\bar{u} = \frac{1}{2}\iint_A (E\varepsilon_x^2 + G\gamma_{x\theta}^2)\,dA \tag{8.71a}$$

$$= \frac{1}{2}EI\kappa^2 + \frac{1}{2}GJ\left(\frac{d\phi}{dx}\right)^2 \tag{8.71b}$$

$$= \frac{1}{2EI}M^2 + \frac{1}{2GJ}T^2 \tag{8.71c}$$

These equations are strictly applicable to a straight rod.

Consider now the rod of Fig. 8.15, wherein the centroidal line has circular form in the unloaded state. The curvature $(1/R)$ is *relatively* small, $r \ll R$, where r denotes the radius of the circular cross section. Then the behavior of any small segment $(ds = R\,d\beta)$ is nearly that of a straight rod, and with little error Eqs. (8.68), (8.69), and (8.71) are applicable.

The curved rod is fixed at O and subject to the couple C at end A. We require the displacement of end A, which is normal to the plane. To obtain that displacement by Castigliano's theorem, we place the dummy load P at A normal to the plane. To express the total internal energy in terms of couple C and force P, we consider the free-body diagram of Fig. 8.15(b). Equilibrium provides the three equations

$$M = -C\cos\beta - PR\sin\beta \tag{8.72a}$$

$$T = C\sin\beta + PR(1 - \cos\beta) \tag{8.72b}$$

$$Q = P \tag{8.72c}$$

Fig. 8.15

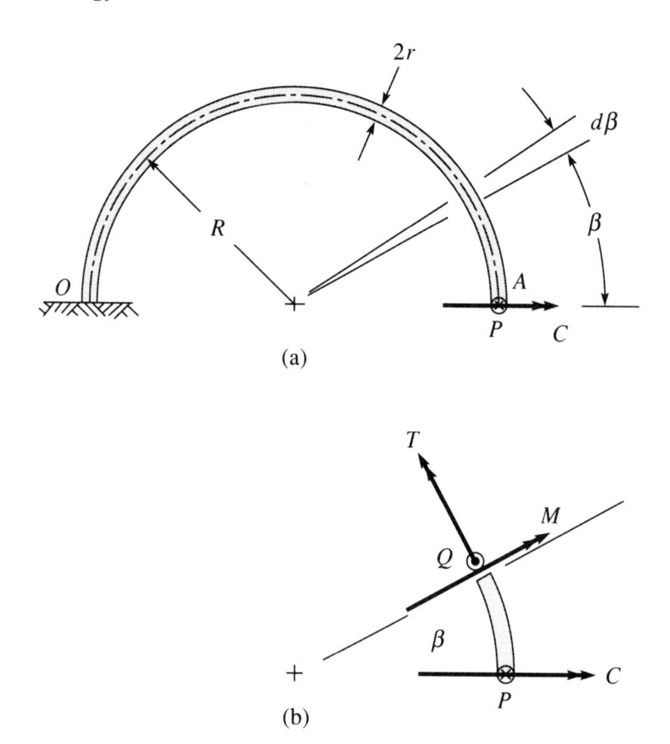

By our theory of beams the deformation and energy caused by the shear force Q is neglected. The total internal energy is obtained by substituting the couples M and T of Eqs. (8.72a, b) into Eq. (8.71c) and integrating throughout the curved rod $(0 \leq \beta \leq \pi)$:

$$V_I = \frac{R}{2} \int_0^{\pi} \left[\frac{1}{EI}(C \cos \beta + PR \sin \beta)^2 + \frac{1}{GJ}(C \sin \beta + PR - PR \cos \beta)^2 \right] d\beta$$

The required displacement is obtained according to Eq. (8.57a):

$$v = \left. \frac{\partial V_I}{\partial P} \right|_{P=0} = R^2 \int_0^{\pi} \left[\frac{1}{EI} C \cos \beta \sin \beta + \frac{1}{GJ} C \sin \beta (1 - \cos \beta) \right] d\beta$$

$$= \frac{2R^2 C}{GJ}$$

Note that the deflection at end A is entirely caused by twisting. The bending couple $M(\beta)$ is odd with respect to $\beta = \pi/2$, negative in the region $0 < \beta < \pi/2$, and positive in $\pi/2 < \beta < \pi$.

The rotation of the end A in the sense of couple C is simply $\partial V_I / \partial C$. ◆

PROBLEMS

Assume HI-HO behavior unless otherwise stated. Assume also that curved members are relatively slender.

8.21 Solve the following problems by means of Castigliano's theorem:

5.65	Truss	6.77(d)	Beam
5.68	Truss	7.24	Beam
5.71	Truss	7.35	Torsion + flexure
8.12	Truss	7.36	Torsion + flexure
6.59	Beam	7.37(b), (c)	Torsion + flexure
6.71	Beam	7.40(a), (b)	Torsion + flexure
6.77(b)	Beam	7.49	Torsion + flexure

8.22 Obtain the displacement of joint O in Prob. 8.4 if the modulus of both cables is 2.50 kN/cm.

8.23 The L-shaped member in Fig. P8.23 is made of steel with a uniform circular cross section. The member is clamped at A and simply supported at C; it forms a right angle at B. It is stress-free when the normal force P is applied. Express (in terms of the load P, cross-sectional diameter D, and modulus $E = 8G/3$) (a) the reaction at C and (b) the deflection at B.

Fig. P8.23

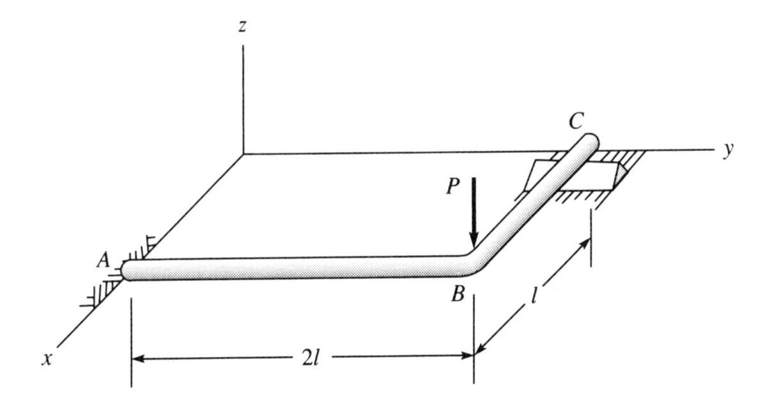

8.24 The split circular ring of Fig. P8.24 is made of steel wire (HI-HO). The diameter of the wire (d) is small compared to the mean diameter (D) of the ring; $d \ll D$. Compute the relative displacement of the adjacent ends for the loadings shown. Neglect transverse shear deformations; consider only the dominant effects of flexure and torsion. Give your answer in terms of the dimensions (d, D), properties (E, ν) and load: couple C, force P, or pressure p (force length).

Fig. P8.24

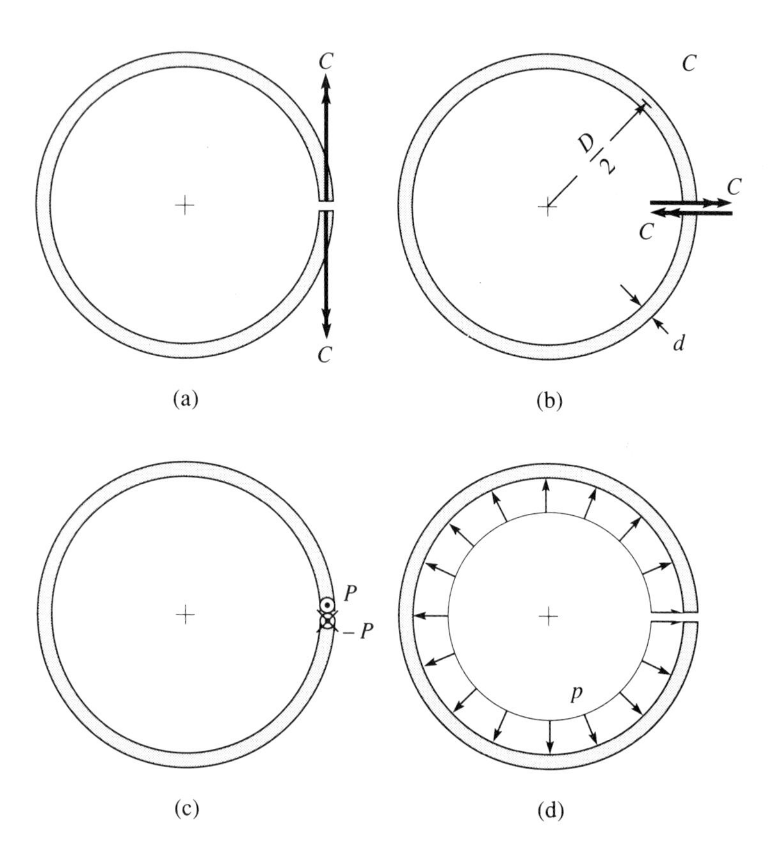

8.25 The simple chain link of Fig. P8.25 is made of steel wire (HI-HO); the form consists of straight and semicircular segments. The diameter of the wire is relatively small ($d \ll R$) compared with the mean radius R, so that deformation is essentially attributed to flexure. $P = 5.5$ N, $d = 0.200$ cm, $R = 2.00$ cm, and $E = 200 \times 10^3$ MPa. Determine (a) the maximum normal stress (neglect concentration at the load), (b) the deflection at the load, and (c) the change in width w.

Fig. P8.25

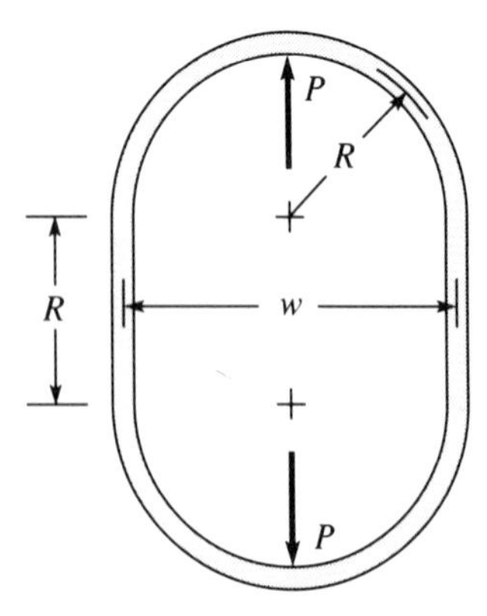

8.26 The semicircular member of Fig. P8.26 is made of HI-HO stock with a circular cross section ($d \ll R$). It is clamped at A and B under no loads or stresses. The

Fig. P8.26

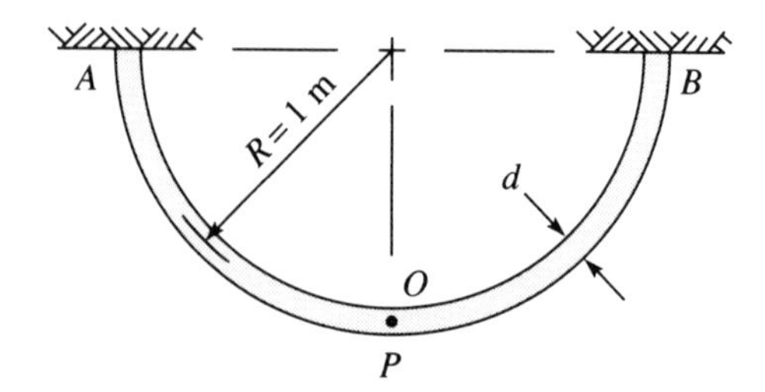

transverse load $P = 1000$ N is applied at O, perpendicular to the plane AOB. $E = 2.5G$. Calculate the reactions of the supports at A and B.

8.27 The helical coil spring of Fig. P8.27 is made of high-strength steel stock that has a circular cross section; diameter d is small compared to the mean diameter D of the coil. Load P is applied axially. The deflections can be traced to torsion. Obtain the formula for the axial deflection in terms of the dimensions (d, D), the number of coils (N), the modulus (G), and the load P. Neglect the deformations of the end loops, note that the helix angle is small $(d/D \ll 1)$, and assume that the spring is initially stress-free.

Fig. P8.27

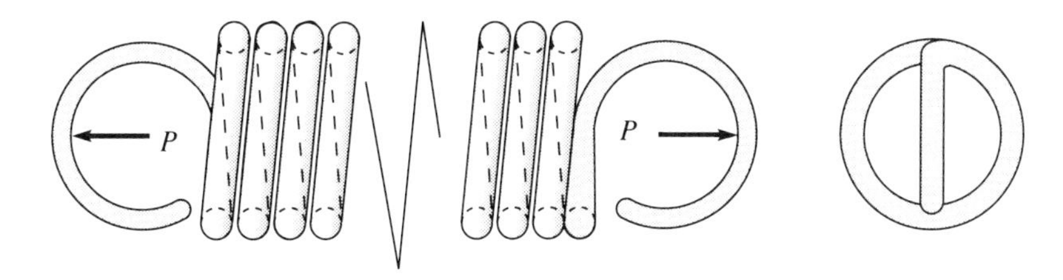

8.28 Consider the simply supported beam (HI-HO) of Fig. P8.28, which is subjected to the loads P_1 and P_2 at $x = x_1$ and $x = x_2$. Use Castigliano's theorem to show that

$$v_1(P_1 = 0, P_2 = P) = v_2(P_1 = P, P_2 = 0)$$

In words, the deflection at x_1 caused by a load $P_2 = P$ at x_2 is equal to the deflection at x_2 caused by a like load $P_1 = P$ at x_1. The result is a special case of Maxwell's reciprocity theorem, which holds for any similar circumstance of linear elastostatics.

Fig. P8.28

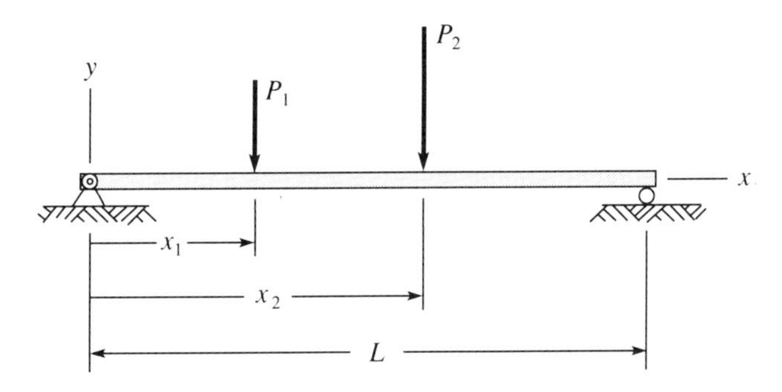

8.6 Stability of Equilibrium: Minimum Potential

A state of equilibrium need not be stable. Stated otherwise, although all conditions of equilibrium may be satisfied, the actual mechanical system may not sustain the configuration. To illustrate instability of equilibrium, consider the rollers resting upon the curved surface of Fig. 8.16. Each is resting upon a smooth surface at a location with a horizontal tangent; each has the free-body diagram of Fig. 8.16(b); each is in equilibrium; $N = W$ (the weight). However, we know that the roller at A cannot stay in this position. (The rollers and surface are smooth.) If, indeed, one could place the roller at A with the aid of a little friction, a slight disturbance would cause it to roll down the

Fig. 8.16

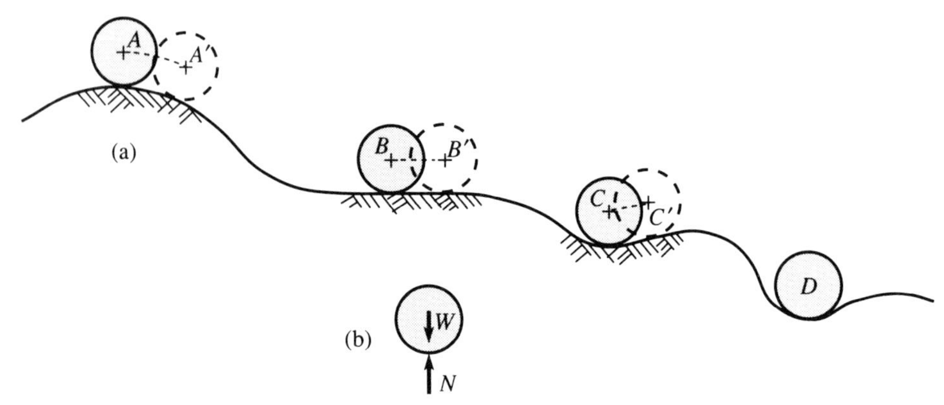

hill. On the other hand, the roller in the valley C is stable; if it were displaced slightly to C', it would roll back. What distinguishes the unstable from the stable states?

The stable state C requires that we do *positive* work to move the roller along the surface to the adjacent position C'. The unstable state A requires no work to dislodge the roller; indeed, if we imagine that we move the roller from point A along the surface to the adjacent point A', then we do *negative* work. The state B on the plateau is distinguished by the fact that zero work is performed in transporting the roller to the adjacent position B'.

The simple system of Fig. 8.16 is a conservative mechanical system. The roller is acted upon by the gravitational force W and the workless normal force N. Any work that we perform upon the roller (against the weight) as we transport it along the surface constitutes a change in potential. If we must increase the potential (location C), then the system is stable. If we decrease the potential (location A), then the system is unstable. If we do not change the potential (location B), then the system is in a state of neutral equilibrium. It is important to note that our test compares the potentials in the immediate neighborhood of the state. In reality, a mechanical system may be strictly stable, as is the roller at C, but a sufficiently large disturbance could move the system to another state of lower potential with a finite change of the configuration (i.e., kick the roller from location C to D).

Based upon our experiences, such as the example of Fig. 8.16, we assert the criterion for stability: *A conservative mechanical system is in a state of equilibrium if and only if the potential is a relative minimum.*

Let us now apply the criterion to a mechanical system, one that possesses many attributes of structural instabilities. The system depicted in Fig. 8.17(a) consists of a rigid bar that is constrained by a smooth pin at end O, by a linear torsional spring at end O, and by a horizontal linear spring at end B. For simplicity, we suppose that the end of the lateral spring is attached at a great distance so that it remains (practically) horizontal as the bar rotates and end B displaces, as shown in Fig. 8.17(b). The bar is subjected to an axial force P, which remains vertical and constant. This "dead" load

Fig. 8.17

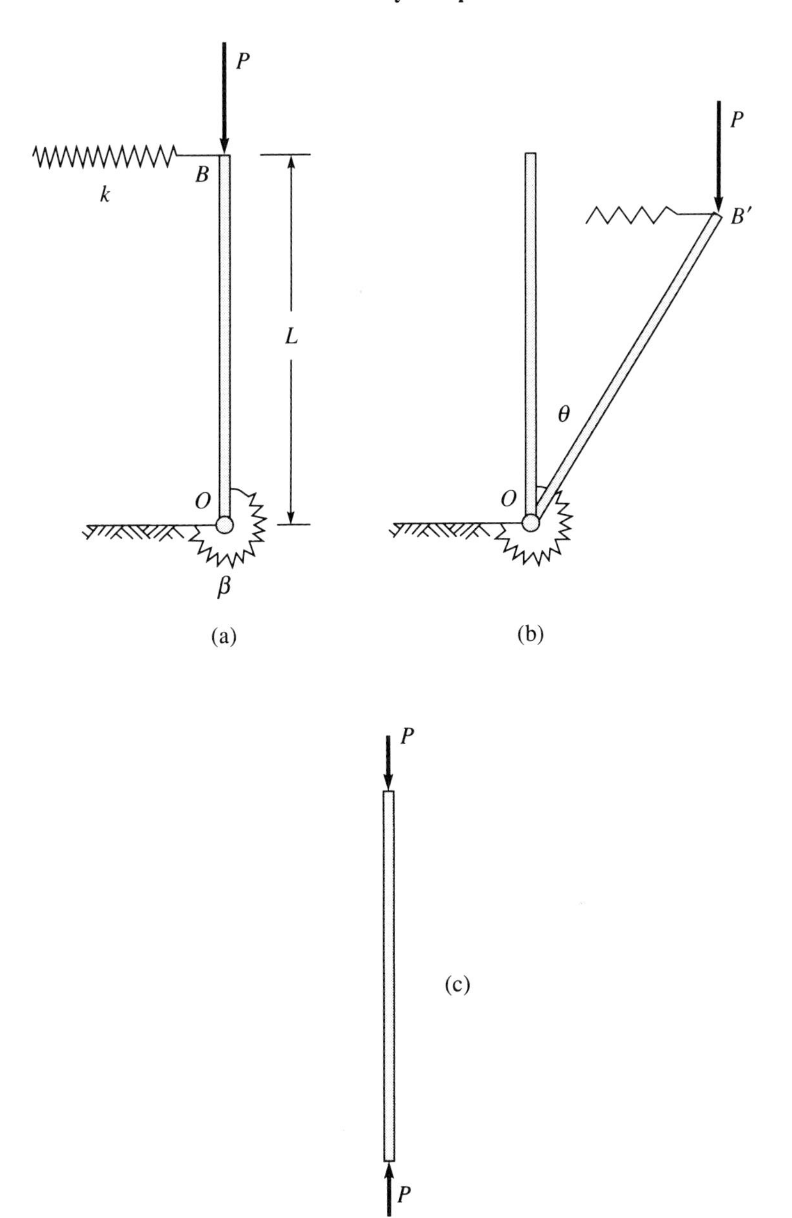

acts just as a weight; relative to the point O it possesses the potential

$$\overline{V}_E = PL \tag{8.73}$$

The torsional and lateral springs have moduli β and k, respectively. The lateral spring resists equally the movement to the left or right; in other words, the system is symmetric about the initial vertical position OB. If the system is displaced to the configuration OB', then the potential V_E of the load P and the internal potential V_I of the springs are

$$V_E = PL\cos\theta \tag{8.74}$$

$$V_I = \tfrac{1}{2}k(L\sin\theta)^2 + \tfrac{1}{2}\beta\theta^2 \tag{8.75}$$

We know that the initial configuration ($\theta = 0$) of Fig. 8.17(a) is an equilibrium configuration for *any* value of the load P. From the free-body diagram of Fig. 8.17(c), we see that the pin exerts an opposing force $-P$ and the undeflected springs exert no actions (couple or force) upon the rod. The question of stability requires that we examine the change of potential as the system is moved from the initial position ($\theta = 0$) of Fig. 8.17(a) to the adjacent position ($\theta \neq 0$) of Fig. 8.17(b). The change in the potential of the external force P is the difference between the values of Eqs. (8.73) and (8.74). The change in potential of the internal forces (strain energy of the springs) is given by Eq. (8.75). The total change of potential is

$$\Delta V = \Delta V_E + \Delta V_I = PL(\cos\theta - 1) + \tfrac{1}{2}kL^2\sin^2\theta + \tfrac{1}{2}\beta\theta^2 \tag{8.76}$$

We are concerned with the change only in the immediate neighborhood of the equilibrium state ($\theta = 0$). Accordingly, we explore ΔV for small angles $\theta \ll 1$; then, we need examine only the smallest nonzero terms. To this end we use the series expansions:

$$\cos\theta = 1 - \frac{\theta^2}{2} + \frac{\theta^4}{4!} - \cdots$$

$$\sin\theta = \theta - \frac{\theta^3}{3!} + \cdots$$

Eq. (8.76) has the expansion

$$\Delta V = PL\left(-\frac{\theta^2}{2} + \frac{\theta^4}{4!} - \cdots\right) + \frac{1}{2}kL^2\left(\theta^2 - \frac{\theta^4}{3} + \cdots\right) + \frac{1}{2}\beta\theta^2 \tag{8.77}$$

Again, we are concerned only with the smallest change. We can always choose $\theta \ll 1$ so that the change ΔV is dominated by the quadratic terms. Let us denote that *approximation* by ΔV_2.

$$\Delta V \doteq \Delta V_2 = (-\tfrac{1}{2}PL + \tfrac{1}{2}kL^2 + \tfrac{1}{2}\beta)\theta^2 \tag{8.78}$$

ΔV_2 is *positive* and the system is stable if

$$P < kL + \frac{\beta}{L}$$

ΔV_2 is *negative* and the system is unstable if

$$P > kL + \frac{\beta}{L}$$

Therefore, the condition of instability is reached if the load reaches or exceeds the *critical* value:

$$P_{CR} = kL + \frac{\beta}{L} \tag{8.79}$$

We might ask about the state if the load is precisely at the critical value. Then the second-degree terms (ΔV_2) of Eq. (8.77) vanish and

$$\Delta V = \left(\frac{PL}{4!} - \frac{kL^2}{3!} \right) \theta^4 + O(\theta^6)$$

Now, the change is dominated by the fourth-degree term; let's denote that dominant term by ΔV_4. At the critical load ($P = kL + \beta/L$)

$$\Delta V \doteq \Delta V_4 = \frac{1}{3!} \left(\frac{\beta}{4} - \frac{3}{4} kL^2 \right) \theta^4$$

We see that ΔV is positive and the system is stable if

$$\beta > 3kL^2$$

ΔV is negative and the system is unstable if

$$\beta < 3kL^2$$

The nature of the system is significantly changed if the relative magnitude of the modulus k exceeds the value $k = \beta/3L^2$. When that occurs, the response at the critical load is very different. Structural systems that possess this characteristic, instability at the critical load, tend to snap abruptly to a severely deformed state, just as the roller at the top of the hill plunges off to a distant valley. Further discussion of such behavior is presented in Chapter 9.

The example of Fig. 8.17 is a very simple case wherein the system has but one degree of freedom; that is, one variable θ defines the configuration. Let us now apply the concepts to other systems.

EXAMPLE 11

Buckling of a Discrete Mechanical System The system of Fig. 8.18 consists of two rigid links, OA and AB, pinned at O and A. Relative rotations at O and A are resisted by linear torsional springs with moduli β. The system is subjected to the constant force P acting at B in the vertical direction BO. The springs are undeformed—that is, they exert no resisting couples—when the links are aligned.

The displaced configuration of Fig. 8.18(b) is defined by the two angles θ_1 and θ_2. The system has two degrees of freedom. The total potential of the system is the sum of the potentials of the external force P and internal forces (strain energy of the springs):

$$\begin{aligned} V &= V_E + V_I \\ &= PL(\cos\theta_1 + \cos\theta_2) + \tfrac{1}{2}\beta\theta_1^2 + \tfrac{1}{2}\beta(\theta_2 - \theta_1)^2 \end{aligned}$$

Fig. 8.18

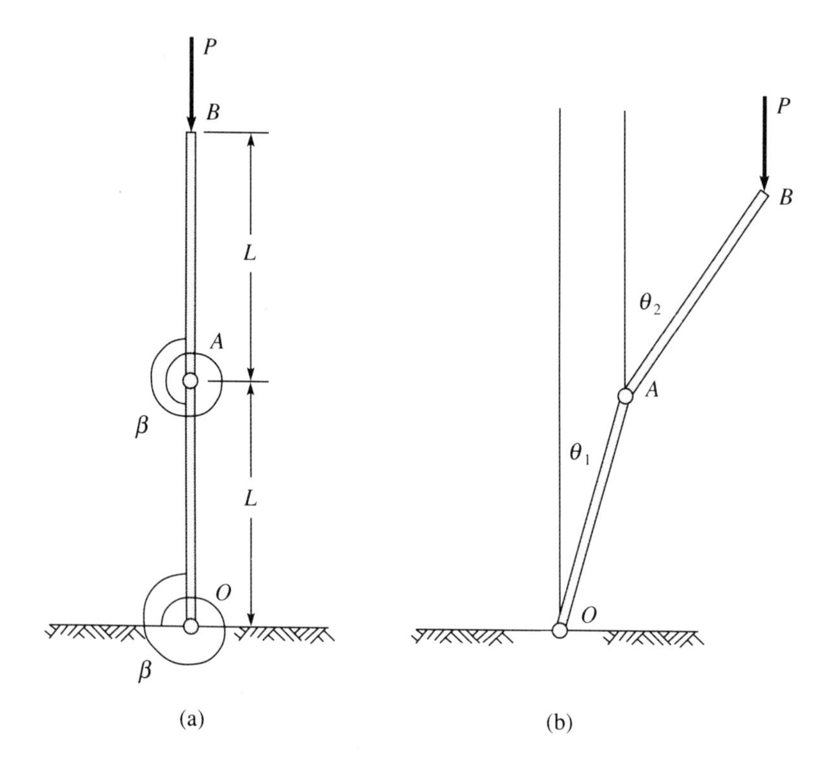

The potential in the initial state ($\theta_1 = \theta_2 = 0$) is the potential of the load,

$$V_0 = 2PL$$

The change of potential is

$$\begin{aligned}\Delta V &= V - V_0 \\ &= PL(\cos\theta_1 + \cos\theta_2 - 2) + \tfrac{1}{2}\beta(2\theta_1^2 - 2\theta_1\theta_2 + \theta_2^2)\end{aligned}$$

As before, we expand the function in powers of the small angles:

$$\Delta V = PL\left(-\frac{\theta_1^2}{2} - \frac{\theta_2^2}{2} + \frac{\theta_1^4}{4!} + \frac{\theta_2^4}{4!} + \cdots\right) + \frac{1}{2}\beta(2\theta_1^2 - 2\theta_1\theta_2 + \theta_2^2) \tag{8.80}$$

For sufficiently small values of these angles the change is dominated by the quadratic term. Therefore, we examine the approximation

$$\Delta V \doteq \Delta V_2 = \left(-\frac{PL}{2}\theta_1^2 - \frac{PL}{2}\theta_2^2 + \beta\theta_1^2 - \beta\theta_1\theta_2 + \frac{1}{2}\beta\theta_2^2\right) \tag{8.81}$$

By the principle of stationary potential, the first-order variation must vanish if the system is in equilibrium.

$$\delta(\Delta V) = \frac{\partial(\Delta V)}{\partial\theta_1}\delta\theta_1 + \frac{\partial(\Delta V)}{\partial\theta_2}\delta\theta_2 = 0 \tag{8.82}$$

However, the variations, $\delta\theta_1$ and $\delta\theta_2$, are independent; Eq. (8.82) is satisfied for all admissible variations if but only if

$$\frac{\partial(\Delta V)}{\partial\theta_1} = (-PL + 2\beta)\theta_1 - \beta\theta_2 = 0 \tag{8.83a}$$

$$\frac{\partial(\Delta V)}{\partial\theta_2} = -\beta\theta_1 + (-PL + \beta)\theta_2 = 0 \tag{8.83b}$$

These are two linear homogeneous algebraic equations in the variables θ_1 and θ_2. They have a nontrivial solution ($\theta_1 \neq 0, \theta_2 \neq 0$) if and only if the determinant of the coefficient vanishes:

$$\begin{vmatrix} (-PL + 2\beta) & -\beta \\ -\beta & (-PL + \beta) \end{vmatrix} = 0 \tag{8.84a}$$

or

$$(PL - 2\beta)(PL - \beta) - \beta^2 = 0 \tag{8.84b}$$

$$P^2 - \frac{3\beta}{L}P + \frac{\beta^2}{L^2} = 0 \tag{8.84c}$$

or

$$P = \frac{3}{2}\frac{\beta}{L} \pm \sqrt{\left(\frac{3\beta}{2L}\right)^2 - \left(\frac{\beta}{L}\right)^2}$$

$$P = \frac{3}{2}\frac{\beta}{L}\left(1 \pm \sqrt{\frac{5}{9}}\right)$$

We are interested in the smallest real value that admits an adjacent ($\theta_1 \neq 0, \theta_2 \neq 0$) configuration, namely,

$$P_{CR} = \frac{3}{2}\frac{\beta}{L}\left(1 - \sqrt{\frac{5}{9}}\right)$$

$$= 0.382\frac{\beta}{L}$$

This is the buckling load that admits a slightly deflected position of equilibrium. Then, according to Eq. (8.83a) or (8.83b),

$$\frac{\theta_2}{\theta_1} = \left(2 - \frac{P_{CR}L}{\beta}\right) = \frac{1}{\left(1 - \frac{P_{CR}L}{\beta}\right)}$$

$$= 1.62 \tag{8.85}$$

Let us note some features of the foregoing analyses that are typical of such problems. Firstly, the adjacent buckled configuration is governed by a system of linear homogeneous equations such as Eqs. (8.83). Secondly, the critical load is determined by the requirement that the determinant of the coefficients must vanish, as in Eq. (8.84). Finally, the buckled mode (shape of the buckled system) is determined by linear equations such as Eq. (8.83), but only the shape is determined, not the amplitude—for example, here, only the ratio θ_2/θ_1. ◆

EXAMPLE 12

Buckling of a Symmetric Euler Column The beam of Fig. 8.19 is hookean, fixed at one end ($x = 0$), and free at the other. It is initially straight and subjected to the constant axial load P. Because it supports axial load, it is now called a column. The change in potential of the force P as the column passes from the straight prebuckled configuration to the adjacent, *slightly* buckled configuration is

$$\Delta V_E = P(L_* - L_0) \tag{8.86}$$

Fig. 8.19

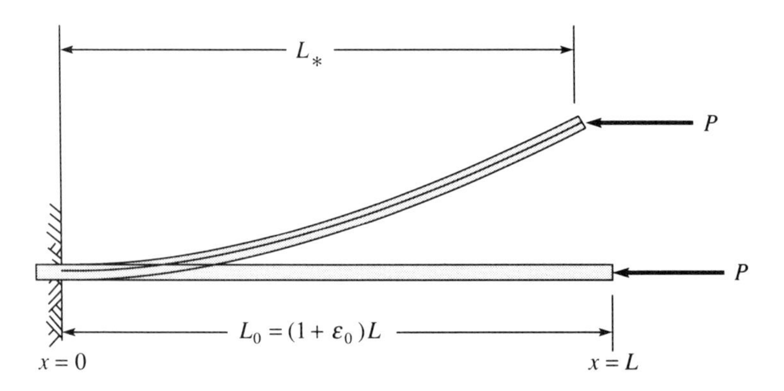

If ε_0 denotes the extensional strain immediately before the onset of buckling, then

$$L_0 = (1 + \varepsilon_0)L \tag{8.87}$$

To determine the distance L_* we consider a segment of the deformed (buckled) axis, as shown in Fig. 8.20. Here, θ is a small angle, since we are concerned with configurations *near* the straight prebuckled configuration. In the limit as $dx \to 0$, we have

$$dx_* = (1 + \varepsilon_*)\cos\theta\, dx$$

Fig. 8.20

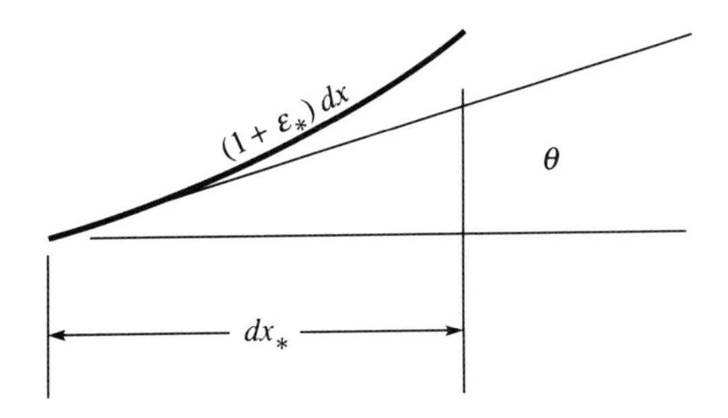

and, therefore,

$$L_* = \int_0^L (1 + \varepsilon_*) \cos\theta \, dx \tag{8.88a}$$

Here, we introduce an approximation. We assume that little extensional strain accompanies the deformation that carries the column to the *slightly* buckled position; we set $\varepsilon_* = \varepsilon_0$, a constant. Since the rotation θ is very small,

$$\cos\theta \doteq 1 - \frac{\theta^2}{2}$$

Then, by Eq. (8.88a) we have

$$L_* \doteq \int_0^L (1 + \varepsilon_0)\left(1 - \frac{\theta^2}{2}\right) dx \tag{8.88b}$$

Substituting Eqs. (8.87) and (8.88b) into Eq. (8.86), we obtain

$$\begin{aligned}\Delta V_E &= P \int_0^L (1 + \varepsilon_0)\left(1 - \frac{\theta^2}{2}\right) dx - P(1 + \varepsilon_0)L \\ &= -\frac{P}{2}(1 + \varepsilon_0) \int_0^L \theta^2 \, dx\end{aligned}$$

Since $\varepsilon_0 \ll 1$, we accept the approximation

$$\Delta V_E = -\frac{P}{2} \int_0^L \theta^2 \, dx \tag{8.89}$$

The assumption of inextensionality ($\varepsilon_* = \varepsilon_0 \ll 1$) embodies the very plausible view that the movement $(L_* - L_0)$ of the end $(x = L)$ is largely due to bending, since the force P remains constant.

During the buckling, the internal potential changes, also by virtue of the bending. As before, subscripts 0 and * denote the pre- and postbuckled states:

$$\Delta V_I = V_* - V_0$$

We assume that the column is hookean and behaves according to the Bernoulli theory, so that

$$\Delta V_I = \int_0^L \left(\frac{EA}{2}\varepsilon_*^2 + \frac{EI}{2}\kappa^2\right) dx - \int \frac{EA}{2}\varepsilon_0^2\, dx$$

As before, we assume the inextensionality $\varepsilon_* = \varepsilon_0$ and accept the approximation

$$\Delta V_I = \frac{EI}{2}\int_0^L \kappa^2\, dx \tag{8.90}$$

In accordance with our previous theory [see Eqs. (6.26) and (6.34)],

$$\theta \doteq \frac{dw}{dx} \tag{8.91}$$

$$\kappa = \frac{d\theta}{dx} = \frac{d^2w}{dx^2} \tag{8.92}$$

Substituting Eqs. (8.91) and (8.92) into Eqs. (8.89) and (8.90), respectively, we obtain the total change of potential:

$$\begin{aligned}\Delta V &= \Delta V_E + \Delta V_I \\ &= \int_0^L \left[-\frac{P}{2}\left(\frac{dw}{dx}\right)^2 + \frac{EI}{2}\left(\frac{d^2w}{dx^2}\right)^2\right] dx \end{aligned} \tag{8.93}$$

Note that this approximation is second-order in the variables, much as Eq. (8.81) is for the discrete system of Example 11.

As in the preceding example, we examine an adjacent equilibrium state. The potential must be stationary; that is, the first-order variation must vanish. A change δw in the function w produces the following changes in the derivatives (dw/dx) and (d^2w/dx^2):

$$\left[\frac{d}{dx}(w + \delta w)\right]^2 - \left[\frac{dw}{dx}\right]^2 = 2\frac{dw}{dx}\frac{d\delta w}{dx} + \left(\frac{d\delta w}{dx}\right)^2$$

$$\left[\frac{d^2(w + \delta w)}{dx^2}\right]^2 - \left[\frac{d^2w}{dx^2}\right]^2 = 2\frac{d^2w}{dx^2}\frac{d^2\delta w}{dx} + \left(\frac{d^2\delta w}{dx^2}\right)^2$$

In each equation the second term on the right is quadratic in the change δw. We are interested only in the first-order approximation, since that term dominates. Then the first variation of potential is

$$\delta(\Delta w) = \int_0^L \left[-P\frac{dw}{dx}\frac{d}{dx}\delta w + EI\frac{d^2w}{dx^2}\frac{d^2}{dx}\delta w\right] dx \tag{8.94}$$

We need to recast the equation, putting the integrand in terms of the variation δw rather than the first and second derivatives of that function. To that end the terms are integrated by parts:

$$\int_0^L \left(\frac{dw}{dx}\frac{d}{dx}\delta w\right)dx = \frac{dw}{dx}\delta w\Big|_0^L - \int_0^L \left(\frac{d^2w}{dx^2}\delta w\right)dx$$

$$\int_0^L \left(\frac{d^2w}{dx^2}\frac{d^2}{dx^2}\delta w\right)dx = \frac{d^2w}{dx^2}\frac{d}{dx}\delta w\Big|_0^L - \frac{d^3w}{dx^3}\delta w\Big|_0^L + \int_0^L \left(\frac{d^4w}{dx^2}\delta w\right)dx$$

Substituting these forms into Eq. (8.94), we obtain

$$\delta(\Delta V) = -\left(P\frac{dw}{dx} + EI\frac{d^3w}{dx^3}\right)\delta w\Big|_0^L + EI\frac{d^2w}{dx^2}\frac{d}{dx}\delta w\Big|_0^L$$

$$+ \int_0^L \left[P\frac{d^2w}{dx^2} + EI\frac{d^4w}{dx^4}\right]\delta w\,dx \tag{8.95}$$

This first variation must vanish for arbitrary variation δw (virtual displacement). That variation is subject only to the requirements of admissibility. In the context of our theory, the function δw (as w) must be continuous with continuous derivatives ($dw/dx = 0, d^2w/dx^2 = M/EI, d^3w/dx^3 = -V/EI$). The variations at the ends are subject also to the constraints of this problem, namely,

$$w|_0 = 0, \qquad \delta w|_0 = 0, \tag{8.96a}$$

$$\frac{dw}{dx}\Big|_0 = 0, \qquad \frac{d\delta w}{dx}\Big|_0 = \delta\theta_0 = 0 \tag{8.96b}$$

Otherwise, the values at the ends can be varied independently of the variation in the interior ($0 < x < L$); that is, $\delta w(L)$ and $d\delta w/dx(L)$ are arbitrary. Therefore, the variation $\delta(\Delta V)$ can vanish only if

$$\left(EI\frac{d^2w}{dx^3} + P\frac{dw}{dx}\right)\Big|^L = 0 \tag{8.96c}$$

$$EI\frac{d^2w}{dx^2}\Big|^L = 0 \tag{8.96d}$$

Because the variation $\delta w(x)$ is arbitrary—for example, some positive function over an arbitrarily small interval—the variation of potential $\delta(\Delta V)$ can vanish only if the bracketed expression of the integrand vanishes everywhere:

$$\frac{d^4w}{dx^4} + \frac{P}{EI}\frac{d^2w}{dx^2} = 0, \qquad 0 < x < L \tag{8.97}$$

If the end conditions of Eq. (8.96) and the differential equation (8.97) are satisfied, then an adjacent (buckled) equilibrium configuration exists. The shape (buckled mode) $w(x)$ and the corresponding load P are determined by the solution of the problem governed by the differential equation and the end conditions. It remains only to give physical meanings to these equations.

We recall the equations governing a symmetrical hookean beam, namely, the equilibrium equations (8.10b, c), the moment-curvature relation (6.33), and the kinematic equations (8.91) and (8.92):

$$\frac{dV}{dx} + \kappa F = \bar{p} \tag{8.98a}$$

$$\frac{dM}{dx} = -V \tag{8.98b}$$

$$\kappa = \frac{d\theta}{dx} = \frac{M}{EI} \tag{8.98c}$$

$$\frac{dw}{dx} = \theta \tag{8.98d}$$

Eliminating θ in Eqs. (8.98c, d) we obtain

$$\frac{M}{EI} = \kappa = \frac{d^2w}{dx^2} \tag{8.99a}$$

Substituting Eqs. (8.99a) and (8.98b), we have

$$V = -EI\frac{d^3w}{dx^3} \tag{8.99b}$$

Substituting Eqs. (8.99a) and (8.99b) into Eq. (8.98a) and noting that $\bar{p} = 0$ and $F \doteq -P$, we obtain Eq. (8.97):

$$\frac{d^4w}{dx^4} + \frac{P}{EI}\frac{d^2w}{dx^2} = 0$$

In other words, the fourth-order differential equation governing the deflection $w(x)$ replaces the four first-order equations governing the four functions (w, θ, M, V). Equations (8.98d) and (8.99a, b) express the variables θ, M, and V in terms of the deflection $w(x)$. Equations (8.96c, d) express the dynamic conditions at the free end ($x = L$), namely,

$$V = P\frac{dw}{dx}, \qquad M = 0 \; \blacklozenge$$

In Sec. 8.4 we describe the Rayleigh method to obtain approximations of equilibrium states. At present we are concerned with adjacent states that signal a bifurcation point and the onset of buckling. As before, we choose a displacement that satisfies geometric constraints and, we hope, similates the actual buckled configuration. Unspecified constants and then the approximation of the critical load are determined by the criteria of stationary potential. A simple example serves to illustrate this approach.

EXAMPLE 13

A Fixed-Free Column The symmetric HI-HO column of Fig. 8.18 is subject to geometric constraints at one end, namely, at $x = 0$, $w = dw/dx = 0$. An approximation that meets minimal conditions is the quartic

$$w(x) = Ax^2 + Bx^3$$

Our approximation is differentiated once and then twice; those derivatives are substituted into Eq. (8.93) to obtain an approximation of the potential:

$$\begin{aligned}\Delta V &= \int_0^L \left[-\frac{P}{2}(2Ax + 3Bx^2)^2 + \frac{EI}{2}(2A + 6Bx)^2 \right] dx \\ &= -\frac{P}{2}\left(\frac{4}{3}A^2L^3 + 3ABL^4 + \frac{9}{5}B^2L^5\right) + \frac{EI}{2}(4A^2L + 12ABL^2 + 12B^2L^3)\end{aligned}$$

The stationary criteria follow:

$$\frac{\partial \Delta V}{\partial A} = 4\left(1 - \frac{PL^2}{3EI}\right)A + 3L\left(2 - \frac{PL^2}{2EI}\right)B = 0$$

$$\frac{\partial \Delta V}{\partial B} = 3\left(2 - \frac{PL^2}{2EI}\right)A + 3L\left(4 - \frac{3}{5}\frac{PL^2}{EI}\right)B = 0$$

Again, the two linear homogeneous equations have a nontrivial solution ($A \neq 0, B \neq 0$) if and only if the determinate of the coefficients vanishes. With $PL^2/EI = p$, we have

$$\begin{vmatrix} 4(1 - \frac{1}{3}p) & 3(2 - \frac{1}{2}p) \\ 3(2 - \frac{1}{2}p) & 3(4 - \frac{3}{5}p) \end{vmatrix} = 0$$

This gives the quadratic equation

$$p^2 - \tfrac{104}{3}p + 80 = 0,$$

The roots are

$$p = +\tfrac{52}{3} \pm \tfrac{1}{3}\sqrt{52^2 - 720} = \tfrac{4}{3}(52 \pm 8\sqrt{31})$$

The smaller root is our approximation of the critical value, $p_{cr} \doteq 2.4860$. The exact solution is $p_{cr} = \pi^2/4 = 2.4674$. Our estimate is high by 0.75%. Recall that such assumed configurations are effectively constrained and hence are stiffer than the actual system. ◆

PROBLEMS

8.29 The mechanical systems in Fig. P8.29 are all made up of rigid links joined by frictionless pins and constrained by linear extensional and/or torsional springs with moduli k and β, respectively. Determine the critical load (P, p, or M) for each case in terms of the given properties and parameters (L, k, β).

Fig. P8.29

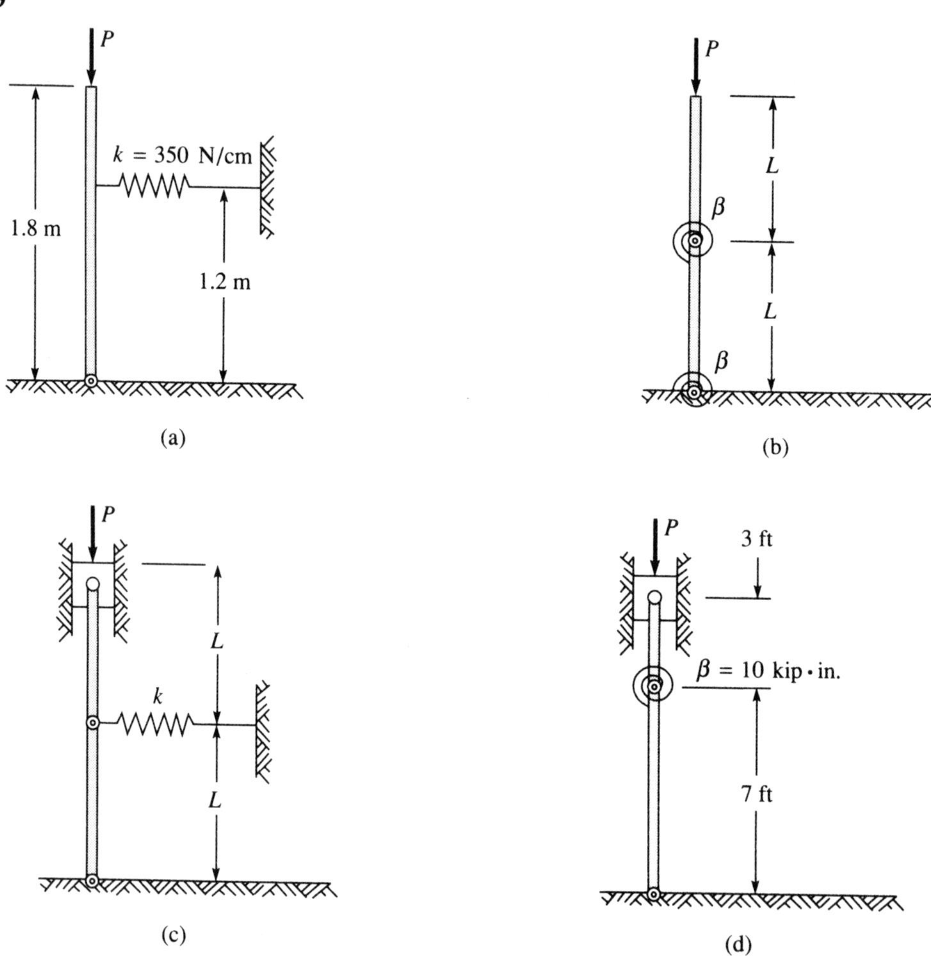
P
k = 350 N/cm
1.8 m
1.2 m
(a)
P
L
β
L
β
(b)
P
L
k
L
(c)
P
3 ft
β = 10 kip · in.
7 ft
(d)

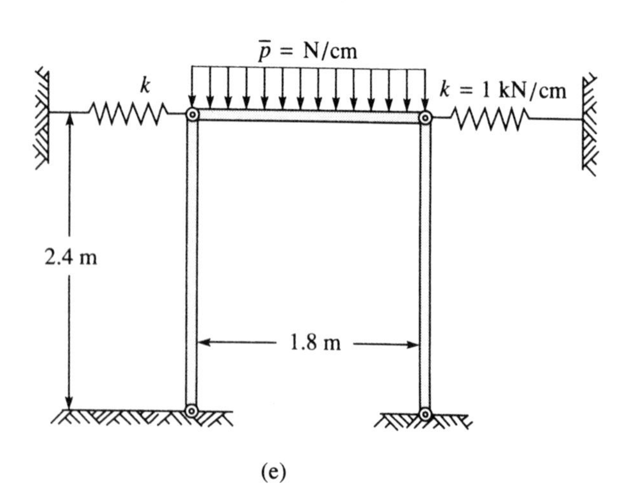
$\bar{p}$ = N/cm
k
k = 1 kN/cm
2.4 m
1.8 m
(e)

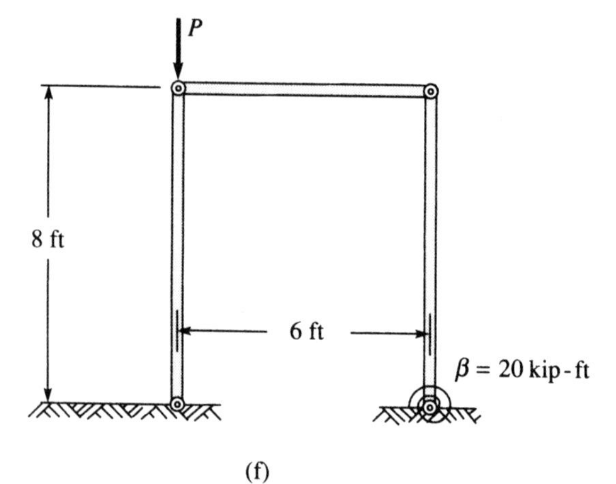
P
8 ft
6 ft
β = 20 kip - ft
(f)

Fig. P8.29
(cont.)

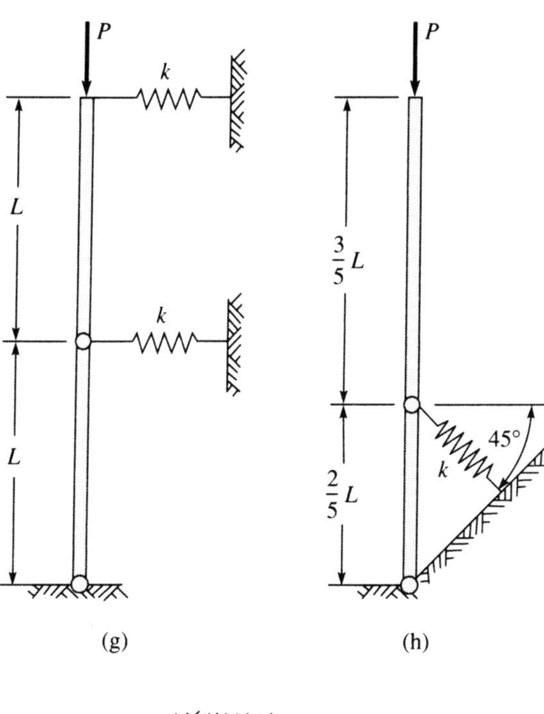

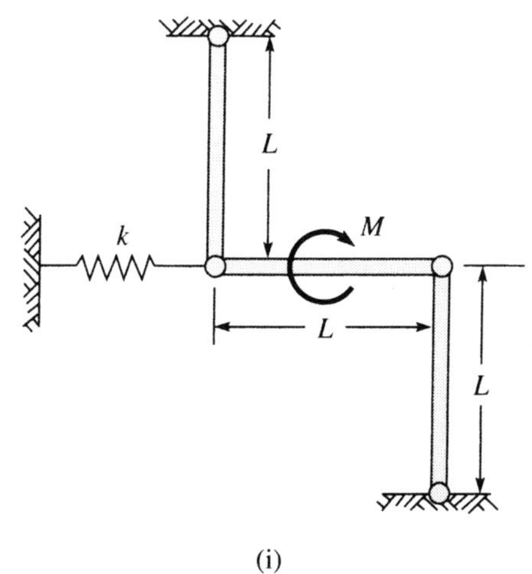

8.30 A homogeneous bar of length $L = 50$ cm is suspended on a pin located 10 cm from the top of the bar. The weight of the bar is 1800 N, and it is loaded by a vertical force P at the top. Find the critical value of the force P.

8.31 A homogeneous bar of length L is suspended on a pin 4 in. from the top of the bar. The bar weighs 20 lb per inch of length. If it is stable under a load $P = 450$ lb placed at the top, find the minimum length L of the bar.

8.32 The mechanical system of Fig. P8.32 consists of two rigid links, a torsional spring that resists relative rotation at A, and an extension-compression spring that resists lateral displacement. In terms of the parameters (L, β, k), determine (a) the critical load P_{CR} and (b) the requirement for stability at the critical load (relative minimum versus maximum potential).

Fig. P8.32

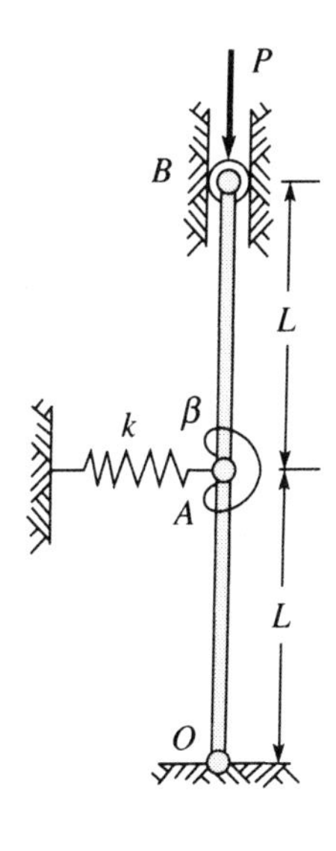

8.33 The system of Fig. P8.33 consists of two similar rigid links, pinned at O and A. Torsional springs (moduli β) resist relative rotation at O and A. The extension-compression spring (modulus k) resists lateral displacement at B. In terms of the given parameters (L, β, k), determine (a) the critical load P_{CR} and (b) the condition for stability at the critical load (relative minimum versus maximum potential).

Fig. P8.33

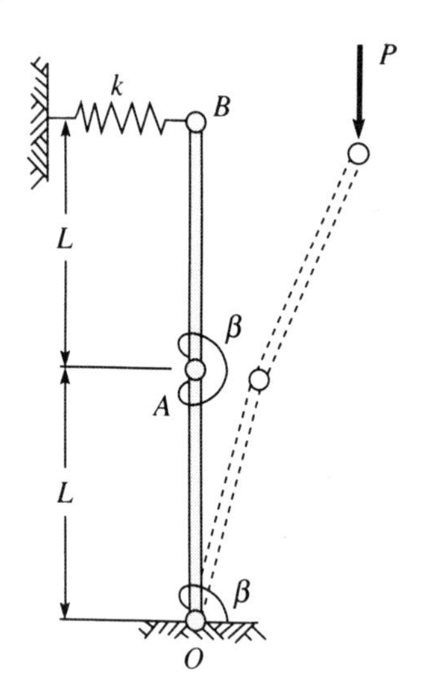

Fig. P8.34

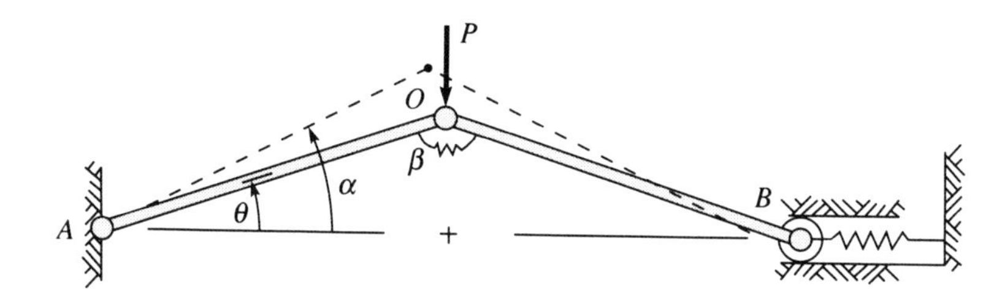

***8.34** The toggle mechanism shown in Fig. P8.34 is made up of the two identical rigid links (length L) pinned at A, O, and B. The end B rolls on the surface; movement is resisted only by the compression of the spring with modulus k. Relative rotation at O is resisted by the torsional spring with modulus β. The mechanism assumes the small angle α when unloaded. Use the energy criteria to obtain (a) the relationship between P and θ and (b) the conditions for instability under a constant load.

8.35 The general solution of Eq. (8.97) has the form

$$w(x) = A\sin\lambda x + B\cos\lambda x + Cx + D$$

Show that this function satisfies Eq. (8.97) if $\lambda^2 = P/EI$.

Substitute the solution into the four end conditions (8.96a, b, c, d) to obtain four linear homogeneous equations in the four constants (A, B, C, D). Impose the necessary and sufficient condition for a nontrivial solution; specifically, the determinant of the coefficients must vanish. Then, from that condition obtain the equation for the critical load:

$$\cos\lambda L = 0, \quad \text{or} \quad P_{CR} = \frac{\pi^2 EI}{4L^2}$$

8.36 Use the general solution of Eq. (8.97), as given in Prob. 8.35, and the end conditions for a column that is simply supported at both ends. Follow the procedure outlined in Prob. 8.35 to show that the critical load is

$$P_{CR} = \frac{\pi^2 EI}{L^2}$$

Show also that $B = C = D = 0$ and that the buckled mode is $w(x) = A\sin\pi x/L$.

8.37 Approximate the buckled mode of a simple supported column with the polynomial

$$w(x) = Ax\left(1 - \frac{x}{L}\right)$$

Use Rayleigh's method to obtain an approximation of the critical load and compare it to the exact value, $P_{CR} = \pi^2 EI/L^2$.

8.38 Repeat Prob. 8.37 but use the approximation

$$w(x) = Ax\left(1 - \frac{x}{L}\right) + Bx^2\left(1 - \frac{x}{L}\right)$$

8.39 A column is fixed at one end ($w = dw/dx = 0$ at $x = 0$) and simply supported at the other ($w = d^2w/dx^2 = 0$ at $x = L$). Use Rayleigh's method to approximate the critical load with the following approximation of the buckled mode:

$$w(x) = A\left(1 - \frac{x^2}{L^2}\right)x^2 + B\left(1 - \frac{x}{L}\right)x^3$$

Compare your result with the exact value $P_{CR} = 2.05\pi^2 EI/L^2$.

8.40 A flagpole is cylindrical, hookean, and homogeneous; let w denote the weight per unit length, E and I be the usual modulus and sectional integral, and L be the length. It is fixed at the base ($x = 0$) and free at the top ($x = L$). Assume the buckling mode $w(x) = A(1 - \cos \pi x/2L)$ and use Rayleigh's method to derive a formula for the maximum height that would forestall buckling.

9 Instability

The term *instability* implies that a situation is precarious and subject to abrupt, perhaps unanticipated, change. Such situations can arise in mechanical systems, notably structures. Unstable states are always undesirable; they usually signal failure and sometimes signal catastrophic collapse.

Several simple examples serve to illustrate such instability: A long, straight, and slender rod, such as a yardstick, can sustain axial thrust; at a certain load the rod bends and, more importantly, continues to bend with little additional resistance. That certain load is the *critical* load. A toggle switch in an electrical circuit requires a force to move it from the on to the off position; again, at a certain force the switch snaps *abruptly* to the off (or on) position.

Our last example suggests a more critical situation, because now the critical force does not initiate a slight movement but causes a *large* movement—that is, from completely on to completely off. The thin rod typifies the most common form of instability; the bending, which accompanies the critical load, is a form of buckling. The very abrupt and large movement of the toggle switch is a form of "snap buckling." Both can occur in various mechanical systems and, especially, structures. The purpose of the present chapter is (1) to introduce and explore the phenomena and (2) to set forth the most common means to ascertain critical loads, as exemplified by the simple column of traditional structures.

9.1 Structural Instability

Throughout our previous analyses, we required equilibrium but not stable equilibrium. To appreciate the implications of instability, let us briefly reconsider our introductory example of the simple truss (Fig. 1.1; Fig. 5.20), redrawn as Fig. 9.1. Recall that our analysis, based on hookean behavior and equilibrium, led to the stresses of Eq. (5.54), namely,

$$\sigma_D = 0.793\frac{W}{A}, \qquad \sigma_C = -0.207\frac{W}{A}, \qquad \sigma_B = 0.293\frac{W}{A}$$

Fig. 9.1

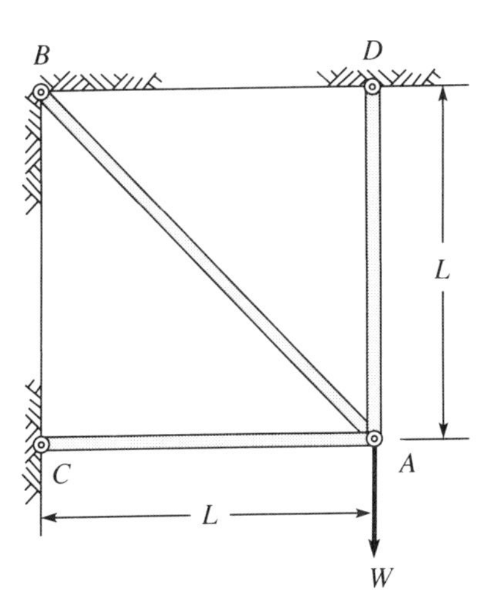

According to that earlier analysis, general yielding is initiated in the member AD when the stress attains the yield stress: $\sigma_D = \sigma_0$. Now, let us consider some possible properties. Suppose that the truss is made of structural steel rods with diameter $d = 0.500$ in. and length $L = 50.0$ in. Note that these are quite slender, that is, $d/L = \frac{1}{100}$. Suppose that the yield stress is $\sigma_0 = 50{,}000$ lb/in^2. Then, according to Eq. (5.55), yielding is initiated at

$$W_0 = 1.26\sigma_0 A = 1.26 \times 50 \times 10^3 \times \pi(0.25)^2$$
$$= 63.0 \times 10^3 A = 12{,}400 \text{ lb}$$

Then the force on the rod AC is

$$F_C = A\sigma_C = -0.207W_0 = -2{,}600 \text{ lb}$$

This is a *compressive* force about equivalent to the weight of a midsized automobile. Recall that the rod is 50 in. long and 0.50 in. in diameter. All our experiences tell us that this very slender member AC would buckle at some compressive load much less than 2,600 lb.[1] The reader can experiment with a slender ruler or common yardstick. The buckling of the slender member is the simplest form of structural instability.

The foregoing example does not necessarily negate our previous analyses. We might stiffen the rod AC by choosing a different cross section. For example, instead of the solid circular cross section, we might utilize a tubular rod, thereby increasing the resistance to flexure. [A hollow cross section with the same area has a greater rigidity (I).] It is clear that equilibrium alone is not sufficient to preclude failures, particularly buckling of thin structural bodies, beams, columns, plates, and shells. We must

1 The stress that causes buckling of the straight member AC is (theoretically) 1850 lb/in.^{2}; the force is 363 lb.

consider criteria for stable equilibrium and the possibility that the body might deform to an alternative state under certain conditions of loading.

9.2 The Stability of Equilibrium

A system is in an equilibrium configuration when the net force and couple acting on the system (and all of its parts) are zero. When any equilibrium configuration is disturbed, the system either returns to the original configuration or it seeks a new equilibrium configuration. Consider, for example, the simple situation illustrated in Fig. 9.2. In both cases the ball is in equilibrium in the position shown. Case I, however, is stable in the sense that if the ball is displaced slightly, it will return to its original equilibrium position. On the other hand, a slight disturbance of the ball in Case II will send it off looking for a new equilibrium position; the configuration shown is one of unstable equilibrium. Now start with Case I and gradually open the angle between the retaining sides until Case II is reached. There will be an intermediate position that marks the change from stable to unstable behavior. In this case the intermediate state occurs when the angle between the sides is π; that is, when the surface is flat.

Fig. 9.2

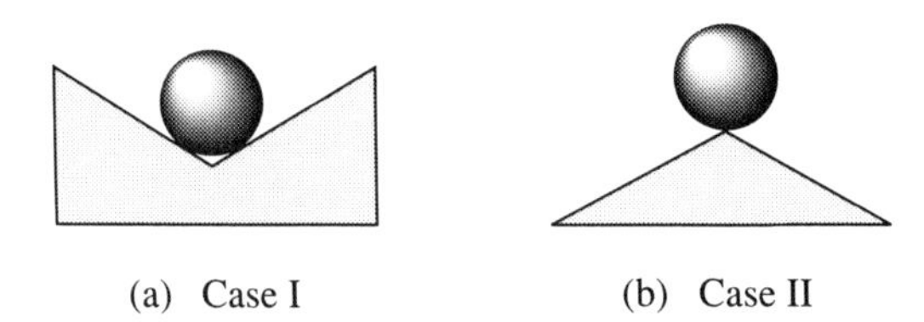

(a) Case I (b) Case II

The circumstances depicted in Fig. 9.2 are extraordinary because the surfaces have a corner at the position of equilibrium; movement along the surface (left to right) does not follow a smooth path. The smooth surface of Fig. 9.3(a) is more typical. Here, we see rollers, each in equilibrium, at the positions A, B, C, and D. The free-body diagram of Fig. 9.3(b) applies to all those positions. We see that the position A is a state of unstable equilibrium, whereas C and D are clearly stable. The latter (stable) states are characterized by the fact that we must do positive work on the rollers (C or D) to move them (along the surface) from their position in a valley. We do negative work to move the roller A (along the surface) to any neighboring position. The roller B is in a special position on a local plain; we need to do no work (positive or negative) to move it to a neighboring position. Roller B is in a state of neutral equilibrium.

The examples of the rollers in Fig. 9.3 are conservative systems. Any work expended to move a roller (along the surface) constitutes a change in the potential. That change of potential is the work expended in lifting or lowering the weight, which represents an increase or decrease in the potential. For such conservative systems, we can state the criterion for stable equilibrium as follows: *The conservative system is in stable equilibrium if the potential energy is a relative minimum.* Our experience tells us

Fig. 9.3

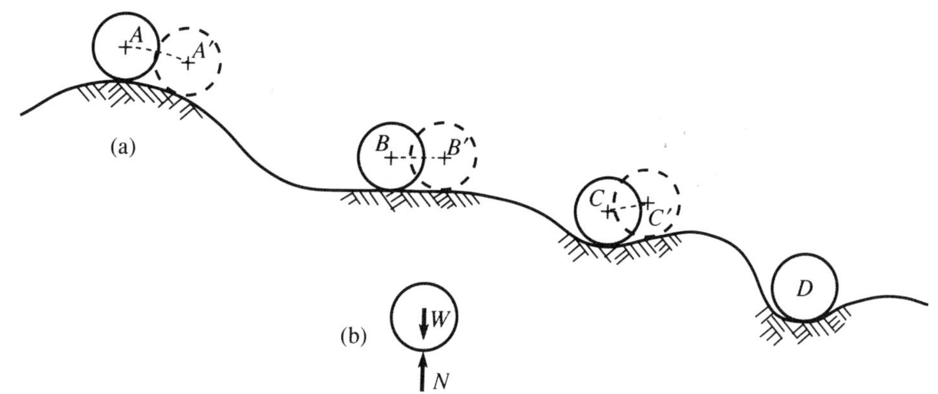

that some positions are more stable than others; for example, the roller C rests in a shallow valley, so a small kick might move it over the hill, sending it to some lower valley of lesser potential.

The rollers upon the terrain of Fig. 9.3 are simple examples of conservative mechanical systems in equilibrium. Elastic bodies and structures are also conservative mechanical systems. Section 8.6 provides a more complete account of the criterion of minimum potential and its applications. Here we adopt a different approach to stability: We examine the conditions that are sufficient to admit equilibrium of a neighboring state.

*9.3 Stability of a Conservative (Elastic) System

The simple discrete system of Sec. 8.6, Fig. 8.17, possesses many attributes of more complicated elastic bodies. That system is redrawn in Fig. 9.4(a). It consists of a rigid link OB, joined at O to a fixed surface, restrained by a torsional spring at O and by a lateral spring at B. Both springs are linear and undeformed when the bar OB is vertical, $\theta = 0$. The torsional spring has modulus β and exerts a restoring couple $\beta\theta$, which inhibits the rotation θ. The lateral spring has modulus k; it is supposedly anchored at some distance so that it remains essentially horizontal and exerts a restoring force $F = kL\sin\theta$. The springs resist movement equally in either direction; for example, the lateral spring resists compression and tension, so that the system is actually symmetric about the undeformed state ($\theta = 0$). The load P is supposed to act in the fixed direction. Clearly, the vertical position OB is an equilibrium state for *all* values of the load P, as evidenced by the free-body diagram of Fig. 9.4(c). We are, however, concerned about the stability of that state and, therefore, consider the possibilities of other *deformed* states of equilibrium, as illustrated by Fig. 9.4(b) and the

Fig. 9.4

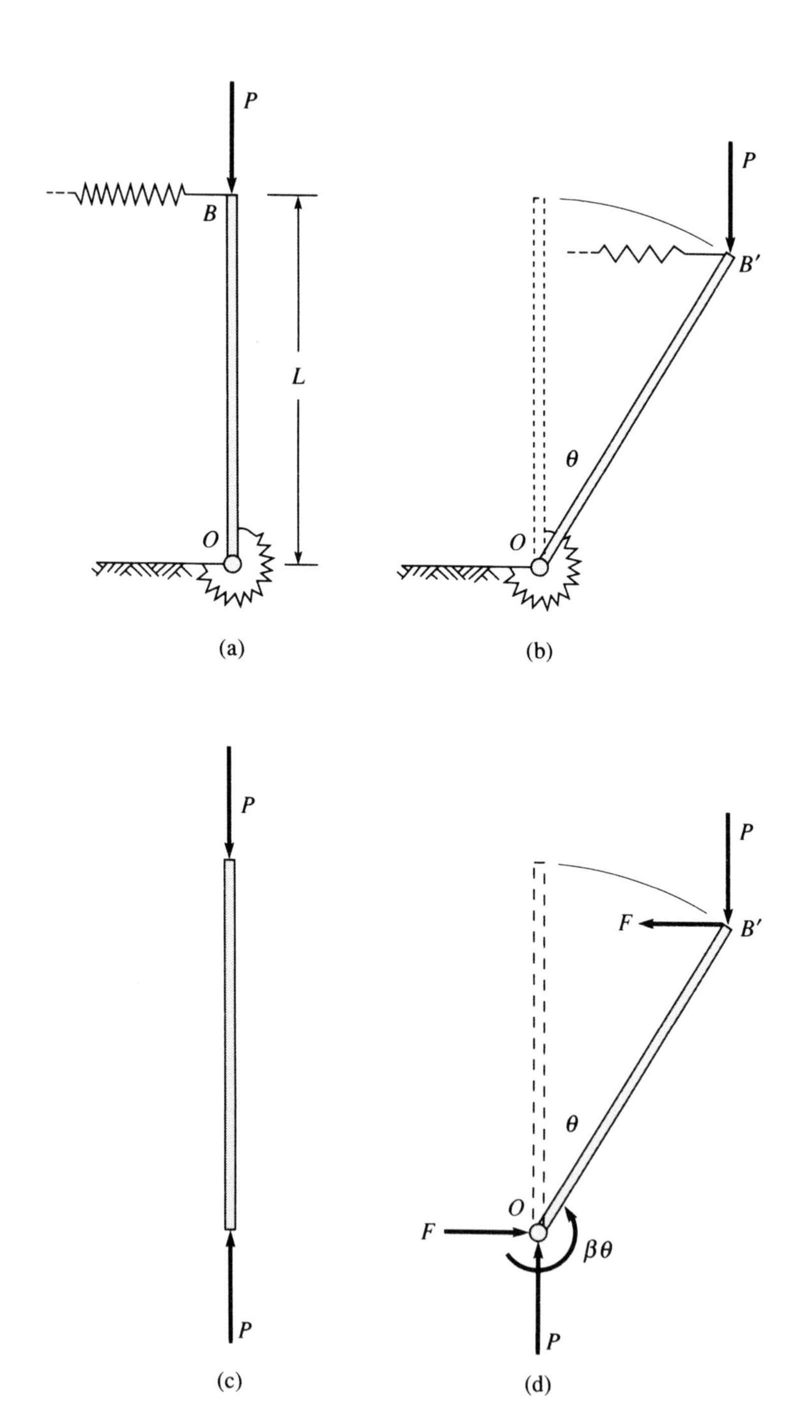

free-body diagram of Fig. 9.4(d). Equilibrium requires that

$$\sum M_0 = FL\cos\theta + \beta\theta - PL\sin\theta = 0$$
$$= kL^2\sin\theta\cos\theta + \beta\theta - PL\sin\theta = 0$$

or

$$P = kL\cos\theta + \frac{\beta}{L}\frac{\theta}{\sin\theta} \tag{9.1}$$

The system is in equilibrium for nonzero values of angle θ and load P that satisfy Eq. (9.1). That relation depends upon the values of the parameters: moduli (k and β) and length (L). We are particularly concerned about states of equilibrium near $\theta = 0$, the vertical position. In the limit $\theta \to 0$, the load approaches

$$\lim_{\theta\to 0} P \equiv P_{CR} = kL + \frac{\beta}{L} \tag{9.2}$$

Now, some simplification of Eq. (9.1) results if we divide by the constant P_{CR} and define a nondimensional "load,"

$$p \equiv \frac{P}{P_{CR}} = \frac{1}{1 + \dfrac{\beta}{kL^2}}\cos\theta + \frac{1}{\dfrac{kL^2}{\beta} + 1}\frac{\theta}{\sin\theta} \tag{9.3}$$

We see that this load p depends upon only one nondimensional parameter, kL^2/β. To appreciate the relation, we plot the equation in Fig. 9.5 for various values of the

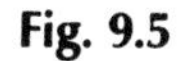
Fig. 9.5

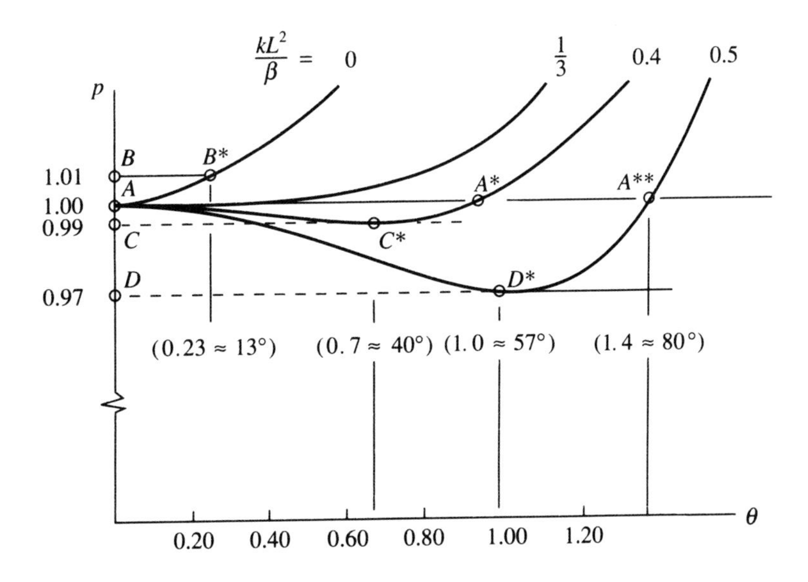

parameter. As noted previously, the system is symmetric about $\theta = 0$, so that the curves can be reflected about that axis; that is, the bar can assume such equilibrium states via movement to the left or right. Our attention focuses on the *bifurcation* point A, where the path ($\theta = 0$) of equilibrium states branches (left or right) along the curve of Eq. (9.3). Our concern is the physical significance of this bifurcation: What is the response of the system when the load attains the value $p = 1$ ($P = P_{CR}$)?

As anticipated, the vertical position ($\theta = 0$) becomes an unstable state at loads greater than the critical value ($p > 1, P > P_{CR}$). The branch of Eq. (9.3) assumes different shapes, depending upon the parameter. All have the horizontal tangent ($dp/d\theta = 0$) at the bifurcation point but slope upward after bifurcation if the parameter is small ($kL^2/\beta < \frac{1}{3}$) or downward if it is large ($kL^2/\beta > \frac{1}{3}$).

In the case of small values, such as $kL^2/\beta = 0$, the equilibrium is *strictly* stable at the critical state, *but* for very small increases the bar OB sways drastically sideward. Note that a 1% increase in load causes the bar to move from A to B^* ($\theta = 13°$). If the device is intended to support the load with little or no movement, then the system has failed in a practical sense.

Now, consider a case for a larger value ($kL^2/\beta = 0.5$): At the critical state A this system is *unstable*; at the constant load ($p = 1$), the bar OB moves to the only available *stable* state A^{**}, where the position is far removed ($\theta = 80°$) from the initial state. In reality that motion would be very abrupt; hence it is called *snap-through*. The latter circumstances are like the behavior at the bottom of a traditional oil can. The can has a slightly conical (convex) bottom; the bottom is quite stiff and resists deflection until we exert a critical pressure, which then causes a snap-buckling, or abrupt inward deflection, and the consequent squirt of oil.

One additional distinction between the two types of response ($kL^2/\beta < \frac{1}{3}$ and $kL^2/\beta > \frac{1}{3}$) has great practical significance. In the first case ($kL^2/\beta < \frac{1}{3}$), little can happen at loads below the critical value; there are no other equilibrium states at lower loads. In the second case—for example, $kL^2/\beta = 0.5$—a state D at the lower load (e.g., $p = 0.973$) also has another equilibrium configuration at D^*. Practically, a sufficient disturbance can cause the bar to snap from state D to D^*. Like the roller C in the shallow valley of Fig. 9.3, little energy is needed to kick it to another state of equilibrium.

Another important distinction between the two types of systems is the effect of initial imperfection. The bar OB of Fig. 9.3 might initially lean to one side or the load P might not act along the axis through the pin O. In cases of such imperfection, any load causes a deflection; bar OB sways sideward with increasing load, as illustrated by the dotted curves of Fig. 9.6. In the first case ($kL^2/\beta = 0$), the effect is not too damaging, since the system can resist load above the critical value. In the second case ($kL^2/\beta = 0.5$), the imperfection can be very serious. Indeed, the imperfect system cannot sustain the critical load but snaps at a lesser load from E to E', a severely deformed state ($\theta \doteq 75°$).

The foregoing examination of the simple system of Fig. 9.4 is intended only to provide insights into the mechanisms of instabilities. It is a simple system with one degree of freedom; hence, a thorough examination is quite simple. Proofs of the

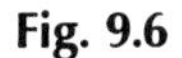

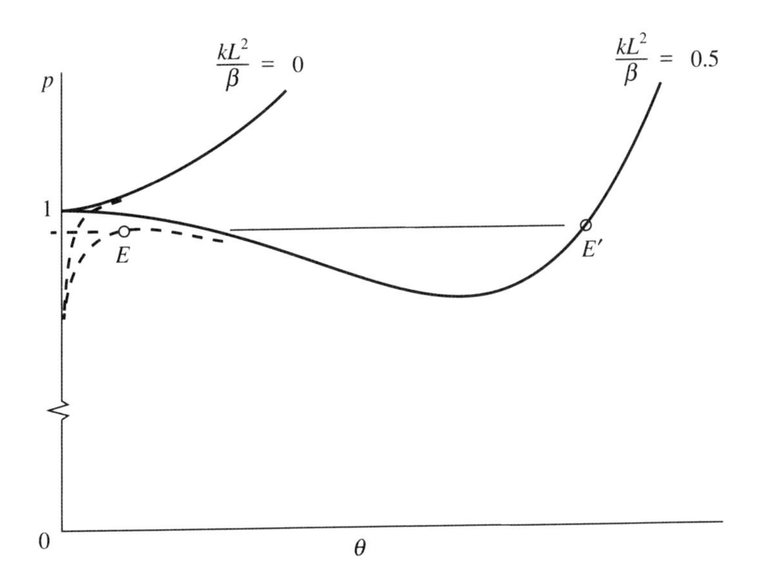

stability and instability of the states require the criterion of the preceding chapter (see Sec. 8.6). A method of determining the critical load, which admits an adjacent state of equilibrium (at point A of Fig. 9.5), is described in the subsequent sections.

To further appreciate the practical significance of the behavior (snap-buckling) of certain structures, we cite the case of a thin cylindrical shell (e.g., a beverage can) under axial load. If the shell is nearly perfect, it can sustain a remarkably great axial load. At some load, near the theoretical critical value at a bifurcation point, a slight disturbance (e.g., an extraneous lateral load) can cause abrupt, catastrophic collapse. We note that the nearly perfect shell resists the load through simple compression, whereas an imperfection or a disturbance induces bending. The wall of a very thin shell (beverage can) has little resistance to bending and folds, much like an accordion. The interested reader can conduct his or her own experiment. A trim, well-coordinated collegian can balance his or her weight on the common beverage can. A slight disturbance, such as a classmate tapping opposing sides of the can lightly with a pencil, will cause the abrupt descent of the loader. Figure 9.7 shows the results of actual tests on 17 aluminum cans; load-deflection curves are shown for six. The average buckling load was 250 lb. The axial displacement was imposed by the steady motion of the head. Note the immediate reduction in load at buckling; the average post-buckled load was 113 lb. (45% of the buckling load). Two typical buckled cans are shown in Fig. 9.8.

Fig. 9.7

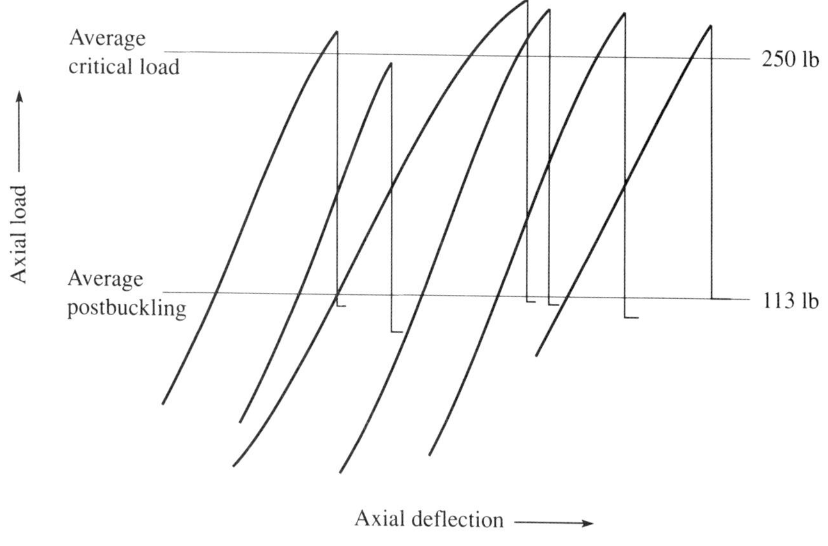

Load-deflection plots

Fig. 9.8

*9.4 General Method of Determining the Critical Load and Buckled Mode

The preceding examples show that buckling is typically characterized by the existence of neighboring states of equilibrium at the same load. That load is *critical* load. The adjacent buckled configuration is typically very different; for instance, a *straight* column assumes a *bent* form. The shape of that adjacent configuration is referred to as a buckled *mode*. We now consider another simple system, but one with two degrees of freedom. With this example, depicted in Fig. 9.9, we can illustrate the nature of the analysis, which determines a critical load and buckled mode.

Fig. 9.9

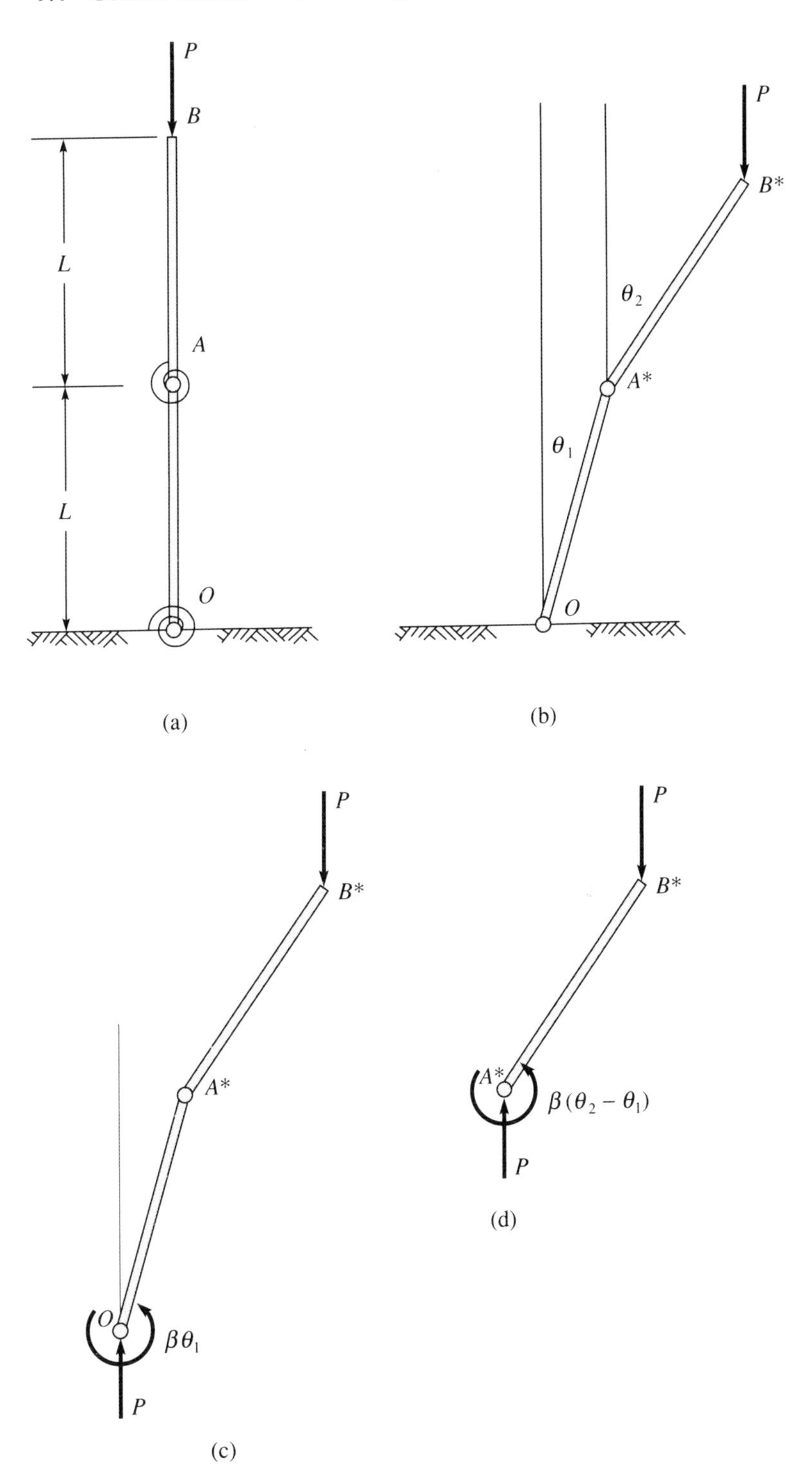

The mechanical system of Fig. 9.9 (the system considered in Sec. 8.6, Fig. 8.18) consists of two rigid links, OA and AB, joined with ideal pins at O and A. In the straight configuration of Fig. 9.9(a), the two identical torsional springs with moduli β at O and A are undeformed. Relative rotations (θ_1 and θ_2) at these joints are resisted by these linear springs, which exert the couples shown on the free-body diagrams of Fig. 9.9(c) and (d).

Equilibrium of the deformed system of Fig. 9.9(c) requires that

$$\sum M_0 = \beta\theta_1 - PL(\sin\theta_1 + \sin\theta_2) = 0 \tag{9.4a}$$

Equilibrium of the deflected link A^*B^* in Fig. 9.9(d) requires that

$$\sum M_{A*} = \beta(\theta_2 - \theta_1) - PL\sin\theta_2 = 0 \tag{9.4b}$$

Values of the angles θ_1 and θ_2 and load P that satisfy the two equations, Eq. (9.4a, b), define equilibrium states. However, we are concerned about adjacent states that exist near the straight (unbuckled) state ($\theta_1 = \theta_2 = 0$). Therefore, we explore solutions for small angles. Hence, we are justified in the approximations ($\sin\theta_1 \doteq \theta_1, \sin\theta_2 \doteq \theta_2$) of Eqs. (9.4a, b):

$$(\beta - PL)\theta_1 - PL\theta_2 = 0 \tag{9.5a}$$

$$-\beta\theta_1 + (\beta - PL)\theta_2 = 0 \tag{9.5b}$$

These are two linear homogeneous equations in the variables θ_1 and θ_2. They have a nontrivial ($\theta_1 \neq 0, \theta_2 \neq 0$) solution if and only if the determinant of the coefficients vanishes; that is,

$$\begin{vmatrix} (\beta - PL) & -PL \\ -\beta & (\beta - PL) \end{vmatrix} = 0 \tag{9.6}$$

Expanding the determinant, we obtain the quadratic equation

$$P^2 - 3\frac{\beta}{L}P + \left(\frac{\beta}{L}\right)^2 = 0$$

The solutions follow:

$$P = \frac{\beta}{L}\left(\frac{3}{2} \pm \sqrt{\frac{5}{4}}\right)$$

We are interested in the *smallest* real value that admits an adjacent state ($\theta_1 \neq 0$, $\theta_2 \neq 0$), that is, the *critical* load,

$$P_{CR} = \frac{\beta}{L}\left(\frac{3}{2} - \sqrt{\frac{5}{4}}\right) = 0.382\frac{\beta}{L} \tag{9.7}$$

Then, according to Eq. (9.5a) [or (9.5b)], the buckled mode has a shape such that

$$\frac{\theta_2}{\theta_1} = 1.62 \tag{9.8}$$

The foregoing example has the essential features of most buckling problems. To determine the critical load and the buckling mode, we proceed as follows:

1. Write the equilibrium equations for an adjacent (deformed) state.
2. Employ the approximations appropriate to a neighboring state; for example, changes of angles are small. The resulting equations are linear equations in the variables.
3. The solution is reduced to a system of linear homogeneous algebraic equations. Such a system has a solution if and only if the determinant of the coefficients vanishes. This last condition has roots (P); the smallest value is the critical load (P_{CR}).

The real problems of deformable bodies and structures are not as simple as our example, wherein all the deformation is embodied in two discrete springs. The following treatment of the continuous column provides further insights to the analyses of structural instability.

9.1 Obtain the critical load for the mechanical system depicted in Fig. P8.29(a)–(i) by examining equilibrium of the free bodies in an adjacent state and imposing the condition for a nontrivial solution of the resulting linear homogeneous equations. Express the critical load (P, p, or M) in terms of the given properties (L, k, β).

9.2 Determine the critical load for the system of Fig. P8.32 in terms of the given parameters (L, k, β). Links OA and AB are rigid, pinned at O, A, and B, and restrained by the linear extension-compression (k) and torsional (β) springs.

9.3 Determine the critical load for the system of Fig. P8.33 in terms of the given parameters (L, k, β). Links OA and AB are rigid, pinned at O and A, and restrained by the linear extension-compression (k) and torsional (β) springs.

9.5 Buckling of a Symmetric Hookean Column

Under compressive loading, thin members are susceptible to buckling, that is, to assuming a quite different bent configuration. When buckling occurs, the member can sustain little, if any, additional load. The thin column is a simple, but striking, example because the initial configuration is straight, and the buckled form is clearly bent. The hookean column of Fig. 9.10 is a particular case, wherein the lower end ($x = 0$) is fixed; the upper end ($x = L$) is free but subject to the axial load P, which remains constant and vertical. The cross section is symmetric about the xy-plane, which is also the plane of least rigidity (EI) and, therefore, the plane of buckling.

Fig. 9.10

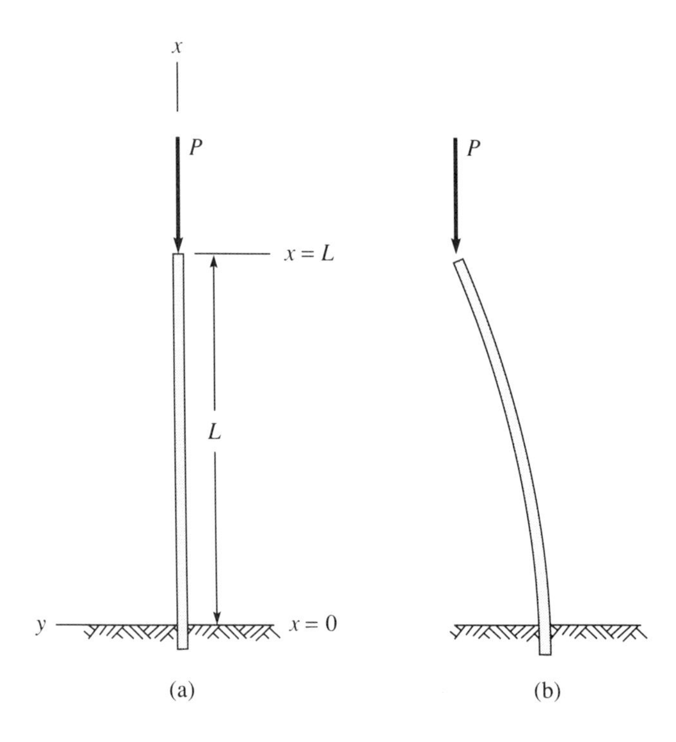

To determine the critical load, which precipitates the buckling, and also the buckled form, we must examine the equilibrium conditions of such *deformed* states. The key word is *deformed.* In our earlier considerations of equilibrium [see Eqs. (6.8) and (6.9)], deformation was neglected (a slice of the beam, the free body of Fig. 6.8, was undeformed). Now, we examine a deformed slice of the column, as depicted in Fig. 9.11. The notations are those in our previous analyses of symmetric hookean beams (see Sec. 6.5, Figs. 6.8 and 6.9). Only the notations n and t are added to signify the directions, normal and tangent, to the *bent* axis x.

Equilibrium of the free body in Fig. 9.11 requires that

$$\sum F_n = -V + (V + \Delta V)\cos\Delta\theta + (F + \Delta F)\sin\Delta\theta - \bar{p}\cos\varepsilon\Delta\theta = 0$$

Here, since the orientation of the lateral (normal) load $\bar{p}$ is intermediate, $0 < \varepsilon < 1$. As before, we divide by Δs, pass to the limit ($\Delta s \to 0$), and obtain

$$\frac{dV}{ds} + F\frac{d\theta}{ds} = p \tag{9.9}$$

Equilibrium also requires that

$$\sum M_0 \doteq -M + (M + \Delta M) + (V + \Delta V)\Delta S\cos\Delta\theta = 0$$

Here, a small approximation (the distance and orientation of the shear force) disappears in the limit, which gives

Fig. 9.11

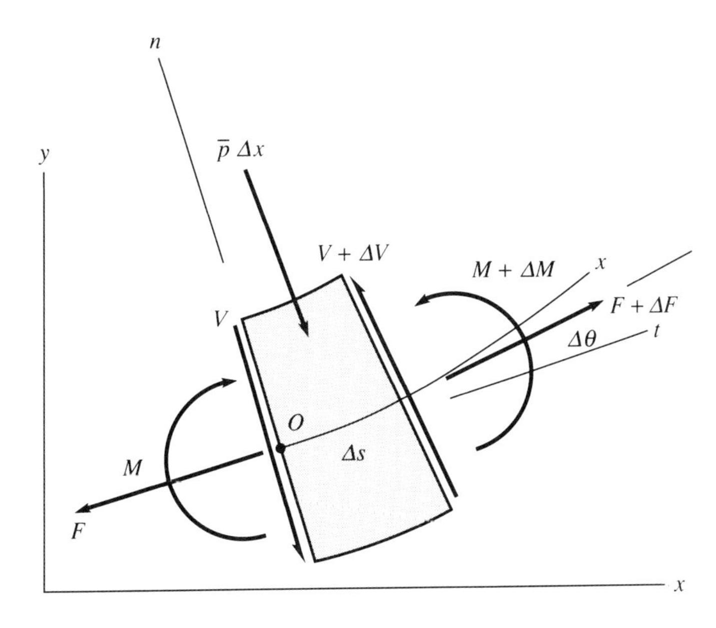

$$\frac{dM}{ds} = -V \tag{9.10}$$

Now, the strain of the axis (ε_0) is small compared to unity, so that we can again justify the approximation

$$ds = (1 + \varepsilon_0)\,dx \doteq dx$$

Then the two equations (9.9) and (9.10) take the forms

$$\frac{dV}{dx} + F\frac{d\theta}{dx} = p \tag{9.11}$$

$$\frac{dM}{dx} = -V \tag{9.12}$$

Only the former, Eq. (9.11), differs from the previous version [see Eqs. (6.31) and (6.32)].

The column is again a symmetric hookean beam, but the loading differs; now $p = 0$, $F \neq 0$. The moment curvature relation, Eq. (6.33), holds, namely,

$$\frac{d\theta}{ds} \doteq \frac{d\theta}{dx} = \frac{M}{EI} \tag{9.13}$$

Again, the axis must be at the centroid of the cross section.

Since we intend to examine only the onset of buckling, the rotation θ remains small and the kinematic relation, Eq. (6.34), prevails:

$$\frac{dw}{dx} = \theta \tag{9.14}$$

A slightly bent form of the column must satisfy four equations: the two equilibrium equations, (9.11) and (9.12), the moment-curvature relation, (9.13), and the kinematic equation, (9.14). Only the first, Eq. (9.11), is strictly nonlinear; it contains the product $F(d\theta/dx)$. However, the force F is constant ($= -P$) at the onset of buckling. Moreover, we intend to explore only equilibrium states adjacent (slightly bent) to the prebuckled (straight) state. Accordingly, the force can be assumed constant, $F = -P$. Then the buckled state is governed by four first-order linear equations in the four variables w, θ, M, V. These must be augmented by four end conditions. Note that our present problem differs from the previous problem of the beam (Sec. 6.5) in two respects: Now, $p = 0$ and one equation, Eq. (9.11), contains an additional, crucial term, $-P\,d\theta/dx$.

Before indulging in the mathematics of our problem, let us complete our formulation with the end conditions for the column of Fig. 9.10. At the lower end the conditions are the kinematic constraints, which prevent rotation and displacement:

$$\theta = 0, \qquad w = 0 \qquad \text{at } x = 0 \tag{9.15a, b}$$

To understand the mechanics at the upper end, consider the displaced end shown in Fig. 9.12. The only action upon the end is the force P; therefore,

$$M = 0$$

$$V = P\sin\theta$$

$$F = -P\cos\theta$$

Fig. 9.12

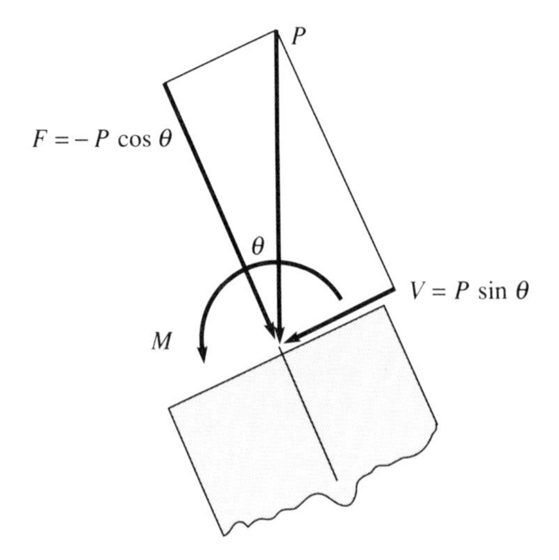

Again, we note that our interests are limited to *small* rotation, so that first-order approximations are justified ($\sin\theta \doteq \theta, \cos\theta \doteq 1$). Also note that we already anticipated the final equation, $F = -P$. The remaining end conditions follow:

$$M = 0 \qquad \text{at } x = L \tag{9.15c}$$

$$V = P\theta \qquad \text{at } x = L \tag{9.15d}$$

Mathematically, our present problem differs from that of a beam under lateral load p. In that case one could successively integrate the four equations and determine directly the four constants. Now, the presence of the additional term $(F\,d\theta/dx)$ in Eq. (9.11) couples these equations. To effect the solution, we rephrase the problem in terms of displacement w as follows: Firstly, we eliminate θ by substituting Eq. (9.14) into Eq. (9.13) and obtain

$$M = EI\frac{d^2w}{dx^2} \tag{9.16}$$

Secondly, we substitute Eq. (9.16) into Eq. (9.12) and obtain

$$V = -EI\frac{d^3w}{dx^3} \tag{9.17}$$

Note that Eqs. (9.14), (9.13), (9.16), and (9.17) express the variables θ, κ, M, and V in terms of displacement w. Finally, we substitute Eqs. (9.17), (9.14), $F = -P$, and $p = 0$ into the remaining differential equation (9.11) and obtain

$$-EI\frac{d^4w}{dx^4} - P\frac{d^2w}{dx^2} = 0$$

or

$$\frac{d^4w}{dx^4} + \left(\frac{P}{EI}\right)\frac{d^2w}{dx^2} = 0 \tag{9.18}$$

In like manner, we express the four end conditions, Eqs. (9.15a, b, c, d) in terms of the displacement w and derivatives:

$$\frac{dw}{dx} = 0, \qquad w = 0 \qquad \text{at } x = 0 \tag{9.19a, b}$$

$$\frac{d^2w}{dx^2} = 0, \qquad \frac{d^3w}{dx^3} + \left(\frac{P}{EI}\right)\frac{dw}{dx} = 0 \qquad \text{at } x = L \tag{9.19c, d}$$

Equilibrium of the slightly buckled column of Fig. 9.10 is governed by the one fourth-order linear differential equation, Eq. (9.18), and the four linear end conditions, Eqs. (9.19a, b, c, d). Certain features are common to the buckling of all symmetric hookean columns that are uniform in their properties (EI is constant) and supported only at their ends. These problems are all governed by the same differential equation, Eq. (9.18); it is homogeneous; that is, the right side is zero ($p = 0$). These problems are all characterized by four homogeneous end conditions, such as Eqs. (9.19a, b, c, d); certain linear combinations of displacement, w, dw/dx, d^2w/dx^2, and d^3w/dx^3, vanish at the ends.

To obtain the critical load and buckling mode $[w(x)]$, we require the solution of Eq. (9.18), which satisfies the end conditions, Eqs. (9.19a, b, c, d). The form of the general solution is

$$w = A\sin\lambda x + B\cos\lambda x + Cx + D \tag{9.20}$$

Derivatives ($dw/dx, d^2w/dx^2, d^3w/dx$) are required in the end conditions. The fourth derivative (d^4w/dx^4) appears in the differential equation. These results follow:

$$\frac{dw}{dx} = \lambda A\cos\lambda x - \lambda B\sin\lambda x + C \tag{9.21a}$$

$$\frac{d^2w}{dx^2} = -\lambda^2 A\sin\lambda x - \lambda^2 B\cos\lambda x \tag{9.21b}$$

$$\frac{d^3w}{dx^3} = -\lambda^3 A\cos\lambda x + \lambda^3 B\sin\lambda x \tag{9.21c}$$

$$\frac{d^4w}{dx^4} = +\lambda^4 A\sin\lambda x + \lambda^4 B\cos\lambda x \tag{9.21d}$$

When Eqs. (9.21b, d) are substituted into the differential equation, we find that the equation is satisfied everywhere if and only if

$$\lambda^2 = \frac{P}{EI}, \qquad \lambda = \sqrt{\frac{P}{EI}} \tag{9.22}$$

In the particular case at hand, Fig. 9.10, the end conditions, Eqs. (9.19a, b, c, d) take the form of four linear homogeneous algebraic equations:

$$(\lambda)A \qquad + C \qquad = 0 \tag{9.23a}$$

$$B \qquad + D = 0 \tag{9.23b}$$

$$-(\lambda^2\sin\lambda L)A - (\lambda^2\cos\lambda L)B \qquad = 0 \tag{9.23c}$$

$$\lambda^2 C \qquad = 0 \tag{9.23d}$$

Every problem of this kind (buckling of a uniform hookean column with end supports) reduces to a linear homogeneous algebraic system, such as Eq. (9.23a, b, c, d), in terms of the four constants A, B, C, and D and the parameter $\lambda(=\sqrt{P/EI})$. The homogeneous equations have a nontrivial solution ($A = B = C = D = 0$ is trivial, the straight configuration) if and only if the determinant of the coefficients vanishes; that is,

$$\begin{vmatrix} \lambda & & 1 & \\ & 1 & & 1 \\ (-\lambda^2\sin\lambda L) & (-\lambda^2\cos\lambda L) & & \\ & & \lambda^2 & \end{vmatrix} = 0 \tag{9.24}$$

Of course, $\lambda = 0$ ($P = 0$) is also trivial; only positive values are relevant. We can, therefore, factor $-\lambda^2$ from the third row and λ^2 from the fourth. We follow the rules for expanding a determinant to obtain an equation in the parameter λ (the load $\sqrt{P/EI}$); here

$$\cos \lambda L = 0 \tag{9.25}$$

The nontrivial roots are

$$\lambda L = \frac{\pi}{2}, \quad \frac{3\pi}{2}, \ldots, \frac{(2n-1)\pi}{2} \tag{9.26}$$

The only important root is the first, which gives the least load that admits a buckled configuration; that root gives the critical load

$$P_{CR} = \frac{\pi^2}{4} \frac{EI}{L^2} \tag{9.27}$$

Other roots characterize unstable states of equilibrium, which cannot be realized.

The buckled mode is determined by the four constants that must satisfy Eqs. (9.23a, b, c, d); in this case

$$A = C = 0, \qquad B = -D$$

Hence the form (mode) of the buckled column is

$$w(x) = D(1 - \cos \lambda x) \tag{9.28}$$

Note that the amplitude of the mode (D) remains unknown. This is always true in the solution of the linearized formulation.

It is never enough to achieve a mathematical result in engineering. We must examine the implications and limitations. Figure 9.13(a) depicts the column and a dial gage arranged to measure lateral deflection $w(L)$. Our theory suggests that the straight column sustains load P in the straight configuration OA $[w(L) = 0]$ in Fig. 9.13(b)

Fig. 9.13

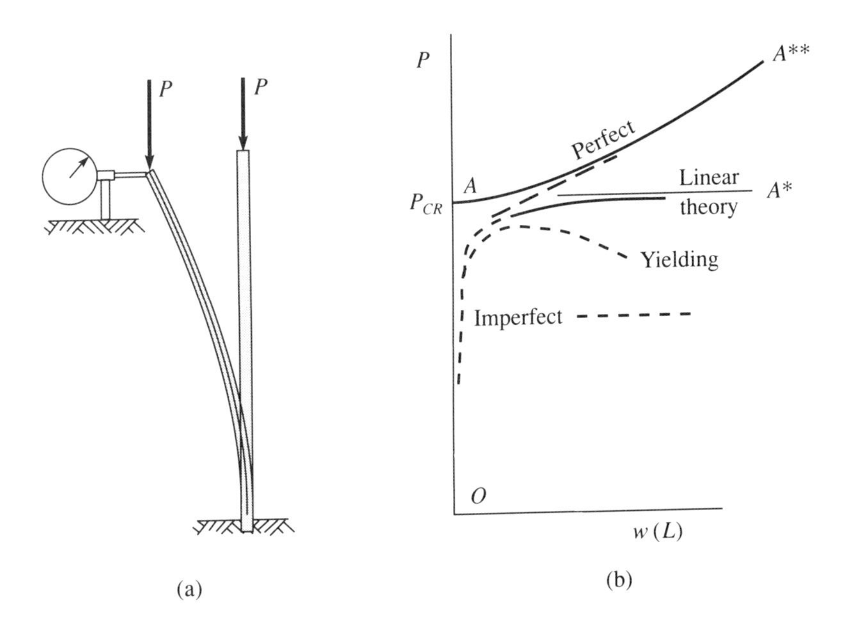

and then encounters the bifurcation A at the critical load and assumes the buckled form with unrestricted amplitude.

In reality, as the deflection increases, nonlinearities must be considered ($\sin\theta \neq \theta$) and *slight* additional load is required to cause further deflection. The trace for a perfect column is AA^{**}. Of course, no column is perfect; all possess some initial crookedness. Also, it is impossible to align the load perfectly. Therefore, the actual imperfect column experiences some bending at the onset of loading. That deflection grows and becomes excessively large at the critical load; this behavior is indicated by the dotted curve of Fig. 9.13(b). From a practical viewpoint the critical load signals the useful limit, since deflection becomes excessive with very little additional loading; P remains practically constant at the critical load P_{CR}. In many instances, buckling is immediately followed by yielding of the material (e.g., structural steel) and collapse ensues.

*9.6 Columns with Intermediate Supports

If a column has an intermediate support (for example, a column is connected to a beam that acts as an elastic support), then the deflection to either side is given by functions $w_-(x)$ and $w_+(x)$, which satisfy Eq. (9.18). Each function has the form of Eq. (9.20); each has four unspecified constants. As in the case of a beam with an intermediate support (see Sec. 6.5, "Support Conditions," p. 415) or a concentrated load (see Example 9 in Chapter 6, p. 422), each support entails four additional conditions of continuity and/or discontinuity in the variables (deflection w, rotation w', moment EIW'', and/or shear $-EIw'''$). Consequently, each support introduces four additional constants *and* also four additional equations. Therefore, the solution for the critical load (and buckled mode) entails a determinant, which is $4(1+n) \times 4(1+n)$, where n is the number of such additional supports. In general, the critical load for most *linearly* elastic systems is characterized by a bifurcation of equilibrium states (point A of Fig. 9.5) and a system of *homogeneous* linear equations. The critical load is the least value that provides a nontrivial solution (buckled mode), specifically, the vanishing of the determinant $[4(1+n) \times 4(1+n)]$.

9.7 Effect of Imperfections

To gain some insight into the consequences of initial imperfection, let us suppose that the column of Fig. 9.10 is initially slightly bent. For simplicity, let us suppose an initial deflection $w_0(x)$ in the form of the buckling mode:

$$w_0(x) = \bar{w}\left(1 - \cos\frac{\pi x}{2L}\right) \tag{9.29}$$

Let $w(x)$ denote any additional deflection, so that the actual bent form has the net deflection

$$w_1(x) = w_0(x) + w(x) \tag{9.30}$$

The deflection w_1 is related to the net rotation θ_1 according to Eq. (9.14):

$$\frac{dw_1}{dx} = \theta_1 = \frac{dw_0}{dx} + \frac{dw}{dx} \tag{9.31}$$

Neither couple M nor force V act initially. The former is proportional to the *change* of curvature associated with the additional deflection. Therefore, relations of couple and force are like Eqs. (9.16) and (9.17):

$$M = EI\frac{d^2w}{dx^2}, \qquad V = -EI\frac{d^3w}{dx^3} \tag{9.32a, b}$$

As before, we substitute Eqs. (9.31) and (9.32b) into the equilibrium equation (9.11) and set $p = 0$ and $F = -P$ to obtain

$$-EI\frac{d^4w}{dx^4} - P\frac{d^2w}{dx^2} = +P\frac{d^2w_0}{dx^2} \tag{9.33a}$$

Now, our initial deflection w_0 has the form of Eq. (9.29); therefore, Eq. (9.33a) takes the following form:

$$\frac{d^4w}{dx^4} + \frac{P}{EI}\frac{d^2w}{dx^2} = -\frac{\pi^2}{4L^2}\frac{P}{EI}\overline{w}\cos\frac{\pi x}{2L} \tag{9.33b}$$

This equation differs from Eq. (9.18) by virtue of the "loading" on the right side. We require the appropriate particular solution to augment the homogeneous solution Eq. (9.20). The particular solution, in this special case, also satisfies the end conditions and, therefore, constitutes the complete solution:

$$w(x) = w_p(x) = -\frac{\overline{w}}{1 - \dfrac{\pi^2}{4L^2}\dfrac{EI}{P}}\cos\frac{\pi x}{2L} \tag{9.34}$$

Note that the deflection becomes unbounded as $P \to P_{CR} = \pi^2 EI/4L^2$. Again, as the deflection grows, the error of our linear theory increases ($\sin\theta \neq \theta$); then the nonlinearities must be accommodated, and the refined theory shows that some additional load can be sustained *but* only as deflection becomes excessive. Such severe deflection is usually accompanied by yielding or other failure of the material.

*9.8 Significance of Higher Modes

Our solution to the problem of Fig. 9.10 indicates an infinity of equilibrium states [$w = D(1 - \cos\lambda x)$] under the loads $\lambda = (2n - 1)\pi/2L$. As noted, only the first mode at the least load ($P_{CR} = EI\lambda^2, n = 1$) can be realized; the higher modes are unstable. To examine the question, let us consider yet another example, the column of Fig. 9.14, which is simply supported at both ends:

$$w = \frac{d^2w}{dx^2} = 0 \qquad \text{at } x = 0, L$$

Fig. 9.14

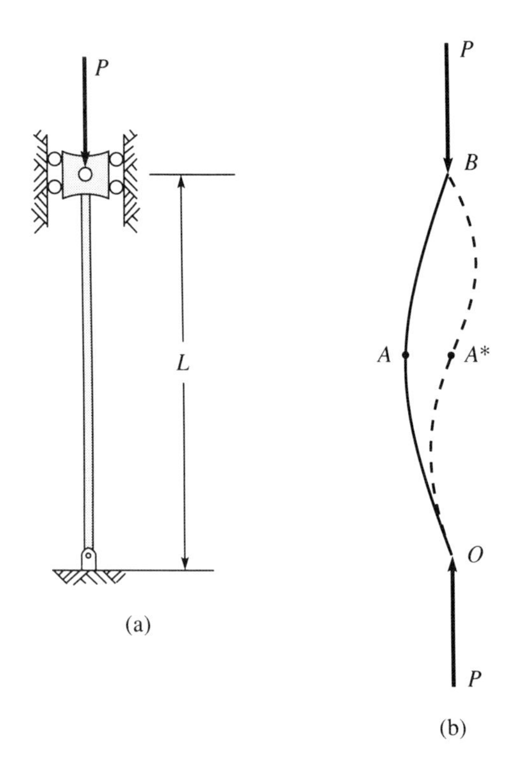

The solutions in this case are

$$w_n = A \sin \lambda_n x, \qquad \lambda_n = n\pi/L, \qquad n = 1, 2, \ldots \tag{9.35}$$

The critical load is again the smallest ($n = 1$):

$$P_{CR} = P_1 = \lambda_1^2 EI = \pi^2 EI/L^2 \tag{9.36}$$

The buckling mode follows:

$$w(x) = A \sin \pi x/L \tag{9.37}$$

However, all solutions satisfy equilibrium conditions. The second mode ($n = 2$) is shown dotted in Fig. 9.14(b). Indeed, with a simple support at the midpoint, that second mode is the buckling mode, which occurs at the load

$$P_2 = \lambda_2^2 EI = 4\pi^2 EI/L^2 \tag{9.38}$$

The midpoint is a point of inflection ($d^2w/dx^2 = M/EI = 0$), and each half of our original column behaves as a simply supported column of length $L/2$.

Let us now consider the implications of the column with an intermediate support. Theoretically, that support need exert no lateral force; a very weak support ought to suffice. We might imagine that we manually hold the midpoint and, therefore, attain the higher load $P_2 = 4P_1$. What would happen if we suddenly removed our support or, in a practical circumstance, a weak support failed? We then have our original column (without benefit of intermediate support) subjected to $P_2 = 4P_1 = 4P_{CR}$. In this state a perfect column *is* in equilibrium—*but unstable* equilibrium. It is in a state such as the one of roller A at the top of the hill in Fig. 9.3. Like the roller, the column tends toward a stable state. Because the load is far in excess of the sustainable load, abrupt collapse ensues. There is an important practical lesson in the example. A weak intermediate support, such as an adjoining beam that supports a column, can greatly increase the critical load, but can also create a state of instability upon failure (see, for example, Figs. 9.4 and 9.5).

PROBLEMS

9.4 A symmetric HI-HO column is simply supported at both ends. Following the procedure of Sec. 9.5, obtain the 4 × 4 determinant and the critical load.

9.5 A symmetric HI-HO column is fixed at one end ($x = 0$) and simply supported at the other ($x = L$). Following the procedure of Sec. 9.5, obtain the 4 × 4 determinant and the equation that determines the critical load, $\tan \lambda L = \lambda L$. Determine the buckled mode $w(x)$, that is, the constants B/A, C/A, and D/A.

9.6 A symmetric HI-HO column is fixed at both ends, but only with respect to transverse displacement w and slope w'. Use the procedures of Sec. 9.5 to determine the critical load and the buckled configuration $w(x)$.

9.7 A HI-HO column has a small initial deflection prior to loading:

$$w_0 = A_0 \sin \pi x/L$$

The bending moment is proportional to the change of curvature; that is,

$$M = EI\left(\frac{d^2 w_1}{dx^2} - \frac{d^2 w_0}{dx^2}\right)$$

where w_1 is the total deflection. Otherwise the basic governing equations are unchanged (see Sec. 9.6). The column is simply supported at both ends. When the column is subjected to axial loading, show that the additional deflection $w = w_1 - w_0$ is governed by the differential equation

$$\frac{d^4 w}{dx^4} + \frac{P}{EI}\frac{d^2 w}{dx^2} = -\frac{P}{EI}\frac{d^2 w_0}{dx^2}$$

Show that the complete solution to the equation and the end condition is

$$w(x) = \frac{A_0 \sin \pi x/L}{\pi^2 EI/L^2 P - 1}$$

What occurs as $P \to P_{CR} = \pi^2 EI/L^2$?

9.8 A symmetric HI-HO beam-column is simply supported at both ends ($x = 0, L$). It is simultaneously subjected to the transverse load

$$p(x) = \bar{p} \sin \pi x/L$$

and the axial thrust P. The transverse loading appears on the right sides of Eqs. (9.11) and (9.18). Show that the solution is

$$w(x) = -\frac{\bar{p}L^4}{\pi^4 EI}\left(\frac{1}{1 - PL^2/\pi^2 EI}\right)\sin \pi x/L$$

Note the influence of the axial thrust and the implications as $P \to \pi^2 EI/L^2$.

9.9 A symmetric HI-HO column is simply supported at both ends ($x = 0, L$). The column lies within an elastic medium, which resists lateral deflection $w(x)$; specifically, the deflection causes the lateral load $p(x) = \alpha w(x)$, where α is the effective modulus of the medium. The load appears on the right sides of Eqs. (9.11) and (9.18). Instead of the latter, the deflection is governed by the linear differential equation

$$\frac{d^4w}{dx^4} + \frac{P}{EI}\frac{d^2w}{dx^2} + \frac{\alpha}{EI}w = 0$$

Verify the solution

$$w = A \sin n\pi x/L$$

and show that

$$P_{CR} = \text{minimum}\frac{\pi^2 EI}{L^2}\left[n^2 + \frac{\gamma}{n^2}\right]$$

where

$$\gamma = \frac{\alpha L^4}{\pi^4 EI} \qquad \text{(nondimensional modulus)}$$

At what value of γ and P_{CR} does the buckled mode pass from one loop ($n = 1$) to two ($n = 2$)?

9.9 Euler's Formula and End Conditions

The analysis of thin hookean columns, the determination of the critical load, bifurcation, and also postbuckling behavior were first investigated by Leonhard Euler.[2] Euler's formula for the critical load has the general form

$$P_{CR} = C\frac{\pi^2 EI}{L^2} \tag{9.39}$$

The constant C is determined by the conditions of support at the ends. It has been called the *end-fixity factor*. Four cases and the critical loads are shown in Fig. 9.15.

The situations depicted in Fig. 9.15 are ideal; the ends are either totally fixed, completely free, or ideally pinned. In reality columns are connected to adjoining members; for example, a vertical column might be joined to a horizontal beam that resists lateral deflection and rotation as it bends and twists but does not completely prevent the movement. The adjoining member acts as a spring; the lateral and torsional moduli depend upon the properties of the member. Such a circumstance is illustrated in Fig. 9.16, where a torsional spring resists rotation at the lower end. Here, the conditions at the lower end ($x = 0$) are

$$w(0) = 0,$$

$$M(0) = EI\left.\frac{d^2w}{dx^2}\right|_0 = \beta\left.\frac{dw}{dx}\right|_0$$

The solution is obtained according to the procedure of Sec. 9.5. The critical load is determined by the smallest root of the equation

$$\tan \lambda L = \frac{\alpha\lambda L}{\alpha + \lambda^2 L^2} \tag{9.40}$$

where

2 Euler obtained his formula for the critical load in a memoir of 1757. The works of Euler are chronicled by Todhunter and Pearson, *Elasticity and Strength of Materials*, vol. 1 (New York: Cambridge University Press, 1886): 32–57.

Fig. 9.15

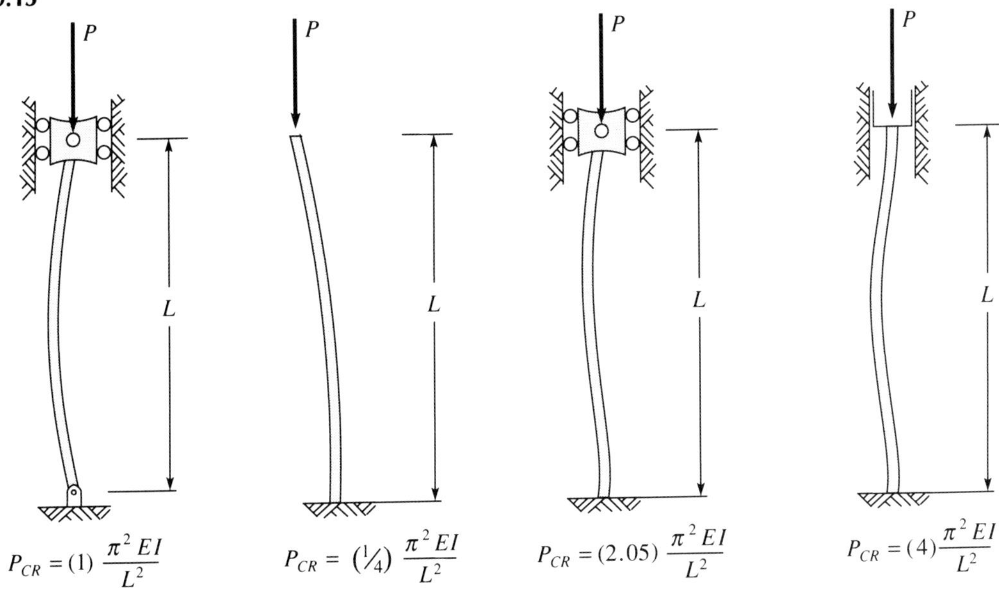

Fig. 9.16

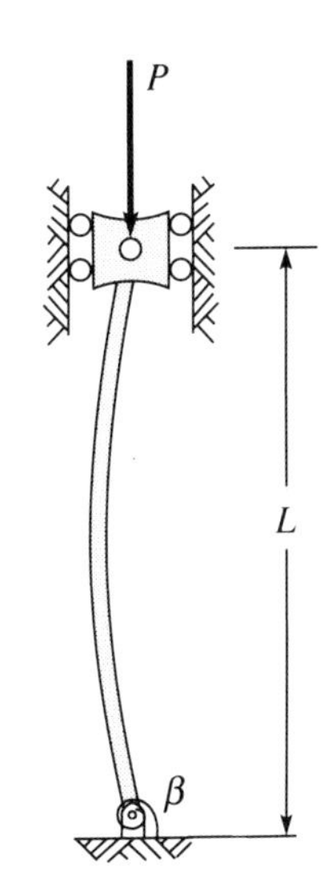

$$\lambda^2 L^2 = \frac{PL^2}{EI}, \qquad \alpha = \frac{\beta L}{EI}$$

If $\beta = \alpha = 0$, the lower end is simply supported and $\lambda L = \pi$ [see Fig. 9.15(a)]. If $\beta \to \infty$, $\alpha \to \infty$, the lower end is fixed and $\lambda L = \sqrt{2.05}\pi$ [see Fig. 9.15(c)].

Sometimes Eq. (9.39) is written as

$$P_{CR} = \frac{\pi^2 EI}{(kL)^2} \tag{9.41}$$

where $k = 1/\sqrt{C}$. In this form kL is called the *reduced length* of the column. The significance of this terminology can be seen in Fig. 9.17. For the pin-connected column, $k = 1$. In this case the column buckles into one arch of a sine wave. The equivalent length of a column is the portion of length that assumes the same buckled shape as the pin-connected column. For the clamped-free column, we would need twice the actual length; hence, $k = 2$. For the clamped-clamped column, half its actual length buckled into one arch of a sine wave; hence, in this case, $k = \frac{1}{2}$. For the clamped-pinned column, the equivalent length turns out to be approximately 0.7 of the actual length.

Fig. 9.17

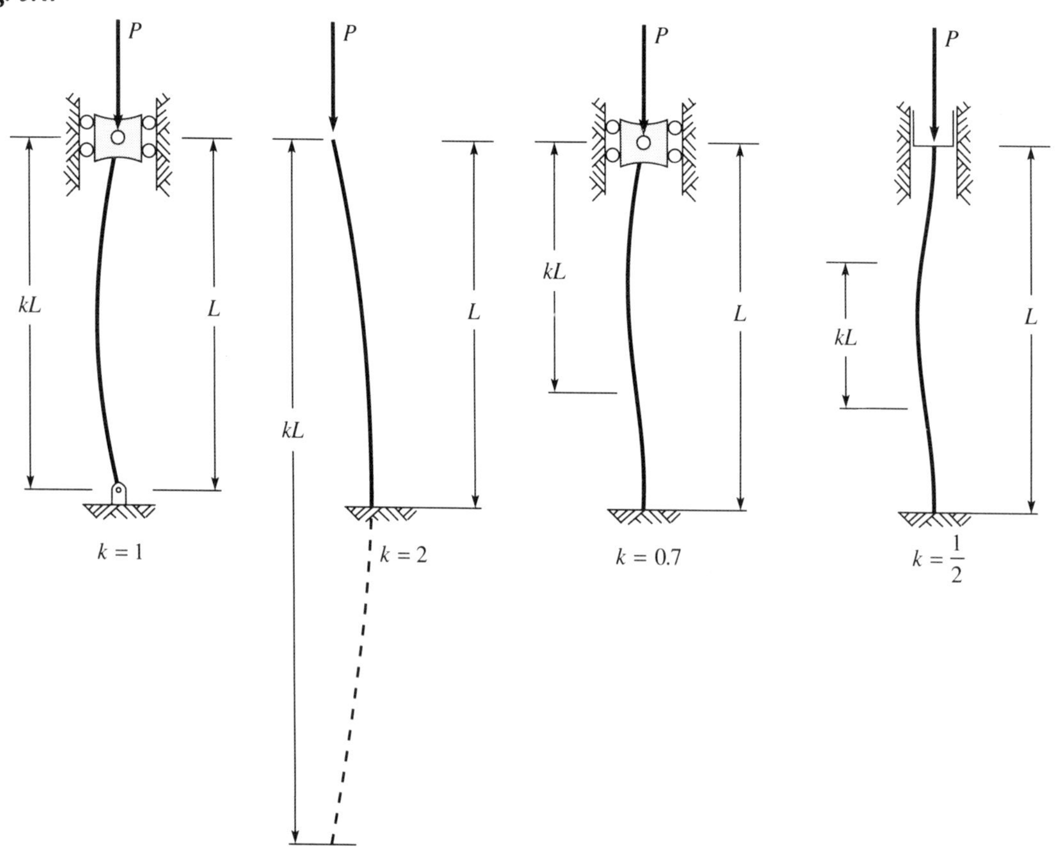

Slenderness Ratio

One further modification is usually made in Eq. (9.41) by dividing through by the cross-sectional area of the column. Then

$$\sigma_{CR} = \frac{P_{CR}}{A} \tag{9.42}$$

is called the *critical stress*, whereas

$$r^2 = \frac{I}{A} \tag{9.43}$$

defines a parameter r, the radius of gyration of the cross section. With this notation, Eq. (9.42) can be written

$$\sigma_{CR} = \frac{\pi^2 E}{\left(\frac{kL}{r}\right)^2} \tag{9.44}$$

The quantity kL/r is called the *slenderness ratio* of the column. (Some authors call L/r the slenderness ratio.)

Additional Remarks

A plot of critical stress versus slenderness ratio as determined by Eq. (9.44) is shown in Fig. 9.18 (The scales are not indicated, since they depend on the value of E for any particular column.) As the slenderness ratio decreases, the critical stress grows quite large. When it exceeds the proportional limit for the material, the Euler formula is no

Fig. 9.18

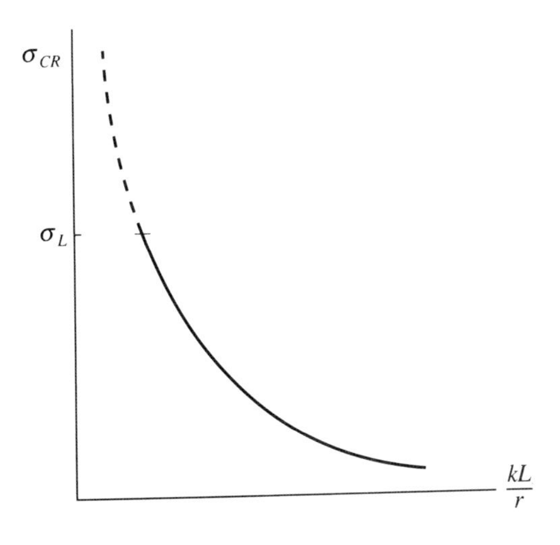

longer valid, since it is based on linearly elastic behavior. The portion of the curve that is not valid is shown dashed.

For very small values of slenderness ratio, the column becomes a squat compression block. In this case buckling is no longer a problem, since failure is the result of compression (fracture for some very brittle materials or excessive deformation for ductile materials). However, there is an intermediate range of slenderness for which the Euler formula gives values above the proportional limit, but failure can still occur as the result of buckling. This situation is considered in the next section.

One final point must be emphasized concerning the elastic buckling of columns. The critical load (or stress) is the *smallest* nonzero value for which a buckled equilibrium configuration is possible. Extreme care must, therefore, be exercised in choosing the values of k and r that apply in a particular case. These quantities in general can vary widely if the plane of bending is changed. For example, a thin rectangular cross section will bend much more easily in one direction than another because of the difference in moments of inertia or, equivalently, the difference in radii of gyration. Also, an end connection can act as a pin with respect to one plane of bending but as a clamp with respect to another. To give the smallest critical load, the plane of buckling must be chosen so that k/r is a maximum.

PROBLEMS

Give your answers in terms of the indicated parameters (L, k, β, λ) when numerical values are not specified.

9.10 Find the critical load for a steel column that is 150 cm long and 2.5 cm in diameter if the ends are pin-connected.

9.11 A yardstick made from wood has a modulus of elasticity equal to 1.6×10^6 psi and cross-sectional dimensions of $\frac{1}{4}$ in. by 1 in. Find the critical load for the stick if it is (a) pin-connected at each end and (b) clamped at one end and free at the other.

9.12 If a pin-ended column is prevented from deflection at the center, as shown in Fig. P9.12, by what factor is the critical load increased?

Fig. P9.12

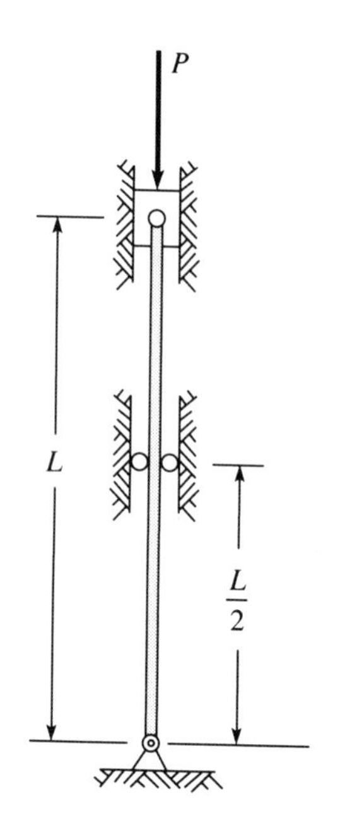

9.13 Bar AB in Fig. P9.13 is an aluminum bar with a 2.5-cm diameter and a modulus of elasticity equal to 70×10^3 MPa. Find the value of the load P that causes bar AB to buckle.

Fig. P9.13

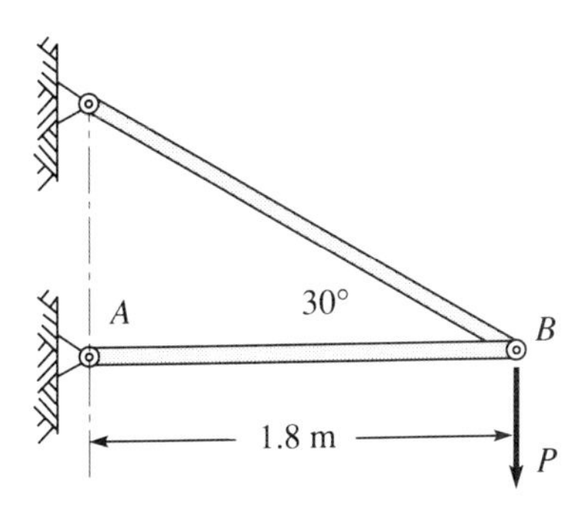

9.14 A steel tubular column is 20 ft long and has a 3-in. outside diameter and a 2.5-in. inside diameter. Find the critical load for the column if (a) both ends are clamped and (b) one end is clamped and the other is pinned.

9.15 A steel pipe used as a column has a cross-sectional area of 25 cm^2 and a radius of gyration of 7.5 cm. It is pinned at both ends and supports a load of 180 kN. (a) What is the maximum length the column can have to ensure it will not buckle? (b) if the column is designed to carry twice the anticipated load (360 kN), what is its maximum length?

9.16 Each member of the equilateral-triangle truss in Fig. P9.16 is made of aluminum ($E = 10^7$ lb/in.^{2}), is 10 ft long, and has a cross-sectional area of 2 in.^{2} and a radius of gyration of 1 in. Find the greatest value of P the frame can support without buckling.

Fig. P9.16

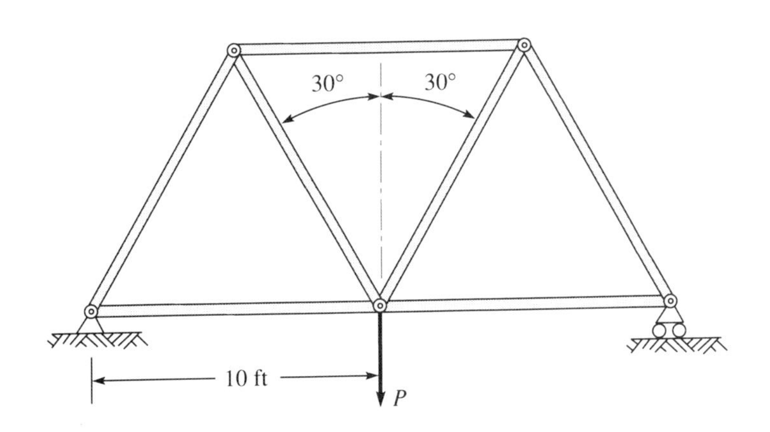

9.17 The column in Fig. P9.17 is pin-supported at the base and restrained from lateral motion at a distance λL above the base. If the column has a length L and a stiffness EI, determine the characteristic equation.

Fig. P9.17

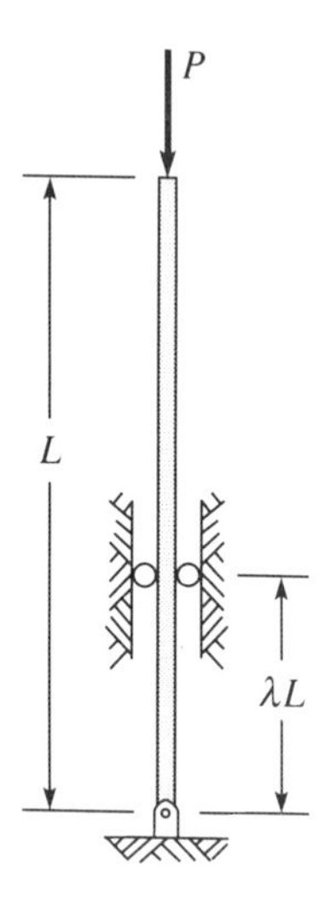

9.18 Determine the characteristic equation for the elastic column shown in Fig. P9.18.

Fig. P9.18

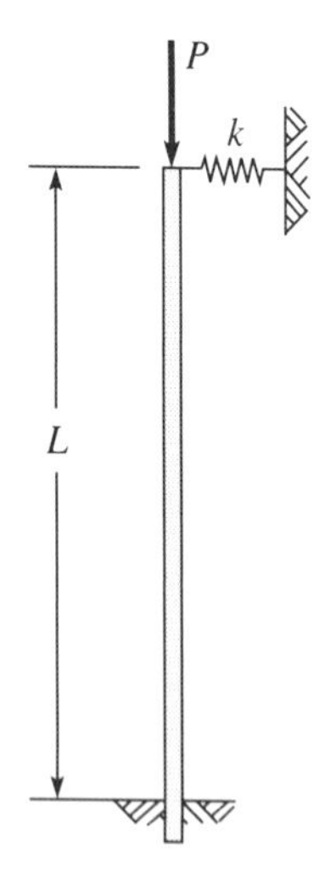

9.19 A slender, round-ended elastic column of length L has a rectangular cross section $a \times 2a$. If the critical load for this column is P, express the new critical load as a multiple of P if (a) the length of the column is halved, (b) the modulus E is doubled, and (c) the cross section is reduced to $a \times a$.

9.20 A uniform bar (with rounded ends) just fits between two rigid walls at a temperature Q_0. If the coefficient of thermal expansion is α and the length is L, determine the temperature at which the straight form of the bar becomes unstable.

9.21 In Fig. P9.21, AB and CD are slender round-ended aluminum columns, each with a 1.9-cm diameter. Compute the critical load P. (Assume the bars EF and GC are rigid.)

Fig. P9.21

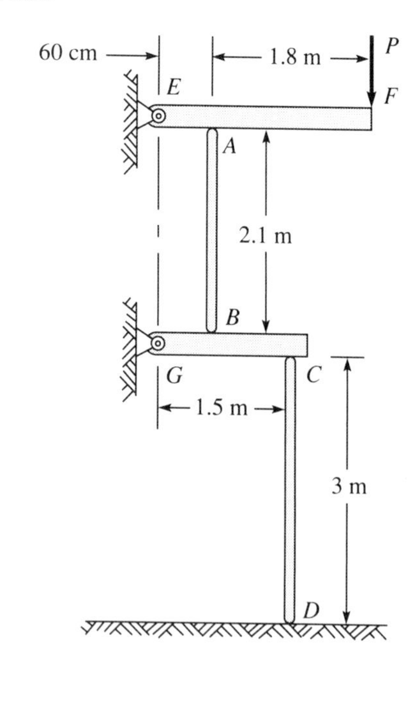

9.22 The plane truss of Fig. P9.22 is composed of circular steel tubes ($E = 30 \times 10^3$ ksi). The yield stress in tension and compression is $\sigma_0 = \pm 50$ ksi. Inner and outer diameters are 1.90 in. and 2.10 in., respectively. Failure is initiated if a member buckles or yields. Calculate the load P that initiates failure.

Fig. P9.22

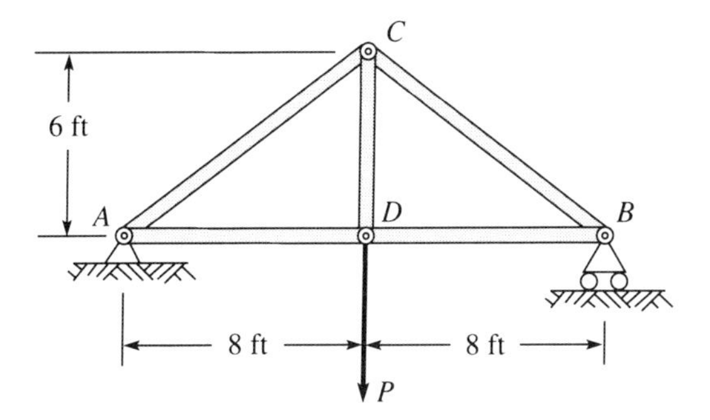

9.23 A pin-connected truss is loaded as shown in Fig. P9.23. Determine the critical load P. All the members have the same cross section and are made of the same material.

Fig. P9.23

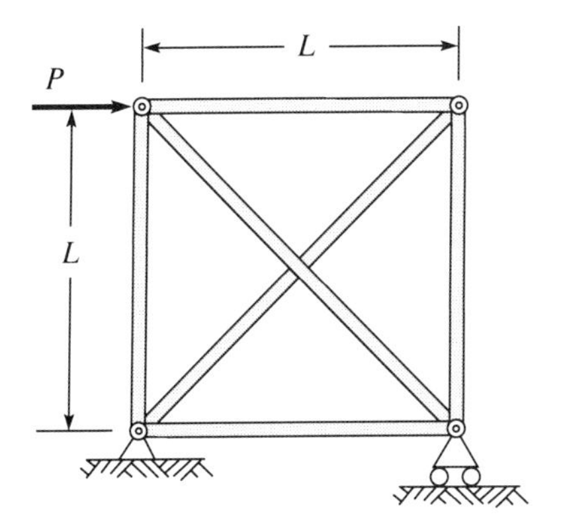

9.24 A tripod consists of three identical steel legs that can rotate freely at O, A, B, and C in Fig. P9.24. The base ABC forms an equilateral triangle on the xy surface. Apex O is at (0, 0, 60 in.), where each leg forms a 30° angle with the z axis. Each rod is homogeneous with a diameter of 0.400 in. The modulus and yield stress of the steel are

$$E = 30 \times 10^3 \text{ ksi} \quad \text{and} \quad \sigma_0 = \pm 50 \text{ ksi}$$

Determine the practical limit of the vertical load P.

Fig. P9.24

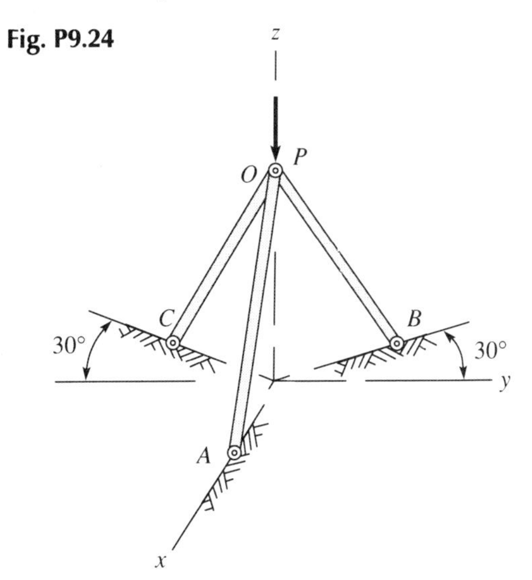

9.25 A slender steel rod, clamped rigidly at its ends, is stress-free at 22°C. It is 3 m long and 5 cm in diameter. If $\sigma_0 = 200$ MPa for the steel, determine the range of temperature within which the rod will neither yield nor buckle.

9.26 A uniform homogeneous column is clamped at the bottom and free at the top. It has a total weight W and a length L and supports its own weight. Show that any buckled configuration must satisfy

$$EI \frac{d^2}{dx^2} w(x) = -W\left(1 - \frac{x}{L}\right) w(x) + \int_x^L \frac{W}{L} w(\xi)\, d\xi$$

Hence, the critical weight is the least value of W for which the preceding equation has a nontrivial solution satisfying boundary conditions.

9.10 Buckling of Columns with Nonlinear (Inelastic) Behavior

There are circumstances in which the Euler formula predicts a critical stress beyond the proportional limit of the material. In this range the Euler formula is not valid; it gives values that are much higher than those actually attainable. Figure 9.19 shows

Fig. 9.19

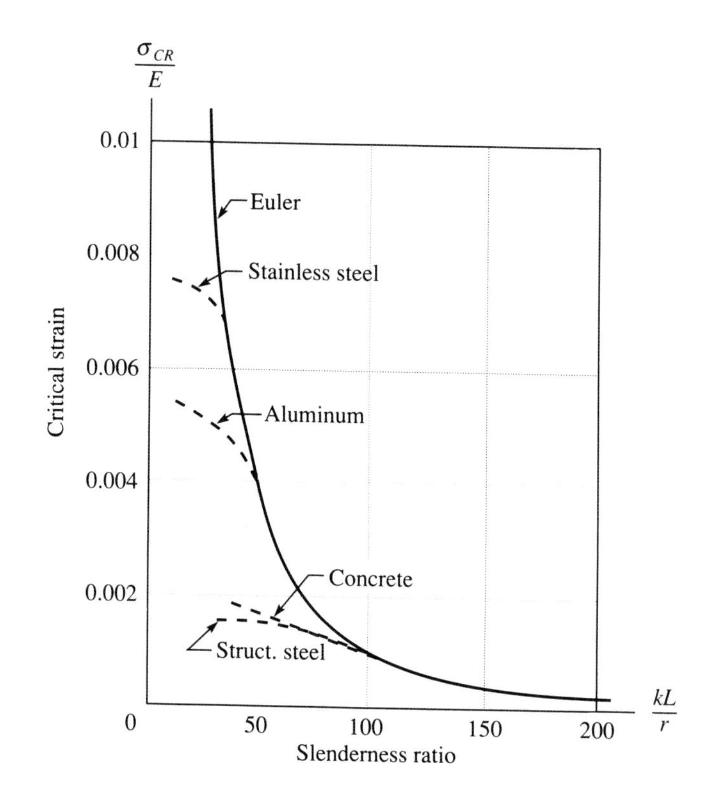

experimental results for some typical engineering materials. The ordinate is the critical stress divided by the material's modulus of elasticity, so that all materials follow the same curve in the linear range of their stress-strain diagram.

Empirical Relations

Many empirical formulas have been devised for the design of columns in the intermediate range. Widely used by engineers, these are often found in building codes, where their use is mandatory. Such empirical formulas are simple algebraic relations between critical stress and slenderness ratio, which are fitted to experimental data. One of the earliest formulas used was of the form

$$\sigma_{CR} = \frac{a}{1 + b\left(\frac{kL}{r}\right)^2}$$

where a and b are constants determined from test data. The simplest approximation is obtained by expressing the critical stress as a linear function of the slenderness ratio. Very widely used, it has the form

$$\sigma_{CR} = a - b\left(\frac{kL}{r}\right)$$

The parabolic approximation

$$\sigma_{CR} = a - b\left(\frac{kL}{r}\right)^2$$

is occasionally used.

The constants in all these empirical formulas must be determined from experimental data, since their values are different for different materials. It is customary to adjust these values to ensure against the uncertainties of actual service conditions. For example, if a and b are experimentally determined values for the coefficients in one of the formulas, then the a and b actually used in the formula are those that give a curve somewhat lower than the experimental one.

These empirical formulas are restricted to a range of slenderness ratio that depends both on the particular formula and the material. By a suitable choice of constants, some of the empirical formulas can be extended well into the range of high slenderness, obviating the need for Euler's formula in these cases.

Engesser's Tangent Modulus Theory

Another approach to the intermediate range of buckling is to modify the calculation of critical stresses to obtain reliable results for all engineering materials in the nonlinear stress-strain range. One of the simplest and most successful attacks was proposed by F. Engesser in 1889.

Let Fig. 9.20 represent the stress-strain diagram for the material as obtained from a compression test. Suppose that σ_{CR} is the actual critical stress for the column we are considering, and ε_{CR} is the corresponding strain. The slope of the stress-strain diagram at this point is E_T. Then, for stresses and strains very close to σ_{CR} and ε_{CR}, we

Fig. 9.20

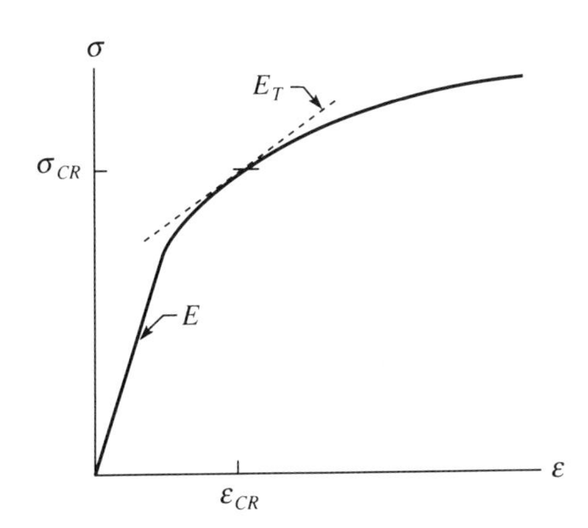

can approximate the stress strain diagram by the straight line

$$\sigma - \sigma_{CR} = E_T(\varepsilon - \varepsilon_{CR})$$

At the critical load, the column is axially compressed and the extensional strain is constant everywhere on a cross section; its value is ε_{CR}. If cross sections remain plane when the column bends, the extensional strain due to bending varies linearly over the cross section. The states of strain due to compression and due to bending represent infinitesimal deformations. Therefore, the resulting state of strain can be obtained by adding the components of strain for the two states.

If the centroidal axis of the column (the neutral axis with respect to the bending deformation alone) remains at the critical stress during buckling, the strain there will be ε_{CR}; then measuring y from this axis, the extensional strain anywhere in the cross section is

$$\varepsilon = \varepsilon_{CR} - \frac{y}{\rho}$$

where ρ is the radius of curvature of the centroidal axis. Since the curvature is very small, the strains throughout each cross section will be close to ε_{CR}; consequently, we may use the approximate stress-strain relation developed here. The stresses are then given by

$$\sigma = \sigma_{CR} - E_T \frac{y}{\rho}$$

The bending couple due to these stresses is

$$M = -\iint_A y\sigma \, dA = -\sigma_{CR} \iint_A y \, dA + \frac{E_T}{\rho} \iint_A y^2 \, dA$$

The first integral is zero, since y is measured from the centroid of the cross section. The second is the moment of inertia, I. Thus

$$M = \frac{E_T I}{\rho}$$

or

$$\frac{1}{\rho} = \frac{M}{E_T I}$$

This is exactly the same formula used in analysis of the hookean column except that E_T replaces E. By following the same procedure, we obtain the Euler-Engesser formula

$$\sigma_{CR} = \frac{\pi^2 E_T}{\left(\dfrac{kL}{r}\right)^2} \tag{9.45}$$

analogous to Eq. (9.44). If σ_{CR} is in the linearly elastic range, then E_T is the slope in this range, namely, E, and Eq. (9.45) reduces to Eq. (9.44). The quantity E_T is called the *tangent modulus*, and Eq. (9.44) is sometimes referred to as the *tangent modulus formula*. Although Engesser's theory does not account for inelastic unloading during buckling, it provides good agreement with experimental results for many materials.

Strain Hardening Material

The following example illustrates the use of the Euler-Engesser formula. Suppose a column is made of an aluminum alloy with the stress-strain diagram in compression, as shown in Fig. 9.21. In the same figure, the dotted curve is a plot of the tangent modulus versus stress with the tangent-modulus scale at the top.

Fig. 9.21

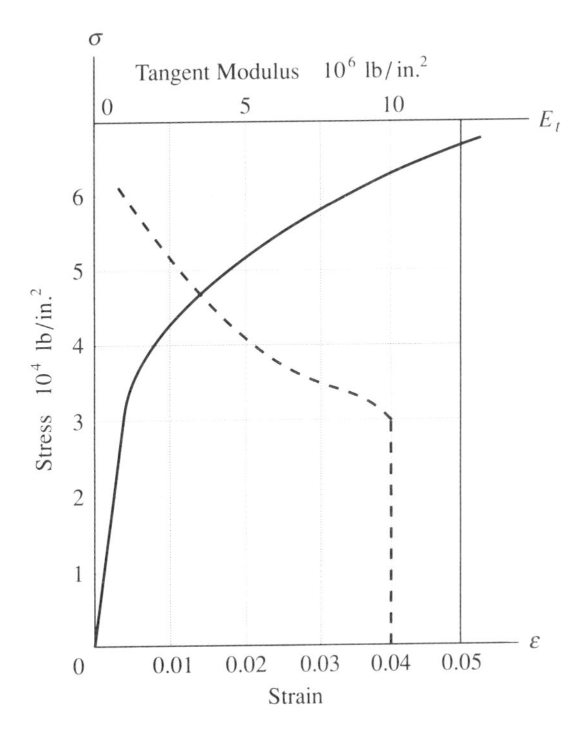

Note that, to the proportional limit, the tangent modulus is constant and equal to the modulus of elasticity. Then, as the slope of the stress-strain diagram decreases, so does the tangent modulus. If the column is designed to have a given critical stress, we enter the curve at that value of stress and find the corresponding tangent modulus. The slenderness ratio is then determined from

$$\frac{kL}{r} = \sqrt{\frac{\pi^2 E_T}{\sigma_{CR}}}$$

If, on the other hand, the slenderness ratio is known and we want the critical stress, a trial-and-error procedure can be used. Guess at the critical stress (try the Euler value, for instance) and find the corresponding value of E_T from the tangent modulus curve. If the guess is correct, these values will satisfy

$$\frac{\sigma_{CR}}{E_T} = \frac{\pi^2}{\left(\frac{kL}{r}\right)^2}$$

If not, adjust σ_{CR} up or down as indicated and repeat the calculation with the new guess until the desired accuracy is achieved.

9.27 Determine the critical load for the column shown in Fig. P9.27 if the material is linear hardening according to the diagram in the figure.

Fig. P9.27

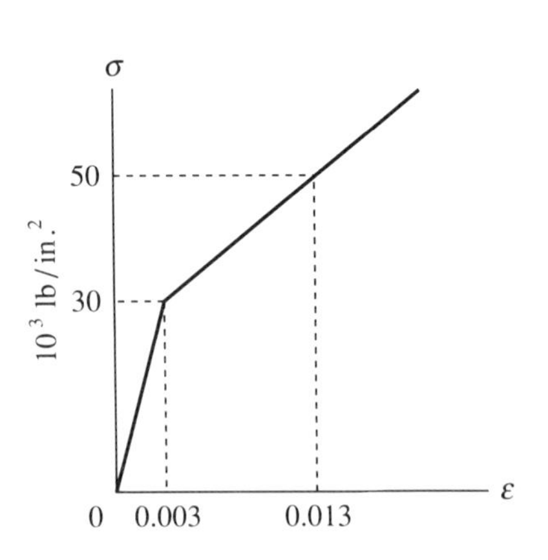

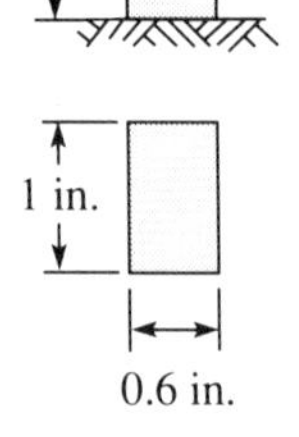

9.28 If the column of Prob. 9.27 consists of a material for which the compressive stress-strain law is $\sigma^2 = 8 \times 10^9 \varepsilon$ (σ is in psi), determine the critical load.

9.29 A pin-ended column 25 cm long has a rectangular cross section 10 cm by 5 cm. If the tangent modulus E_T versus stress σ is as shown in Fig. P9.29, what is the critical load for the column?

Fig. P9.29

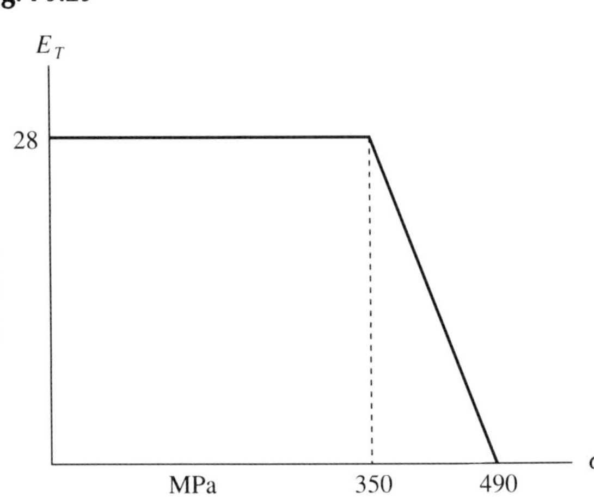

9.30 A material has the stress-strain curve shown in Fig. P9.30. Sketch the tangent-modulus curve.

Fig. P9.30

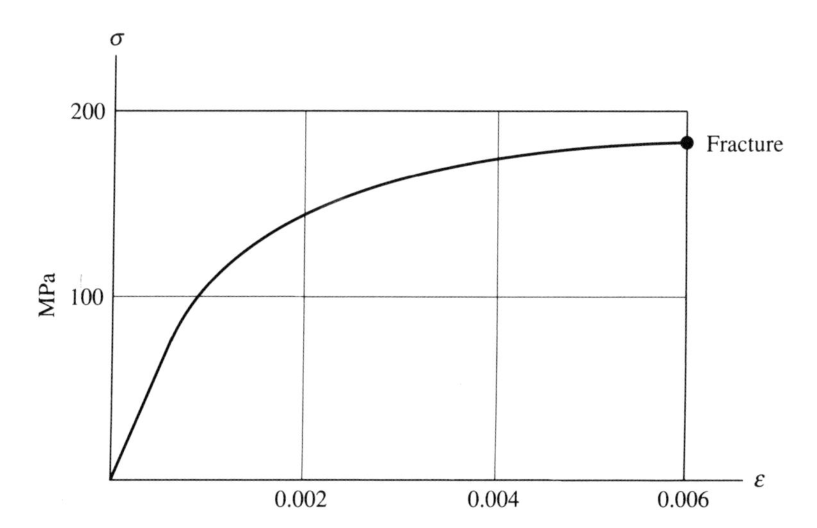

9.31 A linear hardening material has the approximate stress-strain curve shown in Fig. P9.31. Write expressions for the Euler-Engesser critical stress in terms of slenderness ratio only, and indicate the range of slenderness ratio for which each expression is valid.

Fig. P9.31

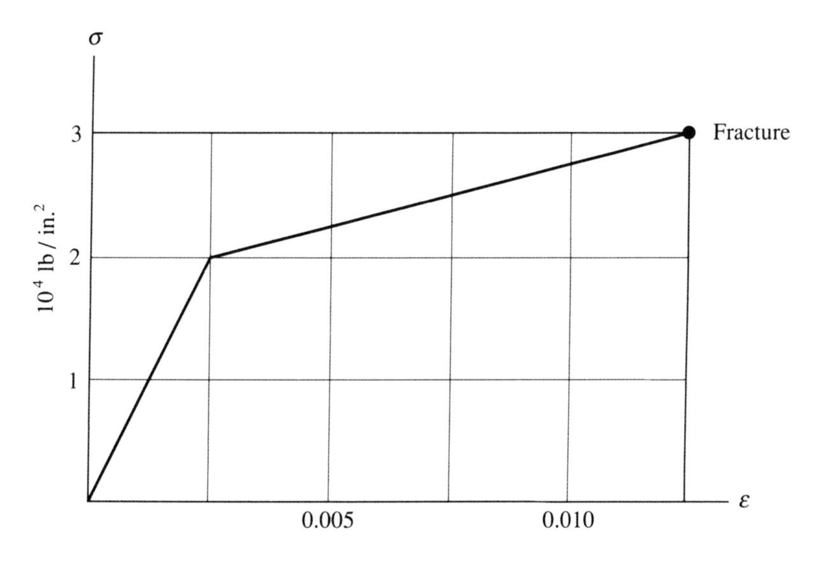

9.32 The compressive stress-strain relation for a material is approximated by the expression

$$\sigma = E\frac{\varepsilon}{1+\varepsilon}$$

Show that the Euler-Engesser critical stress satisfies

$$\sigma_{CR} = \frac{\pi^2 E\left[1 - \dfrac{\sigma_{CR}}{E}\right]^2}{\left(\dfrac{kL}{\rho}\right)^2}$$

9.33 A column with a solid circular cross section is made of the material with the compression stress-strain diagram of Fig. 9.21. It is clamped at one end and free at the other. The cross section has a diameter of 4 in. and the column is 3 ft long. (a) What is the Euler-Engesser critical load? (b) If the length of the column were cut in half, how much would the critical load be increased?

9.34 A pin-connected column is made of the material with the compression stress-strain diagram of Fig. 9.21. The column must be stable under a load of 640,000 lb and the maximum permissible compressive stress in the column cannot exceed 40,000 psi. If the length of the column is 4 ft and the cross section is square, determine the cross-sectional dimensions.

9.35 For an inelastic material under a uniform compressive stress σ, the loading stress-strain curve has a slope E_T, whereas the unloading curve has the initial slope E, as indicated in Fig. P9.35. Assume that the centroidal axis of a column remains at the critical stress during buckling, and show that the critical stress is given by Euler's formula with E replaced by an average modulus (Assume a doubly symmetric cross section.):

$$E_A = \frac{E + E_T}{2}$$

Fig. P9.35

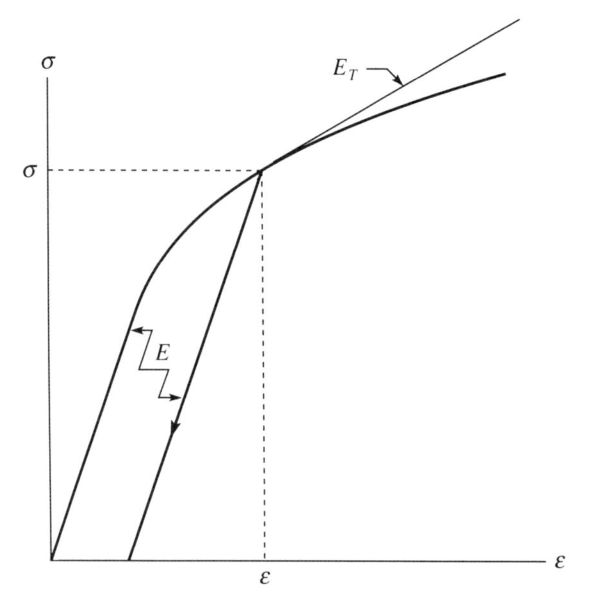

9.36 Repeat Problem 9.35, assuming that the average stress on a cross section is constant at σ_{CR} during buckling. Then show that, for a rectangular cross section, E in Euler's formula is replaced by a reduced modulus:

$$E_R = \frac{4EE_T}{(\sqrt{E} + \sqrt{E_T})^2}$$

9.11 Other Forms of Instability

The buckling of a column is by no means the only instability of interest to the engineer. Numerous situations occur in which the usual equilibrium configuration of a loaded member can become unstable; in most cases this is a very undesirable situation. The analysis of such problems is often quite difficult and will, therefore, not be pursued here. The purpose of this section is merely to describe such situations.

Localized Buckling

Localized buckling, or *wrinkling*, is a stability failure that might occur in members that have thin sections. A thin-walled tube subjected to compression could fail in this manner. Diamond-shaped dents appear in the tube wall, as shown in Fig. 9.22. This reduces the compressive strength of the tube, leading to more localized buckling and, usually, complete collapse.

Fig. 9.22

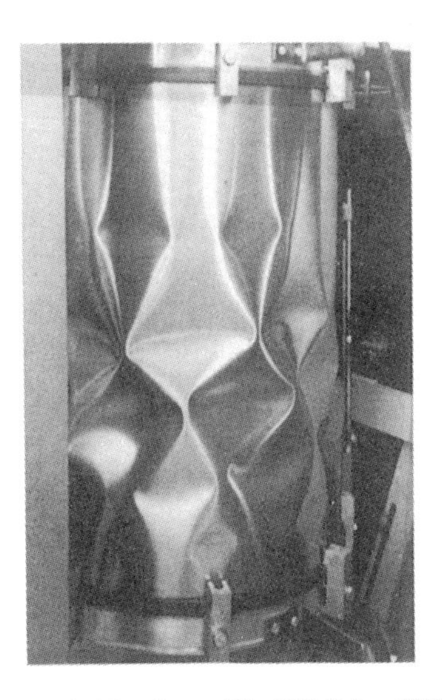

From H.S. Suer et al., *J. Aero. Sci.*, 25, No. 5, pp 281–287 (May 1958); reproduced in D. Bushnell, *Computerized Buckling Analysis of Shells* (Kluwer Academic Publishers, 1985). Reprinted by Permission of Kluwer Academic Publishers.

Another form of localized buckling occurs when beams with thin sections are subjected to bending. In an I-beam, for example, the inner flange is compressed during bending. If the stresses are large enough, the flange might buckle into a shape such as the one indicated in Fig. 9.23. This reduces the beam's resistance to bending, and again collapse might follow. A thin plate subjected to compressive loads in the plane of the plate could buckle in a similar manner.

Fig. 9.23

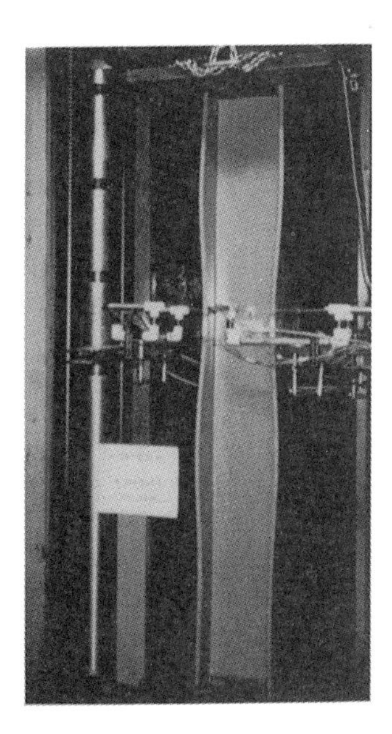

Shear Buckling

Shear buckling can occur when a thin plate or shell is subjected to shear loads, as in Fig. 9.24. The skin on a panel of an airplane wing is loaded in this manner. At 45° to the shear load, the panel experiences compressive stresses. It is this compressive stress that accounts for the buckling of the panel. The same type of situation occurs in the torsion of thin-walled tubes. The localized buckling is oriented with the compressive stress at 45° to the axis of the tube.

Fig. 9.24

From W.H. Horton et al., *An Introduction to Stability*, Stanford University Report Version of Paper No. 219 presented at ASTM Annual Meeting, Atlantic City, N.J. (June 1966), reproduced in D. Bushnell, *Computerized Buckling Analysis of Shells* (Kluwer Academic Publishers, 1985). Reprinted by Permission of Kluwer Academic Publishers.

Lateral Buckling

Lateral buckling of beams is another common stability failure. Such buckling occurs in instances where the transverse bending resistance of the beam is much larger than the lateral bending resistance. For large-enough loads the transversely bent configuration can become unstable, so that the beam twists and bends laterally, as shown in Fig. 9.25. This is prevented by lateral supports.

Fig. 9.25

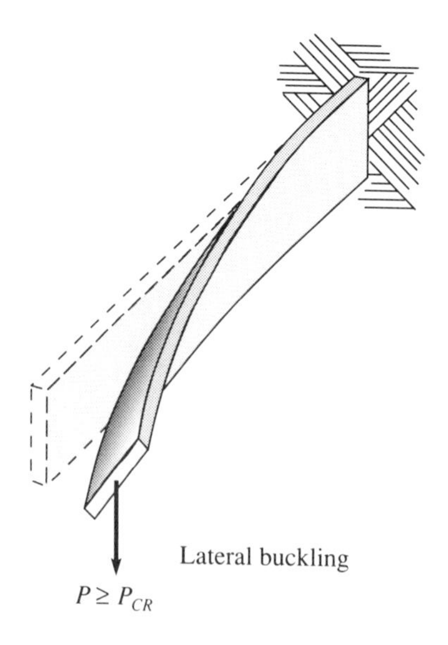

Snap-Through Buckling

Snap-through buckling can be catastrophic, since it occurs abruptly and carries the structure to a very different deformed configuration. For example, the critical load upon the curved bar of Fig. 9.26 causes the bar to snap through to the position shown in the bottom figure. The bottom of a conventional oil can is so constructed that this snap-through occurs for small pressures. For obvious reasons, snap-through buckling is often referred to as "oil canning." The safety lid on an evacuated jar is a familiar usage of snap-buckling: When the seal is broken, the pressure is equalized, and the abrupt snap-through is accompanied by a distinct "pop."

An important application of snap-through buckling is the use of bimetallic caps as thermostatic control elements. These caps consist of two bonded materials with different coefficients of thermal expansion. There is an initial curvature in the cap. As the temperature rises to a critical value, the cap pops through, making or breaking an electrical contact. When the cap cools to a certain temperature, it snaps back, reactivating the control device.

Fig. 9.26

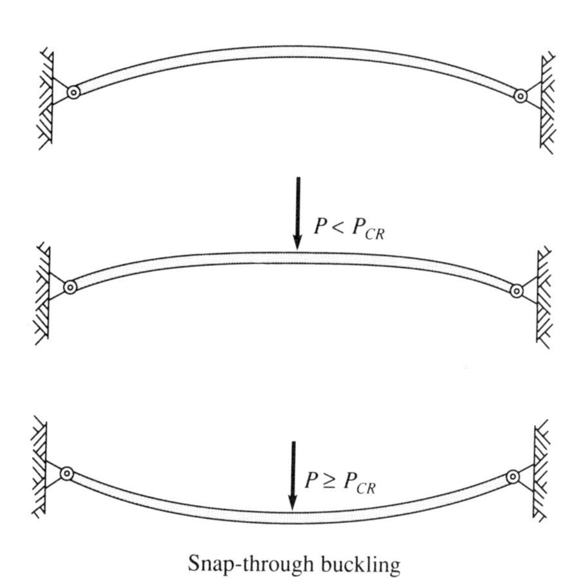

Snap-through buckling

Centroid of an Area

In the study of HI-HO prismatic members under extension and flexure, the centroid of the cross-section plays a useful role. Specifically, it is convenient to represent the resultant by a force *at* the centroid and a couple (moment about the centroid); consequently, the centroid is a convenient origin of the coordinates.

If the origin of a Cartesian system (y, z) is at the centroid, then

$$\iint_A y\,dA = \iint_A z\,dA = 0 \qquad \text{(A.1a, b)}$$

Equations A.1 *define* the centroid and enable us to locate the centroid.

Consider an arbitrary section A, depicted in Fig. A.1, and the arbitrary coordinates (y, z). We require the location of the centroid C. It is located by the two unknowns, Y and Z. Let $\bar{y}$ and $\bar{z}$ be Cartesian coordinates with axes parallel to y and z and originating at C. By definition (A.1)

Fig. A.1

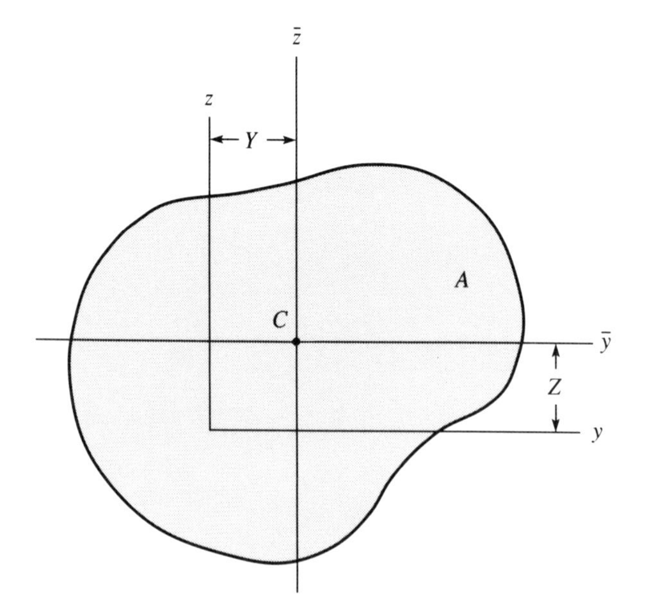

$$\iint_A \bar{y}\, dA = \iint_A \bar{z}\, dA = 0$$

or

$$\iint_A (y - Y)\, dA = \iint_A (z - Z)\, dA = 0$$

It follows that

$$Y = \frac{1}{A} \iint_A y\, dA, \qquad Z = \frac{1}{A} \iint z\, dA \tag{A.2a, b}$$

EXAMPLE 1

To illustrate the use of Eqs. (A.2) in order to locate the centroid, let us apply them to the section depicted in Fig. A.2. The region consists of the rectangular strip A_1 and the semicircular region A_2. We choose the apparently convenient system (y, z) with origin at the center of the circle. This lies on a line of symmetry; centroid C lies on the line of symmetry by definition (A.1). Then unknown Z is determined by (A.2b); the integral throughout the area consists of the integrals over area A_1 and A_2:

$$\iint_A z\, dA = \iint_{A_1} z\, dA + \iint_{A_2} z\, dA \tag{A.3}$$

The differential areas are depicted by the shaded strips. In A_1 and A_2,

Fig. A.2

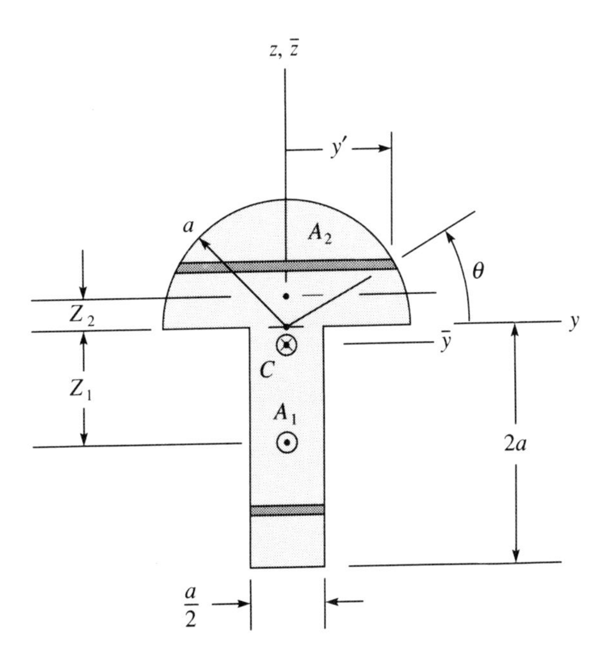

$$dA = \frac{a}{2} dz, \qquad dA = 2y'\, dz$$

respectively. We note that on the circular boundary,

$$z = a \sin \theta, \qquad y = y' = a \cos \theta$$

Therefore,

$$\iint_A z\, dA = \frac{a}{2} \int_{-2a}^{0} z\, dz + 2a^3 \int_0^{\pi/2} \sin \theta \cos^2 \theta\, d\theta$$

$$= -a^3 + 2\frac{a^3}{3} = -\frac{a^3}{3}$$

By Eq. (A.2b)

$$Z = \frac{-\frac{1}{3}a^3}{\left(\frac{\pi}{2}a^2 + a^2\right)} = -\frac{2a}{3(\pi + 2)}$$

Note that Eqs. (A.2b) and (A.3) give

$$Z = \frac{1}{A}\iint_{A_1} z\, dA + \frac{1}{A}\iint_{A_2} z\, dA \qquad \text{(A.4)}$$

If Z_1 and Z_2 are coordinates to the centroids of A_1 and A_2, respectively, then by Eqs. (A.2) and (A.4),

$$Z = \frac{Z_1 A_1 + Z_2 A_2}{A} \qquad \text{(A.5)}$$

In short, if we know the positions Z_1 and Z_2 of the centroids for the individual parts, we can obtain the result by the algebraic equation A.5. Values have been tabulated for simple geometric forms. The centroid of the rectangle (or any doubly symmetric area) is at the midpoint; $Z_1 = -a$. The centroid of the semicircular region is at $Z_2 = 4a/3\pi$. Then, by Eq. (A.5)

$$Z = \frac{-a^3 + \frac{4a}{3\pi} \times \frac{\pi a^2}{2}}{a^2 + \frac{\pi}{2}a^2} = -\frac{2a}{3(2 + \pi)} \quad \blacklozenge$$

B Integrals I_{yy}, I_{zz}, and I_{yz}

Rotation of Axes

In the Bernoulli-Euler theory of beams, we encounter three integrals (in alternative notations):

$$I_y \equiv I_{yy} \equiv \iint_A y^2 \, dA \tag{B.1a}$$

$$I_z \equiv I_{zz} \equiv \iint_A z^2 \, dA \tag{B.1b}$$

$$I_{yz} \equiv I_{zy} \equiv \iint_A yz \, dA \tag{B.1c}$$

The reader must be forewarned that many books use another notation: $I_y = I_{yy}$ is often denoted by I_z. Their notation can be traced to a physical anlogue, namely, the "moment of inertia," which arises in dynamics of rotating rigid bodies. A body rotating about axis z has a moment of inertia about that axis:

$$I_z \equiv \iiint_v (x^2 + y^2)\rho \, dv$$

where ρ denotes mass density and v denotes volume. If the body is *very* thin in the x direction (thickness t) and the density ρ is constant, then

$$I_z \doteq t\rho \iint_A y^2 \, dA$$

Hence, the term *area moment of inertia* is attached to the integral and the subscript z persists. We adopt the notation of Eq. (B.1) for two reasons. (1) The subscript signals the integrand: I_{yy} has integrand y^2, I_{yz} has integrand yz, and I_{zz} has integrand z^2. (2) More importantly, the notations of Eq. (B.1) conform completely to those employed previously for the *mathematically* similar components of stress (σ) and strain (ε). Let us show that these "components" (I_{yy}, I_{yz}, I_{zz}) transform via the same linear equations.

Figure B.1 shows an arbitrary cross section of area A and rectangular coordinates (y, z) and (n, t). Both originate at the common point O, which is arbitrary. From the geometry of the figure

$$n = a + b = y\cos\theta + z\sin\theta$$

$$t = c - d = z\cos\theta - y\sin\theta$$

Fig. B.1

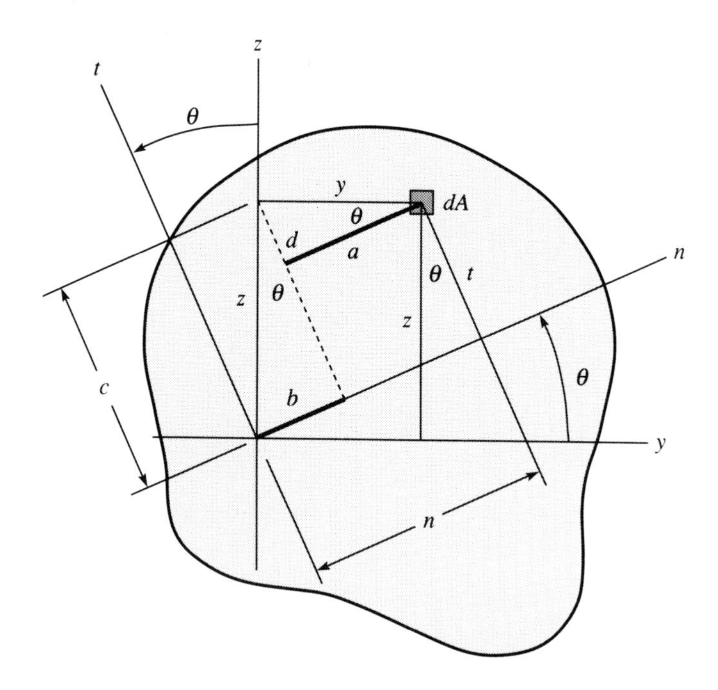

From the definitions:

$$I_{nn} = \iint_A n^2 \, dA$$

$$= \cos^2\theta \iint_A y^2 \, dA + 2\cos\theta\sin\theta \iint yz \, dA + \sin^2\theta \iint z^2 \, dA$$

$$= I_{yy}\cos^2\theta + 2I_{yz}\cos\theta\sin\theta + I_{zz}\sin^2\theta \tag{B.2a}$$

Likewise,

$$I_{nt} = -(I_{yy} - I_{zz})\cos\theta\sin\theta + I_{yz}(\cos^2\theta - \sin^2\theta) \tag{B.2b}$$

In the notation of Eqs. (2.55), (2.57), (3.19), and (3.20),

$$l_n = \cos\theta, \qquad m_n = +\sin\theta$$

$$l_t = -\sin\theta, \qquad m_t = \cos\theta$$

$$I_{nn} = l_n l_n I_{yy} + l_n m_n I_{yz} + m_n l_n I_{zy} + m_n m_n I_{zz} \tag{B.3a}$$

$$I_{nt} = l_n l_t I_{yy} + l_n m_t I_{yz} + m_n l_t I_{yz} + m_n m_t I_{zz} \tag{B.3b}$$

Of course, here the coordinates y and z replace x and y in keeping with our notations for the cross section of a beam.

With the notations of Eqs. (B.1), the linear transformations of Eqs. (B.2) or (B.3) are the same as those for the components of stress (σ) and strain (ε). It follows by the same arguments that principal directions $(\bar{y}, \bar{z})$ always exist, such that $I_{\bar{y}\bar{z}} = 0$ and $I_{\bar{y}\bar{y}}$ and $I_{\bar{z}\bar{z}}$ are extrema (maximum or minimum). As before, their orientation is determined by setting Eq. (B.2b) to zero. In terms of the double angle 2θ,

$$\tan 2\theta = \frac{2I_{yz}}{I_{yy} - I_{zz}}$$

Translation of Axes

In the Bernoulli-Euler theory of beams, the centroid of the cross section is often the convenient origin of coordinates (y, z). Then we need the integrals I_{yy}, I_{yz}, and I_{zz} with respect to the centroidal axes $(\bar{y}, \bar{z})$. In such cases, one can transfer axes as follows: We refer to Fig. A.1 and the definitions (B.1a) and (B.1c):

$$\begin{aligned}
I_{yy} &= \iint_A y^2\, dA = \iint_A (\bar{y} + Y)^2\, dA \\
&= \iint_A \bar{y}^2\, dA + 2Y \iint_A \bar{y}\, dA + Y^2 A \\
I_{yz} &= \iint_A yz\, dA = \iint (\bar{y} + Y)(\bar{z} + Z)\, dA \\
&= \iint_A \bar{y}\bar{z}\, dA + Z \iint_A \bar{y}\, dA + Y \iint_A \bar{z}\, dA + YZA
\end{aligned}$$

By the definition of a centroid, the middle terms on the right sides are zero. Hence, we obtain the "transfer of axes" theorem:

$$I_{yy} = I_{\bar{y}\bar{y}} + Y^2 A \tag{B.4a}$$

$$I_{yz} = I_{\bar{y}\bar{z}} + YZA \tag{B.4b}$$

Likewise,

$$I_{zz} = I_{\bar{z}\bar{z}} + Z^2 A \tag{B.4c}$$

These Eqs. (B.4) collectively constitute the *parallel-axis theorem.* This can be useful if the section is composed of parts that are geometrically simple.

EXAMPLE 1

Consider again the section shown in Fig. A.2. Recall that the centroid of the semicircular part is at $z = 4a/3\pi$. We know that the integral I''_{zz} (the double prime signifies part 2) for this part is

$$I''_{zz} = \frac{\pi}{8} a^4$$

With respect to the centroid of A_2,

$$\bar{I}''_{zz} = I''_{zz} - \left(\frac{4a}{3\pi}\right)^2 A_2$$
$$= \frac{\pi a^4}{8} - \left(\frac{4a}{3\pi}\right)^2 \left(\frac{\pi a^2}{2}\right)$$
$$= \left(\frac{\pi}{8} - \frac{8}{9\pi}\right) a^4$$

The integral I'_{zz} (the prime signifies part 1) with respect to the centroid of A_1 is

$$\bar{I}'_{zz} = \frac{a^4}{3}$$

The integral for the cross section with respect to the centroid of the entire section is

$$I_{\bar{z}\bar{z}} = I''_{\bar{z}\bar{z}} + I'_{\bar{z}\bar{z}} \tag{B.5}$$

The two integrals on the right can be obtained by transfer of axes according to Eq. (B.4c):

$$I''_{\bar{z}\bar{z}} = \bar{I}''_{zz} + (\bar{z}'')^2 A_2 \tag{B.6a}$$
$$I'_{\bar{z}\bar{z}} = \bar{I}'_{zz} + (\bar{z}')^2 A_1 \tag{B.6b}$$

Note: The distances $\bar{z}'$ and $\bar{z}''$ are the distances between the centroidal axis of the cross section and the centroidal axes of the parts 1 and 2, respectively. The result follows from Eqs. (B.5) and (B.6):

$$I_{\bar{z}\bar{z}} = \bar{I}''_{zz} + \bar{I}'_{zz} + (\bar{z}'')^2 A_2 + (\bar{z}')^2 A_1$$
$$I_{\bar{z}\bar{z}} = \left(\frac{\pi}{8} - \frac{8}{9\pi}\right) a^4 + \left(\frac{4a}{3\pi} - Z\right)^2 \frac{\pi a^2}{2} + \frac{a^4}{3} + (a + Z)^2 a^2 \tag{B.7a}$$

By direct integration

$$I_{\bar{z}\bar{z}} = 2a^2 \int_0^{\pi/2} (Z + a\sin\theta)^2 \cos^2\theta \, d\theta + \frac{a}{2} \int_{-(2a+z)}^{-Z} \bar{z}^2 \, dA \tag{B.7b}$$

where $Z = -2a/3(\pi + 2)$, as previously determined. Both Eqs. (B.7a) and (B.7b) give $I_{\bar{z}\bar{z}} = 1.683a^4$, the integral throughout the cross section with respect to the centroidal axes. ◆

C Hookean Beams under Concentrated Loads

In Chapter 6, we employed the differential equations for a symmetric hookean beam, specifically, the equilibrium equations (6.31) and (6.32) and the curvature-moment equation (6.33) and the kinematic equation (6.34). We reiterate these for convenience.

$$\frac{dV}{dx} = p(x) \tag{C.1}$$

$$\frac{dM}{dx} = -V(x) \tag{C.2}$$

$$\frac{d\theta}{dx} = \frac{M(x)}{EI} \tag{C.3}$$

$$\frac{dw}{dx} = \theta(x) \tag{C.4}$$

Recall that a concentrated force, or couple, can*not* be represented by a function that is continuous. Consequently, the integrations of the equations are *not* straightforward. Accordingly, we consider a suitable representation of such loads and the integration of the discontinuous functions.

From a practical point of view the concentrated load is actually distributed about a point ($x = a$) over some small region; let us say that the width is 2ε. Indeed, the concept of the concentrated load is often employed because the actual distribution is unknown and/or because the behavior near the load is not required. What is relevant is the total load and the fact that the region is small ($2\varepsilon \ll L$). Mathematically, if the resultant load is just a force of *unit* magnitude, then

$$\int_{x<a-\varepsilon}^{x>a+\varepsilon} p(x)\,dx = 1 \tag{C.5}$$

Likewise, if the resultant is only a (counterclockwise) couple of *unit* magnitude, then

$$\int_{x<a-\varepsilon}^{x>a+\varepsilon} p(x)\,dx = 0 \tag{C.6a}$$

and

$$\int_{x<a-\varepsilon}^{x>a+\varepsilon} (a - x)p(x)\,dx = 1 \tag{C.6b}$$

Any distributions that satisfy these requirements are therefore acceptable. Distributions that satisfy the requirements for the concentrated couple and force, respectively, are plotted in Fig. C.1(a) and C.1(b).

Fig. C.1

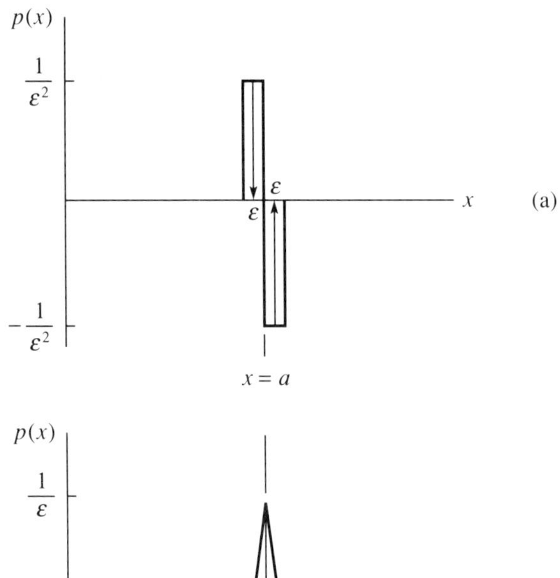

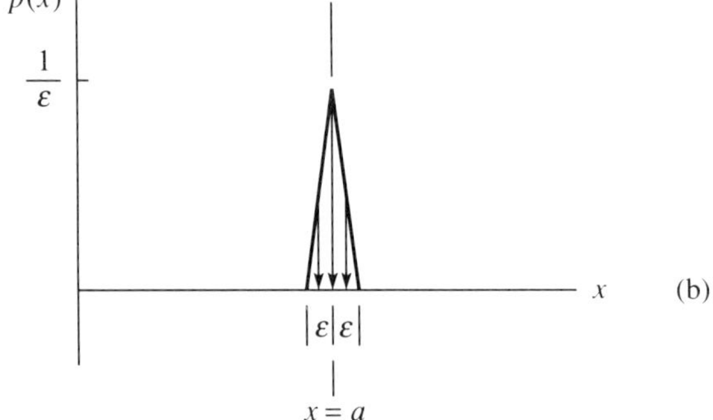

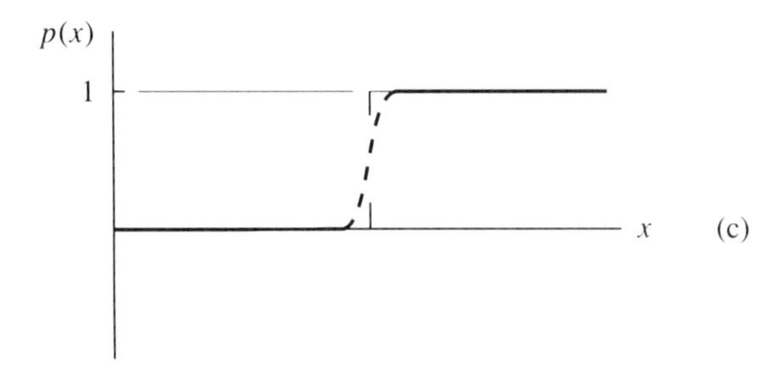

The constant force $(+1/\varepsilon^2)$ over the region $a - \varepsilon < x < a$ and the constant force $(-1/\varepsilon^2)$ over the region $a < x < a + \varepsilon$ in Fig. C.1(a) have a resultant that is a simple couple (one-unit magnitude). The net force is zero.

The pyramidlike distribution of Fig. C.1(b) has a resultant that is a simple force (1-unit magnitude). The net moment is zero. Clearly, these distributions satisfy our requirements for a concentrated couple and force, respectively. Moreover, each is represented by a simple, albeit discontinuous, function. We can integrate them. To simplify the matter, bear in mind that the integral of a function can be interpreted graphically as the area under the plotted function.

Let us begin with the function plotted in Fig. C.1(a), which represents a concentrated couple of unit magnitude. Let's call it $C\langle x - a\rangle$. The bracketed argument $\langle x - a\rangle$ is simply a signal that the action is at $x = a$, where the argument $(x - a)$ vanishes; the letter C helps us remember that it is a couple. For all $x < a - \varepsilon$, the integral is zero. From $x = a - \varepsilon$ to $x = a$, the integral (area under the curve) grows linearly, *exactly* as the function plotted in Fig. C.1(b). Moreover, at $x = a$ the integral reaches the value $1/\varepsilon$, *exactly* the value plotted in Fig. C.1(b). From $x = a$ to $x = a + \varepsilon$, the integral of $C\langle x - a\rangle$ decreases linearly to zero at $x = a + \varepsilon$, *exactly* as the function of Fig. C.1(b). Subsequently, for all $x > a + \varepsilon$, the integral of $C\langle x - a\rangle$ remains zero; the area under the entire plot is zero. Indeed, we have just observed that the integral of $C\langle x - a\rangle$ is the pyramidlike function of Fig. C.1(b). Let's call it $D\langle x - a\rangle$ (D for Δ).

The delta function $D\langle x - a\rangle$ of Fig. C.1(b) represents our concentrated force acting about $x = a$. Again, the bracketed argument $\langle x - a\rangle$ is merely a signal to indicate that the action is at $x = a$. Now, the integral of the function $D\langle x - a\rangle$ is zero for all $x < a - \varepsilon$. On integrating to any $x > a + \varepsilon$, the value of the integral climbs to unity, and the area under the delta is the triangular area $\frac{1}{2} \times (2\varepsilon) \times (1/\varepsilon) = 1$. For all $x > a + \varepsilon$, the value remains unity. The plot of that function is the step shown in Fig. C.1(c). Let's call this function $\langle x - a\rangle^0$. Again, the bracketed argument signals the location of the step. For $x > a$ the function is $(x - a)^0 = 1$.

The functions $C\langle x - a\rangle$, $D\langle x - a\rangle$, and $\langle x - a\rangle^0$ are listed in Table C.1, along with their plots. This is a table of integrals. We have shown that $\langle x - a\rangle^0$ is the integral of $D\langle x - a\rangle$, which is the integral of $C\langle x - a\rangle$. Now, let's extend that table. But first note that the magnitude of ε is irrelevant for our purposes, so that in our further examination we can neglect ε compared with distances such as a. For example, the step of Fig. C.1(c) can be treated as a sharp jump at $x = a$ rather than an ascent from $x = a - \varepsilon$ to $x = a + \varepsilon$.

The function $\langle x - a\rangle^0$ is zero for all $x < a$, and its integral is also zero. For all $x > a$,

$$\int_0^{x>a} \langle x - a\rangle^0 \, dx = \int_a^{x>a} 1 \, dx = (x - a)$$

Let us denote the integral for all x by the symbol $\langle x - a\rangle^1$. This function is zero for all $x < a$, and it is the function $(x - a)^1$ for all $x > a$. Again, the bracket indicates that the change occurs where $x = a$ (the argument $x - a$ vanishes).

Now, the reader can easily extend the table. All subsequent integrals are also zero for $x < a$. For all $x > a$ the integrals are of the form $1/n!(x - a)^n$. Indeed, the appropriate notation follows:

$$\frac{1}{n!}\langle x - a\rangle^n = \begin{cases} 0, & x < a \\ \dfrac{1}{n!}(x - a)^n, & x > a \end{cases}$$

With the representation of concentrated couples and forces by the functions $C\langle x - a\rangle$ and $D\langle x - a\rangle$, respectively, Eqs. (C.1), (C.2), (C.3), and (C.4) can be integrated. The correct continuity or discontinuity of the integrals or their derivatives follow. The determination of shear force, bending couple, rotation, and deflection requires no more effort than the analyses for the more familiar continuous functions. We need only use the representations, notations, and integrals of Table C.1. An example illustrates this.

Table C.1

Label	Plot	Integral
$C\langle x-a\rangle$	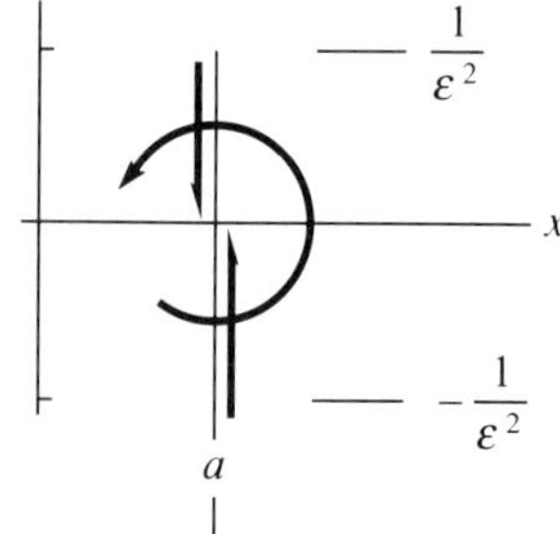	
$D\langle x-a\rangle$	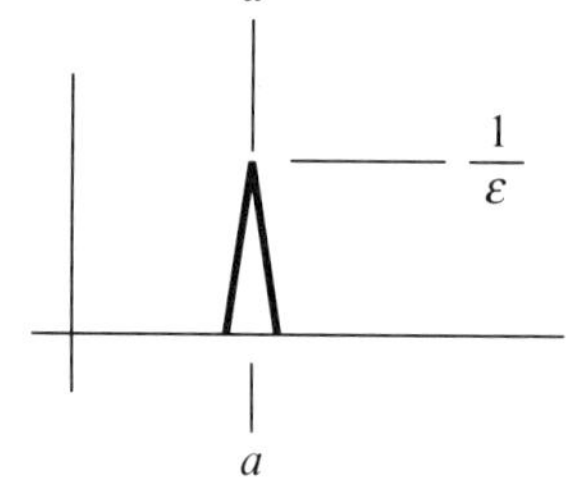	$D\langle x-a\rangle = \int^{x} C\langle x-a\rangle\,dx$
$\langle x-a\rangle^0$	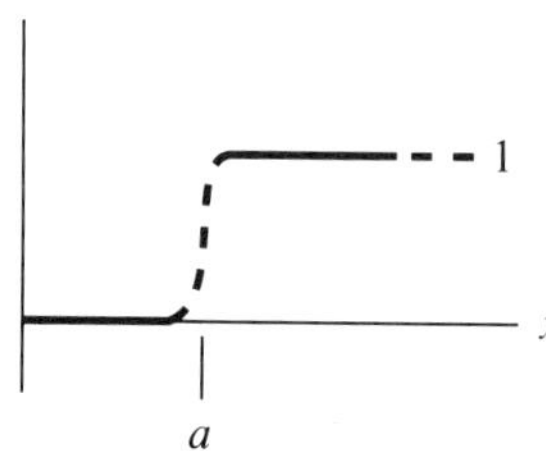	$\langle x-a\rangle^0 = \int^{x} D\langle x-a\rangle\,dx$
$\langle x-a\rangle^1$	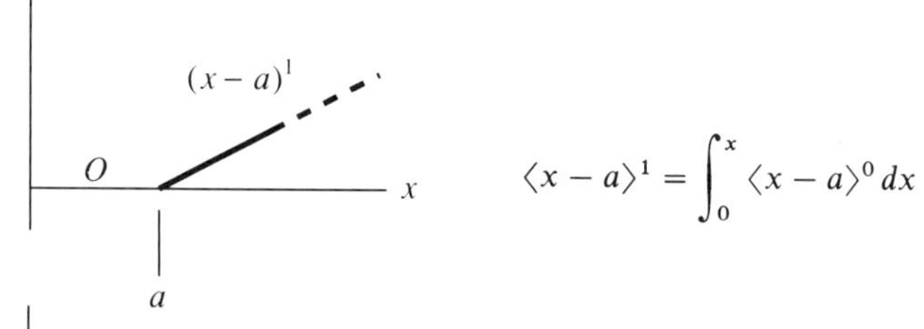	$\langle x-a\rangle^1 = \int_0^{x} \langle x-a\rangle^0\,dx$
$\frac{1}{2}\langle x-a\rangle^2$	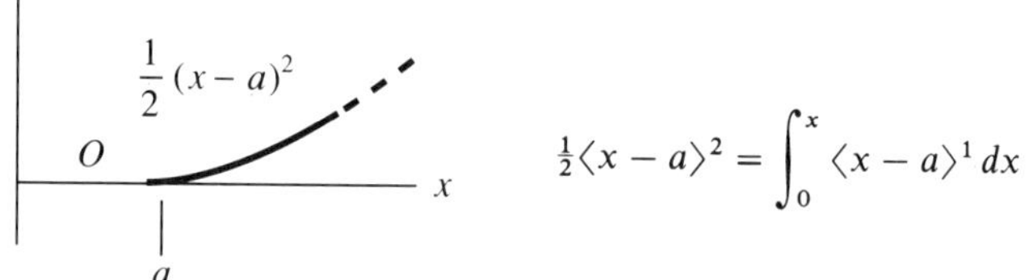	$\frac{1}{2}\langle x-a\rangle^2 = \int_0^{x} \langle x-a\rangle^1\,dx$
$\frac{1}{6}\langle x-a\rangle^3$	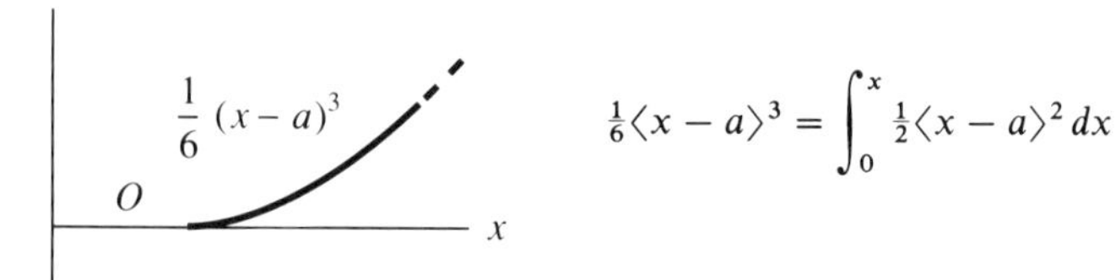	$\frac{1}{6}\langle x-a\rangle^3 = \int_0^{x} \frac{1}{2}\langle x-a\rangle^2\,dx$

EXAMPLE 1

To illustrate the foregoing representation and integration, let us consider the beam depicted in Fig. C.2. The beam is fixed at the left end, $x = 0$, and simply supported at the right, $x = L$. For simplicity, the total of the uniformly distributed load ($\bar{p}$) is taken equal to the concentrated load ($\bar{p}L = P$); the latter (P) acts at midspan, $x = L/2$. The loading $p(x)$ consists of two functions, a constant and a delta. The latter must be multiplied by the factor P, since our delta has unit magnitude.

$$p(x) = P/L + PD\langle x - a \rangle$$

Fig. C.2

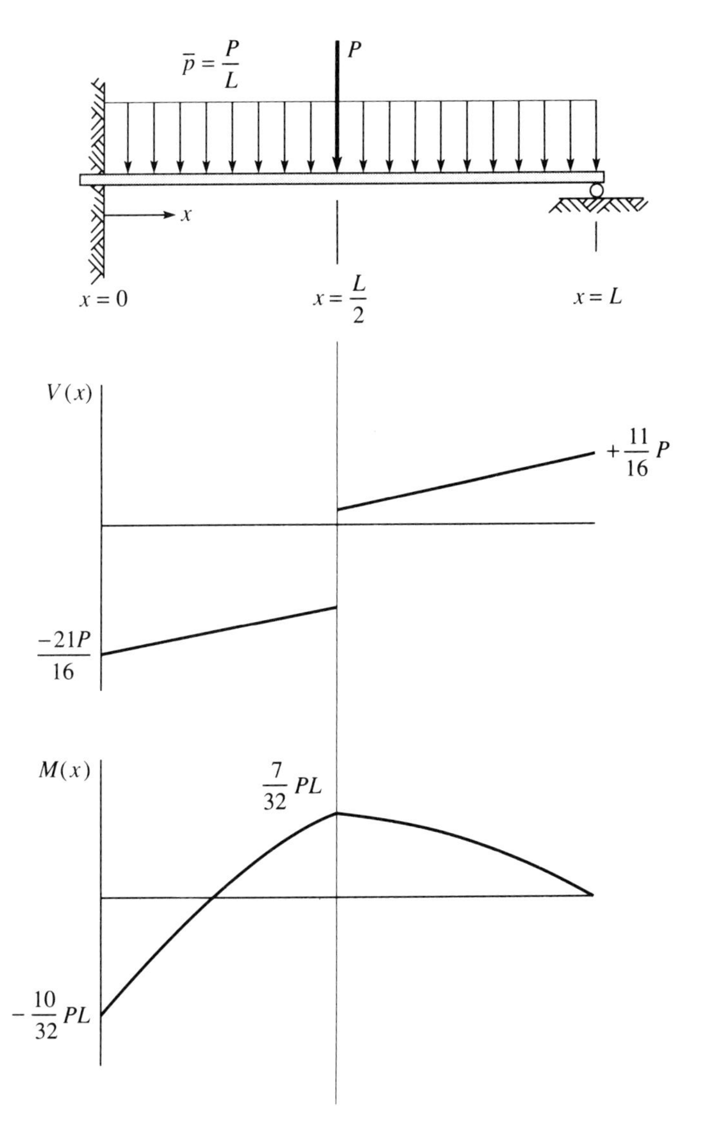

By Eq. (C.1)

$$dV = (P/L + PD\langle x - a \rangle)\,dx$$

Using our table of integrals, we obtain

$$\int_{V_0}^{V(x)} dV = \int_0^x \left(\frac{P}{L} + PD\langle x - a \rangle\right) dx$$

$$V - V_0 = \frac{P}{L}x + P\langle x - a \rangle^0$$

Substituting this into the right side of Eq. (C.2) and integrating, we obtain

$$M = M_0 - V_0 x - \frac{P}{2L}x^2 - P\langle x - a \rangle^1$$

Substituting this result into Eq. (C.3), using the end condition $\theta(0) = 0$, and integrating, we obtain

$$EI\theta(x) = M_0 x - \frac{V_0}{2}x^2 - \frac{P}{6L}x^3 - \frac{P}{2}\langle x - a \rangle^2$$

Using this, the end condition $w(0) = 0$, and Eq. (C.4), we arrive at the deflection,

$$EIw(x) = \frac{M_0}{2}x^2 - \frac{V_0}{6}x^3 - \frac{P}{24L}x^4 - \frac{P}{6}\langle x - a \rangle^3$$

Note that the shear force $V(x)$ exhibits the anticipated jump of magnitude P at the concentrated load. Note, too, that the couple $M(x)$, rotation $\theta(x)$, and deflection $w(x)$ are all continuous everywhere ($0 \le x \le L$). To complete the solution, we need only enforce the conditions at the simply supported end $x = L$:

$$M(L) = M_0 - V_0 L - \frac{PL}{2} - P(L - a) = 0$$

$$EIw(L) = \frac{M_0}{2}L^2 - \frac{V_0}{6}L^3 - \frac{PL^3}{24} - \frac{P}{6}(L - a)^3 = 0$$

The solution of these two algebraic equations determines the unknown constants M_0 and V_0. Note that the concentrated load is at $x = a = L/2$:

$$V_0 = -\frac{21}{16}P, \qquad M_0 = -\frac{5}{16}PL$$

The plots of the shear $V(x)$ and couple $M(x)$ are displayed in Fig. C.2(b) and (c).

The solution by means of the special functions and integrals of Table C.1 requires no more than end conditions and the same steps as needed with any other continuous loading.

Note that these representations of concentrated loadings are useful when the loads are placed at intermediate locations ($0 < x < L$) but *not* at the ends. Concentrated forces, or couples, applied at ends appear simply as end conditions. ◆

D

Simple Problems of Symmetric HI-HO Beams

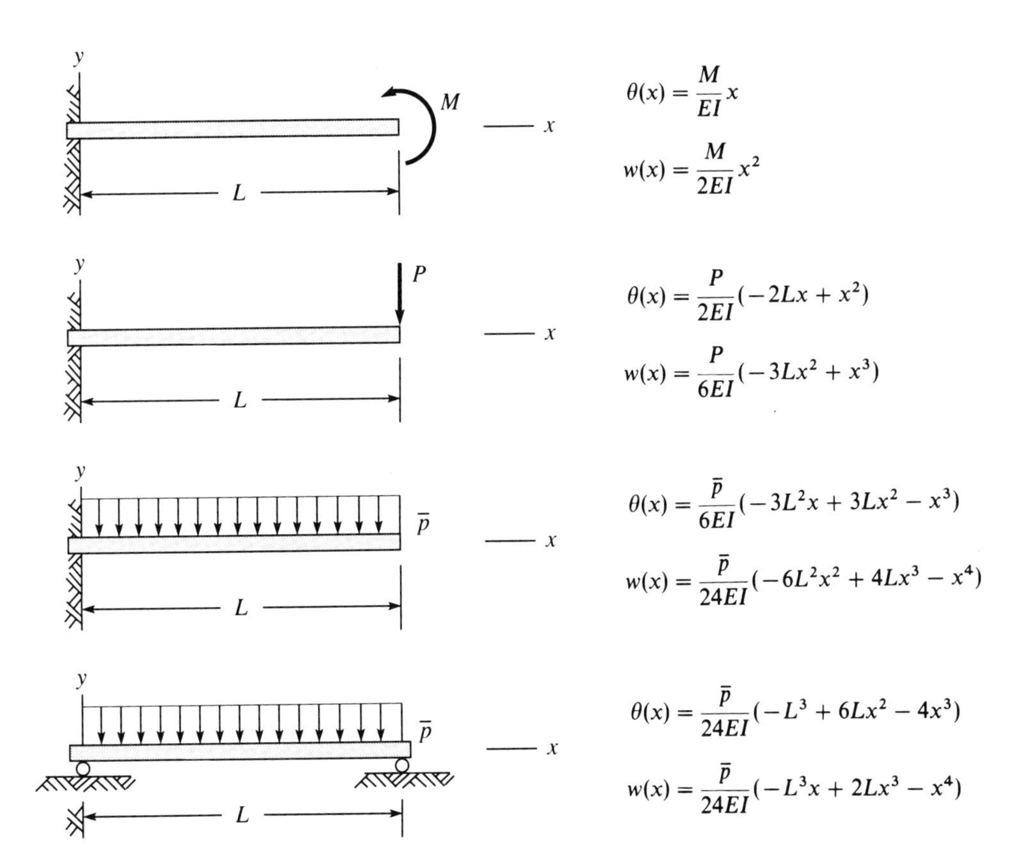

$$\theta(x) = \frac{M}{EI}x$$

$$w(x) = \frac{M}{2EI}x^2$$

$$\theta(x) = \frac{P}{2EI}(-2Lx + x^2)$$

$$w(x) = \frac{P}{6EI}(-3Lx^2 + x^3)$$

$$\theta(x) = \frac{\bar{p}}{6EI}(-3L^2x + 3Lx^2 - x^3)$$

$$w(x) = \frac{\bar{p}}{24EI}(-6L^2x^2 + 4Lx^3 - x^4)$$

$$\theta(x) = \frac{\bar{p}}{24EI}(-L^3 + 6Lx^2 - 4x^3)$$

$$w(x) = \frac{\bar{p}}{24EI}(-L^3x + 2Lx^3 - x^4)$$

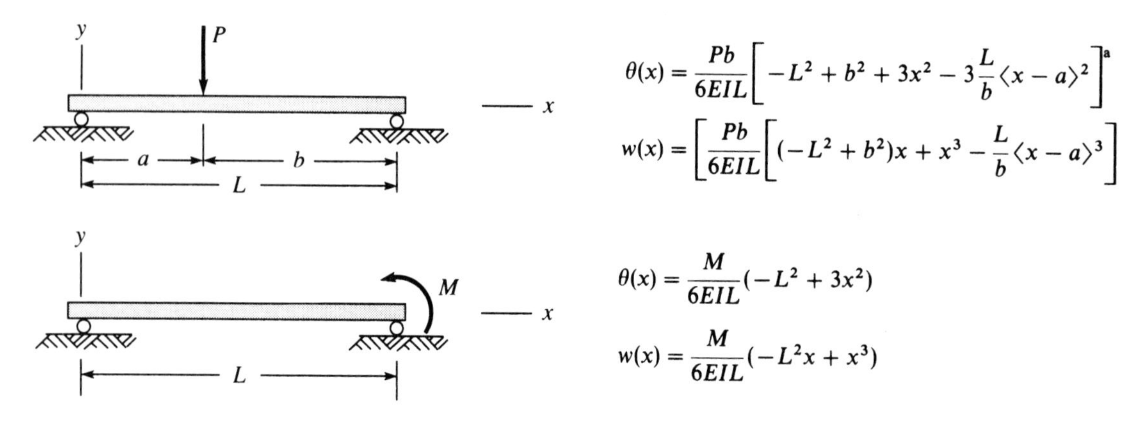

$$\theta(x) = \frac{Pb}{6EIL}\left[-L^2 + b^2 + 3x^2 - 3\frac{L}{b}\langle x - a\rangle^2\right]^{a}$$

$$w(x) = \left[\frac{Pb}{6EIL}\left[(-L^2 + b^2)x + x^3 - \frac{L}{b}\langle x - a\rangle^3\right]\right.$$

$$\theta(x) = \frac{M}{6EIL}(-L^2 + 3x^2)$$

$$w(x) = \frac{M}{6EIL}(-L^2x + x^3)$$

[a] See Table C.1 for definitions of special functions $\langle x - a\rangle^n$.

E Typical Cross-sectional Properties of Wide-Flange I-Beams

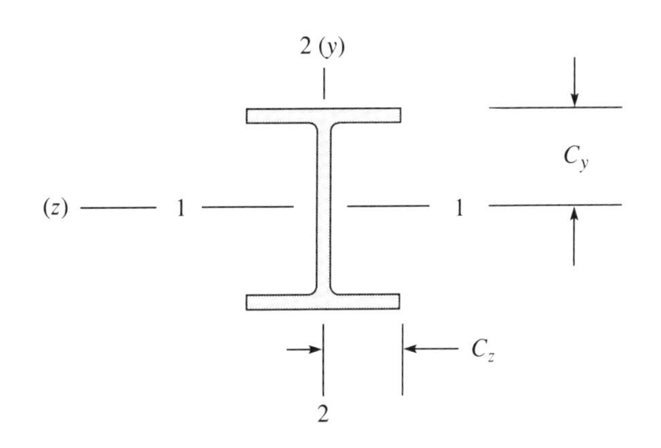

Nominal Size in.	Weight per Foot lb	Area in.^{2}	Depth in.	Flange Width in.	Flange Thick-ness in.	Web Thick-ness in.	Axis 1-1 I_{yy} in.^{4}	Axis 1-1 $\frac{I_{yy}}{C_y}$ in.^{3}	Axis 1-1 $\sqrt{\frac{I_{yy}}{A}}$ in.	Axis 2-2 I_{zz} in.^{4}	Axis 2-2 $\frac{I_{zz}}{C_z}$ in.^{3}	Axis 2-2 $\sqrt{\frac{I_{zz}}{A}}$ in.
36[a]	300	88.17	36.72	16.655	1.680	0.945	20,290.2	1,105.1	15.17	1,225.2	147.1	3.73
12	190	55.86	14.38	12.670	1.736	1.060	1,892.5	263.2	5.82	589.7	93.1	3.25
				Note:	A wide range of 12-in WF beams exists, from 190 to 27 (lb/ft).							
12	27	7.97	11.96	6.500	0.400	0.240	204.1	34.1	5.06	16.6	5.1	1.44
8	17	5.00	8.00	5.250	0.308	0.230	56.4	14.1	3.36	6.72	2.6	1.16

[a] Designation = 36 WF 300 [Depth–Wide-Flange–Weight].
Numerical values are taken from *Manual of Steel Construction*, American Institute of Steel Construction.

This excerpt is intended to indicate the kinds of available information on standard rolled-steel I-beams. The interested reader should consult the manuals for similar data on a wide variety of forms: I, C (channels), L (angles), T (tees), and Z (zees).

Index